CODING AND SIGNAL PROCESSING FOR MAGNETIC RECORDING SYSTEMS

Computer Engineering Series

Series Editor: Vojin Oklobdzija

Low-Power Electronics Design
Edited by Christian Piguet

Digital Image Sequence Processing,
Compression, and Analysis
Edited by Todd R. Reed

Coding and Signal Processing for
Magnetic Recording Systems
Edited by Bane Vasic and Erozan M. Kurtas

CODING AND SIGNAL PROCESSING FOR MAGNETIC RECORDING SYSTEMS

EDITED BY

Bane Vasic
University of Arizona
Tucson, AZ

Erozan M. Kurtas
Seagate Technology
Pittsburgh, PA

CRC PRESS

Boca Raton London New York Washington, D.C.

Library of Congress Cataloging-in-Publication Data

Coding and signal processing for magnetic recording systems / edited by Bane Vasic and Erozan M. Kurtas.
    p. cm. — (Computer engineering; 2)
    Includes bibliographical references and index.
    ISBN 0-8493-1524-7 (alk. paper)
      1. Magnetic recorders and recording. 2. Signal processing. 3. Coding theory.
      I. Vasic, Bane II. Kurtas, M. Erozan III. Title IV. Series: Computer engineering (CRC Press); 2.

TK7881.6C62 2004
621.39—dc22

2004050269

Visit the CRC Press Web site at www.crcpress.com

Preface

A steady increase in recording densities and data rates of magnetic recording systems in the last 15 years is mostly due to advances in recording materials, read/write heads and mechanical designs. The role of signal processing and coding has been to make the best use of the capacity and speed potentials offered by these advances. As the recording technology matures, the hard disk drive "read channel" is becoming more and more advanced, reaching the point where it uses equally or even more complicated signal processing, coding and modulation algorithms than any other telecommunication channel and where, due to the speed, power consumption and cost requirements, the challenges in implementing new architectures and designs have been pushed to today's integrated circuit manufacturing technology limits.

This book reviews advanced coding and signal processing techniques, and architectures for magnetic recording read channels. In the most general terms, the *read channel* controls reading and writing the data to/from recording medium. The operations performed in the data channel include: *timing recovery, equalization, data detection, modulation coding/decoding* and limited *error control*. Besides this so-called *data channel*, a read channel also has a *servo channel* whose role is to sense head position information, and to regulate a proper position of the head above the track. The *error control* functions of a hard drive reside in a *controller,* a separate system responsible for a diverse set of electronic and mechanical functions to provide the user with a data storage system that implements the high-level behavior described by the magnetic hard drive's user interface. A trend in hard drive systems is to merge the functionalities of a read channel and controller into a so-called superchip. This book gives an in-depth treatment of all of these subsystems, with an emphasis on coding and signal processing aspects.

The book has six sections. Each section begins with a review of the underlying principles and theoretical foundations, describes the state-of-the-art systems, and ends with novel and most advanced techniques and methodologies.

The *first* section gives an introduction to recording systems. After a brief history of magnetic storage, we give basic principles of physics of longitudinal and perpendicular magnetic recording, and the physics of optical recording.

A modern hard disk drive comprises a recording medium in the form of a thin film on a surface of a disk, an inductive write head and a giant magneto-resistive read head. We describe and compare two types of recording mechanisms: (i) longitudinal recording in which the media magnetic anisotropy is oriented in the thin film plane, and (ii) perpendicular recording, where the magnetic anisotropy is aligned perpendicular to the film plane. We discuss a pulse response, the media noise powers, and the signal-to-noise ratio calculation for both types of recording. In a recording system, the playback noise originates from the head electronics and the media magnetization random patterns. Generally the medium noise is decomposed into the direct current (DC) remanent and the transition components. However, the transition jitter noise is dominant. In longitudinal recording, due to the random anisotropy dispersion, there always exist some levels of the DC remanent noise. However, in perpendicular recording, the loop full squareness is required to maintain thermal stability. Therefore, the DC remanent noise vanishes.

The principles of optical recording are becoming increasingly important in hard drives because of a promise of a dramatic increase in the recording density that can be achieved by applying heat assisted magnetic recording (HAMR). In this approach, a laser beam at the spot where data are being recorded heats the magnetic medium. Heating the medium results in a reduction of the coercivity required to write the data to a level accessible by the recording head, while rapid subsequent cooling stabilizes the written data.

We give an introduction and history of magnetic recording heads, their evolution and importance, followed by a description of the write head for both longitudinal and perpendicular recording. We describe how various write head design parameters affect the written and read-back waveforms.

It is well understood that one of the key challenges to increasing areal density in magnetic recording systems is media noise suppression. Reducing the size of magnetic domains written to the medium, which has been the conventional approach to this problem, has a side effect whereby the magnetic domains become thermally unstable. To avoid this, so-called super paramagnetic effects media with increased coercivity are being used. However, the magnetic materials from which the head is made limit the fields that can be applied, and these limits are being approached. When the bit size is reduced, the signal-to-noise ratio is decreased, making the detection problem much more difficult.

The first section should equip the reader with a solid understanding of the physical principles of write and read back processes and constraints in designing a data storage system.

The *second* section gives communication and information theory tools necessary for a design and analysis of coding and signal processing techniques. We begin with modeling the recording channel. Design and analysis of coding and signal processing techniques require a suitable communications channel model for magnetic storage systems. Such a model should correctly reflect the essential physics of the read and write processes of magnetic recording, but must also provide a system level description that allows convenient design, analysis and simulation of the communications and signal processing techniques under study. We introduce common signal and noise models for recording channel models, statistical analysis tools, and partial response (PR) signaling as a method of controlling the intersymbol interference (ISI). We give models for medium noise, nonlinear distortion of magnetic transitions, and jitter. We show the effect of these noises on a number of conventional data detectors.

The first chapter on error control codes introduces finite fields and error correction capabilities of algebraically constructed and decoded codes. We introduce finite fields, define linear codes over finite fields, discuss the relation between minimum distance and error correction capability of a code, introduce cyclic codes and Reed-Solomon (RS) codes and explain in detail their encoding and decoding algorithms. The second chapter gives a theoretical foundation of message-passing algorithms and linear time sub-optimal probabilistic decoding algorithms that achieve near optimum performance. Recently such algorithms have attracted tremendous interest and have been investigated as an alternative decoding scheme for new generation of read channels.

The next chapter reviews sofic systems, the theoretical foundation of constrained (or modulation) codes that are used to translate an arbitrary sequence of user data to a channel sequence with special properties required by the physics of the recording medium. Modulation coding in a read channel serves a variety of important roles. Generally speaking, modulation coding eliminates those sequences from a recorded stream that would degrade the error performance, for example, long runs of consecutive symbols that impact the timing recovery or sequences that result in signals on a small Euclidian distance.

The section ends with a view of recording channels from the information theory standpoint, and a chapter summarizing techniques for bounding the capacity of a PR channel. Such results are very important in that they provide theoretical limits on the performance of practical coding/decoding schemes. Since recording channels can be modeled as ISI channels with binary inputs, we first present the capacity of the general ISI channels with additive white Gaussian noise, which is achieved by correlated Gaussian inputs. When taking the constraint of binary inputs into account, no closed-form solutions exist; however, several upper and lower bounds are derived. Monte-Carlo simulation techniques to estimate the achievable information rates of ISI channels are also described. The simulation-based techniques can be extended further to compute the achievable information rates of the magnetic recording channels with media noise, present in high density recording systems.

The *third* section begins with a description of physical and logical organization of data in various recording systems and methods of increasing recording density. Then we give a cross section of the state-of-the-art read channels and explain their subsystems. We explain organization of data on the disc tracks, servo sectors and data sectors, seeking and tracking operations, and phase and frequency acquisition. The section on servo information detection explains sensing radial information and read channel subsystem used to perform this operation.

The image of a magnetic storage device presented to a host computer by a standard interface is an abstraction quite different from the reality of the actual mechanisms providing the means for storing data. The device's presentation via the interface is that of a linear and contiguous array of data blocks that are trivially read from and stored in a defect-free space of logical blocks. This image belies the elaborate choreography of signal processing, coding, data structure, control systems, and digital electronics technologies that is exercised with every block of data moved to and from the magnetic media. The function of a hard drive controller is to implement and coordinate the operation of these disparate technologies so as to map the behavior of the physical storage device into the abstract storage model defined by the drive's interface to the host computer. The chapter on hard drive controllers describes some of the architectures implemented by current hard drive controllers to serve this function.

The *fourth* section is concerned with modulation and error control coding for read channels. It starts with an introduction of modulation coding techniques. The first class of constraints discussed is the runlength constraint. It is the most common constraint in hard drives and is imposed to bound the minimal or maximal lengths of consecutive like channel symbols in order to improve timing recovery, reduce intersymbol interference in channels with excess bandwidth and reduce transition noise in nonlinear media. We also discuss other important classes of codes: namely maximum transition run (MTR) coding and spectrum shaping codes.

Maximum transition run constraint, which limits the number of consecutive transitions, improves minimum distance properties of recorded sequences for a variety of channel responses applicable to recording systems. The channel is characterized by types of error events and their occurrence probability, and pairs of coded sequences that produce these errors are determined. This ambiguity is resolved by simply enforcing a constraint in the encoder, which prevents one or both of the coded sequences.

The first type of spectrum shaping codes considered here is codes with higher order spectral zero at zero frequency. Another class of spectrum shaping codes was invented to support the use of frequency multiplexing technique for track following. Both techniques require the existence of spectral nulls at nonzero frequencies. The third class of spectrum shaping codes is those that give rise to spectral lines. Their purpose is to give the reference information to the head positioning servo system.

We continue with modulation codes with error correcting capability and convolutional codes for PR channels designed to increase the minimum Euclidean distance. The section continues with an overview of the research in new classes of modulation codes and detection techniques: capacity approaching codes for partial response channels, coding and detection for multitrack systems and two-dimensional PR equalization and error control.

The *fifth* section gives an in depth treatment of the signal processing techniques for read channels. In PR channels a clock is used to sample the analog waveform to provide discrete samples to symbol-by-symbol and sequence (Viterbi) detectors. Improper synchronization of these discrete samples with respect to those expected by the detectors for a given partial response will degrade the eventual bit error rate (BER) of the system. The goal of adaptive timing recovery is to produce samples for a sequence detector that are at the desired sampling instances for the partial response being used. We review the basics of timing recovery as well as commonly used algorithms for timing recovery in magnetic recording channels, and we introduce a novel technique called interpolated timing recovery. We also introduce adaptive equalization architectures for partial response channels.

Data in a disk drive is stored in concentric tracks, and when a given track needs to be accessed, the head assembly moves the read/write head to the appropriate radial location. This positioning of the head on top of a given track and maintaining a proper position is achieved by the use of a feedback servo system. The servo system analysis includes a chapter on head position estimation and a chapter on servo signal processing. The first chapter reviews methods to estimate the head position, which is uniquely determined from the radial and angular position of the head with respect to a disk surface.

The second chapter gives an overview of servo channel signal processing using examples of techniques employed in practical servo channels. It reviews how detectors for PR channels can be used to improve the detection of the servo data with time varying Viterbi detector matched to a servo code.

The first article on data detection gives the basic detection principles and techniques. The second one describes signal dependent detectors. It introduces the concepts and tools necessary for performing optimal detection for ISI channels when the noise is not AWGN and signal dependent. We describe the techniques to design both hard and soft decision detectors, in particular the so-called K-step, noise prediction and signal dependent noise prediction detectors.

We end this section with an overview of traditional architectures for signal processing in magnetic read-write channels. Today's dominant architecture, where the majority of signal processing is performed in the digital domain is analyzed in detail. The main challenges and alternatives for implementation of major building blocks are discussed. As the previous chapters had covered the theoretical aspects of operation of these blocks in considerable detail, most of the discussion in this chapter focuses on architectures and techniques that are used in practical read channels. The techniques for implementing iterative decoders as possible future detectors are presented at the end of this chapter.

We conclude this book by the review of new trends in coding, namely iterative decoding, given in the *sixth* section. Iterative coding techniques that improve the reliability of input-constrained ISI channels have recently driven considerable attention in magnetic recording applications.

It has been shown that randomly selected codes of very large block lengths can achieve channel capacity. One way of obtaining a large block length code is concatenating two simple codes so that the encoding and the decoding of the overall code are less complex. Turbo codes represent a way of concatenating two simple codes to obtain codes that achieve the channel capacity. In turbo coding, two systematic recursive constituent convolutional encoders are concatenated in parallel via a long interleaver. For decoding, a practical suboptimal iterative decoding algorithm is employed. The first chapter in this section describes the turbo coding principle in detail.

Drawing inspiration from the success of turbo codes, several authors have considered iterative decoding architectures for coding schemes combining concatenation of outer block, convolutional or turbo encoder with a rate one code representing the channel. Such an architecture is equivalent to a serial concatenation of codes with the inner code being the ISI channel. The decoding of such concatenated codes is facilitated using the concept of codes on graphs.

We continue with an introduction to low-density parity check (LDPC) codes, and describe single-parity check turbo product codes and well-structured LDPC codes. We describe several classes of combinatorially constructed LDPC codes along with their performance PR channels. Due to their mathematical structure, and unlike random codes, these LDPC codes can lend themselves to very low complexity implementations. We describe constructions of regular Gallager LDPC codes based on combinatorial designs, finite geometries and finite lattices.

Two constructions of single-parity check (SPC) codes are considered, one in the form of turbo codes where two SPC branches are concatenated in parallel using a random interleaver, and the other in the form of product codes where multiple SPC codewords are arranged row-wise and column-wise in a two-dimensional array. Despite their small minimum distances, concatenated SPC codes, when combined with a precoded PR channel, possess good distance spectra.

The last chapter introduces turbo coding for multi-track recording channels. It describes a modified *maximum a posteriori* (MAP) detector for the multi-track systems with deterministic or random ITI.

The turbo equalization and the iterative decoding are performed by exchanging the soft information between the channel MAP detector and the outer soft-input soft-output decoder that corresponds to the outer encoder. The resulting system performance is very close to the information theoretical limits.

We would like to end this preface by acknowledging the excellent work done by the contributors. Also it is a pleasure to acknowledge the financial support from the National Science Foundation (Grant CCR 020859) and the continuous support from the Information Storage Industry Consortium and Seagate Technologies.

Bane Vasic
Erozan M. Kurtas

Contributor Listing

Viswanath Annampedu
Agere Systems
Storage Products
Allentown, PA

Pervez Aziz
Agere Systems
Storage Products
Dallas, TX

Dragana Bajic
Faculty of Technical Sciences
University of Novi Sad
Serbia and Montenegro

John R. Barry
School of Electrical and
    Computer Engineering
Georgia Institute of
    Technology
Atlanta, GA

Mario Blaum
Hitachi Global Storage
    Technologies
San Jose, CA

Eric D. Boerner
Seagate Technology
Pittsburgh, PA

Barrett J. Brickner
Bermai, Inc.
Minnetonka, MN

Bruce Buch
Maxtor Corporation
Shrewsbury, MA

William A. Challener
Seagate Technology
Pittsburgh, PA

Roy W. Chantrell
Seagate Technology
Pittsburgh, PA

Willem A. Clarke
Department of Electrical and
    Electronic Engineering
Rand Afrikaans University
Auckland Park, South Africa

Stojan Denic
School of Information
    Technology and Engineering
University of Ottawa
Ottawa, ON

Miroslav Despotović
Department of Electrical
    Engineering and
    Computer Science
Faculty of Engineering
University of Novi Sad,
Serbia and Montenegro

Dusan Drajic
Faculty of Electrical
    Engineering
University of Belgrade
Serbia and Montenegro

Tolga M. Duman
Department of Electrical
    Engineering
Arizona State University
Tempe, AZ

Mehmet Fatih Erden
Seagate Technology
Pittsburgh, PA

John L. Fan
Flarion Technologies
Bedminster, NJ

Hendrik C. Ferreira
Department of Electrical and
    Electronic Engineering
Rand Afrikaans University
Auckland Park, South Africa

Kiyoshi Fukahori
TDK Semiconductor
    Corporation
Mountain View, CA

Roy W. Gustafson
Seagate Technology
Pittsburgh, PA

Mark A. Herro
Department of Electrical and
   Electronic Engineering
University College
Dublin, Ireland

Kees A. Schouhamer Immink
University of Essen
Essen, Germany
Turing Machines Inc.
Rotterdam, Netherlands

Aleksandar Kavčić
Division of Engineering and
   Applied Sciences
Harvard University
Boston, MA

Mustafa Kaynak
Department of Electrical
   Engineering
Arizona State University
Tempe, AZ

Brian M. King
Electrical and Computer
   Engineering Department
Worcester Polytechnic Institute
Worcester, MA

Piya Kovintavewat
School of Electrical and
   Computer Engineering
Georgia Institute of
   Technology
Atlanta, GA

Erozan M. Kurtas
Seagate Technology
Pittsburgh, PA

Alexander Kuznetsov
Seagate Technology
Pittsburgh, PA

Michael Leung
SolarFlare Communications
Irvine, CA

Jing Li
Department of Electrical and
   Computer Engineering
Lehigh University
Bethlehem, PA

Michael Link
Seagate Technology
Pittsburgh, PA

Xiao Ma
Department of Electronic
   Engineering
City University of Hong Kong
Kowloon, Hong Kong

Brian Marcus
Department of Mathematics
University of British Columbia
Vancouver, BC

Terry W. McDaniel
Seagate Technology
Pittsburgh, PA

Olgica Milenkovic
Electrical and Computer
   Engineering Department
University of Colorado
Boulder, CO

Jaekyun Moon
Department of Electrical and
   Computer Engineering
University of Minnesota
Minneapolis, MN

Krishna R. Narayanan
Texas A&M University
College Park, TX

Mark A. Neifeld
Electrical and Computer
   Engineering Department
University of Arizona
Tucson, AZ 85721

Borivoje Nikolic
Department of Electrical
   Engineering and Computer
   Sciences
University of California
Berkeley, CA

Travis Oenning
IBM Corporation,
Rochester, MI

Dean Palmer
Seagate Technology
Minneapolis, MN

Jongseung Park
Seagate Technology
Pittsburgh, PA

Ara Patapoutian
Maxtor Corporation
Shrewsbury, MA

John Proakis
Department of Electrical and
   Computer Engineering
University of California
   San Diego
San Diego, CA

William Radich
Seagate Technology
Pittsburgh, PA

Thomas A. Roscamp
Seagate Technology
Pittsburgh, PA

Robert E. Rottmayer
Seagate Technology
Pittsburgh, PA

William E. Ryan
Department of Electrical and
   Computer Engineering
University of Arizona
Tucson, AZ

Sundararajan Sankaranarayanan
Department of Electrical and
  Computer Engineering
University of Arizona
Tucson, AZ

Necip Sayiner
Agere Systems
Allentown, PA

Vojin Šenk
Department of Electrical
  Engineering and
  Computer Science
Faculty of Engineering
University of Novi Sad
Novi Sad, Yugoslavia

Emina Soljanin
Bell Laboratories
Lucent Technologies
Murray Hill, NJ

Bartolomeu F. Uchôa-Filho
Department of Electrical
  Engineering
Federal University of Santa
  Catarina
Florianopolis, Brazil

Nedeljko Varnica
Division of Engineering and
  Applied Sciences
Harvard University
Boston, MA

Bane Vasic
Department of Electrical and
  Computer Engineering
Department of Mathematics
University of Arizona
Tucson, AZ

Shaohua Yang
Division of Engineering and
  Applied Sciences
Harvard University
Boston, MA

Xueshi Yang
Seagate Technology
Pittsburgh, PA

Engling Yeo
Department of Electrical
  Engineering and Computer
  Sciences
University of California
Berkeley, CA

Zheng Zhang
Department of Electrical
  Engineering
Arizona State University
Tempe, AZ

Hong J. Zhou
Seagate Technology
Pittsburgh, PA

Contents

Section I: Recording Systems

1   A Brief History of Magnetic Storage
Dean Palmer 1-1

2   Physics of Longitudinal and Perpendicular Recording
Hong Zhou, Tom Roscamp, Roy Gustafson, Eric Boernern, and Roy Chantrell 2-1

3   The Physics of Optical Recording
William A. Challener and Terry W. McDaniel 3-1

4   Head Design Techniques for Recording Devices
Robert E. Rottmayer 4-1

Section II: Communication and Information Theory of Magnetic Recording Channels

5   Modeling the Recording Channel
Jaekyun Moon 5-1

6   Signal and Noise Generation for Magnetic Recording Channel Simulations
Xueshi Yang and Erozan M. Kurtas 6-1

7   Statistical Analysis of Digital Signals and Systems
Dragana Bajic and Dusan Drajic 7-1

8   Partial Response Equalization with Application to High Density
Magnetic Recording Channels
John G. Proakis 8-1

9   An Introduction to Error-Correcting Codes
Mario Blaum 9-1

10 Message-Passing Algorithm
    Sundararajan Sankaranarayanan and Bane Vasic **10**-1

11 Modulation Codes for Storage Systems
    Brian Marcus and Emina Soljanin **11**-1

12 Information Theory of Magnetic Recording Channels
    Zheng Zhang, Tolga M. Duman, and Erozan M. Kurtas **12**-1

13 Capacity of Partial Response Channels
    Shaohua Yang and Aleksandar Kavčić **13**-1

Section III: Introduction to Read Channels

14 Recording Physics and Organization of Data on a Disk
    Bane Vasic, Miroslav Despotović, and Vojin Šenk **14**-1

15 Read Channels for Hard Drives
    Bane Vasic, Pervez M. Aziz, and Necip Sayiner **15**-1

16 An Overview of Hard Drive Controller Functionality
    Bruce Buch **16**-1

Section IV: Coding for Read Channels

17 Runlength Limited Sequences
    Kees A. Schouhamer Immink **17**-1

18 Maximum Transition Run Coding
    Barrett J. Brickner **18**-1

19 Spectrum Shaping Codes
    Stojan Denic and Bane Vasic **19**-1

20 Introduction to Constrained Binary Codes with Error Correction Capability
    Hendrik C. Ferreira and Willem A. Clarke **20**-1

21 Constrained Coding and Error-Control Coding
    John L. Fan **21**-1

22 Convolutional Codes for Partial-Response Channels
    *Bartolomeu F. Uchôa-Filho, Mark A. Herro, Miroslav Despotović,
    and Vojin Šenk* **22**-1

23 Capacity-Approaching Codes for Partial Response Channels
    Nedeljko Varnica, Xiao Ma, and Aleksandar Kavčić **23**-1

24 Coding and Detection for Multitrack Systems
Bane Vasic and Olgica Milenkovic 24-1

25 Two-Dimensional Data Detection and Error Control
Brian M. King and Mark A. Neifeld 25-1

Section V: Signal Processing for Read Channels

26 Adaptive Timing Recovery for Partial Response Channels
Pervez M. Aziz and Viswanath Annampedu 26-1

27 Interpolated Timing Recovery
Piya Kovintavewat, John R. Barry, M. Fatih Erden, and Erozan M. Kurtas 27-1

28 Adaptive Equalization Architectures for Partial Response Channels
Pervez M. Aziz 28-1

29 Head Position Estimation
Ara Patapoutian 29-1

30 Servo Signal Processing
Pervez M. Aziz and Viswanath Annampedu 30-1

31 Evaluation of Thermal Asperity in Magnetic Recording
M. Fatih Erden, Erozan M. Kurtas, and Michael J. Link 31-1

32 Data Detection
Miroslav Despotović and Vojin Šenk 32-1

33 Detection Methods for Data-dependent Noise in Storage Channels
*Erozan M. Kurtas, Jongseung Park, Xueshi Yang, William Radich,
and Aleksandar Kavčić* 33-1

34 Read/Write Channel Implementation
Borivoje Nikolić, Michael Leung, Engling Yeo, and Kiyoshi Fukahori 34-1

Section VI: Iterative Decoding

35 Turbo Codes
Mustafa N. Kaynak, Tolga M. Duman, and Erozan M. Kurtas 35-1

36 An Introduction to LDPC Codes
William E. Ryan 36-1

37 Concatenated Single-Parity Check Codes for High-Density
Digital Recording Systems
Jing Li, Krishna R. Narayanan, Erozan M. Kurtas, and Travis R. Oenning 37-1

38 Structured Low-Density Parity-Check Codes
Bane Vasic, Erozan M. Kurtas, Alexander Kuznetsov, and Olgica Milenkovic 38-1

39 Turbo Coding for Multitrack Recording Channels
Zheng Zhang, Tolga M. Duman, and Erozan M. Kurtas 39-1

Index I-1

I

Recording Systems

1 **A Brief History of Magnetic Storage**   *Dean Palmer* **1**-1
Introduction • The Early Days of Magnetic Recording • Tape Drives • Disk
Drives • Acknowledgements

2 **Physics of Longitudinal and Perpendicular Recording**   *Hong Zhou,*
Tom Roscamp, Roy Gustafson, Eric Boerner, and Roy Chantrell **2**-1
Introduction • Transition Parameter of an Isolated Transition • Playback Signal from the
Reciprocity Principle • Playback Media Noise Spectra • SNR Calculations • Nonlinear
Transition Shift • Summary

3 **The Physics of Optical Recording**   *William A. Challener and Terry W. McDaniel* ... **3**-1
Introduction • Optical Recording Media • Optical Recording Systems • Conclusion

4 **Head Design Techniques for Recording Devices**   *Robert E. Rottmayer* **4**-1
Introduction • History of Magnetic Recording Transducers • Air Bearings and Head to
Media Spacing • Write Heads • Longitudinal Write Heads • Perpendicular Heads • CIP
Spin Valve Sensors • CPP Sensor Designs • The Future of Read/Write Heads

1

A Brief History of Magnetic Storage

Dean Palmer

Seagate Technology
Minneapolis, MN

1.1 Introduction **1-1**
1.2 The Early Days of Magnetic Recording **1-1**
1.3 Tape Drives **1-5**
1.4 Disk Drives **1-9**
1.5 Acknowledgements **1-15**

1.1 Introduction

With the advent of the digital computer in the 1950s, there arose a parallel, seemingly insatiable need for digital storage and memory. Early computer architects turned to existing forms of storage such as punched Hollerith cards derived from loom machinery and magnetic recording developed by the audio recording industry. Other forms were quickly devised: paper tape, magnetic core, then semiconductor memory and optical recording. These technologies evolved into the current hierarchy of computer storage: fast semiconductor memory in and attached to the processor, magnetic disks providing quick access to volumes of data, optical and magnetic disks as input/output devices, and magnetic tape for back-up and archival purposes. But in terms of sheer volume of data stored and manipulated on a continuous basis, it is the magnetic media that provide the foundation of this storage pyramid.

Data stored on magnetic recording devices are found everywhere. We carry magnetic data on our credit cards and transportation tickets, we use magnetic recording devices to store our computer programs, e-mails, Internet transactions, and personal photos, and we find more and more of our entertainment in a variety of magnetic recording formats. Having recently celebrated its 100th anniversary, magnetic recording could be considered a very mature technology, except for the rapid pace of innovation that continues today. And the trends in magnetic storage continue to be toward smaller, faster, cheaper and denser devices, with no sure end in sight.

The following brief history of digital storage is necessarily incomplete. For an excellent history of magnetic recording of all types, the reader is encouraged to read *Magnetic Recording: The First 100 Years*, edited by E.D. Daniel, C.D. Mee and M.H. Clark [1] and published on the 100th anniversary of this venerable technology. In this chapter we will examine only magnetic storage and of that, only tape and hard disk drives. These two technologies adequately illustrate the development of signal processing techniques and the various innovations that were devised to solve the problems raised in the base technologies.

1.2 The Early Days of Magnetic Recording

An inspired Oberlin Smith published the first known drawings in 1878, 20 years later Valdemar Poulson made it work, and Kurt Stille made a commercial success of it in the 1920s. Thus began 105 years of magnetic recording development in a business that bloomed into a multibillion dollar industry. This fledgling business was to become the backbone of the modern entertainment, computer, and communication industries. For the first 60 years magnetic recording developed at a slow but steady pace that was accelerated by several events along the way: the discoveries of electronic amplification and AC bias, the technology boost caused by World War II, and the advent of the mainframe and the personal computers.

Magnetic recording incubated in a time of great technological innovation. Alexander Graham Bell had uttered his famous request into the prototype telephone receiver and introduced his telephone to the world in 1876. Thomas Edison had set up his laboratory and was developing a host of new products, including the universal stock ticker, the carbon telephone transmitter, and the automatic telegraph system. In 1877, Edison patented the device that was to have a long-lived effect on the entertainment industry, his cylinder phonograph. This early phonograph consisted of a cylinder covered with metal foil to record sound vibrations concentrated by a trumpet, and on playback it converted those wavy tracks back into sound vibrations. The basic technology evolved over a 100 year period with new media, sound transducers and electronic assistance, until magnetic and optical recording finally relegated it to the museum.

In 1878 a visitor to Edison's laboratory took a look at the phonograph, and like the true engineer that he was, decided to improve upon the mechanical approach to sound reproduction. The visitor was Oberlin Smith. A successful businessman, Smith was concerned about the noisy signal that hampered the early phonograph. He wanted to avoid the mechanical sources of noise and distortion and find a new medium and method for recording altogether. Magnetism and electromagnetism had found their way into new transmission devices, the telegraph in 1835 and the telephone 41 years later; could these same techniques be applied to the storage of sound?

The answer, of course, was a resounding yes, but the path to commercial success for magnetic recording was circuitous. Oberlin Smith reasoned that a coil of wire could magnetize a magnetic medium such as steel wire, so that the magnetic pattern along the wire would mimic the sound vibrations of the spoken word. The recorded pattern could be played back using the same coil passing over the wire to induce a voltage in the coil, which would, in turn, create sound in the speaker. In 1878 Smith published a memorandum outlining his idea for magnetic recording, and in 1888 he discussed magnetic recording alternatives in *The Electrical World* [2]. Unfortunately, he was never able to commercialize his groundbreaking idea; indeed, it is not even known if he succeeded in magnetizing sound waves in the laboratory. (See Figure 1.1.)

It was left to Valdemar Poulsen, a Danish engineer, to produce the first working magnetic recording device, which he did in 1898. In his early years Poulson worked as a telephone engineer and was familiar with

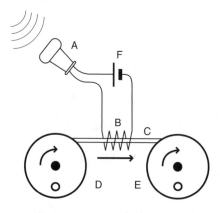

FIGURE 1.1 The magnetic recording apparatus described by Oberlin Smith in the journal *Electrical World* in 1888. The wire, or impregnated string, would be transported through the coil for both the write and read back process. He also mentioned the possibility of using an electromagnet on the outside of the wire. Here it is shown in record mode: A is the microphone, B is the recording coil, C is the steel wire or other recording media, D and E are the transport reels, and F is a battery. (Adapted from Smith, O., *Electrical World*, 116, Sept., 1888.)

telegraph techniques. His motivation for building a sound recorder was to allow people to leave messages on the telephone, a forerunner of the telephone answering machine. His laboratory prototype was a steel wire stretched between two pegs on which he recorded the sound patterns with an electromagnetic recording head containing a magnetic core. Speaking into the microphone produced electrical waveforms that were transformed into magnetic patterns on the wire as he moved the recording head along the wire. On playback, the recorded magnetization induced a voltage in the head, which was converted to sound in a telephone earpiece [3]. Poulsen applied for his first patent for the more practical design of his magnetic recorder in 1898; other patents followed [4]. Instead of a single stretched steel wire, his medium was a length of wire wound in a spiral fashion around a cylinder. In some ways it resembled the magnetic drum storage device, developed much later in the 1950s, which began with a magnetic medium of iron oxide particles coated onto the drum.

But Poulsen's Telegraphone was not destined to be a commercial success. Poulsen joined forces with Danish businessmen to form a company, eventually partnering with a German firm that helped to build a prototype that was the star of the 1900 Paris Exhibition. The Telegraphone attracted a lot of attention, but a commercial version was slow in coming. Poulsen moved on to other things, and the Telegraphone was stalled.

Various American investor groups tried to commercialize the Telegraphone idea, but none was successful. The technology was still immature, and in spite of the discovery of DC-bias recording, more innovations and refinements were necessary to make magnetic recording viable.

By the 1920s a number of commercially successful ventures in Germany, England and the United States began to manufacture magnetic recording equipment for dictation, motion picture sound, radio broadcasting and other uses. Kurt Stille, a German inventor, was the first to enjoy commercial success. Technical innovations such as electronic amplifiers and cassettes for the magnetic media helped to improve the quality of the recorders and make them more convenient. In 1938 three Japanese engineers discovered the benefits of AC-bias recording in reducing the noise and signal distortion present with DC-bias recording. The input signal was added to a high-frequency tone, which resulted in a purer magnetic copy of the sound in the medium. The medium used in all of these devices was steel wire or better yet steel tape, which would not twist like wire and so reduce the sound quality. In England the steel tape machines developed by Ludwig Blattner were adapted for use by the British Broadcasting Company for radio broadcasting. The enormous reels of steel tape weighed nearly 80 pounds, and changing one was a two-person job.

In 1928 Fritz Pfleumer, an Austrian inventor, developed a new type of recording medium that would revolutionize the recording process. Pfleumer pioneered a process for coating paper with thin layers of metal powder and applied this process to the manufacture of magnetic tape. Particulate iron was the first magnetic material used on a paper base. The German company AEG, in partnership with BASF, produced a number of media innovations: cellulose acetate as the tape base and magnetic coatings of carbonyl iron, then magnetite, and finally in 1939, the γ-iron oxide that would still be in use decades later. AEG, for its part, introduced a new tape recorder in 1935, the Magnetophon, which incorporated these media developments, a modern ring recording head, and eventually AC-bias recording. The result was a recorder with significantly improved performance that would launch the modern tape recording industry. (See Figure 1.2.)

In the United States, there were several manufacturers with an interest in the magnetic recording business, notably the Armour Research Foundation and the Brush Development Company. Brush partnered with the Battelle Memorial Institute to develop better magnetic coatings, an effort that resulted in magnetite particles with much better uniformity and performance. In 1944 Brush requested the help of several companies to coat tapes with this new material, including a Minnesota sandpaper manufacturer. Over the next few years the Minnesota Mining and Manufacturing Company provided to Brush samples of the Battelle particles coated on paper tape. In 1947, 3M introduced its own first audio tape, Scotch™ 100, which was a paper tape coated with acicular magnetite particles. A year later 3M brought out its best-selling Scotch™ 111 audio tape, which used a cellulose-acetate backing coated with acicular γ-iron oxide, the same tape that would be used in the first IBM digital tape drive in 1953 [5].

The rise of the Ampex Corporation is an example of the close link between the early magnetic recording industry and the entertainment industries. In 1946 the founder of Ampex, Alexander M. Poniatoff, was

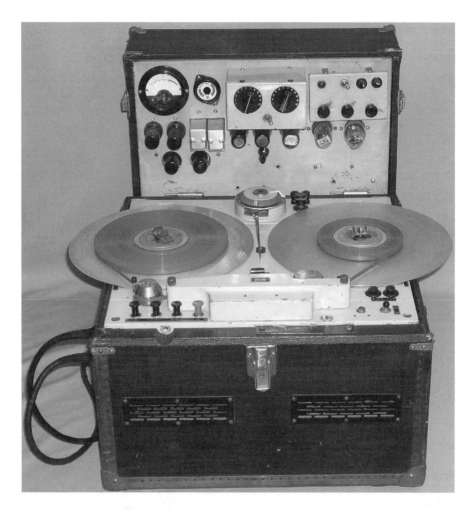

FIGURE 1.2   An original Magnetophon built by AEG and one of two brought back to the U.S. by Jack Mullin after World War II. The Magnetophon was instrumental in the renaissance of the American audio recording industry. (From the Jack Mullin Collection at the Pavek Museum of Broadcasting, St. Louis Park, MN. With permission.)

searching for a new product line for his company to replace his wartime contracts [6]. He came into contact with Jack Mullin, who gave him a demonstration of one of the original Magnetophons. Impressed, Poniatoff assembled a development group and immediately began work on a professional model to serve the radio broadcast and music recording industry. After much development work on the ring recording heads, the company demonstrated the Model 200 audio recorder to a group of broadcasters in Hollywood, using 3M's Type 111 audio tape.

Bing Crosby became associated with Ampex because of his aversion to performing live on his weekly Philco Radio Show. He preferred to record his programs in advance but was dissatisfied with the quality of sound recording available at the time. Because this problem threatened to end his broadcast career, Crosby's producers called together several manufacturers to demonstrate their newest recording products. When Bing Crosby heard the improved sound coming from the cabinet-sized Ampex Model 200, the search was over, and Ampex had its big break into the audio business. Ampex was to build 20 units at $4000 apiece for his company, Bing Crosby Enterprises, and Crosby wrote a personal check for the down payment, which was a boon to the cash-strapped company. A total of 60 units were built for Bing Crosby Enterprises, which acted as a distributor in Hollywood and eventually sold hundreds of Ampex recorders. (See Figure 1.3.)

FIGURE 1.3   The Ampex Model 200 professional audio tape recorder launched the Ampex Corporation in a new direction and filled a need for high quality sound reproduction for the entertainment industry. Bing Crosby used the Model 200 recorder to tape his Philco Radio Hour. (From the Jack Mullin Collection at the Pavek Museum of Broadcasting, St. Louis Park, MN. With permission.)

Ampex followed with improved models of the professional audio series and even got into the consumer market. In the late 1940s and early 1950s, consumer interest was growing and a large number of companies in the United States, Europe, and Japan hurried to satisfy the demand for home tape recorders. The postwar period was a time for expansion of the industry, refinement of the technology and development of new formats and uses for magnetic recording. What started as reel-to-reel tape recording would evolve into cartridges, cassettes and mini-cassettes on the audio side, and magnetic recording would eventually penetrate into video recording. Already more than 50 years old in the 1950s, the magnetic recording industry had finally come of age.

1.3   Tape Drives

In the early 1950s the fast-growing computer sector looked to magnetic recording tape, so successfully developed for the entertainment industry, in order to provide memory and data storage and to allow buffering for the relatively slow input and output devices of the time. Punched card storage was replaced by the much faster tape drives, though the processing was still in batch mode. Tape storage brought the promise of higher transfer rates and faster access to desired blocks of data. But the required improvements in speed could not be met using existing analog audio recorders. Innovation was required in both mechanical tape handling and recording techniques. The end result was a series of high-speed, multi-track, digital tape recorders.

Unlike the analog audio recorders the digital tape devices use a recording technique that imitate the digital nature of the data used in the early computers. The digital information to be used in the computer, in strings of 0's and 1's, was stored on the magnetic tape by reversing a current through the write head at specific intervals. The magnetic field of alternating polarity from the write head resulted in a series of tiny magnets of varying length that were oriented North-pole to North-pole and South-pole to South-pole along the track. This method of recording is termed longitudinal magnetic recording because the magnetization is almost entirely in the plane of the medium.

The information was read back some time later by sensing the passage of the magnetic transitions, that is, the change from one magnet to the next along the track. Unlike the arbitrary analog waveform generated by an audio read back head, the digital read back waveform consists of a series of discrete pulses with alternating polarities. Each of the pulses signifies that a magnetic transition has passed under the read back head.

The user data from the computer or input/output device is fed into the storage device as a series of 1's and 0's. This string is usually modified by a modulation code in order to best match the pattern written on the disk to the characteristics of the recording system. In the simplest embodiment of a recording code, the string of 1's and 0's is written onto the tape by creating a magnetic transition for 1's and no transition for 0's. This simple and most efficient of all codes is the NRZI, or Non-Return to Zero Inverted (or inhibit or IBM) code devised by Byron Phelps of IBM [7].

What system characteristics are the codes designed to assist? One of them is the timing of the read-back waveform. In order for a recorded bit to be identified as a 1-bit, the pulse that marks the passage of the corresponding transition has to fall within a prescribed timing window. This timing window can be created in several ways, including the use of a dedicated clock track. To avoid wasting valuable real estate with dedicated clock tracks, it is preferable to use codes that can provide clocking information in the data tracks. But the NRZI code is not self-clocking because of the possibility of long strings of 0's in the input data that result in long intervals of time between updates to the clocks. Modifications to the NRZI code were soon created that gave extra timing bits to guarantee good clock synchronization, but at the expense of code rate.

Another constraint placed on the recording code was the size of the timing window. All else being equal, a larger window gave more tolerance for peak shifting due to intersymbol interference (ISI) from adjacent pulses and to noise and distortion of the signal. The window size is directly related to the rate of the code, which is the ratio of user bits to encoded bits. All else being equal, the larger the rate, the better. NRZI has a code rate of 1, the highest possible for binary codes.

Another important characteristic of a code is the minimum spacing allowed between adjacent pulses (or transitions along the track) since this is one of the determining factors for intersymbol interference, ISI, between pulses. Again, the largest minimum spacing is the best, all else being equal. Unfortunately, not all these characteristics can be optimized simultaneously, so the progress in code construction for pulse detectors has involved a series of different compromises that seek to optimize the total recording performance.

Several other recording considerations have influenced codes, for example, the build-up of DC magnetism or "charge" can be detrimental to a system that cannot support DC. This constraint can be satisfied with special charge-constrained codes. Unfortunately, the more constraints that are added to a code, the more difficult it is to find a code with a high code rate. Some of these constraints are characterized in the run-length limitation (RLL) formalism. A code can be classified as belonging to an "$r(d, k)$" RRL group where r is the code rate, the d-constraint is the minimum number of encoded 0's between two encoded 1's, and the k-constraint is the maximum. The c-constraint is occasionally specified, which gives the maximum magnetization or "charge" that can be accumulated along the track.

The pulses in the read-back stream, which signify 1's, were initially detected by simple threshold circuitry. The threshold detector was soon succeeded by more advanced peak detectors that examined the pulse characteristics from several angles in order to discriminate true data pulses from the ever-present noise and signal distortions.

The first of the digital tape recorders was the Uniservo I, which was introduced in 1951 by an arm of the Remington Rand Corporation. This initial product used eight tracks recorded on a removable reel of 0.5 in. wide metal tape. A throwback to the early days of magnetic recording, solid metal tape was used to get the mechanical durability needed for the high-speed tape transport system. The weight of the metal tape made it difficult to handle and store. It was also hard to accelerate in the tape handling system and required special measures to prevent tribology issues with the recording heads. But in spite of the unfortunate choice of media, the Uniservo I incorporated many of the digital tape recording features that have been carried forward to the present day.

In 1953 the IBM Corporation shipped the Model 726 tape drive, the first in a series of tape drives that set the standard for digital tape for many years. The most obvious difference was the adoption of plastic based tape as the recording medium, which opened up possibilities but demanded invention to protect the durability of the tape. The biggest challenge was the design of a method to buffer the movement of the tape from the high inertia take-up reels to the fast accelerating drive capstans on either side of the recording head array. The Uniservo I had used a pair of spring-loaded tension pulleys to allow for the different acceleration rates, which worked well for the durable metal tape. For the Model 726 the IBM engineers developed a pair of vacuum columns that extended to nearly the full height of the drive and provided a long, free, but perfectly tensioned loop of tape on either side of the recording head array. This arrangement gave the high inertia tape reels a few moments to catch up with the drive capstans. The two vertical vacuum columns and the reel-to-reel technology were a distinctive feature of IBM tape drives until 1984 [8]. (See Figure 1.4.)

The write and read heads used in the IBM 726 were 7-head arrays preceded by a two-gap erase head designed to completely demagnetize the tape. The tape was provided by 3M, which initially used the Scotch^{TM} 111 audio tape introduced in 1948. This tape was hand selected for low defects and no splices and consisted of an acicular γ-iron oxide particle coating on a 0.5 in. wide cellulose acetate backing from Cellanese [5].

The coding scheme chosen for the Model 726 was the NRZI code, patented in 1956. The NRZI code is an $r = 1 (0, \infty)$ RRL code and has the largest possible timing window. Since this code is not self-clocking,

FIGURE 1.4 The IBM Model 726 was the first digital tape recorder to use twin vacuum columns to buffer the magnetic tape, which set a standard for reel-to-reel recorders lasting many years. (From the IBM Corporation Archives. With permission.)

additional tracks were added to guarantee clocking bits. Initially the tracks were doubled, with one track recorded with normal NRZI and the other with the complement of the normal track to ensure that one of the two tracks contained a 1-bit for clocking. Later the six "character" tracks were joined by an odd parity track, which ensured both clocking information and error-detecting redundancy across the six data tracks.

Other manufacturers joined in the development of the tape drive, and many improvements were made over the next 20 years. Reel sizes went from 12 to 10.5 in., which became the standard for years. In 1953, 3M started offering a more durable tape based on DuPont's Mylar^{TM} film in order to solve the problem of tape breakage [5]. Pulse slimming circuits were introduced to reduce intersymbol interference. The cyclic redundancy check (CRC) was added to the parity checking already in place to give a boost in error detection and limited correction. Mechanical and electronic tape skewing improvements minimized track-to-track timing problems. A two-gapped read/write head made it possible to read-verify while writing, thus improving reliability. Timing in the channel was improved with a shift to an FM code, which introduces a clocking bit between each pair of data bits.

The next major change in tape recording came in 1984, when IBM shipped the 3480 tape device with the $4 \times 5 \times 1$ in. Square Tape Cartridge containing 200 MB of information, followed by the 3480E Cartridge that could hold 400 MB. The previous generation of 10-in. tape reels held only 150 MB. The IBM 3480 was the answer to the pressing problem of where to store the tape; users by then had thousands of tape reels in huge storage racks to keep track of. The small cartridges had a density increase of a factor of 6 due to some remarkable developments in the head, tape and signal processing. The magnetic material on the tape was chromium dioxide, which had a high coercivity and could support a much higher data density.

The 18 track head used the conventional inductive writer scheme, but readback was achieved with an entirely new device. Back in 1971, Robert Hunt of Ampex had invented the magnetoresistive (MR) read head to detect magnetic fields from the media. Using thin films of permalloy he found that the resistance of the film changed by a few percent in response to the magnetic fields found in magnetic recording [9]. A sense current applied through the permalloy film developed a voltage across the stripe and produced a pulse whenever a transition passed by the head. Although the resistance change was a small percentage of the total, the absolute change in voltage could be made large by increasing the sense current up to the limits of the film's reliability.

An advantage of the MR head is that the output is proportional to the magnetic flux and not to the rate of flux change; so the MR head gives a uniform signal independent of tape velocity. The MR head found its first use in a tape device and not in a disk drive because of the relatively wide track widths of the tape system. It was far easier to make the MR head magnetically stable with large track widths; it would be another 8 years before the MR head was introduced in a commercially available disk drive.

The MR head had the disadvantage of limited dynamic signal range, which resulted in a different approach to the signal processing of the IBM 3480 tape drive. A series of coding structures were applied to the channel to get the desired result. First, an $r = 8/9$ (0, 3) RLL code was chosen to assure adequate timing updates. Additional closely spaced transitions were added to break up the longer runs of 0's allowed by the (0, 3) code. These transitions were too crowded to be resolved into separate pulses by the readback system. The deliberately introduced intersymbol interference from the closely spaced transitions serves to equalize the waveform and reduces the dynamic range of the waveform pulses. These codes, together with $1 - D$ equalization on the read side, produce a three-level waveform that is detected using an amplitude-type peak sensor [10, 11].

The error characteristics of the tape drive, long bursts of errors along the track but little correlation from track to track, dictated an error correction strategy based on dedicated parity tracks, called adaptive cross parity, or AXP. Four tracks of the 18 were used for parity checks across the width or at diagonals of the tape. Up to three long burst errors could be corrected if the error bursts were sufficiently separated.

In the same year that the IBM 3480 was introduced, Ampex announced an innovation that would improve the data densities of their tape recorders and pave the way for density increases in disk drives by allowing more bits per unit bandwidth. The advance came in the form of a partial response maximum likelihood channel (PRML), which had been used already in communication channels. The first use of PRML in storage products was in the Ampex digital cassette recording system (DCRS). [12]

Many other improvements would follow those made in the Model 3480, such as cartridges of various sizes and formats offered by numerous companies; Philips, Ampex, Exabyte, 3M, and Digital Equipment Corporation. For large data operations there were automated tape storage libraries with devices that could seek to racks of cartridges and produce archived data within minutes. Densities and data rates were improved by using track-following servo methods, and reliability was enhanced by using redundancy across a single cartridge or spread across many cartridges.

The niches that tape storage has occupied in recent years are inexpensive back up for small and personal computers and archival storage for large computer installations. As the capacity of hard disk drives continues to grow and the cost per Megabyte continues to drop, there will be increasing pressure on tape storage in both these market segments.

1.4   Disk Drives

The long and successful reign of the magnetic disk drive began in part as an effort to lighten the burden of keypunchers. The data processing industry wanted to streamline the method by which punched card input was prepared for use in accounting machines. If repetitive information were kept in magnetic storage, the operators would only have to add new data and so could reduce data input time. As the development of the first disk drive proceeded, designers realized that random access magnetic storage would make it possible to switch from sequential batch processing of data transactions to a system that would give immediate access to individual transactions. Magnetic tape memories were used at that time to store the batch processing data. This method was cumbersome, and it precluded many desirable applications, such as real-time computing of individual transactions. Applications that we take for granted, such as airline ticketing information and instant bank account balance information, were not possible [13].

These were the motivations for one of the first projects tackled by Reynold Johnson, when he was asked to establish an experimental laboratory for IBM in the early 1950s. What IBM wanted was a new development laboratory to be located in California, in proximity to their innovative customers in the aircraft industry who were located in Los Angeles and Seattle. The city of San Jose was a natural choice because a punched card plant was already established there. In 1952 IBM asked Johnson to start up the laboratory with a small staff of 30 to 50 people and to select projects that emphasized innovation. After completing several projects in support of unique customer applications, the laboratory started focusing on the means to provide nonvolatile storage of repetitive data for keypunch operators and eventually to allow the random access of selected data [14].

Reviewing the possible technologies, the design team decided that magnetic tape did not provide the access times needed for such a device and that magnetic drums had poor volumetric density. They chose rotating magnetic disks in spite of the fact that significant innovation would be required to turn the idea into a product. A magnetic disk drive would be more economical in terms of the surface area available and would allow rapid random access to any piece of data in storage.

Since the recording head would be in constant proximity to the disk over long periods of time, a method was needed for maintaining a precise separation between the two. A constant separation during the lifetime of the drive would avoid wear without excessively degrading the magnetic performance. The engineers achieved this compromise by injecting compressed air into the recording head to create a fluid bearing that maintained an 800 μin. head-to-disk spacing [15]. By contrast, current products are designed to operate with less than 1 μin. of spacing in order to achieve high linear densities.

Although the data density of the first drive was only 2000 bits/square inch, this was 4000 times the data density of an 80-column punched card. In order to achieve an acceptable capacity for the device, the available surface area was maximized by using fifty disks, each 24-in. in diameter. The disks were oriented in a vertical column rotating at 1200 RPM. The head positioner moved a single pair of recording heads to one disk at a time in order to access any of the hundred tracks on each disk surface. As large as a refrigerator and weighing over 500 lb, the total capacity of the drive was only 5 MB, miniscule by

FIGURE 1.5 The IBM 350 Disk Drive was the first commercially available hard drive and was leased as part of the random access method of accounting and control (RAMAC) system. With 50 24-in. disks and weighing in at over 500 lb, it boasted a capacity of 5 MB. (From the Hitachi Global Storage Technologies Archives, San Jose. With permission.)

today's standards but large enough at the time to make a significant difference in the speed at which data processors and computers manipulated data. (See Figure 1.5.)

The magnetic disks used in the drive were made of aluminum coated with particulate γ-iron oxide in an epoxy base. The oxide paint was filtered in a silk stocking and applied to the aluminum substrate by means of a spin process in which the disks were rotated and the coating was poured from a paper cup onto the inner radius of the disk. To assure coating uniformity, an entire tray of paper cups was filled at the same time. This method, with many improvements, was used for particulate media for decades until the advent of thin film media.

Signal processing was the only area where no major innovations were needed or attempted. The NRZI code and amplitude detector combination was borrowed from existing tape drive devices. The amplitude detector operated by registering a 1-bit whenever a magnetic transition passed and the head voltage exceeded a specified level. Timing was achieved with two oscillators, triggered by alternate 1-bits. The NRZI code stream was augmented with odd parity bits, which assured that the oscillators would be resynchronized periodically even if the data were all 0's. The oscillators ran open-loop, powered by vacuum tube circuitry. There was no ECC, no address marking, and no flagging for spare tracks [15]. Life was simple.

In February of 1954, the following words were written and read back from the disk drive: "This has been a day of solid achievement" [14]. Sober and perhaps not so dramatic as the poetic words spoken through the first telephone or tapped on the first telegraph, they are nevertheless significant in that they triggered the avalanche of information — the musings, music, pictures, video, e-mails, drawings and data — that have been written onto disk drives ever since.

The IBM 350 disk drive was announced in 1956 and shipped in 1957 as part of the IBM RAMAC (Random Access Method of Accounting and Control) system. This was the birth of the first commercially available disk drive, and it marked the beginning of a long history of new technologies and applications for what turned out to be a very robust means of providing nonvolatile and inexpensive storage. In the meantime efforts were underway to continue the steady series of improvements occurring throughout the disk drive industry, which led to a staggering increase in storage density of a factor of more than 50 million by 2004.

The next generation of disk drive that was developed in the IBM San Jose Laboratory was a radical departure from the first. In order to put a head on every surface and thus improve the access time, it was necessary to find a different method for suspending the head above the disk surface. The system of air compressors and complicated pressure pistons and bearings used in the IBM 350 did not lend itself to easy scaling. Instead, a new hydrodynamic air bearing was developed which automatically created a pressurized film of air under the recording head by taking advantage of the relative motion of the head and disk to compress the air.

Another radical innovation was the use of perpendicular recording, that is, the magnetization of the disk was oriented perpendicular to the plane of the disk. The medium consisted of an oxidized steel surface laminated onto an aluminum core, but the surface could not be made sufficiently smooth to sustain the densities required [13]. Even though the advanced disk file (ADF) was eventually terminated because of these surface problems, the recording method is significant in that perpendicular recording is currently a hot prospect for replacing longitudinal recording.

Fortunately IBM had pursued an alternate program based on longitudinal recording and the new hydrodynamic air bearing. In 1962 IBM shipped the Model 1301 disk drive with a total capacity of 56 MB. The signal processing elements were much the same as in the IBM 350, an NRZI code and an amplitude detector, but a dedicated clock track was added for timing in the data channel.

There followed a series of technical advances, starting with a line of drives with removable media. The smaller 13 and 14-in. disks were mounted in packs of 10 to 20 disks that could be removed and archived to effectively increase the storage capacity and versatility of the disk drive system. The IBM 1311, introduced in 1963 and followed by the 2314 and 3330, gave small businesses and institutions the benefit of random access memory at a reduced cost. The fixed disk design returned in 1973 in the IBM 3340 drive, which incorporated a new hydrodynamic head design, the tri-rail "Winchester" recording head that was to become an industry standard for years.

The first removable disk drive, the IBM 1311, used the same coding and detection schemes as previous fixed disk drives, but the follow-on IBM 2314 switched to frequency modulation (FM) encoding, which was self-clocking, coupled with a peak detection channel. The peak detector senses the readback pulses that mark the passage of a magnetic transition by differentiating the recording head signal and looking for the presence or absence of a zero-crossing within the timing tolerance or "window."

Both the IBM 3330 and 3340 drives used the modified frequency modulation (MFM) code, also self-clocking, and an improved peak detector, the delta-V detector. This detector added logic in order to distinguish spurious noise pulses from valid signal pulses by measuring both the leading and trailing slopes of the pulse and the pulse amplitude. No extra clock tracks or added transitions were needed with the FM and MFM codes, since the encoded bits ensured adequate clock synchronization within the data stream. A phase locked loop (PLL) was introduced, which tied the timing windows to the data more accurately and permitted higher data rates.

Both the FM and MFM codes are half rate codes, that is, each user bit cell lasted for two clock periods. The FM code, also known as the bi-phase or Manchester code, is identical to the NRZI code with an additional clock bit inserted between each pair of data bits. The additional bits limit the linear density that can be obtained because of the intersymbol interference that is caused by the consecutive, closely spaced clock and data bits. The MFM code resolved this problem by systematically eliminating clock bits that would be immediately adjacent to data 1-bits, leaving all pulses at least two clock periods apart. The move from FM to MFM enabled a doubling of the data density with little additional hardware cost.

A host of technology enhancements fueled the exponential increase in areal density at a steady 30% compound annual growth rate (CAGR) for the first 35 years: ferrite inductive heads, metal-in-gap (MIG) heads and finally thin film inductive heads introduced by IBM in 1979.

A new run-length-limited code, the half rate (2, 7) RLL code, was introduced in the IBM 3370 (1979) and 3380 (1981) disk drives [16] that further increased the distance between adjacent 1-bits to allow higher density data to be recorded. George Jacoby at Sperry Univac developed a similar code, the 3PM code [17]. In combination with the rate of the code, the d-constraint controls the intersymbol interference while the k-constraint limits the maximum time between clocking updates. The FM and MFM codes can be categorized as $r = \frac{1}{2}(0,1)$ and $r = \frac{1}{2}(1,3)$ RLL codes, respectively. The $r = \frac{1}{2}(2,7)$ RLL code allowed for at most eight bit cells between adjacent 1's, which was sufficient to enable the improved PLL's to maintain synchronization [18].

The detector used in the 3370 drive was the delta clipper, an enhanced peak detector, which had more sophisticated qualification circuits to assure that only true peaks were reported. The delta clipper checked for alternating polarity of the pulses, minimum amplitude, correct timing and the proper fall-off of the signal after a peak.

The growing popularity of personal and mid-sized computers in the late 1970s started a movement toward smaller drive form factors, with a variety of disk sizes. In 1979 IBM introduced an 8-inch diameter disk drive, the IBM 3310, with an MFM encoding scheme and a compact rotary actuator for use in mid-sized computers. But the personal computer required an even smaller form factor in order to be compatible with a diskette drive.

In 1980 the newly formed Seagate Technology, headed by Al Shugart, started shipping the ST506 disk drive with 5.25-in. disks. The capacity of the drive was 5 MB, which was the same capacity as the IBM 350 built 23 years before, but had 14 times the capacity of and was many times faster than the 5.25-in. floppy disk drive that it was designed to replace. Many other manufacturers joined in the success enjoyed by the first 5.25-in. disk drives, and in most of them the code of choice was the simple MFM code. In 1983 Rodime introduced the 3.5-in. form factor, and in 1988 Prairietek followed with a 2.5-in. drive, which was needed for the growing notebook computer market. In 1991 Hewlett Packard opted for a 1.3-in. disk in the Kitty Hawk drive, but it was cancelled due to poor demand. Integral shipped a 1.8-in. disk drive in 1992, and IBM introduced the Microdrive in 1998, with a miniature disk the size of a quarter. (See Figure 1.6.)

Thin film media made their debut in the mid-80s and were adopted by small form factor disk drives. They quickly replaced the particulate disk and after many revisions remain the standard technology for all existing drive applications.

The most important advancement in head technology was the switch from inductive readback heads to magnetoresistive (MR) heads. MR heads had the advantage that the signal output was large compared

FIGURE 1.6 In 1998, IBM Corporation Storage Division (now Hitachi Global Storage Technologies) introduced the Microdrive with a 1-in. diameter disk. The initial storage capacity of 170 MB has grown to 4 GB. Small form factor disk drives are finding their way into handheld and portable electronic devices like cameras and palm PC's. (From the Hitachi Global Storage Technologies Archives, San Jose. With permission.)

to inductive readers and the signal was not dependent on velocity, which made the heads ideal for smaller disks with lower linear velocities. The first storage application of MR heads was in the IBM 3480 tape drive because the wider track widths made it easier to resolve some of the inherent difficulties with MR heads. The first use of the MR head in a commercial disk drive was in the Model 0663 Corsair drive developed by IBM Rochester in 1992.

The smaller drives remained loyal to the peak detect channel and to a succession of run-length-limited codes, starting with MFM and progressing to the $r = \frac{1}{2}$ (2, 7) RLL code, which was very popular for an extended time, and finishing with an $r = 2/3$ (1, 7) RLL code. Compared to the (2, 7) RLL code, the (1, 7) RLL code had the disadvantage of a closer physical spacing between 1-bits but the advantage of a larger timing window to counteract the effects of ISI and all noise sources.

However, the shipment of the IBM 0681 in 1990 marked the beginning of the end of the 34-year reign of the peak detection channel. An entirely new method of data detection was introduced, which promised higher data densities and the ability to perform with a significantly reduced signal-to-noise ratio. The ideas of partial response signaling and maximum likelihood detection were not new to signal processors; they had been in use for years in communication devices and found their way into tape recorders. Communication systems used these techniques separately, together, and in combination with others. Only in the disk drive have we locked them together without any new permutations. The IBM 0681 was the first commercially available disk drive that incorporated the partial response maximum likelihood (PRML) channel.

The peak detector is useful only as long as the intersymbol interference is low enough to keep the readback pulses relatively separate. This separation was accomplished from 1966 to 1990 by using improved recording components: lower flying heights, better run-length-limited codes and high-powered equalization methods to slim the individual pulses. But this strategy was becoming less effective as the drive for higher densities continued. The PRML strategy toward ISI differs completely; instead of struggling to eliminate intersymbol interference, the PR channel allows a large degree of interference, but in a controlled manner so that the signal can be unraveled into its original form. To paraphrase the subtitle of the 1960s movie about the bomb, "Dr. Strangelove," PRML enables us to explain "How I learned to stop worrying and love intersymbol interference."

As mentioned before, the first storage product utilizing PRML was a digital cassette tape recording system (DCRS), introduced in 1984 by Ampex for high data rate instrumentation. Subsequently at the 1984 Intermag Conference Roger Wood from Ampex reported on an experimental disc recording system with a PRML channel that achieved 56 Mb/in^2 using a PRML channel [19].

The history of the PRML channel within IBM goes back to the 1970s, when theoretical studies by Kobayashi suggested that the maximum likelihood sampling detector and partial response equalization could be applied to the recording channel [20]. In the mid-70s interest in the application to a disk drive was developing in the IBM Zurich Research Lab under the leadership of Francois Dolivo [21] and in the late-70s the IBM San Jose Research team became involved. A collaboration of the two IBM Research teams led by Dolivo and Tom Howell resulted in the development of PRML channel prototypes in order to investigate the potential and pitfalls of the new detection scheme. This same channel, in a new prototype based on a chip developed by Fritz Wiedmer's team, was later used to demonstrate an areal density of 1 Gb/in^2 using the newly developed MR head coupled with a thin film disk [22].

In the 1980s the San Jose Development Group built their own prototype channel and coined the acronym PRML, but did not fully embrace the advantages of the channel. It was not until a collaboration was formed between teams at IBM Zurich and IBM Rochester that the momentum was begun to put the channel into a product. The collaboration's first accomplishment was the completion of a prototype PRML channel in 1984. An ungainly device the size of a college student's fridge, the discrete channel was used to evaluate the potential of the channel and to begin some of the integration experiments that were necessary to incorporate PRML into a drive. One of the problems to be overcome was the distortion caused by nonlinear transition shift, which was a consequence of the $d = 0$ code used with PR channels [23]. The flux density on the disk was considerably higher than it had been with peak detection codes, and closely spaced transitions were interfering magnetically with each other. The solution to this signal distortion was write precompensation, taking advantage of the deterministic nature of the transition shift. In order to

FIGURE 1.7 The shipment of the IBM 0681 Redwing in 1990 marked the first use of the PRML channel in a commercially available disk drive. The channel had both analog and digital equalization sections coupled with a digital detector. (From the Hitachi Global Storage Technologies Archives, San Jose. With permission.)

write the transitions at the proper physical position, the timing of closely spaced transitions was altered so that the transitions ended up in the correct position along the track [19].

Availability of a new mixed mode IC technology, which promised a low cost, high-performance implementation, made possible the development of single-chip PRML channels. The earliest implementations used an 8/9-rate (0, 4/4) code. This first-generation channel had both analog and digital equalization sections and a digital detector; the three-tap digital section allowed adjustments to account for head-disk variations and radius effects [24]. When coupled with ferrite MIG inductive heads and 5.25 in. thin film disks, the IBM 0681 Redwing disk drive provided 1 GB of unformatted data. The Quantum Corporation shipped its first PRML channel with thin film inductive heads; MR heads were not required for PRML operation. However, the advent of MR heads made the use of PRML easier and more advantageous [26]. (See Figure 1.7.)

The initial PRML drive was followed by an all-digital implementation of the equalizer, dubbed the PRDF channel, for "digital filter." This channel was used in the IBM 0664 2-GB 3.5 in. disk drive. The all-digital implementation gave much tighter control over the equalization, and permitted many auxiliary functions such as fly height measurement and equalization optimization. Quantum used a digital implementation to ship a drive with an adaptive digital equalizer in 1994. At the beginning of the 1990's only a few IBM drives had the PRML channel, and the rest of the drive industry was using peak detection. By the end of the decade the entire industry had switched to some form of partial response channel.

There are several useful classes of partial response equalization, each of which defines the desired time and frequency response of the pulse [27]. The frequency response of the early Class IV PRML implementation is symmetric with nulls at both zero and the Nyquist frequency. The Class IV response was chosen initially because the spectral response matched that of the recorded signal rather well. But there were other partial response equalization targets that matched even better. These targets minimized the high frequency noise boost. A scheme known as extended partial response (EPRML) equalization shifts the response maximum to lower frequencies and results in less noise enhancement; this was used in the mid-90s by Quantum, Seagate and others with and without postprocessing. EEPRML equalization, or E^2PRML, aligns the equalization with the response of the recording components even more closely, but it did not enjoy the same popularity as earlier versions. Instead, several drive and component companies developed proprietary equalization targets and incorporated them into channel IC's.

Since those early days the targets for equalization have become more sophisticated, using generalized PR polynomials to better reduce noise enhancement at the equalizer output. The detectors have become more complex and contain many more decision-making branches to match the complexity of the equalization targets. Postprocessors can be added after the Viterbi detectors to reduce the probability of certain errors even further and/or to minimize the complexity of the Viterbi detector [28]. For example, a Class IV partial response system with a postprocessor can be made equivalent to an EPRIV system in performance but with reduced complexity. Noise predictive maximum likelihood (NPML) detectors were introduced in the late-90s, which use noise whitening filters and more sophisticated Viterbi detectors to wring more advantage out of the signal processing [29]. Both single-bit and two-bit parity codes were added to the channel to assist the postprocessor and improve error rates in the face of high media noise. And finally, efforts are currently underway to further reduce the effects of the transition noise by including signal-dependent information in the detector.

These and other signal processing enhancements and the use of the MR head and improved thin film disks essentially doubled the rate of increase in the areal storage density from 30% CAGR for the first 35 years of magnetic recording to 60% CAGR after 1992. This rate of increase rose to 100% CAGR in the late 1990s and peaked at greater than 130% CAGR at the turn of the century, primarily due to the introduction of the giant magnetoresistive (GMR) head. But there are signs that the areal density growth rate is returning to historic trends. The slowdown in growth rate for longitudinal recording is largely a consequence of the onset of physical effects in the recording medium. As the recording density grows, the media magnetic features, or grains, must shrink to maintain the signal-to-noise ratio. However this reduction cannot continue forever, because the magnetic grains are reaching the superparamagnetic limit. If the grains become superparamagnetic, the recorded data will spontaneously erase after a short period of time. Efforts to work around this limit have so far proven successful. But the industry, with its insatiable appetite for density growth, may have to settle for a slower, more sustainable pace. New types of media and new recording techniques, such as perpendicular recording, patterned media, and heat-assisted writing are under development to keep the areal density growing and satisfy the demands of the ever-expanding storage industry.

1.5   Acknowledgements

The author would like to thank the following for recounting their experiences in the development of signal processing as applied to magnetic recording: Thomas Howell, Roger Wood, Bill Abbott, Bill Radich, Pablo Ziperovitch, John Mallinson, and Jim Weispfenning. The author is grateful to the Ampex Corporation, Hitachi Global Storage Technologies, IBM Corporation, Imation Corporation, Magnetic Disk Heritage Center, 3M Company, Pavek Museum of Broadcasting, and Seagate Technology for information and materials relating to the development of magnetic recording.

References

[1] Daniel, E.D., Mee, C.D., and Clark, M.H., *Magnetic Recording: The First 100 Years*, IEEE Press, Piscataway, 1999.

[2] Smith, O., Some possible forms of phonograph, *The Electrical World*, 12, 116, 1888.

[3] Camras, M., *Magnetic Recording Handbook*, Van Nostrand Reinhold, New York, 1988.

[4] Poulsen, V., U.S. Patent 661, 619, 1899.

[5] Richards, D., *A short history of 3M's involvement in magnetic recording*, Imation Corporation.

[6] Poniatoff, A.M., *History of Ampex*, unpublished talk, Ampex Corporation, 1952.

[7] Phelps, B.E., U.S. Patent 2,774,646, 1956.

[8] Harris, J.P. et al., Innovations in the design of magnetic tape subsystems, *IBM J. Res. Develop.*, 25, 691, 1981.

[9] Mallinson, J.C., *Magneto-Resistive Heads*, Academic Press, San Diego, 1996, 57.

[10] Schneider, R.C., Write equalization in high-linear-density magnetic recording, *IBM J. Res. Develop.*, 29, 563, 1985.

[11] Wood, R., Magnetic and optical storage systems: opportunities for communications technology, *Proc. ICC '89*, 1605, 1989.

[12] Wood, R., Magnetic megabits, *IEEE Spectrum*, 27, 32, 1990.

[13] Hoagland, A.S., History of magnetic disk storage based on perpendicular magnetic recording, *IEEE Trans. Mag.*, 39, 1871, 2003.

[14] Johnson, R.B., Dinner speech at DataStorage '89 Conference, San Jose, 1989.

[15] Harker et al., A quarter century of disk file innovation, *IBM J. Res. Develop.*, 25, 677, 1981.

[16] Franaszek, P.A., U.S. Patent 3,689,899, 1972.

[17] Jacoby, G.V., U.S. Patent 4,323,931, 1977.

[18] Siegel, P.H., Recording codes for digital magnetic storage, *IEEE Trans. Mag.*, 21, 1344, 1985.

[19] Wood, R. et al., An experimental eight-inch disc drive with one-hundred megabytes per surface, *IEEE Trans. Mag.*, 20, 698, 1984.

[20] Kobayashi, H., Application of probabilistic decoding to digital magnetic recording systems, *IBM J. Res. Develop.*, 15, 64, 1971.

[21] Dolivo, F., The PRML recording channel, *IBM Res. Magazine*, Summer, 11, 1990.

[22] Howell, T.D. et al., Error rate performance of experimental gigabit per square inch recording components, *IEEE Trans. Mag.*, 26, 2298, 1990.

[23] Palmer, D.C. et al., Identification of nonlinear write effects using pseudorandom sequences, *IEEE Trans. Mag.*, 24, 2377, 1987.

[24] Coker, J.D. et al., Implementation of PRML in a rigid disk drive, *IEEE Trans. Mag.*, 27, 4538, 1991.

[25] Kerwin, G.J., Galbraith, R.L., and Coker, J.D., Performance evaluation of the disk drive industry's second-generation partial response maximum likelihood data channel, *IEEE Trans. Mag.*, 29, 4005, 1993.

[26] Howell, T., Abbott, W., and Fisher, K., Advanced read channels for magnetic disk drives, *IEEE Trans. Mag.*, 30, 3807, 1994.

[27] Kabal, P. and Pasupathy, S., Partial-response signaling, *IEEE Trans. Comm.*, 23, 921, 1975.

[28] Wood, R., Turbo-PRML, a compromise EPRML detector, *IEEE Trans. Mag.*, 29, 4018, 1993.

[29] Eleftheriou, E. and Hirt, W., Noise-predictive maximum likelihood (NPML) detection for the magnetic recording channel, *IEEE Intl. Conf. On Commun.*, 556, 1996.

2

Physics of Longitudinal and Perpendicular Recording

Hong Zhou

Tom Roscamp

Roy Gustafson

Eric Boerner

Roy Chantrell
Seagate Research
Pittsburgh, PA

2.1   Introduction **2**-1
2.2   Transition Parameter of an Isolated Transition **2**-3
2.3   Playback Signal from the Reciprocity Principle **2**-7
       Playback Signal of an Isolated Magnetization
       Transition • Playback Signal of a Square-wave Magnetization
       Pattern • Playback Signal Power Spectrum of a Pseudo-random
       Bit Sequence
2.4   Playback Media Noise Spectra **2**-13
       Media Noise Spectra of an Isolated Magnetization
       Transition • Transition Noise Spectrum of a Square-wave
       Pattern • Transition Noise Spectrum of a PRBS
2.5   SNR Calculations **2**-17
       SNR of the Square-wave Pattern • SNR of the PRBS
2.6   Nonlinear Transition Shift **2**-21
2.7   Summary **2**-22

2.1   Introduction

Since the first hard disk drive was invented at IBM in 1957, substantial progress has been achieved in both the recording areal density and the data rate [1–3]. A modern hard disk drive, as illustrated in Figure 2.1, comprises a recording medium, an inductive write head and a giant magneto-resistance (GMR) playback sensor. The inductive head is connected with the GMR sensor through a shared soft magnetic pole. The disk drive system is typically classified into two types: (1) in the longitudinal recording, the media magnetic anisotropy is oriented in the thin film plane; and (2) in the perpendicular recording, the magnetic anisotropy is aligned perpendicular to the film plane [4,5]. In addition, the perpendicular recording is also divided into two groups, with and without the inclusion of a soft magnetic underlayer (SUL) [6]. The discussion here will only cover perpendicular recording system with a single pole head and an SUL. The latest laboratory areal density demonstrations were about 140 Gbit/in^2 for longitudinal recording [7] and 169 Gbit/in^2 for perpendicular recording [8].

0-8493-1524-7/05/$0.00+$1.50
© 2005 by CRC Press, LLC

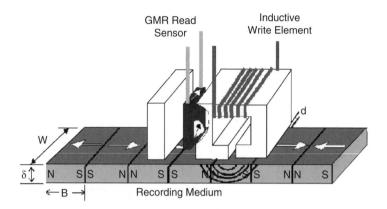

FIGURE 2.1 An illustration of the magnetic recording system (permission from the website of San Jose Research Center of Hitachi Global System Technologies).

Perpendicular recording exhibits several advantages over longitudinal recording:

1. Due to the imaging effect of the SUL, writing and playback processes are less dependent on the film thickness. Thus, higher areal densities can be achieved using a perpendicular medium with a smaller grain diameter but a larger film thickness to maintain thermal stability.
2. At the same linear density, the perpendicular recording exhibits a larger playback signal.
3. The write head field gradient can be larger in the perpendicular recording, which yields a smaller transition jitter and no DC particulate noise.
4. The demagnetization field decreases with increasing linear density in the perpendicular recording [6,9].

Though there exist technical challenges to commercialize the perpendicular recording technology [10–12], it is a primary candidate to achieve areal densities up to Tbit/in^2 [13–16]. However, because the recording industry is currently completely dominated by longitudinal recording, both longitudinal and perpendicular recordings will be given equal focus in this chapter.

To achieve an ultra-high density recording system, the primary medium requirements are: a high coercivity, a sufficiently large value of $K_u \langle V \rangle$ to maintain thermal stability and a narrow grain size distribution. In addition, an inductive head delivering sufficient writing fields and the GMR playback head with a large magnetoresistance ratio are required [6,17]. However, the ultimate requirement is that the recording system exhibits good signal-to-noise ratio (SNR) and small nonlinear effects to satisfy the bit error rate (BER) requirement [18,19].

Even though there are some architectural differences between the longitudinal and perpendicular recording technologies, they are based on the same recording physics principles [2,20]. To study the complicated recording processes, two approaches have been applied: the recording bubble model [21–25] and micromagnetic simulations [2,26–28]. Assuming a continuous medium structure, the bubble model has been widely utilized to understand and study the fundamental recording phenomena. However, to examine the effect of intergranular interactions in the granular media on recording performance, micromagnetic simulations are required [28].

In this chapter, determinations of the transition parameter, the nonlinear transition shift (NLTS) and the playback processes are studied briefly using the recording bubble model. Formulas to calculate the playback signal, noise power and SNR are derived. The chapter is organized as follows: in Section 2.2, the Williams-Comstock model is applied to calculate the transition parameter of an isolated transition; in Section 2.3, the playback signal based on the reciprocity principle is discussed; in Section 2.4, calculations of the noise powers are introduced, and the corresponding SNR discussions are covered in Section 2.5. Section 2.6 is devoted to a brief discussion of the nonlinear transition shift. A final summary is included in Section 2.7.

It should be emphasized that the electromagnetic CGS units are applied throughout this chapter. In all the discussions, effects of the AC coupling and low-pass and high-pass filters on the playback processes are not considered. In addition, studies of perpendicular recording will focus on the undifferentiated playback, because the differentiated playback is similar to the Gaussian pulse in longitudinal recording.

2.2 Transition Parameter of an Isolated Transition

A typical written transition pattern in longitudinal recording is illustrated in Figure 2.2 where the grains form a zig-zag structure at the transition location. To study the writing and playback processes, the transition shape and the transition parameter of an isolated transition are essential [23]. When a transition is recorded, both the head field profile and the media hysteresis properties determine the transition parameter and the transition shape.

To simplify the discussions, the write head is assumed to be infinite in the cross-track direction so that only two-dimensional head fields are considered. In addition, in perpendicular recording studies, a semi-infinite single pole head is utilized. These simplifications are very useful to study the on-track recording performance.

The coordinate systems are illustrated in Figure 2.3. The head to medium spacing is denoted as d, the film thickness is t and the medium to SUL spacing is s. In a longitudinal recording system (Figure 2.3(a)), for an infinite track-width head, the Karlqvist equations for the head fields are the most widely applied (x is the down-track direction and y is the perpendicular direction) [29]:

$$H_x(x, y) = \frac{H_0}{\pi} \left[\tan^{-1} \left(\frac{g_W/2 + x}{y} \right) + \tan^{-1} \left(\frac{g_W/2 - x}{y} \right) \right]$$

$$H_y(x, y) = -\frac{H_0}{2\pi} \ln \frac{(x + g_W/2)^2 + y^2}{(x - g_W/2)^2 + y^2}$$

(2.1)

where g_W is the deep gap width and H_0 is the deep gap field that is determined by the head magnetization moment and the head efficiency [30,31]. The zero point of the coordinate system is set at the gap center on the air-bearing surface (ABS). In perpendicular recording (Figure 2.3(b)), assuming the SUL acts

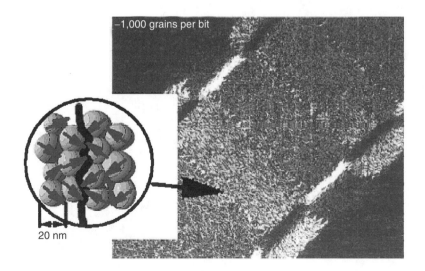

FIGURE 2.2 An illustration of the written transition pattern in longitudinal recording (permission from the website of IBM Almaden Research Center).

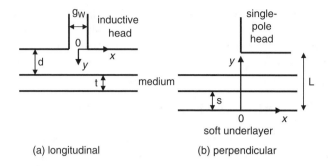

FIGURE 2.3   Recording dimensions and coordinate systems.

as a perfect image plane, Westmijze obtained the following field expressions by applying the conformal mapping method [32]:

$$Z = -\frac{1}{\pi}\left[\frac{2i}{H^*} + \ln\left(\frac{1 + iH^*}{1 - iH^*}\right)\right] \tag{2.2}$$

where the complex numbers are $Z = x + iy$ and $H = H_x + i\,H_y$. It should be noted that all the lengths are scaled to the head-to-SUL spacing $L\,(L = d + t + s)$ and the fields are normalized to the deep gap field H_0. Note that the zero point of the coordinate system is fixed at the SUL top surface. In perpendicular recording, a simple formula for the deep gap field has been suggested [33]:

$$\frac{H_0}{B_s} = \frac{RW_{\text{write}}}{RW_{\text{write}} + (R + W_{\text{write}})L\frac{4}{\pi}\ln\left(1 + \frac{F}{L}\right)} \tag{2.3}$$

where B_s is the head saturation magnetization, R is the down-track pole length, W_{write} is the cross-track pole width, and F is the head flare distance from the ABS surface. For a given head geometry, Equation 2.2 must be solved numerically to obtain the fields.

A typical hysteresis loop in a longitudinal recording medium with the external field applied in the film plane is plotted in Figure 2.4. The loop is usually characterized by the remanent magnetization M_r, the

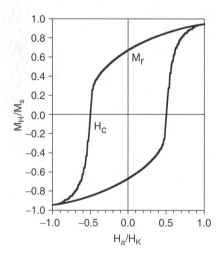

FIGURE 2.4   A typical hysteresis loop of longitudinal recording medium with in-plane applied field.

coercivity H_c, and the loop squareness S^* defined in

$$\left.\frac{dM}{dH}\right|_{H_c} = \frac{M_r}{H_c(1 - S^*)} \tag{2.4}$$

Micromagnetic simulations have demonstrated that these parameters depend primarily on the intergranular interactions and the media microstructures (anisotropy orientation, grain size distribution, etc.) [28]. Hysteresis loops can also be measured for perpendicular media. However, the perpendicular loops tend to exhibit larger loop squareness due to the high degree of orientation of the magnetic anisotropy axes, and a characteristic shearing of the loop caused by the demagnetizing field.

When both the write and the read head widths are wide, the playback voltage is related to the cross-track averaged magnetization. Here, variation of the averaged magnetization with the down-track location x (x_0 is the transition center) is assumed to follow a hyperbolic tangent (tanh) function [34]:

$$M(x) = M_r \tanh\left[\frac{2(x - x_0)}{\pi a}\right] \tag{2.5}$$

which satisfies

$$\left.\frac{dM(x)}{dx}\right|_{x=x_0} = \frac{2M_r}{\pi a} \tag{2.6}$$

Here, a is the transition parameter that is a measure of the transition width. It should be emphasized that in longitudinal recording, only the down-track magnetization component is considered, while in perpendicular recording the perpendicular component is studied. Both arctangent and error functions have also been used for the transition shape [34].

To determine the transition parameter a, Williams and Comstock analyzed the transition slope at the transition center:

$$\left.\frac{dM(x)}{dx}\right|_{x=x_0} = \left.\frac{dM}{dH}\right|_{H=H_c} \times \left[\frac{dH_a}{dx} + \frac{dH_d}{dx}\right]_{x=x_0} \tag{2.7}$$

where H_a is the applied head field and H_d is the demagnetization field in the medium [35]. The transition center location x_0 is determined from

$$H_a(x_0) = H_c \tag{2.8}$$

The transition location and the head field gradient can be calculated from the field expressions in Equation 2.1 and Equation 2.2. In longitudinal recording, only the down-track head field component is considered, while in perpendicular recording, the perpendicular head field is used (the effect of the longitudinal field component is neglected). Effect of head field angle on the transition parameter has been studied elsewhere [36,37]. The head fields are evaluated at the thin film center plane. In perpendicular recording, since there is no explicit expression for the fields and the distances, both the transition location and the head field gradient have to be obtained numerically. In longitudinal recording, the Karlqvist head field $H_x(x, y)$ in Equation 2.2 leads to the transition location ($y = d + t/2$):

$$x_0 = \frac{g_W}{2}\sqrt{1 - \left(\frac{2y}{g_W}\right)^2 + 4y/[g_W \tan(\pi H_c/H_0)]} \qquad (H_0 > 2H_c)$$
$$\tag{2.9}$$
$$x_0 = \frac{g_W}{2}\sqrt{1 - \left(\frac{2y}{g_W}\right)^2 - 4y/[g_W \tan(\pi(1 - H_c/H_0))]} \quad (H_0 \le 2H_c)$$

The head field gradient at the transition center can be easily obtained:

$$\left.\frac{dH_a}{dx}\right|_{x=x_0} = -\frac{QH_c}{y}, \tag{2.10}$$

where

$$Q = \frac{2x_0 H_0}{\pi g_W H_c} \sin^2 \frac{\pi H_c}{H_0}$$

Determination of the media demagnetization field entails the convolution of the demagnetization field of a perfect transition and the derivative of the transition shape [38]. In the discussions here, the demagnetization fields from the head imaging are neglected in both the longitudinal and the perpendicular modes. In perpendicular recording, the perpendicular demagnetization field component at the medium center plane is [39]:

$$H_d(x)/M_r = - \int\limits_0^\infty \frac{\sin kx}{k} [8e^{-kt/2} - 4e^{-k(2s+t/2)} + 4e^{-k(2s+3t/2)}] \frac{\pi^2 ak/4}{\sinh(\pi^2 ak/4)} \, dk \qquad (2.11)$$

In longitudinal recording, the hyperbolic tangent transition shape (Equation 2.5) induces a complicated longitudinal demagnetization field expression in the medium center plane [40]:

$$H_d(x) = M_r G(x, t/2)$$

where

$$G(x, u) = \sum_{\substack{m=1 \\ m=\text{odd}}}^\infty \frac{\pi^3 a^2 x m u / 2}{(x^2 - \pi^4 a^2 m^2 / 16 + u^2)^2 + \pi^4 x^2 a^2 m^2 / 4}$$

$$+ \tanh\left(\frac{2x}{\pi a}\right) - \frac{\sinh(2x/(\pi a)) \cosh(2x/(\pi a))}{\cos^2(2u/(\pi a)) + \sinh^2(2x/(\pi a))} \qquad (2.12)$$

with the following gradient at the transition center:

$$\left.\frac{d H_d(x)}{dx}\right|_{x=x_0} = -\frac{4M_r t I}{a^2} \qquad (2.13)$$

where I depends on the transition shape function: specially, $I = 0.691$ for the hyperbolic tangent function [34]. A combination of the above expressions yields the commonly used Williams-Comstock transition parameter in longitudinal recording [34]:

$$a_{WC} = \frac{(1 - S^*)y}{\pi Q} + \sqrt{\left[\frac{(1 - S^*)y}{\pi Q}\right]^2 + \frac{4M_r t y I}{Q H_c}} \qquad (2.14)$$

The transition parameter in perpendicular recording has to be obtained numerically.

As stated above, the Williams-Comstock model is based on the continuous medium assumption. However, the recording medium is composed of finite size grains. When the media granular structure is taken into account, micromagnetic simulations have found that Equation 2.14 needs to be modified according to (with $\langle D \rangle$ the average grain size):

$$a = \sqrt{\left(\frac{\langle D \rangle}{\pi}\right)^2 + 0.35 a_{WC}^2} \qquad (2.15)$$

for both advanced longitudinal and perpendicular recording systems [41–45]. As can be seen, a sufficiently large head field gradient exhibits the smallest transition parameter of $\langle D \rangle / \pi$.

The finite head field rise time and medium velocity also modify the transition parameter [46–48]. If there exist multiple transitions in the trailing direction, the demagnetization fields from these transitions also alter the transition parameter [49,50].

2.3 Playback Signal from the Reciprocity Principle

When a GMR sensor is used in the playback, the weak magnetic field from the recording medium slightly rotates the magnetization in the free layer of the GMR sensor, leading to the following output voltage:

$$V = IR_{sq} \frac{W_{read}}{2h} \frac{\Delta R}{R} [\langle \sin \theta \rangle - \langle \sin \theta_{eq} \rangle] \tag{2.16}$$

where I is the current and R_{sq} is the ohms in the device [51–53]. W_{read} and h are the width and height of the free layer, respectively. The MR ratio $\Delta R/R$ is about 15% in the advanced recording system. The equilibrium magnetization orientation θ_{eq} is usually chosen to be zero [51]. In order to calculate the playback signal, in principle it is necessary to calculate the field from the medium integrated over the read sensor, which involves a six-fold integral over the medium and the sensor. The reciprocity principle (e.g., [52]) is most easily described for an inductive head, where it reflects a mutual inductance. In using reciprocity one assumes the sensor to be uniformly magnetized and calculates the magnetic field at any point in the medium, and the resulting output voltage can be shown to be proportional to the integral of the magnetization and the field over the medium. Although the use of reciprocity simplifies the calculation of playback voltages it should be noted that it does not take into account the complex interactions between the head and the medium. Based on the reciprocity principle and linear signal analysis, the playback voltage (Equation 2.16) can be rewritten as

$$V(x) = IR_{sq} \frac{W_{read}}{h} \frac{\Delta R}{R} E \frac{(M_r t)_{media}}{(M_s \delta)_{MR}} \frac{1}{t} \int_d^{d+t} dy_1 \int_{-\infty}^{\infty} dx_1$$

$$\times \Phi(x + x_1, y_1) \left[\frac{1}{2M_r} \nabla \cdot \vec{M}(x_1, y_1) \right]_{media} \tag{2.17}$$

where δ is the GMR free layer thickness and $\Phi(x, y)$ expresses the virtual magnetic potential in the medium from the MR sensor [51,54]. The efficiency E expresses the average flux in the free layer (the active part of the GMR sensor device) divided by the input flux at the surface:

$$E = \frac{\tanh(h/2l)}{h/l} \tag{2.18}$$

where l is the characteristic length in the GMR sensor given by

$$l = \sqrt{\pi M_s \delta (G - \delta)/H_K} \tag{2.19}$$

Here, G is the GMR head's shield-to-shield spacing. A very small height h yields the largest efficiency $E = 0.5$ [51].

Calculation of Equation 2.17 is more conveniently carried out in Fourier space. Hereafter, the superscript (L) denotes longitudinal recording, and (P) is used for perpendicular recording. The magnetic potential is generally evaluated at the film center plane ($y = d + t/2$ for longitudinal recording). Fourier transformation of Equation 2.17 yields

$$V^{(L)}(x) = \frac{V_0^{(L)} W_{read}}{\pi} \int_0^{\infty} k\Phi^{(L)}(k, y) M(k)\cos(kx)\, dk$$

$$V^{(P)}(x) = \frac{V_0^{(P)} W_{read}}{\pi} \int_0^{\infty} [\Phi^{(P)}(k, d) - \Phi^{(P)}(k, d + t)] M(k)\cos(kx)\, dk \tag{2.20}$$

with the constants:

$$V_0^{(L)} = IR_{sq}\frac{t}{2h}\frac{\Delta R}{R}E\frac{1}{(M_s\delta)_{MR}}$$

$$V_0^{(P)} = IR_{sq}\frac{1}{2h}\frac{\Delta R}{R}E\frac{1}{(M_s\delta)_{MR}} \tag{2.21}$$

If the magnetic potential at the GMR ABS surface $\Phi_s(k)$ is known, the potentials in the medium can be easily obtained:

$$\Phi^{(L)}(k,y) = \Phi_s(k)e^{-ky},$$

$$\Phi^{(P)}(k,y) = \Phi_s(k)e^{-ky}\frac{1-e^{-2k(L-y)}}{1-e^{-2kL}} \tag{2.22}$$

The surface potential is usually approximated from Potter's linear variation $(g = (G-\delta)/2)$ [55]:

$$\Phi_s(x) = \frac{g+\delta/2-|x|}{g} \qquad \begin{cases} 0 & (|x| > g+\delta/2) \\ (\delta/2 \le |x| \le g+\delta/2) \\ 1 & (|x| < \delta/2) \end{cases} \tag{2.23}$$

Here, the linear potential drop is also assumed to be applicable to perpendicular recording without further modification due to the presence of the SUL [56]. Simple calculations show that the potential difference in perpendicular recording in Equation 2.20 can be further simplified as

$$\Phi^{(P)}(k,d) - \Phi^{(P)}(k,d+t) \approx \Phi_s(k)\frac{t}{L}e^{-k^2W_P} \tag{2.24}$$

where $W_P = [d^2 + 2d(t+s) - s(2s+t)]/6$ [56]. Fourier transformation of Equation 2.23 yields

$$\Phi_s(k) = \frac{4}{k^2g}\sin\left(\frac{kg}{2}\right)\sin\left(\frac{k(g+\delta)}{2}\right) \tag{2.25}$$

In recording characterization and channel design, the playback pulse shape of an isolated magnetization transition is essential. Based on linear superposition, the playback signal of a square-wave pattern and the signal power spectrum of a pseudo-random bit sequence (PRBS) can be easily obtained from the isolated pulse.

2.3.1 Playback Signal of an Isolated Magnetization Transition

For the isolated magnetization transition shape in Equation 2.5, Fourier transformation leads to

$$M(k) = \frac{2M_r}{k}\frac{\pi^2ka/4}{\sinh(\pi^2ka/4)} \tag{2.26}$$

Combining Equation 2.20, Equation 2.22, and Equation 2.24 through Equation 2.26 yields

$$V_{sp}^{(L)}(x) = \frac{2M_rV_0^{(L)}(g+\delta)W_{read}}{\pi}\int_0^\infty e^{-ky}\frac{\pi^2ka/4}{\sinh(\pi^2ka/4)}\frac{\sin(kg/2)}{kg/2}\frac{\sin(k(g+\delta)/2)}{k(g+\delta)/2}\cos kx\,dk$$

$$\tag{2.27}$$

$$V_{sp}^{(P)}(x) = \frac{2M_rV_0^{(P)}(g+\delta)W_{read}t}{\pi L}\int_0^\infty \frac{1}{k}e^{-k^2W_P}\frac{\pi^2ka/4}{\sinh(\pi^2ka/4)}\frac{\sin(kg/2)}{kg/2}\frac{\sin(k(g+\delta)/2)}{k(g+\delta)/2}\cos kx\,dk$$

The subscript "sp" is used to specify the output of an isolated magnetization transition. In general there is no solution in closed analytical form for Equation 2.27. Since the primary contributions to the integration come from very small k values, the following approximations have been used [57]:

$$\frac{\sin x}{x} \sim e^{-x^2/6} \quad \text{and} \quad \frac{x}{\sinh x} \sim e^{-x^2/6} \tag{2.28}$$

Thus, Equation 2.27 can be reevaluated as

$$V_{\text{sp}}^{(L)}(x) \approx \frac{2 M_r V_0^{(L)}(g+\delta) W_{\text{read}}}{\pi} \int_0^\infty e^{-\alpha_L^2 k^2 - ky} \cos kx \, dk$$

where

$$\alpha_L^2 = (g^2 + g\delta + \delta^2/2 + \pi^4 a^2/8)/12 \tag{2.29}$$

$$V_{\text{sp}}^{(P)}(x) \approx \frac{2 M_r t V_0^{(P)}(g+\delta) W_{\text{read}}}{\pi L} \int_0^\infty \frac{1}{k} e^{-\alpha_P^2 k^2} \cos kx \, dk$$

where

$$\alpha_P^2 = \alpha_L^2 + W_P$$

It is easy to show that the integration in the perpendicular pulse yields an error function:

$$V_{\text{sp}}^{(P)}(x) = \frac{M_r t V_0^{(P)}(g+\delta) W_{\text{read}}}{L} \text{erf}\left(\frac{x}{2\alpha_P}\right) \tag{2.30}$$

Thus, in perpendicular recording, the pulse shape of an isolated magnetization transition can be expressed as an error function [56]. From Equation 2.29 and Equation 2.30, the output peak voltages are

$$V_{\text{peak}}^{(L)} = \frac{M_r V_0^{(L)}(g+\delta) W_{\text{read}}}{\alpha_L \sqrt{\pi}} e^{y^2/(4\alpha_L^2)} \left(1 - \text{erf}\left(\frac{y}{2\alpha_L}\right)\right)$$

$$= IR_{\text{sq}} \frac{W_{\text{read}}}{2h} \frac{\Delta R}{R} E \frac{(g+\delta)_{\text{MR}}}{(M_s \Delta)_{\text{MR}}} \frac{(M_r t)_{\text{media}}}{\alpha_L \sqrt{\pi}} e^{y^2/(4\alpha_L^2)} \left(1 - \text{erf}\left(\frac{y}{2\alpha_L}\right)\right) \tag{2.31}$$

$$V_{\text{peak}}^{(P)} = \frac{M_r V_0^{(P)}(g+\Delta) W_{\text{read}} t}{L} = IR_{\text{sq}} \frac{W_{\text{read}}}{h} \frac{\Delta R}{R} E \frac{(g+\delta)_{\text{MR}}}{(M_s \Delta)_{\text{MR}}} \frac{(M_r t)_{\text{media}}}{2L}$$

As can be seen, the output voltage depends on both the medium and the GMR sensor parameters. For the same medium and GMR dimensions, because typically α_L is much larger than L, the output peak signal in perpendicular recording is generally larger than that in longitudinal recording. For the condition of $y < \alpha_L$, the longitudinal playback in Equation 2.29 can be evaluated using an infinite summation:

$$V_{\text{sp}}^{(L)}(x) = \frac{2 M_r V_0^{(L)}(g+\delta) W_{\text{read}}}{\pi \alpha_L} \sum_{n=0}^\infty \left[\frac{(y/\alpha_L)^{2n}}{(2n)!} \int_0^\infty k^{2n} \cos\left(\frac{kx}{\alpha_L}\right) e^{-k^2} dk \right.$$

$$\left. - \frac{(y/\alpha_L)^{2n+1}}{(2n+1)!} \int_0^\infty k^{2n+1} \cos\left(\frac{kx}{\alpha_L}\right) e^{-k^2} dk \right] \tag{2.32}$$

From Equation 2.30, the differentiated perpendicular pulse is a Gaussian:

$$V_{\text{sp}}^{'(P)}(x) = \frac{V_{\text{peak}}^{(P)}}{\alpha_P \sqrt{\pi}} e^{-x^2/(4\alpha_P^2)} \tag{2.33}$$

For the longitudinal type playback pulse shape, PW_{50} is used to denote the width of the half peak. However, in perpendicular recording, since PW_{50} is only defined in the differentiated pulse, dPW_{50} is used instead. From $V(PW_{50}/2) = V_{peak}/2$, the PW_{50} values are [56,57]:

$$PW_{50}^{(L)} \approx \sqrt{g^2 + g\delta + \delta^2/2 + \pi^4 a^2/8} + 1.1y$$

$$dPW_{50}^{(P)} \approx 0.96\sqrt{g^2 + g\delta + \delta^2/2 + \pi^4 a^2/8 + 12W_P}$$

(2.34)

In perpendicular recording, for the undifferentiated pulse (Equation 2.30), $T_{50}^{(P)}$ is also used to describe the width of the half maximum $V_{sp}^{(P)}(T_{50}^{(P)}/2) = V_{peak}^{(P)}/2$ [56]. It is easy to show that

$$dPW_{50}^{(P)} \approx 1.73 T_{50}^{(P)}$$

(2.35)

Using dPW_{50}, the perpendicular pulses in Equation 2.30 and Equation 2.33 can be rewritten as

$$V_{sp}^{(P)}(x) = V_{peak}^{(P)} \text{erf}\left(\frac{x}{\beta_P}\right)$$

where

$$\beta_P = dPW_{50}^{(P)}/2\sqrt{\ln 2}$$

$$V_{sp}^{'(P)}(x) = \frac{2V_{peak}^{(P)}}{\beta_P \sqrt{\pi}} e^{-x^2/\beta_P^2}$$

(2.36)

In longitudinal recording, the playback pulse expression in Equation 2.32 is complicated. In the signal processing, a Lorentzian function is normally assumed [19]:

$$V_L^{(L)}(x) = \frac{V_{peak}^{(L)}}{1 + \left(2x/PW_{50}^{(L)}\right)^2}$$

(2.37)

while a Gaussian shape has also been suggested:

$$V_G^{(L)}(x) = V_{peak}^{(L)} e^{-(x/\beta_L)^2} \left(\beta_L = \frac{PW_{50}^{(L)}}{2\sqrt{\ln 2}}\right)$$

(2.38)

A combination of Lorentzian and Gaussian functions with a weighting parameter η (Voigt function) has been found to fit the measured shape more accurately [58]:

$$V_V^{(L)}(x) = \eta V_L^{(L)}(x) + (1 - \eta)V_G^{(L)}(x)$$

(2.39)

In the following discussions, the subscript "L" denotes the Lorentzian function, "G" is for the Gaussian function and "V" specifies the Voigt shape. In Figure 2.5, these expressions are compared to the exact result (Equation 2.27) using the parameters $g = 50$ nm, $\delta = 5$ nm, $a = 5$ nm, and $y = 15$ nm. As can be seen, major differences exist when the head moves far away from the transition center; the Lorentzian shape is always larger than the exact value but the Gaussian function is always smaller. By fitting (Equation 2.39) to the exact result, a weighting parameter $\eta = 0.43$ is obtained and excellent agreement is found between the exact value and the Voigt function. Thus, a Voigt function can be applied to express the actual pulse shape in longitudinal recording. It can be easily understood that η depends primarily on the GMR head's shield-to-shield spacing and the net magnetic spacing $y = d + t/2$. Changing the transition parameter does not change η significantly. In Figure 2.6(a), for the fixed magnetic spacing of 15 nm, a transition parameter of 5 nm, and a GMR element thickness of 5 nm, η decreases with increasing the shield-to-shield spacing. However, as shown in Figure 2.6(b), for a fixed shield-to-shield spacing of 100 nm, increasing

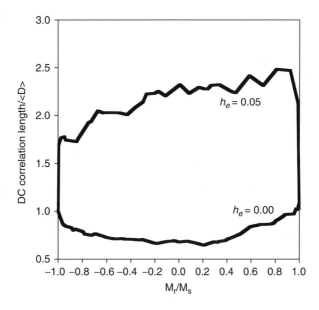

FIGURE 2.8 DC correlation length of perpendicular media.

where θ_c is determined from

$$\frac{M_r}{M_s} = \frac{\sin\theta_c}{\theta_c}$$

The remanent magnetization variance decreases substantially with increasing remanent squareness M_r/M_s and vanishes, as expected, for squareness of unity.

Micromagnetic simulations have shown that $S_{\text{down}}^{(DC)} \approx S_{\text{cross}}^{(DC)}$ for isotropic longitudinal media and perpendicular media. The correlation lengths depend on the magnetization state and the switching mechanism [69]. As an example, micromagnetic simulations have been performed on perpendicular media with $t/\langle D \rangle = 1.0, s/\langle D \rangle = 0.5$ and $M_s/H_K = 0.03$. Initially the magnetizations were saturated in the negative direction. In Figure 2.8, with varying the magnetization state, the DC correlation lengths were compared at $h_e = 0.00$ and $h_e = 0.05$ (h_e denotes the strength of the intergranular exchange coupling [27]). Without intergranular exchange, the correlation length initially decreases with increasing the applied field and reaches a minimum at about $0.68\langle D \rangle$ at the coercivity point, then increases at higher applied field. A symmetrical curve with respect to the coercivity is exhibited. However, the trend changes in the presence of exchange coupling. With increasing applied field, the correlation length continues to increase and reaches at the maximum value of $2.5\langle D \rangle$ at $M_r/M_s \sim 0.82$, then decreases very quickly to the initial level. The differences are from different reversal mechanisms in the presence of exchange coupling [69]. Therefore, for a given M_r/M_s level, the intergranular exchange couplings increase the DC noise power. Similar conclusions have also been found in longitudinal media [69].

It should be emphasized that the grain volume dispersion is characteristic of the media microstructure. The grain volume dispersion $\sigma_V/\langle V \rangle$ contributes an additional term to the noise spectrum [68]. On applying the same approximations as those in the Section 2.3 the DC remanent noise spectrum in Fourier space can be approximated as

$$\text{PSD}^{(DC)}(k) \approx \frac{2}{1.85\sqrt{6\pi}} \frac{V_{\text{peak}}^2 S_{\text{cross}}^{(DC)} S_{\text{down}}^{(DC)} (1 + (\sigma_V/\langle V \rangle)^2)(\langle M^2 \rangle - M_r^2)}{W_{\text{read}} M_r^2} e^{-k^2 \text{PW}_{50}^2/6} \qquad (2.57)$$

Note that $d\text{PW}_{50}$ needs to be substituted for perpendicular recording. The DC remanent noise spectrum is independent of the recorded pattern.

2.4.1.2 Transition Noise Spectrum

In the transition region, only the cross-track correlation length can be defined. It should be noted that the cross-track correlation length differs in the transition region and the DC remanent region [68]. The evaluation of Equation 2.51 in the transition region is reduced to

$$\Delta V_{(tr)}^2 = V_0^2 S_{cross}^{(tr)} W_{read} \int\limits_{-\infty}^{\infty} dx_1 \int\limits_{-\infty}^{\infty} dx_2 h(x + x_1, y) h(x + x_2, y)$$

$$\times \left[\langle M(x_1)M(x_2) \rangle - \langle M(x_1) \rangle \langle M(x_2) \rangle \right] \tag{2.58}$$

Since the down-track magnetization correlation is nonstationary in the transition region, the following approximation was suggested [66,67]:

$$\langle M(x_1)M(x_2) \rangle - \langle M(x_1) \rangle \langle M(x_2) \rangle = (M_r + M(x^<))(M_r - M(x^>)) \tag{2.59}$$

where

$$(x^<, x^>) = \begin{cases} (x_1, x_2) & \text{for } M(x_1) \le M(x_2) \\ (x_2, x_1) & \text{for } M(x_2) \le M(x_1) \end{cases}$$

Similarly, the transition noise spectra including the grain volume dispersion contribution are

$$PSD_{(tr)}^{(L)}(k) = \frac{V_0^{(L)2} S_{cross}^{(tr)} W_{read} (g + \delta)^2 4 M_r^2}{\pi} \left[1 + \left(\frac{\sigma_V}{\langle V \rangle} \right)^2 \right]$$

$$\times e^{-2ky} \left[\frac{\sin(kg/2)}{kg/2} \right]^2 \left[\frac{\sin(k(g+\delta)/2)}{k(g+\delta)/2} \right]^2 \left[1 - \left(\frac{\pi^2 ka/4}{\sinh(\pi^2 ka/4)} \right)^2 \right]$$

$$PSD_{(tr)}^{(P)}(k) = \frac{V_0^{(P)2} S_{cross}^{(tr)} W_{read} (g + \delta)^2 4 M_r^2 t^2}{\pi L^2} \left[1 + \left(\frac{\sigma_V}{\langle V \rangle} \right)^2 \right]$$

$$\times \frac{e^{-2k^2 W_P}}{k^2} \left[\frac{\sin(kg/2)}{kg/2} \right]^2 \left[\frac{\sin(k(g+\delta)/2)}{k(g+\delta)/2} \right]^2 \left[1 - \left(\frac{\pi^2 ka/4}{\sinh(\pi^2 ka/4)} \right)^2 \right] \tag{2.60}$$

Since the transition noise is the dominant noise source in both the longitudinal and perpendicular recordings, experiments have demonstrated that the primary noise modes are the transition location fluctuation (jitter mode) and the transition shape variance [70]. When only these two modes are considered, the noise can be expressed as

$$\delta V(x) = \frac{\partial V(x)}{\partial x} \delta x + \frac{\partial V(x)}{\partial a} \delta a \tag{2.61}$$

where δx and δa denote the change of transition location and transition shape, respectively. A comparison of Equation 2.61 and Equation 2.60 determines the variance of the jitter mode σ_j and the shape variance σ_a (assuming both obey Gaussian distributions) [71,72]:

$$\sigma_j = \sqrt{\frac{\pi^4}{48} \frac{a^2 S_{cross}^{(tr)}}{W_{read}}}$$

$$\sigma_a/\sigma_j = \sqrt{\frac{\pi^4 a^2}{120}} \tag{2.62}$$

2.4.2 Transition Noise Spectrum of a Square-wave Pattern

For a square-wave pattern with bit length B, based on the linear superposition, the transition noise spectrum is (Equation 2.60) divided by B [62]:

$$
\begin{aligned}
\mathrm{PSD}^{(L)}_{(\mathrm{tr})}(k) = {} & \frac{V_0^{(L)2} S^{(\mathrm{tr})}_{\mathrm{cross}} W_{\mathrm{read}}(g+\delta)^2 4 M_r^2}{\pi B} \left[1 + \left(\frac{\sigma_V}{\langle V \rangle} \right)^2 \right] \\
& \times e^{-2ky} \left[\frac{\sin(kg/2)}{kg/2} \right]^2 \left[\frac{\sin(k(g+\delta)/2)}{k(g+\delta)/2} \right]^2 \left[1 - \left(\frac{\pi^2 ka/4}{\sinh(\pi^2 ka/4)} \right)^2 \right]
\end{aligned}
$$

$$
\begin{aligned}
\mathrm{PSD}^{(P)}_{(\mathrm{tr})}(k) = {} & \frac{V_0^{(P)2} S^{(\mathrm{tr})}_{\mathrm{cross}} W_{\mathrm{read}}(g+\delta)^2 4 M_r^2 t^2}{\pi B L^2} \left[1 + \left(\frac{\sigma_V}{\langle V \rangle} \right)^2 \right] \\
& \times \frac{e^{-2k^2 W_P}}{k^2} \left[\frac{\sin(kg/2)}{kg/2} \right]^2 \left[\frac{\sin(k(g+\delta)/2)}{k(g+\delta)/2} \right]^2 \left[1 - \left(\frac{\pi^2 ka/4}{\sinh(\pi^2 ka/4)} \right)^2 \right]
\end{aligned}
\tag{2.63}
$$

2.4.3 Transition Noise Spectrum of a PRBS

Similarly, based on linear superposition, the transition noise spectrum of a PRBS is (if only the dominant jitter and shape fluctuation modes are considered) [60]:

$$
\mathrm{PSD}^{(\mathrm{tr})}(k) = \frac{1}{2\pi} \frac{N+1}{NB} \sigma_j^2 k^2 V_{\mathrm{sp}}^2(k) \left[1 + \left(\frac{\sigma_V}{\langle V \rangle} \right)^2 \right] \left(1 + \sigma_a^2 k^2 \right)
\tag{2.64}
$$

2.5 SNR Calculations

After obtaining both the playback signal and the noise spectra, the SNR can be easily calculated. However, the SNR depends on the choice of the signal. In the literature, for a square-wave pattern, the signal is generally chosen as the peak playback signal of an isolated magnetization transition [73–75]. For the PRBS pattern, both the signal and the noise power are defined in the PRBS frequencies [76].

2.5.1 SNR of the Square-wave Pattern

For the square-wave pattern, the noise spectrum is generally extended to the wide-band (infinite frequency). The DC remanent noise power $\mathrm{NP}^{(\mathrm{DC})}$, the transition noise power $\mathrm{NP}^{(\mathrm{tr})}$, and the total noise power TNP are defined as

$$
\mathrm{NP}^{(\mathrm{DC})} = \int_0^\infty \mathrm{PSD}^{(\mathrm{DC})}(k)\, dk
$$

$$
\mathrm{NP}^{(\mathrm{tr})} = \int_0^\infty \mathrm{PSD}^{(\mathrm{tr})}(k)\, dk
\tag{2.65}
$$

$$
\mathrm{TNP} = \mathrm{NP}^{(\mathrm{DC})} + \mathrm{NP}^{(\mathrm{tr})}
$$

Hereafter, the superscript "(DC)" denotes the DC remanent noise and the superscript "(tr)" is used for the transition noise. The DC noise power can be easily obtained from Equation 2.57. For the transition noise, though Equation 2.60 can be used, Equation 2.61 is usually applied to obtain the noise power [71]:

$$\mathrm{NP}^{(\mathrm{tr})} = \frac{\pi^3 a^2 S_{\mathrm{cross}}^{(\mathrm{tr})}}{48\,B\,W_{\mathrm{read}}} \left[1 + \left(\frac{\sigma_V}{\langle V \rangle}\right)^2\right] \int_0^\infty k^2 V_{\mathrm{sp}}^2(k) \left(1 + \frac{\pi^4 a^2}{60} k^2\right) dk \tag{2.66}$$

where the k^2 term is due to the jitter mode and the k^4 term comes from the shape fluctuation. In longitudinal recording, the noise power depends on the pulse shape. From Equation 2.43, we have

$$\mathrm{NP}_L^{(L,\mathrm{tr})} = \frac{\pi^5 S_{\mathrm{cross}}^{(\mathrm{tr})} a^2 V_{\mathrm{peak}}^{(L)2}}{96\,B\,W_{\mathrm{read}}\mathrm{PW}_{50}^{(L)}} \left(1 + \frac{\pi^4 a^2}{10\,\mathrm{PW}_{50}^{(L)}}\right) \left[1 + \left(\frac{\sigma_V}{\langle V \rangle}\right)^2\right]$$

$$\mathrm{NP}_G^{(L,\mathrm{tr})} = \frac{\pi^{9/2}\sqrt{2\ln 2}\,S_{\mathrm{cross}}^{(\mathrm{tr})} a^2 V_{\mathrm{peak}}^{(L)2}}{48\,B\,W_{\mathrm{read}}\mathrm{PW}_{50}^{(L)2}} \left(1 + \frac{\pi^4 a^2 \ln 2}{10\,\mathrm{PW}_{50}^{(L)2}}\right) \left[1 + \left(\frac{\sigma_V}{\langle V \rangle}\right)^2\right],$$

$$\mathrm{NP}_V^{(L,\mathrm{tr})} = \eta^2 NP^{(L)} + (1-\eta)^2 \mathrm{NP}^{(G)} + \frac{\pi^{9/2}\ln 2\,S_{\mathrm{cross}}^{(\mathrm{tr})} a^2 V_{\mathrm{peak}}^{(L)2}\eta(1-\eta)}{3\,B\,W_{\mathrm{read}}\mathrm{PW}_{50}^{(L)}} \tag{2.67}$$

$$\times \left[1 + \left(\frac{\sigma_V}{\langle V \rangle}\right)^2\right] \int_0^\infty k^2 e^{-(k+\sqrt{\ln 2})^2}\left(1 + \frac{4\ln 2\pi^4 a^2}{30\,\mathrm{PW}_{50}^{(L)2}}\right) dk$$

In perpendicular recording, from Equation 2.36 we have

$$\mathrm{NP}^{(P,\mathrm{tr})} = \frac{\pi^3 a^2 S_{\mathrm{cross}}^{(tr)} V_{\mathrm{peak}}^{(P)2}}{12\,W_{\mathrm{read}} B}\frac{\sqrt{2\pi}}{2\beta_P}\left(1 + \frac{\pi^4 a^2}{120\,\beta_P^2}\right)\left[1 + \left(\frac{\sigma_V}{\langle V \rangle}\right)^2\right] \tag{2.68}$$

Therefore, when including the grain volume dispersion, the SNR for each noise source in perpendicular recording is

$$\mathrm{SNR}^{(P,\mathrm{tr})} = \frac{0.185\,d\mathrm{PW}_{50}^{(P)} B\,W_{\mathrm{read}}}{a^2 S_{\mathrm{cross}}^{(\mathrm{tr})}(1 + (\sigma_V/\langle V \rangle)^2)}\left(1 + \frac{18\,a^2}{d\mathrm{PW}_{50}^{(P)2}}\right)^{-1} \approx \frac{0.37\,d\mathrm{PW}_{50}^{(P)} B}{\sigma_j^2(1 + (\sigma_V/\langle V \rangle)^2)}$$

$$\mathrm{SNR}^{(P,\mathrm{DC})} = 1.85\frac{d\mathrm{PW}_{50}^{(P)} W_{\mathrm{read}} M_r^2}{S_{\mathrm{cross}}^{(\mathrm{DC})} S_{\mathrm{down}}^{(\mathrm{DC})}\left(M_s^2 - M_r^2\right)(1 + (\sigma_V/\langle V \rangle)^2)} \tag{2.69}$$

$$\frac{1}{\mathrm{SNR}^{(P,\mathrm{tot})}} = \frac{1}{\mathrm{SNR}^{(P,\mathrm{tr})}} + \frac{1}{\mathrm{SNR}^{(P,\mathrm{DC})}}$$

However, in longitudinal recording, the SNR depends on the pulse shape. Because the jitter mode is dominant, the noise from the transition shape fluctuation is neglected here. For the Lorentzian and Gaussian functions, the SNR can be easily obtained:

$$\mathrm{SNR}_L^{(L,\mathrm{tr})} \approx \frac{0.31\,\mathrm{PW}_{50}^{(L)} B\,W_{\mathrm{read}}}{a^2 S_{\mathrm{cross}}^{(\mathrm{tr})}(1 + (\sigma_V/\langle V \rangle)^2)} = \frac{2}{\pi}\frac{\mathrm{PW}_{50}^{(L)} B}{\sigma_j^2(1 + (\sigma_V/\langle V \rangle)^2)}$$

$$\mathrm{SNR}_G^{(L,\mathrm{tr})} \approx \frac{0.24\,\mathrm{PW}_{50}^{(L)} B\,W_{\mathrm{read}}}{a^2 S_{\mathrm{cross}}^{(\mathrm{tr})}(1 + (\sigma_V/\langle V \rangle)^2)} \approx 0.76\,\mathrm{SNR}_L^{(L,tr)}$$

$$\mathrm{SNR}^{(L,\mathrm{DC})} = 1.85\frac{\mathrm{PW}_{50}^{(L)} W_{\mathrm{read}} M_r^2}{S_{\mathrm{cross}}^{(\mathrm{DC})} S_{\mathrm{down}}^{(\mathrm{DC})}(\langle M^2 \rangle - M_r^2)(1 + (\sigma_V/\langle V \rangle)^2)} \tag{2.70}$$

$$\frac{1}{\mathrm{SNR}^{(L,\mathrm{tot})}} = \frac{1}{\mathrm{SNR}^{(L,\mathrm{tr})}} + \frac{1}{\mathrm{SNR}^{(L,\mathrm{DC})}}$$

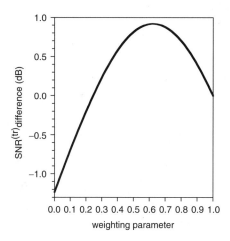

FIGURE 2.9 SNR$^{(tr)}$ difference with the weighting parameter in longitudinal recording.

Thus, the SNR due to transition noise from a Gaussian playback waveform is about 1.2 dB smaller than the result from a Lorentzian function. As discussed above, a Voigt function is found to be a better fit to the measured pulse shape than both Lorentzian and Gaussian functions. In Figure 2.9, the transition noise SNR difference between a Voigt and a Lorentzian function is plotted versus the weighting parameter η. The result shows that when $\eta < 0.24$, the Voigt function leads to a smaller SNR but with increasing η, the SNR from the Lorentzian function is always smaller. The largest SNR difference of 0.92 dB occurs at $\eta = 0.62$. For the above example with $\eta = 0.43$, a higher SNR of 0.66 dB is found for the Voigt function.

2.5.2   SNR of the PRBS

For a PRBS, the signal and noise powers are typically evaluated up to the Nyquist frequency π/B [60]. The SNR of a PRBS is usually called the auto-correlation signal-to-noise ratio (ACSN) [76]. From Equation 2.64, the signal power is

$$\mathrm{E} = 2\frac{N+1}{N^2 B^2} \sum_{l=1}^{(N-1)/2} \sin^2\left(\frac{\pi l}{N}\right) V_{\mathrm{sp}}^2\left(\frac{2\pi l}{NB}\right) \tag{2.71}$$

The noise powers are

$$\mathrm{NP}^{(tr)} = \frac{\pi^3}{96}\frac{N+1}{NB}\frac{a^2 S_{\mathrm{cross}}^{(tr)}(1+(\sigma_V/<V>)^2)}{W_{\mathrm{read}}} \times \int_0^{\pi/B} k^2\left(1+\frac{\pi^4 a^2 k^2}{120}\right) V_{\mathrm{sp}}^2(k)\,dk$$

$$\mathrm{NP}^{(DC)} = \int_0^{\pi/B} \mathrm{PSD}^{(DC)}(k)dk \tag{2.72}$$

$$\mathrm{TNP} = \mathrm{NP}^{(tr)} + \mathrm{NP}^{(DC)}$$

In perpendicular recording, we have

$$E^{(P)} = 2\frac{N+1}{N^2} V_{\text{peak}}^{(P)2} \sum_{l=1}^{(N-1)/2} \frac{\sin^2(\pi l/N)}{(\pi l/N)^2} \exp\left[-2\left(\frac{1}{2\sqrt{\ln 2}}\frac{\pi l}{N}\frac{d\text{PW}_{50}^{(P)}}{B}\right)^2\right]$$

$$\text{NP}^{(P,\text{tr})} = \frac{\pi^3\sqrt{2\ln 2}}{12}\frac{N+1}{N}\frac{a^2 S_{\text{cross}}^{(\text{tr})} V_{\text{peak}}^{(P)2}\left[1+\left(\frac{\sigma_V}{<V>}\right)^2\right]}{d\text{PW}_{50}^{(P)} B\, W_{\text{read}}} \int_0^{\frac{\pi}{2\sqrt{2\ln 2}}\frac{d\text{PW}_{50}^{(P)}}{B}} \left(1+\frac{\pi^4\ln 2\, a^2 k^2}{15 d\text{PW}_{50}^{(P)2}}\right) e^{-k^2}\, dk \quad (2.73)$$

$$\text{NP}^{(P,\text{DC})} = \frac{2}{1.85\sqrt{\pi}}\frac{V_{\text{peak}}^{(P)2} S_{\text{cross}}^{(\text{DC})} S_{\text{down}}^{(\text{DC})}\left(M_s^2 - M_r^2\right)\left[1+\left(\frac{\sigma_V}{<V>}\right)^2\right]}{d\text{PW}_{50}^{(P)} W_{\text{read}} M_r^2} \int_0^{\frac{\pi}{\sqrt{6}}\frac{d\text{PW}_{50}^{(P)}}{B}} e^{-k^2}\, dk$$

The corresponding SNRs are

$$\text{SNR}^{(P,\text{tr})} = \frac{0.33 d\text{PW}_{50}^{(P)} B\, W_{\text{read}}}{a^2 S_{\text{cross}}^{(\text{tr})}\left[1+\left(\frac{\sigma_V}{\langle V\rangle}\right)^2\right]} \frac{\frac{2}{N}\sum_{l=1}^{(N-1)/2}\frac{\sin^2(\pi l/N)}{(\pi l/N)^2}\exp\left[-2\left(\frac{1}{2\sqrt{\ln 2}}\frac{\pi l}{N}\frac{d\text{PW}_{50}^{(P)}}{B}\right)^2\right]}{\int_0^{\frac{\pi}{2\sqrt{2\ln 2}}\frac{d\text{PW}_{50}^{(P)}}{B}}\left(1+\frac{\pi^4\ln 2\, a^2 x^2}{15 d\text{PW}_{50}^{(P)2}}\right)e^{-x^2}\, dx}$$

$$\approx \frac{0.67 d\text{PW}_{50}^{(P)} B}{\sigma_j^2\left[1+\left(\frac{\sigma_V}{\langle V\rangle}\right)^2\right]} \frac{\frac{2}{N}\sum_{l=1}^{(N-1)/2}\frac{\sin^2(\pi l/N)}{(\pi l/N)^2}\exp\left[-2\left(\frac{1}{2\sqrt{\ln 2}}\frac{\pi l}{N}\frac{d\text{PW}_{50}^{(P)}}{B}\right)^2\right]}{\int_0^{\frac{\pi}{2\sqrt{2\ln 2}}\frac{d\text{PW}_{50}^{(P)}}{B}} e^{-x^2}\, dx}$$

$$(2.74)$$

$$\text{SNR}^{(P,\text{DC})} = 1.64\frac{N+1}{N}\frac{d\text{PW}_{50}^{(P)} W_{\text{read}} M_r^2}{S_{\text{cross}}^{(\text{DC})} S_{\text{down}}^{(\text{DC})}\left(M_s^2 - M_r^2\right)\left[1+\left(\frac{\sigma_V}{<V>}\right)^2\right]}$$

$$\times \frac{\frac{2}{N}\sum_{l=1}^{(N-1)/2}\frac{\sin^2(\pi l/N)}{(\pi l/N)^2}\exp\left[-2\left(\frac{1}{2\sqrt{\ln 2}}\frac{\pi l}{N}\frac{d\text{PW}_{50}^{(P)}}{B}\right)^2\right]}{\int_0^{\frac{\pi}{\sqrt{6}}\frac{d\text{PW}_{50}^{(P)}}{B}} e^{-x^2}\, dx}$$

In longitudinal recording, if only Lorentzian and Gaussian shapes are considered and only jitter noise mode is included, the signal and noise powers are

$$E_L^{(L)} = \frac{N+1}{N^2}\frac{\pi^2 V_{\text{peak}}^{(L)2}\text{PW}_{50}^{(L)2}}{2B^2}\sum_{l=1}^{(N-1)/2}\sin^2(\pi l/N)\exp\left[-\frac{2\pi l}{N}\frac{\text{PW}_{50}^{(L)}}{B}\right]$$

$$E_G^{(L)} = \frac{N+1}{N^2}\frac{\pi V_{\text{peak}}^{(L)2}\text{PW}_{50}^{(L)2}}{2\ln 2\, B^2}\sum_{l=1}^{(N-1)/2}\sin^2(\pi l/N)\exp\left[-2\left(\frac{1}{2\sqrt{\ln 2}}\frac{\pi l}{N}\frac{\text{PW}_{50}^{(L)}}{B}\right)^2\right] \quad (2.75a)$$

$$\text{NP}_L^{(L,\text{tr})} = \frac{\pi}{8}\frac{N+1}{N}\frac{\sigma_j^2\left[1+\left(\frac{\sigma_V}{\langle V\rangle}\right)^2\right]V_{\text{peak}}^{(L)2}}{\text{PW}_{50}^{(L)} B}\int_0^{\frac{\pi\text{PW}_{50}^{(L)}}{B}} x^2 e^{-x}\, dx$$

$$\mathrm{NP}_G^{(L,\mathrm{tr})} = \sqrt{8\ln 2}\,\frac{N+1}{N}\,\frac{\sigma_j^2\left[1+\left(\frac{\sigma_V}{\langle V\rangle}\right)^2\right]V_{\mathrm{peak}}^{(L)2}}{\mathrm{PW}_{50}^{(L)}B}\int_0^{\frac{\pi}{2\sqrt{2\ln 2}}\frac{\mathrm{PW}_{50}^{(L)}}{B}} x^2 e^{-x^2}\,dx \tag{2.75b}$$

$$\mathrm{NP}^{(L,\mathrm{DC})} = \frac{2}{1.85\sqrt{\pi}}\,\frac{V_{\mathrm{peak}}^{(L)2}\,S_{\mathrm{cross}}^{(\mathrm{DC})}\,S_{\mathrm{down}}^{(\mathrm{DC})}\left[1+\left(\frac{\sigma_V}{\langle V\rangle}\right)^2\right](\langle M^2\rangle - M_r^2)}{\mathrm{PW}_{50}^{(L)}\,W_{\mathrm{read}}\,M_r^2}\int_0^{\frac{\pi}{\sqrt{6}}\frac{\mathrm{PW}_{50}^{(L)}}{B}} e^{-x^2}\,dx$$

It is easy to obtain the SNR for both Lorentzian and Gaussian pulse shapes:

$$\mathrm{SNR}_L^{(L,\mathrm{tr})} = \frac{2\pi\,\mathrm{PW}_{50}^{(L)3}}{\sigma_j^2 B\left[1+\left(\frac{\sigma_V}{\langle V\rangle}\right)^2\right]}\,\frac{\frac{2}{N}\displaystyle\sum_{l=1}^{(N-1)/2}\sin^2(\pi l/N)\exp\left[-\frac{2\pi l}{N}\frac{\mathrm{PW}_{50}^{(L)}}{B}\right]}{\int_0^{\frac{\pi\,\mathrm{PW}_{50}^{(L)}}{B}} x^2 e^{-x}\,dx} \tag{2.76a}$$

$$\mathrm{SNR}_L^{(L,\mathrm{DC})} = 4.0\,\frac{N+1}{N}\,\frac{\mathrm{PW}_{50}^{(L)3}\,W_{\mathrm{read}}\,M_r^2}{S_{\mathrm{cross}}^{(\mathrm{DC})}\,S_{\mathrm{down}}^{(\mathrm{DC})}\,B^2\left[1+\left(\frac{\sigma_V}{\langle V\rangle}\right)^2\right](M_s^2 - M_r^2)}$$

$$\times\,\frac{\frac{2}{N}\displaystyle\sum_{l=1}^{(N-1)/2}\sin^2(\pi l/N)\exp\left[-\frac{2\pi l}{N}\frac{\mathrm{PW}_{50}^{(L)}}{B}\right]}{\int_0^{\frac{\pi}{\sqrt{6}}\frac{\mathrm{PW}_{50}^{(L)}}{B}} e^{-x^2}\,dx}$$

$$\mathrm{SNR}_G^{(L,\mathrm{tr})} = \frac{0.68\mathrm{PW}_{50}^{(L)3}}{\sigma_j^2 B\left[1+\left(\frac{\sigma_V}{\langle V\rangle}\right)^2\right]}\,\frac{\frac{2}{N}\displaystyle\sum_{l=1}^{(N-1)/2}\sin^2(\pi l/N)\exp\left[-2\left(\frac{1}{2\sqrt{\ln 2}}\frac{\pi l}{N}\frac{\mathrm{PW}_{50}^{(L)}}{B}\right)^2\right]}{\int_0^{\frac{\pi}{2\sqrt{2\ln 2}}\frac{\mathrm{PW}_{50}^{(L)}}{B}} x^2 e^{-x^2}\,dx}$$

$$\mathrm{SNR}_G^{(L,\mathrm{DC})} = \frac{N+1}{N}\,\frac{1.87\,\mathrm{PW}_{50}^{(L)3}\,W_{\mathrm{read}}\,M_r^2}{S_{\mathrm{cross}}^{(\mathrm{DC})}\,S_{\mathrm{down}}^{(\mathrm{DC})}\,B^2\left[1+\left(\frac{\sigma_V}{\langle V\rangle}\right)^2\right](M_s^2 - M_r^2)} \tag{2.76b}$$

$$\times\,\frac{\frac{2}{N}\displaystyle\sum_{l=1}^{(N-1)/2}\sin^2(\pi l/N)\exp\left[-2\left(\frac{1}{2\sqrt{\ln 2}}\frac{\pi l}{N}\frac{\mathrm{PW}_{50}^{(L)}}{B}\right)^2\right]}{\int_0^{\frac{\pi}{\sqrt{6}}\frac{\mathrm{PW}_{50}^{(L)}}{B}} e^{-x^2}\,dx}$$

Clearly the SNR value depends on the pulse shape in the longitudinal recording.

2.6 Nonlinear Transition Shift

In the above discussions, linear superposition was assumed to be applicable for all recording patterns. However, in the measured playback waveform, this assumption can be violated due to nonlinear effects in the recording system. Nonlinear effects exist in both the writing and playback processes [23,25] and degrade the BER level. The strength of the different nonlinear effects can be obtained from the dipulse extraction method [77–80]. Here, one of the most important nonlinear effects, the nonlinear transition shift (NLTS), is studied. From Equation 2.8, the transition location x_0 is solely determined by the head field since the demagnetization field is zero at the transition center. However, this analysis only applies when an isolated transition is written; for a multitransition pattern, the magnetostatic field from the previously recorded transitions shifts the transition location. Since the magnetostatic field from the nearest transition is the largest, the transition shift in a dibit pattern is considered to demonstrate the NLTS. At a bit length

B, the magnetostatic field from the nearest neighbor $H_d^{dibit}(B)$ shifts the transition location from x_0 to $x_0 + \Delta x$. Similar to Equation 2.8, we have

$$H_c = H_d^{dibit}(B) + H_a(x_0 + \Delta x) \tag{2.77}$$

A combination of Equation 2.8 and Equation 2.77 yields the transition location shift:

$$\Delta x = -\frac{H_d^{dibit}(B)}{\partial H_a(x_0)/\partial x} \tag{2.78}$$

For longitudinal recording, if $B > 2a$, the magnetostatic field can be approximated as [23]:

$$H_d^{dibit}(B) \approx -\frac{16 M_r t y^2}{B^3} \tag{2.79}$$

and the transition location shift Δx is

$$\Delta x = \frac{16 M_r t y^3}{Q H_c B^4} \tag{2.80}$$

However, the exact shift is more complicated due to the head imaging effect. In general, the transition shift can be written as

$$\Delta x = \frac{K}{B^q} \tag{2.81}$$

where K is related to medium parameters and head geometry [40]. The power factor q depends on the recording density: q is approximately 3 at low densities and drops to 2 at high densities [40].

In perpendicular recording, when head imaging is neglected, the magnetostatic field from the nearest transition can be easily obtained from Equation 2.11:

$$H_d^{dibit}(B) = 4\pi M_r - H_d(B) \quad \text{(Equation 2.11)} \tag{2.82}$$

and the amount of transition shift can be calculated from Equation 2.77.

It should be emphasized that the direction of the transition location shift is different for different recording modes: the transition center location is shifted towards the previous transition in longitudinal recording, while in perpendicular recording, the shift is away from the previous transition [50,81,82]. In order to reduce the NLTS, write precompensation is implemented [23,83].

2.7 Summary

In this chapter, for both longitudinal and perpendicular recording modes, calculations of the transition parameter, the playback pulse, the media noise powers, and the SNR are discussed. Assuming a hyperbolic tangent shape for an isolated transition, the playback pulse can be well described using a Voigt function in longitudinal recording. For idealized perfect soft magnetic underlayer design in perpendicular recording, an error function expresses the pulse shape. In addition, differentiating the perpendicular pulse returns to the Gaussian longitudinal pulse.

In a recording system, the playback noise originates from the head electronics and the media magnetization random patterns. Generally the medium noise is decomposed into the DC remanent and the transition components. However, the transition jitter noise is dominant. In longitudinal recording, due to the random anisotropy dispersion, there always exist some levels of the DC remanent noise. However, in perpendicular recording, the loop full squareness is required to maintain thermal stability. Therefore, the DC remanent noise vanishes.

Acknowledgments

The authors wish to thank Bogdan Valcu, H. Neal Bertram, Gregory J. Parker, Dieter Weller, Sharat Batra, Hans J. Richter, and Eric Champion for useful discussions.

References

[1] For the history of magnetic recording, please refer to Daniel, E.D., Mee, C.D., and Clark, M.H., Magnetic Recording: The First 100 Years, IEEE Press, 1999.

[2] Mee, C.D. and Daniel, E.D., *Magnetic Storage Handbooks,* 2nd ed., McGraw-Hill, New York, 1996.

[3] A review of the disk drive evolution can be found in the special issues of *IBM J. Res. Devel.,* 44, 1–3, 2000.

[4] Iwasaki, S.I. and Nakamura, Y., An analysis for the magnetization mode for high density magnetic recording, *IEEE Trans. Magn.,* 13, 1272, 1977.

[5] For a history of perpendicular recording, please refer to Iwasaki, S., Perpendicular magnetic recording focused on the origin and its significance, *IEEE Trans. Magn.,* 38, 1609, 2002; Iwasaki, S., History of perpendicular magnetic recording — focused on the discoveries that guided the innovation, *J. Magn. Soc. Jpn.,* 25, 1361, 2001.

[6] Thompson, D.A., The role of perpendicular recording in the future of hard disk storage, *J. Magn. Soc. Jpn.,* 21, Suppl. S2, 9, 1997.

[7] Chen, Y. et al., Inductive write heads using high moment pole materials for ultra-high areal density demonstrations, Boston, March 28–April 3, 2003, paper CE-07, to be published in *IEEE Trans. Magn,* 39, 2368, 2003.

[8] A commentary review of Fuji Electric's 169 Gbit/in^2 perpendicular recording demonstration can be found in *Nikkei Electronics,* April 28th, 30, 2003.

[9] Thompson, D.A. and Best, J.S., The future of magnetic data storage technology, *IBM J. Res. Devel.,* 44, 311, 2000.

[10] Cain, W. et al., Challenges in the practical implementation of perpendicular magnetic recording, *IEEE Trans. Magn.,* 32, 97, 1996.

[11] Nakamura, Y., Technical issues for realization of perpendicular magnetic recording, *J. Magn. Soc. Jpn.,* 21, Suppl. S2, 125, 1997.

[12] Richter, H.J., Champion, E., and Peng, Q., Theoretical analysis of longitudinal and perpendicular recording potential, *IEEE Trans. Magn.,* 39, 697, 2003.

[13] Wood, R., The feasibility of magnetic recording at 1 terabit per square inch, *IEEE Trans. Magn.,* 36, 36, 2000.

[14] Wood, R., Miles, J., and Olson, T., Recording technologies for terabit per square inch systems, *IEEE Trans. Magn.,* 38, 1711, 2002.

[15] Victora, R.H., Senanan, K., and Xue, J., Areal density limits for perpendicular magnetic recording, *IEEE Trans. Magn.,* 38, 1886, 2002.

[16] Mallary, M., Torabi, A., and Benakli, M., One terabit per square inch perpendicular recording conceptual design, *IEEE Trans. Magn.,* 38, 1719, 2002.

[17] Weller, D. and Moser, A., Thermal effect limits in ultra high-density magnetic recording, *IEEE Trans. Magn.,* 35, 423, 1999.

[18] Bertram, H.N., Zhou, H., and Gustafson, R., Signal to noise ratio scaling to density density limit estimates in longitudinal magnetic recording, *IEEE Trans. Magn.,* 34, 1846, 1998.

[19] Howell, T.D., McEwen, P.A., and Patapoutian, A., Getting the information in and out: the channel, *J. Appl. Phys.,* 87, 5371, 2000.

[20] Bromley, D.J., A comparison of vertical and longitudinal magnetic recording based on analytic models, *IEEE Trans. Magn.,* 19, 2239, 1983.

[21] White, R., *Introduction to Magnetic Recording,* IEEE Press, 1985.

[22] Mallinson, J.C., *The Foundations of Magnetic Recording,* 2nd ed., Academic Press, New York, 1993.

[23] Bertram, H.N., *Theory of Magnetic Recording*, Cambridge University Press, 1994.

[24] Ashar, K.G., *Magnetic Disk Drive Technology*, IEEE Press, 1996.

[25] Wang, S.X. and Taratorin, A.M., *Magnetic Information Storage Technology*, Academic Press, New York, 1999.

[26] Hughes, G.F., Magnetization reversal in cobalt-phosphorus films, *J. Appl. Phys.*, 54, 5306, 1983.

[27] Zhu, J. and Bertram, H.N., Micromagnetic studies of thin metallic films, *J. Appl. Phys.*, 63, 3248, 1988.

[28] Bertram, H.N. and Zhu, J., Fundamental magnetization processes in thin-film recording media, *Solid State Physics*, edited by H. Ehrenreich and D. Turnbull, 36, 271, 1992.

[29] Karlqvist, O., Calculation of the magnetic field in the ferromagnetic layer of a magnetic drum, *Trans. Roy. Inst. Tech. Stockholm*, 86, 3, 1954.

[30] Williams, M.L. and Grochowski, E., Design of magnetic inductive write heads for high-density storage, *Datatech*, 2, 53, 1999.

[31] Bertram, H.N., *Theory of Magnetic Recording*, Cambridge University Press, 1994, Chap. 3.

[32] Westmijze, W.K., Studies on magnetic recording, *Philips Res. Rep.*, Part II, 8, 161, 1953.

[33] Williams, M.L. et al., Perpendicular write process and head design, *IEEE Trans. Magn.*, 38, 1643, 2002.

[34] Bertram, H.N., *Theory of Magnetic Recording*, Cambridge University Press, 1994, Chap. 8.

[35] Williams, M.L. and Comstock, R.L., An analytical model of the write process in digital magnetic recording, *AIP Conf. Proc.*, 5, 738, 1971.

[36] Richter, H.J., A generalized slope model for magnetization transitions, *IEEE Trans. Magn.*, 31, 1073, 1997.

[37] Zhong, L.P. et al., Head field angle-dependent writing in longitudinal recording, *IEEE Trans. Magn.*, 39, 1851, 2003.

[38] Bertram, H.N., *Theory of Magnetic Recording*, Cambridge University Press, 1994, Chap. 4.

[39] Nakamura, K., Analytical model for estimation of isolated transition width in perpendicular magnetic recording, *J. Appl. Phys.*, 87, 4993, 2000.

[40] Zhang, Y. and Bertram, H. N., A theoretical study of nonlinear transition shift, *IEEE Trans. Magn.*, 34, 1955, 1998.

[41] Zhou, H. and Bertram, H.N., Scaling of hysteresis and transition parameter with grain size in longitudinal thin film media, *J. Appl. Phys.*, 85, 4982, 1999.

[42] Bertram, H.N. and Williams, M.L., SNR and density limit estimates: a comparison of longitudinal and perpendicular recording, *IEEE Trans. Magn.*, 36, 4, 2000.

[43] Richter, H.J., Recent advances in the recording physics of thin-film media, *J. Phys. D (Appl. Phys.)*, 32, R147, 1999.

[44] Yen, E.T. et al., Case study of media noise mechanism in longitudinal recording, *IEEE Trans. Magn.*, 35, 2730, 1999.

[45] Richter, H.J., Longitudinal recording at 10-20 Gbit/in^2 and beyond, *IEEE Trans. Magn.*, 35, 2790, 1999.

[46] Xing, X. and Bertram, H.N., Analysis of the effect of head field rise time on signal, noise, and nonlinear transition shift, *J. Apply. Phys.*, 85, 5861, 1999.

[47] Thayamballi, P., Modeling the effects of write field rise time on the recording properties in thin film media, *IEEE Trans. Magn.*, 32, 61, 1996.

[48] Nakamoto, K. and Bertram, H.N., Head-field rise-time effect in perpendicular recording, *IEEE Trans. Magn.*, 38, 2069, 2002.

[49] Barany, A.M. and Bertram, H.N., Transition noise model for longitudinal thin-film media, *IEEE Trans. Magn.*, 23, 1776, 1987.

[50] Nakamoto, K. and Bertram, H.N., Analytic perpendicular-recording model for the transition parameter and NLTS, *J. Magn. Soc. Jpn.*, 26, 79, 2002.

[51] Bertram, H.N., Linear signal analysis of shielded AMR and spin valve heads, *IEEE Trans. Magn.*, 31, 2573, 1995.

[52] Mallinson, J.C., *Magneto-resistive Heads: Fundamentals and Applications,* Academic Press, New York, 1996.

[53] Williams, E.M., *Design and Analysis of Magnetoresistive Recording Heads,* John Wiley & Sons, 2001.

[54] Bertram, H.N., *Theory of Magnetic Recording,* Cambridge University Press, 1994, Chaps. 5 and 7.

[55] Potter, R.I., Digital magnetic recording theory, *IEEE Trans. Magn.,* 10, 502, 1974.

[56] Valcu, B., Roscamp, T., and Bertram, H.N., Pulse shape, resolution, and signal-to-noise ratio in perpendicular recording, *IEEE Trans. Magn.,* 38, 288, 2002.

[57] Zhang, Y. and Bertram, H.N., PW_{50}, D_{50} and playback voltage formulae for shielded MR heads, *J. Appl. Phys.,* 81, 4897, 1997.

[58] Stupp, S.E., Approximating the magnetic recording step response using a Voigt function, *J. Appl. Phys.,* 87, 5010, 2000.

[59] Bertram, H.N., *Theory of Magnetic Recording,* Cambridge University Press, 1994, Chap. 6.

[60] Slutsky, B. and Bertram, H.N., Transition noise analysis of thin film magnetic recording media, *IEEE Trans. Magn.,* 30, 2808, 1994.

[61] Arnoldussen, T.C. and Nunnelley, L.L., Eds., *Noise in digital magnetic recording,* World Scientific, 1992.

[62] Bertram, H.N., *Theory of magnetic recording,* Cambridge University Press, 1994, Chaps. 10–12.

[63] Lin, G.H. et al., Texture induced noise and its impact on system performance, *IEEE Trans. Magn.,* 33, 950, 1997.

[64] Chen, J. et al., Texture noise and its impact on recording performance at high recording density, *IEEE Trans. Magn.,* 35, 2727, 1999.

[65] Seagle, D.J. et al., Transition curvature analysis, *IEEE Trans. Magn.,* 35, 619, 1999.

[66] Bertram, H.N., Beardsley, I.A., and Che, X., Magnetization correlations in thin film media in the presence of a recorded transition, *J. Appl. Phys.,* 73, 5545, 1993.

[67] Bertram, H.N. and Che, X., General analysis of noise in recorded transitions in thin film recording media, *IEEE Trans. Magn.,* 29, 201, 1993.

[68] Zhou, H. and Bertram, H.N., Effect of grain size distribution on recording performance in longitudinal thin film media, *IEEE Trans. Magn.,* 36, 61, 2000.

[69] Bertram, H.N. and Arias, R., Magnetization correlations and noise in thin-film recording media, *J. Appl. Phys.,* 71, 3439, 1992.

[70] Yuan, S.W. and Bertram, H.N., Statistical data analysis of magnetic recording noise mechanisms, *IEEE Trans. Magn.,* 28, 84, 1992.

[71] Xing, X. and Bertram, H.N., Analysis of transition noise in thin film media, *IEEE Trans. Magn.,* 33, 2959, 1997.

[72] Caroselli, J. and Wolf, J.K., Applications of a new simulation model for media noise limited magnetic recording channels, *IEEE Trans. Magn.,* 32, 3917, 1996.

[73] Tarnopolsky, G.J. and Pitts, P.R., Media noise and signal-to-noise ratio estimates for high areal density recording, *J. Appl. Phys.,* 81, 4837, 1997.

[74] Arnoldussen, T.C., Bit cell aspect ratio: an SNR and detection perspective, *IEEE Trans. Magn.,* 34, 1851, 1998.

[75] Bertram, H.N., Zhou, H., and Gustafson, R., Signal to noise ratio scaling to density limit estimates in longitudinal magnetic recording, *IEEE Trans. Magn.,* 34, 1846, 1998.

[76] Mian, G. and Howell, T.D., Determining a signal to noise ratio for an arbitrary data sequence by a time domain analysis, *IEEE Trans. Magn.,* 29, 3999, 1993.

[77] Palmer, D. et al., Identification of nonlinear write effects using pseudorandom sequence, *IEEE Trans. Magn.,* 23, 2377, 1987.

[78] Palmer, D. et al., Characterization of the read/write process for magnetic recording, *IEEE Trans. Magn.,* 31, 1071, 1995.

[79] Muraoka, H. and Nakamura, Y., Experimental study of nonlinear transition shift in perpendicular magnetic recording with single-pole head, *IEICE Trans. Electron.,* E80-C, 1181, 1997.

[80] Muraoka, H., Wood, R., and Nakamura, Y., Nonlinear transition shift measurement in perpendicular magnetic recording, *IEEE Trans. Magn.*, 32, 3926, 1996.

[81] Senanan, K. and Victora, R.H., Theoretical study of nonlinear transition shift in double-layer perpendicular media, *IEEE Trans. Magn.*, 38, 1664, 2002.

[82] Chen, J. et al., Effect of reader saturation on the measurements of NLTS in perpendicular recording, *J. Appl. Phys.*, 93, 6534, 2003.

[83] Che, X., Nonlinearity measurements and write precompensation studies for a PRML recording channel, *IEEE Trans. Magn.*, 31, 3021, 1995.

3

The Physics of Optical Recording

William A. Challener

Terry W. McDaniel

Seagate Research
Pittsburgh, PA

3.1 Introduction 3-1
3.2 Optical Recording Media 3-3
    Introduction • Phase Change, Organic Dye, and
    Ablative/Deformative Media • Magneto-Optic Media
    • Dielectrics • Heat Sink/Reflector • Film Stack Design
    • Substrate • Direct Overwrite • Super-Resolution
3.3 Optical Recording Systems 3-15
    Light Source • Beamsplitter and Waveplate • Objective
    • Focus and Tracking Servo • Read Signal • Noise
3.4 Conclusion 3-19

3.1 Introduction

Optical data storage was first proposed in 1969 by Klass Compaan at Philips Research, was initially commercialized in the late 1970s, expanded rapidly through the next two decades, and today enjoys a preeminent position as the underlying technology supporting audio and video content distribution (CD, DVD) and removable media data storage (read-only, write-once, and rewritable disk formats). In this chapter, we will explore the scientific and engineering basis for this commercially important quarter-century old technology.[1]

The simplest class of optical data storage system is read-only or "read-only-memory" (ROM). Here, the user device merely reads a prewritten medium, but cannot induce any physical change (i.e., record) in the medium. This class is historically and technologically very important because from the late 1970s it offered a means to inexpensively replicate and distribute vast amounts of digital information on a highly portable medium. A ROM medium is mass-produced by replicating multiple copies from a master source by a very low cost process (capital costs are amortized over huge volumes of output, thus driving unit cost very low). The ubiquitous CD-ROM (and now DVD-ROM) disks have become common as free advertising distribution, while the materials and processing cost of even relatively expensive entertainment content distribution (music or movie DVD disks) are but a tiny fraction of the product cost to the consumer. This is data recording at an enormous rate in which a medium unit of 10^{11} bytes can be replicated from a master on a single machine in a few seconds. ROM readout, on the other hand, is much slower, and is more similar in execution to the end-user optical recording process described below.

Conventional optical recording by an end-user means bit-by-bit digital recording (writing) in which a focused light beam interacts with a medium to effect some local physical change in the material, and this local property modification is then detected with an optical probe beam. The physical change is optically or thermally driven, and can be a structural phase change in a solid film material, a photochromic/thermal reaction in a dye material, ablation or deformation of a film material, or a magnetic state change (often induced with an external magnetic bias field). Clearly, the mode of detection of such induced physical changes differs depending on the type of local change. In phase change (PC) recording, one usually observes a simple modulation of reflectance, while a local topographical alteration due to ablation or deformation induces a light scattering that may be more simply understood as a modulation in the diffractive properties of the film surface. If the magnetization direction of a magnetic film is varied, this is conveniently detected optically with a magneto-optic (MO) effect (Kerr effect in reflection; Faraday effect in transmission). In each of these cases, it is normal to work with a diffraction-limited focused light spot to maximize the spatial resolution of the system. As the data storage medium moves beneath the light beam, the detection process is in every way analogous to that of scanning microscope, and some mode of diffractive analysis (in amplitude, phase, or both) is applicable.

For engineering simplicity, and associated cost minimization, the recording and readout optical systems have as much commonality as possible. This definitely applies to the beam focusing optics, and thus the optical properties of the diffraction limited spot. Only the beam power is modulated, high for writing and low for reading. Readout clearly must be performed in a nonperturbative regime so that data detection is nondestructive, and this limits the readout power that can be employed so that medium heating is constrained. It is also clear that the light detection optics for reading will differ depending upon the type of physical change induced in the medium. For example, modulation of reflectivity or beam scattering may require detection of a variation of the zeroth-order and first-order diffraction, while MO detection implies some analysis of the polarization state of the scattered light, and hence the presence of polarization inducing and detection optics.[2,3] This makes it quite apparent that MO recording systems have the potential for elevated system costs relative to other forms of optical recording.

A great deal of effort has been devoted to increasing optical storage densities. By switching to shorter wavelengths as inexpensive laser diodes become available, the diffraction limit is proportionally reduced. Objectives with higher numerical apertures are also used to reduce the optical spot size. Super-resolution techniques, which overcome the diffraction limit, have been addressed from both the media and the drive design. Marks as small as 20 nm have been recorded in MO media using exchange-coupled multilayers.[4] Thin layers of AgO_x, PtO_x, Sb, or Te in the medium film stack have also been used to create small apertures or scattering centers, and a resultant super-resolution effect, within the recording medium itself during the recording process.[5,6] In the design of the disk drive, solid immersion lenses and physical apertures in the optical read path have enabled recording and reading marks smaller than the conventional diffraction limit. Nevertheless, laboratory demonstrations of small, recorded marks have not yet been translated into commercial products and there is still a great deal of development to be done in this area.

The detected raw bit error rate for optical recording depends on many factors in both the recording medium and the disk drive. Noise sources in the recording medium include defects in the substrate or thin films, roughness in the edges of the recorded marks, groove edge roughness, birefringence in the substrate, and thermal cycling degradation of the films. Noise in the disk drive originates in the laser, in the detector, and in the electronics. A great deal of effort has been devoted to understanding and minimizing these noise sources as well as designing encoding schemes and read channels to provide error detection and correction with minimal overhead, although this is the subject of other chapters.

In this chapter, we will present an overview of the storage medium and recording hardware design issues for a variety of optical recording systems that have been commercialized. We will draft the evolution in these component systems from the introduction of the technology to the present, and to glimpse ahead to the advances that might be expected in the coming decade. Optical data storage shares with other forms of data storage (notably magnetic) a common technological goal of relentlessly advancing data storage density and reducing the cost per unit of capacity. Both trends are intimately linked, and the historical exponential variation is well documented.[7] These metrics have provided consumers of stored information (entertainment and data) with perhaps the most striking example of the fruits of technological advance that we know.

3.2   Optical Recording Media

3.2.1   Introduction

In this section, we consider the constituents of thin film materials and structures that comprise the important types of recordable optical storage media: phase change, organic dye, ablative, surface deformation, and magneto-optical. In each of these media structures, a "memory layer" is usually encapsulated in a sandwich of protective film layers. The attendant layers may also serve a central role in the optical/thermal design of the thin film stack. These films reside on a relatively rigid *disk* substrate for rotary memory devices — for example, CD and DVD drives, various *write once, read many* (WORM) and rewritable disk drives. A primary attribute of optical disk drives is the removability of their storage medium, and the requirement that the disk medium survive repeated exposure to handling and potentially dirty environments has meant that the memory layer be protected with a relatively thick cover plate. Consequently, the structure shown in Figure 3.1 is prototypical for optical storage media — the substrate may double as an optical cover plate to keep contaminants physically separated from the focal plane. Such an arrangement is sometimes called "second-surface illumination." A recent design trend has been to reduce the thickness of the protective cover plate so as to enable employment of higher numerical aperture (NA) optical systems for increased spatial resolution, since diffraction limited optical spot sizes scale as λ/NA. In the extreme, as the cover layer thickness shrinks to zero, one has moved to "first-" or "front-surface illumination." Obviously, this architectural change reduces and eventually eliminates the possibility of keeping disk contaminants away from the optical focal plane (memory layer plane), thus increasing the difficulty of engineering a media removability option.

3.2.2   Phase Change, Organic Dye, and Ablative/Deformative Media

3.2.2.1   GeSbTe and AgInSbTe Phase Change

This chalcogenide alloy system comprises inorganic compounds containing an element from column VIA of the periodic table ("chalcogens"). This system has a propensity for forming amorphous glasses, and phase transitions between amorphous and crystalline states can be induced thermally. Alloys such as the binary Sb_2Te_3, the ternary $Ge_2Sb_2Te_5$, and the quaternary AgInSbTe are prototypical materials forming the basis for most recordable phase change optical storage media, both write once and rewritable. A ternary phase diagram of the representative material GeSbTe (GST) in Figure 3.2 indicates the nearly complete miscibility of the system at all compositions. Of course, physical properties vary considerably over this space. The essential characteristic of interest for optical storage is the ready control of the solid phase structure (crystalline, amorphous) of the material with accessible temperatures (Figure 3.3). For optical detection of phase variation, one utilizes the marked change in refractive indices, and hence reflectivity.

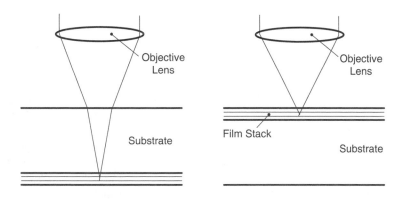

FIGURE 3.1   First- and second-surface illumination film stack architecture of optical media.

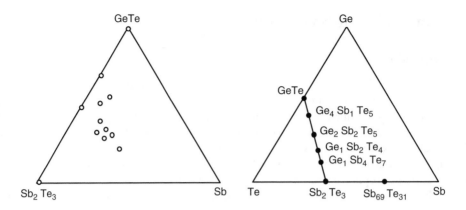

FIGURE 3.2　Ternary phase diagram of GeSbTe.

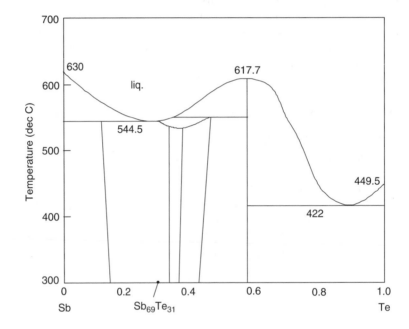

FIGURE 3.3　A binary phase diagram of a chalcogenide compound (SbTe) showing the role of temperature in the setting of the phase state.

Because the physical mechanism of writing in phase change materials is atomic structure rearrangement driven by ambient thermal energy, one is mainly concerned with rates of thermal diffusion and the associated temporal evolution of some atomic order parameter, a measure of the relative degree of crystallization or amorphization. Generally, as one might expect of thermally-driven entropy changes, the production of an amorphous (disordered) atomic state can be induced quickly with sufficiently high temperature followed by rapid cooling (quenching), since it is a high entropy (high probability) configuration. Conversely, annealing the solid into an ordered, crystalline state is relatively slow due to its low entropy (low probability) nature. It is precisely through control of peak temperatures and cooling rates that one controllably sets a "final" metastable amorphous state, or a highly stable crystalline state. Through careful joint engineering of the PC material and the drive laser power control, both WORM and rewritable PC realizations have been extensively commercialized.

3.2.2.2   Organic Dye

Several organic dye materials have been developed for write-once recordable optical media. The three main ones are cyanine, phthalocyanine, or metal-azo. A reflective layer, either silver alloy or gold, is coated over the dye. Phthalocyanine (golden color) and metal-azo dyes appear to have become the preferred memory layer choices for WORM CD or DVD media (CD-R or DVD-R). Phthalocyanine is attractive due to its high transparency and greater stability than cyanine. Metal-azo dye[8] has a blue color and is used with a silver reflective layer. It is more stable against ultraviolet radiation than cyanine. Cyanine has a green color and is generally less reflective, resulting in lower contrast signal between marks and no marks. In the writing process, incident photons induce an irreversible change in the local properties of the dye, resulting in features that are detectible optically via a reflected readout beam. The actual physical change in the dye seems not to be very well understood. It is usually described as a darkening, which changes the local transmission to the reflective layer. However, some dyes absorb enough infrared or visible laser light that they deform from the heat, and so a local topographical feature ("pit") may form. In any case, the feature created by the writing laser power is such that it modulates the readout light returning to the optical detectors via some combination of reflectivity, diffraction, or light scattering. The lifetime of organic dye CD-R or DVD-R media is claimed to exceed ten years, and in some cases even 100 years.

3.2.2.3   Ablative/Deformative/Other

In these WORM media, a thermally-induced topographical change is created on the memory layer, which is read out by a modulated diffractive signal. One of the early forms of recordable optical media used ablative Te metal or Te-based alloys to create pits or holes in a continuous metal film. A hole opened by melting is usually surrounded by a raised rim of frozen metal upon cooling, and so the hole interacts with a reading beam with a combination of diffraction and modulated reflection (between the inside and outside of the hole).

In "bubble-forming" optical media, heat can form dome-shaped marks in a polymer film material. The marks scatter scanned light in readout.

Plasmon once commercialized a form of WORM media employing "texture change," and it was called *moth-eye* media. Here, the surface texture (or roughness) media material was locally smoothed following heating with the writing laser beam. This resulted in a subtle topographical change yielding a large contrast in effective surface reflectivity. "Alloy-forming" media is another WORM type with similarities to phase change media. In this case, thermally-induced mixing of atomic species results in the formation of a metallic alloy exhibiting a reflectivity contrast from the surrounding, unwritten media. This mixing is another physical process with entropic increase, which is practically irreversible.

3.2.2.4   Media Noise Sources

Readout noise sources for the media can be traced fundamentally to stochastic (random) processes associated with light reflection, diffraction, or scattering from the written features in the memory layer. Medium readout noise inevitably accompanies the reproduction of signal, which is the deterministic modulation of the light intensity returned to the optical detectors of the optical head from the written information. Medium noise is observed to be additive or multiplicative, which means that its power level may be independent of, or scale with, signal level.

It is apparent that background media noise arises from light scattering or diffraction from imperfections in the medium reflecting film. Such imperfections could be surface roughness on a spectrum of length scales, or material inhomogeneity in the film's physical properties, such as optical property variation or physical phase variation (which could be expressed as a variable atomic order parameter).

When information is written on the medium, additional sources of media noise can arise. Ensemble variation in the 3D physical profile of the transition from one binary state to the inverse state is a fundamental kind of writing noise in optical or magnetic recording. This transition noise is commonly called *jitter*. Such variation can arise due to changes in writing parameters extrinsic to the medium (temperature, laser power, medium speed, focus and tracking servo), but also due to material property variation in the same sense as background media noise. It is understandable that noise associated with written signal (variance on that signal) can scale in magnitude with that signal level (proportional to the volume of material recorded), and hence be multiplicative. Regions between transitions reflect the background noise level of each of the binary states of the medium. And track edges can have a 50%-level of transition zone (on average, depending on the intertrack media state) which is also a potential noise source.

A polycrystalline material might be expected to exhibit more inherent variation in light reflectivity, diffraction, or scattering than amorphous material, since polycrystalline structure will have variation of grain size, shape, and orientation. This is the very reason that amorphous materials became the preferred vehicle for magneto-optical media.

3.2.3 Magneto-Optic Media

Magneto-optic media is an important form of rewritable optical media. As its name implies, it is *magnetic* media, and the term magneto-optic refers to the physical effect utilized in reading out information recorded magnetically. The MO effects were discovered in the 19th century by Faraday and Kerr as a rotation of the plane of polarization of transmitted or reflected light upon interaction with magnetized material, and the corresponding effects are named for the discoverers.

However, the MO effects are only half the story, since magnetic media must be written in an MO data storage device. As we know, an optical drive is fundamentally an *optical* system, and so a means of high-resolution recording of magnetic media with the simplest possible magnetic components is needed. The answer turns out to be *thermomagnetic recording*, a process employing the focusing optical head as a means of heating the recording medium. Thermomagnetic recording exploits the fact that all permanent magnet materials have temperature-dependent magnetic properties, and by modulating the properties locally by changing the medium temperature, one can enhance and simplify a magnetic recording process. Figure 3.4 illustrates how it is done.

A simple external source of a modest bias magnetic field is placed in the vicinity of the recording zone of the medium, and the optical heating induces a striking reduction in the magnetic switching field

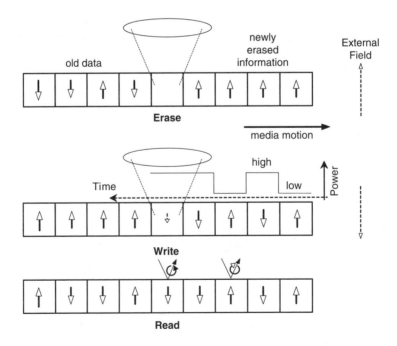

FIGURE 3.4 Thermomagnetic recording schematic: erasing, writing, and reading.

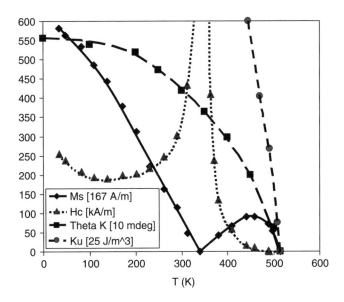

FIGURE 3.5   Temperature-dependent magnetic properties of MO media.[72]

(coercivity H_c) of the magnetic medium (see Figure 3.5). In the medium where the coercivity is reduced below the level of the external bias field, the local magnetization will switch to the bias field direction. Usually the direction of the medium magnetization and the bias field are set perpendicular to the plane of the medium because the MO effects used for readout are strongest in this configuration. As the moving medium passes from under the heating optical spot, the medium rapidly cools, restoring the medium's high coercivity at the writing location. Consequently, at ambient temperature, the written medium is exceedingly stable against undesired reversal due to any cause.

Perpendicular permanent magnetization is a reliable physical realization of a binary state system — therefore magnetization up or down represents the binary digits 0 or 1. We can imagine this local modulation of magnetization being done by switching the direction of the bias magnetic field. However, it is practically difficult to write by modulating the magnetic field, as the self-inductance of the simple bias coil precludes the field switching rates required by the disk speed and recorded density. Instead, the medium may be uniformly erased along a track in a first disk rotation prior to recording, and for the next rotational pass of the disk, the bias field direction is reversed requiring a switching rate of several hundred Hertz. Then, reversed magnetic bits can be written by simply switching the light source on and off at data rates, a much simpler task. It *is* possible, however, to switch the field source at data rates (as is done in conventional magnetic recording), and thus avoid the lost time of one disk revolution to perform a preerase step. This process is called *magnetic field modulation* (MFM) *direct overwrite* (DOW) recording.

3.2.3.1   RE-TM Alloys

When practical thermomagnetic recording was first considered in the 1960s (due to the appearance of laser diodes), the motivation in identifying a medium for MO recording was to find a highly reflective magnetic material with a strong Kerr effect. An early material was the intermetallic compound MnBi, which could be deposited in the form a polycrystalline film on a suitable substrate. Although this metal had both a considerable Kerr response and a reasonably high reflectivity (both important, since the electrical signal derived from an optical detection system designed to measure the Kerr rotation angle of the plane of polarized light gives a voltage $V \propto R \cdot \Theta_K$, where R is the mirror reflectance, and Θ_K is the Kerr rotation angle), it suffered from rather high media noise. The reason is the high variance in reflectance that a polycrystalline metallic film exhibits. A solution was found in the early 1970s when sputter processing of an amorphous film alloy system with rare earth (RE) and transition metal (TM) components was

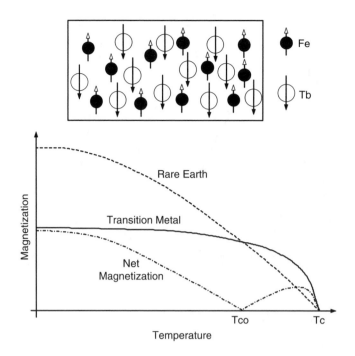

FIGURE 3.6 Two antiferromagnetically coupled magnetic moment subnetworks in an amorphous binary RE-TM alloy.

developed. A prototypical binary alloy is $Tb_x Fe_{1-x}$, where $0.15 < x < 0.35$ yields properties of interest in MO recording systems.

These RE-TM alloys are ferrimagnets over wide composition ranges. A ferrimagnet is a strong cooperatively coupled magnetic material similar to a ferromagnet, except that the ferromagnetically aligned magnetic moments on each atomic species are coupled antiferromagnetically to one another. This is shown in Figure 3.6, where two subnetworks of each alloy component are antialigned. In the amorphous solid, the atoms are arranged quite randomly beyond some short, interatomic correlation length, and yet there exists a definite polarization direction for the magnetization on each amorphous subnetwork, a consequence of the strong uniaxial magnetic anisotropy in these materials.

Figure 3.5 shows schematically the very important temperature dependence of the RE-TM ferromagnetic alloys. While the magnetization versus temperature behavior of each species subnetwork is that of a typical ferromagnet, the difference in magnetization magnitude of the two antiferromagnetically coupled species often gives rise to a special condition at a so-called *compensation temperature* T_{co}, whereby the magnetization magnitudes of the two components crossover and cancel vectorially. At the temperature where the net magnetization vanishes, the coercivity appears to diverge, since an infinite applied field is needed to apply a finite reversal torque to zero magnetization. Two more aspects of the RE-TM alloy system are very important. One observes experimentally that the amorphous RE-TM films have extremely low media noise in comparison to polycrystalline films. This results in superior media signal-to-noise ratios (SNR) for RE-TM materials, even though the Kerr signal itself is only modest, with $\Theta_K < 0.5°$ typically. Second, the RE-TM alloys are chemically very reactive (due to the RE component), and thus environmentally unstable unless special care is taken in media construction. The unprotected alloy will oxidize in the atmosphere very rapidly, and this destroys the magnetism.

The MO Kerr effect is strongest when the light propagation vector $\vec{k}$ is parallel to the mirror magnetization $\vec{M}$. For the optical disk, this implies that the memory layer magnetization should lie perpendicular to the disk (film) plane, that is, we desire perpendicular magnetic anisotropy (PMA). PMA is energetically *unfavorable* from the viewpoint of film shape anisotropy, due to the much stronger demagnetizing fields when $\vec{M}$ is normal to the film rather than parallel to it. For PMA to persist overall requires that the so-called magnetocrystalline anisotropy energy density K_u dominate the shape anisotropy term $K_s = 2\pi M_s^2$ for the perpendicular orientation.

The MO Kerr effect was introduced above, and many excellent references discuss the fundamentals.[9-12] Two practical matters that impact the usable media SNR in MO recording systems are that (a) the optical penetration into the film upon reflection is limited by the skin depth of the metal at the wavelength of operation (in the 20–40 nm range), and (b) the spectral response of the MO effects. Both of these effects diminish the achievable SNR using the Kerr effect. Currently, most drive products operate at $\lambda \sim 650$ nm, and are now evolving toward $\lambda \sim 405$ nm. While the MO response of RE-TM materials is reasonable in the near infrared, it falls off markedly toward the blue. This is a weak point for the application of RE-TM MO materials in the future, unless they can be modified successfully.[13]

3.2.3.2 Co/Pt and Co/Pd Multilayers

This class of MO materials has not yet been commercialized to nearly the extent of the RE-TM alloys. Nevertheless, they have been the focus of research interest for many years due to their potential for short wavelength MO recording in the future.[14-16] We will now summarize the important characteristics of these multilayer (ML) systems in comparison to the RE-TM materials.

Essentially, these are artificial superlattice film structures in which one deposits alternate thin layers (typically $\sim 2 - 20$ Å) of elemental species. These materials exhibit strong PMA and Kerr response when the Co layers are in the 3–6 Å range, and the Pd or Pt layers are 8–20 Å. The films are polycrystalline, but with grains small enough[17] (<20 nm) to provide optical readout without overwhelming media noise. Another attractive feature of the Co/X (X = Pd, Pt) is their chemical stability. A multilayer capped with Pd or Pt can be safely exposed to the atmosphere without fear of chemical reactions that seriously degrade the magnetic properties.

Co is the ionic species mainly responsible for the formation of strong PMA in these films, although an unambiguous identification of the mechanism responsible has been elusive. The Pd and Pt contribute to the sizeable MO effect at short wavelength, probably due to partial magnetic polarization of these layers by the ferromagnetic Co.[18] The Co/X films exhibit transitional behavior between 2D and 3D ferromagnetics, depending on the thickness of the Co layers. As such, the $M_s(T)$ and $H_c(T)$ curves are more or less convex, dropping monotonically to zero at the Curie temperature.[2,3] T_{Curie} for Co/Pt can be adjusted over a wide range from a value close to that of pure Co (1404 K) downward to ~ 500 K for Co(2.4 Å)/Pt(13.1 Å), thus illustrating the dilution of the exchange coupling in the material. The Co/X ML's can offer two to three times higher MO figure of merit ($R\Theta_K \propto$ MO signal) around $\lambda = 400$ nm compared to the RE-TM materials.[19] This alone will give the ML's a large advantage for MO media products of the future.

3.2.4 Dielectrics

Practical optical media structures require materials beyond the memory layer(s). In virtually all recordable media, the element of protection of the active memory layer is paramount, for the memory materials alone may be mechanically fragile and/or chemically reactive.

Dielectrics for optical media application are usually required to have an optical, thermal, chemical, and mechanical function. A common usage is to encapsulate the memory material, providing it mechanical support, environmental protection, and proper thermal and optical interaction with the light source and the remainder of the medium structure. Figure 3.7 shows examples of complete film stacks for phase change and MO disk media, and we see how dielectric materials surround the memory layer. In order for light to reach the memory layer, it must traverse one or more dielectrics to perform its function of heating or information interrogation.

In optical storage media, dielectrics are a class of electrically insulating materials that are typically highly transparent oxides, nitrides, sulfides, or carbides. Examples are the chemically inert compounds silicon nitride SiN, silicon carbide SiC, zinc sulfide ZnS, aluminum nitride AlN, and silicon dioxide SiO_2. High transparency implies that the imaginary part of their refractive indices is very small at the operating wavelength (consistent with the insulating property). Ideally, the dielectric material is optically homogeneous so that optical scattering (a noise source) is minimal. Associated with the property of electrical insulation

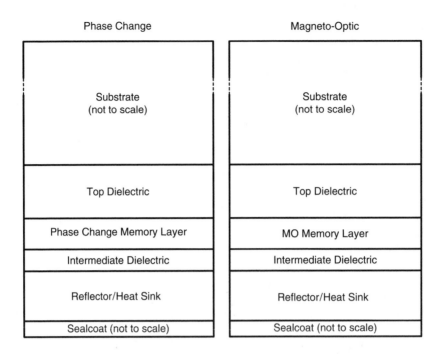

FIGURE 3.7   Typical film stacks for phase change and RE-TM MO media.

is a thermal resistivity generally much higher than that of metals. Consequently, dielectrics often play an important role as thermal insulators in film stacks.

Mechanically, Figure 3.7 suggests that dielectrics should exhibit strong film adhesion to maintain high integrity encapsulation throughout the lifetime of optical media that is repeatedly thermally cycled. Perhaps less apparent is the issue of film stress. Films can be prepared in states of high or low tension or compression. High stress is undesirable, as it can be a precursor to film delamination, or stress can have detrimental effects on substrate (e.g., birefringence in plastics) or memory materials (e.g., magnetomechanical effects in MO films).

3.2.5   Heat Sink/Reflector

Another film stack constituent is a nonmagnetic metallic layer that serves as a reflector, heat sink, protective cap, or some combination of these. Again, in Figure 3.7 we see examples of this in phase change or MO media. Metals provide the opposite optical and thermal function of dielectrics — metals are opaque, but highly reflective (depending on surface roughness), and highly thermally conductive. We note in passing that thermal conductivity of solids can have important components due to electronic transport and lattice vibrations. The former is coincident with electrical conductivity — in fact electrical and thermal conductivity are proportional to one another via the Wiedemann-Franz law (from the framework of free electrons in a metal):

$$\kappa_e = \frac{\pi^2 k_B^2 T \sigma}{3e} = LT\sigma \tag{3.1}$$

where κ_e is the electrical component of thermal conductivity, k_B is Boltzmann's constant, T is absolute temperature, σ is electrical conductivity, e is electronic charge, and L is the Lorentz number, a fundamental constant. Lattice vibrations are present in both insulators and conductors.

The physical function of the heat sink/reflector in optical disks is vital. The placement of a reflector in an optical film stack is clearly helpful if one is attempting to achieve optical cavity function in which light energy is recycled within the film stack. And the reflector blocks the light from escaping from the media system and returns it to interact further with the memory layer. A heat sink function implies that cooling rates can be manipulated through placement of the sink through its thermal conductivity and its heat capacity. Finally, not only is this metallic layer a barrier to light, but it is also an effective barrier to atomic diffusion. This is why a metallic protective cap is effective in sealing a reactive film stack against environmental degradation (corrosion).

3.2.6 Film Stack Design

A layered film (stack) can be optimally engineered and configured for the intended function, and there are in fact common design objectives among phase change, organic dye, and MO disks. , Fundamentally, we want to utilize the incident light energy efficiently in both writing and reading. This means we do not want to generate wasted heat in writing, for excess heat is always detrimental to material systems. The disk must be designed to function reliably for thousands to millions of writing cycles, so the heat management requirements are severe. Similarly, on reading we want to use limited light energy to extract the maximum possible signal with a minimum of detected noise.

3.2.6.1 Optical Figure of Merit

Optical recording engineers have devised optical figures of merit (FOM) primarily as a quality metric for the readout process. Nevertheless, one cannot design for reading solely while ignoring the requirements for writing. A central reason for this is that writing requires light *absorption in* the medium, while reading requires primarily *reflection from* the medium. (Reading may involve absorption as well — this is the case in MO readout.) Since most optical disks do *not* transmit any light, and because the simple general relation $R + A + T = 1$ (where R = reflectance, A = absorbance, and T = transmittance) expresses conservation of the flux of light energy (or power) entering and leaving the medium, we have in practice the simpler result $R + A = 1$. This tells us that reflectance and absorbance are complementary — as one increases, the other decreases, and vice versa. The obvious meaning of this for optical media design is that writing and reading are complementary.

The first, and simplest, MO system FOM is that for MO signal alone which is proportional to $R\Theta_K$. Taking the next step toward SNR, but recognizing that system noise can be complicated (having many potential sources, with different character), we simplify system noise by supposing that our laser, media, and electronics are perfect, but that we cannot escape a fundamental quantum noise of light detectors — shot noise amplitude $\propto \sqrt{I_{det}} \propto \sqrt{R}$. In this case, SNR (power ratio) $\propto R\Theta_K^2$. This is an upper bound on system SNR, and is a second important MO FOM.

A third MO FOM was developed by Mansuripur.[20] For a single thin MO film in an arbitrary film stack, he derived the fundamental relation

$$r_\perp = \left(|\varepsilon'| / 2\text{Im}\,\varepsilon \right) \left(P_{abs} / P_{inc} \right) \tag{3.2}$$

where $r_\perp$ is Kerr reflection coefficient (E_y/E_x), ε' is the off-diagonal term in the MO film dielectric function, ε is the diagonal term of the dielectric function, and P_{abs} and P_{inc} are the absorbed (in the MO film) and incident reading light power, respectively. $r_\perp$ is proportional to Θ_K, and hence MO signal. This expresses an upper bound on the MO signal that can be expected from any film stack design, given the intrinsic MO and optical properties of the medium.

One final MO FOM is worth noting. This recognizes that the physics giving rise to the MO Faraday effect is really the same as that for the MO Kerr effect — only the positioning of the detection optics is different to handle transmission or reflection, respectively. However, another class of MO media material that we did not discuss is "hybrid" in the sense that it is a highly transparent rare earth oxide — called garnets, crystalline solids with the general formulae $R_3Fe_5O_{12}$ or $3R_2O_3 \bullet 5Fe_2O_3$, where R is a rare earth species. Many of their properties are nearly ideal for short wavelength MO application.[21-23] Due to the low absorption, rather thick garnet films can be built into a reflective film stack by putting a reflective layer at the back. In such a situation, an MO FOM proportional to MO signal can be written as[19]

$$2d\Theta_F \exp\left(-2\alpha d\right) \qquad\qquad (3.3)$$

where d is the film thickness, Θ_F is the Faraday rotation, and α is the film absorption coefficient.

FOM's for phase change (or other reflective readout optical media) are possible.[24] They turn out to be similar to the basic concept for any optical media in that one is seeking an optimal combination of reflectance and absorbance to accommodate both the writing and reading function on the same medium.

A more practical means of optimizing optical design for media film stacks involves computer simulations, usually utilizing analytic and/or empirical expressions for thin film optical interaction and system noise components[2,25,26]. Much more elaborate FOM's can be constructed which consist of arbitrary combinations various optical/system responses, such as signal, SNR, reflectance, Kerr rotation, phase shift, etc.

3.2.6.2   Thermal Properties

Analogous to optical stack design optimization is stack thermal design. Conventionally, engineers have used classical heat diffusion analysis in continuum materials for practical thin film structures. Such analysis ignores additional effects on the transport of thermal energy such as phonon or electron scattering by interfaces, boundaries, and material defects, inhomogeneities and imperfections. Microscale or nanoscale heat transport analysis is the remedy, but it increases the complexity of analysis considerably.[27,28] To some extent, this shortcoming can overcome by adjusting the continuum bulk parameters of thermal conductivity and specific heat to effective values, which may be derived either from experiment or rigorous simulation of actual geometries via the Boltzmann transport equation.[29]

Recently, PC recording has matured to the point that detailed thermal management has become essential to control heating and cooling rates, and these approaches have long been standard practice in MO recording.[30] In rewritable PC recording, the physical processes are essentially annealing (for crystallization) and quenching (for amorphization), so it is apparent that thermal management is absolutely central to this technology.

3.2.6.3   Magnetic Properties: Exchange-Coupled and Field-Coupled Multilayers

Returning to MO media considerations, more complex multilayers can be utilized to construct a composite memory layer.[3] This adds design flexibility for extending the optimization of thermomagnetic recording. The general concept is to deposit two or more adjacent magnetic alloy films that are coupled together magnetically. One may employ either a very strong interatomic exchange coupling (either ferro- or antiferromagnetic), or a weaker, longer-range magnetostatic coupling. Three leading motivations for this strategy are to achieve extended media function by (a) using separate layers for information storage (memory layer) and readout, (b) establishing a media direct overwrite solution, or (c) achieving a media magnetic super-resolution (MSR) function.[31,32]

3.2.7   Substrate

A thin film stack is always deposited on a thick supporting substrate.[1,3] In removable optical disk systems, the substrate is normally transparent and doubles as an optical cover plate in "second-surface" illumination. This arrangement keeps dust, dirt, and other contamination far from the focal plane of the media. Further, the substrate is a carrier for topographical formatting information to define tracks and data addresses. In this application, the optical properties of the substrate are critical.

3.2.7.1   Optical Properties: Transparency, Birefringence, Roughness

When the substrate is an optical cover sheet, it must be highly transparent and should contribute minimal optical aberration to the transmitted beam. Low cost optical storage systems have utilized plastic or polymer substrate materials (polycarbonate or polymethyl methacrylate — PMMA). These are convenient for format replication through a molding process (see below), but these organic solids tend to exhibit much larger birefringence than, for example, glass. Birefringence is an optical aberration (anisotropy in refractive indices) imparted by the molecular structure of the material, or by stress-induced strains. Surface roughness is critical when the surface in question is near the optical focal plane. This is the surface on which the thin films are deposited. Surface irregularities are replicated in the growing films, and this can become another source of light scattering, which contributes readout medium noise.

Glass substrates are preferable from the optical quality standpoint. However, glass is relatively expensive, and is difficult to impart with replicated format structure. Nevertheless, glass is used in some high-end optical storage systems.

3.2.7.2   Mechanical Properties

The substrate should be very smooth, flat, and rigid. Flatness is an issue for the focusing and tracking servos, since a rotating disk presents a dynamic surface for the optical system to follow. Similarly, mechanical rigidity is important to minimize dynamic surface vibrations and fluctuations for the servo systems to follow.

Additionally, it is useful if the substrate material can act as a mechanical barrier to contaminants and corrosion agents, since thin film stacks are usually protected on at least one surface by the substrate. This function can be mitigated if a metallic cap is deposited on the substrate prior to deposition of the remainder of the film stack, but this "underlayer" may complicate film growth and property control.

3.2.7.3   Thermal Properties

Because the substrate is in intimate contact with the thin film stack, it inevitably is a part of the thermal environment of the medium. Plastics are poor thermal conductors, and glasses are only moderately so, which means reliance on a nonmetallic substrate to function as a heat sink may be ineffective. On the other hand, the heat capacity of the substrate is huge by virtue of its volume relative to the thin films, so it will have moderate effectiveness as a heat sink if it is not purposely insulated from the film stack.

3.2.7.4   Formatting

This is a process of creation of topographical features (or at least optically contrasting ones) on the surface of the substrate. Spiral or concentric grooves comparable to the dimensions of the intended track pitch (approximately equal to the optical spot size) are formed in plastic by an injection molding process using a hard metallic stamper. The grooves are accompanied by (or sometimes interrupted by) small pits placed on the "lands" which separate the grooves (see Figure 3.8). As described in the Optical Head section, servo signals for the focusing and tracking systems are derived from light diffraction from these features on the rotating disk.

One of the cost advantages of optical storage over hard disk drive magnetic storage is that this hard-formatting can be replicated inexpensively to thousands of disks with the use of a single format master.

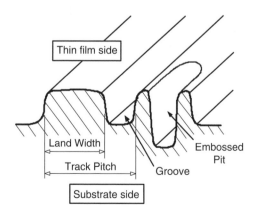

FIGURE 3.8   Schematic of embossed optical disk substrate surface topography, showing tracking grooves and pits on the land.

Formatting of glass can be done with etching through masks, and while not as inexpensive as molding or stamping of plastic, it can still be low cost relative to servo-writing of each medium.

3.2.7.5 Noise Sources

Birefringence and light scattering from defects or surface roughness and contamination (dust, dirt, fingerprints, etc.) are the main contributors of substrate noise. Clearly, this is only an issue in second-surface configurations, but this arrangement is still dominant in commercial, removable media optical storage products. The recent trend toward reduced thickness substrate cover sheets is helpful in diminishing such noise contributions. Because substrate noise arises when a converging focused beam traverses the cover sheet, the light energy is sampling a relatively large area of the substrate material. Also, birefringence due to internal stress/strain patterns in the material is a relatively long-range physical perturbation. For these reasons, noise due to substrate imperfections tends to be a low frequency phenomenon, and depending on the details of the signal detection channel, one may have an opportunity to separate this noise spectrally from the main information band.

3.2.8 Direct Overwrite

We earlier introduced DOW as a special function necessitated in MO recording by the restriction on bias field source switching rate. While it is possible to adopt a direct solution for DOW using a fast switching bias field source as in magnetic recording, this may prove more expensive at the drive level than an alternate media solution.

The first form of DOW proposed for MO media was the simplest, in which erasure of previously written information proceeded under the action of thermally-induced self-demagnetization.[33] That is, an existing magnetic domain has an internal demagnetizing field proportional to its magnetization, and if the local temperature is elevated such that the coercivity drops, then the demagnetizing field may be sufficient in strength to form a reversed domain or collapse and erase an existing domain without application of any external field. Unfortunately, this form of recording with automatic overwrite was never perfected to a level enabling it to appear in commercial products.

Through clever multilayer engineering using RE-TM alloys, intricate DOW function has been achieved.[3,31,32] In one approach, use of an "initialization layer" coupled with a two-level DC bias magnetic field and a two-level writing power sequence enabled one-pass DOW with good performance. The engineering price for this achievement is more complex media thin film structure (and higher media cost), coupled with tighter tolerances on media production and drive operating margins (including temperature-power control). Nevertheless, media DOW solutions have been commercialized, perhaps proving that this engineering tradeoff can be made viable.

DOW function is built into rewritable phase change media designs. Sometimes, however, perfecting the rewritable function, particularly at high data rates, requires adding one or more thermal management layers to even an existing quadrilayer PC disk structure.

3.2.9 Super-Resolution

In general, super-resolution [SR] implies an attempt to extend the available effective spatial resolution beyond that allowed by the optical system. Because optical system resolving power is hard-limited by parameters that may be nontrivial to adjust — wavelength and numerical aperture, there is important payoff in terms of areal density and data capacity in finding alternate means of achieving SR, both in PC and MO systems.

3.2.9.1 AgO_x or PtO_x and Sb or Te

These materials have been recently introduced into PC media as SR approaches. The acronym put forward by the developers is super-RENS, meaning "super-resolution near-field structure."[34,35] Recently, more understanding of the phenomenon has come forth.[36,37] Super-RENS structures have a thin, nonlinear optical material such as AgO_x, PtO_x, Sb, or Te deposited just above a PC memory layer. Apparently incident radiation for readout can dissociate Ag or Pt particles from the oxide, and a near-field re-radiation from the metallic particles (involving surface and bulk plasma polaritons) mediates radiation reflection from the memory material with significantly enhanced spatial resolution. The process with AgO_x or PtO_x appears to be reversible to some degree, so reasonable readout cycling looks to be possible. The process with Sb (or Te) metal is less clear. Apparently, the thin Sb or Te film can melt at the hottest location in the medium, and a dynamic aperture is created. The level of SNR achieved in these PC SR systems is improving, but so far it is not believed adequate for commercialization.

3.2.9.2 Magnetic Multilayers (MSR, MAMMOS, DWDD)

Magnetic super-resolution has been commercialized in many MO drive designs. Early research demonstrations led the way for these product solutions.[38-41] A multilayer MSR solution in MO media relies on a form of "Kerr readout aperturing." If the MO readout layer in the drive has in-plane magnetization, there is zero polar Kerr effect in this configuration. Through use of a readout layer material (RE-TM) that transitions between in-plane magnetic anisotropy at $T_{ambient}$ and PMA at elevated temperature, one can effect a "masking" of memory layer information under the "cold" media zones beyond the light spot heating zone. This effectively opens a thermal aperture of controllable size through which information is read. While this can enhance optical resolving power (at the expense of received light intensity), additional readout noise can arise from the mask, and it can suffer from an SNR penalty.

Two other forms of super-resolution in MO media are magnetic amplification MO system (MAMMOS) and domain wall displacement detection (DWDD). In both of these multilayer systems, a super-resolution readout function is combined with signal amplification for SNR enhancement. In both cases, information is recorded at a linear density exceeding the nominal spatial resolution of the readout optical system. This is done by recording with MFM, possibly combined with laser pulse modulation, to form short domains in the memory layer. Upon readout at the lower power levels appropriate for this function, information from the memory layer is copied upward to a readout layer, usually through at least one intermediate coupling layer. Finally, the replicated domain in the readout layer expands to a size comparable to the thermal profile imparted by the readout beam, which is fully compatible with readout at maximum signal level by the diffraction-limited beam.

In MAMMOS, the domain expansion mechanism enlarges a replicate of the small memory domain in each lateral direction. Two types of currently available MAMMOS are called RF-MAMMOS ("radio-frequency" applied, modulated magnetic field) or ZF-MAMMOS ("zero field"). The DWDD method involves a readout domain expansion driven mainly by the advancing thermal gradient at the rear of the readout beam. Both of these schemes have been perfected through clever attention to thermal and magnetic material engineering detail, and both are expected to appear in commercial products in the foreseeable future.

3.3 Optical Recording Systems

There are a variety of optical recording systems corresponding to the different types of optical media. All recording and playback systems, however, may be divided into discrete units with similar functions. The optical layout for a generic optical recording system is pictured in Figure 3.9. The layout is divided into a light source with beam shaping optics, a beamsplitter, an objective, the storage medium, the focus and tracking optics, and the read optics. Several of these functions are often combined into a single optical device within a commercial product, but each of these sections will be considered separately.

3.3.1 Light Source

The diffraction limit is a well-known criterion for determining the smallest optical spot size for a focused beam.[42-45] The minimum spot size for a focused beam at the full width - half maximum (FWHM) is

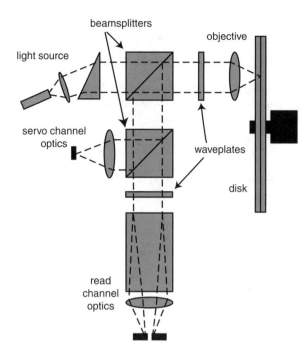

FIGURE 3.9   Generic optical disk drive layout.

approximately $0.51 \lambda/(NA)$ where λ is the wavelength in free space and NA is the numerical aperture of the objective. The optical spot size is directly related to the smallest mark which can be recorded or read back in an optical storage system. A source of coherent light is necessary to obtain the diffraction limit. Semiconductor lasers, also called laser diodes (LD's), are low cost, robust sources of coherent light of sufficient power for optical storage. LD's operating in the near infrared at 780 or 830 nm are used for CD's and low density MO disk drives. Visible LD's at wavelengths of 635 to 680 nm are used in DVD and high capacity MO disk drives. LD's which operate at the blue/violet wavelength of 405 nm are now appearing in DVD Blu-Ray disk drives. A LD must provide a power of 1 to 2 mW at the disk for read back, while the typical power required at the disk for recording is about 8 to 10 mW.

Index-guided laser diodes are primarily used in optical data systems where high beam quality and focusing to the diffraction limit is essential. The astigmatism for this type of laser is very small (typically 5 to 15 microns) and the divergence angles are very consistent from laser to laser. The polarization extinction ratio is about 100:1.

Diffraction from the rectangular output facet of the laser diode gives rise to a diverging output beam with an elliptical cross section. The divergence is typically 2 to 3 times larger for the narrow dimension of the facet than for the wide dimension and the point from which the diverging beam appears to originate is not the same in the vertical and horizontal directions. A combination of a spherical and cylindrical lens or a single molded aspheric lens can be used to correct for this astigmatism and collimate the beam. To obtain the smallest focused spot size, the cross section of the beam must also be circularized either by an anamorphic prism pair or a pair of cylindrical lenses.

3.3.2   Beamsplitter and Waveplate

The collimated beam passes through a beamsplitter (BS) on its way to the objective lens. The BS splits off some of the light reflected from the disk for the read signal and the focus and tracking servos. In CD and DVD systems a polarizing BS (PBS) can prevent light from returning to the laser and causing feedback noise. The PBS transmits p-polarized light and reflects s-polarized light. A quarter wave plate (QWP) after the PBS and before the objective is oriented at 45° to the incident polarization so that the beam transmitted to the objective becomes circularly polarized. The reflected beam upon passing back through the QWP becomes linearly polarized with the orthogonal polarization to the incident beam and is entirely reflected by the PBS to the detectors.

Circularly polarized light also minimizes the effect of birefringence in the plastic substrate on the reflected light intensity. Unfortunately the light incident on MO media must be linearly polarized to generate a read signal by a small change in polarization angle. The PBS for MO systems is "leaky." It transmits most of the incident p-polarized light and reflects nearly all of the incident s-polarized light. The light reflected from the medium has its plane of polarization rotated slightly by the MO Kerr effect of the medium. On the return pass through the leaky PBS about 15% of the unrotated p-polarized component is reflected to the read optics and all of the rotated s-polarized signal component is reflected.

3.3.3   Objective

The objective lens brings the light to a diffraction-limited spot on the active layer of the recording medium. This lens is generally a single molded glass part with two aspheric surfaces.[46] Increasing the NA of the objective reduces the spot size. CD's use objectives with an NA of 0.45. The objectives in MO drives have an NA of 0.55, and in DVD drives an NA of 0.60. The new DVD Blu-Ray drives use objectives with an NA of 0.85.

Vector diffraction theory must be used for objectives with large NA's to calculate the correct size and polarization of the focused spot.[47,48] The diffraction-limited spot for a uniformly illuminated, aberration-free objective with an NA of 0.85 at a wavelength of 405 nm corresponding to DVD Blu-Ray is graphed in Figure 3.10. The spot size is also calculated for an objective with an NA of 0.45 at a wavelength of 780 nm, corresponding to CD. The difference in spot size is striking and explains in part the 30 × increase in storage capacity of DVD Blu-Ray over CD.

3.3.4   Focus and Tracking Servo

The dependence of the depth of focus is quadratic in the NA of the objective.[49] For a CD player with $\lambda = 780$ nm and an NA of 0.45, the depth of focus is about 2 μm. For a Blu-Ray DVD player with $\lambda = 405$ nm and an NA of 0.85, the depth of focus is only 0.28 μm. Disk warpage causes axial runout.

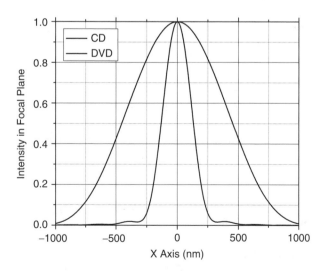

FIGURE 3.10   Theoretical intensity in the focal plane of an aberration-free objective in a typical CD and DVD Blu-Ray system.

The maximum allowable warpage for CD's is ± 500 μm. For DVD's it is ± 300 μm.[50] As a general rule of thumb, the focus servo of the disk player/recorder must keep the focused spot on the reading/recording layer to within about 10% of the depth of focus as the disk spins and the optical head changes tracks. The objective is attached to a voice coil motor which actuates the lens motion to maintain focus. A large variety of techniques have been invented to generate the focus error signal for feedback to the servo.[51-53]

As the disk spins, eccentricity of the disk or improper centering of the disk on the axle causes radial runout. For CD's this is specified to be no more than 140 μm peak-to-peak. The allowable radial runout for DVD's is only 100 μm peak-to-peak. Vibrations add nonrepeatable runout. There are also effects from temperature variations and stretching of the substrate as it spins that cause the data track to shift under the focused spot. Again, a variety of techniques have been invented to generate an error signal from the reflected beam for feedback to a tracking actuator attached to the objective to maintain the focused spot on the desired data track.[54-62]

3.3.5   Read Signal

The read signal is due to a variation in light intensity on the read detector(s). For prerecorded CD's and DVD's and write-once CD-R and DVD$\pm$R media the light intensity variation is caused by an interference or diffraction effect from the light reflected from the pits (or bubbles) in the disk. For CD$\pm$RW and DVD$\pm$RW rewritable media the light variation is due to a reflectance change in the recorded mark. Recorded marks in magneto-optical media use the polar MO Kerr effect to change the state of the reflected polarization. The change in polarization state is converted into an intensity variation at the detector by a PBS with or without an associated phase retarder.

In the generic layout for an optical disc player in Figure 3.9, the objective captures only a portion of the light reflected from the disk which is scattered by the pits. This scattered light is focused on the read signal detector. The difference in signal intensity between a pit and the surrounding land is due to an interference or diffractive effect. When the focused spot overlaps a pit, the light reflected from the "land" around the edges of the pit arrives at the detector with a phase shift relative to the light reflected from the pit. If the pit is a quarter wavelength deep, the relative phase shift is 180° and destructive interference between the two parts of the beam causes a reduction in the reflected light intensity. In actual practice the pits for a CD have a depth of $\lambda/6$ which is a compromise to enable the push-pull tracking technique. The pit depth for DVD's is optimized for the read signal at $\lambda/4$ and push-pull tracking is not used.[63]

The optical film stack for CD-RW media is designed so that when the active layer, the phase change material, is in its amorphous, written state the reflectance is low. In the crystalline, erased state, the reflectance is higher. The read signal is due to the reflected intensity change at the detector.

The film stack for MO media is designed so that the polar MO Kerr effect causes a small (typically less than 1°) rotation of the plane of polarization of the reflected beam. The direction of rotation depends on the magnetic state of the recorded mark. The reflectance of the two magnetic states, however, is equal. A waveplate may be inserted into the optical path after the leaky PBS to adjust the polarization state of the reflected beam for optimum detection. A second PBS is oriented at 45° to the incident polarization so that it splits the reflected beam nearly in half. Each half beam is then detected, and the read signal is the difference of the two signals. Jones matrices can be used to analyze the read path and signal.[64]

3.3.6   Noise

There are a variety of noise sources in optical recording, including laser noise, disk (media) noise, shot noise, and post-detection electronic noise.[65-67] The common mode noise is due to both the laser and the disk. A well-balanced differential detection system as used in MO read back systems can reduce the common mode noise by typically 5 to 10 dB.[68] MO differential detection can be unbalanced by birefringence in the substrate and unmatched or nonlinear detectors and amplifiers.[67] The rotation of the polarization due to the Kerr signal also slightly unbalances the differential channel. The sum channel read back for CD and DVD systems does not reject the common mode noise.

When a LD is operated at lower power levels as is typical during read back, laser emission generally occurs for several different longitudinal modes.[69] Small changes in temperature affect the cavity length via thermal expansion, and cause a redistribution in laser power for each mode which is one source of laser noise. In addition, the individual longitudinal modes differ in wavelength by about 0.2 to 1 nm depending on the length of the laser cavity. When the power is redistributed in wavelength even if there is no change in total power emission, if the optical system is not achromatic there can be variations in the focus and/or tracking channels as well as variations in the focused spot size and position along the track, causing additional noise.

Optical feedback into the laser cavity generates noise by varying the amplification of different longitudinal modes. If optical isolation with a QWP is not used, the feedback noise can be minimized by impressing upon the laser drive current a high frequency (e.g. 0.5 GHz) modulation that drives the laser between cut off and the operating power level. This forces the laser to rapidly switch between different longitudinal modes, thereby washing out the noise from optical feedback. This technique can reduce the laser noise by 2 to 6 dB, although a stable, temperature-controlled laser with no feedback is always quieter.

Shot noise is due to the inherent randomness in the times at which incident photons create electron-hole pairs within a semiconductor photodetector. Shot noise is not usually the dominant noise source. Electronic noise is that noise component present at the output of the detector amplifiers when there is no light on the detectors. It can be due to several factors including shot noise from dark current in the detector, bias current, thermal noise sources, and $1/f$ noise.[70]

The optical spot focused on the data medium is not entirely confined to the track that is being read. The Gaussian intensity profile of the focused spot allows some of the light to reach the neighboring tracks and be reflected into the read channel. Aberrations in the focused spot increase the spot size and the amount of unwanted reflected light from neighboring tracks as well as the next transition down track. The high frequency performance of the optical channel is degraded and intersymbol interference increased. Therefore, the spot size must be carefully controlled to allow only a minimal amount of unwanted light into the reflected beam.

3.4 Conclusion

Optical data storage has developed over the past 30 years into a robust medium for information interchange and archiving. It has found an even larger application as a distribution medium for audio and video. Many technological achievements were required to enable this storage technique. Continuing improvements in these technologies have enabled storage capacities to increase from 700 MB/disc for CD's to 23 GB/disc for single layer DVD Blu-Ray. Further increases in storage density are expected by continuing to decrease wavelength, increase the numerical aperture (NA) of the objective lens, and by beginning to employ various superresolution techniques like solid immersion lenses, Super-RENS, or MAMMOS. At the same time there are many challenges to overcome. Larger NA's require thinner substrates or cover layers to minimize the effects of aberrations, but this means that surface contamination will become more of a problem. At UV wavelengths plastics are not transparent and near field optics will require flying optical heads like hard disk drives. Nevertheless, the capacity per disk is expected to reach 200 Gb/in^2.[71]

Acknowledgments

Useful conversations with Walt Eppler, Ed Gage, Chubing Peng, and Tim Rausch are gratefully acknowledged.

References

[1] Marchant, A.B., The optical stylus, *Optical Recording, a Technical Overview*, Addison-Wesley, Reading, MA, 1990.

[2] Mansuripur, M., *The Physical Principles of Magneto-optical Recording*, Cambridge University Press, Cambridge, UK, 1995.

[3] McDaniel, T.W. and Victora, R. H., Eds. *Handbook of Magneto-Optical Data Recording*, Noyes Publications, Westwood, NJ, 1997.

[4] Awano, H., 20 nm domain expansion readout by magnetic amplifying MO System (MAMMOS), *IEEE Trans. Magn.* **36**, 2261, 2000.

[5] Tominaga, J., Nakano, T., and Atoda, N., An approach for recording and readout beyond the diffraction limit with an Sb thin film, *Appl. Phys. Lett.* **73**, 2078, 1998.

[6] Tominaga, J. et al., The characteristics and the potential of super resolution near-field structure, *Jpn. J. Appl. Phys.* **39**, 957, 2000.

[7] *Optical Disk Storage Roadmap*, National Storage Industry Consortium, San Diego, CA, 2000.

[8] Park, H. et al., Synthesis of metal-azo dyes and their optical and thermal properties as recording materials for DVD-R, *Bull. Chem. Soc. Jpn.*, **75**, 2067, 2002.

[9] Argyres, P.N., Theory of the Faraday and Kerr effects in ferromagnetics, *Phys. Rev.* **97**, 334, 1955.

[10] Hunt, R.P., Magneto-optic scattering from thin solid films, *J. Appl. Phys.* **38**, 1652, 1967.

[11] Pershan, P.S., Magneto-optical effects, *J. Appl. Phys.* **38**, 1482, 1967.

[12] Freiser, M.J., A survey of magnetooptic effects, *IEEE Trans. Magn.* **4**, 152, 1968.

[13] Le Gall, H., Sbiaa, R., and Pogossian, S., Present and future magnetooptical recording materials and technology, *J. Alloys Compounds,* **275–277**, 677, 1998.

[14] Mes, M.H. et al., CoNi/Pt multilayers for magneto-optical recording, *J. Magn. Soc. Jpn.*, **17**, Supplement S1 44, 1993.

[15] Meng, Q. et al., Curie temperature dependence of magnetic properties of CoNi/Pt multilayers, *J. Magn. Magn. Mater.* **156**, 296–298, 1996.

[16] Van Drent, W.P. et al., Spectroscopic Kerr investigations of CoNi/Pt multilayers, *J. Appl. Phys.* **79**(8), 6190, 1996.

[17] Weller, D. et al., Thickness dependent coercivity in sputtered Co/Pt multilayers, *IEEE Trans. Magn.* **28**, 2500, 1992.

[18] T.R. McGuire, J.A. Aboaf, and E. Klokholm, Magnetic and transport properties of Co-Pt thin films, *J. Appl. Phys.* **55**, 1951, 1984.

[19] Suzuki, T., Magnetic and magneto-optic properties of rapid thermally crystallized garnet films, *J. Appl. Phys.* **69**, 4756, 1991.

[20] Mansuripur, M., Figure of merit for magneto-optic media based on the dielectric tensor, *Appl. Phys. Lett.* **49** 19, 1986.

[21] Hansen, P. and Krumme, J-P., Magnetic and magneto-optical properties of garnet films, *Thin Solid Films* **114**, 69, 1984.

[22] Krumme, J.P. et al., Optical recording aspects of rf magnetron-sputtered iron-garnet films, *J. Appl. Phys.* **66**, 4393, 1989.

[23] Hansen, P., Krumme, J.P., and Mergel, D., *J. Magn. Soc. Jpn.* **15**, 219, 1991.

[24] Challener, W.A., Figures of merit for recordable optical media, *Proc. SPIE* **3109**, 52, 1997.

[25] Grove, S.L. and Challener, W.A., *Jpn. J. Appl. Phys.* **28**, 51, 1989.

[26] Atkinson, R., Salter, I.W., and Xu, J., Angular Performance of Phase-optimised magneto-optic quadrilayers, *Opt. Eng.* **32**, 3288, 1993; Design, fabrication and performance of enhanced magneto-optic quadrilayers with controllable ellipticity, *Appl. Opt.* **31**, 4847, 1992.

[27] Majumdar, A., Microscale heat conduction in dielectric thin film, *J. Heat Transfer, Trans. ASME* **115**, 7, 1993.

[28] Asheghi, M., Leung, Y.K., Wong, S.S., and Goodson, K.E., Phonon-boundary scattering in thin silicon layers, *Appl. Phys. Let.*, **71**, 1798, 1997.

[29] K. Banoo, Direct solution of the Boltzmann transport equation in nanoscale Si devices, Ph.D. thesis, School of Electrical and Computer Engineering, Purdue University, December 2000.

[30] See proceedings of recent optical storage conferences, such as ISOM, ODS, MORIS for many papers on thermal design.

[31] Saito, J. et al., *Jpn. J. Appl. Phys.* **26**, 155, 1987.

[32] Aratani, K. et al., Overwriting on a magneto-optical disk with magnetic triple layers by means of the light intensity modulation method, *Proc. SPIE* **1078**, 258, 1989.

[33] H-P.D. Shieh and M. Kryder, Magneto-optic recording materials with direct overwrite capability, *Appl. Phys. Lett.*, **49**(8), 473, 1986.

[34] Tominaga, *op. cit.*, Ref. 5.

[35] Tominaga, J. et al., Antimony aperture properties on super-resolution near-field structure using different protection layers, *Jpn. J. Appl. Phys.* **38**, 4089, 1999.

[36] J. Tominaga, "High density recording and plasmon technologies in phase change films," presented at E*PCOS 01, to be published by *SPIE* (2001); http://www.epcos.org/Papers_PDF/Tominaga.pdf.

[37] http://w3.opt-sci.arizona.edu/ODSCsponsors/02-04-11IABSpringMeeting/O1 - Butz-Milster - Super-RENS- April 2002.PDF

[38] Aratani, K. et al., Magnetically induced super resolution in a novel magneto-optical disk, *Proc. SPIE* **1499**, 209, 1991.

[39] Fukumoto, A. et al., Super resolution in a magneto-optical disk with an active mask, *Proc. SPIE* **1499**, 216, 1991.

[40] Murakami, Y. et al., *J.Magn. Soc. Jpn.*, **17**, Supplement S1 201, 1993.

[41] Ohta, N., ISOM/ODS Digest, p. 63, 1993.

[42] Abbe, E., Betrage zur theorie der microscope und der microscopischen wahrehmung, *Arch. Mikrosk. Anat.* **9**, 413, 1873.

[43] Lord Rayleigh, On the theory of optical images with special reference to the microscope, *Philos. Mag.* **5**, 167, 1896.

[44] Vigoureux, J.M. and Courjon, D., Detection of nonradiative fields in light of the Heisenberg uncertainty principle and the Rayleigh criterion, *Appl. Opt.* **31**, 3170, 1992.

[45] Courjon, D. and Bainier, C., Near-field microscopy and near-field optics, *Rep. Prog. Phys.* **57**, 989, 1994.

[46] Haisma, J., Hugues, E., and Babolat, C., Realization of a bi-aspherical objective lens for the Philips video long play system, *Opt. Lett.* **42**, 70, 1979.

[47] Wolf, E., Electromagnetic diffraction in optical systems, I. An integral representation of the image field, *Proc. Roy. Soc. London Ser. A* **253**, 349, 1959.

[48] Richards, B., and Wolf, E., Electromagnetic diffraction in optical systems, II. Structure of the image field in an aplanatic system, *Proc. Roy. Soc. London Ser. A* **253**, 358, 1959.

[49] Braat, J., Read-out of optical discs, *Principles of Optical Disc Systems*, Adam Hilger Ltd, Bristol, 1986, Chap. 2.

[50] Physical format of read-only discs, *Pioneer DVD Technical Guide*, Pioneer Corporation, http://www.pioneer.co.jp/crdl/tech/index-e.html, 2003, Chap. 2.

[51] Mansuripur, M. and Pons, C., Diffraction modeling of optical path for magneto-optical disk systems, *Proc. SPIE* **899**, 56, 1988.

[52] Mansuripur, M., Computer modeling of the optical path, *op. cit.*, Ref. 2, pp. 264–282.

[53] Bricot, C. et al., Optical readout of videodisc, *IEEE Trans. Consum. Electron.* **CE-22**, 304, 1976.

[54] Braat, J., Read-out of optical discs, *Principles of Optical Disc Systems*, Adam Hilger Ltd, Bristol, 1986, Chap. 2.

[55] Pasman, J., Read-out of optical discs, *Principles of Optical Disc Systems*, Adam Hilger Ltd, Bristol, 1986, Chap. 3.

[56] Chandezon, J. et al., Multicoated gratings: a differential formalism applicable in the entire optical region, *J. Opt. Soc. Am.* **72**, 839, 1982.

[57] Challener, W.A., Vector diffraction of a grating with conformal thin films, *J. Opt. Soc. Am. A* **13**, 1859, 1996.

[58] Pasman, J., Read-out of optical discs, *Principles of Optical Disc Systems*, Adam Hilger Ltd, Bristol, 1986, Chap. 3.

[59] Chandezon, J. et al., Multicoated gratings: a differential formalism applicable in the entire optical region, *J. Opt. Soc. Am.* **72**, 839, 1982.

[60] Challener, W.A., Vector diffraction of a grating with conformal thin films, *J. Opt. Soc. Am. A* **13**, 1859, 1996.

[61] Bouwhuis, G. and Burgstede, P., The optical scanning system of the Philips 'VLP' record player, *Philips Tech. Rev.* **33**, 186, 1973.

[62] Bouwhuis, G. and Burgstede, P., The optical scanning system of the Philips 'VLP' record player, *Philips Tech. Rev.* **33**, 186, 1973.

[63] Braat, J., Differential time detection for radial tracking of optical disks, *Appl. Opt.* **37**, 6973, 1998.

[64] Challener, W.A. and Rinehart, T.A., Jones matrix analysis of magneto-optical media and read-back systems, *Appl. Opt.* **26**, 3974, 1987.

[65] Mansuripur, M., Connell, G.A.N., and Goodman, J.W., Signal and noise in magneto-optical readout, *J. Appl. Phys.* **53**, 4485, 1982.

[66] Treves, D. and Bloomberg, D., Signal, noise, and codes in optical memories, *Opt. Eng.* **25**, 881, 1986.

[67] Finkelstein, B.I., and Williams, W.C., Noise sources in magnetooptic recording, *Appl. Opt.* **27**, 703, 1988.

[68] Marchant, A.B., The optical stylus, *op. cit.*, Ref. 1, Chap. 8.

[69] Marchant, A.B., The optical stylus, *op. cit.*, Ref. 1, Chap. 6.

[70] Heemskerk, J.P.J., Noise in a video disk system: experiments with an (AlGa)As laser, *Appl. Opt.* **17**, 2007, 1978.

[71] Bechevet, B., New trends and technical challenges in optical storage, *Trans. Magn. Soc. Jpn.* **2**, 126, 2002.

[72] Greidanus, F.J.A.M. and Zeper, W.B., Magneto-optical storage materials, *Mater. Res. Soc. Bull.* **15**(4), 31, 1990.

4

Head Design Techniques for Recording Devices

4.1 Introduction **4**-1
4.2 History of Magnetic Recording Transducers **4**-1
4.3 Air Bearings and Head to Media Spacing **4**-4
4.4 Write Heads **4**-6
4.5 Longitudinal Write Heads **4**-7
4.6 Perpendicular Heads **4**-8
    The Read Head
4.7 CIP Spin Valve Sensors **4**-10
4.8 CPP Sensor Designs **4**-13
4.9 The Future of Read/Write Heads **4**-15

Robert E. Rottmayer

Seagate Technology
Pittsburgh, PA

4.1 Introduction

Magnetic Recording heads are a key part of any recording system. The recording head takes the information from the medium (tape or disk) and converts it to an electrical signal that is then processed by the channel. Conversely, the head also takes an electrical signal from the channel and writes that signal on the medium. The process of interaction between the head and the medium is complex, generally nonlinear and subject to several sources of noise and errors. In the early days of recording heads one inductive transducer was used to both read and write. In today's magnetic recording heads the read and write function is generally separated with inductive heads doing the writing and flux sensing heads doing the reading (see Figure 4.1). Almost all flux sensing heads today are based on magnetoresistance, the change of resistance in a material due to an applied field. Almost all recording up to now has been longitudinal, that is, the magnetization lies at the plane of the medium.

With the exception of some MO recording all magnetic recording today depends critically on the proximity of the head to the media. Much of the advance in areal density has been driven by the ability to put heads and media in close proximity to each other. In today's drives a head-to-media spacing (HMS) of less than 1 μin. (25 nm) is common.

4.2 History of Magnetic Recording Transducers

Oberlin Smith, an American, patented the first magnetic recording head in 1888. It featured a wire recording medium with an electromagnet consisting of a coil wound around a rod of iron. Valdemar Poulsen of

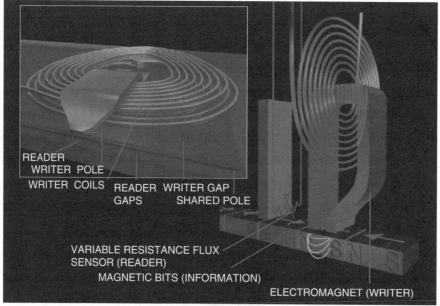

READER
WRITER POLE
WRITER COILS   READER   WRITER GAP
GAPS   SHARED POLE

VARIABLE RESISTANCE FLUX
SENSOR (READER)
MAGNETIC BITS (INFORMATION)
ELECTROMAGNET (WRITER)

[Courtesy of Seagate Technology LLC]

FIGURE 4.1   Merged MR Head.

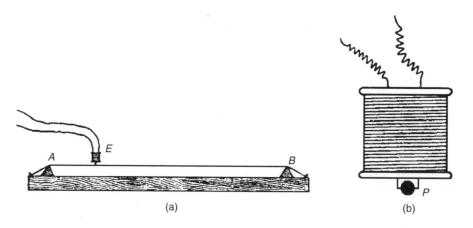

FIGURE 4.2   The apparatus used by Poulsen for his first experiments. A steel wire is stretched between A and B. The electromagnet E was moved along the wire by hand to record or replay sound. In (b) the wire is shown partially surrounded by the pole piece of the electromagnet.

Denmark built the first functioning recorder using this head design in 1898 (see Figure 4.2). In Poulsen's device a wire was used as a medium for the coil wound with iron. In this particular design the field from the head was vertical instead of horizontal as in today's commercial disk and tape drives. However, in spite of the vertical field of the head the writing was still longitudinal due to media properties.

In Germany, the Stabltone-Bandmachine was first offered for sale in 1935. In England the Blattnerphone was first offered for lease in 1931. In both of these more advanced machines a steel tape replaced the wire.

In 1933 the rights to the Blattnerphone were sold to Marconi's Wireless Telegraph Company, Marconi, in cooperation with Stille, one of the developers of Stabltone-Bandmachine, developed a second-generation machine. The head on this machine consisted of a dual pole head, one on each side of the tape. The flux was concentrated by tapering the tips of each pole. In 1935 Marconi modified the head design to a single pole design. However, the recording was still longitudinal due again to the medium properties.

In 1933 Eduard Schuller of AEG in Germany took out a patent on the ring head. This design produced a substantially longitudinal field at the medium and was a much more efficient transducer. At AEG, tape recorders were built using ring heads for recording, playback and erasure. Ring Heads are still used today for virtually all writing on magnetic recording media.

For the next 50 years ring heads were used for reading and writing in most recording applications. The basic design remained the same but improvements in materials and the manufacture of smaller heads with tighter tolerances extended them to higher areal densities and frequencies. The original steel heads were replaced by mu metal and laminated mu metal. NiZn and then MnZn ferrite followed this. The last evolution in this batch-fabricated technology was the metal in gap head (MIG), which featured a ferrite core with high moment metal deposited in the gap. The purpose of this high moment metal was to concentrate the flux from the rest of the core in gap area and produce a higher writing field.

The first commercial thin film single turn heads were made by Burroughs in 1975 on the B9470 head/track disk file. In 1979 IBM introduced thin film heads. These heads were made by a semiconductor type process with hundreds and then thousands of heads fabricated on a single wafer. The wafers were then sliced up and made into sliders and mounted on head gimbal assemblies (HGAs) as shown in Figure 4.3. This batch process eventually reduced the cost and improved the performance of the magnetic recording heads until MIG ferrite heads were driven from the marketplace for disk drives around 1995.

As the areal density increased in the 1980s the smaller tracks required more and more turns to be put on the heads to make up for the signal lost as the track width was decreased. This was not entirely without cost since the inductance increased with the square of the number of turns and the increasing inductance reduced the frequency response of the heads by decreasing the resonant frequency of the coil. As the number of turns increased the process complexity increased and the yields went down and the costs up. The first IBM thin film heads had a single layer 8 turn coil. Two layer coils followed these. In the early 1990s 3 and 4 layer coils were developed. Inductive thin film heads reached their limit with heads manufactured

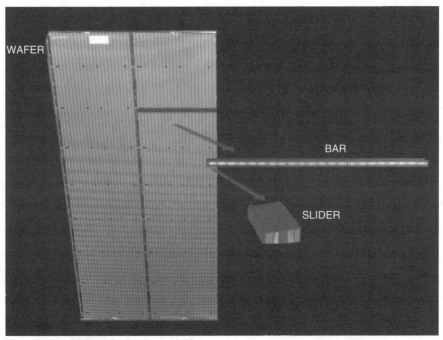

[Courtesy of Seagate Technology LLC]

FIGURE 4.3   Thin Film Head Process — Wafer to Slider.

by Dastek Corp. consisting of 4 layers and 54 coil turns. Western Digital Corporation launched the last read/write inductive head drive in 1995.

In 1991 IBM introduced the magnetoresistive (MR) read sensor that used the anisotropic magnetoresistive (AMR) effect in which the resistance of the element was determined by the angle between the sense current and the magnetization. In 1997 IBM introduced a read sensor based on the giant magnetoreistive (GMR) effect in which the resistance was determined by the relative angle between two magnetic layers. This sensor was capable of over an order of magnitude increase in the % change of resistance and was a key enabler of the increase in the rate of areal density growth in hard disk drives (HDDs) during the late 1990s and early 2000s.

4.3   Air Bearings and Head to Media Spacing

Hand in hand with the improvements in the transducer, the increases in areal density over the past four decades were enabled by the decrease in head to media spacing usual referred to as *fly height* (FH). In 1953, Bill Goddard of IBM designed and flew an externally pressured hydrostatic air bearing as part of the development of the IBM RAMAC disk drive. The heads were mounted on a gimbal spring that allowed the head to pitch and roll so that it could align with the surface of the rotating disk underneath. All subsequent flying heads used a gimbal assembly to allow alignment of the head and disk surfaces. This is referred to as a *head gimbal assembly* (HGA).

The first self-acting air bearing for magnetic recording was developed at IBM beginning in 1955. Jake Hagopian developed the first concept and Bill Gross, Ken Haughton and Russ Brunner refined it. The self-acting air bearing was introduced in the IBM 1301 disk drive in 1962. This air bearing and the following generations had a cylindrical or spherical crown and featured a large mass and load pressure (several hundred grams force). The use of these cylindrical self-acting air bearings reduced the fly height from 800 μin. on the RAMAC to 50 μin. on the IBM 3330. They were of the load/unload type, which means they were lifted from the disk surface before the disk stopped rotating and loaded when the disk was up to speed.

In the early 1970s IBM developed and shipped the 3340 "Winchester" disk drive, which featured a two-rail taper flat air bearing with a low-low mass slider flying over a lubricated Ferric Oxide disk in a contact/start stop mode. The head flew at about 20 μin. when the disk was rotating but was in contact when the disk was stopped. At about the same time Digital Equipment Corporation designed and shipped the RS04 disk drive featuring a taper flat, contact/start stop, air bearing flying on a lubricated thin film plated disk.

IBM in the Corsair Drive used these taper-flat, contact/start-stop sliders on many drives until the introduction of the negative pressure air bearings in 1991. In 1995, Read-Rite Corporation introduced the Tri-Pad air bearing featuring a pseudo-contact air bearing where the head and disk were almost in contact. This approach only worked with inductive heads due to the sensitivity of the MR and GMR heads to contact problems such as electro-static discharge (ESD) and thermal noise as a result of the contact with asperities on the disk surface.

Today's air bearings of the negative pressure type fly between 7 to 10 nm above the recording medium. A negative pressure air bearing has regions of both above and below ambient pressure as shown in Figure 4.4. The resultant pressure is positive and balanced by the head load. These negative pressure air bearings with their many design parameters can be made to give improved dynamic response, insensitivity to skew, altitude and radius. They can be designed to be relatively insensitive to manufacturing tolerances.

The negative pressure contours are formed by photo patterning after the bar is cut from the slider. The cavities are usually formed by ion etching. To be useful in a disk drive, the slider must be mounted on an HGA and assembled into a positioner in the drive as shown in Figure 4.5. Sliders have become increasingly smaller resulting in more sliders/wafer and decreased slider cost. The sliders have less mass and loads are lighter thus reducing the wear and increasing the reliability of the interface when the head touches the medium. The discrete wires of the early heads have given way to miniature flex circuits for connecting the head to the electronics. The drives themselves have gotten smaller and the disk speed has increased. This moved the electronics closer to the head.

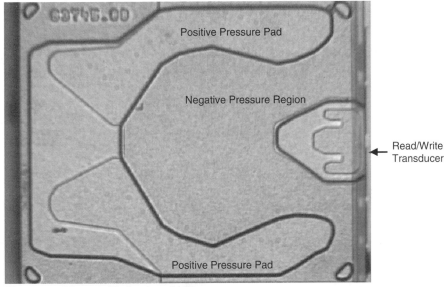

Positive Pressure Pad

Negative Pressure Region

Read/Write
Transducer

Positive Pressure Pad

[Courtesy of Seagate Technology LLC]

FIGURE 4.4 Negative Pressure Air Bearing.

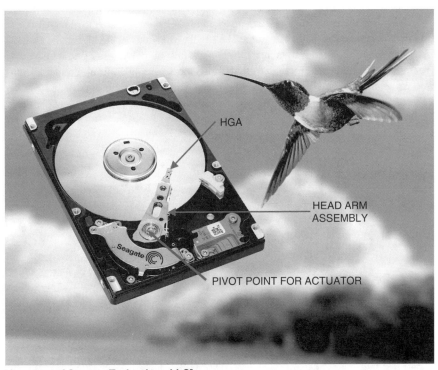

HGA

HEAD ARM
ASSEMBLY

PIVOT POINT FOR ACTUATOR

[Courtesy of Seagate Technology LLC]

FIGURE 4.5 Seagate Momentus Mobile Hard Disk Drive.

4.4   Write Heads

In today's commercial drives longitudinal write heads are used exclusively. However, in the near future, there is a growing consensus in the industry that the technology will switch from longitudinal recording (medium magnetization lies in the plane of the film) to perpendicular recording (magnitization lies perpendicular to the plane of the film). The comparison of the two technologies is illustrated in Figure 4.6. The main difference in the two technologies is that with a ring head the fringing flux of the gap is used to write, whereas with a perpendicular head, the medium is in the gap formed by the write pole and the soft underlayer (SUL) of the medium. This results in increased write field for the perpendicular head. The important parameters of the head with the magnetic recording system are:

1. The magnitude of the write field in the recording layer on the written track.
2. The gradient of the write field at the point at which a transition is written.
3. The magnitude of the field impinging on the adjacent track.
4. The frequency response of the head field.

If everything were ideal, the maximum write field that could be obtained from a conventional write head would be

2piMs (longitudinal ring head)
4piMs (perpendicular pole head with soft underlayer)

Various geometrical parameters of the head design can be optimized to get as close to the maximum field as possible. However, the magnetization of the head pole material determines the amount of field available from either a longitudinal or perpendicular head. The write field from the head places a fundamental limit on how high a coercivity can be written. The coercivity, in turn, is directly related to the anisatotropy, Ku, of the grains. It is the product of the anisotropy and the volume of a grain Ku that determines the thermal stability of the grains in the medium. The smaller the volume the higher the Ku (Hc) needed. However, smaller grains are needed to increase the Signal/Noise at high densities, so we have what is called the superparamagnetic limit (i.e., KuV $\geq$ 40) that limits our density by limiting the stable grain size. The only way to write higher Ku media is with a higher write field and thus the Bs of the pole material is one of the key factors in limiting size of the grains in magnetic medium and thus the areal density.

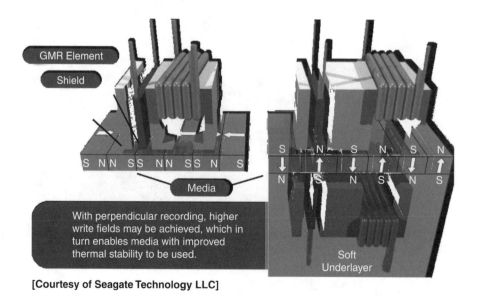

[Courtesy of Seagate Technology LLC]

FIGURE 4.6   Longitudinal vs. Perpendicular Recording.

The materials used in head poles are moving to higher saturation moments with the common electroplated composition of $Ni_{45}Fe_{55}$ (Bs = 1.6T) of a few years ago being replaced by higher moment electroplated alloys. Most of these heads also use higher moment sputtered alloys such as FeN x (x = Ta, Al, Zr etc.) or FeCo x (x = Zr, Nb, B, Ni). These materials give effective moments greater than 2.0T. The high moment material is placed close to the gap where the high flux is needed. Lower moment material guides the flux to the gap.

4.5   Longitudinal Write Heads

The trends in longitudinal recording heads (see Figure 4.7 and Figure 4.8) are towards smaller yokes and fewer turns (3–6). This gives better frequency response enabling drives to be made with higher data rates. Laminated magnetic materials and magnetic materials with better high frequency permeability will also help increase the frequency response. The smaller yokes also give less nonlinear transition shift (NLTS). The write gaps in these heads have become smaller as a consequence of going to higher linear densities. The smaller write gaps combined with the lower HMS also help to contain adjacent track erasure (ATE) where the side fringing field of the head writes on the next track as well as the one being written. Even with this scaling of head dimensions, ATE is an increasing problem at the higher track densities of today's drives.

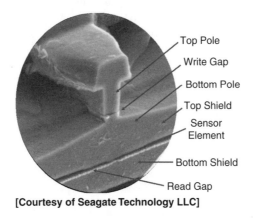

[Courtesy of Seagate Technology LLC]

FIGURE 4.7   Longitudinal Read/Write Head (ABS View).

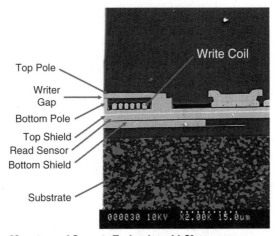

[Courtesy of Seagate Technology LLC]

FIGURE 4.8   Longitudinal Read/Write Head (Cross Section).

Although the longitudinal ring head has served the magnetic recording industry well for 70 years, it now appears that the need for more field will cause the industry to shift to perpendicular recording in near future. As of this writing, the maximum density achieved in a 3.5 in. commercial drive (Seagate Barracuda 7200.7) is 100 GB/3.5 in. disk (65 Gb/in.^{2}) and in a 2.5 in. commercial drive (Fujitsu MHT2060AH), 40 GB/2.5 in. disk (69 Gb/in.^{2}). Both these drives use longitudinal recording.

4.6   Perpendicular Heads

Perpendicular recording offers, at least theoretically, a factor of 2 increase in write field over longitudinal recording. This is because perpendicular recording utilizes a so called pole head to write on the magnetic media that has a soft magnetic underlayer (SUL) as shown in Figure 4.9. This effectively places the medium in the narrow gap formed between the write pole and the SUL. The vertical field produced in this configuration is used to write the media. In perpendicular recording the SUL in the media is really part of the head flux path providing the return path for the flux from the head. Bertram has shown that perpendicular recording can have a 4 to 5 times advantage in areal density over longitudinal recording.

The basic considerations of write head design listed above are similar for perpendicular and longitudinal recording. The write field is higher and the write efficiency tends to be higher for perpendicular. However, as the pole size decreases at higher densities, as shown in Figure 4.10, the amount of field from the head drops off. Also, if the length of the pole is reduced to help with skew the field will drop even more. Skew is the angle the head makes wrt to the track because of the rotary positioner. The write fringing field can also

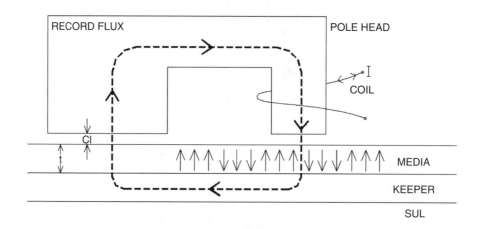

FIGURE 4.9   Perpendicular write head.

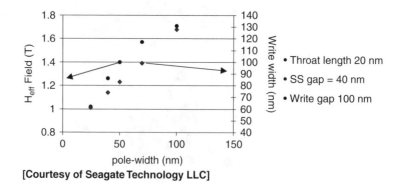

[Courtesy of Seagate Technology LLC]

FIGURE 4.10 H_{eff} and Write Width.

be a problem in perpendicular recording. It has been suggested by Mallary et al. that side shields will be needed at high areal density to prevent ATE. It is also necessary that the medium have magnetic properties that require a high fringing field to write (high nucleation field and a square loop). However, there are important considerations that are unique for perpendicular recording. These are:

1. Skew sensitivity
2. Neighborhood induced transition shift (NITS)
3. Stray field sensitivity
4. SUL magnetic properties

Since modern disk drives have rotary actuators where the head-arm assembly is pivoted about a rotation point outside of the disk (see Figure 4.5), as the head moves from inner to outer track the angle at which the write pole makes to the track center changes. This means that the edge of the pole protrudes over the edge of the track and the pole can write a wider track or erase an adjacent track. Skew angles can be as high as 20° or more. Shortening the pole is one solution but this also reduces the write field. The pole can be formed into a trapezoidal shape to contain the side wall writing but this has drawbacks and is limited to small angles less than 20° at high tpi (tracks per inch).

The write pole of a perpendicular head is susceptible to the pick-up of field from adjacent tracks. This field can add or subtract from the write field supplied by the coil and induce transition shift (NITS). There are two types of NITS: Direct and Indirect (see Figure 4.11). Indirect NITS is caused by the coupling of fields through the SUL and return pole in the case of the write head or through the SUL and the shields in the case of the read head. Direct NITS is the result of direct coupling of the flux from adjacent tracks into the write pole can lead to transition shift and/or incomplete saturation of the media. The worst case occurs when the medium around the transition has the same polarity as the bit. The higher the Ms of the media the higher the NITS. DC free codes with small k values may be needed to reduce the effect of NITS to an acceptable level but this would result in a data rate penalty.

Due to their geometry, perpendicular heads are more susceptible to flux concentration in the pole from stray fields. This problem can be dealt with by shielding the heads.

The SUL is really part of the head flux path. Its orientation, permeability and saturation magnetization are important. Magnetic domain walls in the SUL will create noise spikes in the read-back signal. Magnetic nonuniformities such as ripple will create extra noise.

Although there are new design problems with perpendicular recording there also appear to be solutions to these problems. This will require some changes to the heads, media, channel and perhaps the

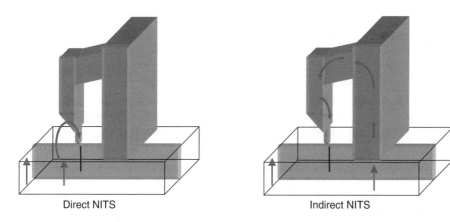

Direct NITS   Indirect NITS

The effect of Stray field from the media effects the write-ability and position of written transition.

FIGURE 4.11 Neighborhood induced transition shift (NITS).

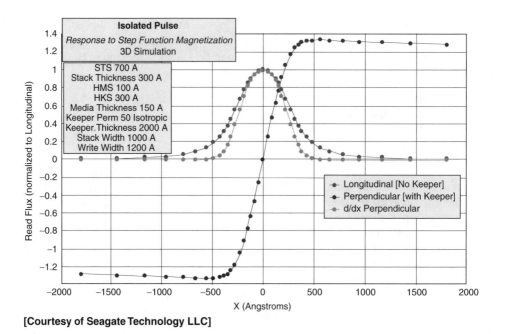

[Courtesy of Seagate Technology LLC]

FIGURE 4.12   Perpendicular and Longitudinal Wave Forms.

drive mechanics. On the other hand, the basic processes and equipment used to make both the heads and media are applicable to perpendicular recording, thus avoiding major new process development and capitalization.

4.6.1   The Read Head

The basic shielded read head design is the same whether the recording mode is perpendicular or longitudinal, however, the wave shape is different as illustrated in Figure 4.12. A Hilbert Transformation will convert an ideal perpendicular waveform to a longitudinal waveform. There is an effect in perpendicular that increases the flux in the transducer due to the reflection of the transition in the SUL. Perpendicular also tends to have a thicker recording layer that can increase the flux in the read head. Perpendicular recording can also have a large DC component in the signal for a large bit spacing. This means that the free layer can be in a state of maximum rotation between transitions especially if the distance between transitions is long. Thus, the main effect for the read head in switching from longitudinal to perpendicular is a change in the flux that the head sees.

In 1988, the giant magnitoresistive (GMR) effect was discovered by Baibich et al. in Fe/Cr multilayers. In this effect, layers of ferromagnetic (FM) material alternate with layers of nonmagnetic metals. These structures exhibit a large change in resistance when switched from the low resistance state where the FM layers are parallel to each other to a high resistance state where the layers are antiparallel.

4.7   CIP Spin Valve Sensors

There are two types of architectures of potential use in read head geometry, the current in plane (CIP) and current perpendicular to the plane (CPP) as shown in Figure 4.13. The vast majority of today's hard disk drive heads are in a mutilayer CIP configuration referred to as a spin valve. Today's heads have separate read and write elements as shown in Figure 4.1. An inductive-style writer is usually built on top of a spin valve reader. As shown in Figure 4.14 and Figure 4.15 the GMR sensor consists of many layers. In the bottom spin valve shown here there is a seed layer, an antiferromagnetic (AF) pinning layer that holds the pinned

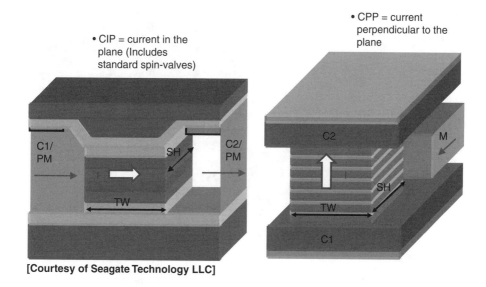

[Courtesy of Seagate Technology LLC]

FIGURE 4.13 ABS View of CIP and CPP Sensors.

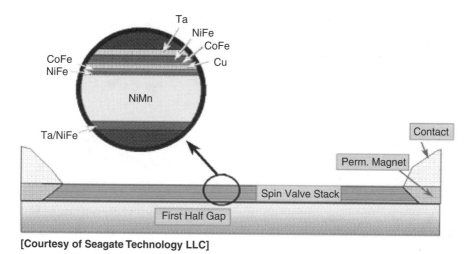

[Courtesy of Seagate Technology LLC]

FIGURE 4.14 GMR Bottom Spin Valve Sensor Stack.

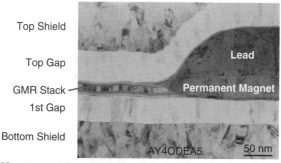

[Courtesy of Seagate Technology LLC]

FIGURE 4.15 Cross-section of a GMR Reader.

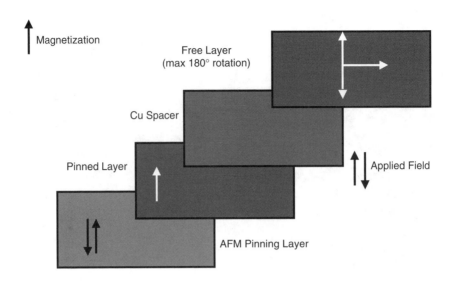

FIGURE 4.16   Rotation of free layer in spin valve.

layer magnetization in a constant direction through exchange coupling with the pinning layer which has no net moment. Between the pinned layer and the free layer is a thin layer of Cu. The free layer is free to rotate under an applied field as shown in Figure 4.16. The structure is protected with a capping layer.

The change in voltage of the spin valve is given by

$$\Delta V = -\tfrac{1}{2} \Delta R \cos \Theta$$

where I is the bias current flowing through the sensor, Θ is the angle between the magnetizations of the two layers, and ΔR is the maximum change in resistance when the two layers change from parallel to antiparallel.

The response of a spin valve to a transition on the disk is shown in Figure 4.17. In a real spin valve the magnetization is not uniform and the $\cos \Theta$ must be averaged over the active area of the spin valve.

$\Delta R / R$'s of over 20% have been observed in spin valves but values of 7 to 12% are typical of today's production. Typical sensor resistances are 50 Ω. Bias currents are typically about 3 mA. Output voltages of over 1 mV have been achieved.

The sensing structure described above is placed between two soft magnetic layers called shields as shown in Figure 4.14 and Figure 4.15. These shields prevent the sensor from picking up flux from bits both down track and up track from the sensor and limit the pick up to the flux immediately under the device. The sensor must be insulated from the shields to prevent shorting of the leads.

These heads require longitudinal stabilization, which keeps the magnetization in these devices more coherent and prevents the break-up into smaller domains. Insufficient stabilizing bias is one cause of unstable heads in which the output voltage can fluctuate as the magnetic state of the sensor changes. Increasing bias can, however, reduce the sensitivity of the device and thus a trade-off must be made between stability and sensitivity.

The longitudinal bias is provided by attaching a permanent magnet (PM)/lead structure placed adjacent to the sides of the GMR sensor. The PM is magnetized in a direction across the track and combined with a lead structure to bring the sense current to the device. The change in resistance is then typically read out as a change in voltage with a constant sense current.

In addition to the shield-to-shield spacing and sensor width, the height of the sensor is important. This is referred to as the stripe height and is produced by precision lapping of the sensor after it is cut from the wafer in the form of a bar. Separate electronic lapping guides (ELGs) are formed at the same step as

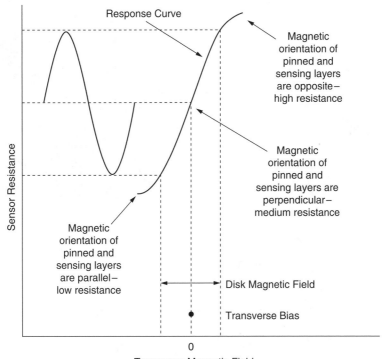

Response Curve

Magnetic orientation of pinned and sensing layers are opposite— high resistance

Magnetic orientation of pinned and sensing layers are perpendicular— medium resistance

Magnetic orientation of pinned and sensing layers are parallel— low resistance

Disk Magnetic Field

Transverse Bias

Sensor Resistance

0

Transverse Magnetic Field

[Courtesy of IBM Corporation]

FIGURE 4.17   GMR sensor response to disk magnetic field.

the sensor and provide a reference to the back of the stripe, which can be used in lapping. It is possible to lap to the correct stripe height by measuring the resistance of the sensor while lapping, but in most cases, practical considerations and the ability to make very accurate, separate guides results in the guides being used to determine final stripe height.

The modern spin valve head is made up of several layers of very thin films (5 - 200Å common) called "stacks" and small feature sizes (submicron track widths). (See Figure 4.14.) These sensors require ultraclean vacuum processing and thickness control better than 1 atomic layer in many cases. New vacuum system designs, utilizing a planetary design where the substrates rotate under the sputtering targets while spinning to average out variations in thickness, have become the tools of choice for these stacks. Precise photo processing and etch techniques are also needed. In spite of these exacting requirements, spin valve heads are manufactured by the millions using advanced techniques unavailable a few years ago. In spite of the precision, sophistication and investment required, the cost of sliders in today's disk drive a small fraction of the total cost of the drive.

4.8   CPP Sensor Designs

There are three main CPP designs under development at the present time to replace the CIP spin valves currently in use. These are the CPP spin valve and the magnetic tunnel junction (MTJ) and the CPP — Multilayer (CPP-ML). All of these devices still use the GMR effect but have the current flowing from top to bottom in the device. The CPP spin valve and CPP-ML have a Cu layer separating the FM layers, and the MTJ has a thin insulating layer separating the free and pinned layer. The MTJ uses the tunneling of polarized electrons between the two layers to create the magnetoresistance. The concept of a head architecture using the shields as leads on the top and bottom of a sensor in the CPP mode was proposed (Rottmayer and Zhu) in 1995.

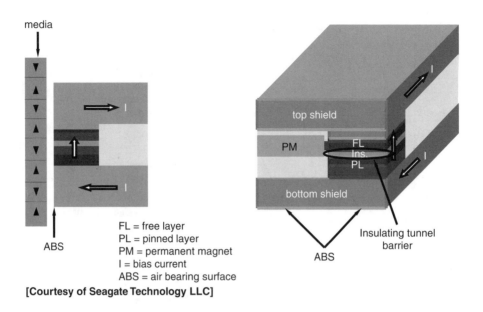

FL = free layer
PL = pinned layer
PM = permanent magnet
I = bias current
ABS = air bearing surface

[Courtesy of Seagate Technology LLC]

FIGURE 4.18   Magnetic Tunnel Junction Read Head.

CPP spin valve structures have obtained MR ratios of 5% with a RA product of $1\ \Omega - \mu m^2$ (Oshima et al.). This is produced using a conventional spin valve structure with nano oxide layers (NOLs). This is claimed by the authors to be adequate for 150 Gb/in.^{2}. Saito et al. obtained a low frequency voltage of 0.6 mV on a sensor with a magnetic track width of 101nm (approx. 100 Gb/in.^{2}). These authors conclude that this is not yet enough for recording. There needs to be more work done on it for commercialization of CPP spin valves.

The MTJ device is made by sandwiching a thin insulating layer of alumina between two ferromagnetic metal layers, one of which is pinned as in a spin valve as shown in Figure 4.18. Electrons can tunnel through the thin (<1 nm) barrier. Since they are spin polarized, the current is larger when the magnetization of the two FM layers is parallel. TMR heads have been reported (Kuwashima et al., 2003) with an RA product of $3\ \Omega - \mu m^2$ and an MR ratio of 18%. These authors claim this is a 100 Gb/in.^{2} class TMR head. Mao et al. have obtained 890 Kbpi linear density and 171 Kbpi trial density (152 Gb/in^2) with a MTJ read/write head.

A CPP multilayer device consists of a number of layer pairs stacked one on top of the other as shown in Figure 4.19. The structure is biased by an external magnet instead of an antiferromagnetic layer used in the above cases. Seigler et al. have achieved a MR ratio of 23% and an RA product of $.2\ \Omega - \mu m^2$ in a structure consisting of 15 layers of Cu/CoFe pairs.

- $R_{min} = 4\ \Omega$
- $J = 1 \times 10^8$ A/cm^2
- (CoFe 10 Å/Cu 19 Å)×15
- DR/R_{min} = 30%
- 90 nm wide physical width
- Shield-to-shield spacing ~ 70 nm

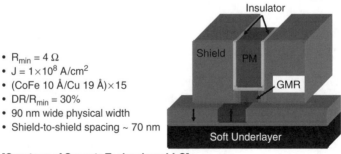

[Courtesy of Seagate Technology LLC]

FIGURE 4.19   CPP ML Reader.

These three types of CPP structures are the subject of active research and development at the leading universities and storage companies. There has been steady progress in all devices over the last 5 years. It is reasonable to expect that CPP devices will replace CIP devices because of the fabrication advantages. It is much easier to control physical track width with the stack etch used in CPP sensors. In addition, the MTJ and CPP-ML have no magnetic material on the sides allowing for better control of side reading. The CPP spin valve and CPP-ML are much more electrostatic discharge (ESD) resistant than CIP spin valves, which have thin oxide spacers in the gap. The MTJ has a thin oxide layer in the gap that may result in ESD or other reliability problems. Both the MTJ and and the CPP-ML have higher MR ratios than today's spin valves. However, both devices also exhibit additional noise, especially at low frequencies, which is a concern. The RA products of CPP-MLs are lower than spin valves and those of MTJs are higher. This gives an advantage to the CPP-MLs as densities increase because the resistance of all devices will increase due to the smaller area of the sensor. Although problems remain to be solved with these devices, they show great promise in delivering the manufacturability and high sensitivity needed as we advance to higher areal densities and data rates.

4.9   The Future of Read/Write Heads

Magnetic recording heads have answered the challenges of the data storage market place for over 100 years. For the hard disk drive industry, heads have been produced for drives from the IBM Ramac (2.6 Kb/in.^{2}) to today's drives of approximately 70 Gb/in.^{2}. This is an increase in areal density of over 7 orders of magnitude in 47 years. The obvious question is whether this rate of progress will continue into the future or not.

The key problem facing recording at the present time is the lack of higher write fields. Recently the number and rate of increase of areal density in demos has slowed. The difficulty in increasing the areal density in commercial products was illustrated by the longer than usual product introduction times for the 80 GB/3.5 in. disk families. This is evidence that at least in the near term the problems associated with areal density increase have become tougher.

Most of the gains over the past 50 years have been due to three factors:

1. Scaling — reducing the size of the key parameters of the recording system as bits became smaller.
2. Introduction of new materials and processes.
3. Better understanding of the physics of recording and the ability to model it.

The write field can be slightly increased by lowering the fly height. Today, however, we are at fly height of 7 to 10 nm with an additional spacing from head overcoat and media overcoat and lube of 7 to 10 nm. It is a daunting challenge to reduce these spacings further and while some progress is expected it will be slow. The HMS is the scaling parameter with the most leverage. The lithographic and deposition techniques available at the present time, while difficult, do not appear to yet impose limits on the scaling of density. In the case of write head materials, most write heads are already made with high Bs material (approaching 2.4T). 2.4T is the limit of moment known ferromagnetic alloys and it is unlikely that new materials with significantly higher moments will be found in the near future.

Read sensors seem to pose less of a problem than write heads since there appear to be a number of approaches to improve the read sensitivity at higher densities.

For these reasons, the magnetic recording industry is looking at new designs to increase the areal densities. As mentioned above, perpendicular recording by virtue of its more favorable geometry, can increase the write field. It has the advantage of being able to use existing head and media fabrication processing and materials. However, the problems mentioned earlier of skew, NITS, stray field sensitivity, and reader bias require significant revisions to the drive and with a new technology, such as perpendicular, much testing still needs to be done. There is also the issue of customer acceptance of a new technology, how long this acceptance will take and what the customer will need to be convinced. There is nevertheless, a growing consensus that the next step for HDD magnetic recording will be perpendicular. Maxtor's media division has announced that they have a process to produce perpendicular media using existing production processes.

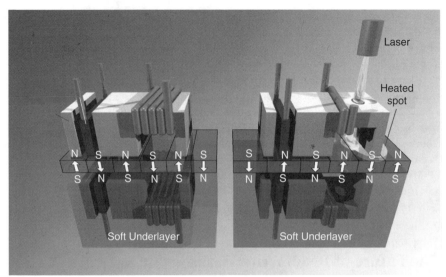

[Courtesy of Seagate Technology LLC]

FIGURE 4.20   Perpendicular vs. HAMR Recording.

The EE Times (September 15, 2003) report that "A consensus appears to be emerging that the shift will come for server and notebook drives at the 160-GB/platter generation in late 2005, followed by a move for desktop drives in the 200 GB/platter generation."

It seems that the industry is on the threshold of introducing perpendicular recording. The marketplace will determine how rapidly it replaces longitudinal.

In the longer term, heat assisted magnetic recording (HAMR) promises to increase densities even further. HAMR uses a focused beam of intense light to heat the media. (See Figure 4.20.) This lowers the coercivity to a point where the fields produced by the head can switch the media. After writing, the medium cools and the higher coercivity insures that it is thermally stable. The cooled media is read in the same manner as a conventional recording system. A number of companies and universities are working on this concept that promises to extend magnetic recording even further. Even though we may not see the compounded areal density gains of the late 1990s, there are many new technologies that promise to overcome our present problems and extend the life of magnetic recording. As in the past, head design and implementation will play an important role.

Acknowledgments

Portions of this work (including figures) were previously published in Magnetic Recording Heads: Historical Perspective and Background, *Encyclopedia of Materials: Science and Technology*, Elsevier, 2001, pp. 4879–4889.

References

Baibich et al. (1988). Giant magnetoresistance of (001) Fe/(001) Cr magnetic superlattices, *Phys. Rev. Lett.* **61**, 2472–2475.

Batra S., Hannay J., Zhou H., and Goldberg J. (2003). Investigations of Perpendicular Write Head Design for 1 Tb/in.^{2} 14th Annual Magnetic Recording Conf. **E7**.

Bhushan B. (1999). *Micro/Nano Tribology*. 2nd ed., CRC Press, Boca Raton FL.

Betram H.N. (1994). *Theory of Magnetic Recording*. Cambridge University Press, London.

Betram N.H., Williams M. (2000). SNR and density limit estimates: a comparison of longitudinal and perpendicular recording. *IEEE Trans. Mag*. **36**, 4–9.

Daniel E.C., Mee C.D., and Clark M.H., (1999). *Magnetic Recording The First 100 Years*. IEEE Press.

Hoagland A.S. (1983). *Digital Magnetic Recording*. Robert E. Krieger Publishing.

Mallary M., Torabi A., and Benakli M. (2002). One Terabit per square inch perpendicular recording conceptual design, *IEEE Trans. Mag*. **38**, No. 4, 1719–1724.

Mao S. et al. (2003). TMR Recording Heads beyond 100 Gb/in^2. 14th Ann Mag Recording Conf. **E4**.

Mee C.D., Daniel E.D. (1996). *Magnetic Storage Handbook*. 2nd ed., McGraw-Hill, New York.

Oshima H. et al. (2003). Current-perpendicular spin valves with partially oxidized magnetic layers for ultrahigh-density magnetic recording. *IEEE Trans. Mag*. **39**, No. 5, 2377–2380.

Rottmayer R. and Zhu J. (1995). A new design for an ultra-high density magnetic recording head using a GMR sensor in the CPP mode. *IEEE Trans. Mag*. **31**, No. 6, 2597–2599.

Rottmayer R. (1994). Magnetic Head Assembly with MR sensor. U.S. Patent #5, 446.613

Rottmayer R., Spash J.L. (1976). Transducer Assembly for a Disc Drive. U.S. Patent #3, 975, 570

Saito et al. (2003). Narrow Track Current-Perpendicular-to-Plane Spin Valve GMR Heads 2003. 14th Annual Magnetic Recording Conf. **A7**.

Seigler, M. van der Heijden P. Parker G., and Rottmayer R. (2002). CPP-GMR Multilayer Read Head for >100 Gbit/in^2. 47th Magnetism and Magnetic Materials Conf. **CA-05**.

Van der Heijden et al. (2002). The effect of media background on reading and writing in perpendicular recording. *J. App. Phy*. **92**, No. 10, 8372–8374.

White R.M., (1984). *Introduction to Magnetic Recording*, IEEE Press.

Wang S.X., Taratorin A.M. (1999). *Magnetic Information Storage Technology*, Academic Press.

II

Communication and Information Theory of Magnetic Recording Channels

5 **Modeling the Recording Channel** *Jaekyun Moon* **5**-1
Introduction • Basic Communication Channel Model for Magnetic
Storage • Signal-Dependent Medium Noise • SNR Definition in Signal-Dependent
Noise • Nonlinearity Characterization • Other Channel Impediments • Applications

6 **Signal and Noise Generation for Magnetic Recording Channel Simulations**
Xueshi Yang and Erozan M. Kurtas **6**-1
Introduction • Simulating the Recording Channel with Electronics Noise Only
• Simulating the Recording Channel with Electronics and Transition Noise • Conclusions

7 **Statistical Analysis of Digital Signals and Systems** *Dragana Bajic*
and Dusan Drajic **7**-1
Introduction • Signals • Averages • Autocorrelation • Power Density Spectrum
of Digital Signals

8 **Partial Response Equalization with Application to High Density Magnetic
Recording Channels** *John G. Proakis* **8**-1
Introduction • Characterization of Intersymbol Interference in Digital Communication
Systems • Partial Response Signals • Detection of Partial Response Signals • Model of a
Digital Magnetic Recording System • Optimum Detection for the AWGN
Channel • Linear Equalizer and Partial Response Targets • Maximum-Likelihood
Sequence Detection and Symbol-by-Symbol Detection • Performance Results from
Computer Simulation • Concluding Remarks

9 **An Introduction to Error-Correcting Codes** *Mario Blaum* **9**-1
Introduction • Linear Codes • Syndrome Decoding, Hamming Codes, and Capacity of
the Channel • Codes over Bytes and Finite Fields • Cyclic Codes • Reed Solomon
Codes • Decoding of RS codes: the key equation • Decoding RS Codes with Euclid's
Algorithm • Applications: Burst and Random Error Correction

10 **Message-Passing Algorithm** *Sundararajan Sankaranarayanan and Bane Vasic* ... **10**-1
Introduction • Iterative Decoding on Binary Erasure Channel • Iterative Decoding
Schemes for LDPC Codes • General Message-Passing Algorithm

11 **Modulation Codes for Storage Systems**   *Brian Marcus and Emina Soljanin* **11**-1
Introduction • Constrained Systems and Codes • Constraints for ISI Channels
• Channels with Colored Noise and Intertrack Interference • An Example
• Future Directions

12 **Information Theory of Magnetic Recording Channels**   *Zheng Zhang,*
Tolga M. Duman and Erozan M. Kurtas **12**-1
Introduction • Channel Capacity and Information Rates for Magnetic Recording
Channels • Achievable Information Rates for Realistic Magnetic Recording Channels
with Media Noise • Conclusions

13 **Capacity of Partial Response Channels**   *Shaohua Yang and Aleksandar Kavčić* **13**-1
Introduction • Channel Model • Information Rate and Channel Capacity • Computing
the Information Rate for a Markov Source • Tight Channel Capacity Lower
Bounds — Maximal Information Rates for Markov Sources • A Tight Channel Capacity
Upper Bound — Delayed Feedback Capacity • Vontobel-Arnold Upper Bound
• Conclusion and Extensions Beyond Partial Response Channels

5

Modeling the Recording Channel

Jaekyun Moon

University of Minnesota
Minneapolis, MN

5.1 Introduction 5-1
5.2 Basic Communication Channel Model for Magnetic
    Storage 5-1
5.3 Signal-Dependent Medium Noise 5-4
5.4 SNR Definition in Signal-Dependent Noise 5-5
5.5 Nonlinearity Characterization 5-8
5.6 Other Channel Impediments 5-9
5.7 Applications 5-9

5.1 Introduction

Design and analysis of coding and signal processing techniques require a suitable communications channel model for magnetic storage systems. Such a model should correctly reflect the essential physics of the read and write processes of magnetic recording, but must also provide a system level description that allows convenient design, analysis and simulation of the communications and signal processing techniques under study. This article provides an overview of such communications channel modeling of magnetic storage. The main focus is on describing intersymbol interference (ISI) and medium noise that plague modern high-density disk drive systems, but nonlinear distortion of magnetic transitions is also discussed. The medium noise model leads to an E_b/N_0-like signal-to-noise ratio (SNR) definition suitable for magnetic recording that can be used to compare different read channel strategies in the presence of signal-dependent medium noise. The usefulness of the model is seen from its application to capacity analysis and design and evaluation of codes and optimized detection schemes.

5.2 Basic Communication Channel Model for Magnetic Storage

Consider a general data storage system depicted in Figure 5.1. Like in other communications channels, error correction coding is used to protect information bits from random noise and other unpredictable disturbances. In commercial magnetic storage devices, the Reed-Solomon (RS) codes are used invariably for this purpose. Modulation coding is then applied to the original bits and the RS parity bits. In storage systems, modulation coding is employed to control minimum and maximum distances between consecutive magnetic transitions [1]. The minimum distance constraint mitigates local medium noise and nonlinearity associated with crowded magnetic transitions. In magnetic recording, the signal rises only in response to a written transition and the absence of a transition over a long period will hamper a proper operation of the gain control and timing recovery circuits. The maximum distance constraint ensures that

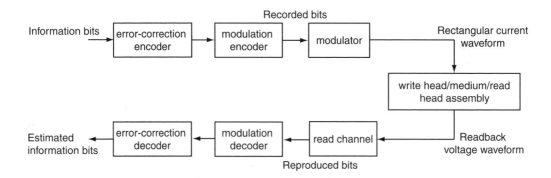

FIGURE 5.1   Data storage system.

the signal arises often enough at readback. Constrained codes can also be designed to improve distance properties at the detector output [2-3].

The modulated bit stream is converted into a rectangular current waveform and then stored in the medium in the form of a magnetization waveform. The corresponding read waveform available at the pre-amplifier output is roughly a linear superposition of shifted transition responses. The transition response is a signal pulse that can be measured with a single written transition. For traditional longitudinal recording, this pulse is a bell-shaped waveform whose polarity alternates between consecutive transitions. In perpendicular recording, the transition response is more or less proportional to the shape of the transition, other than a blurring effect due to some frequency-dependent signal loss terms associated with the read process.

The read waveform is then passed through what is generally called the read channel in the data storage community. The read channel consists of some type of band-limiting filter, a sampler driven by a phase locked loop, an equalizer and a symbol detector (in a broader definition the read channel also includes the modulation encoder/decoder and, possibly, some auxiliary inner error correction or error detection encoder/decoder). The detected bit sequence is then applied to the modulation decoder and finally to the error correction decoder.

A simple pulse-amplitude-modulation (PAM)-like description of the read waveform can be obtained based on the linear superposition of isolated transition responses. Figure 5.2 shows a current waveform whose amplitude levels represent written bits, according to the non-return-to-zero (NRZ) signaling convention, and the corresponding read waveform in longitudinal recording. Letting b_k denote the input data, $a_k = b_k - b_{k-1}$ indicates the polarity (or absence) of the transition and it is easy to see that the read waveform can be expressed as

$$z(t) = \sum_k \underbrace{(b_k - b_{k-1})}_{a_k} h(t - kT) + n(t) \tag{5.1}$$

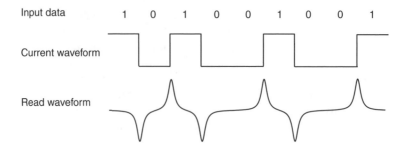

FIGURE 5.2   Input data, current waveform, and read signal in longitudinal recording.

where $h(t)$ is the isolated transition response and $n(t)$ is the additive white Gaussian noise (AWGN) due to the read head and electronics. The shape of $h(t)$ is governed by various frequency-dependent signal loss terms such as the head-medium spacing in the read process, the read head gap effect, the medium thickness effect and imperfect writing of magnetic transitions. The read signal in perpendicular recording can also be described in a similar way, except that the shape of $h(t)$ is quite different, as mentioned earlier. For convenience $h(t)$ is often modeled as Lorentzian for longitudinal recording. In perpendicular recording, $h(t)$ tends to follow the shape of the magnetization and is more appropriately modeled by a function that exhibits odd symmetry such as the hyperbolic tangent or the error function. The Lorentzian model of the transition response is given by

$$h(t) = \sqrt{\frac{4E_t}{\pi PW50}} \cdot \frac{1}{1 + (2t/PW50)^2} \tag{5.2}$$

where E_t is the energy of the transition response, that is, $\int |h(t)|^2 \, dt = E_t$, and $PW50$ is the width of the pulse at half the peak height. Figure 5.3(a) shows a more realistic transition response from a practical longitudinal recording system. For comparison, the transition response of perpendicular recording is also shown in Figure 5.3(b).

Equation 5.1 can be rewritten as

$$z(t) = \sum_k b_k \underbrace{[h(t - kT) - h(t - (k-1)T)]}_{p(t-kT)} + n(t) \tag{5.3}$$

where $p(t)$ represents the "dibit" response, that is, the channel's response to a pair of transitions separated by T. This representation is more in line with the traditional PAM signal representation in digital communication, but in the presence of nonlinearity and media noise that arise in transitions, the model of Equation 5.1 provides a more meaningful tool, as nonlinearity and medium noise should be viewed as distortions in $h(t)$ rather than in $p(t)$. In typical commercial magnetic recording systems $h(t)$ and $p(t)$ extend over several symbol periods. As such, a typical magnetic recording channel can be characterized as a severe ISI channel. The dibit response $p(t)$ matches, often to a good approximation, a partial response of the form $(1 - D)(1 + D)^n$ in discrete-time, where D denotes a unit symbol delay and n is a nonnegative integer that controls the amount of ISI [4]. The factor $(1 - D)$ apparently arises due to the "difference" operation used in converting b_k to a_k. The ratio $PW50/T$ is often referred to as the symbol density and is also a normalized measure of the extent of ISI.

In nonreturn-to-zero-inverse (NRZI) convention, a binary 1 in the data pattern produces a transition in the magnetization waveform whereas a binary 0 results in the absence of transition. The basic signal model

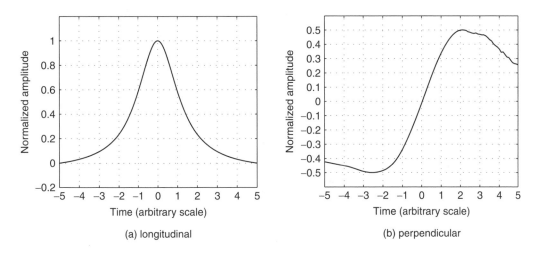

FIGURE 5.3 Transition responses in disk drive systems.

of Equation 5.1 or Equation 5.3 can still be used to describe the input/output relationship of a recording system based on the NRZI convention if it is understood that b_k in Equation 5.1 or Equation 5.3 in this case is really a precoded sequence, that is, a bit sequence obtained by passing the original bit sequence b_k through a precoder with transfer function $1/(1 \oplus D)$, where $\oplus$ is the modulo-2 addition.

5.3   Signal-Dependent Medium Noise

As there exist local variations in the magnetic properties of a disk as well as time-varying fluctuations in the field gradient of the writing head, written transitions generally are not identical in shape from one transition to next [5]. In fact, the geometry of each written transition is subject to highly statistical variations. In a given transition, however, there exists considerable averaging, during the read process, of fluctuations that occur across the track in small scales. For example, the exact shape of each tooth in a zig-zag wall that forms a magnetic transition is not important in the final shape of the corresponding transition response, as the read head outputs what is effectively an across-track average of its responses over a number of "micro-tracks." The statistical variations in the transition geometry do, however, result in fluctuations from one transition to another of some large-scale parameters that control the functional shape of the transition response.

To be more specific, a simple but fairly realistic model for transition noise is obtained by introducing random variations to the width and position parameters of the transition response [6]. Let $h(t, w)$ be the read waveform corresponding to a noise-free transition located at $t = 0$, where w is a parameter that represents the physical width of the written transition. While it is clear that w is a certain fraction of the parameter PW50, the exact relationship between the two parameters depends on physical characteristics of the particular head/disk combination. Now, let $h_k(t, w)$ denote the transition response for the k-th symbol interval. The subscript k emphasizes the underlying assumption that the transition response is in general different from one transition to next. Then, the read response to a noisy transition in the k-th symbol interval can be written as

$$h_k(t, w) = h(t - kT + j_k, w + w_k) \tag{5.4}$$

where j_k and w_k are random parameters representing deviations in the position and width, respectively, from the nominal values. Taking an n-th order Taylor series expansion, the above expression can be approximated as a linear sum of the noise-free response and residual responses due to deviations around the nominal position and width of the pulse. Figure 5.4 shows the second order model in both continuous-time and discrete-time, assuming only the position jitter component. The width variation components can be included in a similar fashion. The model in Figure 5.4 also includes additive noise. As for the accuracy of series expansion relative to Equation 5.4, simulation results show that the first order model is sufficient

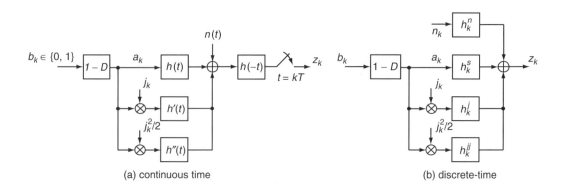

(a) continuous time   (b) discrete-time

FIGURE 5.4   Channel model with second order position jitter.

in evaluating partial response maximum likelihood (PRML) detection techniques. However, to analyze detection techniques optimized for medium noise, the second order components must be included [7].

Note that the matched filter, although not necessarily optimum in the presence of medium noise, does provide sufficient statistics for all practical purposes since at reasonably high densities both the signal and medium noise are effectively band-limited. Assuming a Lorentzian transition response and a matched filter, tedious but straightforward algebraic manipulations yield the following convenient closed-form expressions for discrete-time channel representation [8]:

$$h_k^s = E_t \frac{D_s^2}{D_s^2 + k^2} \tag{5.5}$$

$$h_k^j = -\frac{E_t}{PW50} \frac{2D_s^3 k}{\left(D_s^2 + k^2\right)^2} \tag{5.6}$$

$$h_k^w = -\frac{E_t}{PW50} \cdot D_s^2 \frac{D_s^2 - k^2}{\left(D_s^2 + k^2\right)^2} \tag{5.7}$$

$$h_k^n = \sqrt{\frac{E_t D_s}{2\pi}} \tanh\left(\frac{D_s \pi}{2}\right) \frac{k + D_s/2}{(D_s/2)^2 + k^2} \tag{5.8}$$

$$h_k^{ww} = h_k^{jj} = \frac{-2E_t}{PW50^2} D_s^4 \frac{D_s^2 - 3k^2}{\left(D_s^2 + k^2\right)^3} \tag{5.9}$$

where D_s denotes the symbol density (defined as $PW50/T$), and h_k^w and h_k^{ww} represent the discrete-time filters corresponding to the first and second order width variation components. With N_0 denoting the single-sided power spectrum of the additive noise $n(t)$, the variance of the discrete-time noise sample n_k is given by $\sigma_n^2 = N_0/2$. To relate the variance of j_k and w_k to the continuous-time medium noise spectrum, first define M_0 as the single-sided, equivalent band-limited white spectrum of the medium noise for the single frequency transition pattern (so called $1T$ pattern). That is, M_0 is the height of a box-car power spectral density (PSD) of width $1/2T$ that integrates to the same value as the $1T$ pattern's medium noise PSD. Further, let λ denote the fraction of the medium noise power due to jitter. Then, it can be shown that for the first order model, the jitter and width variances are given by $\sigma_j^2 = PW50^2(\lambda M_0/4E_t)$ and $\sigma_w^2 = PW50^2[(1 - \lambda)M_0/4E_t]$, respectively [8]. For the second order model, it can be shown that $\sigma_j^2 = (PW50^2/18E_t)(-E_t + \sqrt{E_t^2 + 9\lambda E_t M_0})$ and $\sigma_w^2 = (PW50^2/18E_t)(-E_t + \sqrt{E_t^2 + 9(1 - \lambda)E_t M_0})$ [8].

In practice, a low pass filter is often used as the front-end receive filter. A similar set of discrete-time filters can be derived for the noise modeling purposes with a low pass filter as the front-end filter [8].

Other transition noise modeling techniques exist. Some rely on more elaborate reasoning based on recording physics [9]. Some others deal with efficient ways of fitting to experimental data the parameters of a generic model that does not require any knowledge of the underlying recording physics [10].

5.4 SNR Definition in Signal-Dependent Noise

One of the difficulties experienced by coding and signal processing researchers working on data storage is that there is no universal signal-to-noise ratio (SNR) definition that is being used to readily compare the performance merits of different schemes that potentially have different code rates and operate at different linear densities. In digital communication, the information bit energy to noise spectral height ratio, universally denoted by E_b/N_o, has been used extensively to compare the performance merits of different coding and modulation techniques that make use of a given bandwidth. In recording channels with additive noise, a similar definition can be made based on the energy of the transition response $h(t)$ that is the inherent characteristic of the given head-medium assembly [11]. Recalling that E_t denotes the energy in an isolated transition response, the required level of E_t/N_o in achieving a particular bit error rate (BER) of, say, 10^{-5} can be compared among read channel systems employing different sets of coding and signal processing schemes with possibly varying rates and operating densities. This definition is a simple measure of the level of noise that can be tolerated by a given read channel technique in achieving a prescribed target BER, for some head-medium assembly whose goodness is characterized by the single parameter E_t. Apparently, E_t does not change as a function of the operating density or the code rate, and thus the E_t/N_o definition of SNR can be used to compare read channel strategies with different code rates.

However, a difficulty arises when the noise depends on the written data pattern as in real recording systems. The statistical properties of medium noise depend on the written bit pattern as well as the written bit (symbol) density, and finding a universal definition of SNR in this case is not trivial. Consider the following modified SNR definition [12]:

$$\text{SNR} = \frac{E_t}{N_0 + M_0} = \frac{E_t}{N_\alpha} \tag{5.10}$$

where M_0, as defined above, is the single-sided equivalent box car spectrum of the medium noise for the $1T$ pattern and $N_\alpha = N_0 + M_0$ with α denoting the medium noise power expressed as a percentage of the total in-band noise power, that is,

$$\alpha = \frac{M_0}{N_0 + M_0} \times 100 \tag{5.11}$$

We can further write $N_0 = [(100 - \alpha)/100]N_\alpha$ and $M_0 = (\alpha/100)N_\alpha$ with $0 \leq \alpha \leq 100$. With $\alpha = 0$, the SNR reduces to E_t/N_0.

From the definition of M_0, it follows that $M_0/2T$ represents the total integrated medium noise power for the $1T$ pattern. On the other hand, the total medium noise power arising from the $1T$ pattern is easily shown to be $\int_{-\infty}^{\infty} \overline{n_m^2(t)} \, dt/T$, assuming medium noise is statistically independent among transitions and that $n_m(t)$ denotes the medium noise voltage waveform associated with each transition. The statistical average (denoted by the over-bar operation) is taken over transitions written at different positions in the disk. This gives rises to an equivalent, alternative definition for M_0, namely, twice the energy in the noise voltage waveform associated with each transition:

$$M_0 = 2 \int_{-\infty}^{\infty} \overline{n_m^2(t)} \, dt \tag{5.12}$$

The SNR definition of (10) can also be rewritten as

$$\text{SNR} = \frac{E_t}{N_0 + M_0} = \frac{1}{2} \cdot \frac{E_t/T}{\frac{N_0}{2T} + \frac{M_0}{2T}} \tag{5.13}$$

giving rise to an interpretation:

$$\text{SNR} = \frac{1}{2} \cdot \frac{E_t/T}{\text{(total inband noise power with } 1T \text{ pattern)}} \tag{5.14}$$

which can be viewed as the ratio of the isolated transition pulse power to the overall in-band noise power.

For the first-order position jitter and width variation ($1PW$) noise model, M_0 can be written as

$$M_0 = 2\sigma_j^2 I_t + 2\sigma_w^2 I_w \tag{5.15}$$

where σ_j^2 and σ_w^2 are the variances of position jitter and width variation, respectively, and

$$I_t = \int_{-\infty}^{\infty} [\partial h/\partial t]^2 \, dt \tag{5.16}$$

$$I_w = \int_{-\infty}^{\infty} [\partial h/\partial w]^2 \, dt \tag{5.17}$$

With the above definitions, specifying E_t/N_α with a particular value of α determines both N_o and M_0 (assuming $h(t)$ is given as well). Once M_0 is set, the noise parameters necessary for BER simulation or analysis are specified. For example, both σ_j^2 and σ_w^2 are specified once M_0 is known, provided that λ, the fraction of the medium noise due to jitter, is also known. Once N_o, σ_j^2 and σ_w^2 as well as the symbol period are specified, one can proceed with performance analysis or BER simulations to compare codes/detectors operating at different symbol or user densities. With the microtrack noise model of [9], fixing M_0 amounts to determining the number of microtracks and the magnetization width parameter, which are required to simulate the recording channel output. Thus, assuming $h(t)$ and the noise model are predetermined, specifying E_t/N_α with a particular value of α completely characterizes the channel for analysis and simulation purposes.

Note that the key assumption here is that M_0 is independent of the symbol period T so that the noise composition parameter α is also independent of T. This means that the E_t/N_α definition is free of symbol density, as desired. This assumption will hold for any recording systems where the total integrated medium noise power increases linearly with $1/T$ (i.e., where the intrinsic medium noise power per transition remains the same as the symbol pattern or density changes), a condition that is met with well-designed media. For the Lorentzian channel it can be shown that

$$I_t = I_w = V_o^2 \cdot \pi \cdot \frac{1}{2PW50} = \frac{2E_t}{PW50^2} \tag{5.18}$$

(where, for the purpose of computing I_w, $w = PW50/2$ is assumed) and thus

$$E_t/N_\alpha = \frac{E_t}{N_o + 4\frac{\left(\sigma_j^2 + \sigma_w^2\right)}{PW50^2} E_t} = \frac{E_t/N_o}{1 + 4\frac{\left(\sigma_j^2 + \sigma_w^2\right)}{PW50^2} \frac{E_t}{N_o}} \tag{5.19}$$

It is useful to understand the relationship between the E_t/N_α SNR definition and other SNR definitions recording engineers are familiar with. Consider, for example, the peak isolated pulse amplitude to the total in-band noise ratio for a pattern written at some density D_s. This SNR can be expressed as

$$SNR_p(D_s, \gamma) = \frac{V_o^2}{NP(D_s, \gamma)} \tag{5.20}$$

where V_o is the peak-zero amplitude of $h(t)$, $NP(D_s, \gamma)$ is the total $1/2T$-band noise power measured with a data pattern whose symbol density is D_s and average transition density (total number of transitions divided by the total number of written bits) is γ. We can further write

$$SNR_p(D_s, \gamma) = \frac{V_o^2}{\frac{N_0}{2T} + \frac{\gamma M_0}{2T}} = \frac{8}{\pi \cdot D_s} \cdot \frac{E_t}{N_0 + \gamma M_0} \tag{5.21}$$

where the last equality holds for the Lorentzian pulse. Comparing Equation 5.21 and Equation 5.10, the two SNRs can be related via

$$SNR_p(D_s, \gamma = 1) = \frac{8}{\pi \cdot D_s} \cdot \frac{E_t}{N_\alpha} \tag{5.22}$$

Another SNR definition of interest is the mean-square (ms) SNR ratio defined as

$$SNR_{ms}(D_s, \gamma) = \frac{\frac{1}{2\Delta} \int_{-\Delta}^{\Delta} s^2(t)\, dt}{\frac{1}{2\Delta} \int_{-\Delta}^{\Delta} n^2(t)\, dt} \tag{5.23}$$

where Δ is some large interval ($\Delta \gg T$), and $s(t)$ and $n(t)$ are overall signal and noise (band-limited to $1/2T$) voltages, respectively, corresponding to a written pattern with symbol density D_s and average transition density γ. For Lorentzian, we have

$$SNR_{ms}(D_s, \gamma) = \frac{E_t}{\left(D_s^2 + 1\right)(\gamma M_0 + N_0)} \tag{5.24}$$

This SNR can be relate to E_t/N_α via

$$\text{SNR}_{ms}(D_s, \gamma = 1) = \frac{1}{D_s^2 + 1} \cdot \frac{E_t}{N_\alpha} \tag{5.25}$$

5.5 Nonlinearity Characterization

The signal and noise modeling discussed above assumes that the signal portion of the read waveform can be constructed as a linear superposition of isolated transition responses. But, in reality, magnetic interactions of closely located transitions give rise to nonlinear distortion [13]. A well-known form of nonlinearity in longitudinal recording is the shift in transition position to the direction of the previously written transition, which occurs as the demagnetizing field of the previous transition aids the head field writing the current transition. In addition, a transition gets broadened as the head field gradient currently writing the transition is reduced by the demagnetizing field from the previous transition. As a result, a transition is both shifted earlier in time and broadened when there exists a nearby transition written earlier in time. As transition spacing becomes even smaller at higher densities, adjacent transitions tend to partially erase each other, resulting in a substantial reduction of the signal amplitude. While the position shift can be eliminated to a large extent by precompensating the write current so that the actual written position of a transition coincides with the intended position, the transition broadening and partial erasure effects are difficult to avoid as linear density increases. Nonlinear distortion of transition response also exists in perpendicular recording, but nonlinearity there manifests itself in different ways. The interested reader should see, for example, [14].

Whether longitudinal or perpendicular, unlike transition noise, nonlinearity is largely deterministic once the local written transition patterns are specified. Accordingly, the read waveform of a nonlinear magnetic channel can be described as a superposition of pattern-dependent transition responses. Let the vector $\mathbf{a}_k$ denote the local sequence of transition symbols that contribute to nonlinearity for the transition in time interval k. Emphasizing the dependence of a given transition response on the local transition pattern, we write the read waveform arising from a nonlinear magnetic channel as

$$z(t) = \sum_k a_k h(t - kT, \mathbf{a}_k) + n(t) \tag{5.26}$$

Expressing the nonlinear distortion as a deviation around the linear step response, we can write

$$z(t) = \sum_k a_k [h(t - kT) + \delta(t - kT, \mathbf{a}_k)] + n(t) \tag{5.27}$$

where $\delta(t - kT, \mathbf{a}_k)$ represents the pattern-dependent deviation from the linear response in the k-th symbol interval.

The nonlinear distortion terms can be described by a finite Volterra functional series [15]. The Volterra series model constructs the nonlinear portion of the signal as the sum of the outputs of nonlinear kernels. These nonlinear kernels are driven by a product of the present symbol and some combination of the neighboring symbols contributing to the nonlinearity in the present symbol. As an example, assume that the nonlinear distortion in the kth symbol interval is affected by three preceding transition symbols a_k, a_{k-1} and a_{k-2}. Then, the Volterra model describes the nonlinear distortion with three different kernels: two driven by the second-order symbol products, $a_k a_{k-1}$ and $a_k a_{k-2}$, and one driven by the third-order product $a_k a_{k-1} a_{k-2}$. This is depicted in Figure 5.5, where $h_1^{(2)}(t)$ and $h_2^{(2)}(t)$ represent the two second-order kernels and $h^{(3)}(t)$ the third-order kernel. The kernels in Figure 5.5 can be estimated by training the assumed structure to measured data using the traditional least-mean-square type of criterion. Other nonlinear characterization methods exist that range from systematic identification of Volterra-like nonlinear filters to modeling that relies heavily on physical understanding of the write process [16–18].

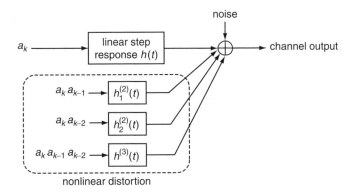

FIGURE 5.5   Volterra series model for nonlinearity.

5.6   Other Channel Impediments

Mainly the medium noise and nonlinearity have been discussed here, but there exist other types of channel impediments that read channel designers must be aware of. They include offtrack interference, channel response fluctuation, especially in tape recording, and thermal asperity effects. Signal processing techniques can be designed specifically to combat these channel impediments, if the hardware complexity is justified given the nature of application.

5.7   Applications

The simple medium noise model and the SNR definition discussed here provides a convenient tool for comparing different read channel strategies in terms of their noise tolerance in mixed noise environments. As a point of reference, to meet a 10^{-5} target BER for the 1PW Lorentzian channel at user density (defined as $PW50$ over the use bit interval) 2.5, the extended class IV PRML (EPR4ML) detector with a rate 16/17 runlength-limited code requires an E_t/N_o slightly above 24 dB and an E_t/N_{90} approximately equal to 23 dB when the 90% medium noise comprises only of independent position jitter [8]. The E_t/N_α definition allows a nice quantitative characterization of the read channel performance that enables direct comparison of the noise tolerances among recording systems with different read channel strategies and even different operating user densities.

Pattern-dependent ML detectors optimized for medium noise have been tested using the simple medium noise model and the results are presented in BER versus E_t/N_α curves for different values of α [7]. The simple medium noise model has also been used in computing capacity of the medium noise dominant channels [19]. Iterative code performance has also been analyzed using this type of medium noise model [20–21].

References

[1] Siegel, P.H., "Recording codes for digital magnetic storage," *IEEE Trans. Mag.*, 21, 1344, 1985.
[2] Moon, J. and Brickner, B., "Maximum transition run codes for data storage systems," *IEEE Trans. Magn.*, 32, 3992, 1996.
[3] Siegel, P.H., Karabed R., and Siegel, P.H., "Coding for high-order partial response channels," *SPIE Conference*, 1995.
[4] Thapar, H. and Patel, A.M., A class of partial response systems for increasing storage density in magnetic recording, *IEEE Trans. Mag.*, 23, 3666, 1987.
[5] Arnoldussen, T.C. and Tong, H.C., "Zigzag transition profiles, noise and correlation statistics in highly oriented longitudinal film media," *IEEE Trans. Mag.*, 22, 889, 1986.

[6] Moon, J., Carley, L.R., and Katti R.R., "Density dependence of noise in thin metallic longitudinal media," *J. Appl. Phys.*, 63, 3254, 1988.

[7] Moon, J. and Park, J., "Pattern-dependent noise prediction in signal-dependent noise," *IEEE J-SAC*, 19, 730, 2001.

[8] Oenning, T. and Moon, J., "Modeling the Lorentzian magnetic recording channels with transition noise," *IEEE Trans. Mag.*, 37, 583, 2001.

[9] Caroselli, J. and Wolf, J.K., "Application of a new simulation model for media noise limited magnetic recording channels," *IEEE Trans. Mag.*, 32, 3917, 1996.

[10] Kavcic, A. and Patapoutian, A., "A signal-dependent autoregressive channel model," *IEEE Trans. Magn.*, 35, 2316, 1999.

[11] Ryan, W., "Optimal code rates for concatenated codes on a PR4-equalized magnetic recording channel", *CISS '99*.

[12] Moon, J., "SNR definition for magnetic recording channels with transition noise," *IEEE Trans. Magn.*, 36, 3881, 2000.

[13] Palmer, D., Ziperovich, P., Wood, R., and Howell, T.D., "Identification of nonlinear write effects using pseudorandom sequences," *IEEE Trans. Mag.*, 23, 2377, 1987.

[14] Senanan, K. and Victora, R.H., "Theoretical study of nonlinear transition shift in double-layer perpendicular media," *IEEE Trans. Mag.*, 38, 1664, 2002.

[15] Hermann, R., "Volterra modeling of digital magnetic saturation recording channels," *IEEE Trans. Mag.*, 26, 2125, 1990.

[16] Yamauchi, T. and Cioffi, J.M., "A nonlinear model for thin film disk recording systems," *IEEE Trans. Mag.*, 29, 3993, 1993.

[17] Zeng, W. and Moon, J., "A practical nonlinear model for magnetic recording channels" *IEEE Trans. Mag.*, 31, 4233, 1994.

[18] Barndt, R.D., Armstrong, A.J., Bertram, N., and Wolf, J.K., "A simple statistical model of partial erasure in thin film disk recording systems," *IEEE Trans. Mag.*, 27, 4978, 1991.

[19] Zhang, Z., Duman, T.M., and Kurtas E., "Information rates of binary-input intersymbol interference channels with signal-dependent media noise," *IEEE Trans. Mag.*, 39, 599, 2003.

[20] Duman, T. and Kurtas, E., "Performance of turbo codes over magnetic recording channels," *MILCOM '99*.

[21] Oenning, T. and Moon, J., "A low density generator matrix interpretation of parallel concatenated single bit parity codes," *IEEE Trans. Magn.*, 37, 737, 2001.

6

Signal and Noise Generation for Magnetic Recording Channel Simulations

Xueshi Yang

Erozan M. Kurtas
Seagate Technology
Pittsburgh, PA

6.1 Introduction **6**-1
6.2 Simulating the Recording Channel with Electronics
    Noise Only **6**-4
    Communications Channel Analogy • Generation of the Noisy
    Readback Signal for a given SNR$_e$ • Numerical Examples
6.3 Simulating the Recording Channel with Electronics and
    Transition Noise **6**-9
    The Direct Approach • Numerical Examples • Remarks on
    Direct Approach • The Indirect Approach • Simulation Methods
    • Numerical Examples • Remarks on the Indirect Approach
6.4 Conclusions **6**-20

6.1 Introduction

The magnetic recording system can be regarded as a communication channel, where the data are recorded on a disk and then read out at a later time. The *communication* is instantiated in the time domain instead of in the space domain where normal communications occur.

In magnetic recording channels, user data bits are encoded and then recorded on the disk by magnetizing the storage media into two opposite directions, representing either bit 0 or bit 1. Subsequently, reading of the data recorded is achieved by sensing the change of the magnetic field emanating from the disk. The overall write and read process can be approximated by a simple linear system model, as illustrated in Figure 6.1. Since during the reading process, the read-head produces an output only when there is a magnetization flux change, the system can be viewed as a differential system, where the encoded user bits are differentiated by the $1 - D$ unit shown in Figure 6.1. Here, D is the delay unit in terms of the channel clock. The read out process then can be characterized by the *transition response* $h(t, w)$, which corresponds to a channel input of $+1$, that is, positive transition from bit 0 to bit 1, ($-h(t, w)$ for input -1, that is, negative transition from bit 1 to bit 0, respectively). For longitudinal recording, the transition response is often modeled as the Lorenzian pulse:

$$h(t, w) = \frac{V_p}{1 + \left(\frac{2t}{w}\right)^2} \tag{6.1}$$

0-8493-1524-7/05/$0.00+$1.50

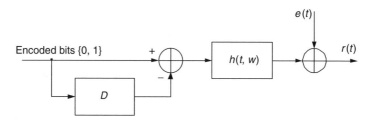

FIGURE 6.1　Linear model for a magnetic recording channel.

where $w = PW_{50}$ is the width of $h(t, w)$ at half of its peak value V_p. For perpendicular recording, the transition response $h(t, w)$ can be expressed as

$$h(t, w) = V_p \cdot \text{erf}\left(\frac{2\sqrt{\ln 2}}{w} t\right) \tag{6.2}$$

where $\text{erf}(x)$ is the error function, defined as

$$\text{erf}(t) = \frac{2}{\sqrt{\pi}} \int_0^t e^{-x^2} dx \tag{6.3}$$

Again, V_p is the peak value of the transition response. In Equation 6.2, $w = PW_{50}$ is defined as the width of the derivative of $h(t, w)$ (a.k.a. impulse response) at half of its peak amplitude.

The readback signal $r(t)$ can thus be represented by

$$r(t) = \sum_k (a_k - a_{k-1}) h(t - kT, w) + e(t) \tag{6.4}$$

where $\{a_k\}$ represent the encoded user bits (0 or 1), T is the bit interval and $e(t)$ is the electronics noise, assumed to be additive white Gaussian (AWGN). The readback signal can also be written as

$$r(t) = \sum_k a_k p(t - kT, w) + e(t) \tag{6.5}$$

where $p(t) = h(t) - h(t - T)$ is the *dibit response* of the system. The recording linear density (or normalized density) is defined as

$$D_s = w/T \tag{6.6}$$

which indicates how many bits are recorded in the interval w. Clearly, the overlapping between the output from transitions aggravates as the normalized density D_s increases, and hence more inter-symbol interference (ISI). The ISI effect is particular evident by considering the dibit response $p(t)$ as a function of the normalized density D_s. The dibit response $p(t)$ essentially depicts the channel response of a positive transition followed immediately by a negative transition. Figure 6.2(a) shows the dibit response of a perpendicular recording channel when D_s is 2 and 3 respectively, while Figure 6.2(b) shows the cases for longitudinal channel. From these two figures, we observe that the amplitude of the dibit response decreases as the normalized density increases, due to more severe ISI. Consequently, the readback signal power reduces when D_s increases, entailed by Equation 6.5.

In magnetic recording channels, the readback signal is corrupted by not only electronics noise and ISI as indicated by Equation 6.4 but by a variety of of noise sources due to imperfections in magnetic media and the recording head and other parts of the physical system. In this chapter, we will consider a very simple medium noise model and ignore all other noise and nonlinearity sources. How well such a simple model correlates with reality however is beyond the scope of the subject of this chapter. We assume that

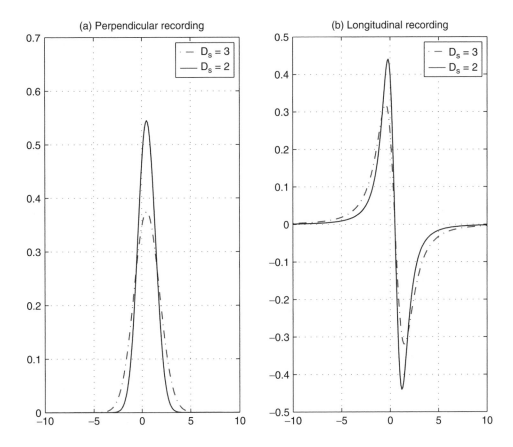

FIGURE 6.2   Dibit responses for perpendicular and longitudinal recording channels at different normalized densities.

transition noise is the only medium noise source and it can be characterized by the position jitter $\triangle t_k$ and pulse broadening effect $\triangle w_k$. The transition noise can be formulated in terms of $\triangle t_k$ and $\triangle w_k$ through the response of the k-th transition as $h(t - kT - \triangle t_k, w + \triangle w_k)$. Consequently, the readback signal that includes the transition noise can be rewritten as

$$r(t) = \sum_k b_k h(t - kT + \triangle t_k, w + \triangle w_k) + e(t) \tag{6.7}$$

where, b_k is the transition sequence, that is, $b_k = a_k - a_{k-1}$ and $\{a_k\}$ is the user data sequence (0 or 1). The position jitter $\triangle t_k$ can be modeled by a truncated Gaussian random variable, that is, $\triangle_k$ is Gaussian distributed but limited to $|\triangle t_k| < T/2$. The pulse-broadening $\triangle w_k$ is often modeled a single-sided Gaussian, that is, its probability density function can be written as

$$f_{\triangle w_k}(x) = \frac{1}{\pi \sigma_w} e^{-\frac{x^2}{2\sigma_w^2}}, \quad x > 0 \tag{6.8}$$

where σ_w is the standard deviation of $\triangle w_k$.

Our aim in this chapter can be summarized as follows:

- To detail the procedures to simulate the signal and noise for a given physical channel (head, media and other physical parts that make up the physical channel).
- To explain how one can compare two recording channels with different coding, detection and signal processing algorithms but the same physical channel.

Clearly the following are **not** among our goals:

- Claim a *uniquely correct* signal-to-noise-ratio (SNR) definition for recording channels.
- Find an *ultimate* SNR definition that relates analytically to system metrics such as bit error rate (BER) or sector failure rate (SFR).

We also hope that the readers of this chapter will be able to make their informed judgment on the relevance of the last two items by the use of the techniques presented here.

The rest of the Chapter is organized as follows. In Section 6.2, we start with an analogy to the classical communication channel and introduce an SNR definition that includes only electronics noise for magnetic recording channels. We then detail how to generate the noisy readback signals according to the SNR definition and subsequently present some numerical examples for reference purposes. To account for the transition noise, in Section 6.3, we present a direct approach where the transition noise is not included in the SNR definition but specified through the corresponding statistical parameters. The detail signal and noise generation methods are given along with some numerical examples and remarks. We then introduce an SNR definition that includes both the electronics noise and transition noise. We elaborate the methods in simulating the signal and noise according to this SNR definition. Finally, we conclude this Chapter in Section 6.4

6.2   Simulating the Recording Channel with Electronics Noise Only

In this section we consider generating noisy readback signals when the only source of noise is electronics noise, which is considered to be additive white Gaussian like in a classical AWGN communications channel. Therefore it makes sense to relate such a recording channel to its counterparts used in communication theory.

6.2.1   Communications Channel Analogy

In the general communication context, E_b/N_0 is defined as the SNR for power-constraint AWGN channel to evaluate the performance of different codes, different modulation schemes and detection algorithms. Here E_b is the average information bit energy and N_0 is the noise spectral density height (single-sided). To see why E_b/N_0 is applicable for comparisons between different codes with different code rates for communication channels, consider an uncoded system versus a coded system where the code rate $R < 1$. A fair comparison between the two system will be to compare the BER observed under the assumption that the same amount of energy is used to transmit the same amount of *information bits* due to the power-constraint (hence, the same E_b), through the same AWGN channel (hence, the same noise power-spectral density N_0). In other words, we shall compare the BER for the two systems under the same SNR $= E_b/N_0$. The often referred code-rate loss or SNR penalty is reflected through the *detection SNR*, defined as the signal-to-noise ratio observed by the receiver. Assuming the signal bandwidth is $1/T$ for both systems (which can be achieved through different modulation methods and/or channel signaling), the receiver for the uncoded system will observe a signal power E_b/T and noise power $N_0/2T$, which results in a detection SNR

$$\text{Detection SNR}_{\text{uncoded}} = \frac{\text{signal power}}{\text{noise power}} = \frac{E_b/T}{N_0/2T} = \frac{2E_b}{N_0} \tag{6.9}$$

In contrast, the receiver for the coded system observes a detection SNR

$$\text{Detection SNR}_{\text{coded}} = \frac{\text{signal power}}{\text{noise power}} = \frac{R \cdot E_b/T}{N_0/2T} = R \cdot \frac{2E_b}{N_0} \tag{6.10}$$

where the SNR penalty associated with the code rate is obvious. Therefore, the codes used in the coded system must provide large enough gains to compensate for this SNR loss in order to furnish performance improvements.

Comparing to other communication channels, magnetic recording channels possess many unique characteristics. In particular, unlike other communication channels where more channel bits can be transmitted in a certain fixed time period without incurring bandwidth expansion, adding redundant bits resulted from coding to the recording channel can only be realized by increasing the recording density, which entails bandwidth expansion. This is owing to the fact that positive/negative transitions are the only signaling scheme available for recording channels. Moreover, unlike other communication channels where there is power-constraint, magnetic recording channels are constrained by the transition response $h(t, w)$, which is *fixed once the head/media and the associated electronic circuits are defined* (see Equation 6.1 and Equation 6.2), that is, the physical channel is fixed.

Given these differences, among many, between communication and recording channels, how can we proceed then if the only noise source considered is the electronics noise in generating noisy readback signals? From the earlier discussion it should be clear that for a fixed physical channel N_0 does not change. Therefore the question boils down to how we should represent the signal term while defining the SNR. The answer to this question is not unique and one can come up with a variety of other meaningful answers different than presented in this chapter. We choose a specific approach to emphasize the link between the communication and recording channels in a natural way. In particular, we shall represent the signal term in such a way that it is independent of the code rate (or equivalently the normalized linear density) just as the E_b term in the classical E_b/N_0 SNR definition. Since in recording channels, the *transition response constraint* plays a similar role as the *power constraint* does in other communication channels, we can represent the signal by replacing E_b with an energy term associated with the transition response of the channel. Specifically, we use the term E_i defined by

$$E_i \overset{\triangle}{=} \int_{-\infty}^{\infty} i(t)^2 dt \qquad (6.11)$$

where $i(t)$ is the *normalized impulse response* of the recording channel defined as

$$i(t) \overset{\triangle}{=} w \cdot \frac{\partial h(t, w)}{\partial t} \qquad (6.12)$$

The normalization factor w in Equation 6.12 is in place primarily to make the SNR definition dimensionless, as will be seen shortly. The SNR for recording channels which suffers from only electronics noise can thus be defined as

$$\boxed{\text{SNR}_e = 10 \log 10 \frac{E_i}{N_0}} \qquad (6.13)$$

where N_0 is the power-spectral density height of electronics noise.[1] The subscript e in Equation 6.13 is used to signify that only electronics noise has been included in the SNR definition.

Let us examine the unit of SNR definition given in Equation 6.13. Since the transition response $h(t, w)$ can be measured in practice and has a unit of *volt*, then the unit of $i(t)$ defined in Equation 6.12 is also *volt*, since w has a unit of *time*. Thus, E_i and N_0 both have units of energy and SNR$_e$ is dimensionless as desired.

---

[1] In [1] and [2], the authors use E_t, the energy of the isolated transition response instead of E_i. Note that, however, E_t cannot be well defined for perpendicular recording, if the isolated transition response of perpendicular recording is modeled as the error function.

For longitudinal recording, the transition response of Equation 6.1 implies that

$$E_i = \frac{\pi w V_p^2}{2} \tag{6.14}$$

and for perpendicular recording,

$$E_i = 4w V_p^2 \sqrt{\frac{2 \ln 2}{\pi}} \tag{6.15}$$

For convenience, we shall normalize the isolated transition response by $\sqrt{E_i}$ such that the energy of the impulse response becomes unity, that is, $E_i = 1$. For longitudinal recording, $E_i = 1$ implies that

$$V_p = \sqrt{2/\pi w} \tag{6.16}$$

and the transition response is given by

$$h(t, w) = \frac{\sqrt{2/\pi w}}{1 + \left(\frac{2t}{w}\right)^2} \tag{6.17}$$

similarly, for perpendicular recoding, the peak value of the transition response becomes

$$V_p = \frac{1}{2} \left(\frac{\pi}{2w^2 \ln 2}\right)^{1/4} \tag{6.18}$$

and the isolated transition response is

$$h(t, w) = \frac{1}{2} \left(\frac{\pi}{2w^2 \ln 2}\right)^{1/4} \cdot \text{erf}\left(\frac{2\sqrt{\ln 2}}{w} t\right) \tag{6.19}$$

Note that the SNR defined by Equation 6.13 can be rewritten as

$$\begin{aligned}
10^{\text{SNR}_e/10} &= \frac{E_i}{N_0} \\
&= \frac{1}{2} \cdot \frac{E_i/T}{N_0/2T} \\
&= \frac{1}{2} \cdot \frac{E_i/T}{\text{In-band electronics noise power}}
\end{aligned} \tag{6.20}$$

which can be interpreted as the ratio of the impulse response power to the overall in-band electronics noise power.

6.2.2 Generation of the Noisy Readback Signal for a given SNR$_e$

In this section, we answer the following question: for the SNR definition given by Equation 6.13, how to set the appropriate simulation parameters, i.e., the electronics noise variance, for a desired SNR at a designated normalized density D_s?

Before we proceed, we should observe that for a fixed head/media combination, PW_{50} is fixed and independent of operating normalized density. Normalized density is adjusted by changing T. Therefore we assume that

$$w = PW_{50} \equiv 1 \tag{6.21}$$

Note that such an assumption will not incur any loss of generality, since in simulations we are dealing with discrete time signals and the *time t* is always normalized by PW_{50} (see Equation 6.17 and Equation 6.19).

FIGURE 6.3   A typical system model of magnetic recording channels.

Consider the recording system model in Figure 6.3. We wish to calculate the variance of the additive white Gaussian noise $e(t)$ for a desired $SNR_e = z$dB. For simplicity, we shall assume that the low-pass filter in Figure 6.3 is an ideal low-pass filter[2] with cut-off frequency $1/2T$. By Equation 6.20 we have

$$\frac{N_0}{2T} = 10^{-z/10} \cdot \frac{E_i}{2T} \tag{6.22}$$

which implies that

Total in-band electronics noise power

$$= \frac{N_0}{2T}$$
$$= 10^{-z/10} \cdot \frac{E_i}{2T}$$
$$= 10^{-z/10} \cdot \frac{E_i/PW_{50}}{2T/PW_{50}}$$
$$= 10^{-z/10} \cdot \frac{E_i D_s}{2} \tag{6.23}$$

where D_s is the normalized density, that is, $D_s = PW_{50}/T$. In Equation 6.23, we have used the assumption that $PW_{50} = 1$.

In practice, continuous-time waveforms are simulated through over-sampled discrete-time waveforms (sampling period $T_s \ll T$). An oversampling ratio of $OSR = T/T_s$ entails a scaling factor of OSR for the electronics noise power in the oversampling domain. This is because the analog frequencies corresponding to the same discrete-time frequency in the oversampling domain and baud-rate sampling domain, respectively, differ by a ratio of OSR, as illustrated in Figure 6.4. Consequently, for the oversampled signal, the variance of AWGN becomes

$$\boxed{\sigma_e^2 = OSR \cdot 10^{-z/10} \cdot \frac{E_i D_s}{2}} \tag{6.24}$$

If the isolated transition responses are given by Equation 6.17 and Equation 6.19, for longitudinal and perpendicular recording, respectively, then E_i can be simply substituted by 1 in Equation 6.24.

As for the signal, when no jitter and pulse-broadening noise are simulated, it can be computed by using Equation 6.17 for longitudinal recording channel and Equation 6.19 for perpendicular recording channel, respectively. It should be noted that the variable t in Equation 6.17 and Equation 6.19 in fact are not *time* if $w = PW_{50}$ is interpreted to have the unit of length. However, since we implicitly assume a constant rotational speed, the relationship between time and distance becomes obvious.

---

[2]By assuming the transition response is perfectly bandlimited in $1/2T$, it can be shown that the systems resulted from matched filtering and ideal low-pass filtering are equivalent. For a proof, see, for example, [3]. By using different low-pass filters or receivers, the overall system performance will vary, due to the variations in detection SNR seen by the receiver. However, this will not affect the calculation here, since our purpose is to establish a reference channel (excluding the effect of different receivers), and to compare the performance of different codes, detectors, etc.

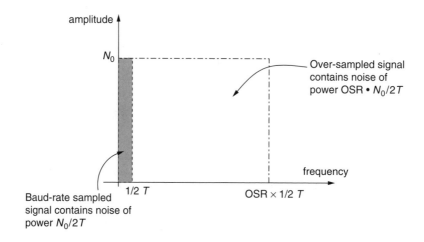

FIGURE 6.4   If the AWGN noise PSD height is N_0, the oversampled (oversampling ratio is OSR) signal contains OSR times more noise power than the baud-rate sampled signal.

In order to generate signals in the discrete-time domain, where the basic time unit is the sampling period $T_s = T/\text{OSR}$, the variable t needs to be normalized by T_s. We shall denote the i-th sample in the oversampling domain by s_i, then, for longitudinal recording, we have

$$s_i = h(t, w)|_{t=iT_s}$$

$$= \frac{\sqrt{2/\pi}}{1 + \left(\frac{2t/T_s}{w/T_s}\right)^2}\Bigg|_{t=iT_s}$$

$$= \frac{\sqrt{2/\pi}}{1 + \left(\frac{2i}{D_s \cdot \text{OSR}}\right)^2}, \quad i = 0, 1, 2, \ldots \tag{6.25}$$

and for perpendicular recording, the signal is calculated via

$$s_i = h(t, w)|_{t=iT_s}$$

$$= \frac{1}{2}\left(\frac{\pi}{2\ln 2}\right)^{1/4} \cdot \text{erf}\left(\frac{2\sqrt{\ln 2}}{\text{OSR} \cdot D_s} i\right), \quad i = 0, 1, 2, \ldots. \tag{6.26}$$

Note that, in the above equations, we have used the assumption that $PW_{50} = 1$.

6.2.3   Numerical Examples

For references purposes, we provide a set of simulation results here both for longitudinal and perpendicular recording channels. Readback signals are simulated through the linear superposition model of Equation 6.27, as shown in Figure 6.3. The transition responses are given in Equation 6.17 for longitudinal and Equation 6.19 for perpendicular channel, respectively. An 7-th order elliptic low-pass filtering (with cut-off frequency $1/2T$) and baud-rate sampling is assumed. The equalizer is a 21-tap transversal linear filter, optimized by the minimum mean-square error criterion (see e.g., [4]). For each SNR, at least 200 bit errors are collected to compute the bit-error-rate (BER).

Figure 6.5 shows the BER results for longitudinal recording channel of normalized density 2 and 2.5, with AWGN only. A maximum likelihood (ML) detector implemented via Viterbi algorithm is used. We assumed $E_i = 1$ and $PW_{50} = 1$ for signal generation. For each density, both $PR4 = [1 \ 0 \ -1]$ and $EPR4 = [1 \ 1 \ -1 \ -1]$ targets are used. The SNR loss associated with the density increase is obvious from the figure.

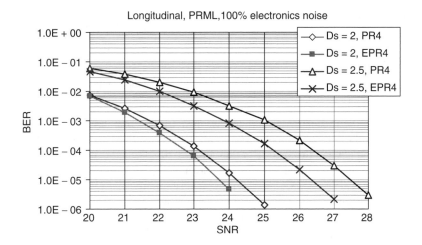

FIGURE 6.5   Bit-error-rate versus SNR_e for longitudinal channel of normalized density 2 and 2.5, with targets PR4 [1 0 −1] and EPR4 [1 1 −1 −1], respectively. Only electronics noise is simulated.

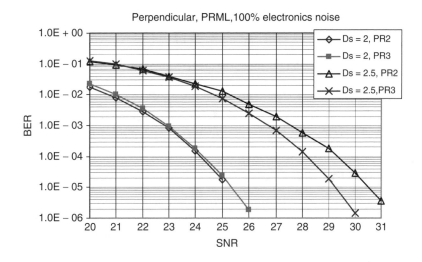

FIGURE 6.6   Bit-error-rate versus SNR_e for perpendicular channel at normalized density 2 and 2.5, with targets PR2 [1 2 1] and PR3 [1 3 3 1], respectively. Only electronics noise is simulated.

In Figure 6.6, we plot the BER curves versus SNR_e defined in Equation 6.13 for perpendicular recording channel of normalized density 2 and 2.5, for PR2 = [1 2 1] and PR3 = [1 3 3 1] targets, respectively. Similarly, we observe that as density increases, significant BER deterioration is incurred.

6.3   Simulating the Recording Channel with Electronics and Transition Noise

In this section we will take our analysis one step further by including transition noise into our noisy readback signal generation. We will describe two different approaches, the direct approach and the indirect approach, and discuss the pros and cons for both.

6.3.1 The Direct Approach

Consider the recording system model in Figure 6.3. The readback signal $r(t)$ is given in Equation 6.7, which is rewritten here for convenience:

$$r(t) = \sum_k b_k h(t - kT + \triangle t_k, w + \triangle w_k) + e(t) \tag{6.27}$$

where, b_k is the transition sequence; $h(t, w)$ denotes the transition response; $\triangle t_k$ and $\triangle w_k$ represent the position variation and width variation of the k-th transition respectively. In the direct approach we do not include the transition noise in the SNR definition. Instead, we specify the distributions of $\triangle t_k$ and $\triangle w_k$. As discussed in Section 6.1, $\triangle t_k$ and $\triangle w_k$ are often assumed to be truncated Gaussian and single-sided Gaussian respectively. Therefore, we can specify the the distributions of $\triangle t_k$ and $\triangle w_k$ through their standard deviations, to be denoted by σ_t and σ_w respectively. For convenience, we further designate σ_t and σ_w as the percentages of PW_{50}. For example, we say jitter noise is 5%, which is equivalent to $\sigma_t/PW_{50} = 0.05$.

When jitter and pulse-broadening noise are simulated, the readback signal is given by Equation 6.27. If the position jitter $\triangle t$ and pulse-broadening $\triangle w$ are random values with unit $w = PW_{50}$,[3] the resulted discrete-time signal corresponding to a single transition is given by

$$
s_i = h(t - w \cdot \triangle t, w - w \cdot \triangle w)|_{t=iT_s}
$$
$$
= \frac{\sqrt{2/(\pi(1 + \triangle w))}}{1 + \left(\dfrac{2(i - \mathrm{OSR} \cdot D_s \cdot \triangle t)}{\mathrm{OSR} \cdot D_s \cdot (1 + \triangle w)}\right)^2}, \quad i = 0, 1, 2, \dots. \tag{6.28}
$$

Similarly, for perpendicular recording channel, the discrete-time signal can be computed as

$$
s_i = h(t, w)|_{t=iT_s}
$$
$$
= \frac{1}{2}\left(\frac{\pi}{2(1 + \triangle w)^2 \ln 2}\right)^{1/4} \cdot \mathrm{erf}\left(\frac{2\sqrt{\ln 2}(i - \mathrm{OSR} \cdot D_s \cdot \triangle t)}{\mathrm{OSR} \cdot D_s \cdot (1 + \triangle w)}\right), \quad i = 0, 1, 2, \dots \tag{6.29}
$$

6.3.2 Numerical Examples

Figure 6.7 shows the BER results for longitudinal recording channel of normalized density 2 (EPR4 [1 1 −1 −1] target) with 0%, 2.5% and 5% jitter noise (in terms of PW_{50}) respectively. Clearly, the jitter noise results in system performance loss and the loss increases as the jitter noise level. In particular, notice that when the jitter noise is severe, that is, 5% of PW_{50}, the performance of the system appears to be "saturated" by the jitter noise. This is expected since fixed amount of jitter noise is always present in the system even as the SNR goes to infinity and is due to that fact that the jitter noise is not included in the SNR calculation.

In Figure 6.8, we plot the BER curves versus SNR_e defined in Equation 6.13 for perpendicular recording channel with different jitter noise levels. The target is the PR2 [1 2 1] target and the operating linear density is 2.0. Again, substantial performance loss is observed due to the presence of jitter noise and the loss increases as the jitter noise level escalates.

6.3.3 Remarks on Direct Approach

As it is outlined in the previous subsections, we observe that the main difficulty of dealing with transition noise stems from the fact that a true SNR can not be a scalar quantity. It should instead have enough dimensions to capture relevant noise models such as electronics, transition noise, nonlinearity, among many others. For example, by considering electronics and transition noise only, the BER versus noise

---

[3]In other words, the actual position jitter corresponding to a value of $\triangle t$ is $\triangle t \cdot PW_{50}$; the actual pulse broadening corresponding to a value of $\triangle w$ is $\triangle w \cdot PW_{50}$.

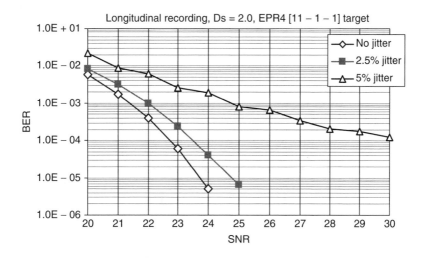

FIGURE 6.7 Bit-error-rate versus SNR_e for longitudinal recording channel of normalized density 2, with EPR4 [1 1 −1 −1] target for 0, 2.5 and 5% jitter noise, respectively.

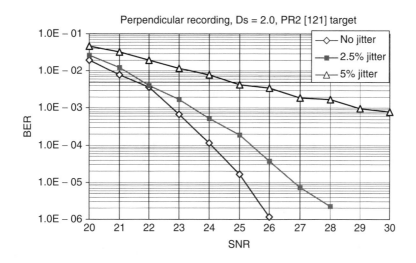

FIGURE 6.8 Bit-error-rate versus SNR_e for perpendicular channel of normalized density 2, with PR2 [1 2 1] target for 0, 2.5 and 5% jitter noise, respectively.

should be an 3-D plot, as illustrated in Figure 6.9. The direct approach essentially obtains a slice of the 3-dimensional plot at a certain fixed transition noise level. We summarize the advantages of including the transition noise using the direct approach as follows:

- Since all the relevant parameters (jitter and pulse width variation) and their distributions can be measured using real signals or complicated physical models conveniently, we can apply these parameters directly to the simulations.
- The direct approach adds fixed amount of transition noise on top of the electronics noise and does not allow exchanging one for the other, such that one can decouple the different effects of transition noise and electronics noise on system performance.

The main issue that has been raised with the direct approach, to our knowledge, has classically been the *difficulty* in comparing two systems with different code rates. We believe this is superficial and one

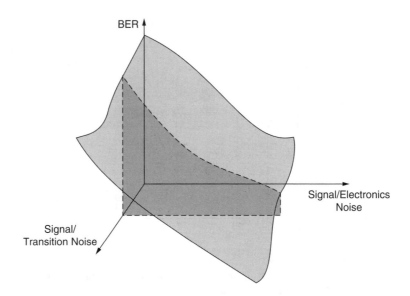

FIGURE 6.9   An 3-D view of the BER versus electronics and transition noise. Note the direct approach essentially bisects the 3-D plot at a fixed transition noise level.

can evaluate the coding gain by either measuring horizontally the different electronics level required to meet a certain BER target or vertically the BER difference at a fixed electronics noise level, as shown in Figure 6.10. These measurements can be performed at different transition noise levels, for example, 5% or 2.5% jitter noise, and in general one would expect that the specific gains varies for different transition noise levels, a phenomenon also observed in reality.

We should also mention that for certain recording media, the transition noise parameters scale with the normalized density, possibly in a nonlinear fashion. For such cases, it is relevant to adjust the transition noise parameters with the linear density while evaluating coding gains or system performance at different linear densities. A more detail discussion on this subject is given in the next Section.

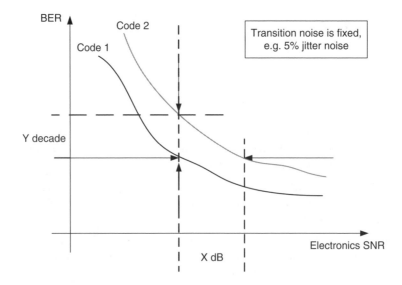

FIGURE 6.10   Evaluation of different codes with the direct approach. The SNR is defined in Equation 6.13 and does not include the transition noise, which is fixed at certain amount.

6.3.4   The Indirect Approach

In contrast to the direct approach, in this section we focus on modifying the SNR definition of Equation 6.13 to take the transition noise into account. A major difficulty in dealing with transition noise is that, unlike AWGN, transition noise is correlated and varies with recording densities, possibly in a nonlinear fashion. In the following, we show that, under certain reasonable assumptions, we can use a parameter M_0 to characterize the transition noise, similar to the parameter N_0 for AWGN (see also [2]).

Let us consider the linear superposition model of Equation 6.7. The k-th transition noise waveform can be written as

$$n_m(k,t) = b_k[h(t - kT, w) - h(t - kT + \Delta t_k, w + \Delta w_k)] \tag{6.30}$$

By assuming that the transition jitter Δt_k and width variation Δw_k are independent between transitions, which holds for well-designed media where the transition noise power spectral density does not exhibit a supralinear behavior as a function of recording density [5], the average transition noise power becomes (for all-transition pattern)

Average transition noise power for all-transition pattern

$$= \lim_{K \to \infty} \frac{1}{KT} \int_{-\infty}^{\infty} \left[\sum_{k=1}^{K} n_m(k,t) \right]^2 dt$$

$$= \frac{1}{T} \int_{-\infty}^{\infty} \mathrm{E}\left\{ b_k^2[h(t - kT, w) - h(t - kT + \Delta t_k, w + \Delta w_k)]^2 \right\} dt$$

$$+ \lim_{K \to \infty} \frac{1}{K} \sum_{k_1} \sum_{k_2 \neq k_1} \int_{-\infty}^{\infty} n_m(k_1, t) n_m(k_2, t)\, dt$$

$$= \frac{1}{T} \int_{-\infty}^{\infty} \mathrm{E}\{[h(t - kT, w) - h(t - kT + \Delta t_k, w + \Delta w_k)]^2\}\, dt$$

$$+ \lim_{K \to \infty} \frac{1}{K} \sum_{k_1} \sum_{k_2 \neq k_1} \int_{-\infty}^{\infty} n_m(k_1, t) n_m(k_2, t)\, dt \tag{6.31}$$

If we further assume that $n_m(k,t)$ only has non-negligible values when $t \in [(k-L)t, (k+L)t]$, where L is some large positive integer, we can calculate the second term in Equation 6.31 as

$$\lim_{K \to \infty} \frac{1}{K} \sum_{\substack{k_1=1 \\ }}^{K} \sum_{\substack{k_2=1 \\ k_2 \neq k_1}}^{K} \int_{-\infty}^{\infty} n_m(k_1, t) n_m(k_2, t)\, dt$$

$$= \lim_{K \to \infty} \int_{-\infty}^{\infty} \frac{1}{K} \sum_{\substack{k_1=1 \\ }}^{K} \sum_{\substack{k_2=k_1-L \\ k_2 \neq k_1}}^{k_1+L} n_m(k_1, t) n_m(k_2, t)\, dt$$

$$= \int_{-\infty}^{\infty} \mathrm{E}\left\{ \sum_{\substack{k_2=k_1-L \\ k_2 \neq k_1}}^{k_1+L} n_m(k_1, t) n_m(k_2, t) \right\} dt$$

$$= \int_{-\infty}^{\infty} \sum_{\substack{k_2=k_1-L \\ k_2 \neq k_1}}^{k_1+L} \mathrm{E}\{n_m(k_1, t) n_m(k_2, t)\}\, dt \tag{6.32}$$

It is not difficult to see that for independent (mutually and individually) and zero-mean sequences $\{b_k\}$, $\{\triangle t_k\}$, and $\{\triangle w_k\}$,

$$E\{n_m(k_1, t)n_m(k_2, t)\} = 0, \quad \text{for } k_1 \neq k_2 \tag{6.33}$$

Now, let us define[4]

$$M_0 \overset{\triangle}{=} \int_{-\infty}^{\infty} E\{[h(t, w) - h(t + \triangle t_k, w + \triangle w_k)]^2\} \, dt \tag{6.34}$$

and assume that the transition probability (i.e., total number of transitions divided by the total number of written bits) for identically independent distributed (i.i.d.) random bit sequence $\{a_k\}$ is 1/2. Then, it holds that

$$\text{Average transition noise power for i.i.d. random data} = \frac{1}{2} \cdot \frac{1}{T} M_0 \tag{6.35}$$

Note that the M_0 defined in Equation 6.34 can be interpreted as the *average transition noise energy associated with an isolated transition*. Moreover, if we assume that the distribution of $\{\triangle t_k\}$ and $\{\triangle w_k\}$ do not change with normalized density, M_0 is completely determined by the transition response and is independent of the normalized density. Thus, M_0 can be deemed as an analogy of N_0 for AWGN, and represents the equivalent power spectral density height for transition noise. In other words, the integral of M_0 over the in-band frequencies from 0 to $1/2T$ is the same as the integral of the PSD of the transition noise.

We can thus define the SNR for recording channels with transition noise by extending Equation 6.13 as

$$\boxed{\text{SNR} = 10 \log 10 \frac{E_i}{N_0 + M_0}} \tag{6.36}$$

where E_i, as defined before, is the energy of the normalized channel impulse response; N_0 is the single-sided PSD height of AWGN; and M_0 is defined as in Equation 6.34. Note that the SNR definition of Equation 6.36 is free of normalized densities and code rates, as desired. Equation 6.36 can be interpreted as

$$10^{\text{SNR}/10} = \frac{1}{2} \cdot \frac{E_i/T}{\frac{N_0}{2T} + \frac{M_0}{2T}} \tag{6.37}$$

$$= \frac{1}{2} \cdot \frac{E_i/T}{\text{In-band electronics noise power+ transition noise power}}$$

$$= \frac{1}{2} \cdot \frac{E_i/T}{\text{Total in-band noise power for i.i.d. random data pattern}} \tag{6.38}$$

The notion of M_0 can be further confirmed by numerical simulations. In Figure 6.11, we show the measured transition noise power, for random data, at different normalized densities for longitudinal recording channel. The readback signal is calculated through the linear superposition model of Equation 6.27 with transition jitter noise only. The transition jitter $\triangle t_k$ is drawn from an i.i.d truncated Gaussian sequence with zero mean, and fixed standard deviation of 2.5% $\times$ PW_{50} across different normalized densities. The transition noise is measured after the readback signal is low-pass filtered and baud-rate sampled. At each normalized density, 200 data sectors are simulated, and each sector contains 4096 bits. The measured noise power are indicated at the corresponding clustered points along with the mean and the approximate 95% confidence intervals (mean$\pm$ 2* standard deviation). We can see that the linear scaling between the transition noise power and the normalized density holds very well. Similar observations can be made for perpendicular channel, for which the jitter noise power versus normalized density for fixed 2.5% position jitter is shown in Figure 6.12.

---

[4]Note that in [2], M_0 is defined as twice the energy of the transition noise waveform, a difference of scale of 2 from the definition used here.

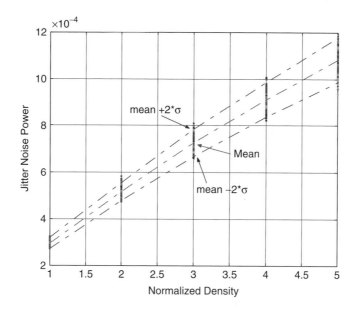

FIGURE 6.11 Jitter noise power as a function of normalized density for longitudinal recording channel. The standard deviation of $\triangle t_k$ is 2.5% of PW_{50}, assumed to to i.i.d Gaussian with zero-mean. Noise power is measured at the baud-rate sampler output.

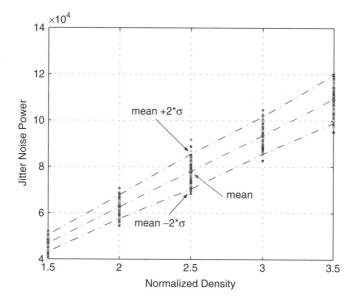

FIGURE 6.12 Jitter noise power as a function of normalized density for perpendicular recording channel. The standard deviation of $\triangle t_k$ is 2.5% of PW_{50}, assumed to to i.i.d Gaussian with zero-mean. Noise power is measured at the baud-rate sampler output.

In practice, it is of interest to evaluate the performance of a coding scheme or a detection method under different electronics and transition noise mixture environment. Let us denote the percentage of the transition noise power of the total in-band noise power by α, that is,

$$\alpha = \frac{M_0}{N_0 + M_0} \times 100 \qquad (6.39)$$

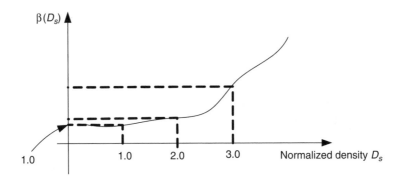

FIGURE 6.13 An example of scaling function $\beta(D_s)$. The function is defined with reference to the M_0 value at normalized density 1, for which $\beta(1) = 1$ and $M_0 = M_{0,1}$. The M_0 value for other normalized density D_s can be obtained by multiplying $M_{0,1}$ by $\beta(D_s)$.

For a given SNR and α, the electronics noise N_0 and transition noise M_0 can then be determined by Equation 6.36 and Equation 6.39.

The definition of Equation 6.36 assumes that the position jitter $\{\triangle t_k\}$ and pulse broadening $\{\triangle w_k\}$ do not scale with the linear density, which holds for well designed and low transition noise media. However, for certain media with high transition noise, the transition noise does not scale linearly with the normalized density and exhibits supralinear behavior at high linear densities. In the following analysis, for simplicity, we shall assume the independence properties of $\{b_k\}$, $\{\triangle t_k\}$ and $\{\triangle w_k\}$ still holds so that we can use M_0 to characterize the transition noise. More specifically, let us assume that for a given head/media, the scaling of M_0 with the normalized density can be described by the formula

$$M_0(D_s) = \beta(D_s) \cdot M_{0,1} \tag{6.40}$$

where $M_0(D_s)$ and $M_{0,1}$ are the M_0 values at normalized density D_s and 1 respectively, and $\beta(D_s)$ is the scaling function. Clearly, in the definition of Equation 6.36, $\beta(D_s) \equiv 1$. Now, let us consider the case when the scaling function $\beta(D_s)$ is not a constant. Such an example is shown in Figure 6.13. Recall that the purpose of SNR definition is to make a fair comparison between systems operating at different normalized densities for the same head/media combination and electronics. It is thus desirable to make the SNR independent of operating densities. For the case illustrated in Figure 6.13, we can define the SNR as follows

$$\text{SNR} = 10 \log 10 \frac{E_i}{N_0 + M_{0,1}} \tag{6.41}$$

where $M_{0,1}$ is the M_0 value at normalized density 1. It should be noted that the reference point of normalized density 1 is chosen for convenience. One can always define the SNR at other normalized densities, for example, $D_s = 2$, with slight modifications to the scaling function $\beta(D_s)$.

6.3.5 Simulation Methods

In this section, we first present the methods for generating the signal and noise for a given SNR defined in Equation 6.36. We then discuss the case when the media exhibits supralinear effect, for which we shall use the SNR definition of Equation 6.41.

For the SNR definition of Equation 6.36 and a desired SNR $= z$ dB, the noise parameter can be calculated as follows. We assume that the transition noise accounts for α-percent of the total noise. In light of Equation 6.37, We have

$$\frac{N_0}{2T} + \frac{M_0}{2T} = 10^{-z/10} \cdot \frac{E_i}{2T} \tag{6.42}$$

Since the electronics noise is responsible for $(100 - \alpha)$-percent of the total in-band noise, the in-band electronics noise power can be calculated as

$$\text{In-band electronics noise power} = \frac{N_0}{2T}$$

$$= (1 - \alpha/100) \cdot 10^{-z/10} \cdot \frac{E_i}{2T}$$

$$= (1 - \alpha/100) \cdot 10^{-z/10} \cdot \frac{E_i/PW_{50}}{2T/PW_{50}}$$

$$= (1 - \alpha/100) \cdot 10^{-z/10} \cdot \frac{E_i D_s}{2} \tag{6.43}$$

where D_s is the channel normalized density, that is, $D_s = PW_{50}/T$. Referring to Figure 6.3, if we simulate the signal/noise in the oversampling domain prior to the low-pass filter and baud-rate sampler, the AWGN power or variance should be scaled by the oversampling ratio OSR:

$$\sigma_e^2 = \text{OSR} \times (1 - \alpha/100) \cdot 10^{-z/10} \cdot \frac{E_i D_s}{2} \tag{6.44}$$

Similarly, the transition noise can be calculated as

$$\text{In-band transition noise power} = \frac{M_0}{2T}$$

$$= \frac{\alpha}{100} \cdot 10^{-z/10} \cdot \frac{E_i}{2T}$$

$$= \frac{\alpha}{100} \cdot 10^{-z/10} \cdot \frac{E_i D_s}{2} \tag{6.45}$$

In practice, it is necessary to convert the calculated transition noise power to the parameters associated with the transition position jitter $\triangle t_k$ and pulse width variation $\triangle w_k$ in order to synthesize the desired transition noise level. By assuming that $\triangle t_k$ and $\triangle w_k$ are zero-mean Gaussian distributed, we shall only need to compute the standard deviations of $\triangle t_k$ and $\triangle w_k$, which will be denoted by σ_j and σ_w respectively. In [2], the first-order position jitter and width variation (1PW) model is used to find σ_j and σ_w. More specifically, for the 1PW model, M_0 is related to σ_j^2 and σ_w^2 through

$$M_0 = \sigma_j^2 \int_{-\infty}^{\infty} \left[\frac{\partial h(t, w)}{\partial t} \right]^2 dt + \sigma_w^2 \int_{-\infty}^{\infty} \left[\frac{\partial h(t, w)}{\partial w} \right]^2 dt \tag{6.46}$$

Once the ratio between the position jitter noise power to the width variation noise power is given, σ_j^2 and σ_w^2 can then be obtained by solving Equation (6.46).

It should be pointed out that, however, the 1PW model is only a rough approximation, particularly for large σ_j and/or σ_w. One could possibly use higher-order approximation models than the 1PW model to calculate σ_j and σ_w. However, this often leads to very complex expressions. In the following, we present a more practical and convenient approach. The approach uses the gradient method (c.f. [6] pp. 639) to calculate the noise parameter iteratively. The following pseudo code describes the steps to calculate σ_j when only Gaussian position jitter is present:

```
1. Calculate the desired position jitter noise power using
   Equation 6.45
2. Set σj = 0.05 (or other initial values, e.g., 0).
3. Simulate the readback signal using σj and i.i.d. random data
   sequence; compute the simulated transition noise power mns.
```

4. While $|\frac{mn_s - mn_d}{mn_d}| < $ precision threshold

    (a) Update $\sigma_j = \sigma_j + \lambda(mn_s - mn_d)$

    (b) Simulate the readback signal using σ_j and random data sequence compute the simulated transition noise power mn_s.

5. End while

In the above pseudo-code, mn_s and mn_d represent the simulated transition noise power and desired transition noise power respectively. In Step (4), the parameter *precision threshold* is the convergence criterion, and can be set by the user according to the precision requirement. For example, one may terminate the iteration when the $(mn_s - mn_d)/mn_d < 0.01$. The iteration step size λ in Step 4(a) is a positive number, and can be chosen appropriately to ensure fast convergence. Through numerical experiments, it is found that for a typical value of $\lambda = 2$, the iteration often terminates in less than 20 iterations.

The above procedure can also be used to calculate σ_w if width variation noise is the only transition noise source. However, when both position jitter noise and width variation noise are present, it is impossible to separate them due to the fact that jitter noise and width variation interact with each other (see Equation 6.30). Nevertheless, if such interactions are ignored, the values of σ_j and σ_w can be independently obtained by using the above procedure once the ratio of position jitter noise power to the width variation noise power is given. Finally, it should be noted that the above procedure is only applicable when the position jitter or pulse broadening can be specified by a single parameter and the relation between the parameter and the transition noise is monotonic.

After the parameters of $\{\triangle t_k\}$ and $\{\triangle w_k\}$ are obtained, the readback signal can then be calculated by Equation 6.28 and Equation 6.29 respectively.

If instead, the SNR definition of Equation 6.41 is used, care must be taken when specifying the noise mixture levels. In specific, for the same head/media and electronics, different normalized densities imply different noise mixtures, as the transition noise scales differently from the electronics noise with the normalized density. Consequently, to compare the performance of different codes with different code rate, it is necessary to specify the noise mixture at a fixed normalized density. For example, if we want to evaluate the coding gains provided two different codes with code rate R_1 and R_2 respectively, given a system with transition noise scaling function $\beta(D_s)$, we can specify the noise mixture (electronics/transition noise) at normalized density 1, say 50% electronics noise and 50% transition noise. For a desired SNR, we then can calculate N_0 and $M_{0,1}$ according to Equation 6.41. If, for the first code, the system is operating at normalized density 2, we have

$$M_0 = M_{0,1} \cdot \beta(2.0) \tag{6.47}$$

and the corresponding parameters of position jitter and/or pulse broadening can then be obtained as previously described. As for the second code, the system needs to be evaluated at normalized density $2R_1/R_2$ and for the same SNR the associated M_0 is given by

$$M_0 = M_{0,1} \cdot \beta(2R_1/R_2) \tag{6.48}$$

Then, the corresponding σ_j and σ_w can be found by the above procedures.

6.3.6   Numerical Examples

For references purposes, we provide a set of simulation results here for both longitudinal and perpendicular recording channels when the SNR definition in Equation 6.36 is used. Unless otherwise specified, we use the same simulation settings as in Section 6.2.3.

Figure 6.14 shows the BER results for Viterbi detector at normalized density 2 and 2.5 with 50% electronics noise and 50% transition jitter (width variation noise is zero). For each density, PR4 $= [1 \ 0 \ -1]$ and EPR4 $= [1 \ 1 \ -1 \ -1]$ targets are selected. From the figure, the SNR loss associated with the density increase is obvious.

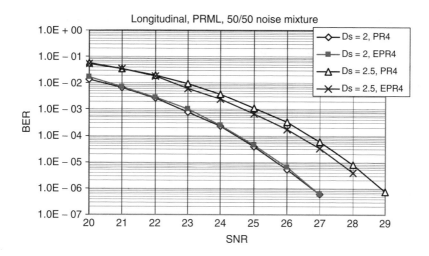

FIGURE 6.14 Bit-error-rate versus SNR for longitudinal channel at normalized density 2 and 2.5 with targets PR4 [1 0 −1] and EPR4 [1 1 −1 −1], respectively.

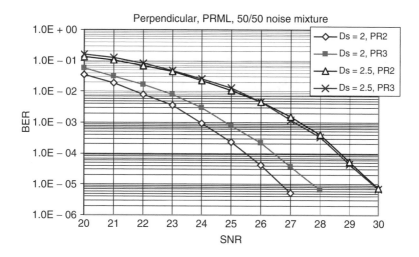

FIGURE 6.15 Bit-error-rate versus SNR for perpendicular channel at normalized density 2 and 2.5 with targets PR2 [1 2 1] and PR3 [1 3 3 1], respectively.

In Figure 6.15, we plot the BER curves versus SNR for perpendicular recording channel at normalized density 2 and 2.5, for PR2 = [1 2 1] and PR3 = [1 3 3 1] targets, respectively. Similarly, we observe that as density increases, significant BER deterioration is incurred.

6.3.7 Remarks on the Indirect Approach

The key difference between the direct approach and the indirect approach lies in whether the transition noise is included in the SNR definition. In the direct approach, the transition noise is not included in the SNR calculation and is specified directly by the noise statistical parameters (e.g., σ_j and σ_w); while for the indirect approach, the transition noise is taken into account by the SNR definition and is calculated indirectly via the parameter M_0. It should be noted however once the distribution of the transition noise and the channel transition response are given, there exists an one-to-one mapping between M_0 and the transition noise parameters. Hence, both approaches can be used in simulations.

We further remark that when the SNR definition of Equation 6.36 is used for simulations, the ratio between electronics noise and transition noise must be specified. Otherwise, infinitely many noise combinations can give rise to the same SNR number. In addition, we should also point out that even for the same values of N_0 and M_0, the simulated electronics noise and transition noise ratio may become different for differently coded data sequence, since coding in general changes the statistical properties of the data sequence, in particular the transition probability. As a result, the often-used notion of α-percent of electronics noise and $(1 - \alpha)$-percent of transition noise must be associated with a specific data pattern, based on which the comparisons of different codes can be made. For example, one can specify the noise mixture for uncoded random data sequence and subsequently calculate the associated noise parameters N_0 and M_0. Once the N_0 and M_0 are given, the corresponding physical channel is then fixed and fair comparisons can be made for different coding and signal processing schemes using the SNR definitions of Equation 6.36.

6.4 Conclusions

We addressed the problem of signal and noise generation for magnetic recording channels simulations. We focused on a simple system with electronics and transition noise. We looked at different ways of adding these noises to generate readback signals and outlined both the advantages and disadvantages of the different approaches we presented.

References

[1] Ryan, W.E. "Optimal code rates for concatenated codes on a pr4-equalized magnetic recording channel," *IEEE Trans. Mag.*, 36, no. 6, pp. 4044–4049, 2000.

[2] Moon, J., "Signal-to-noise ratio definition for magnetic recording channels with transition noise," *IEEE Trans. Mag.*, 36, no. 5, pp. 3881–3883, 2000.

[3] Moon, J., "Performance comparison of detection methods in magnetic recording," *IEEE Trans. Mag.*, 26, no. 6, pp. 3155–3172, 1990.

[4] Moon J. and Zeng, W., "Equalization for maximum likelihood detectors," *IEEE Trans. Mag.*, 31, no. 2, pp. 1083–1088, 1995.

[5] Wang S.X. and Taratorin, A. M., *Magnetic Information Storage Technology*, Adademic Press, San Diego, USA, 1999.

[6] Proakis, J.G., *Digital Communications*, McGraw-Hill, New York, 1995.

7

Statistical Analysis of Digital Signals and Systems

Dragana Bajic
University of Novi Sad
Novi Sad, Yugoslavia

Dusan Drajic
University of Belgrade
Serbia and Montenegro

7.1   Introduction 7-1
7.2   Signals 7-2
7.3   Averages 7-3
     Time Average (mean time value) • Statistical Average
7.4   Autocorrelation 7-5
     Time Average Autocorrelation • Statistical Autocorrelation
     • Phase Randomising
7.5   Power Density Spectrum of Digital Signals 7-8

7.1   Introduction

Enormous technological development of components, circuits and systems offered transmission and recording rates that could envisage only the most optimistic adherent of Moore's Law (that predicts the doubling of technological performances within constant period of time). The achievements have formed a positive feedback towards consumer's appetites, demanding all the better performances. This implies exclusive usage of digital signals within contemporary systems.

Theory of digital signals transmission has been developed over the previous decades and it can be applied nowadays, adapting to the occasions (higher data rates and new media).

The notions of stationarity and ergodicity play important role in the analysis of random signals (processes), either continuous or discrete. Considering corresponding *ensemble* (i.e. all the realizations — sample functions — of a random process) it is said that it is *stationary* if the joint probability density (joint distribution function) of any set of amplitudes (random variables) obtained by observing the random process is invariant with respect to the location of the origin [1]. For some random processes almost every member (sample function) of the ensemble exhibits the same statistical behavior as the whole ensemble, allowing the determination of the whole ensemble behavior by examining only one typical member. Such process is said to be *ergodic* [2]. For an ergodic random process all its averages (mean value, autocorrelation function, higher order moments) are the same for every ensemble member, "along the process" (called *time averages*), and equal to the *ensemble (statistical) averages* (obtained by statistical averaging — "across the process" — using joint probability densities). The stationarity is not the sufficient condition for ergodicity.

The fulfilment of given conditions results in the *strict sense stationarity (ergodicity)*, sometimes very difficult to assert. However, *wide sense stationarity (ergodicity)* is sufficient for the most cases of practical interest. The random process is wide sense stationary if its mean value (obtained upon the first-order probability density) does not depend on time and its autocorrelation function (obtained upon the second-order probability density) depends only on time interval between the observed signal amplitudes. Assuming further the equalities of time and ensemble averages, the corresponding *mean value ergodic* and *autocorrelation function ergodic* random process is defined. According to Wiener-Hinchine theorem, the average *power density spectrum* (PDS) of a wide sense stationary (and ergodic) random process is a Fourier transform of its autocorrelation function. PDS, appropriately normalized, can be considered as a kind of probability density function.

7.2   Signals

Signals are used to transmit or to store information. Within this process the following signal components can be clearly distinguished:

- Information content — the sequence of information symbols a_l from finite L-ary alphabet $a(l) = 1, \ldots, L$, considered to be a discrete-time random wide sense ergodic (implying stationarity) process; *for clarity* of further exposition, subscript denotes time epoch $(\ldots a_{n-1}, a_n, a_{n+1}, \ldots)$, while index in parenthesis denote set member $(a(1), a(2), \ldots, a(n), \ldots, a(L))$;
- Basic pulse shapes — the random waveforms taking values from a set of M deterministic (finite energy) signals $x_m(t)$ $(m = 1, \ldots, M)$;
- Transmission rate corresponding to the uniform time intervals T between successive pulses (synchronous transmission, necessary for high data rates);
- System memory — sequence of discrete random variables $S(i)$, $i = 1, \ldots, I$, referred to as the states of the coder (modulator); again, subscript denotes time epoch (S_n) and index in parenthesis denotes member of the set of states.

At the output the coder generates signal known as digitally modulated random signal, or, shortly, digital signal [3]:

$$\xi(t) = \sum_{n=-\infty}^{\infty} x(t - n \cdot T; a_n, S_n) \tag{7.1}$$

Pulse shapes are usually nonoverlapping, confined to one digit interval T.

If the coder states (S_i) do not appear in Equation 7.1, the corresponding scheme is *memoryless*, yielding

$$\xi(t) = \sum_{n=-\infty}^{\infty} x(t - n \cdot T; a_n) \tag{7.2}$$

Further, all the waveforms can be of the same shape $x(t)$, the signals being its scalar multiples. The corresponding signal (waveform) generator (modulator) is called *linear* [3], the signal at its output being represented by

$$\xi(t) = \sum_{n=-\infty}^{\infty} a_n \cdot x(t - n \cdot T) \tag{7.3}$$

This type of signal is known as synchronous pulse amplitude modulation [4].

7.3 Averages

The assumption of wide sense ergodicity of information sequence implies equality between the corresponding time and statistical averages:

$$\mu_a = E\{a_n\} = \sum_{l=1}^{L} a(l) \cdot \Pr\{a_n = a(l)\} = \overline{a_n} = \lim_{N \to \infty} \frac{1}{2 \cdot N + 1} \sum_{n=-N}^{N} a_n \tag{7.4}$$

$$R_a(k) = E\{a_n \cdot a_{n+k}\} = \sum_{j=1}^{L} \sum_{l=1}^{L} a(j) \cdot a(l) \cdot \Pr\{a_n = a(j), a_{n+k} = a(l)\}$$

$$= \overline{a_n \cdot a_{n+k}} = \lim_{N \to \infty} \frac{1}{2 \cdot N + 1} \sum_{n=-N}^{N} a_n \cdot a_{n+k}. \tag{7.5}$$

It is reasonable to suppose that "distant" symbols a_n and a_{n+k} $(k \to \infty)$ are statistically independent, yielding:

$$\lim_{k \to \infty} R_a(k) = R_a(\infty) = \mu_a^2 \tag{7.6}$$

7.3.1 Time Average (mean time value)

Time average of digital signal can be considered as its DC component and expressed as:

$$\overline{\xi(t)} = \lim_{N \to \infty} \frac{1}{2 \cdot N \cdot T} \int_{-N \cdot T}^{N \cdot T} \xi(t) \cdot dt = \lim_{N \to \infty} \frac{1}{2 \cdot N \cdot T} \sum_{n=-N}^{N} a_n \cdot \lim_{N \to \infty} \left[\int_{-N \cdot T}^{N \cdot T} x(t - n \cdot T) \cdot dt \right] \tag{7.7}$$

The last term of the Equation 7.7 has been obtained by interchanging the order of integration and summation.

It is obvious that, for an isolated signal pulse shape independent of n, it holds:

$$\lim_{N \to \infty} \left[\int_{-N \cdot T}^{N \cdot T} x(t - n \cdot T) \cdot dt \right] = \int_{-\infty}^{\infty} x(t) \cdot dt \tag{7.8}$$

and therefore

$$\overline{\xi(t)} = \lim_{N \to \infty} \frac{1}{2 \cdot N \cdot T} \sum_{n=-\infty}^{\infty} a_n \cdot \int_{-\infty}^{\infty} x(t) \cdot dt$$

$$= \frac{1}{T} \cdot \int_{-\infty}^{\infty} x(t) \cdot dt \cdot \lim_{N \to \infty} \frac{2 \cdot N + 1}{2 \cdot N} \cdot \lim_{N \to \infty} \frac{1}{2 \cdot N + 1} \cdot \left[\sum_{n=-\infty}^{-N-1} a_n + \sum_{n=-N}^{N} a_n + \sum_{n=N+1}^{\infty} a_n \right]$$

$$= \frac{1}{T} \cdot \int_{-\infty}^{\infty} x(t) \cdot dt \cdot \lim_{N \to \infty} \frac{1}{2 \cdot N + 1} \cdot \sum_{n=-N}^{N} a_n = \frac{\mu_a}{T} \cdot \int_{-\infty}^{\infty} x(t) \cdot dt = \frac{\mu_a}{T} \cdot X(j \cdot 0) \tag{7.9}$$

where $X(j \cdot \omega)$ denotes a Fourier transform of $x(t)$:

$$X(j\omega) = \int_{-\infty}^{\infty} x(t) \cdot e^{-j\omega t} \cdot dt \tag{7.10}$$

Value of $X(j \cdot 0)$ equals to the algebraic value of area under the basic pulse.

7.3.2 Statistical Average

The statistical average of digital signal is defined as:

$$E\{\xi(t)\} = E\left\{ \sum_{n=-\infty}^{\infty} a_n \cdot x(t - n \cdot T) \right\} = \sum_{n=-\infty}^{\infty} E\{a_n\} \cdot x(t - n \cdot T)$$

$$= \mu_a \cdot \sum_{n=-\infty}^{\infty} x(t - n \cdot T) \tag{7.11}$$

From the previous equation it is obvious that the statistical average depends upon time.

This can be further illustrated if a basic shape is rectangular (return-to-zero, RZ shape):

$$x(t) = \begin{cases} 1, & 0 \le t < T/2 \\ 0, & \text{elsewhere} \end{cases} \tag{7.12}$$

as shown in Figure 7.1, where several signal realizations $\xi_{(.)}$, as well as their statistical average and time averages (over finite interval) are shown. From Equation 7.11 and Figure 7.1 it can be clearly seen that $E\{\xi(t)\}$ is a periodical function:

$$E\{\xi(t)\} = E\{\xi(t + T)\} \tag{7.13}$$

This property enables signal decomposition into two components (Figure 7.2) — the first is periodical equalling to the statistical average, while the second is random and equals:

$$\xi_R(t) = \xi(t) - E\{\xi(t)\} = \sum_{n=-\infty}^{\infty} (a_n - \mu_a) \cdot x(t - n \cdot T) \quad \Rightarrow \quad E\{\xi_R(t)\} = 0 \tag{7.14}$$

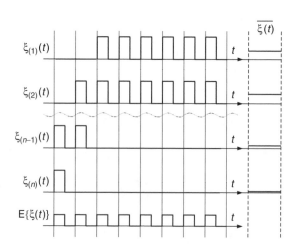

FIGURE 7.1   Signal realisations and their statistical average and time averages (over finite interval!).

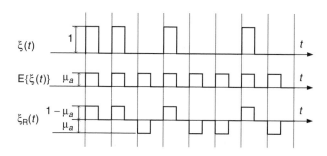

FIGURE 7.2 Periodic ($E\{\xi(t)\}$) and random ($\xi_R(t)$) component of random discrete signal ($\xi(t)$).

Using the Poisson sum formula, [4]:

$$\sum_{n=-\infty}^{\infty} x(t - n \cdot T) = \frac{1}{T} \cdot \sum_{n=-\infty}^{\infty} X\left(j \cdot 2 \cdot \pi \cdot \frac{n}{T}\right) \cdot e^{-jn\omega_T t}, \quad \omega_T = 2 \cdot \pi \cdot f_T, \quad f_T = \frac{1}{T} \quad (7.15)$$

so

$$E\{\xi(t)\} = \frac{\mu_a}{T} \cdot \sum_{n=-\infty}^{\infty} X(j \cdot n \cdot \omega_T) \cdot e^{-jn\omega_T t} \quad (7.16)$$

Spectrum of this periodical signal consists of discrete components at the multiples of f_T, with amplitudes proportional to the Fourier transform of an isolated basic shape.

By comparing Equation 7.9 and Equation 7.16 it can be concluded that signal would be mean value ergodic either if $\mu_a = 0$ (depending upon the information content) or if amplitudes of all harmonics equal to zero:

$$X(j \cdot n \cdot \omega_T) = 0, \quad n \neq 0 \quad (7.17)$$

Therefore, our illustrative example of RZ basic shapes is not mean value ergodic (not to be confused with the ergodicity of information content!), but the ergodicity could be achieved by suitable low-pass filtering, thus eliminating all unnecessary components.

Further comparison of Equation 7.9 and Equation 7.16 shows that the time average is a DC component of the ensemble average and consequently can be obtained by averaging $E\{\xi(t)\}$ over a period (T).

7.4 Autocorrelation

7.4.1 Time Average Autocorrelation

Time average autocorrelation function is defined as:

$$\overline{R_\xi(\tau)} = \overline{\xi(t) \cdot \xi(t+\tau)} = \lim_{N \to \infty} \frac{1}{2 \cdot N \cdot T} \cdot \int_{-N \cdot T}^{N \cdot T} \xi(t) \cdot \xi(t+\tau) \cdot d\tau \quad (7.18)$$

From Equation 7.3 it follows:

$$\xi(t) \cdot \xi(t+\tau) = \left[\sum_{n=-\infty}^{\infty} a_n \cdot x(t - n \cdot T)\right] \cdot \left[\sum_{l=-\infty}^{\infty} a_l \cdot x(t + \tau - l \cdot T)\right]$$

$$= \sum_{n=-\infty}^{\infty} \sum_{l=-\infty}^{\infty} a_n \cdot a_l \cdot x(t - n \cdot T) \cdot x(t + \tau - l \cdot T) \quad (7.19)$$

By substitution

$$l = n + k$$
$$x(t) \cdot x(t + \tau) = w(t, \tau) \tag{7.20}$$

and

$$\xi(t) \cdot \xi(t + \tau) = \sum_{n=-\infty}^{\infty} \sum_{k=-\infty}^{\infty} a_n \cdot a_{n+k} \cdot w(t - n \cdot T, \tau - k \cdot T) \tag{7.21}$$

into Equation 7.19, it is obtained:

$$\overline{R_\xi(\tau)} = \lim_{N \to \infty} \frac{1}{2 \cdot N \cdot T} \cdot \sum_{n=-\infty}^{\infty} \sum_{k=-\infty}^{\infty} a_n \cdot a_{n+k} \cdot \lim_{N \to \infty} \int_{-N \cdot T}^{N \cdot T} w(t - n \cdot T, \tau - k \cdot T) \cdot d\tau \tag{7.22}$$

$$\lim_{N \to \infty} \int_{-N \cdot T}^{N \cdot T} w(t - n \cdot T, \tau - k \cdot T) \cdot d\tau = \int_{-\infty}^{\infty} w(t - n \cdot T, \tau - k \cdot T) \cdot d\tau \tag{7.23}$$

therefore:

$$\overline{R_\xi(\tau)} = \frac{1}{T} \sum_{k=-\infty}^{\infty} \int_{-\infty}^{\infty} w(t, \tau - k \cdot T) \cdot \left[\lim_{N \to \infty} \frac{1}{2 \cdot N} \cdot \sum_{n=-\infty}^{\infty} a_n \cdot a_{n+k} \right] \cdot d\tau \tag{7.24}$$

Taking into account that:

$$\lim_{N \to \infty} \frac{1}{2 \cdot N} \cdot \sum_{n=-\infty}^{\infty} a_n \cdot a_{n+k}$$

$$= \lim_{N \to \infty} \frac{2 \cdot N + 1}{2 \cdot N} \cdot \lim_{N \to \infty} \frac{1}{2 \cdot N + 1} \cdot \left[\sum_{n=-\infty}^{-N-1} a_n \cdot a_{n+k} + \sum_{n=-N}^{N} a_n \cdot a_{n+k} + \sum_{n=N+1}^{\infty} a_n \cdot a_{n+k} \right]$$

$$= \lim_{N \to \infty} \frac{1}{2 \cdot N + 1} \cdot \sum_{n=-N}^{N} a_n \cdot a_{n+k} \tag{7.25}$$

and having in mind Equation 7.5, time average autocorrelation function is finally expressed as:

$$\overline{R_\xi(\tau)} = \frac{1}{T} \sum_{k=-\infty}^{\infty} R_a(k) \cdot \int_{-\infty}^{\infty} w(t, \tau - k \cdot T) \cdot d\tau$$

$$= \frac{1}{T} \sum_{k=-\infty}^{\infty} R_a(k) \cdot W(j \cdot 0, \tau - k \cdot T) \tag{7.26}$$

with $W(j \cdot \omega, \tau)$ Fourier transform of signal $w(t, \tau)$:

$$W(j \cdot \omega, \tau) = \int_{-\infty}^{\infty} X(j \cdot \lambda) \cdot X(j \cdot (\omega - \lambda)) \cdot e^{j \tau \lambda} \cdot d\lambda \tag{7.27}$$

7.4.2 Statistical Autocorrelation

Statistical autocorrelation function is defined as:

$$R_\xi(t,\tau) = E\{\xi(t) \cdot \xi(t+\tau)\} = \sum_{n=-\infty}^{\infty} \sum_{k=-\infty}^{\infty} E\{a_n \cdot a_{n+k}\} \cdot w(t-n\cdot T, \tau - k\cdot T)$$

$$= \sum_{n=-\infty}^{\infty} \sum_{k=-\infty}^{\infty} R_a(k) \cdot w(t-n\cdot T, \tau - k\cdot T) \tag{7.28}$$

Therefore, statistical autocorrelation exhibits periodicity, as $R_\xi(t+T,\tau) = R_\xi(t,\tau)$. Since $E\{\xi(t)\}$ is a periodical function as well, this class of signals is said to be cyclostationary in the wide sense [5].

In spite of the supposed ergodicity of the information content, the observed digital signal is not even stationary!

Applying Poisson sum formula over Fourier series of periodical function $R_\xi(t,\tau)$:

$$R_\xi(t,\tau) = \frac{1}{T} \sum_{k=-\infty}^{\infty} \sum_{n=-\infty}^{\infty} R_a(k) \cdot W(j\cdot n\cdot \omega_T, \tau - k\cdot T) \cdot e^{-j\cdot n\omega_T t} \tag{7.29}$$

Observing the expression for $\overline{R_\xi(\tau)}$ (Equation 7.26) it could be noticed that time autocorrelation function corresponds to DC component of statistical autocorrelation. In the previous section, we had a similar case of time and statistical average. Analogously, signal would be ergodic in autocorrelation sense if equality of time and statistical average holds, that is, if

$$W(j\cdot n\cdot \omega_T, \tau - k\cdot T) = 0 \quad \text{for } |n| \geq 1 \tag{7.30}$$

However, as signal $w(t,\tau)$ is product of shifted versions of basic shape pulse $x(t)$, its spectrum $W(j\cdot n\cdot \omega_T, \tau)$ equals to the convolution of spectra of basic pulses. For the sake of ergodicity (in autocorrelation sense!), it is sufficient that the Fourier transform of basic shape has no spectral components for $f \geq f_T/2$. This condition is equivalent to the one given by Equation 7.17, resulting in the equality of time and statistical averages, that is, in the wide sense ergodicity!

Statistical autocorrelation function of random component of digital signal $\xi_R(t)$ is

$$R_{\xi_R}(t,\tau) = \frac{1}{T} \sum_{k=-\infty}^{\infty} \sum_{n=-\infty}^{\infty} \left[R_a(k) - \mu_a^2\right] \cdot W(j\cdot n\cdot \omega_T, \tau - k\cdot T) \cdot e^{-jn\omega_T t} \tag{7.31}$$

leading to the conclusion that statistical autocorrelation of the digital signal is a periodic time function even if it does not posses the periodical component! This property enables digital clock extraction even if the periodic signal component is eliminated by appropriate line coding.

Similarly, as for the mean values, by averaging the statistical autocorrelation over a period, time autocorrelation function is obtained

$$\overline{R_\xi(\tau)} = \frac{1}{T} \cdot \int_0^T R_\xi(t,\tau) \cdot dt = \frac{1}{T} \cdot \sum_{k=-\infty}^{\infty} R_a(k) \cdot \int_{-\infty}^{\infty} w(t-n\cdot T, \tau - k\cdot T) \cdot dt = \overline{R_\xi(\tau)} \tag{7.32}$$

7.4.3 Phase Randomising

So far, it is proven that digital signals are cyclostationary. If the suitable filtering is applied, signal becomes wide sense ergodic, indicating wide sense stationarity as well, the time dependence being removed (Equation 7.26).

This approach implies that the observation of the received signal starts always at fixed time related to local oscillator. However, the duration of the signal is not of infinite. The beginning of the reception

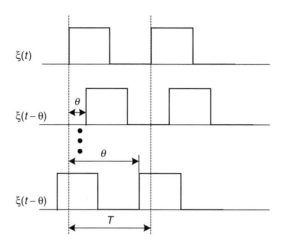

FIGURE 7.3   Various time shifts of the received signal.

(observation) is independent of the local clock and it is arbitrarily shifted in time, as shown in Figure 7.3. This time shifts can be regarded as an uniformly distributed independent random variable θ within a time period T. Starting from Equation 7.28, after averaging over a period [4]:

$$\overline{R_\xi(t,\tau)}^\theta = E\{\xi(t-\theta)\cdot\xi(t-\theta+\tau)\} = \sum_{n=-\infty}^{\infty}\sum_{k=-\infty}^{\infty} E\{a_n\cdot a_{n+k}\}\cdot\int_0^T w(t-\theta-n\cdot T,\tau-k\cdot T)\cdot\frac{1}{T}\cdot d\theta$$

$$= \frac{1}{T}\cdot\sum_{k=-\infty}^{\infty} R_a(k)\cdot\int_{-\infty}^{\infty} w(t-n\cdot T,\tau-k\cdot T)\,dt \tag{7.33}$$

the same result as in Equation 7.32 is obtained, but the physical insight may be easier.

7.5   Power Density Spectrum of Digital Signals

According to the Wiener-Hinchine theorem, PDS and autocorrelation function of a wide-sense stationary random process (wide-sense ergodicity is assumed as well) form a Fourier transform pair [1, 2, 6]. For digital signals under consideration, time autocorrelation function will be used (starting from ensemble autocorrelation function, the two-dimensional "cyclic spectrum" interesting for signal interception and synchronization would be obtained [7]).

Starting from

$$\overline{R_\xi(\tau)} = \frac{1}{2\cdot\pi}\cdot\int_{-\infty}^{\infty}\Phi_\xi(\omega)\cdot e^{j\omega\tau}\cdot d\omega \tag{7.34}$$

and comparing to

$$\overline{R_\xi(\tau)} = \frac{1}{2\cdot\pi}\cdot\int_{-\infty}^{\infty}\left[\frac{X(j\omega)\cdot X(-j\omega)}{T}\cdot\sum_{k=-\infty}^{\infty} R_a(k)\cdot e^{-j\omega kT}\right]\cdot e^{j\omega\tau}\cdot d\omega \tag{7.35}$$

obtained by the corresponding manipulations after substituting Equation 7.27 in Equation 7.26, assuming the real signal $x(t)$:

$$X(j\omega)\cdot X(-j\omega) = |X(j\omega)|^2 \tag{7.36}$$

PDS of digital signal is obtained in the form:

$$\Phi_\xi(\omega) = \frac{|X(j\omega)|^2}{T} \sum_{k=-\infty}^{\infty} R_a(k)e^{-j\omega kT} \tag{7.37}$$

Therefore, PDS is product of two factors — one depending on the pulse shape ($|X(j\omega)|^2$), and the other — depending on the information content. The "*information content spectrum*"

$$K_a(\omega) = \frac{1}{T} \sum_{k=-\infty}^{\infty} R_a(k)e^{-j\omega kT} \tag{7.38}$$

can be further divided into two parts

$$K_a(\omega) = \frac{R_a(\infty)}{T} \sum_{k=-\infty}^{\infty} e^{-j\omega kT} + \frac{1}{T} \sum_{k=-\infty}^{\infty} [R_a(k) - R_a(\infty)] \cdot e^{-j\omega kT} = K_{a_p}(\omega) + K_{a_c}(\omega) \tag{7.39}$$

The first one has discrete structure with the components at the signalling frequency multiples, according to Poisson sum formula:

$$\sum_{k=-\infty}^{\infty} \delta(\omega - k\omega_T) = T \cdot \sum_{k=-\infty}^{\infty} e^{-j\omega kT} \tag{7.40}$$

There are no discrete components if $\mu_a = 0$.

The second one is continuous, because $R_a(k) - R_a(\infty) = R_a(k) - \mu_a^2$ has no discrete components. Of course, this part is periodic in frequency with period f_T.

Finally,

$$\Phi_{\xi_p}(\omega) = |X(j\omega)|^2 \cdot \frac{\mu_a^2}{T^2} \cdot \sum_{k=-\infty}^{\infty} \delta(\omega - k\omega_T) \tag{7.41}$$

$$\Phi_{\xi_c}(\omega) = |X(j\omega)|^2 \cdot \frac{1}{T} \cdot \sum_{k=-\infty}^{\infty} \left[R_a(k) - \mu_a^2\right] \cdot e^{-j\omega kT} \tag{7.42}$$

is obtained.

Example 7.1

Statistically independent (binary) information content [4, 8]

Let the information content be formed from statistically independent bits (denoted by 0 and 1) having the probability $\Pr\{a_n = 1\} = p$ and $\Pr\{a_n = 0\} = 1 - p$. Joint probabilities are

$$\Pr\{a_n, a_{n+k}\} = \begin{cases} \Pr\{a_n\}, & k = 0 \\ \Pr\{a_n\} \cdot \Pr\{a_{n+k}\}, & k \neq 0 \end{cases} \tag{7.43}$$

Mean value and autocorrelation are

$$\mu_a = E\{a_n\} = p$$
$$R_a(k) = E\{a_n a_{n+k}\} = \begin{cases} p, & k = 0 \\ p^2, & k \neq 0 \end{cases} \tag{7.44}$$

Easily verifying $R_a(\infty) = \mu_a^2 = p^2$, PDS is obtained in the form

$$\Phi_x(\omega) = \frac{|X(j\omega)|^2}{T^2} \cdot p^2 \cdot \sum_{n=-\infty}^{\infty} \delta(\omega - n \cdot \omega_T) + \frac{|X(j\omega)|^2}{T} \cdot p \cdot (1 - p) \tag{7.45}$$

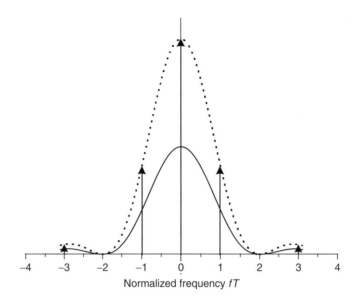

FIGURE 7.4   PDS of statistically independent binary signal, $p = 2/3$; full line — continuous part; dashed line — envelope of the discrete part.

The corresponding PDS for $p = 2/3$ is shown in Figure 7.4. Basic pulse is a RZ type while duration is $T/2$. If $p = 1/3$, the envelope of the discrete part and the continuous part exchange their positions; if $p = 1/2$, they overlap.

From the Equation 7.37 it can be concluded that the power density spectrum can be shaped (modified) either by basic pulse choice, or by the corresponding information content, or by combination of both. In the previous example it is supposed that information content is represented by statistically independent bits, thus giving predominant influence to the pulse shape. However, even if the original information content is formed by statistically independent bits (symbols), some type of coding is very often used before recording or transmission. Therefore the statistical dependence must be introduced, modifying the PDS shape.

Example 7.2

Differential binary coding [4]

The differential binary coding is sometimes more precisely called "differential coding of ones." Bit one ("1") at the input changes the output signal to its complementary value, while the input bit "0" leaves the output signal unchanged.

Let the output signals $b_n \in \{0, 1\}$ correspond to the input statistically independent bits $a_n \in \{0, 1\}$ ($\Pr\{a_n = 1\} = p$ and $\Pr\{a_n = 0\} = 1 - p$). Then $a_n = 0 \Rightarrow b_n = b_{n-1}, a_n = 1 \Rightarrow b_n \neq b_{n-1} = 1 - b_n$.

The output signal is the *first-order Markov chain* with *state diagram* shown in Figure 7.5. The corresponding probabilities are denoted along the branches.

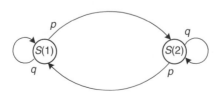

FIGURE 7.5   State-transition diagram for differential binary code.

$\Pr\{b_n = 1\} = \Pr\{b_n = 0\} = 1/2$, regardless the input bits probability, as symmetrical state diagram also suggests, (trivial case $p = 0$ is an exception). The mean value is

$$E\{b_n\} = \frac{1}{2} = \overline{b_n}. \tag{7.46}$$

After some more calculation, the correlation coefficients are obtained from difference equation [4]

$$R_b(k) - (1 - 2p) \cdot R_b(k - 1) = \frac{1}{2}p \tag{7.47}$$

Starting from $R_b(0) = 1/2$, the following expression is obtained

$$R_b(k) = \frac{1}{4}[(1 - 2p)^{|k|} + 1] \tag{7.48}$$

Finally, PDS of differentially encoded binary signal is

$$\Phi_\xi(\omega) = \frac{|X(j\omega)|^2}{4T} \cdot K_b(\omega) + \frac{|X(j\omega)|^2}{4T^2} \sum_{k=-\infty}^{\infty} \delta(\omega - k\omega_d) \tag{7.49}$$

where

$$K_b(\omega) = \sum_{k=-\infty}^{\infty} (1 - 2p)^{|k|} e^{j\omega kT} = \frac{p \cdot (1 - p)}{p^2 + (1 - 2p) \cdot \sin^2(\omega T/2)} \tag{7.50}$$

The corresponding PDS is shown for some values of p (Figure 7.6).

As shown within the previous example, even for a relatively simple kind of coding, certain amount of calculations is necessary. For more sophisticated coding, this amount can be quite overwhelming. However, a fairly simple expression for the PDS in a matrix form can be derived [9, 10] if a coder is considered as a finite automat. It is often supposed that the coder can be considered as a first-order Markov source, that is, producing a first-order Markov chain. (A higher order Markov chain can be reduced to the first-order Markov chain. A very short overview of Markov chains terminology can be found at the end of this chapter).

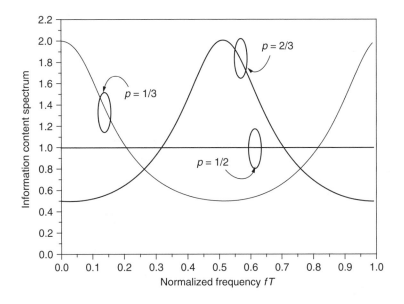

FIGURE 7.6   Information content spectrum for differential binary coding.

Let us now return to the digital signal described at the beginning of the chapter (Equation 7.1). It is supposed that it is generated by the first-order Markov source (either coder or modulator). Very often, it is considered that the coder is driven by the sequence of statistically independent symbols (information content) and that the dependence is introduced by the coder (modulator) only. However, this assumption is not of great significance, since the information content can be higher order Markov chain as well [11].

Let the information content be a sequence of statistically independent symbols a_n (not necessarily equally probable). Let the coder (finite automat) "wonders" through states according to the following rule

$$S_{n+1} = f(a_n, S_n) \tag{7.51}$$

where $f(.)$ is defined for all the states and all the input symbols. Accordingly, when coder is in state S_n, the next state (S_{n+1}) depends on present state and on the input symbol a_n. Therefore, the state *transition probabilities* equal to the input symbol probabilities. For every pair (a_n, S_n) the signal $x_m(t, a_n, S_n)$ appears at the output. The visualization is easier by using the corresponding state diagram, shown earlier in Figure 7.5. for differential binary coding. The "output signals" are 0 and 1. Transition probabilities between states form the *transition probability matrix*, where element p_{ij} corresponds to the probability of transition from the state $S(i)$ to the state $S(j)$. This is $I \times I$ matrix Π. For the previous example:

$$\Pi = \begin{bmatrix} q & p \\ p & q \end{bmatrix} \tag{7.52}$$

The corresponding output signals can also be presented in the matrix form. But, if the same signal is always "emitted" from a specific state, regardless the previous state, it is sufficient to define *output vector* $\mathbf{A}$ $(1 \times I)$. Accepting the basic pulse shape $x(t)$:

$$\mathbf{A} = [0 \quad x(t)] = x(t) \cdot [0 \quad 1] \tag{7.53}$$

Assuming further that the Markov source is *homogenous* (its transition probabilities do not depend on time n, denoted as a *stationary* source by some authors), as well as *irreducible and ergodic,* usually valid for cases under consideration, the following expression for autocorrelation function of the output signal is obtained $((\cdot)^T$ denotes transposition) [8]:

$$R_a(k) = \mathbf{A} \cdot \mathbf{P_0} \cdot \Pi^k \cdot \mathbf{A}^T \tag{7.54}$$

where $\mathbf{P_0} = \mathrm{diag}(\pi)$ and π is a *vector of state selection probabilities* (equalling, if the duration of each state is equal, to a *vector of stationary state probabilities* [12]), obtained by solving the matrix equations:

$$\pi = \pi \cdot \Pi; \quad \pi \cdot 1^T = 1; \quad 1 = [1 \quad 1 \quad \cdots \quad 1] \tag{7.55}$$

For the previous example

$$\pi = \begin{bmatrix} \frac{1}{2} & \frac{1}{2} \end{bmatrix} \tag{7.56}$$

If Markov source is *fully regular*, the continuous part of the power spectrum density is:

$$T \cdot K_a(\omega) = \mathrm{Re}\{2 \cdot z \cdot \mathbf{A} \cdot \mathbf{P_0} \cdot (z \cdot \mathbf{I} - \Pi + \Pi_\infty)^{-1} \cdot \mathbf{A}^T - \mathbf{A} \cdot \mathbf{P_0} \cdot (\mathbf{I} + \Pi_\infty) \cdot \mathbf{A}^T\} \tag{7.57}$$

where

$$\Pi_\infty = \lim_{k \to \infty} \Pi^k = 1^T \cdot \pi \tag{7.58}$$

and $z = e^{-j\omega T}$ and $\mathbf{I}$ is the identity matrix (dimensions $(I \times I)$). The inverse matrix is obtained by the summation of infinite matrix geometric series comprising $R_a(k)$ values. It is interesting to note that discrete part remains the same ($K_{a_p}(\omega)$), where

$$R_a(\infty) = \mathbf{A} \cdot \mathbf{P_0} \cdot \Pi_\infty \cdot \mathbf{A}^T = \mu_a^2 \tag{7.59}$$

Of course, for our example (differential binary coding), the using of matrix notation will give the equivalent result. However, for more complex cases, the above explained approach is practically always used and very often the matrix inversion can be obtained in suitable closed form. If it is not possible, matrix inversion can be done numerically for each frequency (f).

Example 7.3

Delay modulation (precoded Miller code) [3, 13]

This type of coding has been used very often for digital magnetic recording and for carrier communication systems employing PSK (Phase Shift Keying). State diagram with signals corresponding to states is shown in Figure 7.7. The output vector is

$$\mathbf{A} = [x_1(t) \quad x_2(t) \quad -x_2(t) \quad -x_1(t)] \tag{7.60}$$

The correlation matrix is

$$\mathbf{R} = \begin{bmatrix} 1 & 0 & 0 & -1 \\ 0 & 1 & -1 & 0 \\ 0 & -1 & 1 & 0 \\ -1 & 0 & 0 & 1 \end{bmatrix} \tag{7.61}$$

Therefore, signal waveforms are biorthogonal. The complete procedure can be envisaged as 1B-2B coding where input bit 1 is coded by $+-$ or $-+$, while input bit 0 is coded by $++$ or $--$. The other coding description is as follows: for the input bit 1 there is a transition (positive or negative) in the middle of the output signal interval, for the input bit 0 there is no transition in the middle of the interval, but when the next input bit (after 0) is also 0, there is a transition between the intervals. When applied to PSK modulator, input bit 1 will cause the phase shift in the middle of the signal interval.

The corresponding transition probability matrix is

$$\mathbf{\Pi} = \begin{bmatrix} 0 & p & 0 & q \\ 0 & 0 & p & q \\ q & p & 0 & 0 \\ q & 0 & p & 0 \end{bmatrix} \tag{7.62}$$

For $\Pr\{1\} = \Pr\{0\} = 1/2$, the state (signal) selection probabilities are all equal to 1/4. In this case, the following relation can be easily verified [3, 13]

$$\mathbf{\Pi}^{k-4} \cdot \mathbf{R} = -\frac{1}{4} \cdot \mathbf{\Pi}^k \cdot \mathbf{R} \quad (k \geq 1). \tag{7.63}$$

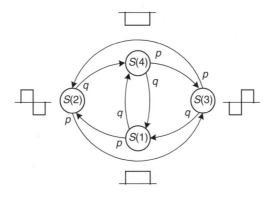

FIGURE 7.7 State-transition diagram for delayed modulation with corresponding pulse shapes.

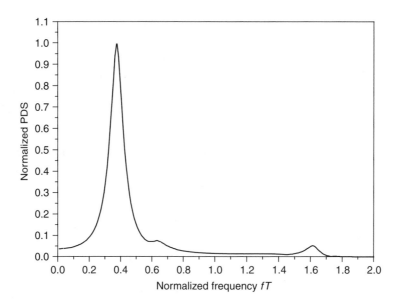

FIGURE 7.8   Normalized PDS of delayed modulation.

Using this relation, the summation of the infinite series in Equation 7.54 can be carried out using geometrical progression, yielding finally [13]

$$\Phi_\xi(\omega) = \frac{23 - 2\cos x - 22\cos 2x - 12\cos 3x + 5\cos 4x + 12\cos 5x + 2\cos 6x - 8\cos 7x + 2\cos 8x}{2x^2(17 + 8\cos 8x)}$$

(7.64)

where $x = (\omega T)/2$.

The corresponding PDS is shown in Figure 7.8. The existence of a relatively narrow high peak, as well as a very little power in the vicinity of $f = 0$, should be noted.

For a specific code types, some special methods to obtain PDS are developed [14]. Some basis for synthesis of a code having prescribed PDS can be found in [15].

Appendix: A Short Overview of Markov Chains Terminology

Markov chain nomenclature is still not unique [6, 16–19]. A short and incomplete list of some basic notions follows.

Finite Markov chain has a finite number of states (there are also countable Markov chains having a countable number of states).

For the *first-order Markov chain*, the conditional probability of the next state depends only on the present state, that is

$$\Pr\{S_{n+1}/S_n, S_{n-1}, \ldots\} = \Pr\{S_{n+1}/S_n\}$$

In the following, the first-order Markov chains are considered.

Markov source (chain) is *homogenous* (*stationary* for some authors [18]) if the transition (conditional) probabilities do not depend on time.

A Markov chain is *irreducible* if, after some steps, every state can be reached from every other state.

A state is *periodic* if it can be entered only in certain periodic intervals. Otherwise it is *aperiodic*. For the chains having periodic states, the state selection probabilities can be found as Cesaro averages.

A Markov chain is *ergodic* if every state is aperiodic and irreducible.

A Markov chain is *regular* if all the roots of the characteristic equation of the transition probability matrix of unit modulus are identically 1. Chain is *fully regular* if 1 is a simple root of the characteristic equation.

References

[1] Haykin, S., *Communication Systems*, 3rd ed., John Wiley & Sons, Inc., New York, 1994, chap. 4.

[2] Cooper, G.R. and Mc Gillem, C.D., *Probabilistic Methods of Signal and System Analysis*, 3rd ed., Oxford University Press, Oxford 1999, chap. 7.

[3] Benedetto, S., Biglieri, E., and Castellani, V., *Digital Transmission Theory*, Prentice-Hall, Inc., Englewood Cliffs, N.J., 1999, chap. 2.

[4] Franks, L.E., *Signal Theory*, Prentice-Hall, Inc., Englewood Cliffs, N.J., 1969, chap. 8.

[5] Bennett, W.R., Statistics of regenerative digital transmission, *The Bell Sys. Tech. Journal*, 37, 1501, 1958.

[6] Thomas, J.B., *An Introduction to Statistical Communication Theory*, John Wiley & Sons, Inc., New York, 1969, chap. 3.

[7] Gardner, W.A., *Introduction to Random Processes*, Macmillan Publishing Company, New York, 1986, chap. 10.

[8] Lukatela, G., Drajic, D., Petrovic, G, and Petrovic, R., *Digital Communications* (in Serbian), 2nd ed., Gradjevinska knjiga, Beograd, 1984, chap. 14.

[9] Cariolaro, G.L. and Tronca, G.P., Spectra of block coded digital signals, *IEEE Trans. on Commun.* COM-22, 1555, 1974.

[10] Cariolaro, G.L., Pierobon, G.L., and Pupolin, S.G., Spectral analysis of variable-length coded digital signals, *IEEE Trans. on Info. Theory*, IT-28, 473, 1982.

[11] Bilardi, G., Padovani, R., and Pierobon, G.L., Spectral analysis of functions of Markov chains with applications, *IEEE Trans. on Commun.* COM-31, 853, 1983.

[12] Wolff, R.W., *Stochastic Modelling and the Theory of Queues*, Prentice-Hall International Inc, Englewood Cliffs, NJ, 1989, chap. 3.

[13] Proakis, J.G., *Digital Communications*, 3rd ed., McGraw-Hill Inc., New York, 1995, chap. 4.

[14] Gallopoulos, A., Heegard, C., and Siegel, P.H., The power spectrum of run-length-limited codes, *IEEE Trans. on Commun.* 37, 906, 1989.

[15] Justensen, J., Information rates and power spectra of digital codes, *IEEE Trans. on Info. Theory*, IT-28, 457, 1982.

[16] Galko, P. and Pasupathy, S., The mean power spectral density of Markov chain driven signals, *IEEE Trans. on Info. Theory*, IT-27, 746, 1981.

[17] Kemeny, J.G., and Snell, J.L., *Finite Markov Chains*, D. van Nostrand Company, Inc., Princeton, N.J., 1960. chap. 2.

[18] Isaacson, D.L., and Madsen, R.W., *Markov Chains Theory and Applications*, John Wiley & Sons, Inc., New York, 1976. chap. 2.

[19] Chung, K.L., *Markov Chains with Stationary Transition Probabilities*, 2nd ed., Springer, Berlin, 1967, chap. 1.

8

Partial Response Equalization with Application to High Density Magnetic Recording Channels

8.1 Introduction **8**-1
8.2 Characterization of Intersymbol Interference in Digital Communication Systems **8**-3
8.3 Partial Response Signals **8**-6
8.4 Detection of Partial Response Signals **8**-9
8.5 Model of a Digital Magnetic Recording System **8**-10
8.6 Optimum Detection for the AWGN Channel **8**-13
8.7 Linear Equalizer and Partial Response Targets **8**-17
8.8 Maximum-Likelihood Sequence Detection and Symbol-by-Symbol Detection **8**-18
    Symbol-by-Symbol Detector • Maximum-Likelihood Sequence Detector
8.9 Performance Results from Computer Simulation **8**-20
8.10 Concluding Remarks **8**-21

John G. Proakis

University of California San Diego
San Diego, CA

8.1 Introduction

Most communication channels, including data storage channels, may be generally characterized as band-limited linear filters. Consequently, such channels are described by their frequency response $C(f)$, which may be expressed as

$$C(f) = A(f)e^{j\theta(f)}$$

where $A(f)$ is called the *amplitude response* and $\theta(f)$ is called the *phase response*. Another characteristic that is sometimes used in place of the phase response is the *envelope delay* or *group delay*, which is defined as

$$\tau(f) = -\frac{1}{2\pi}\frac{d\theta(f)}{df}$$

0-8493-1524-7/05/$0.00+$1.50
© 2005 by CRC Press, LLC

A channel is said to be nondistorting or ideal if, within the bandwidth W occupied by the transmitted signal, $A(f) = $ const and $\theta(f)$ is a linear function of frequency [or the envelope delay $\tau(f) = $ const]. On the other hand, if $A(f)$ and $\tau(f)$ are not constant within the bandwidth occupied by the transmitted signal, the channel distorts the signal. If $A(f)$ is not constant, the distortion is called *amplitude distortion* and if $\tau(f)$ is not constant, the distortion on the transmitted signal is called *delay distortion*.

As a result of the amplitude and delay distortion caused by the nonideal channel frequency response characteristic $C(f)$, a succession of pulses transmitted through the channel at rates comparable to the bandwidth W are smeared to the point that they are no longer distinguishable as well-defined pulses at the receiving terminal. Instead, they overlap and, thus, we have **intersymbol interference (ISI)**. As an example of the effect of delay distortion on a transmitted pulse, Figure 8.1(a) illustrates a band-limited pulse having zeros periodically spaced in time at points labeled $\pm T, \pm 2T$, etc. If information is conveyed by the pulse amplitude, as in pulse amplitude modulation (PAM), for example, then one can transmit

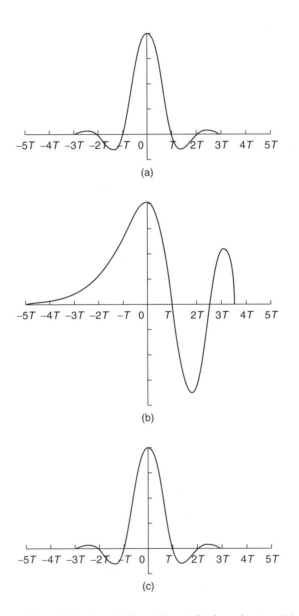

FIGURE 8.1   Effect of channel distortion: (a) channel input, (b) channel output, (c) equalizer output.

a sequence of pulses, each of which has a peak at the periodic zeros of the other pulses. Transmission of the pulse through a channel modeled as having a linear envelope delay characteristic $\tau(f)$ [quadratic phase $\theta(f)$], however, results in the received pulse shown in Figure 8.1(b) having zero crossings that are no longer periodically spaced. Consequently, a sequence of successive pulses would be smeared into one another, and the peaks of the pulses would no longer be distinguishable. Thus, the channel delay distortion results in intersymbol interference. As will be discussed in this chapter, it is possible to compensate for the nonideal frequency response characteristic of the channel by use of a filter or equalizer at the demodulator. Figure 8.1(c) illustrates the output of a linear equalizer that compensates for the linear distortion in the channel.

8.2 Characterization of Intersymbol Interference in Digital Communication Systems

As illustrated above, channel distortion causes intersymbol interference. To develop a mathematical model that characterizes the ISI, let us consider the transmission of a binary sequence $\{a_k\}$, where $a_k = \pm 1$ with equal probabilities, through a baseband communication channel having frequency response $C(f)$ and impulse response $c(t)$. The transmitted signal may be expressed as

$$s(t) = \sum_{n=-\infty}^{\infty} a_n g_T(t - nT) \tag{8.1}$$

where $1/T$ is the symbol rate (in this case the symbols are bits) and $g_T(t)$ is a basic pulse shape that is selected to control the spectral characteristics of the transmitted signal and to ensure that the spectrum of the transmitted signal falls within the bandwidth limitations of the channel.

The received signal may be expressed as

$$r(t) = \sum_{n=-\infty}^{\infty} a_n h(t - nT) + w(t) \tag{8.2}$$

where $h(t) = g_T(t) * c(t)$, where the asterisk denotes convolution, and $w(t)$ represents the additive noise in the received signal, which is the noise introduced by the channel and by the front end of the receiver.

To characterize the ISI, suppose that the received signal is passed through a receiving filter and sampled at the rate $1/T$ samples. When the additive noise $w(t)$ is a sample function of a stationary white Gaussian noise process, the optimum filter at the receiver is matched to the received signal pulse $h(t)$. Hence, the frequency response of this filter is $H^*(f)$, where the superscript asterisk denotes the complex conjugate of $H(f)$. Assuming that this filter is used at the receiver, its output may be expressed

$$y(t) = \sum_{n=-\infty}^{\infty} a_n x(t - nT) + v(t) \tag{8.3}$$

where $x(t)$ is the signal pulse response of the receiving filter, that is, $X(f) = H(f)H^*(f) = |H(f)|^2$, and $v(t)$ is the response of the receiving filter to the noise $w(t)$. Now, if $y(t)$ is sampled at times $t = kT$, $k = 0, 1, 2, \ldots$, we have

$$y(kT) \equiv y_k = \sum_{n=-\infty}^{\infty} a_n x(kT - nT) + v(kT)$$

$$= \sum_{n=-\infty}^{\infty} a_n x_{k-n} + v_k, \quad k = 0, 1, 2, \ldots \tag{8.4}$$

The sample values can be expressed as

$$y_k = x_0 a_k + \sum_{\substack{n=-\infty \\ n \neq k}}^{\infty} a_n x_{k-n} + v_k \tag{8.5}$$

The term x_0 is an arbitrary scale factor, which we arbitrarily set to unity for convenience. Then,

$$y_k = a_k + \sum_{\substack{n=-\infty \\ n \neq k}}^{\infty} a_n x_{k-n} + v_k \tag{8.6}$$

The term a_k represents the desired information symbol at the kth sampling instant, the term

$$\sum_{\substack{n=-\infty \\ n \neq k}}^{\infty} a_k x_{n-k}$$

represents the ISI, and v_k is the additive noise variable at the kth sampling instant.

The effect of the ISI and noise in a digital communication system can be viewed on an oscilloscope. We display the received signal $y(t)$ on the vertical input and set the horizontal sweep rate at $1/T$. The resulting oscilloscope display is called an *eye pattern* because of its resemblance to the human eye. Figure 8.2 illustrates the eye pattern for binary modulation. The effect of the ISI is to cause the eye to close, thereby reducing the margin for additive noise to cause errors. Figure 8.3 graphically illustrates the effect of ISI in reducing the opening of the eye. Note that ISI distorts the position of the zero crossings and causes a reduction in the eye opening. As a consequence, it also causes the system to be more sensitive to any timing error in the clock that provides timing for the sampler.

To eliminate ISI, it is necessary and sufficient that $x(kT - nT) = 0$ for $k \neq n$ and $x(0) \neq 0$, where without loss of generality we can assume that

$$x(kT) = \begin{cases} 1, & k = 0 \\ 0, & k \neq 0 \end{cases} \tag{8.7}$$

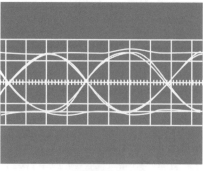

BINARY

FIGURE 8.2   Example of eye pattern for binary signal.

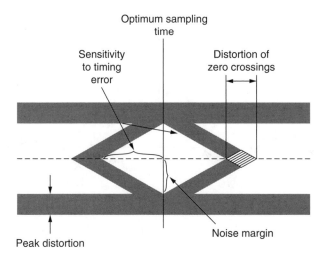

FIGURE 8.3 Effect of intersymbol interference on eye opening.

Nyquist [1] has proved that a necessary and sufficient condition for zero ISI is that the Fourier transform $X(f)$ of the received signal pulse $x(t)$ satisfy the condition

$$\sum_{m=-\infty}^{\infty} X\left(f + \frac{m}{T}\right) = T \tag{8.8}$$

If the available channel bandwidth is WHz, Nyquist proved that the maximum binary transmission rate that is possible with zero ISI is $1/T = 2W$ symbols/sec. This rate, called the *Nyquist rate* is obtained with the signal pulse whose Fourier transform is

$$X(f) = \begin{cases} T, & |f| \leq W \\ 0, & \text{otherwise} \end{cases} \tag{8.9}$$

However, such a frequency response characteristic cannot be achieved with physically realizable filters. On the other hand, if the transmission rate $1/T$ is reduced below the Nyquist rate of $2W$, it is possible to design physically realizable filters that satisfy Equation 8.8 and, thus, result in zero ISI at the sampling instants.

A particular pulse spectrum that has desirable spectral properties and has been widely used in practice is the raised cosine spectrum. The raised cosine frequency characteristic is given as

$$X_{rc}(f) = \begin{cases} T, & 0 \leq |f| \leq (1-\alpha)/2T \\ \frac{T}{2}\left[1 + \cos\frac{\pi T}{\alpha}\left(|f| - \frac{1-\alpha}{2T}\right)\right], & \frac{1-\alpha}{2T} \leq |f| < \frac{1+\alpha}{2T} \\ 0, & |f| > \frac{1+\alpha}{2T} \end{cases} \tag{8.10}$$

where α is called the *rolloff* factor, which takes values in the range $0 \leq \alpha \leq 1$, and $1/T$ is the symbol rate. The frequency response $X_{rc}(f)$ is illustrated in Figure 8.4(b) for $\alpha = 0, 1/2$, and 1. Note that when $\alpha = 0$, $X_{rc}(f)$ reduces to an ideal brick wall physically nonrealizable frequency response with bandwidth occupancy $1/2T$. The frequency $1/2T$ is called the *Nyquist frequency*. For $\alpha > 0$, the bandwidth occupied by the desired signal $X_{rc}(f)$ beyond the Nyquist frequency $1/2T$ is called the *excess bandwidth* and is usually expressed as a percentage of the Nyquist frequency. For example, when $\alpha = 1/2$, the excess bandwidth

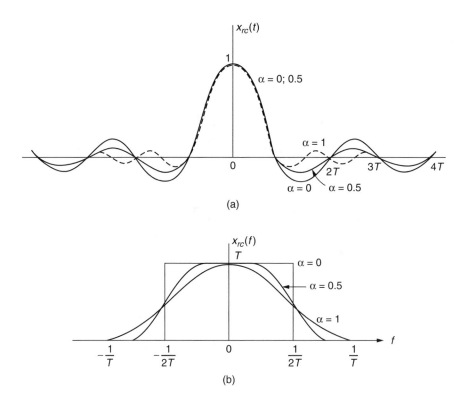

FIGURE 8.4 Pulses having a raised-cosine spectrum.

is 50%, and when $\alpha = 1$, the excess bandwidth is 100%. The signal pulse $x_{rc}(t)$ having the raised-cosine spectrum is

$$x_{rc}(t) = \frac{\sin \pi t/T}{\pi t/T} \frac{\cos(\pi \alpha t/T)}{1 - 4\alpha^2 t^2/T^2} \tag{8.11}$$

Figure 8.4(a) illustrates the pulse shape $x_{rc}(t)$ for $\alpha = 0, 1/2$, and 1. Note that $x_{rc}(t) = 1$ at $t = 0$ and $x_{rc}(t) = 0$ at $t = kT, k = \pm 1, \pm 2, \ldots$. Consequently, at the sampling instants $t = kT, k \neq 0$, there is no ISI from adjacent symbols. In order to satisfy the condition of zero ISI with physically realizable filters, the transmission rate $1/T$ was reduced below the Nyquist rate of $2W$. For example, with $\alpha = 1$, the transmission rate $1/T = W$ symbols/sec. Signal pulses that are designed to have zero ISI, such as $x_{rc}(t)$ given in Equation 8.11, are called *full response signals*.

8.3   Partial Response Signals

From our discussion of signal design for zero ISI, we have observed that it is necessary to reduce the transmission rate $1/T$ below the Nyquist rate of $2W$ symbols/sec in order to design practical transmitting and receiving filters. Now, suppose we relax the condition of zero ISI and, thus, achieve the transmission rate of $2W$ symbols/sec. By allowing a controlled amount of ISI, we can achieve the rate of $2W$ symbols/sec, as described below.

Let us consider the design of a bandlimited signal pulse $x(t)$ that results in ISI at one time instant. That is, we allow one additional nonzero value in the samples $\{x(nT)\}$. The ISI that we introduce is deterministic or "controlled," hence, it can be taken into account in the signal detection that is performed at the receiver.

For example, suppose we select the following signal pulse samples:

$$x(nT) = \begin{cases} 1, & n = 0, 1 \\ 0, & \text{otherwise} \end{cases} \tag{8.12}$$

where $T = 1/2W$. In this case, the ISI is limited to two adjacent symbols. Since the channel frequency response $C(f)$ is bandlimited to $W Hz$, the signal pulse $x(t)$ is also bandlimited to $W Hz$, that is,

$$X(f) = 0, \quad |f| > W$$

By applying the sampling theorem for bandlimited signals, the signal pulse $x(t)$ can be expressed as

$$x(t) = \sum_{n=-\infty}^{\infty} x\left(\frac{n}{2W}\right) \left(\frac{\sin 2\pi W(t - n/2W)}{2\pi W(t - n/2W)}\right) \tag{8.13}$$

where $\{x(n/2W)\}$ are samples of $x(t)$ taken at $t = nT = n/2W$. Therefore, the signal pulse $x(t)$ for the sample values given by Equation 8.12 is

$$x(t) = \frac{\sin 2\pi Wt}{2\pi Wt} + \frac{\sin 2\pi W(t - 1/2W)}{2\pi W(t - 1/2W)} \tag{8.14}$$

Its spectrum is

$$X(f) = \begin{cases} \frac{1}{2W}\left[1 + e^{-j\pi f/W}\right], & |f| \le W \\ 0, & \text{otherwise} \end{cases}$$

$$= \begin{cases} \frac{1}{W} e^{-j\pi f/2W} \cos\left(\frac{\pi f}{2W}\right), & |f| \le W \\ 0, & \text{otherwise} \end{cases} \tag{8.15}$$

This pulse and its amplitude spectrum $|X(f)|$ are illustrated in Figure 8.5. Note that the spectrum $|X(f)|$ decays smoothly to zero at $|f| = W$. This implies that transmitting and receiving filters can be readily implemented. The signal pulse given by Equation 8.14 is called a *duobinary signal pulse*.

Another special case of Equation 8.13 that results in physically realizable transmitting and receiving filters is specified by the samples

$$x\left(\frac{n}{2W}\right) = x(nT) = \begin{cases} 1, & n = -1 \\ -1, & n = 1 \\ 0, & \text{otherwise} \end{cases} \tag{8.16}$$

The corresponding signal pulse, obtained from Equation 8.13 is

$$x(t) = \frac{\sin 2\pi W(t + 1/2W)}{2\pi W(t + 1/2W)} - \frac{\sin 2\pi W(t - 1/2W)}{2\pi W(t - 1/2W)} \tag{8.17}$$

and its spectrum is

$$X(f) = \begin{cases} \frac{1}{2W}(e^{j\pi f/W} - e^{-j\pi f/W}) = \frac{j}{W}\sin\frac{\pi f}{W}, & |f| \le W \\ 0, & |f| > W \end{cases} \tag{8.18}$$

This pulse and its amplitude spectrum are illustrated in Figure 8.6. It is called a *modified duobinary signal pulse*. It is interesting to note that the spectrum of this signal pulse is zero at $f = 0$, making it suitable for transmission over a channel that does not pass dc.

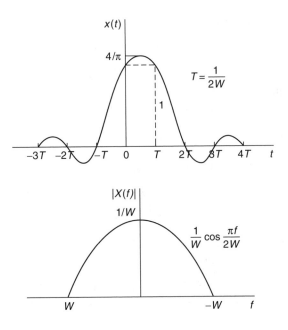

FIGURE 8.5   Time domain and frequency-domain characteristics of a duobionary signal.

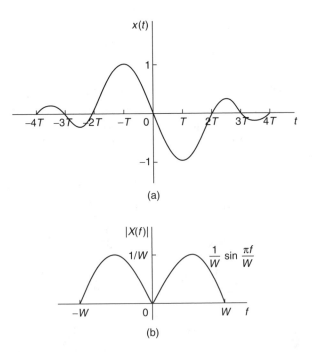

FIGURE 8.6   Time domain and frequency domain characteristics of a modified duobinary signal.

From Equation 8.13 one can obtain other interesting and physically realizable filter characteristics by selecting different values for $\{x(n/2W)\}$ with two or more nonzero samples. The signal pulses obtained in this manner in which controlled ISI is purposely introduced by selecting two or more nonzero samples from the set $\{x(n/2W)\}$ are called *partial response signals*. The resulting signal pulses allow us to transmit data at the Nyquist rate of 2W symbols/sec.

8.4 Detection of Partial Response Signals

The optimum detection method for recovering the information symbols at the demodulator when the received signal is a partial response signal is based on the maximum-likelihood criterion for detecting a sequence of symbols. This detection method, called a *maximum-likelihood sequence detection* (MLSD), minimizes the probability of error for a received sequence of bits [2,6]. An efficient algorithm for implementing MLSD is the Viterbi algorithm [2,6].

An alternative detection method for recovering the information symbols when the received signal is a partial response signal is based on symbol-by-symbol (SBS) detection. SBS detection is relatively easy to implement, but the average bit error probability is higher than MLSD. To describe the SBS method, let us consider the detection of the duobinary and the modified duobinary partial response signals. In both cases, we assume that the desired spectral characteristic $X(f)$ for the partial response signal is split evenly between the transmitting and receiving filters, that is, $|G_T(f)| = G_R(f) = |X(f)|^{1/2}$.

For the duobinary signal pulse, $x(nT) = 1$, for $n = 0, 1$ and zero otherwise. Hence, the samples at the output of the receiving filter have the form

$$y_m = b_m + v_m = a_m + a_{m-1} + v_m \tag{8.19}$$

where $\{a_m\}$ is the transmitted sequence of amplitudes and $\{v_m\}$ is a sequence of additive Gaussian noise samples. Let us ignore the noise for the moment and consider the binary case where $a_m = \pm 1$ with equal probability. Then, b_m takes on one of three possible values, namely, $b_m = -2, 0, 2$ with corresponding probabilities $1/4, 1/2, 1/4$. If a_{m-1} is the detected symbol from the $(m-1)st$ signaling interval, its effect on b_m, the received signal in the mth signaling interval, can be eliminated by subtraction, thus allowing a_m to be detected. This process can be repeated sequentially for every received symbol.

The major problem with this procedure is that errors arising from the additive noise tend to propagate. For example, if a_{m-1} is in error, its effect on b_m is not eliminated but, in fact, it is reinforced by the incorrect subtraction. Consequently, the detection of a_m is also likely to be in error.

Error propagation can be avoided by *precoding* the data at the transmitter instead of eliminating the controlled ISI by subtraction at the receiver. The precoding is performed on the binary data sequence prior to modulation. From the data sequence $\{d_n\}$ of 1s and 0s that is to be transmitted, a new sequence $\{p_n\}$, called the *precoded sequence* is generated. For the duobinary signal, the precoded sequence is defined as

$$p_m = d_m \ominus p_{m-1}, \quad m = 1, 2, \ldots \tag{8.20}$$

where the symbol $\ominus$ denotes modulo-2 subtraction.[1] Then, we set $a_m = -1$ if $p_m = 0$ and $a_m = 1$ if $p_m = 1$, that is, $a_m = 2p_m - 1$.

The noise-free samples at the output of the receiving filter are given as

$$\begin{aligned} b_m &= a_m + a_{m-1} \\ &= (2p_m - 1) + (2p_{m-1} - 1) \\ &= 2(p_m + p_{m-1} - 1) \end{aligned} \tag{8.21}$$

---

[1]Although this is identical to modulo-2 addition, it is convenient to view the precoding operation for duobinary in terms of modulo-2 subtraction.

Consequently,

$$p_m + p_{m-1} = \frac{b_m}{2} + 1 \tag{8.22}$$

Since $d_m = p_m \oplus p_{m-1}$, it follows that the data sequence d_m is obtained from b_m by using the relation

$$d_m = \frac{b_m}{2} + 1 (\text{mod } 2) \tag{8.23}$$

Consequently, if $b_m = \pm 2$, $d_m = 0$ and if $b_m = 0$, $d_m = 1$. In the presence of additive noise the sampled outputs from the receiving filter are given by Equation 8.19. In this case $y_m = b_m + v_m$ is compared with the two thresholds set at $+1$ and -1. The data sequence $\{d_m\}$ is obtained according to the detection rule

$$d_m = \begin{cases} 1, & \text{if } -1 < y_m < 1 \\ 0, & \text{if } |y_m| \geq 1 \end{cases} \tag{8.24}$$

In the case of the modified duobinary pulse, the controlled ISI is specified by the values $x(n/2W) = -1$, for $n = 1$, $x(n/2W) = 1$, for $n = -1$ and zero otherwise. Consequently, the noise-free sampled output from the receiving filter is given as

$$b_m = a_m - a_{m-2} \tag{8.25}$$

where the sequence $\{a_m\}$ is obtained by mapping a precoded sequence

$$p_m = d_m \oplus p_{m-2} (\text{mod } 2) \tag{8.26}$$

as described above, that is, $a_m = 2p_m - 1$. From these relations, it is easy to show that the detection rule for recovering the data sequence $\{d_m\}$ from $\{b_m\}$ in the absence of noise is

$$d_m = \frac{b_m}{2} (\text{mod } 2) \tag{8.27}$$

In the presence of noise, the received signal-plus-noise $y_m = b_m + v_m$ is compared with the two thresholds set at ± 1, and the data sequence $\{d_m\}$ is obtained according to the detection rule

$$d_m = \begin{cases} 0, & \text{if } -1 < y_m < 1 \\ 1, & \text{if } |y_m| \geq 1 \end{cases} \tag{8.28}$$

As demonstrated above, the precoding of the data at the transmitter makes it possible to detect the received data on a symbol-by-symbol basis without having to look back at previously detected symbols. Thus, error propagation is avoided.

8.5 Model of a Digital Magnetic Recording System

Figure 8.7 illustrates a block diagram of a digital magnetic recording system. The binary data sequence to be stored is fed to the write current driver, which generates a two-level signal waveform called the write current, as illustrated in Figure 8.8. NRZI (non-return-to-zero inverse) is the method used for mapping the binary data sequence into the write current. In NRZI recording, a change in polarity in the write current corresponds to the binary digit 1, while no change in polarity corresponds to the binary digit 0.

The write current is fed to the write head, which is basically the transducer that magnetizes the medium. Thus, the write current determines the direction of the magnetization on the magnetic medium, as shown in Figure 8.9. The positive polarity of the write current magnetizes the medium to saturation in one direction and the negative polarity magnetizes the medium in the opposite direction.

In the readback process, the flux-to-voltage conversion is performed by the readback head, which may be either inductive or magnetoresistive. The inductive readback head senses the derivative of the flux through

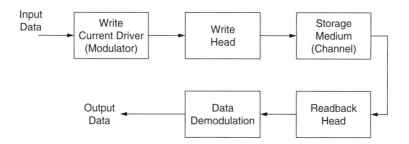

FIGURE 8.7 Block diagram of a magnetic storage read/write system.

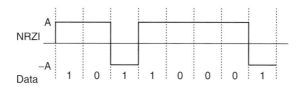

FIGURE 8.8 NRZI signal.

FIGURE 8.9 Magnetization pattern for the write current in Figure 8.8.

the magnetic core, while the magnetoresistive head responds to the flux emanating from the magnetic transition directly. For an isolated positive transition, the output voltage waveform of the readback head is a pulse, $p(t)$. For an isolated negative transition, the output voltage waveform of the readback head is a pulse, $-p(t)$. We call $p(t)$ the *isolated transition response* of the magnetic recording system. The pulse, $p(t)$, is well-modeled mathematically as a *Lorentzian pulse* (normalized to unit peak amplitude), defined as [3]

$$p(t) = \frac{1}{1 + \left(\frac{2t}{T_{50}}\right)} \tag{8.29}$$

where T_{50} denotes the width of the pulse at its 50% amplitude level, as illustrated in Figure 8.10. The value of T_{50} is determined by the characteristics of the magnetic medium, the head, and the distance of the head to the medium. This parameter basically determines the resolution of the recording process and ultimately becomes the limiting factor in the recording density of the system.

Now, suppose we write a positive transition followed by a negative transition. Let us vary the time interval between two transitions, which we denote as T (the bit time interval). Figure 8.11 illustrates the readback signal pulses, which are obtained by superposition of $p(t)$ with $-p(t - T)$. The parameter $D = T_{50}/T$ is defined as the normalized density. The closer the bit transitions (T small), the larger the value of the normalized density and, hence, the larger the density of the system. We observe that as D is increased, the peak amplitudes of the readback signal are reduced and are also shifted in time from the desired time instants. In other words, the pulses interfere with one another, thus limiting the density with which we can write. Note that for $D = 1$, the interference effects are negligible, but as D is increased, this is no longer the case. The reduction in amplitude level for $D > 1$ results in a decrease in the signal-to-noise ratio in the data detection process. Furthermore, the overlapping pulses in a sequence of data bits result in intersymbol interference.

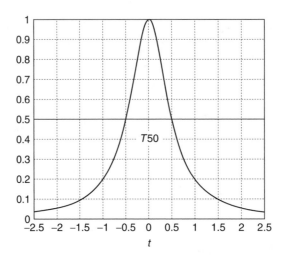

FIGURE 8.10   Readback pulse in a magnetic recording system.

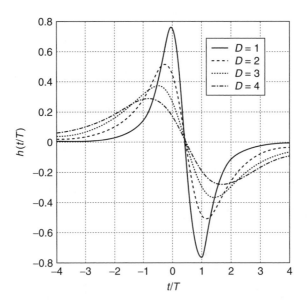

FIGURE 8.11   Readback signal response to a pulse.

Our treatment is focused on the problem of achieving normalized densities of $D > 1$, where ISI is present in the readback signal. In such a case, the recovery of the information sequence from the readback signal requires the use of an equalization technique that compensates for the presence of ISI.

Mathematically, we may express the readback signal as

$$r(t) = \sum_{k=-\infty}^{\infty} a_k h(t - kT) + w(t) \qquad (8.30)$$

where $\{a_k\}$ is the recorded binary data sequence representing the magnetization state with values $\{+1, -1\}$; $h(t) = p(t) - p(t - T)$ is the pulse response of the readback head and medium, frequently called the

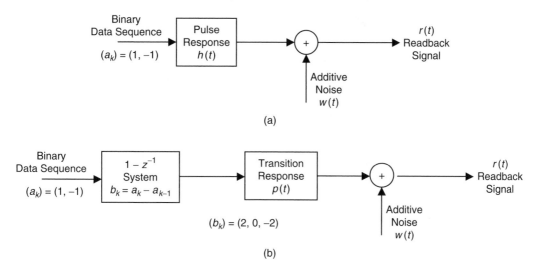

FIGURE 8.12   Model for the readback signal in a magnetic recording system: (a) model for Equation 8.30 (b) model for Equation 8.31.

magnetic recording channel; $p(t)$ is the Lorentzian pulse defined in Equation 8.29; and $w(t)$ represents the additive noise in the system. Note that $h(t)$ may be viewed as a partial response signal. Alternatively, we may express the readback signal in the form

$$r(t) = \sum_{k=-\infty}^{\infty} b_k p(t - kT) + w(t) \tag{8.31}$$

where $b_k = a_k - a_{k-1}$ may be called the *transition sequence*. Note that $\{b_k\}$ is a three-level sequence with possible values taken from the set $\{2, 0, -2\}$. Figure 8.12 illustrates the two representations of the readback signal as expressed in Equation 8.30 and Equation 8.31.

In the frequency domain, the Lorentzian pulse, $p(t)$, may be expressed as

$$P(f) = \frac{\pi T_{50}}{2} \exp\{-\pi T_{50}|f|\}$$
$$= \frac{\pi T}{2} D \exp\{-\pi D|fT|\} \tag{8.32}$$

and the pulse response, $h(t)$, may be expressed as

$$H(f) = P(f)[1 - e^{-j2\pi fT}]$$
$$= j2 \sin(\pi fT) P(f) e^{-j\pi fT} \tag{8.33}$$
$$= j\pi T D \sin(\pi fT) \exp\{-\pi D|fT|\} e^{-j\pi fT}$$

The frequency response characteristics $|P(f)|$ and $|H(f)|$ are illustrated in Figure 8.13, plotted as a function of the normalized frequency, fT, with D as a parameter. We observe that $20 \log |P(f)|$ falls off linearly as a function of the normalized frequency, fT. As D increases, the bandwidth of $P(f)$ and $H(f)$ decreases. We also note that $H(f)$ exhibits a spectral null at $f = 0$.

8.6   Optimum Detection for the AWGN Channel

The noise in a magnetic recording system is a combination of media noise, head noise, and thermal noise generated in the preamplifier. These three noise sources are generally mutually uncorrelated. Although media noise is non-Gaussian, the other two noise sources are well modeled as white Gaussian processes.

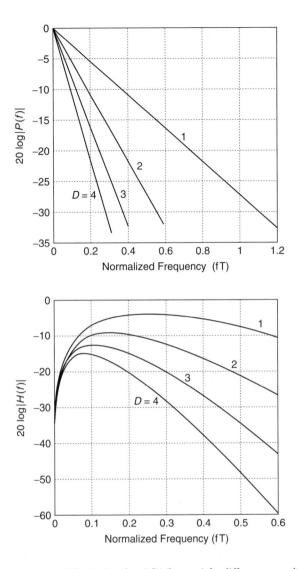

FIGURE 8.13   Graphs of $|P(f)|$ (top) and $|H(f)|$ (bottom) for different normalized density, D.

For the purpose of specifying the optimum signal processing and detection method for the readback signal, we assume that the additive noise $w(t)$ in $r(t)$, is a sample function of a white Gaussian noise process with zero mean and spectral density $N_0/2$ W/Hz. In such a case, the optimum signal processing and detection method consists of a filter matched to the pulse response, $h(t)$, a symbol-rate sampler, and a maximum-likelihood sequence detector [2], as illustrated in Figure 8.14. The output of the symbol-rate sampler is the signal sequence

$$y_n = \sum_{k=-\infty}^{\infty} a_k x_{n-k} + v_n, \quad -\infty < n < \infty \tag{8.34}$$

where the sequence $\{x_n\}$ is the sampled autocorrelation function of $h(t)$, that is,

$$x_n = \int_{-\infty}^{\infty} h(t) h(t + nT) \, dt \tag{8.35}$$

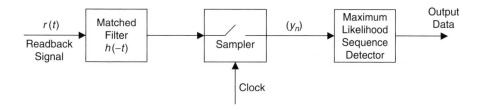

FIGURE 8.14   Block diagram of optimum signal processing and detection of readback signal.

and $\{v_k\}$ is the additive noise sequence, defined as,

$$v_k = \int_{-\infty}^{\infty} w(t)h(t - kT)\,dt \tag{8.36}$$

The noise sequence $\{v_k\}$ is zero mean and Gaussian. Its variance is $\sigma_n^2 = N_0 x_0/2$. The sampled signal sequence $\{y_n\}$ may be expressed as

$$y_n = a_n x_0 + \sum_{\substack{k=-\infty \\ k \neq n}}^{\infty} a_k x_{n-k} + v_n \tag{8.37}$$

where the term $a_n x_0$ represents the desired signal component, and the second term on the right-hand-side is the ISI term.

The sampled sequence $\{y_n\}$ given in Equation 8.37 is passed to the MLSD, which yields an estimate of the binary information sequence $\{a_n\}$. Although the form in Equation 8.37 implies that the ISI affects an infinite number of bits on either side of the desired bit a_n, from a practical viewpoint the summation in Equation 8.37 can be truncated to a finite number of terms. Figure 8.15 illustrates the autocorrelation sequence $\{x_n\}$ for different values of D, with x_0 normalized to unity for convenience [4, 5]. We observe that for $D = 0.5$, the ISI is basically limited to two bits on either side of the desired bit. On the other hand, when $D = 3$, the ISI extends to about 10 bits on either side of the desired bit. Consequently, the digital magnetic recording system including the matched filter and sampler can be modeled as an equivalent discrete-time FIR system with system function $X(z)$, where

$$X(z) = \sum_{n=-L}^{L} x_n z^{-n} \tag{8.38}$$

whose output is corrupted by the additive noise sequence $\{v_n\}$, as illustrated in Figure 8.16. The value of the parameter L is determined by the value of D. The system with transfer function $X(z)$ will be called the *equivalent discrete-time channel model* for the magnetic recording system. The frequency response of this channel model is

$$X(\omega) = \sum_{n=-L}^{L} x_n e^{-j\omega T n} \tag{8.39}$$

It is interesting to note that frequency response $X(\omega)$ of the equivalent discrete-time channel model is related to frequency response $H(\omega)$ given in Equation 8.33 as follows [2]:

$$X(\omega) = \frac{1}{T} \sum_{n=-\infty}^{\infty} \left| H\left(\omega + \frac{2\pi n}{T}\right) \right|^2 \tag{8.40}$$

The expression in the right-hand-side of Equation 8.40 is called the *folded power spectrum* of the analog channel characteristic.

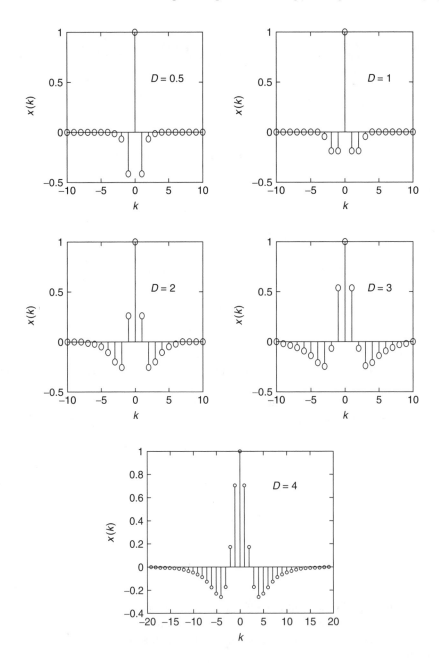

FIGURE 8.15   Sampled autocorrelation sequence (x_k).

The MLSD that follows the sampler determines the most probable sequence of bits based on observation of the sequence $\{y_n\}$ at its input. It is well known [2, 6] that the MLSD is efficiently implemented by use of the Viterbi algorithm (VA) based on Euclidean distance metrics.

In spite of the computational efficiency of the VA in the implementation of MLSD, its complexity is exponentially dependent on the span of the ISI. Hence, with an ISI span of $2L$ bits, the computational complexity of the VA is 2^{2L}. It is apparent that the optimum detector is too complex to implement for $D > 2$. In such cases, simpler types of equalization techniques may be employed to combat ISI. In particular, a linear equalizer may be used to reduce the span of ISI and, thus, to reduce the complexity of the MLSD.

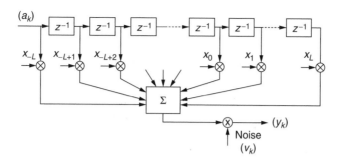

FIGURE 8.16   Equivalent discrete-time FIR channel model.

Since the equalizer is designed to yield some specified or controlled ISI pattern its output is a partial response signal [2] and the corresponding equalizer is called a partial response equalizer [7–10].

8.7   Linear Equalizer and Partial Response Targets

The most common type of equalizer used in practice to reduce ISI is a linear filter with adjustable coefficients $\{c_k\}$, as shown in Figure 8.17. Its input is the sampled sequence $\{y_k\}$ from the output of the matched filter. Let us assume for the moment that the filter has an infinite number of taps. The equalizer coefficients are adjusted so that the overall channel and equalizer response result in a specified partial response, $Q(z)$. That is

$$C(z)X(z) = Q(z)$$

or, equivalently,

$$C(z) = \frac{Q(z)}{X(z)} \tag{8.41}$$

where $Q(z)$ is a low-degree polynomial that serves as the partial response target. For magnetic recording channels, the partial response polynomials of the form [9, 10]

$$Q_n(z) = (1 - z^{-1})(1 + z^{-1})^n, \quad n = 0, 1, 2, \ldots \tag{8.42}$$

have been found to be particularly suitable because their spectral characteristics are closely matched to the channel characteristics. Thus, the linear equalizer with transfer function given by Equation 8.41 shapes the incoming signal sequence $\{y_k\}$ to one of the partial response shapes given by Equation 8.42 and forces the ISI to zero at all other time instants. Such a linear equalizer is called a *zero-forcing equalizer* [2].

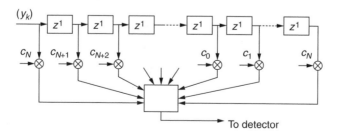

FIGURE 8.17   Linear equalizer with adjustable coeffiecients (c_k).

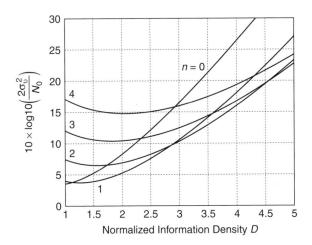

FIGURE 8.18   Normalized variance as a function of normalized density, D.

In effect, the linear zero-forcing equalizer reduces the span of the ISI to a small number of bits and allows us to implement an MLSD for the selected target shape, $Q_n(z)$.

As we have observed, the cascade of the equivalent discrete-time channel with transfer function $X(z)$ and the linear zero-forcing equalizer result in an equivalent FIR system with system function $Q_n(z)$. The noise at the output of the linear equalizer, denoted as $\{v_n\}$, has a power spectral density of

$$\Phi_v(f) = \frac{N_0}{2} \frac{|Q_n(f)|^2}{X(f)} \tag{8.43}$$

where $Q_n(f)$ is simply $Q_n(z)$, evaluated at $z = e^{j2\pi f T}$ and $X(f)$ is given by Equation 8.39 with $\omega = 2\pi f$. Hence, the variance of v_n is

$$\sigma_v^2 = \frac{N_0 T}{2} \int_{-1/2T}^{1/2T} \Phi_v(f)\, df$$

$$= \frac{N_0 T}{2} \int_{-1/2T}^{1/2T} \frac{|Q_n(f)|^2}{X(f)}\, df \tag{8.44}$$

The normalized variance, $2\sigma_v^2/N_0$, is plotted in Figure 8.18 as a function of the normalized density, D, for $n = 0, 1, 2, 3, 4$. We observe that there is an optimum choice of n for any given D. Furthermore, as D is increased, the optimum choice of n also increases. In particular, $n = 0$ is the best choice for $D \le 1, n = 1$ is suitable for $1 < D < 3$, etc. However, in selecting the value of n, we should also consider not only the noise enhancement in the equalizer but also the choice of the detector as described by Tyner et al. [10].

8.8   Maximum-Likelihood Sequence Detection and Symbol-by-Symbol Detection

The output of the zero-forcing linear equalizer that shapes the output of the desired partial response characteristic, $Q_n(z)$, is a multilevel sequence. For example, when $n = 0$ and $n = 1$, the target shapes are

$$Q_0(z) = 1 - z^{-1}$$
$$Q_1(z) = 1 - z^{-2}$$

These two characteristics result in a three-level sequence, with possible values $(2, 0 - 2)$ at the input to the detector. For $n = 2$, the sequence of received symbols is a five-level sequence with possible values $\{4, 2, 0, -2, -4\}$, and for $n = 3$ the received sequence contains seven levels. To detect this multilevel sequence, we may use either the optimum MLSD or the suboptimum symbol-by-symbol (SBS) detector, which does not exploit the inherent memory in the sequence at the input to the detector.

8.8.1 Symbol-by-Symbol Detector

The multilevel signal at the output of the equalizer may be applied to an SBS detector that simply maps the multi-level sequence into a two-level sequence $\{a_n\}$. In SBS detection, the input data sequence $\{a_n\}$ is usually precoded as previously described. For all partial responses shapes of the form of Equation 8.42, the signal levels are equi-spaced, the signal amplitude is unity, and the corresponding SNR at the input to the detector is $1/\sigma_v^2$. As a consequence, the probability of error for the SBS detector is approximately given as [10].

$$P_b \approx 2 \left[1 - \frac{1}{2^{2\lceil (n+1)/2 \rceil}} \right] F \left(\sqrt{\frac{1}{\sigma_v^2}} \right) \tag{8.45}$$

where, by definition,

$$F(x) = \frac{1}{\sqrt{2\pi}} \int_x^\infty e^{-t^2/2} \, dt \tag{8.46}$$

and $\lceil x \rceil$ denotes the smallest integer contained in x. In this case, we may use the graphs in Figure 8.18 to select the best value of n.

8.8.2 Maximum-Likelihood Sequence Detector

As indicated earlier, the optimum detector for the signal sequence is an MLSD that may be efficiently implemented by use of the VA. Its performance for the additive Gaussian noise partial response channels of the form $Q_n(z)$ given by Equation 8.42 may be determined by the technique described in the paper by Forney [6]. To a first approximation, the probability of error for MLSD at high SNRs and white Gaussian noise at the input to the detector is [6]

$$P_b = K F \left(\sqrt{\frac{d_{\min}^2}{4\sigma^2}} \right) \tag{8.47}$$

where $d_{\min}^2$ is the minimum squared Euclidean distance for an error event in the VA, σ^2 is the variance of the additive white Gaussian noise, and K is a constant.

For any partial response channel, $d_{\min}^2$ is upper bounded as [9, 10].

$$d_{\min}^2 < 4 \sum_k q_k^2 \tag{8.48}$$

where $\{q_k\}$ are the coefficients of the polynomial $Q_n(z)$. For $n = 0$ and $n = 1, d_{\min}^2 = 8$ and, hence,

$$P_b = K F \left(\sqrt{\frac{2}{\sigma^2}} \right) \tag{8.49}$$

If we ignore the correlation in the additive noise sequence at the input to the MLSD, and set $\sigma^2 = \sigma_v^2$, the performance obtained from Equation 8.49 is 3 dB better than that of the SBS detector. For $n = 2$, the minimum squared Euclidean distance, $d_{\min}^2 = 16$ and

$$P_b = K F \left(\sqrt{\frac{4}{\sigma^2}} \right) \tag{8.50}$$

which implies that the MLSD yields an additional 3 dB gain compared to $n = 0$ and $n = 1$. This additional factor should be considered in selecting the value of n for a given choice of normalized density D.

From Equation 8.48 we observe that the ratio

$$d_{\min}^2 \big/ 4 \sum_k q_k^2 \leq 1 \tag{8.51}$$

This ratio has the following interpretation. For an isolated pulse, the optimum maximum-likelihood detector results in a bit-error probability of

$$P_k = K_1 F \left(\sqrt{\frac{\sum_k q_k^2}{\sigma^2}} \right) \tag{8.52}$$

If we compare the arguments of $F(x)$ in Equation 8.52 and Equation 8.47, we obtain the ratio in Equation 8.51. Hence, the ratio in Equation 8.51 may be viewed as the loss in SNR in the VA due to the ISI of the channel.

When $n = 0, 1$, and $2, d_{\min}^2 = 4 \sum_k q_k^2$ and there is no loss in the VA due to the channel ISI. For $n \geq 3$, however, $d_{\min}^2 < 4 \sum_k q_k^2$, and the performance of the MLSD degrades by an amount equal to the loss resulting from the ISI in the partial response signal. In [9] and [11], it has been determined that for $n = 3, 10 \log_{10} (4 \sum_k q_{k_2}^2 / d_{\min}^2) = 2.2$ dB and for $n = 4, 10 \log_{10} (4 \sum_k q_k^2 / d_{\min}^2) = 3.7$ dB.

8.9   Performance Results from Computer Simulation

In this section we illustrate the performance of the SBS detector and the MLSD via computer simulation of the Lorentzian channel as given by Equation 8.29. The channel response, $h(t)$, was implemented as an FIR filter with coefficients spaced at $T/4$. A precoder was used on the input symbols when the equalizer is followed by an SBS detector. White Gaussian noise was added to the output of the Lorentzian-shaping FIR filter and the resulting samples were passed through a (unity gain) linear-phase FIR low-pass filter, having an approximate normalized bandwidth of $1/T$ Hz and a peak stopband amplitude of -42.5 dB. This filter eliminates aliasing and limits the bandwidth of the additive noise. The output of this low-pass filter constitutes the band-limited readback signal, which is to be equalized to a partial response of the $Q_n(z)$ and detected either by the SBS detector or the MLSD. In the implementation of the MLSD, the metric computations were based on the assumption that the additive Gaussian noise at the input to the detector is white. In reality, it is colored. Consequently, the performance of the MLSD is degraded to some extent compared to the MLSD for colored noise.

Figure 8.19 illustrates the bit-error probability for the SBS detector for $n = 0, 1, 2$, and 3, and a normalized density of $D = 3$. We observe that the best performance is obtained with $n = 1$ and $n = 2$. For $n = 0$, the performance is about 6 dB poorer compared to $n = 1$ and $n = 2$. For $n = 3$, the performance is about 2 dB poorer compared to $n = 1$ and $n = 2$. These performance results are consistent with the graphs shown in Figure 8.18.

Figure 8.20 illustrates the bit-error probability for the MLSD for $n = 0, 1, 2, 3$ and $D = 3$. We observe that $n = 2$ gives the best performance. For $n = 1$ and $n = 3$, the performance is about 2 dB poorer and for $n = 0$, the performance is about 10 dB poorer at an error probability of 10^{-4}.

When we compare the performance of the MLSD to that of the SBS detector for $D = 3$ and for $n = 2$, we find that the MLSD is about 4 dB better at a bit-error rate of 10^{-4}. If the additive Gaussian noise at

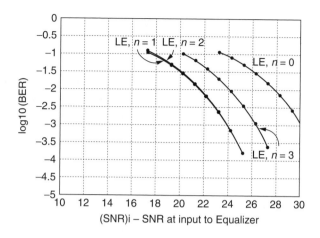

FIGURE 8.19 BER of equalizers with SBS detection when $D = 3$.

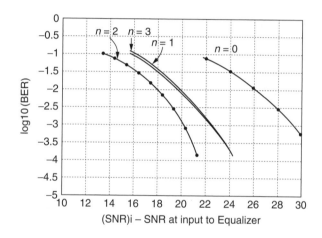

FIGURE 8.20 BER of equalizers with MLSD detection when $D = 3$.

the input to the MLSD were white, the MLSD would perform about 6 dB better than SBS detection. The difference of 2 dB is attributed primarily to the fact that the noise is colored and the MLSD ignores the correlation in the noise sequence.

8.10 Concluding Remarks

We have treated in relatively simple terms the problem of equalization of the magnetic recording channel at high normalized recording densities, where the readback signal is corrupted by ISI. We observed that the optimum detector for the ISI-corrupted signal is an MLSD that can be efficiently implemented by means of the Viterbi algorithm [6]. We also observed that, at high recording densities, the ISI spans tens of data bits and, as a consequence, the implementation complexity of the Viterbi algorithm becomes impractical. To reduce the implementation complexity of the MLSD, we employed a linear zero-forcing-equalizer that reduced and shaped the ISI pattern to a partial response signal that encompasses a small number of bits.

In addition to the optimum MLSD, we also considered an SBS detector for recovering the information from the output of the linear equalizer. We observed that SBS detection is relatively simple compared to MLSD, but the performance of the SBS detector is poorer.

Partial response shapes of the form $Q_n(z) = (1 - z^{-1})(1 + z^{-1})^n$ have been used in magnetic recording because their spectral characteristics match the spectral characteristics of the channel. Furthermore, these partial response shapes result in equally spaced multilevel signals that are easily detected by an SBS detector. However, linear equalizers that shape the output signal to the form $Q_n(z)$ generally result in a noise enhancement compared to an equalizer that shapes the signal to some optimal partial response, where the coefficients of the target response are generally not integers. This problem is treated in papers by Tyner and Proakis [10] and by Fitzpatrick et al. [12]. We note that SBS detection for noninteger coefficients in the partial response shape is generally difficult to implement.

In this article we have treated two types of symbol detectors, namely the SBS detector and the MLSD. We should indicate that other suboptimum detectors may be employed in place of the optimum MLSD [13–19]. In many cases, such detectors come close to the performance of the optimum MLSD, but their complexity is significantly lower.

Other types of equalizers that have been considered for application to magnetic recording systems are decision-feedback equalizers (DFEs) [2, 7, 10]. A DFE may be used as an alternative to the linear equalizer to provide partial response shaping. The primary advantage of the DFE is that it results in a smaller noise enhancement than the linear equalizer. However, because the DFE requires that bit decisions be fed back to cancel the ISI caused by previously detected bits, the DFE is suitable when it is followed by an SBS detector.

The MLSD has an inherent delay in making bit decisions, which is incompatible with the requirement that the DFE employ bit decisions with no delay. Consequently, any performance gain realized by the DFE in reducing the noise enhancement is offset by the loss in the SBS detector relative to MLSD.

An additional performance loss in the DFE/SBS detector is caused by error propagation resulting from the occasional incorrect decisions that are fed back to the feedback filter in the DFE [2, 10]. For these reasons, the combination of the linear equalizer followed by the MLSD generally provides better performance than the DFE followed by the SBS detector. However, for normalized densities of $D > 3$, the DFE/SBS detector may be simpler to implement, because the implementation complexity of the MLSD grows exponentially with the length of the partial response target $Q_n(z)$.

References

[1] Nyquist, Certain topics in telegraph transmission theory, *AIEE Trans.*, vol. 47, pp. 617–644, 1928.

[2] J. G. Proakis, *Digital Communications*, 4th ed., McGraw-Hill, New York, 2001.

[3] C. D. Mee and E. D. Daniel, *Magnetic Recording*, Vol. 1, Technology, McGraw-Hill, New York, 1987.

[4] J. W. M. Bergmans, Discrete-time models for digital magnetic recording, *Phillips J. Res.*, vol. 41, no. 6, pp. 531–558, 1986.

[5] K. A. S. Immink, Coding techniques for the noisy magnetic recording channel: A state of the art report, *IEEE Trans. Commn.*, vol. 37, no. 5, pp. 413–419, May 1989.

[6] G. D. Forney, Jr., Maximum-likelihood sequence estimation of digital sequences in the presence of intersymbol interference," *IEEE Trans. Info. Theory*, vol. IT-18, pp. 363–378, May 1972.

[7] J. W. M. Bergmans, Partial response equalization, *Phillips J. Res.*, vol. 42, no. 2, pp. 209–245, 1987.

[8] J. J. Moon and L. R. Carley, Partial response signaling in a magnetic recording channel," *IEEE Trans. Magn.*, vol. MAG-24, pp. 2973–2975, Nov. 1988.

[9] H. K. Thapar and A. M. Patel, A class of partial response systems for increasing storage density in magnetic recording, *IEEE Trans. Magn.*, vol. MAG-23, pp. 3666–3668, Sept. 1987.

[10] D. J. Tyner and J. G. Proakis, Partial response equalizer performance in digital magnetic recording channels, *IEEE Trans. Magn.*, vol. MAG-29, pp. 4194–4208, November 1993.

[11] G. Vannucci and G. J. Foschini, The minimum distance for digital magnetic recording partial responses, *IEEE Trans. Info. Theory*, vol. IT-37, no. 3, pp. 955–960, May 1991.

[12] J. Fitzpatrick, J. K. Wolf, and L. Barbosa, New equalizer targets for sampled magnetic recording systems, *Conf. Record, Twenty-Fifth Asilomar Conference on Signals, Systems and Computers*, pp. 30–34, Pacific Grove, CA, November 4-6, 1991.

[13] F. L. Vermeulen and M. E. Hellman, Reduced state Viterbi decoders for channels with intersymbol interference, *Proc. 1974 IEEE Int. Commun. Conf.*, pp. 37B-1, 37B-4.

[14] G. J. Foschini, A reduced state variant of maximum likelihood sequence detection attaining optimum performance for high signal-to-noise ratios, *IEEE Trans. Info. Theory*, vol. IT-23, pp. 605–609, September 1977.

[15] M. V. Eyuboglu and S. V. H. Qureshi, Reduced-state sequence estimation with set partitioning and decision-feedback, *IEEE Trans. Commn.*, vol. COM-36, no. 1, pp. 13–20, January 1988.

[16] J. Moon and L. R. Carley, Performance comparison of detection methods in magnetic recording, *IEEE Trans. Magn.*, vol. MAG-26, no. 6, pp. 3155–3172, November 1990.

[17] J. Moon and L. R. Carley, Efficient sequence detection for intersymbol interference channels with run-length constraints, *IEEE Trans. Commn.*, vol. COM-17, no. 9, pp. 2654–2660, September 1994.

[18] A. Duel-Hallen and C. Heegard, Delayed decision-feedback sequence estimation, *IEEE Trans. Commn.*, vol. COM-37, no. 5, pp. 428–436, May 1989.

[19] H. Vinck, Low complexity decoding for intersymbol interference channels, *Proc. IEE Eighth International Conference on Video Audio and Data Processing*, pp. 154–158.

9

An Introduction to Error-Correcting Codes

9.1 Introduction **9**-1
9.2 Linear Codes **9**-3
9.3 Syndrome Decoding, Hamming Codes, and Capacity of the Channel **9**-6
9.4 Codes over Bytes and Finite Fields **9**-8
9.5 Cyclic Codes **9**-10
9.6 Reed Solomon Codes **9**-11
9.7 Decoding of RS codes: the key equation **9**-13
9.8 Decoding RS Codes with Euclid's Algorithm **9**-16
9.9 Applications: Burst and Random Error Correction **9**-18

Mario Blaum
IBM Research Division
San Jose, CA

9.1 Introduction

When digital data are transmitted over a noisy channel, it is important to have a mechanism allowing recovery against a limited number of errors. Normally, a user string of 0s and 1s, called bits, is encoded by adding a number of redundant bits to it. When the receiver attempts to reconstruct the original message sent, it starts by examining a possibly corrupted version of the encoded message, and then makes a decision. This process is called the *decoding*.

The set of all possible encoded messages is called an error-correcting code. The field was started in the late 1940s by the work of Shannon and Hamming, and since then thousands of papers on the subject have been published. There are also several very good books touching different aspects of error-correcting codes, for instance, [1, 3, 4, 5, 7, 8], to mention just a few.

The purpose of this chapter is giving an introduction to the theory and practice of error-correcting codes. In particular, it will be shown how to encode and decode the most widely used codes, Reed Solomon codes.

In principle, we will assume that our information symbols are bits, that is, 0s and 1s. The set $\{0, 1\}$ has a field structure under the exclusive-OR ($\oplus$) and product operations. We denote this field $GF(2)$, which means Galois field of order 2.

Roughly, there are two types of error-correcting codes: codes of block type and codes of convolutional type. Codes of block type encode a fixed number of bits, say k bits, into a vector of length n. So, the information string is divided into blocks of k bits each. Convolutional codes take the string of information bits globally and slide a window over the data in order to encode. A certain amount of memory is needed

0-8493-1524-7/05/$0.00+$1.50

by the encoder. However, in this chapter we concentrate on block codes only. For more on convolutional codes, see [3, 8].

As said above, we encode k information bits into n bits. So, we have a 1-1 function f,

$$f : GF(2)^k \to GF(2)^n$$

The function f defines the encoding procedure. The set of 2^k encoded vectors of length n is called a code of *length n* and *dimension k*, and we denote it as an $[n, k]$ code. We call codewords the elements of the code while we call words the vectors of length n in general. The ratio k/n is called the *rate* of the code.

The error-correcting power of a code is characterized by a parameter called the minimum (Hamming) distance of the code. Formally:

Definition 9.1   Given two vectors of length n, say $\underline{a}$ and $\underline{b}$, we call the Hamming distance between $\underline{a}$ and $\underline{b}$ the number of coordinates in which they differ (notation, $d_H(\underline{a}, \underline{b})$). Given a code $\mathcal{C}$ of length n and dimension k, let

$$d = \min\{d_H(\underline{a}, \underline{b}) : \underline{a} \neq \underline{b}, \ \underline{a}, \underline{b}, \in \mathcal{C}\}$$

We call d the minimum (Hamming) distance of the code $\mathcal{C}$ and we say that $\mathcal{C}$ is an $[n, k, d]$ code.

It is easy to verify that $d_H(\underline{a}, \underline{b})$ verifies the axioms of distance, that is,

1. $d_H(\underline{a}, \underline{b}) = d_H(\underline{b}, \underline{a})$
2. $d_H(\underline{a}, \underline{b}) = 0$ if and only if $\underline{a} = \underline{b}$
3. $d_H(\underline{a}, \underline{c}) \leq d_H(\underline{a}, \underline{b}) + d_H(\underline{b}, \underline{c})$

We call a sphere of radius r and center $\underline{a}$ the set of vectors that are at distance at most r from $\underline{a}$. The relation between d and the maximum number of errors that code $\mathcal{C}$ can correct is given by the following lemma:

Lemma 9.1   *The maximum number of errors that an $[n, k, d]$ code can correct is $\lfloor (d - 1)/2 \rfloor$, where $\lfloor x \rfloor$ denotes the largest integer smaller or equal than x.*

Proof 9.1   Assume that vector $\underline{a}$ was transmitted but a possibly corrupted version of $\underline{a}$, say $\underline{r}$, was received. Moreover, assume that no more than $\lfloor (d - 1)/2 \rfloor$ errors have occurred.

Consider the set of 2^k spheres of radius $\lfloor (d - 1)/2 \rfloor$ whose centers are the codewords in $\mathcal{C}$. By the definition of d, all these spheres are disjoint. Hence, $\underline{r}$ belongs to one and only one sphere: the one whose center is codeword $\underline{a}$. So, the decoder looks for the sphere in which $\underline{r}$ belongs, and outputs the center of that sphere as the decoded vector. As we see, whenever the number of errors is at most $\lfloor (d - 1)/2 \rfloor$, this procedure will give the correct answer.

Moreover, $\lfloor (d-1)/2 \rfloor$ is the maximum number of errors that the code can correct. For let $\underline{a}, \underline{b}, \in \mathcal{C}$ such that $d_H(\underline{a}, \underline{b}) = d$. Let $\underline{u}$ be a vector such that $d_H(\underline{a}, \underline{u}) = 1 + \lfloor (d-1)/2 \rfloor$ and $d_H(\underline{b}, \underline{u}) = d - 1 - \lfloor (d-1)/2 \rfloor$. We easily verify that $d_H(\underline{b}, \underline{u}) \leq d_H(\underline{a}, \underline{u})$, so, if $\underline{a}$ is transmitted and $\underline{u}$ is received (i.e., $1 + \lfloor (d - 1)/2 \rfloor$ errors have occurred), the decoder cannot decide that the transmitted codeword was $\underline{a}$, since codeword $\underline{b}$ is at least as close to $\underline{u}$ as $\underline{a}$.   □

Example 9.1

Consider the following 1-1 relationship between $GF(2)^2$ and $GF(2)^5$ defining the encoding:

$$
\begin{array}{ccc}
00 & \leftrightarrow & 00000 \\
10 & \leftrightarrow & 00111 \\
01 & \leftrightarrow & 11100 \\
11 & \leftrightarrow & 11011
\end{array}
$$

The four vectors in $GF(2)^5$ constitute a $[5, 2, 3]$ code $\mathcal{C}$. From Lemma 9.1, $\mathcal{C}$ can correct one error. For instance, assume that we receive the vector $\underline{r} = 10100$. The decoder looks into the four spheres of radius 1 (each sphere has six elements!) around each codeword, finding that $\underline{r}$ belongs in the sphere with center 11100. If we look at the table above, the final output of the decoder is the information block 01.

Example 9.1 shows that the decoder has to make at most 24 checks before arriving to the correct decision. When large codes are involved, as is the case in applications, this decoding procedure is not practical, since it amounts to an exhaustive search over a huge set of vectors. One of the goals in the theory of error-correcting codes is finding codes with high rate and minimum distance as large as possible. The possibility of finding codes with the right properties is often limited by bounds that constrain the choice of parameters n, k and d. We give some of these bounds in the next section.

Let us point out that error-correcting codes can be used for detection instead of correction of errors. The simplest example of an error-detecting code is given by a parity code: a parity is added to a string of bits in such a way that the total number of bits is even (a more sophisticated way of saying this, is that the sum modulo-2 of the bits has to be 0). For example, 0100 is encoded as 01001. If an error occurs, or, more generally, an odd number of errors, these errors will be detected since the sum modulo 2 of the received bits will be 1. Notice that two errors will be undetected. In general, if an $[n, k, d]$ code is used for detection only, the decoder checks whether the received vector is in the code or not. If it is not, then errors are detected. It is easy to see that an $[n, k, d]$ code can detect up to $d - 1$ errors. Also, we can choose to correct less than $\lfloor (d - 1)/2 \rfloor$ errors, say s errors, by taking disjoint spheres of radius s around codewords, and using the remaining capacity to detect errors. In other words, we want to correct up to s errors or detect up to $s + t$ errors when more than s errors occur.

Another application of error-correcting codes is in erasure correction. An erased bit is a bit that cannot be read, so the decoder has to decide if it was a 0 or a 1. An erasure is normally denoted with the symbol ?. For instance, 01?0 means that we cannot read the third symbol. Obviously, it is easier to correct erasures than to correct errors, since in the case of erasures we already know the location, we simply have to find what the erased bit was. It is not hard to prove that an $[n, k, d]$ code can correct up to $d - 1$ erasures. We may also want to simultaneously correct errors and erasures. In fact, a code $\mathcal{C}$ with minimum distance d can correct s errors together with t erasures whenever $2s + t \leq d - 1$.

9.2 Linear Codes

We have seen in the previous section that a binary code of length n is a subset of $GF(2)^n$. Notice that, being $GF(2)$ a field, $GF(2)^n$ has a structure of vector space over $GF(2)$. We say that a code $\mathcal{C}$ is linear if it is a subspace of $GF(2)^n$, that is,

1. $\underline{0} \in \mathcal{C}$
2. $\forall \, \underline{a}, \underline{b} \in \mathcal{C}, \underline{a} \oplus \underline{b} \in \mathcal{C}$

The symbol $\underline{0}$ denotes the all-zero vector. In general, vectors will be denoted with underlined letters, otherwise letters denote scalars.

In Section 5.1, we assumed that a code had 2^k elements, k being the dimension. However, we can define a code of length n as any subset of $GF(2)^n$.

There are many interesting combinatorial questions regarding nonlinear codes. Probably, the most important question is the following: given the length n and the minimum distance d, what is the maximum number of codewords that a code can have? For more about nonlinear codes, the reader is referred to [4]. From now on, we assume that all codes are linear. Linear codes are in general easier to encode and decode than their nonlinear counterparts, hence they are more suitable for implementation in applications.

In order to find the minimum distance of a linear code, it is enough to find its minimum *weight*. We say that the (Hamming) weight of a vector $\underline{u}$ is the distance between $\underline{u}$ and the zero vector. In other words, the weight of $\underline{u}$, denoted $w_H(\underline{u})$, is the number of nonzero coordinates of the vector $\underline{u}$. The minimum weight of a code is the minimum between all the weights of the nonzero codewords. The proof of the following lemma is left as an exercise.

Lemma 9.2   *Let C be a linear* $[n, k, d]$ *code. Then, the minimum distance and the minimum weight of C are the same.*

Next, we introduce two important matrices that define a linear error-correcting code. Since a code C is now a subspace, the dimension k of C is the cardinality of a basis of C. Consider then an $[n, k, d]$ code C. We say that a $k \times n$ matrix G is a *generator* matrix of a code C if the rows of G are a basis of C. Given a generator matrix, the encoding process is simple.

Explicitly, let $\underline{u}$ be an information vector of length k and G a $k \times n$ generator matrix, then $\underline{u}$ is encoded into the n-vector $\underline{v}$ given by

$$\underline{v} = \underline{u}G \tag{9.1}$$

Example 9.2

Let G be the 2×5 matrix

$$G = \begin{pmatrix} 0 & 0 & 1 & 1 & 1 \\ 1 & 1 & 1 & 0 & 0 \end{pmatrix}$$

It is easy to see that G is a generator matrix of the $[5, 2, 3]$ code described in Example 9.1.

Notice that, although a code may have many generator matrices, the encoding depends on the particular matrix chosen, according to Equation 9.1. We say that G is a *systematic* generator matrix if G can be written as

$$G = (I_k \mid V) \tag{9.2}$$

where I_k is the $k \times k$ identity matrix and V is a $k \times (n - k)$ matrix. A systematic generator matrix has the following advantage: given an information vector $\underline{u}$ of length k, the encoding given by Equation 9.1 outputs a codeword $(\underline{u}, \underline{w})$, where $\underline{w}$ has length $n - k$. In other words, a systematic encoder adds $n - k$ redundant bits to the k information bits, so information and redundancy are clearly separated. This also simplifies the decoding process, since, after decoding, the redundant bits are simply discarded. For that reason, most encoders used in applications are systematic.

A permutation of the columns of a generator matrix gives a new generator matrix defining a new code. The codewords of the new code are permutations of the coordinates of the codewords of the original code. We then say that the two codes are *equivalent*. Notice that equivalent codes have the same distance properties, so their error correcting capabilities are exactly the same.

By permuting the columns of the generator matrix in Example 9.2, we obtain the following generator matrix G':

$$G' = \begin{pmatrix} 1 & 0 & 0 & 1 & 1 \\ 0 & 1 & 1 & 1 & 0 \end{pmatrix} \tag{9.3}$$

The matrix G' defines a systematic encoder for a code that is equivalent to the one given in Example 9.1. For instance, the information vector 11 is encoded into 11 101.

The second important matrix related to a code is the so called *parity check* matrix. We say that an $(n - k) \times n$ matrix H is a parity check matrix of an $[n, k]$ code C if and only if, for any $\underline{c} \in C$,

$$\underline{c}H^T = \underline{0} \tag{9.4}$$

where H^T denotes the transpose of matrix H and $\underline{0}$ is a zero vector of length $n - k$. We say that the parity check matrix H is in systematic from if

$$H = (W \mid I_{n-k}) \tag{9.5}$$

where I_{n-k} is the $(n - k) \times (n - k)$ identity matrix and W is an $(n - k) \times k$ matrix.

Given a systematic generator matrix G of a code C, it is easy to find the systematic parity check matrix H (and conversely). Explicitly, if G is given by Equation 9.2, H is given by

$$H = (V^T \mid I_{n-k}) \tag{9.6}$$

We leave the proof of this fact to the reader.

For example, the systematic parity check matrix of the code whose systematic generator matrix is given by Equation 9.3, is

$$H = \begin{pmatrix} 0 & 1 & 1 & 0 & 0 \\ 1 & 1 & 0 & 1 & 0 \\ 1 & 0 & 0 & 0 & 1 \end{pmatrix} \tag{9.7}$$

We state now an important property of parity check matrices.

Lemma 9.3 *Let C be a linear $[n, k, d]$ code and H a parity-check matrix. Then, any $d - 1$ columns of H are linearly independent.*

Proof 9.2 Numerate the columns of H from 0 to $n - 1$. Assume that columns $0 \leq i_1 < i_2 < \cdots < i_m \leq n - 1$ are linearly dependent, where $m \leq d - 1$. Without loss of generality, we may assume that the sum of these columns is equal to the column vector zero. Let $\underline{v}$ be a vector of length n whose non-zero coordinates are in locations $i_1, i_2, \ldots, i_m$. Then, we have

$$\underline{v} H^T = \underline{0}$$

hence $\underline{v}$ is in C. But $\underline{v}$ has weight $m \leq d - 1$, contradicting the fact that C has minimum distance d. $\quad\square$

Corollary 9.1 For any linear $[n, k, d]$ code, the minimum distance d is the smallest number m such that there is a subset of m linearly dependent columns.

Proof 9.3 It follows immediately from Lemma 9.3. $\quad\square$

Corollary 9.2 (Singleton Bound) For any linear $[n, k, d]$ code,

$$d \leq n - k + 1$$

Proof 9.4 Notice that, since H is an $(n - k) \times n$ matrix, any $n - k + 1$ columns are going to be linearly dependent, so if $d > n - k + 1$ we would contradict Corollary 9.1. $\quad\square$

Codes meeting the Singleton bound are called maximum distance separable (MDS). In fact, except for trivial cases, binary codes are not MDS. In order to obtain MDS codes, we will define codes over larger fields, like the so-called Reed Solomon codes, to be described later in the chapter.

We also give a second bound relating the redundancy and the minimum distance of an $[n, k, d]$ code: the so called Hamming or volume bound. Let us denote by $V(r)$ the number of elements in a sphere of radius r whose center is an element in $GF(2)^n$. It is easy to verify that

$$V(r) = \sum_{i=0}^{r} \binom{n}{i} \tag{9.8}$$

We then have:

Lemma 9.4 (Hamming bound) *Let C be a linear $[n, k, d]$ code, then*

$$n - k \geq \log_2 V(\lfloor (d-1)/2 \rfloor) \tag{9.9}$$

Proof 9.5 Notice that the 2^k spheres with the 2^k codewords as centers and radius $\lfloor (d-1)/2 \rfloor$ are disjoint. The total number of vectors contained in these spheres is $2^k V(\lfloor (d-1)/2 \rfloor)$. This number has to be smaller than or equal to the total number of vectors in the space, that is,

$$2^n \geq 2^k V(\lfloor (d-1)/2 \rfloor) \tag{9.10}$$

Inequality 9.9 follows immediately from Inequality 9.10. □

A *perfect* code is a code for which Inequality 9.9 is in effect equality. Geometrically, a perfect code is a code for which the 2^k spheres of radius $\lfloor (d-1)/2 \rfloor$ and the codewords as centers cover the whole space.

There are not many perfect codes. In the binary case, the only nontrivial linear perfect codes are the Hamming codes (to be presented in the next section) and the $[23, 12, 7]$ Golay code (see [4]).

9.3   Syndrome Decoding, Hamming Codes, and Capacity of the Channel

In this section, we study the first important family of codes, the so called Hamming codes. As we will see, Hamming codes can correct up to one error.

Let C be an $[n, k, d]$ code with parity check matrix H. Let $\underline{u}$ be a transmitted vector and $\underline{r}$ a possibly corrupted received version of $\underline{u}$. We say that the syndrome of $\underline{r}$ is the vector $\underline{s}$ of length $n - k$ given by

$$\underline{s} = \underline{r} H^T \tag{9.11}$$

Notice that, if no errors occurred, the syndrome of $\underline{r}$ is the zero vector. The syndrome, however, tells us more than a vector being in the code or not. Say, as before, that $\underline{u}$ was transmitted and $\underline{r}$ was received, where $\underline{r} = \underline{u} \oplus \underline{e}, \underline{e}$ an error vector. Notice that,

$$\underline{s} = \underline{r} H^T = (\underline{u} \oplus \underline{e}) H^T = \underline{u} H^T \oplus \underline{e} H^T = \underline{e} H^T$$

since $\underline{u}$ is in C. Hence, the syndrome does not depend on the received vector but on the error vector. In the next lemma, we show that to every error vector of weight $\leq (d-1)/2$ corresponds a unique syndrome.

Lemma 9.5 *Let C be a linear $[n, k, d]$ code with parity check matrix H. Then, there is a 1-1 correspondence between errors of weight $\leq (d-1)/2$ and syndromes.*

Proof 9.6 Let $\underline{e}_1$ and $\underline{e}_2$ be two distinct error vectors of weight $\leq (d-1)/2$ with syndromes $\underline{s}_1 = \underline{e}_1 H^T$ and $\underline{s}_2 = \underline{e}_2 H^T$. If $\underline{s}_1 = \underline{s}_2$, then $\underline{s} = (\underline{e}_1 \oplus \underline{e}_2) H^T = \underline{s}_1 \oplus \underline{s}_2 = \underline{0}$, hence $\underline{e}_1 \oplus \underline{e}_2 \in C$. But $\underline{e}_1 \oplus \underline{e}_2$ has weight $\leq d - 1$, a contradiction. □

Lemma 9.5 gives the key for a decoding method that is more efficient than exhaustive search. We can construct a table with the 1-1 correspondence between syndromes and error patterns of weight $\leq (d-1)/2$ and decode by look-up table. In other words, given a received vector, we first find its syndrome and then we look in the table to which error pattern it corresponds. Once we obtain the error pattern, we add it to the received vector, retrieving the original information. This procedure may be efficient for small codes, but it is still too complex for large codes.

Example 9.3

Consider the code whose parity matrix H is given by (7). We have seen that this is a $[5, 2, 3]$ code. We have 6 error patterns of weight ≤ 1. The 1-1 correspondence between these error patterns and the syndromes, can be immediately verified to be

$$
\begin{array}{ccc}
00000 & \leftrightarrow & 000 \\
10000 & \leftrightarrow & 011 \\
01000 & \leftrightarrow & 110 \\
00100 & \leftrightarrow & 100 \\
00010 & \leftrightarrow & 010 \\
00001 & \leftrightarrow & 001
\end{array}
$$

For instance, assume that we receive the vector $\underline{r} = 10111$. We obtain the syndrome $\underline{s} = \underline{r}H^T = 100$. Looking at the table above, we see that this syndrome corresponds to the error pattern $\underline{e} = 00100$. Adding this error pattern to the received vector, we conclude that the transmitted vector was $\underline{r} \oplus \underline{e} = 10011$.

Given a number r or redundant bits, we say that a $[2^r - 1, 2^r - r - 1, 3]$ Hamming code is a code having an $r \times (2^r - 1)$ parity check matrix H such that its columns are all the different nonzero vectors of length r.

A Hamming code has minimum distance 3. This follows from its definition and Corollary 9.1: notice that any two columns in H, being different, are linearly independent. Also, if we take any two different columns and their sum, these three columns are linearly dependent, proving our assertion.

A natural way of writing the columns of H in a Hamming code, is by considering them as binary numbers on base 2 in increasing order. This means, the first column is 1 on base 2, the second columns is 2 and so on. The last column is $2^r - 1$ on base 2, that is, $(1, 1, \ldots, 1)^T$. This parity check matrix, although nonsystematic, makes the decoding very simple.

In effect, let $\underline{r}$ be a received vector such that $\underline{r} = \underline{v} \oplus \underline{e}$, where $\underline{v}$ was the transmitted codeword and $\underline{e}$ is an error vector of weight 1. Then, the syndrome is $\underline{s} = \underline{e}H^T$, which gives the column corresponding to the location in error. This column, as a number on base 2, tells us exactly where the error has occurred, so the received vector can be corrected.

Example 9.4

Consider the $[7, 4, 3]$ Hamming code $\mathcal{C}$ with parity check matrix

$$
H = \begin{pmatrix}
0 & 0 & 0 & 1 & 1 & 1 & 1 \\
0 & 1 & 1 & 0 & 0 & 1 & 1 \\
1 & 0 & 1 & 0 & 1 & 0 & 1
\end{pmatrix}
\tag{9.12}
$$

Assume that vector $\underline{r} = 1100101$ is received. The syndrome is $\underline{s} = \underline{r}H^T = 001$, which is the binary representation of the number 1. Hence, the first location is in error, so the decoder estimates that the transmitted vector was $\underline{v} = 0100101$.

We can obtain 1-error correcting codes of any length simply by shortening a Hamming code. This procedure works as follows: Assume that we want to encode k information bits into a 1-error correcting code. Let r be the smallest number such that $k \leq 2^r - r - 1$. Let H be the parity-check matrix of a $[2^r - 1, 2^r - r - 1, 3]$ Hamming code. Then construct a matrix H' by eliminating some $2^r - r - 1 - k$ columns from H. The code whose parity-check matrix is H' is a $[k + r, k, d]$ code with $d \geq 3$, hence it can correct one error. We call it a shortened Hamming code. For instance, the $[5, 2, 3]$ code whose parity-check matrix is given by Equation 9.7, is a shortened Hamming code.

In general, if H is the parity-check matrix of a code C, H' is a matrix obtained by eliminating a certain number of columns from H and C' is the code with parity-check matrix H', we say that C' is obtained by shortening C.

A $[2^r - 1, 2^r - r - 1, 3]$ Hamming code can be extended to a $[2^r, 2^r - r - 1, 4]$ Hamming code by adding to each codeword a parity bit that is the exclusive-OR of the first $2^r - 1$ bits. The new code is called an extended Hamming code.

So far we have not talked about probabilities of errors. Assume that we have a binary symmetric channel (BSC), that is, the probability of a 1 becoming a 0 or of a 0 becoming a 1 is $p < .5$. Let P_{err} be the probability of error after decoding using a code, that is, the output of the decoder does not correspond to the originally transmitted information vector. A fundamental question is the following: given a BSC with bit error probability p, does it exist a code of high rate that can arbitrarily lower P_{err}? The answer, due to Shannon, is yes, provided that the code has rate below a parameter called the capacity of the channel, as defined next:

Definition 9.2   Given a BSC with probability of bit error p, we say that the capacity of the channel is

$$C(p) = 1 + p \log_2 p + (1 - p) \log_2(1 - p) \tag{9.13}$$

Theorem 9.1 (Shannon)   *For any $\epsilon > 0$ and $R < C(p)$, there is an $[n, k]$ binary code of rate $k/n \geq R$ with $P_{\text{err}} < \epsilon$.*

For a proof of Theorem 9.1 and some of its generalizations, the reader is referred to [5], or even to Shannon's original paper [6].

Theorem 9.1 has enormous theoretical importance: it shows that reliable communication is not limited in the presence of noise, only the rate of communication is. For instance, if $p = 0.01$, the capacity of the channel is $C(0.01) = 0.9192$. Hence, there are codes of rate ≥ 0.9 with P_{err} arbitrarily small. It also tells us not to look for codes with rate 0.92 making P_{err} arbitrarily small.

The proof of Theorem 9.1, though, is based on probabilistic methods and the assumption of arbitrarily large values of n. In practical applications, n cannot be too large. The theorem does not tell us how to construct efficient codes, it just asserts their existence. Moreover, when we construct codes, we want them to have efficient encoding and decoding algorithms. In the last few years, coding methods approaching the Shannon limit have been developed, the so called *turbo codes*. Although great progress has been made towards practical implementations of turbo codes, in applications like magnetic recording their complexity is still a problem. A description of turbo codes is beyond the scope of this introduction. We refer the reader to [2].

9.4   Codes over Bytes and Finite Fields

So far, we have considered linear codes over bits. Next we want to introduce codes over larger symbols, mainly over bytes. A byte of size v is a vector of v bits. Mathematically, bytes are vectors in $GF(2)^v$. Typical cases in magnetic and optical recording involve 8-bit bytes. Most of the general results in the previous sections for codes over bits easily extend to codes over bytes. It is trivial to multiply bits, but we need a method to multiply bytes. To this end, the theory of finite fields has been developed. Next we give a brief introduction to the theory of finite fields. For a more complete treatment, the reader is referred to [4], Chapter 4.

We know how to add two binary vectors: we simply exclusive-OR them componentwise. What we need now is a rule that allows us to multiply bytes while preserving associative, distributive, and multiplicative inverse properties, that is, a product that gives to the set of bytes of length v the structure of a field. To this end, we will define a multiplication between vectors that satisfies the associative and commutative properties, it has a 1 element, each nonzero element is invertible and it is distributive with respect to the sum operation.

Recall the definition of the ring Z_m of integers modulo m: Z_m is the set $\{0, 1, 2, \ldots, m - 1\}$, with a sum and product of any two elements defined as the residue of dividing by m the usual sum or product. It is not difficult to prove that Z_m is a field if and only if m is a prime number. Using this analogy, we will give to $(GF(2))^\nu$ the structure of a field.

Consider the vector space $(GF(2))^\nu$ over the field $GF(2)$. We can view each vector as a polynomial of degree $\leq \nu - 1$ as follows: the vector $a = (a_0, a_1, \ldots, a_{\nu-1})$ corresponds to the polynomial $a(\alpha) = a_0 + a_1\alpha + \cdots + a_{\nu-1}\alpha^{\nu-1}$.

Our goal is to give to $(GF(2))^\nu$ the structure of a field. We will denote such a field by $GF(2^\nu)$. The sum in $GF(2^\nu)$ is the usual sum of vectors in $(GF(2))^\nu$. We need now to define a product.

Let $f(x)$ be an irreducible polynomial (i.e., it cannot be expressed as the product of two polynomials of smaller degree) of degree ν whose coefficients are in $GF(2)$. Let $a(\alpha)$ and $b(\alpha)$ be two elements of $GF(2^\nu)$. We define the product between $a(\alpha)$ and $b(\alpha)$ in $GF(2^\nu)$ as the unique polynomial $c(\alpha)$ of degree $\leq \nu - 1$ such that $c(\alpha)$ is the residue of dividing the product $a(\alpha)b(\alpha)$ by $f(\alpha)$ (the notation $g(x) \equiv h(x) \pmod{f(x)}$) means that $g(x)$ and $h(x)$ have the same residue after dividing by $f(x)$, i.e., $g(\alpha) = h(\alpha)$).

The sum and product operations defined above give to $GF(2^\nu)$ a field structure. The role of the irreducible polynomial $f(x)$ is the same as the prime number m when Z_m is a field. In effect, the proof that $GF(2^\nu)$ is a field when m is irreducible is essentially the same as the proof that Z_m is a field when m is prime. From now on, we denote the elements in $GF(2^\nu)$ as polynomials in α of degree $\leq \nu - 1$ with coefficients in $GF(2)$. Given two polynomials $a(x)$ and $b(x)$ with coefficients in $GF(2)$, $a(\alpha)b(\alpha)$ denotes the product in $GF(2^\nu)$, while $a(x)b(x)$ denotes the regular product of polynomials. Notice that, for the irreducible polynomial $f(x)$, in particular, $f(\alpha) = 0$ in $GF(2^\nu)$, since $f(x) \equiv 0 \pmod{f(x)}$.

So, the set $GF(2^\nu)$ given by the irreducible polynomial $f(x)$ of degree ν, is the set of polynomials of degree $\leq \nu - 1$, where the sum operation is the regular sum of polynomials, and the product operation is the residue of dividing by $f(x)$ the regular product of two polynomials.

Example 9.5

Let us construct the field $GF(8)$. Consider the polynomials of degree ≤ 2 over $GF(2)$. Let $f(x) = 1 + x + x^3$. Since $f(x)$ has no roots over $GF(2)$, it is irreducible (notice that such an assessment can be made only for polynomials of degree 2 or 3). Let us consider the powers of α modulo $f(\alpha)$. Notice that $\alpha^3 = \alpha^3 + f(\alpha) = 1 + \alpha$. Also, $\alpha^4 = \alpha\alpha^3 = \alpha(1 + \alpha) = \alpha + \alpha^2$. Similarly, we obtain $\alpha^5 = \alpha\alpha^4 = \alpha(\alpha + \alpha^2) = \alpha^2 + \alpha^3 = 1 + \alpha + \alpha^2$, and $\alpha^6 = \alpha\alpha^5 = \alpha + \alpha^2 + \alpha^3 = 1 + \alpha^2$. Finally, $\alpha^7 = \alpha\alpha^6 = \alpha + \alpha^3 = 1$.

As we can see, every nonzero element in $GF(8)$ can be obtained as a power of the element α. In this case, α is called a *primitive* element and the irreducible polynomial $f(x)$ that defines the field is called a *primitive* polynomial. It can be proven that it is always the case that the multiplicative group of a finite field is cyclic, so there is always a primitive element. A convenient description of $GF(8)$ is given in Table 9.1.

The first column in Table 9.1 describes the element of the field in vector form, the second one as a polynomial in α of degree ≤ 2, the third one as a power of α, and the last one gives the logarithm (also called Zech logarithm): it simply indicates the corresponding power of α. As a convention, we denote by $-\infty$ the logarithm corresponding to the element 0.

TABLE 9.1 The Finite Field $GF(8)$ Generated by $1 + x + x^3$

Vector	Polynomial	Power of α	Logarithm
000	0	0	$-\infty$
100	1	1	0
010	α	α	1
001	α^2	α^2	2
110	$1 + \alpha$	α^3	3
011	$\alpha + \alpha^2$	α^4	4
111	$1 + \alpha + \alpha^2$	α^5	5
101	$1 + \alpha^2$	α^6	6

It is often convenient to express the elements in a finite field as powers of α: when we multiply two of them, we obtain a new power of α whose exponent is the sum of the two exponents modulo $2^\nu - 1$. Explicitly, if i and j are the logarithms of two element in $GF(2^\nu)$, then their product has logarithm $i + j \pmod{2^\nu - 1}$. In the example, if we want to multiply the vectors 101 and 111, we first look at their logarithms. They are 6 and 5 respectively, so the logarithm of the product is $6 + 5 \pmod 7 = 4$, corresponding to the vector 011.

In order to add vectors, the best way is to express them in vector form and add coordinate to coordinate in the usual way.

9.5 Cyclic Codes

In the same way we defined codes over the binary field $GF(2)$, we can define codes over any finite field $GF(2^\nu)$. Now, a code of length n is a subset of $(GF(2^\nu))^n$, but since we study only linear codes, we require that such a subset is a vector space. Similarly, we define the minimum (Hamming) distance and the generator and parity-check matrices of a code. Some properties of binary linear codes, like the Singleton bound, remain the same in the general case. Others, like the Hamming bound, require some modifications.

Consider a linear code $\mathcal{C}$ over $GF(2^\nu)$ of length n. We say that $\mathcal{C}$ is cyclic if, for any codeword $(c_0, c_1, \ldots, c_{n-1}) \in \mathcal{C}$, then $(c_{n-1}, c_0, c_1, \ldots, c_{n-2}) \in \mathcal{C}$. In other words, the code is invariant under cyclic shifts to the right.

If we write the codewords as polynomials of degree $< n$ with coefficients in $GF(2^\nu)$, this is equivalent to say that if $c(x) \in \mathcal{C}$, then $xc(x) \bmod (x^n - 1) \in \mathcal{C}$. Hence, if $c(x) \in \mathcal{C}$, then, given any polynomial $w(x)$, the residue of dividing $w(x)c(x)$ by $x^n - 1$ is in $\mathcal{C}$. In particular, if the degree of $w(x)c(x)$ is smaller than n, then $w(x)c(x) \in \mathcal{C}$.

From now on, we write the elements of a cyclic code $\mathcal{C}$ as polynomials modulo $x^n - 1$.

Theorem 9.2 $\mathcal{C}$ *is an* $[n, k]$ *cyclic code over* $GF(2^\nu)$ *if and only if there is a (monic) polynomial* $g(x)$ *of degree* $n - k$ *such that* $g(x)$ *divides* $x^n - 1$ *and each* $c(x) \in \mathcal{C}$ *is a multiple of* $g(x)$, *that is*, $c(x) \in \mathcal{C}$ *if and only if* $c(x) = w(x)g(x), \deg(w) < k$. *We call* $g(x)$ *a generator polynomial of* $\mathcal{C}$.

Proof 9.7 Let $g(x)$ be a monic (i.e., lead coefficient is 1) polynomial in $\mathcal{C}$ such that $g(x)$ has minimal degree. If $\deg(g) = 0$ (i.e., $g = 1$), then $\mathcal{C}$ is the whole space $(GF(2^\nu))^n$, so assume $\deg(g) \geq 1$. Let $c(x)$ be any element in $\mathcal{C}$. We can write $c(x) = w(x)g(x) + r(x)$, where $\deg(r) < \deg(g)$. Since $\deg(wg) < n, g \in \mathcal{C}$ and $\mathcal{C}$ is cyclic, in particular, $w(x)g(x) \in \mathcal{C}$. Hence, $r(x) = c(x) - w(x)g(x) \in \mathcal{C}$. If $r \neq 0$, we would contradict the fact that $g(x)$ has minimal degree, hence, $r = 0$ and $c(x)$ is a multiple of $g(x)$.

Similarly, we can prove that $g(x)$ divides $x^n - 1$. Let $x^n - 1 = h(x)g(x) + r(x)$, where $\deg(r) < \deg(g)$. In particular, $h(x)g(x) \equiv -r(x) \bmod (x^n - 1)$, hence, $r(x) \in \mathcal{C}$. Since $g(x)$ has minimal degree, $r = 0$, so $g(x)$ divides $x^n - 1$.

Conversely, assume that every element in $\mathcal{C}$ is a multiple of $g(x)$ and g divides $x^n - 1$. It is immediate that the code is linear and that it has dimension k. Let $c(x) \in \mathcal{C}$, hence, $c(x) = w(x)g(x)$ with $\deg(w) < k$. Also, since $g(x)$ divides $x^n - 1$, $x^n - 1 = h(x)g(x)$. Assume that $c(x) = c_0 + c_1x + c_2x^2 + \cdots + c_{n-1}x^{n-1}$, then, $xc(x) \equiv c_{n-1} + c_0x + \cdots + c_{n-2}x^{n-1} \pmod{x^n - 1}$. We have to prove that $c_{n-1} + c_0x + \cdots + c_{n-2}x^{n-1} = q(x)g(x)$, where $q(x)$ has degree $\leq k - 1$. Notice that

$$
\begin{aligned}
c_{n-1} + c_0x + \cdots + c_{n-2}x^{n-1} &= c_{n-1} + c_0x + \cdots + c_{n-2}x^{n-1} + c_{n-1}x^n - c_{n-1}x^n \\
&= c_0x + \cdots + c_{n-2}x^{n-1} + c_{n-1}x^n - c_{n-1}(x^n - 1) \\
&= xc(x) - c_{n-1}(x^n - 1) \\
&= xw(x)g(x) - c_{n-1}h(x)g(x) \\
&= (xw(x) - c_{n-1}h(x))g(x)
\end{aligned}
$$

proving that the element is in the code. $\square$

Theorem 9.1 gives a method to find all cyclic codes of length n: simply take all the (monic) factors of $x^n - 1$. Each one of them is the generator polynomial of a cyclic code.

Example 9.6

Consider the $[7, 4]$ cyclic code over $GF(2)$ generated by $g(x) = 1 + x + x^3$. We can verify that $x^7 - 1 = g(x)(1 + x)(1 + x^2 + x^3)$, hence, $g(x)$ indeed generates a cyclic code.

In order to encode an information polynomial over $GF(2)$ of degree ≤ 3 into a codeword, we multiply it by $g(x)$.

Say that we want to encode $\underline{u} = (1, 0, 0, 1)$, which in polynomial form is $u(x) = 1 + x^3$. Hence, the encoding gives $c(x) = u(x)g(x) = 1 + x + x^4 + x^6$. In vector form, this gives $\underline{c} = (1\,1\,0\,0\,1\,0\,1)$.

It can be easily verified that the $[7, 4]$ code given in this example has minimum distance 3 and is equivalent to the Hamming code of Example 9.4. In other words, the codewords of the code given in this example are permutations of the codewords of the $[7,4,3]$ Hamming code given in Example 9.4.

The encoding method of a cyclic code with generator polynomial g is then very simple: we multiply the information polynomial by g. However, this encoder is not systematic. A systematic encoder of a cyclic code is given by the following algorithm:

Algorithm 9.1 (Systematic Encoding Algorithm for Cyclic Codes)

Let C be a cyclic $[n, k]$ code over $GF(2^v)$ with generator polynomial $g(x)$. Let $u(x)$ be an information polynomial, $\deg(u) < k$. Let $r(x)$ be the residue of dividing $x^{n-k}u(x)$ by $g(x)$. Then, $u(x)$ is encoded into the polynomial $c(x) = u(x) - x^k r(x)$.

We leave as an exercise proving that Algorithm 9.2 produces indeed a codeword in C.

Example 9.7

Consider the $[7, 4]$ cyclic code over $GF(2)$ of Example 9.6. If we want to encode systematically the information vector $\underline{u} = (1, 0, 0, 1)$(or $u(x) = 1 + x^3$), we have to obtain first the residue of dividing $x^3 u(x) = x^3 + x^6$ by $g(x)$. This residue is $r(x) = x + x^2$. Hence, the output of the encoder is $c(x) = u(x) - x^4 r(x) = 1 + x^3 + x^5 + x^6$. In vector form, this gives $\underline{c} = (1\,0\,0\,1\,0\,1\,1)$.

9.6 Reed Solomon Codes

Throughout this section, the codes considered are over the field $GF(2^v)$. Let α be a primitive element in $GF(2^v)$, that is, $\alpha^{2^v-1} = 1, \alpha^i \neq 1$ for $i \not\equiv 0 \bmod 2^v - 1$. A Reed Solomon (RS) code of length $n = 2^v - 1$ and dimension k is the cyclic code generated by

$$g(x) = (x - \alpha)(x - \alpha^2) \cdots (x - \alpha^{n-k-1})(x - \alpha^{n-k})$$

Since each α^i is a root of unity, $x - \alpha^i$ divides $x^n - 1$, hence g divides $x^n - 1$ and the code is cyclic.

An equivalent way of describing a RS code, is as the set of polynomials over $GF(2^v)$ of degree $\leq n - 1$ with roots $\alpha, \alpha^2, \ldots, \alpha^{n-k}$, that is, F is in the code if and only if $\deg(F) \leq n - 1$ and $F(\alpha) = F(\alpha^2) = \cdots = F(\alpha^{n-k}) = 0$.

This property allows us to find a parity check matrix for a RS code. Say that $F(x) = F_0 + F_1 x + \cdots + F_{n-1}x^{n-1}$ is in the code. Let $1 \leq i \leq n - k$, then

$$F(\alpha^i) = F_0 + F_1\alpha^i + \cdots + F_{n-1}\alpha^{i(n-1)} = 0 \tag{9.14}$$

In other words, Equation 5.14 tells us that codeword $(F_0, F_1, \ldots, F_{n-1})$ is orthogonal to the vectors $(1, \alpha^i, \alpha^{2i}, \ldots, \alpha^{i(n-1)}), 1 \leq i \leq n - k$. Hence these vectors are the rows of a parity check matrix for the

RS code. A parity check matrix of an $[n, k]$ RS code over $GF(2^v)$ is then

$$H = \begin{pmatrix} 1 & \alpha & \alpha^2 & \cdots & \alpha^{n-1} \\ 1 & \alpha^2 & \alpha^4 & \cdots & \alpha^{2(n-1)} \\ \vdots & \vdots & \vdots & \ddots & \vdots \\ 1 & \alpha^{n-k} & \alpha^{(n-k)2} & \cdots & \alpha^{(n-k)(n-1)} \end{pmatrix} \tag{9.15}$$

In order to show that H is in fact a parity check matrix, we need to prove that the rows of H are linearly independent. The next lemma provides an even stronger result.

Lemma 9.6   *Any set of $n - k$ columns in matrix H defined by Equation 9.15 is linearly independent.*

***Proof* 9.8**   Take a set $0 \leq i_1 < i_2 < \cdots < i_{n-k} \leq n-1$ of columns of H. Denote α^{i_j} by α_j, $1 \leq j \leq n-k$. Columns $i_1, i_2, \ldots, i_{n-k}$ are linearly independent if and only if their determinant is nonzero, that is, if and only if

$$\det \begin{pmatrix} \alpha_1 & \alpha_2 & \cdots & \alpha_{n-k} \\ (\alpha_1)^2 & (\alpha_2)^2 & \cdots & (\alpha_{n-k})^2 \\ \vdots & \vdots & \ddots & \vdots \\ (\alpha_1)^{n-k} & (\alpha_2)^{n-k} & \cdots & (\alpha_{n-k})^{n-k} \end{pmatrix} \neq 0 \tag{9.16}$$

Let

$$V(\alpha_1, \alpha_2, \ldots, \alpha_{n-k}) = \det \begin{pmatrix} 1 & 1 & \cdots & 1 \\ \alpha_1 & \alpha_2 & \cdots & \alpha_{n-k} \\ \vdots & \vdots & \ddots & \vdots \\ (\alpha_1)^{n-k-1} & (\alpha_2)^{n-k-1} & \cdots & (\alpha_{n-k})^{n-k-1} \end{pmatrix} \tag{9.17}$$

We call the determinant $V(\alpha_1, \alpha_2, \ldots, \alpha_{n-k})$ a *Vandermonde determinant*: it is the determinant of an $(n - k) \times (n - k)$ matrix whose rows are the powers of vector $\alpha_1, \alpha_2, \ldots, \alpha_{n-k}$, the powers running from 0 to $n - k - 1$. By properties of determinants, if we consider the determinant in Equation 5.16, we have

$$\det \begin{pmatrix} \alpha_1 & \alpha_2 & \cdots & \alpha_{n-k} \\ (\alpha_1)^2 & (\alpha_2)^2 & \cdots & (\alpha_{n-k})^2 \\ \vdots & \vdots & \ddots & \vdots \\ (\alpha_1)^{n-k} & (\alpha_2)^{n-k} & \cdots & (\alpha_{n-k})^{n-k} \end{pmatrix} = \alpha_1 \alpha_2 \ldots \alpha_{n-k} V(\alpha_1, \alpha_2, \ldots, \alpha_{n-k}). \tag{9.18}$$

Hence, by Equation 9.16 and Equation 9.18, since the α_j's are nonzero, it is enough to prove that $V(\alpha_1, \alpha_2, \ldots, \alpha_{n-k}) \neq 0$. A well known result in literature states that

$$V(\alpha_1, \alpha_2, \ldots, \alpha_{n-k}) = \prod_{1 \leq i < j \leq n-k} (\alpha_j - \alpha_i) \tag{9.19}$$

Since α is a primitive element in $GF(2^v)$, its powers $\alpha^l, 0 \leq l \leq n - 1$ are distinct. In particular, the α_i's, $l \leq i \leq n - k$ are distinct, hence, the product in the right hand side of Equation 9.19 nonzero.   □

Corollary 9.3   An $[n, k]$ RS code has minimum distance $d = n - k + 1$.

Proof 9.9   Let H be the parity check matrix of the RS code defined by Equation 9.15. Notice that, since *any $n - k$ columns* in H are linearly independent, $d \geq n - k + 1$ by Lemma 9.3.

On the other hand, $d \leq n - k + 1$ by the Singleton bound (Corollary 9.2), so we have equality.   □

Since RS codes meet the Singleton bound with equality, they are MDS (see Section 9.2).

Example 9.8

Consider the $[7, 3, 5]$ RS code over $GF(8)$, where $GF(8)$ is given by Table 9.1. The generator polynomial is

$$g(x) = (x - \alpha)(x - \alpha^2)(x - \alpha^3)(x - \alpha^4) = \alpha^3 + \alpha x + x^2 + \alpha^3 x^3 + x^4$$

Assume that we want to encode the 3 byte vector $\underline{u} = 101\,001\,111$. Writing the bytes as powers of α in polynomial form, we have $u(x) = \alpha^6 + \alpha^2 x + \alpha^5 x^2$.

In order to encode $u(x)$, we perform

$$u(x)g(x) = \alpha^2 + \alpha^4 x + \alpha^2 x^2 + \alpha^6 x^3 + \alpha^6 x^4 + \alpha^4 x^5 + \alpha^5 x^6$$

In vector form the output of the encoder is given by $001\,011\,001\,101\,101\,011\,111$. If we encode $u(x)$ using a systematic encoder (Algorithm 9.1), then the output of the encoder is

$$\alpha^6 + \alpha^2 x + \alpha^5 x^2 + \alpha^6 x^3 + \alpha^5 x^4 + \alpha^4 x^5 + \alpha^4 x^6$$

which in vector form is $101\,001\,111\,101\,111\,011\,011$.

Next we make some observations:

- The definition given above for an $[n, k]$ Reed Solomon code states that $F(x)$ is in the code if and only if it has as roots the powers $\alpha, \alpha^2, \ldots, \alpha^{n-k}$ of a primitive element α. However, it is enough to state that F has as roots a set of *consecutive* powers of α, say, $\alpha^m, \alpha^{m+1}, \ldots, \alpha^{m+n-k-1}$, where $0 \leq m \leq n - 1$. Although our definition (i.e., $m = 1$) gives the most usual setting for RS codes, often engineering reasons may determine different choices of m. It is easy to verify that with the more general definition of RS codes, the minimum distance remains $n - k + 1$.

- Given an $[n, k]$ RS code, there is an easy way to shorten it and obtain an $[n - l, k - l]$ code for $l < k$. In effect, if we have only $k - l$ bytes of information, we add l zeroes in order to obtain an information string of length k. We then find the $n - k$ redundant bytes using a systematic encoder. When writing, of course, the l zeroes are not written, so we have an $[n - l, k - l]$ code, called a shortened RS code. It is immediately verified that shortened RS codes are also MDS.

We have defined RS codes, proven that they are MDS and showed how to encode them systematically. The next step, to be developed in the next sections, is decoding them.

9.7   Decoding of RS codes: the key equation

Through this section $\mathcal{C}$ denotes an $[n, k]$ RS code (unless otherwise stated). Assume that a codeword $F(x) = \sum_{i=0}^{n-1} F_i x^i$ in $\mathcal{C}$ is transmitted and a word $R(x) = \sum_{i=0}^{n-1} R_i x^i$ is received; hence, F and R are related by an error vector $E(x) = \sum_{i=0}^{n-1} E_i x^i$, where $R(x) = F(x) + E(x)$. The decoder will attempt to find $E(x)$.

Let us start by computing the syndromes. For $1 \leq j \leq n - k$, we have

$$S_j = R(\alpha^j) = \sum_{i=0}^{n-1} R_i \alpha^{ij} = \sum_{i=0}^{n-1} E_i \alpha^{ij} \tag{9.20}$$

Before proceeding further, consider Equation 9.20 in a particular case.

Take the $[n, n - 2]$ 1-byte correcting RS code. In this case, we have two syndromes S_1 and S_2, so, if exactly one error has occurred, say in location i, by Equation 9.20, we have

$$S_1 = E_i \alpha^i \quad \text{and} \quad S_2 = E_i \alpha^{2i} \tag{9.21}$$

Hence, $\alpha^i = S_2 / S_1$, so we can determine the location i in error. The error value is $E_i = (S_1)^2 / S_2$.

Example 9.9

Consider the $[7, 5, 3]$ RS code over $GF(8)$, where $GF(8)$ is given by Table 9.1. Assume that we want to decode the received vector.

$$\underline{r} = (101\,001\,110\,001\,011\,010\,100)$$

which in polynomial form is

$$R(x) = \alpha^6 + \alpha^2 x + \alpha^3 x^2 + \alpha^2 x^3 + \alpha^4 x^4 + \alpha x^5 + x^6$$

Evaluating the syndromes, we obtain $S_1 = R(\alpha) = \alpha^2$ and $S_2 = R(\alpha^2) = \alpha^4$. Thus, $S_2 / S_1 = \alpha^2$, meaning that location 2 is in error. The error value is $E_2 = (S_1)^2 / S_2 = (\alpha^2)^2 / \alpha^4 = 1$, which in vector form is 100. The output of the decoder is then

$$\underline{c} = (101\,001\,010\,001\,011\,010\,100)$$

which in polynomial form is

$$C(x) = \alpha^6 + \alpha^2 x + \alpha x^2 + \alpha^2 x^3 + \alpha^4 x^4 + \alpha x^5 + x^6$$

Let $\mathcal{E}$ be the subset of $\{0, 1, \ldots, n - 1\}$ of locations in error, that is, $\mathcal{E} = \{l : E_l \neq 0\}$. With this notation, Equation 9.20 becomes

$$S_j = \sum_{i \in \mathcal{E}} E_i \alpha^{ij}, \quad 1 \leq j \leq n - k \tag{9.22}$$

The decoder will find the error set $\mathcal{E}$ and the error values E_i when the error correcting capability of the code is not exceeded. Thus, if s is the number of errors and $2s \leq n - k$, the system of equations given by Equation 9.22 has a unique solution. However, this is a nonlinear system, and it is very difficult to solve it directly.

In order to find the set of locations in error $\mathcal{E}$ and the corresponding error values $\{E_i : i \in \mathcal{E}\}$, we define two polynomials. The first one is called the *error locator polynomial*, which is the polynomial that has as roots the values α^{-i}, where $i \in \mathcal{E}$. We denote this polynomial by $\sigma(x)$. Explicitly,

$$\sigma(x) = \prod_{i \in \mathcal{E}} (x - \alpha^{-i}) \tag{9.23}$$

If somehow we can determine the polynomial $\sigma(x)$, by finding its roots, we can obtain the set $\mathcal{E}$ of locations in error. Once we have the set of locations in error, we need to find the errors themselves. We define a second polynomial, called the *error evaluator polynomial* and denoted by $w(x)$, as follows:

$$w(x) = \sum_{i \in \mathcal{E}} E_i \prod_{\substack{l \in \mathcal{E} \\ l \neq i}} (x - \alpha^{-l}) \tag{9.24}$$

Since an $[n, k]$ RS code corrects at most $(n - k)/2$ errors, we assume that $|\mathcal{E}| = \deg(\sigma) \leq (n - k)/2$. Notice also that $\deg(w) \leq |\mathcal{E}| - 1$, since w is a sum of polynomials of degree $|\mathcal{E}| - 1$. Given a polynomial $f(x) = a_0 + a_1 x + \cdots + a_m x^m$ with coefficients over a field F, we define the (formal) derivative of f, denoted f', as the polynomial

$$f'(x) = a_1 + 2a_2x + \cdots + ma_m x^{m-1}$$

For instance, over $GF(8)$, if $f(x) = \alpha + \alpha^3 x + \alpha^4 x^2$, then $f'(x) = \alpha^3$ (since $2 = 0$ over $GF(2)$). The formal derivative has several properties similar to the traditional derivative, like the derivative of a product, $(fg)' = f'g + fg'$. Back to the error locator and error evaluator polynomials, we have the following relationship between the two:

$$E_i = \frac{w(\alpha^{-i})}{\sigma'(\alpha^{-i})} \tag{9.25}$$

Let us prove some of these facts in the following lemma:

Lemma 9.7 *The polynomials $\sigma(x)$ and $w(x)$ are relatively prime, and the error values E_i are given by Equation 9.25*

Proof 9.10 In order to show that $\sigma(x)$ and $w(x)$ are relatively prime, it is enough to observe that they have no roots in common. In effect, if α^{-j} is a root of $\sigma(x)$, then $j \in \mathcal{E}$. By Equation 9.24,

$$w(\alpha^{-j}) = \sum_{i \in \mathcal{E}} E_i \prod_{\substack{l \in \mathcal{E} \\ l \neq i}} (\alpha^{-j} - \alpha^{-l}) = E_j \prod_{\substack{l \in \mathcal{E} \\ l \neq j}} (\alpha^{-j} - \alpha^{-l}) \neq 0 \tag{9.26}$$

Hence, $\sigma(x)$ and $w(x)$ are relatively prime.

In order to prove Equation 9.25, notice that

$$\sigma'(x) = \sum_{i \in \mathcal{E}} \prod_{\substack{l \in \mathcal{E} \\ l \neq i}} (x - \alpha^{-l})$$

hence,

$$\sigma'(\alpha^{-j}) = \prod_{\substack{l \in \mathcal{E} \\ l \neq j}} (\alpha^{-j} - \alpha^{-l}) \tag{9.27}$$

By Equation 9.26 and Equation 9.27, Equation 9.25 follows. $\qquad\square$

The decoding methods of RS codes are based on finding the error locator and the error evaluator polynomials. By finding the roots of the error locator polynomial, we determine the locations in error, while the errors themselves can be found using Equation 9.25. We will establish a relationship between $\sigma(x)$ and $w(x)$, but first we need to define a third polynomial, the syndrome polynomial. We define the syndrome polynomial as the polynomial of degree $\leq n-k-1$ whose coefficients are the $n-k$ syndromes. Explicitly,

$$S(x) = S_1 + S_2x + S_3x^2 + \cdots + S_{n-k}x^{n-k-1} = \sum_{j=0}^{n-k-1} S_{j+1}x^j \tag{9.28}$$

Notice that $R(x)$ is in $\mathcal{C}$ if and only if $S(x) = 0$.

The next theorem gives the so called *key equation* for decoding RS codes, and it establishes a fundamental relationship between $\sigma(x), w(x)$ and $S(x)$.

Theorem 9.3 *There is a polynomial $\mu(x)$ such that the error locator, the error evaluator, and the syndrome polynomials verify the following equation:*

$$\sigma(x)S(x) = -w(x) + \mu(x)x^{n-k} \tag{9.29}$$

Alternatively, Equation 9.29 can be written as a congruence as follows:

$$\sigma(x)S(x) \equiv -w(x)(\bmod x^{n-k}) \tag{9.30}$$

***Proof* 9.11**   By Equation 9.28 and Equation 9.22, we have

$$S(x) = \sum_{j=0}^{n-k-1} S_{j+1}x^j$$

$$= \sum_{j=0}^{n-k-1} \left(\sum_{i \in \mathcal{E}} E_i \alpha^{i(j+1)} \right) x^j$$

$$= \sum_{i \in \mathcal{E}} E_i \alpha^i \sum_{j=0}^{n-k-1} (\alpha^i x)^j$$

$$= \sum_{i \in \mathcal{E}} E_i \alpha^i \frac{(\alpha^i x)^{n-k} - 1}{\alpha^i x - 1}$$

$$= \sum_{i \in \mathcal{E}} E_i \frac{(\alpha^i x)^{n-k} - 1}{x - \alpha^{-i}} \tag{9.31}$$

since $\sum_{l=0}^{m} a^l = (a^{m+1} - 1)/(a - 1)$ for $a \neq 1$. Multiplying both sides of Equation 9.31 by $\sigma(x)$, where $\sigma(x)$ is given by Equation 9.23, we obtain

$$\sigma(x)S(x) = \sum_{i \in \mathcal{E}} E_i ((\alpha^i x)^{n-k} - 1) \prod_{\substack{l \in \mathcal{E} \\ l \neq i}} (x - \alpha^{-l})$$

$$= -\sum_{i \in \mathcal{E}} E_i \prod_{\substack{l \in \mathcal{E} \\ l \neq i}} (x - \alpha^{-l}) + \left(\sum_{i \in \mathcal{E}} E_i \alpha^{i(n-k)} \prod_{\substack{l \in \mathcal{E} \\ l \neq i}} (x - \alpha^{-l}) \right) x^{n-k}$$

$$= -\omega(x) + \mu(x)x^{n-k}$$

since $\omega(x)$ is given by Equation 9.24. This completes the proof.   □

The decoding methods for RS codes concentrate on solving the key equation. In the next section we describe an efficient decoder based on Euclid's algorithm for polynomials. Another efficient decoding algorithm is the so-called Berlekamp-Massey decoding algorithm [1].

9.8   Decoding RS Codes with Euclid's Algorithm

Given two polynomials or integers A and B, Euclid's algorithm provides a recursive procedure to find the greatest common divisor C between A and B, denoted $C = \gcd(A, B)$. Moreover, the algorithm also finds two polynomials or integers S and T such that $C = SA + TB$.

Recall that we want to solve the key equation

$$\mu(x)x^{n-k} + \sigma(x)S(x) = -\omega(x)$$

In the recursion, x^{n-k} will play the role of A and $S(x)$ the role of B; $\sigma(x)$ and $\omega(x)$ will be obtained at a certain step of the recursion.

Let us describe Euclid's algorithm for integers or polynomials. Consider A and B such that $A \geq B$ if they are integers and $\deg(A) \geq \deg(B)$ if they are polynomials. We start from the initial conditions $r_{-1} = A$ and $r_0 = B$.

We perform a recursion in steps $1, 2, \ldots, i, \ldots$. At step i of the recursion, we obtain r_i as the residue of dividing r_{i-2} by r_{i-1}, that is, $r_{i-2} = q_i r_{i-1} + r_i$, where $r_i < r_{i-1}$ for integers and $\deg(r_i) < \deg(r_{i-1})$ for polynomials. The recursion is then given by

$$r_i = r_{i-2} - q_i r_{i-1} \tag{9.32}$$

We also obtain values s_i and t_i such that $r_i = s_i A + t_i B$. Hence, the same recursion is valid for s_i and t_i as well:

$$s_i = s_{i-2} - q_i s_{i-1} \tag{9.33}$$
$$t_i = t_{i-2} - q_i t_{i-1} \tag{9.34}$$

Since $r_{-1} = A = (1)A + (0)B$ and $r_0 = B = (0)A + (1)B$, we set the initial conditions $s_{-1} = 1, t_{-1} = 0, s_0 = 0$ and $t_0 = 1$.

Let us illustrate the process with $A = 124$ and $B = 46$. We will find $\gcd(124, 46)$. The idea is to divide recursively by the residues of the division until obtaining a last residue 0. Then, the last divisor is the gcd. The procedure works as follows:

$$
\begin{aligned}
124 &= & (1)124 + & (0)46 \\
46 &= & (0)124 + & (1)46 \\
32 &= & (1)124 + & (-2)46 \\
14 &= & (-1)124 + & (3)46 \\
4 &= & (3)124 + & (-8)46 \\
2 &= & (-10)124 + & (27)46
\end{aligned}
$$

Since 2 divides 4, 2 is the greatest common divisor between 124 and 46.

The best way to develop the process above, is to construct a table for r_i, q_i, s_i and t_i using the initial conditions and recursions from Equation 9.32 through Equation 9.34.

Let us do it again for 124 and 46.

i	r_i	q_i	$s_i = s_{i-2} - q_i s_{i-1}$	$t_i = t_{i-2} - q_i t_{i-1}$
-1	124		1	0
0	46		0	1
1	32	2	1	-2
2	14	1	-1	3
3	4	2	3	-8
4	2	3	-10	27
5	0	2	23	-62

From now on, let us concentrate on Euclid's algorithm for polynomials. If we want to solve the key equation

$$\mu(x)x^{n-k} + \sigma(x)S(x) = -\omega(x)$$

and the error correcting capability of the code has not been exceeded, then applying Euclid's algorithm to x^{n-k} and to $S(x)$, at a certain point of the recursion we obtain

$$r_i(x) = s_i(x)x^{n-k} + t_i(x)S(x)$$

where $\deg(r_i) \leq \lfloor(n-k)/2\rfloor - 1$, and i is the first with this property. Then, $\omega(x) = -\lambda r_i(x)$ and $\sigma(x) = \lambda t_i(x)$, where λ is a constant that makes $\sigma(x)$ monic. For a proof that Euclid's algorithm gives the right solution, see [1] or [5].

We illustrate the decoding of RS codes using Euclid's algorithm with an example. Notice that we are interested in $r_i(x)$ and $t_i(x)$ only.

Example 9.10

Consider the $[7, 3, 5]$ RS code over $GF(8)$, and assume that we want to decode the received vector

$$\underline{r} = (011\ 101\ 111\ 111\ 111\ 101\ 010)$$

which in polynomial form is

$$R(x) = \alpha^4 + \alpha^6 x + \alpha^5 x^2 + \alpha^5 x^3 + \alpha^5 x^4 + \alpha^6 x^5 + \alpha x^6$$

Evaluating the syndromes, we obtain

$$
\begin{aligned}
S_1 &= R(\alpha) \ \ = \alpha^5 \\
S_2 &= R(\alpha^2) = \alpha \\
S_3 &= R(\alpha^3) = 0 \\
S_4 &= R(\alpha^4) = \alpha^3
\end{aligned}
$$

Therefore, the syndrome polynomial is $S(x) = \alpha^5 + \alpha x + \alpha^3 x^3$.

Next, we apply Euclid's algorithm with respect to x^4 and to $S(x)$. When we find the first i for which $r_i(x)$ has degree ≤ 1, we stop the algorithm and we obtain $\omega(x)$ and $\sigma(x)$. The process is tabulated below.

i	$r_i = r_{i-2} - q_i r_{i-1}$	q_i	$t_i = t_{i-2} - q_i t_{i-1}$
-1	x^4		0
0	$\alpha^5 + \alpha x + \alpha^3 x^3$		1
1	$\alpha^2 x + \alpha^5 x^2$	$\alpha^4 x$	$\alpha^4 x$
2	$\alpha^5 + \alpha^2 x$	$\alpha^2 + \alpha^5 x$	$1 + \alpha^6 x + \alpha^2 x^2$

So, for $i = 2$, we obtain a polynomial $r_2(x) = \alpha^5 + \alpha^2 x$ of degree 1. Now, multiplying both $r_2(x)$ and $t_2(x)$ by $\lambda = \alpha^5$, we obtain $\omega(x) = \alpha^3 + x$ and $\sigma(x) = \alpha^5 + \alpha^4 x + x^2$.

Searching the roots of $\sigma(x)$, we verify that these roots are $\alpha^0 = 1$ and α^5; hence, the errors are in locations 0 and 2. The derivative of $\sigma(x)$ is $\sigma'(x) = \alpha^4$. By Equation 9.25, we obtain $E_0 = \omega(1)/\sigma'(1) = \alpha^4$ and $E_2 = \omega(\alpha^5)/\sigma'(\alpha^5) = \alpha^5$. Adding E_0 and E_2 to the received locations 0 and 2, the decoder concludes that the transmitted polynomial was

$$F(x) = \alpha^6 x + \alpha^5 x^3 + \alpha^5 x^4 + \alpha^6 x^5 + \alpha x^6$$

which in vector form is

$$\underline{c} = (000\ 101\ 000\ 111\ 111\ 101\ 010)$$

If the information is carried in the first 3 bytes, then the output of the decoder is

$$\underline{u} = (000\ 101\ 000)$$

9.9 Applications: Burst and Random Error Correction

In the previous sections we have studied how to encode and decode Reed-Solomon codes. In this section, we will briefly examine how they are used in applications, mainly for correction of bursts of errors. The two main methods for burst and combined burst and random error correction are interleaving and product codes.

In practice, errors often come in bursts. A burst of length l is a vector whose nonzero entries are among l consecutive (cyclically) entries, the first and last of them being nonzero. We consider binary bursts, and we use the elements of larger fields (bytes) to correct them. Below are some examples of bursts of length 4 in vectors of length 15:

$$
\begin{array}{ccccccccccccccc}
0 & 0 & 0 & 1 & 0 & 1 & 1 & 0 & 0 & 0 & 0 & 0 & 0 & 0 & 0 \\
0 & 0 & 0 & 0 & 0 & 0 & 1 & 1 & 1 & 1 & 0 & 0 & 0 & 0 & 0 \\
1 & 0 & 0 & 0 & 0 & 0 & 0 & 0 & 0 & 0 & 0 & 0 & 1 & 0 & 0
\end{array}
$$

Errors tend to come in bursts not only because the channel is bursty. Normally, both in optical and magnetic recording, data is encoded using a so called modulation code, which attempts to match the data to the characteristics of the channel. In general, the ECC is applied first to the random data and then the encoded data is modulated using modulation codes (see the chapter on modulation codes in this book). At the decoding, the order is reversed: when data exits the channel, it is first demodulated and then corrected using the ECC. Now, the demodulator tends to propagate errors, even single-bit errors. Although most modulation codes used in practice tend to control error propagation, nevertheless errors have a bursty character. For that reason, we need to implement a burst-correcting scheme, as we will see next.

A well-known relationship between the burst-correcting capability of a code and its radundancy is given by the Reiger bound, to be presented next, and whose proof is left as an exercise.

Theorem 9.4 (Reiger Bound) *Let C be an $[n, k]$ linear code over a field GF(2^v) that can correct all bursts of length up to l. Then $2l \leq n - k$.*

Cyclic binary codes that can correct bursts were obtained by computer search. A well known family of burst-correcting codes are the so called Fire codes. Here, we concentrate on the use of RS codes for burst correction. There are good reasons for this. One of them is that, although good burst-correcting codes have been found by computer search, there are no known general constructions giving cyclic codes that approach the Reiger bound. Interleaving of RS codes on the other hand, to be described below, provides a burst-correcting code whose redundancy, asymptotically, approaches the Reiger bound. The longer the burst we want to correct, the more efficient interleaving of RS codes is. The second reason for choosing interleaving of RS codes, and probably the most important one, is that, by increasing the error-correcting capability of the individual RS codes, we can correct multiple bursts, as we will see. The known binary cyclic codes are designed, in general, to correct only one burst. Let us start with the use of regular RS codes for correction of bursts. Let C be an $[n, k]$ RS code over $GF(2^b)$ (i.e., b-bit bytes). If this code can correct s bytes, in particular, it can correct a burst of length up to $(s - 1)b + 1$ bits. In effect, a burst of length $(s - 1)b + 2$ bits may affect $s + 1$ consecutive bytes, exceeding the byte-correcting capability of the code. This hapens when the burst of length $(s - 1)b + 2$ bits starts in the last bit of a byte. How good are then RS codes as burst-correcting codes? Given a binary $[n, k]$ code that can correct bursts of length up to l, we define a parameter, called the *burst-correcting efficiency* of the code, as follows:

$$
e_l = \frac{2l}{n - k} \tag{9.35}
$$

Notice that, by the Reiger bound, $e_l \leq 1$. The closer e_l is to 1, the more efficient the code is for correction of bursts. Going back to our $[n, k]$ RS code over $GF(2^b)$, it can be regarded as an $[nb, kb]$ binary code. Assuming that the code can correct s bytes and its redundancy is $n - k = 2s$, its burst-correcting efficiency is

$$
e_{(s-1)b+1} = \frac{(s - 1)b + 1}{bs}
$$

Notice that, for $s \to \infty$, $e_{(s-1)b+1} \to 1$, justifying our assertion that for long bursts, RS codes are efficient as burst-correcting codes (as a comparison, the efficiency of Fire codes, asymptotically, tends to 2/3). However, when s is large, there is a problem regarding complexity. It may not be practical to implement a RS code with too much redundancy. Moreover, the length of a RS code is limited, in the case of 8-bit bytes, it cannot be more than 256 (when extended). An alternative would be to implement a 1-byte correcting RS code interleaved s times.

$c_{0,0}$	$c_{0,1}$	$c_{0,2}$	$\cdots$	$c_{0,m-1}$
$c_{1,0}$	$c_{1,1}$	$c_{1,2}$	$\cdots$	$c_{1,m-1}$
$\vdots$	$\vdots$	$\vdots$	$\ddots$	$\vdots$
$c_{k-1,0}$	$c_{k-1,1}$	$c_{k-1,2}$	$\cdots$	$c_{k-1,m-1}$
$c_{k,0}$	$c_{k,1}$	$c_{k,2}$	$\cdots$	$c_{k,m-1}$
$\vdots$	$\vdots$	$\vdots$	$\ddots$	$\vdots$
$c_{n-1,0}$	$c_{n-1,1}$	$c_{n-1,2}$	$\cdots$	$c_{n-1,m-1}$

FIGURE 9.1 Interleaving m times of code C.

An $[n, k]$ code interleaved m times is illustrated in Figure 9.1. Each column $c_{0,j}, \ldots, c_{n-1,j}$ is a codeword in an $[n, k]$ code. In general, each symbol $c_{i,j}$ is a byte and the code is a RS code. The first k bytes carry information bytes and the last $n - k$ bytes are redundant bytes. The bytes are read in row order, and the parameter m is called the depth of interleaving. If each of the individual codes can correct up to s errors, then the interleaved scheme can correct up to s bursts of length up to m bytes each, or $(m - 1)b + 1$ bits each. This occurs because a burst of length up to m bytes is distributed among m different codewords. Intuitively, interleaving "randomizes" a burst.

The drawback of interleaving is delay: notice that we need to read most of the information bytes before we are able to calculate and write the redundant bytes. Thus, we need enough buffer space to accomplish this.

Interleaving of RS codes has been widely used in magnetic recording. For instance, in a disk, the data are written in concentric tracks, and each track contains a number of information sectors. Typically, a sector consists of 512 information 8-bit bytes (although the latest trends tend to larger sectors). A typical embodiment would consist in dividing the 512 bytes into 4 codewords, each one containing 128 information bytes and 6 redundant bytes (i.e., each interleaved shortened RS codeword can correct up to 3 bytes). Therefore, this scheme can correct up to 3 bursts of length up to 25 bits each.

A natural generalization of the interleaved scheme described above is product codes. In effect, we may consider that both rows and columns are encoded into error-correcting codes. The product of an $[n_1, k_1]$ code C_1 with an $[n_2, k_2]$ code C_2, denoted $C_1 \times C_2$, is illustrated in Figure 9.2. If C_1 has minimum distance d_1 and C_2 has minimum distance d_2, it is easy to see that $C_1 \times C_2$ has minimum distance $d_1 d_2$.

In general, the symbols are read out in row order (although other readouts, like diagonal readouts, are also possible). For encoding, first the column redundant symbols are obtained, and then the row redundant symbols. For obtaining the checks on checks $c_{i,j}, k_1 \leq i \leq n_1 - 1, k_2 \leq j \leq n_2 - 1$, it is easy to see that it is irrelevant if we encode on columns or on rows first. If the symbols are read in row order, normally C_1 is called the outer code and C_2 the inner code. For decoding, there are many possible procedures. The idea

$c_{0,0}$	$c_{0,1}$	$c_{0,2}$	$\cdots$	c_{0,k_2-1}	c_{0,k_2}	c_{0,k_2+1}	$\cdots$	c_{0,n_2-1}
$c_{1,0}$	$c_{1,1}$	$c_{1,2}$	$\cdots$	c_{1,k_2-1}	c_{1,k_2}	c_{1,k_2+1}	$\cdots$	c_{1,n_2-1}
$\vdots$	$\vdots$	$\vdots$	$\ddots$	$\vdots$	$\vdots$	$\vdots$	$\ddots$	$\vdots$
$c_{k_1-1,0}$	$c_{k_1-1,1}$	$c_{k_1-1,2}$	$\cdots$	c_{k_1-1,k_2-1}	c_{k_1-1,k_2}	c_{k_1-1,k_2+1}	$\cdots$	c_{k_1-1,n_2-1}
$c_{k_1,0}$	$c_{k_1,1}$	$c_{k_1,2}$	$\cdots$	c_{k_1,k_2-1}	c_{k_1,k_2}	c_{k_1,k_2+1}	$\cdots$	c_{k_1,n_2-1}
$\vdots$	$\vdots$	$\vdots$	$\ddots$	$\vdots$	$\vdots$	$\vdots$	$\ddots$	$\vdots$
$c_{n_1-1,0}$	$c_{n_1-1,1}$	$c_{n_1-1,2}$	$\cdots$	c_{n_1-1,k_2-1}	c_{n_1-1,k_2}	c_{n_1-1,k_2+1}	$\cdots$	c_{n_1-1,n_2-1}

FIGURE 9.2 Product code $C_1 \times C_2$.

is to correct long bursts together with random errors. The inner code C_2 corrects first. In that case, two events may happen when its error-correcting capability is exceeded: either the code will detect the error event or it will miscorrect. If the code detects an error event (that may well have been caused by a long burst), one alternative is to declare an erasure in the whole row, which will be communicated to the outer code C_1. The other event is a miscorrection, that cannot be detected. In this case, we expect that the errors will be corrected by the error-erasure decoder of the outer code.

Product codes are important in practical applications. For instance, the code used in the DVD (digital video disk) is a product code where C_1 is a [208, 192, 17] RS code and C_2 is a [182, 172, 11] RS code. Both RS codes are defined over $GF(256)$, where $GF(256)$ is generated by the primitive polynomial $1 + x^2 + x^3 + x^4 + x^8$.

References

[1] R. E. Blahut, *Theory and Practice of Error Control Codes*, Addison Wesley, Reading 1983.

[2] C. Heegard and S. B. Wicker, *Turbo Coding*, Kluwer Academic Publishers, Dordrecht, 1999.

[3] S. Lin and D. J. Costello, *Error Control Coding: Fundamentals and Applications*, Prentice Hall, New York, 1983.

[4] F. J. MacWilliams and N. J. A. Sloane, *The Theory of Error-Correcting Codes*, North-Holland, Amsterdam, 1978.

[5] R. J. McEliece, *The Theory of Information and Coding*, Addison-Wesley, Reading 1977.

[6] C. E. Shannon, A mathematical theory of communication, *Bell Syst. Tech. J.*, 27, pp. 379–423 and 623–656, 1948.

[7] W. Wesley Peterson and E. J. Weldon, *Error-Correcting Codes*, MIT Press, Cambridge, 2nd ed., 1984.

[8] S. Wicker, *Error Control Systems for Digital Communications and Storage*, Prentice Hall, New York, 1995.

10

Message-Passing Algorithm

10.1 Introduction **10**-1
10.2 Iterative Decoding on Binary Erasure Channel **10**-2
10.3 Iterative Decoding Schemes for LDPC Codes **10**-7
Gallager's Probabilistic Iterative Decoding Scheme
• Message-Passing Algorithm for Soft-Decision
Iterative Decoding
10.4 General Message-Passing Algorithm **10**-14
Computation of a Marginal Function • Message-Passing
Schedule to Compute a Marginal Function

Sundararajan
Sankaranarayanan

Bane Vasic
University of Arizona
Tucson, AZ

10.1 Introduction

A growing interest in the theory and application of iterative decoding algorithms was ignited by the impressive bit-error rate performance of the turbo-decoding algorithm demonstrated by Berrou et al. [3]. In the last several years, the turbo decoding algorithm has been generalized and mathematically formulated using a graph-theoretic approach. Such iterative decoding algorithms operate by "message-passing" in graphs associated with codes and hence, they are referred to as the *message-passing algorithms*. The theory of codes on graphs has helped to improve the error performance of low-complexity decoding schemes. Also, it has opened new research avenues for investigating alternative suboptimal decoding schemes. One of the key results in the theory of codes on graphs comes from Kschischang et al. [8,9]. They generalized the concept of codes on graphs by introducing the idea of general functions on graphs. They derived a general distributed marginalization algorithm for function described by factor graphs. The theory of factor graphs and the sum-product algorithm generalizes a variety of algorithms developed in computer science and engineering. The belief propagation algorithm, developed by Pearl [14], operating on Bayesian networks is an instance of the sum-product algorithm operating on an appropriate factor graph.

In the early 1960s, Gallager [7] introduced a class of linear block codes called the *low-density parity-check codes*, and devised iterative decoding algorithms for these codes. Mackay and Neal [11], and McEliece et al. [13] showed that Gallager's algorithm for decoding low-density parity-check codes is essentially an instance of the Pearl's belief propagation algorithm. Using extensive simulation results, Mackay and Neal showed that low-density parity-check codes, constructed by Gallager, perform nearly as well as turbo codes.

Wiberg et al. [19] showed that graphs introduced by Tanner [17] provide a natural setting to describe and study iterative soft-decision decoding techniques. Kschischang et al. [9] showed that various graphical models, such as Markov random fields, Tanner graphs, and Bayesian networks, support the basic probability propagation algorithm in factor graphs. In addition, they showed that all compound codes, such as turbo codes, classical serially concatenated codes, Gallager's LDPC codes, and product codes, can be decoded by "message-passing" in graphs.

In recent years, Luby et al. [10], and Richardson and Urbanke [16] have published key results contributing towards better understanding of these iterative decoding algorithms. They devised a method to characterize the average bit-error rate performance of low-density parity-check code (of infinite length) ensembles over various channels in an iterative decodable setting. In addition, they analytically computed thresholds of code ensembles in binary erasure channels, binary symmetric channels and additive white Gaussian noise channels.

The fundamental importance of message-passing algorithms has motivated an in-depth treatment of the theoretical foundation of these algorithms in this chapter. The chapter begins with an introduction to the concept of iterative decoding in binary erasure channels. In addition, we introduce the concept of stopping sets in binary erasure channels and their importance in analytical characterization of the message-passing algorithm. The idea of "message-passing" can be extended to decode low-density parity-check codes that are transmitted in an arbitrary binary-input continuous output channel. In Section 10.2, we study the concepts of a probabilistic iterative decoding scheme introduced by Gallager, and mathematically derive the foundation of the related message-passing algorithm. It is appropriate to conclude the chapter with a discussion on the generalized message-passing algorithm in factor graphs. This algorithm is a generalization of a variety of message-passing algorithms. Also, we present the mathematical foundation of the generalized message-passing algorithm as developed by Kschischang et al.

10.2   Iterative Decoding on Binary Erasure Channel

The principles of a message-passing algorithm can be clearly explained by introducing the concept of iterative decoding in a *binary erasure channel* (BEC). In a BEC channel, the input alphabet is binary, and the output alphabet, besides symbols 0 and 1, has an erasure symbol E (see, Figure 10.1). The channel is symmetric which means that both 0 and 1 are treated alike. Also, an input bit $0(1)$ transmitted through the channel is either received as a $0(1)$ with a probability $1 - \zeta$, or as an erasure E, with a probability ζ. Hence, it is clear that the probability of an error-free transmission of an input bit through the channel is $1 - \zeta$.

To illustrate the message-passing algorithm, assume that input bits to the channel are encoded using a linear binary code, described by the parity-check matrix:

$$\mathbf{H} = \begin{bmatrix} 1 & 0 & 0 & 1 & 0 & 0 & 0 & 1 & 0 & 0 & 0 & 1 & 1 & 0 & 1 & 0 \\ 0 & 0 & 1 & 0 & 0 & 0 & 0 & 0 & 0 & 1 & 1 & 0 & 1 & 1 & 0 & 1 \\ 1 & 0 & 1 & 0 & 0 & 0 & 1 & 1 & 1 & 1 & 0 & 0 & 0 & 0 & 0 & 0 \\ 0 & 0 & 0 & 1 & 1 & 1 & 0 & 0 & 1 & 0 & 0 & 1 & 0 & 1 & 0 & 0 \\ 1 & 1 & 0 & 0 & 1 & 1 & 0 & 0 & 0 & 0 & 1 & 0 & 0 & 0 & 1 & 0 \\ 0 & 1 & 0 & 0 & 1 & 0 & 0 & 0 & 1 & 0 & 0 & 1 & 1 & 0 & 1 & 0 \\ 0 & 0 & 0 & 0 & 0 & 1 & 1 & 1 & 0 & 0 & 1 & 0 & 0 & 1 & 0 & 1 \\ 0 & 1 & 1 & 1 & 0 & 0 & 1 & 0 & 0 & 1 & 0 & 0 & 0 & 0 & 0 & 1 \end{bmatrix} \qquad (10.1)$$

Every codeword $\mathbf{c}$, of the linear code should satisfy the system of linear equations $\mathbf{cH}^T = \mathbf{0}$. The set of constraints defined by the rows of $\mathbf{H}$ on the set of codewords is called the parity-check constraints. Assume that a transmitted codeword $\mathbf{c}$ is transformed into a vector $\mathbf{r}$ at the output of the BEC channel. If the

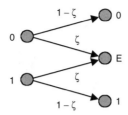

FIGURE 10.1 Binary erasure channel.

transmission was error-free, then **r** satisfies the following system of linear equations:

$$r_1 + r_4 + r_8 + r_{12} + r_{13} + r_{15} = 0$$
$$r_3 + r_{10} + r_{11} + r_{13} + r_{14} + r_{16} = 0$$
$$r_1 + r_3 + r_7 + r_8 + r_9 + r_{10} = 0$$
$$r_4 + r_5 + r_6 + r_9 + r_{12} + r_{14} = 0 \qquad (10.2)$$
$$r_1 + r_2 + r_5 + r_6 + r_{11} + r_{15} = 0$$
$$r_2 + r_5 + r_9 + r_{12} + r_{13} + r_{15} = 0$$
$$r_6 + r_7 + r_8 + r_{11} + r_{14} + r_{16} = 0$$
$$r_2 + r_3 + r_4 + r_7 + r_{10} + r_{16} = 0$$

But, if the transmission was not error-free, then we can try to solve Equation 10.2 in order to correct for erasures. Assume that the received word $r = (1, 0, 0, 1, E, 1, 1, E, 0, E, E, 0, E, 0, 0, 0)$. Now, we can try to solve Equation 10.2 for unknown bits $r_5, r_8, r_{10}, r_{11}, r_{13}$ in **r**. By substituting known values of bits $r_1, r_2, r_3, r_4, r_6, r_7, r_9, r_{12}, r_{14}, r_{15}, r_{16}$ in Equation 10.2, we obtain:

$$r_8 + r_{13} = 0$$
$$r_{10} + r_{11} + r_{13} = 0$$
$$r_8 + r_{10} = 0$$
$$r_5 = 0$$
$$r_5 + r_{11} = 0 \qquad (10.3)$$
$$r_5 + r_{13} = 0$$
$$r_8 + r_{11} = 0$$
$$r_{10} = 0$$

From Equation 10.3, we can solve for bits ($r_5 = 0, r_{10} = 0$) and hence, rewrite the system of linear equations as

$$r_8 + r_{13} = 0$$
$$r_{11} + r_{13} = 0$$
$$r_8 = 0$$
$$r_{11} = 0 \qquad (10.4)$$
$$r_{13} = 0$$

From Equation 10.4, we can solve for bits ($r_8 = 0, r_{11} = 0, r_{13} = 0$) and hence, the estimation of **c**, say $\hat{c}$, is

$$\hat{c} = (1, 0, 0, 1, 0, 1, 1, 0, 0, 0, 0, 0, 0, 0, 0, 0) \qquad (10.5)$$

The above procedure for solving a system of linear equations can be explained as a process of sending messages to and fro between the nodes of a graph that describes the linear code. In [17], Tanner introduced a clever graphical representation of the **H** of a linear code. A simple bipartite graph representation of **H**, introduced by Tanner, is usually referred to as the *Tanner graph*. A Tanner graph consists of two sets of nodes

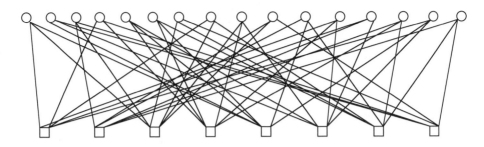

FIGURE 10.2   Tanner graph.

namely, *variable nodes and check nodes*. Each variable node of the graph corresponds to a codeword bit, and each check node of the graph corresponds to a parity-check constraint or a row of **H**. In a Tanner graph, a variable node and a check node are connected by an undirected edge if the variable node participates in the parity-check constraint represented by the check node. The number of edges incident on a variable (check) node is referred to as the *degree* of the variable (check) node. In Figure 10.2, we present the Tanner graph of the parity-check matrix in Equation 10.1.

In Figure 10.2, circular nodes represent the set of variable nodes, and square nodes represent the set of check nodes. There are 16 variable nodes since the code has 16 codeword bits, and there are eight check nodes since **H** has eight parity-check equations.

The graphical interpretation of the message-passing algorithm gives a good insight into the iterative decoding process in general. The parity-check matrix of a linear code, shown in Equation 10.1, and the received word $r = (1, 0, 0, 1, E, 1, 1, E, 0, E, E, 0, E, 0, 0, 0)$ is used to explain the message-passing algorithm. In the initialization step, each variable node of the bipartite graph is assigned its corresponding bit value from r, see Figure 10.3.

In the first step, variable nodes send messages to check nodes over connecting edges. The message sent by a variable node over an edge is nothing but the value assigned to it in the initialization step. Hence, the message sent by a variable node can be a 0, 1 or an E. The number of messages incident on a check node is equal to the degree of the node. Each check node maintains the sum (i.e., modulo 2 addition) of all messages with a value 0 or 1, and disregards all messages with a value E. This process is depicted in Figure 10.4 where only edges emerging from variable nodes having values 0 and 1 are drawn.

In the second step, all edges carrying messages with values 0 and 1 are erased from the Tanner graph. In other words, variables with values 0 and 1 are not required further in the decoding process. The reduced graph, obtained after erasing edges from Figure 10.3, is shown in Figure 10.5. In the reduced graph, check nodes with degree one send messages to variable nodes. A message sent by a check node is equal to the sum stored in it. For example, a message 0 is sent from the fourth check node to the fifth variable node, see Figure 10.5. Similarly, the eighth check node sends a message 0 to the tenth variable node.

Generally, a cycle of sending messages from variable nodes to check nodes, and from check nodes to variable nodes is referred to as a *single iterations*. The first and second steps of the message-passing

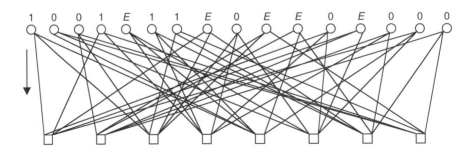

FIGURE 10.3   Message-passing algorithm for BEC — initialization.

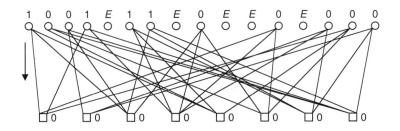

FIGURE 10.4   Message-passing algorithm for BEC — first step.

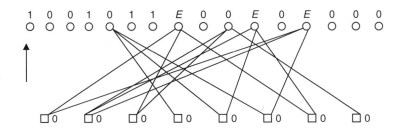

FIGURE 10.5   Message-passing algorithm for BEC — second step.

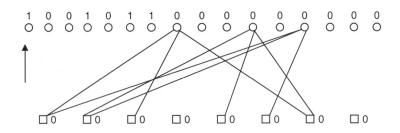

FIGURE 10.6   Message-passing algorithm — final iteration.

algorithm, explained above, constitute a single iteration. Several iterations of the message-passing algorithm are performed until no variable node is in erasure. The decoding procedure is complete when each variable node is associated with a value 0 or 1. In the effort to decode r, another iteration of the message-passing algorithm is required. At the end of the second iteration, each variable has a value 0 or 1, see Figure 10.6.

The complexity of the iterative decoding algorithm is proportional to the number of edges in the Tanner graph. As shown by Gallager [7], the complexity can be reduced to be linear in the codeword length if a code belongs to the class of low-density parity-check codes. The decoding procedure can be implemented using a parallel architecture where, each variable node is a memory element and each check node is a processor.

In the illustration of the message-passing algorithm for binary erasure channel, the decoding procedure is complete because it converges to a codeword. But, this is not necessarily the case for any received word. Assume that the received word r is: $r = (E, 0, 0, 1, E, 1, 1, 0, E, 0, 0, E, 0, 0, E, 0)$. The first and second steps of the iterative decoding algorithm, required to decode r, is illustrated in Figure 10.7 and Figure 10.8 respectively.

In Figure 10.8, none of the check nodes has degree one and hence, the iterative decoding process cannot proceed any further. This situation arises when two or more variable nodes with a value E are connected to each check node. A subset of variable nodes in erasure that has the capacity to stall the message-passing algorithm, as shown in Figure 10.8, is called a *stopping set*.

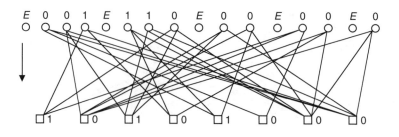

FIGURE 10.7   Message-passing algorithm for BEC channel — first step.

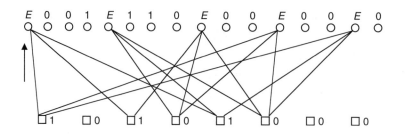

FIGURE 10.8   Message-passing algorithm for BEC — second step.

Since the message-passing algorithm is stalled when all variable nodes of a code are in erasure, a trivial example of a stopping set is the set of all variables. As shown in Figure 10.8, a set of variables $x_1, x_5, x_9, x_{12}, x_{15}$ is a stopping set of the code represented by a parity-check matrix **H**. It is straightforward to show that the cardinality of any stopping set is greater than one. Also, any code has a stopping set of length 2 if and only if it has two variables with two or more like neighbors. In a simple bipartite graph, a check (variable) node is a *neighbor* to a variable (check) node if they are connected by an edge. It is easy to show that the union of two stopping sets is a stopping set.

A *stopping set* of a code is defined as a set S of variable nodes such that all neighbors of S are connected to S at least twice. This definition for stopping set is equivalent to the previous definition. If every variable in S has a value E, then each check node connected to S has at least two neighbors with a value E. The stopping set of the example, shown in Figure 10.8, exhibits this property. The message-passing algorithm does not converge to a codeword when a subset of variable nodes in erasure is a stopping set. C. Di et al. [5] showed that a complete characterization of an ensemble of low-density parity-check (LDPC) codes, a code ensemble is a set of all possible regular bipartite graphs with n variable nodes and nd_v/d_c check nodes. The expected value of the bit-error rate $E[P_b]$, over the regular (n, d_v, d_c)-LDPC code ensemble is given by

$$E[P_b] = \sum_{i=0}^{n} \binom{n}{i} \xi^i (1 - \xi)^{n-i} \sum_{j=0}^{i} \binom{i}{j} \frac{j}{n} \frac{O\left(i, j, n\frac{d_v}{d_c}, 0\right)}{T\left(i, n\frac{d_v}{d_c}, 0\right)} \tag{10.6}$$

where functions O and T

$$T(v, c, d) = \binom{d + cd_c}{vd_v} (vd_v)!$$

$$O(v, s, c, d) = \sum_{k=0}^{c} \binom{c}{k} coef[((1 + x)^{d_c} - 1 - d_c x)^k (1 + x)^d, x^{sd_v}](sd_v)! N(v - s, c - k, d + kd_c - sd_v)$$

$$N(v, c, d) = T(v, c, d) - \sum_{s=0}^{v} \binom{v}{s} O(v, s, c, d)$$

The deviation in bit-error rate performance of any member, say $P_b(G)$, of the ensemble from the average performance of the ensemble decreases exponentially with an increase in length n as shown in Equation 10.7.

$$Pr[\,|P_b(G) - E[P_b]| > \varepsilon] < e^{-\alpha n}, \quad \varepsilon, \alpha > 0 \tag{10.7}$$

For a detailed reading on the analytical characterization of the message-passing algorithm on BEC, see [5].

10.3   Iterative Decoding Schemes for LDPC Codes

In his thesis [7], Gallager introduced a new class of linear codes called the low-density parity-check (LDPC) codes. As the name indicates, an LDPC code has a sparse parity-check matrix, say $\mathbf{H}$. In this article, we deal with binary codes and hence, $\mathbf{H}$ takes values from a Galois field of order 2, GF(2). A generator matrix, say $\mathbf{G}$, of an LDPC code is a matrix that satisfies the equation $\mathbf{GH}^T = \mathbf{0}$. Since the code is assumed to be binary, $\mathbf{G}$ takes values from GF(2). In general, an LDPC code is constructed by first designing its parity-check matrix $\mathbf{H}$. In [7], Gallager proposed a specific method to construct a class of LDPC codes. In recent years, several systematic methods have been proposed to construct different classes of LDPC codes. More details in this regard can be obtained from [1,12,18,20]. An LDPC code of length n with k user of information bits is represented as a (n, k)-LDPC code.

An LDPC code constructed using Gallager's method will have a parity-check matrix $\mathbf{H}$, with uniform column and row weights. For a binary code, the weight of a column (row) of $\mathbf{H}$ is defined as the number of 1s in that column (row). A codeword ($\mathbf{c}$) of a (n, k)-LDPC code is a collection of bits $c_i, 1 \leq i \leq n$, such that $\mathbf{cH}^T = \mathbf{0}$. The set of constraints defined by the rows of $\mathbf{H}$ on the set of codewords of a (n, k)-LDPC code is called the parity-check constraints. The degree of a variable (check) node is equal to the weight of the corresponding column (row) in $\mathbf{H}$. A regular (d_v, d_c)-LDPC code is represented by a Tanner graph with a uniform variable node degree d_v, and a uniform check node degree d_c. Any LDPC code that is not regular is often referred to as an irregular LDPC code. The girth of a graph is defined as the length of the shortest cycle.

For example, we present the Tanner graph, see Figure 10.9, of a $(7, 3)$-LDPC code specified by the following parity-check matrix:

$$\mathbf{H} = \begin{bmatrix} 0 & 1 & 0 & 1 & 0 & 1 & 0 \\ 1 & 0 & 0 & 1 & 1 & 0 & 0 \\ 0 & 0 & 1 & 1 & 0 & 0 & 1 \\ 1 & 1 & 1 & 0 & 0 & 0 & 0 \\ 0 & 1 & 0 & 0 & 1 & 0 & 1 \\ 1 & 0 & 0 & 0 & 0 & 1 & 1 \\ 0 & 0 & 1 & 0 & 1 & 1 & 0 \end{bmatrix} \tag{10.8}$$

In Figure 10.9, circular nodes represent the set of variable nodes, and square nodes represent the set of check nodes. The Tanner graph has a uniform variable and check node degree of 3. It is easy to verify that the girth of this graph is 6.

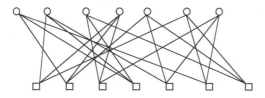

FIGURE 10.9   Tanner graph of a (7, 3)-LDPC code.

10.3.1   Gallager's Probabilistic Iterative Decoding Scheme

In [7], Gallager proposed a probabilistic iterative decoding scheme for LDPC codes that are transmitted over an arbitrary binary-input continuous output channel. In the following discussion, we assume that codewords of a (n, k)-LDPC codes are transmitted over an additive white Gaussian noise (AWGN) channel using antipodal signaling such that a code bit 0 is mapped to a symbol -1, and a code bit 1 is mapped to a symbol 1. Using the AWGN channel model, a transmitted codeword $\mathbf{x}$, is related to the channel observation $\mathbf{y}$, by

$$\mathbf{y} = (2^{*}\mathbf{x} - 1) + \mathbf{n} \tag{10.9}$$

In Equation 10.9, $\mathbf{n}$ is a vector of n independent Gaussian random variables with mean zero and variance σ^2. A Tanner graph of the (n, k)-LDPC code is a (d_v, d_c) regular graph, and it is assumed to be free of cycles. Since the Tanner graph is assumed to be free of cycles, the probabilistic decoding scheme calculates the *a posteriori* probability of each bit based on channel observations.

This iterative decoding scheme involves the technique of passing messages back and forth over the edges of a Tanner graph of the LDPC code [2]. The following derivation is an attempt to calculate the *a posteriori* probability of x_d, the dth bit of the vector $\mathbf{x}$. The *a posteriori* probability of x_d is defined as the probability of x_d being a 1 conditional on $\mathbf{y}$ and on an event S that will be later defined. Since the Tanner graph of the code is free of cycles, it can be represented as a tree rooted on x_d, see Figure 10.10. The tree structure shown in Figure 10.10 is a truncated Tanner graph of the code. In Figure 10.10, x_d in the first tier is connected to d_v check nodes in the second tier. Each parity-check in the second tier is connected to $(d_c - 1)$ variable nodes in the third tier. The event S is defined as an event that the variables in the first and third tier satisfy all the check constraints in the second tier. Using Bayes rule, the *a posteriori* probability of x_d can be written as:

$$\Pr[x_d = 1 \mid \{y\}, S] = \frac{\Pr[S \mid x_d = 1, \{y\}]}{\Pr[\{y\}, S]} \Pr[x_d = 1, \{y\}]$$

$$= \frac{\Pr[S \mid x_d = 1, \{y\}]}{\Pr[\{y\}, S]} \Pr[x_d = 1 \mid \{y\}] \Pr[\{y\}]$$

$$= \frac{\Pr[S \mid x_d = 1, \{y\}]}{\Pr[\{y\}, S]} p_d \Pr[\{y\}] \tag{10.10}$$

In a similar fashion, one can calculate the probability of x_d being a 0 conditional on the set of received symbols $\{y\}$, and on the event S.

$$\Pr[x_d = 0 \mid \{y\}, S] = \frac{\Pr[S \mid x_d = 0, \{y\}]}{\Pr[\{y\}, S]} (1 - p_d) \Pr[\{y\}] \tag{10.11}$$

From Equation 10.10 and Equation 10.11, it is clear that the computation of the probabilities can be simplified by working with a *log-likelihood ratio* (LLR). The log-likelihood ratio of x_d is defined as the

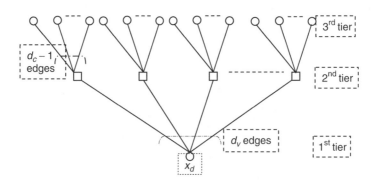

FIGURE 10.10   Truncated tree structure rooted on x_d.

natural logarithm of the ratio of the *a posteriori* probability of x_d being a 1 and x_d being a 0, and this can be written as

$$\ln\left(\frac{\Pr[_{-d} = 1 \,|\, \{y\}, S]}{\Pr[_{-d} = 0 \,|\, \{y\}, S]}\right) = \ln\left(\frac{\Pr[S \,|\, _{-d} = 1, \{y\}]}{\Pr[S \,|\, _{-d} = 0, \{y\}]}\frac{p_d}{(1 - p_d)}\right) \tag{10.12}$$

The ith check node in the second tier is connected to $(d_c - 1)$ variable nodes in the third tier. Let p_{il} be the probability that the lth variable node in the third tier and connected to the ith check node is a 1. Since these $(d_c - 1)$ variables are statistically independent, the probability that an even number of these variables are 1 is given by [7]

$$\frac{1 + \prod_{l=1}^{d_c-1}(1 - 2p_{il})}{2} \tag{10.13}$$

If even number of these variables is 1, then x_d has to be a 0 to satisfy the ith check constraint. Hence, the expression presented in Equation 10.13 can be interpreted as the probability of satisfying the ith check constraint conditional on x_d being a 0. By extending this analogy, the probability that all check constraints in the second tier are satisfied conditional on x_d being a 0 is given by

$$\Pr[S \,|\, x_d = 0, \{y\}] = \prod_{i=1}^{d_v}\frac{1 + \prod_{l=1}^{d_c-1}(1 - 2p_{il})}{2} \tag{10.14}$$

Using Equation 10.14, the expression for the LLR of x_d in Equation 10.12, can be modified as

$$\ln\left(\frac{\Pr[x_d = 1 \,|\, \{y\}, S]}{\Pr[x_d = 0 \,|\, \{y\}, S]}\right) = \ln\left(\frac{p_d}{1 - p_d}\right) + \ln\left(\frac{\Pr[S \,|\, x_d = 1, \{y\}]}{\Pr[S \,|\, x_d = 0, \{y\}]}\right)$$

$$= \ln\left(\frac{p_d}{1 - p_d}\right) + \ln\left(\frac{\prod_{i=1}^{d_v}\left(1 - \frac{1 + \prod_{l=1}^{d_c-1}(1-2p_{il})}{2}\right)}{\prod_{i=1}^{d_v}\frac{1 + \prod_{l=1}^{d_c-1}(1-2p_{il})}{2}}\right)$$

$$= \ln\left(\frac{p_d}{1 - p_d}\right) + \sum_{i=1}^{d_v}\ln\left(\frac{1 - \prod_{l=1}^{d_c-1}(1 - 2p_{il})}{1 - \prod_{l=1}^{d_c-1}(1 - 2p_{il})}\right) \tag{10.15}$$

In order to further simplify Equation 10.15, define a log-likelihood ratio λ_{il}, as

$$\lambda_{il} := \ln\left(\frac{p_{il}}{1 - p_{il}}\right)$$

$$\Rightarrow p_{il} = \frac{e^{\lambda_{il}}}{1 + e^{\lambda_{il}}} = \frac{1}{1 + e^{-\lambda_{il}}} \tag{10.16}$$

Using Equation 10.16, the log-term in Equation 10.15 can be written as

$$\ln\left(\frac{1 - \prod_{l=1}^{d_c-1}(1 - 2p_{il})}{1 + \prod_{l=1}^{d_c-1}(1 - 2p_{il})}\right) = \varphi = \ln\left(\frac{1 - \prod_{l=1}^{d_c-1}\left(1 - \frac{2}{1+e^{-\lambda_{il}}}\right)}{1 + \prod_{l=1}^{d_c-1}\left(1 - \frac{2}{1+e^{-\lambda_{il}}}\right)}\right)$$

$$= \ln\left(\frac{1 - \prod_{l=1}^{d_c-1}\left(\frac{1 - e^{\lambda_{il}}}{1 + e^{\lambda_{il}}}\right)}{1 + \prod_{l=1}^{d_c-1}\left(\frac{1 - e^{\lambda_{il}}}{1 + e^{\lambda_{il}}}\right)}\right)$$

$$= \ln\left(\frac{1 - \prod_{l=1}^{d_c-1}\tanh\left(-\frac{\lambda_{il}}{2}\right)}{1 + \prod_{l=1}^{d_c-1}\tanh\left(-\frac{\lambda_{il}}{2}\right)}\right) \tag{10.17}$$

Using $\tanh\left(\frac{-\varphi}{2}\right) = \frac{1-e^{\varphi}}{1+e^{\varphi}}$, and letting $\Pi = \prod_{l=1}^{d_c-1}\tanh\left(\frac{-\lambda_{il}}{2}\right)$, we get

$$\tanh\left(-\frac{\varphi}{2}\right) = \frac{1 - e^{\ln\left(\frac{1-\Pi}{1+\Pi}\right)}}{1 + e^{\ln\left(\frac{1-\Pi}{1+\Pi}\right)}} = \frac{1 - \left(\frac{1-\Pi}{1+\Pi}\right)}{1 + \left(\frac{1-\Pi}{1+\Pi}\right)} = \Pi = \prod_{l=1}^{d_c-1}\tanh\left(\frac{-\lambda_{il}}{2}\right)$$

$$\Rightarrow \varphi = -2\tanh^{-1}\left(\prod_{l=1}^{d_c-1}\tanh\left(\frac{-\lambda_{il}}{2}\right)\right) \tag{10.18}$$

The expression for φ in Equation 10.18, can be used in Equation 10.15 to obtain the following:

$$\ln\left(\frac{\Pr[x_d = 1 \mid \{y\}, S]}{\Pr[x_d = 0 \mid \{y\}, S]}\right) = \ln\left(\frac{p_d}{1 - p_d}\right) + \ln\left(\frac{\Pr[S \mid x_d = 1, \{y\}]}{\Pr[S \mid x_d = 0, \{y\}]}\right)$$

$$= \ln\left(\frac{p_d}{1 - p_d}\right) + \sum_{i=1}^{d_v}\ln\left(\frac{1 - \prod_{l=1}^{d_c-1}(1 - 2p_{il})}{1 - \prod_{l=1}^{d_c-1}(1 - 2p_{il})}\right)$$

$$= \ln\left(\frac{p_d}{1 - p_d}\right) - 2\sum_{i=1}^{d_v}\tanh^{-1}\left(\prod_{l=1}^{d_c-1}\tanh\left(\frac{-\lambda_{il}}{2}\right)\right) \tag{10.19}$$

The first term in Equation 10.19 represents the contribution from the dth channel observation, and is called the *intrinsic* information, while the second term represents the contribution from channel observations of other bits of the codeword, and is called the *extrinsic* information.

As mentioned above, p_{il} is the probability that the lth variable node in the third tier and connected to the ith check node is a 1. The quantity λ_{il} in Equation 10.16 is a log-likelihood ratio of p_{il} and $(1 - p_{il})$. In order to calculate λ_{il}, it will be helpful to add more tiers to the tree structure shown in Figure 10.10. Such an extended tree structure is shown in Figure 10.11.

In Figure 10.11, the lth variable node in the third tier is shaded black and it is connected to $d_v - 1$ check nodes in the fourth tier. Each check node in the fourth tier is connected $d_c - 1$ variable nodes in the fifth tier. The lth variable node in the third tier is connected to the ith check node in the second tier. The quantity p_{il} can be calculated based on the contributions of $d_v - 1$ check nodes in the fourth tier. Hence, p_{il} can be calculated by using Equation 10.14 but with a modification that the first product is taken over

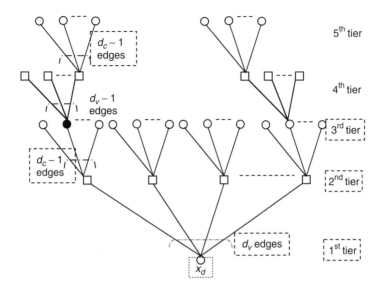

FIGURE 10.11 Extended tree structure rooted on x_d.

only $d_v - 1$ check nodes, see Equation 10.20.

$$p_{il} = \prod_{k=1}^{d_v-1} \frac{1 + \prod_{j=1}^{d_c-1} \left(1 - 2p_{jk}\right)}{2} \tag{10.20}$$

By induction, this iteration process can be used to find the log-likelihood ratio of x_d.

10.3.1.1 Min-Sign Approximation

The function φ, that involves the product of $d_c - 1$ terms in Equation 10.18, can alternatively be expressed as the sum of $d_c - 1$ terms. In hardware implementations, this alternate representation is preferred over the representation shown in Equation 10.18. This alternate representation can be obtained in the following manner:

$$\tanh\left(-\frac{\varphi}{2}\right) = \prod_{l=1}^{d_c-1} \tanh\left(-\frac{\lambda_{il}}{2}\right)$$

$$\text{sign}\left(-\varphi\right)\tanh\left(\left|\frac{\varphi}{2}\right|\right) = \prod_{l=1}^{d_c-1} \text{sign}\left(-\lambda_{il}\right) \prod_{l=1}^{d_c-1} \tanh\left(\left|\frac{\lambda_{il}}{2}\right|\right) \tag{10.21}$$

Since the function $\tanh(\beta)$ is nonnegative if the argument β is nonnegative, we can take logarithm of magnitude terms in Equation 10.21, and hence, we obtain

$$\ln\left(\tanh\left(\left|\frac{\varphi}{2}\right|\right)\right) = \ln\left(\prod_{l=1}^{d_c-1} \tanh\left(\left|\frac{\lambda_{il}}{2}\right|\right)\right)$$

$$= \sum_{l=1}^{d_c-1} \ln\left(\tanh\left(\left|\frac{\lambda_{il}}{2}\right|\right)\right) \tag{10.22}$$

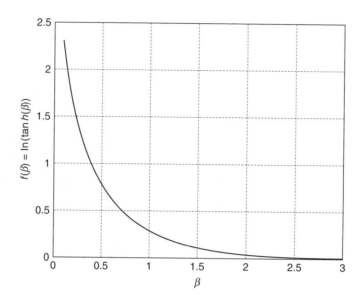

FIGURE 10.12   The function $f(\beta)$.

The function $f(\beta) = -\ln(\tanh(\beta)), \beta \geq 0$, is positive and monotonically decreasing for $\beta > 0$, see Figure 10.12. In Equation 10.23, we show that the function $f(\beta)$, is its own inverse.

$$f(f(\beta)) = -\ln\left(\tanh\left(\frac{f(\beta)}{2}\right)\right) = -\ln\left(\frac{1 - e^{-f(\beta)}}{1 + e^{-f(\beta)}}\right)$$

$$= -\ln\left(\frac{1 - e^{\ln\left(\tanh\left(\frac{\beta}{2}\right)\right)}}{1 + e^{\ln\left(\tanh\left(\frac{\beta}{2}\right)\right)}}\right) = -\ln\left(\frac{1 - \tanh\left(\frac{\beta}{2}\right)}{1 + \tanh\left(\frac{\beta}{2}\right)}\right)$$

$$= -\ln\left(\frac{1 - \left(\frac{1 - e^{-\beta}}{1 + e^{-\beta}}\right)}{1 + \left(\frac{1 - e^{-\beta}}{1 + e^{-\beta}}\right)}\right) = -\ln\left(e^{-\beta}\right)$$

$$= \beta \tag{10.23}$$

Using the result in Equation 10.22, Equation 10.23 can be simplified to

$$|\varphi| = 2f\left(\sum_{l=1}^{d_c-1} f\left(\left|\frac{\lambda_{il}}{2}\right|\right)\right) \tag{10.24}$$

Using the result in Equation 10.24, we can rewrite Equation 10.21 as

$$\varphi = -2\left(\prod_{l=1}^{d_c-1} \text{sign}\left(-\lambda_{il}\right)\right) f\left(\sum_{l=1}^{d_c-1} f\left(\left|\frac{\lambda_{il}}{2}\right|\right)\right) \tag{10.25}$$

As mentioned above, the alternate representation of Equation 10.18, as shown in Equation 10.25, requires multiplication operations to just calculate the sign of φ. Hence, Equation 10.25 is preferred over Equation 10.21 in hardware implementations. From Figure 10.12, it is obvious that the summation term, $\sum_{l=1}^{d_c-1} f\left(\left|\frac{\lambda_{il}}{2}\right|\right)$ in Equation 10.25, is dominated by the minimum of $|\lambda_{il}|, 1 \leq l \leq d_c - 1$. Hence

Equation 10.25 can be approximated as

$$\varphi \approx -2 \left(\prod_{l=1}^{d_c-1} \text{sign}\,(\lambda_{il}) \right) f(f(\min_{1\leq l\leq d_c-1}(|\lambda_{il}|)))$$

$$\approx -2 \left(\prod_{l=1}^{d_c-1} \text{sign}\,(\lambda_{il}) \right) \min_{1\leq l\leq d_c-1}(|\lambda_{il}|) \tag{10.26}$$

10.3.2 Message-Passing Algorithm for Soft-Decision Iterative Decoding

In previous sections, we derived an expression for the log-likelihood ratio of x_d based on the Tanner graph of the LDPC code. Also, a min-sign approximation of the log-likelihood ratio of x_d was derived. This procedure can be extended to iteratively decode all codeword bits based on channel observations. In order to understand this iterative procedure, the Tanner graph of the LDPC code is shown in Figure 10.13.

Each variable node is connected to d_v check nodes, and each check node is connected to d_c variable nodes. The variable nodes and check nodes are serially numbered from left to right. Let M_l be a set of indices of check nodes connected to the lth variable node, and N_m be a set of indices of variable nodes connected to the mth check node.

$$M_l := \{i \mid \mathbf{H}_{i,n} = 1, 1 \leq i \leq n-k\}, \quad 1 \leq l \leq n$$
$$N_m := \{j \mid \mathbf{H}_{m,j} = 1, 1 \leq j \leq n\}, \quad 1 \leq m \leq (n-k) \tag{10.27}$$

The iterative message-passing algorithm involves the process of sending messages, like log-likelihood messages, between variable and check nodes of the Tanner graph. Each variable node acts an independent processor that processes incoming messages from check nodes. Similarly, each check node processes incoming messages from variable nodes. A message sent over an edge from a variable node to a check node is generated based on messages sent by other check node neighbors of the variable node. Similarly, a message sent over an edge from a check node to a variable node is generated based on messages sent by other variable node neighbors of the check node. When this process is iterated for a finite number of times, log-likelihood ratios of variable nodes converges to a true *a posteriori* log-likelihood ratio if the Tanner graph is cycle-free. The cycle-free property of the graph preserves the independence of messages passed on its edges.

Let $\Lambda_{m,l}^p$ be the message sent from the mth check node to the lth variable node in the pth iteration of the message-passing algorithm. Also, let λ_l^p be the log-likelihood ratio of the lth variable in the pth iteration. The message-passing algorithm is summarized as

Initialization

- $\Lambda_{m,l}^0 = 0, \forall m$ and $\forall l \in N_m$
- λ_l^0 is the intrinsic log-likelihood ratio of the lth variable

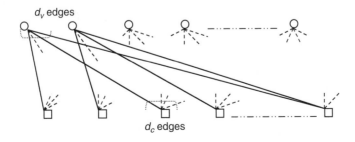

FIGURE 10.13 Tanner graph of the (n, k)-LDPC code.

Iterations

- Check-node update: $\Lambda_{m,l}^{p} = -2\tanh^{-1}\left(\prod_{N_m - l} \tanh\left(\dfrac{-\lambda_i^{p-1}}{2}\right)\right)$

- Bit-node update: $\lambda_l^{p} = \lambda_l^{0} + \sum_{m \in M_l} \Lambda_{m,l}^{p-1}$

For a detailed reading on mathematical characterization of the message-passing algorithm, refer [15].

10.4   General Message-Passing Algorithm

It is clear from the above discussion that each check node of a Tanner graph represents a linear function of variable nodes in GF(2). The global function represented by a Tanner graph is a product of local functions corresponding to the check nodes. Hence, the Tanner graph of a code is a representation of the global function and its factors. The process of sending messages iteratively over the edges of the Tanner graph in order to decode is equivalent to computing the marginal function associated with each variable node. Kschischang et al. [9] introduced this notion, among others, in the paper on factor graphs and the sum-product algorithm. In this paper, the Tanner graph model of a linear code was generalized with an introduction of a *factor graph* model. Unlike the Tanner graph model, *factor nodes* (similar to check nodes of the Tanner graph) of a factor graph represent arbitrary functions of the neighboring variable nodes. The message-passing algorithm on the Tanner graph is a special instance of the *sum-product algorithm* on the factor graph. It is important to note that the applications of these concepts are far reaching and extensive. The sum-product algorithm applied on an appropriately chosen factor graph models a wide variety of seemingly different algorithms developed in computer science and engineering. The following example intends to present such an instance.

Let X_1, X_2, and X_3 be discrete random variables that take values from a sample space $\Phi = \{1, 2, 3, \ldots, M\}$, and their joint probability function $P_{X_1, X_2, X_3}(x_1, x_2, x_3)$ is defined as:

$$P_{X_1 X_2 X_3}(x_1, x_2, x_3) = \Pr[X_1 = x_1, X_2 = x_2, X_3 = x_3] \tag{10.28}$$

A straightforward computation of the marginal probability of X_3 as shown in Equation 10.29 requires M^3 computations.

$$P_{X_3}(x_3) = \sum_{x_1, x_2} P_{X_1, X_2, X_3}(x_1, x_2, x_3) \tag{10.29}$$

To make the discussion interesting, assume that random variables X_1, X_2, and X_3 are random variables at consecutive time instances of a Markov process X. This assumption leads to the following factorization of the joint probability function:

$$P_{X_1, X_2, X_3}(x_1, x_2, x_3) = P_{X_1}(x_1) P_{X_2/X_1}(x_2/x_1) P_{X_3/X_2}(x_3/x_2) \tag{10.30}$$

Similarly, the marginal probability function of X_3 can be modified as

$$P_{X_3}(x_3) = \sum_{x_2} P_{X_3/X_2}(x_3/x_2) \sum_{x_1} P_{X_1}(x_1) P_{X_2/X_1}(x_2/x_1) \tag{10.31}$$

Assuming that the values of inner sums for any given x_2 can be stored, the number of computations required to calculate Equation 10.31 reduces to M^2. Thus the computation complexity of the marginal function is reduced by a factor of M. Note that the inner summation of Equation 10.31 does not depend on x_3. Hence, the inner summation computed for a given x_2 is reused in calculating $P_{X_3}(x_3)$ for different values of x_3. Thus the sum-product algorithm gives an efficient method to calculate marginal functions without recalculating such intermediate values. The above argument can be extended for an arbitrary number of random variables.

In this section, we aim to introduce the reader to the conceptual and mathematical understanding of factor graphs and the sum-product algorithm. For a detailed reading, refer [9].

10.4.1 Computation of a Marginal Function

As mentioned before, a factor graph is a convenient graphical representation of the factors of any global function. Assume that a global function f, of variables $x_1, x_2, \ldots, x_n$, factors into a product of k local functions as shown in Equation 10.32.

$$
\begin{aligned}
f(x_1, x_2, x_3, \ldots, x_n) &= \prod_{i=1}^{k} f_i(y_1, y_2, y_3, \ldots, y_{|F_i|} \mid y_j \in F_i, j = 1, \ldots, |F_i|) \\
&= \prod_{i=1}^{k} f_i(y \mid y \in F_i)
\end{aligned}
\tag{10.32}
$$

In Equation 10.32, a set of arguments of a function f_i is given by F_i, a subset of variables $x_1, x_2, \ldots, x_n$. A *factor graph* of the function f is defined as a bipartite graph that consists of a *variable node* for each x_i, a *factor node* for each f_i, and an edge connecting x_i and f_i iff x_i is an argument of f_i. For example, the factor graph of the global function in Equation 10.30 is shown in Figure 10.14.

In Figure 10.14, circular nodes represent variables, and square nodes represent functions. Also in this example, the factor graph of the global function is free of cycles or closed paths. A connected graph that is free of cycles is called a *tree*. The factor graph of the function f is shown in Figure 10.15.

A generalized message-passing algorithm, called the *sum-product algorithm*, applied on the factor graph of any global function f, is an efficient method to solve for its marginal functions.

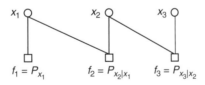

FIGURE 10.14   Factor graph: an example.

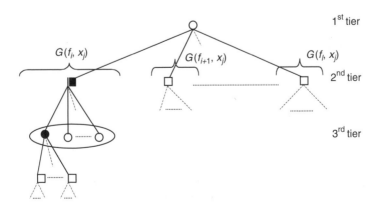

FIGURE 10.15   Factor graph of function f.

Assume that we are interested in computing the marginal function of a variable x_j. A straightforward method to calculate the marginal function $g_{x_j}(x_j)$ is

$$g_{x_j}(x_j) = \sum_{\sim x_j} f(x_1, x_2, x_3, \ldots, x_n) = \sum_{\sim x_j} \prod_i f_i(y \mid y \in F_i) \qquad (10.33)$$

The notation $\sum_{\sim x_j}$ means that the summation is over all variables except x_j. Also, assume that the factor graph of f, shown in Figure 10.15, is a tree. Let $E(x_j)$ be a set of edges (x_j, f_i), incident on x_j. Since the factor graph is a tree, the variable node x_j can be visualized as being connected to a subtree $G(f_i, x_j)$ by an edge (x_j, f_i). The subtree $G(f_i, x_j)$ contains the factor node f_i and all other nodes, except x_j, that are connected to it. Using the above notation, Equation 10.33 can be rewritten as

$$g_{x_j}(x_j) = \sum_{\sim x_j} \left(\prod_{(x_j, f_i) \in E(x_j)} f_i(x_j, y \mid y \in F_i) \prod_{f_m \in \{G(f_i, x_j) - f_i\}} f_m(y \mid y \in F_m) \right) \qquad (10.34)$$

In Equation 10.34, all factor nodes contained in a subtree $G(f_i, x_j)$ are grouped under the outer product for a particular selection of (x_j, f_i). Since the outer product depends only on x_j and the factor functions connected to x_j, the order of summation and outer product in Equation 10.34 can be reversed.

$$
\begin{aligned}
g_{x_j}(x_j) &= \prod_{(x_j, f_i) \in E(x_j)} \sum_{\sim x_j} \left(f_i(x_j, y \mid y \in F_i) \prod_{f_m \in \{G(f_i, x_j) - f_i\}} f_m(y \mid y \in F_m) \right) \\
&= \prod_{(x_j, f_i) \in E(x_j)} g_{x_j, f_i}(x_j) \qquad (10.35)
\end{aligned}
$$

Each factor f_m under the inner product of Equation 10.35 is not a function of the variable x_j because of the underlying assumption that the factor graph is a tree. But, each factor f_m is functionally dependent on one of the variables of f_i. Hence, the summation in $g_{x_j, f_i}(x_j)$ can be split in the following manner:

$$g_{x_j, f_i}(x_j) = \sum_{\sim x_j} \left(f_i(x_j, y \mid y \in F_i) \sum_{\sim y \in F_i} \left(\prod_{f_m \in \{G(f_i, x_j) - f_i\}} f_m(y \mid y \in F_m) \right) \right) \qquad (10.36)$$

Each factor under the inner product of Equation 10.36 has a corresponding factor node contained in one of the subtrees connected to an edge incident on f_i. Each subtree $G(x_p, f_i)$, is connected to f_i by an edge $(x_p \neq x_j, f_i)$. In the fashion similar to that used in arriving at Equation 10.34, the factors under the inner product of Equation 10.36 can be partitioned based on their association with one of the subtrees $G(x_p, f_i)$.

$$
\begin{aligned}
g_{x_j, f_i}(x_j) &= \sum_{\sim x_j} \left(f_i(x_j, y \mid y \in F_i) \sum_{\sim y \in F_i} \left(\prod_{(x_p, f_i) \in \{E(f_i) - x_j\}} \prod_{f_m \in G(x_p, f_i)} f_m(y \mid y \in F_n) \right) \right) \\
&= \sum_{\sim x_j} \left(f_i(x_j, y \mid y \in F_i) \prod_{(x_p, f_i) \in \{E(f_i) - x_j\}} \left(\sum_{\sim y \in F_i} \prod_{f_m \in G(x_p, f_i)} f_m(y \mid y \in F_n) \right) \right) \\
&= \sum_{\sim x_j} \left(f_i(x_j, y \mid y \in F_i) \prod_{(x_p, f_i) \in \{E(f_i) - x_j\}} g_{x_p, \sim f_i} \right) \qquad (10.37)
\end{aligned}
$$

The formulation of $g_{x_p, \sim f_i}$ in Equation 10.37 looks similar to that of $g_{x_j}(x_j)$ in Equation 10.33. Hence, a procedure similar to the one used above can be followed to evaluate $g_{x_p, \sim f_i}$. This procedure is continued until the variable or factor leaf nodes of the subtrees are reached. The quantity $g_{x_p, \sim f_i}$ is set to 1 if the variable node x_p is a leaf node. The quantity $g_{x_j, f_i}(x_j)$ is set to $f_i(x_j)$ if the factor node f_i is a leaf node. Thus a recursive method with an initial condition has been devised to evaluate the marginal function $g_{x_j}(x_j)$.

10.4.2 Message-Passing Schedule to Compute a Marginal Function

A general message-passing algorithm involves the process of sending messages over edges of a graph. It can be shown that the recursive method, designed to evaluate a marginal function, is a message-passing algorithm. This method starts from the leaf nodes in the kth tier of the factor graph sending information (or messages) to the nodes in the $(k-1)$th tier. Thus the marginal function can be computed by continuing to send messages until the first tier is reached.

Let μ_{f_i, x_j} represent the message sent by a factor node f_i to a variable node x_j, and μ_{x_j, f_i} represent the message sent by a variable node x_j to a factor node f_i. A general form of μ_{x_j, f_i} can be obtained by analyzing Equation 10.35. We know that the marginal function $g_{x_j}(x_j)$ is the product of all messages obtained from the second tier. Similarly, a message μ_{x_j, f_i}, generated by a variable node x_j in the kth tier, is equal to the product of messages obtained from the neighboring factor nodes of the $(k-1)$th tier. The quantity $g_{x_j, f_i}(x_j)$ in Equation 10.37 is equal to the message μ_{f_i, x_j} generated by a factor node f_i. The message-passing algorithm on a factor graph can be summarized as

Initialization

- $\mu_{x_p, f_m} = 1 \quad \forall$ leaf node x_p
- $\mu_{f_m, x_p} = f_m(x_p) \quad \forall$ leaf node f_m

Message-passing Algorithm

- $\mu_{x_k, f_l} = \displaystyle\prod_{\{f_m \neq f_l | (x_k, f_m) \in E(x_k)\}} \mu_{f_m, x_k}$
- $\mu_{f_l, x_k} = \displaystyle\sum_{\sim x_k} f_l(x_k, y | y \in F_l - x_k) \prod_{x_p \in F_l, x_p \neq x_k} \mu_{x_p, f_l}$

Marginal function of x_j

- $g_{x_j}(x_j) = \displaystyle\prod_{\{f_m | (x_j, f_m) \in E(x_j)\}} \mu_{f_m, x_j}$

Assume that the message-passing schedule in a factor graph is synchronized with an internal clock. In a clock interval, all variable nodes in the kth tier generate and send messages to factor nodes in the $(k-1)$th tier. In the next clock interval, all factor nodes in the $(k-1)$th tier generate and send messages to variable nodes in the $(k-2)$th tier.

A straightforward approach to calculating the marginal functions of n variables is to repeat the message-passing algorithm n times such that a different variable node forms the root of the factor graph in each repetition. On the other hand, an efficient approach involves a modification to the message-passing algorithm in order to calculate the marginal functions of all variables simultaneously. The efficiency of the approach stems from the fact that several messages generated while evaluating the marginal function of a variable can be reused in evaluating the marginal function of another variable.

Acknowledgement

This work is supported by the NSF under grant CCR-0208597.

References

[1] Assmus, E.F. Jr. and Key, J.D., *Design and their Codes*, Cambridge University Press, The Pitt Building, Trumpington St., Cambridge CB2IRP, 1992.
[2] Barry, J.R., Low-density parity-check codes. [Online]. Available: http://users.ece.gatech.edu/~barry/

[3] Berrou, G., Glavieux, A., and Thitimajshima, P., Near Shannon limit error-correcting coding and decoding: Turbo-codes, in *Proc. IEEE Int. Conf. Commn. (ICC'93)*, Geneva, Switzerland, 2.1064–2.1070, May 1993.

[4] Bahl, L.R., Cocke, J., Jelinek, F., and Raviv, J., Optimal decoding of linear codes for minimizing symbol error rate, *IEEE Trans. Inform. Theory*, IT-20, 284–287, 1974.

[5] Di, C., Proietti, D., Telatar, I.E., Richardson, T.J., and Urbanke, R.L., Finite-length analysis of low-density parity-check codes on the binary erasure channel, *IEEE Trans. Inform. Theory*, 48, 6, 1570–1579, June 2002.

[6] Frey, B.J., *Graphical Models for Machine Learning and Digital Communication*. MIT Press, Cambridge, MA, 1998.

[7] Gallager, R.G., *Low-Density Parity-Check Codes*. MIT Press, Cambridge, MA, 1963.

[8] Kschischang, F.R. and Frey, B.J., Iterative decoding of compound codes by probability propagation in graphical models, *IEEE J. Selected Areas Commn.*, 16, 219–230, February 1998.

[9] Kschischang, F.R., Frey, B.J., and Loeliger, H.-A., Factor graphs and the sum-product algorithm, *IEEE Trans. Inform. Theory*, 47, 2, 498–519, February 2001.

[10] Luby, M., Mitzenmacher, M., Shokrollahi, A., and Spielman, D., Analysis of low-density codes and improved designs using irregular graphs, *Proc. ACM Symp. Theory Comput.*, 249–258, 1998.

[11] MacKay D.J.C. and Neal, R.M., Good codes based on very sparse matrices, in *Cryptography and Coding, 5th IMA Conference, in Lecture Notes in Computer Science*, C. Boyd, Ed., 1995, 1025, 110–111.

[12] MacKay, D.J.C., Good error-correcting codes based on very sparse matrices, *IEEE Trans. Inform. Theory*, 45, 399–431, March 1999.

[13] McEliece, R.J., MacKay, D.J.C., and Cheng, J.-F., Turbo decoding as an instance of Pearl's 'Belief Propagation' algorithm, *IEEE J. Select. Areas Commn.*, 16, 140–152, February 1998.

[14] Pearl, J., *Probabilistic Reasoning in Intelligent Systems: Networks of Plausible Inference*. San Mateo, CA: Morgan Kaufmann, 1988.

[15] Richardson, T. and Urbanke, R., An introduction to the analysis of iterative coding systems, *Codes, Systems and Graphical Models*, IMA Volume in Mathematics and its Application, 1–37, Springer, 2001.

[16] Richardson, T. and Urbanke, R., The capacity of low-density parity-check codes under message-passing decoding, *IEEE Trans. Inform. Theory*, 47, 599–618, February 2001.

[17] Tanner, R.M., A recursive approach to low complexity codes, *IEEE Trans. Inform. Theory*, IT-27, 533–547, September 1981.

[18] Vasic B., and Milenkovic, O., "Combinatorial constructions of lowdensity parity-check codes for iterative decoding," *accepted for publication in IEEE Trans. on Inform. Theory*, 2004.

[19] Wiberg, N., Loeliger, H.-A., and Kötter, R., Codes and iterative decoding on general graphs, *Euro. Trans. Telecommun.*, 6, 513–525, September/October 1995.

[20] Kou, Yu., Lin, S., and Fossorier, M.P.C., Low-density parity-check codes based on finite geometries: a rediscovery and new results, *IEEE Trans. Inform. Theory*, 47:7, 2711–2736, November 2001.

11

Modulation Codes for Storage Systems

11.1	Introduction	**11**-1
11.2	Constrained Systems and Codes	**11**-2
11.3	Constraints for ISI Channels.......................	**11**-4
	Requirements • Definitions	
11.4	Channels with Colored Noise and Intertrack Interference......................................	**11**-6
11.5	An Example.......................................	**11**-7
11.6	Future Directions	**11**-9
	Soft-Output Decoding of Modulation Codes • Reversed Concatenation	

Brian Marcus
University of British Columbia
Vancouver, BC

Emina Soljanin
Lucent Technologies
Murray Hill, NJ

11.1 Introduction

Modulation codes are used to constrain the individual sequences that are recorded in data storage channels, such as magnetic or optical disk or tape drives. The constraints are imposed in order to improve the detection capabilities of the system. Perhaps the most widely known constraints are the runlength limited ($\mathrm{RLL}(d,k)$) constraints, in which 1s are required to be separated by at least d and no more than k 0s. Such constraints are useful in data recording channels that employ peak detection: waveform peaks, corresponding to data ones, are detected independently of one another. The d-constraint helps to increase linear density while mitigating intersymbol interference, and the k-constraint helps to provide feedback for timing and gain control.

Peak detection was widely used until the early 1990s. While it is still used today in some magnetic tape drives and some optical recording devices, most high density magnetic disk drives now use a form of maximum likelihood (Viterbi) sequence detection. The data recording channel is modeled as a linear, discrete-time, communications channel with intersymbol interference (ISI), described by its transfer function and white Gaussian noise. The transfer function is often given by $h(D) = (1 - D)(1 + D)^N$, where N depends on and increases with the linear recording density.

Broadly speaking, two classes of constraints are of interest in today's high density recording channels: (1) constraints for improving timing and gain control and simplifying the design of the Viterbi detector for the channel, and (2) constraints for improving noise immunity. Some constraints serve both purposes.

Constraints in the first class usually take the form of a PRML (G, I) constraint: the maximum run of 0s is G and the maximum run of 0s, within each of the two substrings defined by the even indices and odd indices, is I. The G-constraint plays the same role as the k-constraint in peak detection, while the I-constraint enables the Viterbi detector to work well within practical limits of memory.

0-8493-1524-7/05/$0.00+$1.50
© 2005 by CRC Press, LLC

Constraints in the second class eliminate some of the possible recorded sequences in order to increase the minimum distance between those that remain or eliminate the possibility of certain dominant error events. This general goal does not specify how the constraints should be defined, but many such constraints have been constructed [20] and the references therein for a variety of examples. Bounds on the capacities of constraints that avoid a given set of error events have been given in [26].

Until recently, the only known constraints of this type were the matched-spectral-null (MSN) constraints. They describe sequences whose spectral nulls match those of the channel and therefore increase its minimum distance. For example, a set of DC-balanced sequences (i.e., sequences of ± 1 whose accumulated digital sums are bounded) is an MSN constraint for the channel with transfer function $h(D) = 1 - D$, which doubles its minimum distance [18].

During the past few years, significant progress has been made in defining high capacity distance enhancing constraints for high density magnetic recording channels. One of the earliest examples of such a constraint is the maximum transition run (MTR) constraint [28], which constrains the maximum run of 1s. We explain the main idea behind this type of distance-enhancing codes in Section 11.3.

Another approach to eliminating problematic error events is that of parity coding. Here, a few bits of parity are appended to (or inserted in) each block of some large size, typically 100 bits. For some of the most common error events, any single occurrence in each block can be eliminated. In this way, a more limited immunity against noise can be achieved with less coding overhead [5].

Coding for more realistic recording channel models that include colored noise and intertrack interference are discussed in Section 11.4. We point out that different constraints which avoid the same prescribed set of differences may have different performance on more realistic channels. This makes some of them more attractive for implementation.

For a more complete introduction to this subject, we refer the reader to any one of the many expository treatments, such as [16], [17], or [24].

11.2 Constrained Systems and Codes

Modulation codes used in almost all contemporary storage products belong to the class of constrained codes. These codes encode random input sequences to sequences that obey the constraint of a labeled directed graph with a finite number of states and edges. The set of corresponding constrained sequences is obtained by reading the labels of paths through the graph. Sets of such sequences are called constrained systems or constraints. Figure 11.1 and Figure 11.2 depict graph representations of an RLL constraint and a DC-balanced constraint.

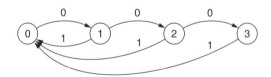

FIGURE 11.1   RLL $(1, 3)$ constraint.

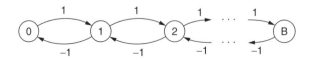

FIGURE 11.2   DC-balanced constraint.

Of special interest are those constraints that do not contain (globally or at certain positions) a finite number of finite length strings. These systems are called systems of finite type (FT). An FT system X over alphabet $\mathcal{A}$ can always be characterized by a finite list of forbidden strings $\mathcal{F} = \{w_1, \ldots, w_N\}$ of symbols in $\mathcal{A}$. Defined this way, FT systems will be denoted by $X_{\mathcal{F}}^{\mathcal{A}}$. The RLL constraints form a prominent class of FT constraints, while DC-balanced constraints are typically not FT.

Design of constrained codes begins with identifying constraints, such as those described in Section 11.1, that achieve certain objectives. Once the system of constrained sequences is specified, information bits are translated into sequences that obey the constraints via an *encoder*, which usually has the form of a finite-state machine. The actual set of sequences produced by the encoder is called a constrained code and is often denoted $\mathcal{C}$. A *decoder* recovers user sequences from constrained sequences. While the decoder is also implemented as a finite-state machine, it is usually required to have a stronger property, called sliding-block decodablility, which controls error propagation [24].

The maximum rate of a constrained code is determined by *Shannon capacity*. The Shannon capacity or simply *capacity* of a constrained system, denoted by C, is defined as

$$C = \lim_{n \to \infty} \frac{\log_2 N(n)}{n}$$

where $N(n)$ is the number of sequences of length n. The capacity of a constrained system represented by a graph G can be easily computed from the *adjacency matrix* (or *state transition matrix*) of G (provided that the labeling of G satisfies some mildly innocent properties). The adjacency matrix of G with r states and a_{ij} edges from state i to state j, $1 \le i, j \le r$, is the $r \times r$ matrix $A = A(G) = \{a_{ij}\}_{r \times r}$. The Shannon capacity of the constraint is given by

$$C = \log_2 \lambda(A)$$

where $\lambda(A)$ is the largest real eigenvalue of A.

The *state-splitting algorithm* [1] (see also [24]) gives a general procedure for constructing constrained codes at any rate up to capacity. In this algorithm, one starts with a graph representation of the desired constraint and then transforms it into an encoder via various graph-theoretic operations including splitting and merging of states. Given a desired constraint and a desired rate $p/q \le C$, one or more rounds of state splitting are performed; the determination of which states to split and how to split them is governed by an approximate eigenvector, that is, a vector $\mathbf{x}$ satisfying $A^q x \ge 2^p x$.

There are many other very important and interesting approaches to constrained code construction — far too many to mention here. One approach combines state-splitting with look-ahead encoding to obtain a very powerful technique which yields competent codes [14]. Another approach involves variable-length and time-varying variations of these techniques [2, 13]. Many other effective coding constructions are described in the monograph [17].

For high capacity constraints, graph transforming techniques, such as the state-splitting algorithm, may result in encoder/decoder architectures with formidable complexity. Fortunately, a block encoder/decoder architecture with acceptable implementation complexity for many constraints can be designed by well-known enumerative [6], and other combinatorial [32] as well as heuristic techniques [25].

Translation of constrained sequences into the channel sequences depends on the modulation method. Saturation recording of binary information on magnetic medium is accomplished by converting an input stream of data into a spatial stream of bit cells along a track where each cell is fully magnetized in one of two possible directions, denoted by 0 and 1. There are two important modulation methods commonly used on magnetic recording channels: *non-return-to-zero* (NRZ) and *modified non-return-to-zero* (NRZI). In NRZ modulation, the binary digits 0 and 1 in the input data stream correspond to 0 and 1 directions of cell magnetizations, respectively. In NRZI modulation, the binary digit 1 corresponds to a magnetic transition between two bit cells, and the binary digit 0 corresponds to no transition. For example, the channel constraint which forbids transitions in two neighboring bit-cells, can be accomplished by either $\mathcal{F} = \{11\}$ NRZI constraint or $\mathcal{F} = \{101, 010\}$ NRZ constraint. The graph representation of these two constraints is shown in Figure 11.3. The NRZI representation is in this case simpler.

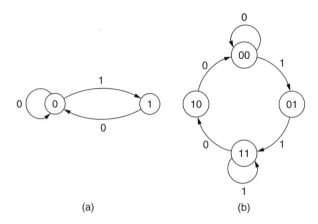

FIGURE 11.3   Two equivalent constraints: (a) $\mathcal{F} = \{11\}$ NRZI, and (b) $\mathcal{F} = \{101, 010\}$ NRZ.

11.3   Constraints for ISI Channels

We discuss a class of codes known as *codes which avoid specified differences*. This is the only class of distance enhancing codes used in commercial magnetic recording systems. There are two main reasons for this: these codes simplify the channel detectors relative to the uncoded channel and even high rate codes in this class can be realized by low complexity encoders and decoders.

11.3.1   Requirements

A number of papers have proposed using constrained codes to provide coding gain on channels with high ISI (see [4, 10, 20, 28]). The main idea of this approach can be described as follows [20]. Consider a discrete-time model for the magnetic recording channel with possibly constrained input $a = \{a_n\} \in \mathcal{C} \subseteq \{-1, 1\}^\infty$, impulse response $\{h_n\}$, and output $y = \{y_n\}$ given by

$$y_n = \sum_m a_m h_{n-m} + \eta_n \tag{11.1}$$

where $h(D) = \sum_n h_n D^n = (1 - D)(1 + D)^3$ (E^2PR4) or $h(D) = \sum_n h_n = (1 - D)(1 + D)^4$ (E^3PR4), η_n are independent Gaussian random variables with zero mean and variance σ^2. The quantity $1/\sigma^2$ is referred to as the signal-to-noise ratio (SNR). The minimum distance of the uncoded channel (Equation 11.1) is

$$d_{\min}^2 = \min_{\epsilon(D) \neq 0} \|h(D)\epsilon(D)\|^2$$

where $\epsilon(D) = \sum_{i=0}^{l-1} \epsilon_i D^i$, $(\epsilon_i \in \{-1, 0, 1\}, \epsilon_0 = 1, \epsilon_{l-1} \neq 0)$ is the polynomial corresponding to a normalized input error sequence $\epsilon = \{\epsilon_i\}_{i=0}^{l-1}$ of length l, and the squared norm of a polynomial is defined as the sum of its squared coefficients. The minimum distance is bounded from above by $\|h(D)\|^2$, denoted by

$$d_{\text{MFB}}^2 = \|h(D)\|^2 \tag{11.2}$$

This bound is known as the *matched-filter bound* (MFB), and is achieved when the error sequence of length $l = 1$, that is, $\epsilon(D) = 1$, is in the set

$$\arg\min_{\epsilon(D) \neq 0} \|h(D)\epsilon(D)\|^2 \tag{11.3}$$

For channels that fail to achieve the MFB, that is, for which $d_{\min}^2 < \|h(D)\|^2$, any error sequences $\epsilon(D)$ for which

$$d_{\min}^2 \leq \|h(D)\epsilon(D)\|^2 < \|h(D)\|^2 \tag{11.4}$$

are of length $l \geq 2$ and may belong to a constrained system $X_{\mathcal{L}}^{\{-1,0,1\}}$, where $\mathcal{L}$ is an appropriately chosen finite list of forbidden strings.

For code $\mathcal{C}$, we write the set of all admissible nonzero error sequences as

$$\mathcal{E}(\mathcal{C}) = \{\epsilon \in \{-1,0,1\}^\infty \mid$$
$$\epsilon \neq \mathbf{0}, \epsilon = (\mathbf{a} - \mathbf{b})/2, \ \mathbf{a}, \mathbf{b} \in \mathcal{C}\}$$

Given the condition $\mathcal{E}(\mathcal{C}) \subseteq X_{\mathcal{L}}^{\{-1,0,1\}}$, we seek to identify the least restrictive finite collection $\mathcal{F}$ of blocks over the alphabet $\{0, 1\}$ so that

$$\mathcal{C} \subseteq X_{\mathcal{F}}^{\{0,1\}} \implies \mathcal{E}(\mathcal{C}) \subseteq X_{\mathcal{L}}^{\{-1,0,1\}} \tag{11.5}$$

11.3.2 Definitions

A constrained code is defined by specifying $\mathcal{F}$, the list of forbidden strings for code sequences. Prior to that one needs to first characterize error sequences that satisfy Equation 11.4 and then specify $\mathcal{L}$, the list of forbidden strings for error sequences. Error event characterization can be done by using any of the methods described by Karabed et al. [20]. Specification of $\mathcal{L}$ is usually straightforward.

A natural way to construct a collection $\mathcal{F}$ of blocks forbidden in code sequences based on the collection $\mathcal{L}$ of blocks forbidden in error sequences is the following. From the above definition of error sequences $\epsilon = \{\epsilon_i\}$ we see that $\epsilon_i = 1$ requires $a_i = 1$ and $\epsilon_i = -1$ requires $a_i = 0$, that is, $a_i = (1 + \epsilon_i)/2$. For each block $\mathbf{w}_{\mathcal{E}} \in \mathcal{L}$, construct a list $\mathcal{F}_{\mathbf{w}_{\mathcal{E}}}$ of blocks of the same length l according to the rule:

$$\mathcal{F}_{\mathbf{w}_{\mathcal{E}}} = \{\mathbf{w}_{\mathcal{C}} \in \{-1,1\}^l \mid$$
$$w_{\mathcal{C}}^i = (1 + w_{\mathcal{E}}^i)/2 \text{ for all } i \text{ for which } w_{\mathcal{E}}^i \neq 0\}$$

Then the collection $\mathcal{F}$ obtained as $\mathcal{F} = \cup_{\mathbf{w}_{\mathcal{E}} \in \mathcal{L}} \mathcal{F}_{\mathbf{w}_{\mathcal{E}}}$ satisfies requirement (Equation 11.5). However, the constrained system $X_{\mathcal{F}}^{\{0,1\}}$ obtained this way may not be the most efficient. (Bounds on the achievable rates of codes which avoid specified differences were found recently in [26].)

We illustrate the above ideas on the example of the E^{2}PR4 channel. Its transfer function is $h(D) = (1 - D)(1 + D)^3$, and its MFB is $\|(1 - D)(1 + D)^3 \cdot 1\|^2 = 10$. The error polynomial $\epsilon(D) = 1 - D + D^2$ is the unique error polynomial for which $\|(1 - D)(1 + D)^3\epsilon(D)\|^2 = 6$, and the error polynomials $\epsilon(D) = 1 - D + D^2 + D^5 - D^6 + D^7$ and $\epsilon(D) = \sum_{i=0}^{l-1}(-1)^i D^i$ for $l \geq 4$ are the only polynomials for which $\|(1 - D)(1 + D)^3\epsilon(D)\|^2 = 8$ (see [20]).

It is easy to show that these error events are not in the constrained error set defined by the list of forbidden error strings $\mathcal{L} = \{+-+00, +-+-\}$, where $+$ denotes 1 and $-$ denotes -1. To see that, note that an error sequence that does not contain the string $+-+00$ cannot have error polynomials $\epsilon(D) = 1 - D + D^2$ or $\epsilon(D) = 1 - D + D^2 + D^5 - D^6 + D^7$, while an error sequence that does not contain string $+-+-$ cannot have an error polynomial of the form $\epsilon(D) = \sum_{i=0}^{l-1}(-1)^i D^i$ for $l \geq 4$. Therefore, by the above procedure of defining the list of forbidden code strings, we obtain the $\mathcal{F} = \{+-+\}$ NRZ constraint. Its capacity is about 0.81, and a rate 4/5 code into the constraint was first given in [19].

In [20], the following approach was used to obtain several higher rate constraints. For each of error strings in $\mathcal{L}$, we write all pairs of channel strings whose difference is the error string. To define $\mathcal{F}$, we look for the longest string(s) appearing in at least one of the strings in each channel pair. For the example above and the $+-+00$ error string, a case-by-case analysis of channel pairs is depicted in Figure 11.4. We can distinguish two types (denoted by A and B in the figure) of pairs of code sequences involved in forming an error event. In a pair of type A, at least one of the sequences has a transition run of length 4. In a pair of type B, both sequences have transition runs of length 3, but for one of them the run starts at an even

		A			B			A			A	
a:	0	1 0 1	0 0	1	1 0 1	0 0	0	1 0 1	0 1	1	1 0 1	0 1
b:	0	0 1 0	0 0	1	0 1 0	0 0	0	0 1 0	0 1	1	0 1 0	0 1

		A			A			B			A	
a:	0	1 0 1	1 0	1	1 0 1	1 0	0	1 0 1	1 1	1	1 0 1	1 1
b:	0	0 1 0	1 0	1	0 1 0	1 0	0	0 1 0	1 1	1	0 1 0	1 1

FIGURE 11.4 Possible pairs of sequences for which error event $+-+00$ may occur.

position and for the other at an odd position. This implies that an NRZI constrained system that limits the run of 1s to 3 when it starts at an odd position, and to 2 when it starts at an even position, eliminates all possibilities shown bold-faced in Figure 11.4. In addition, this constraint eliminates all error sequences containing the string $+-+-$. The capacity of the constraint is about 0.916, and rate 8/9 block codes with this constraint has been implemented in several commercial read channel chips. More about the constraint and the codes can be found in [4, 10, 20, 28].

11.4 Channels with Colored Noise and Intertrack Interference

Magnetic recording systems always operate in the presence of colored noise intertrack interference, and data dependent noise. Codes for these more realistic channel models are studied in [27]. Below, we briefly outline the problem.

Data recording and retrieval process is usually modeled as a linear, continuous-time, communications channel described by its Lorentzian step response and additive white Gaussian noise. The most common discrete-time channel model is given by Equation 11.1. Magnetic recording systems employ channel equalization to the most closely matching transfer function $h(D) = \sum_n h_n D^n$ of the form $h(D) = (1-D)(1+D)^N$. This equalization alters the spectral density of the noise, and a better channel model assumes that the η_n in Equation 11.1 are identically distributed, Gaussian random variables with zero mean, variance σ^2, and normalized cross-correlation $E\{\eta_n \eta_k\}/\sigma^2 = \rho_{n-k}$.

In practice, there is always intertrack interference (ITI), that is, the read head picks up magnetization from an adjacent track. Therefore, the channel output is given by

$$y_n = \sum_m a_m h_{n-m} + \sum_m x_m g_{n-m} + \eta_n \tag{11.6}$$

where $\{g_n\}$ is the discrete-time impulse response of the head to the adjacent track, and $\boldsymbol{x} = \{x_n\} \in \mathcal{C}$ is the sequence recorded on that track. We assume that the noise is white.

In the ideal case (Equation 11.1), the probability of detecting $\boldsymbol{b}$ given that $\boldsymbol{a}$ was recorded is equal to $Q(d(\epsilon)/\sigma)$, where $d(\epsilon)$ is the distance between $\boldsymbol{a}$ and $\boldsymbol{b}$ given by

$$d^2(\epsilon) = \sum_n \left(\sum_m \epsilon_m h_{n-m} \right)^2 \tag{11.7}$$

Therefore, a lower bound, and a close approximation for small σ, to the minimum probability of an error-event in the system is given by $Q(d_{\min,\mathcal{C}}/\sigma)$, where

$$d_{\min,\mathcal{C}} = \min_{\epsilon \in \mathcal{E}_\mathcal{C}} d(\epsilon)$$

is the channel minimum distance of code $\mathcal{C}$. We refer to

$$d_{\min} = \min_{\epsilon \in \{-1,0,1\}^\infty} d(\epsilon) \tag{11.8}$$

as the minimum distance of the uncoded channel, and to the ratio $d_{\min,\mathcal{C}}/d_{\min}$ as the gain in distance of code $\mathcal{C}$ over the uncoded channel.

In the case of colored noise, the probability of detecting b given that a was recorded equals to $Q(\Delta(\epsilon)/\sigma)$, where $\Delta(\epsilon)$ is the distance between a and b given by

$$\Delta^2(\epsilon) = \frac{\left[\sum_n \left(\sum_m \epsilon_m h_{n-m}\right)^2\right]^2}{\sum_n \sum_k \left(\sum_m \epsilon_m h_{n-m}\right) \rho_{n-k} \left(\sum_m \epsilon_m h_{k-m}\right)}$$

Therefore a lower bound to the minimum probability of an error-event in the system is given by $Q\left(\Delta_{\min,\mathcal{C}}/\sigma\right)$, where

$$\Delta_{\min,\mathcal{C}} = \min_{\epsilon \in \mathcal{E}_{\mathcal{C}}} \Delta(\epsilon)$$

In the case of ITI (Equation 11.6), we are interested in the probability of detecting sequence b given that sequence a was recorded on the track being read and sequence x was recorded on an adjacent track. This probability is

$$Q(\delta(\epsilon, x)/\sigma),$$

where $\delta(\epsilon, x)$ is the distance between a and b in the *presence* of x given by [30]

$$\delta^2(\epsilon, x) = \frac{1}{\left[\sum_n \left(\sum_m \epsilon_m h_{n-m}\right)^2\right]} \left[\sum_n \left(\sum_m \epsilon_m h_{n-m}\right)^2 + \sum_n \left(\sum_m x_m g_{n-m}\right)\left(\sum_m \epsilon_m h_{n-m}\right)\right]^2$$

Therefore a lower bound to the minimum probability of an error-event in the system is proportional to $Q(\delta_{\min,\mathcal{C}}/\sigma)$, where

$$\delta_{\min,\mathcal{C}} = \min_{\epsilon \neq 0, x \in C} \delta(\epsilon, x)$$

Distance $\delta_{\min,\mathcal{C}}$ can be bounded as follows [30]:

$$\delta_{\min,\mathcal{C}} \geq (1 - M)d_{\min,\mathcal{C}} \tag{11.9}$$

where $M = \max_{n, x \in C} \sum_m x_m g_{n-m}$, that is, M is the maximum absolute value of the interference. Note that $M = \sum_n |g_n|$. We will assume that $M < 1$. The bound is achieved if and only if there exists an ϵ, $d(\epsilon) = d_{\min,\mathcal{C}}$, for which $\sum_m \epsilon_m h_{n-m} \in \{-1, 0, 1\}$ for all n, and there exists an $x \in C$ such that $\sum_m x_m g_{n-m} = \mp M$ whenever $\sum_m \epsilon_m h_{n-m} = \pm 1$.

11.5 An Example

There are codes that provide gain in minimum distance on channels with ITI and colored noise, but not on the AWGN channel with the same transfer function. This is best illustrated using the example of the partial response channel with the transfer function $h(D) = (1 - D)(1 + D)^2$ known as EPR4. It is well known that for the EPR4 channel $d_{\min}^2 = 4$. Moreover, as discussed in Section 11.3, the following result holds:

Proposition 11.1   Error events $\epsilon(D)$ such that

$$d^2(\epsilon) = d_{\min}^2 = 4$$

take one of the following two forms:

$$\epsilon(D) = \sum_{j=0}^{k-1} D^{2j}, \quad k \geq 1$$

or

$$\epsilon(D) = \sum_{i=0}^{l-1} (-1)^i D^i, \quad l \geq 3$$

Therefore, an improvement of error-probability performance can be accomplished by codes which eliminate the error sequences ϵ containing the strings $-1+1-1$ and $+1-1+1$. Such codes were extensively studied in [20].

In the case of ITI Equation 11.6, we assume that the impulse response to the reading head from an adjacent track is described by $g(D) = \alpha H(D)$, where the parameter α depends on the track to head distance. Under this assumption, the bound (Equation 11.9) gives $\delta_{\min}^2 \geq d_{\min}^2(1 - 4\alpha)^2$. The following result was shown in [30]:

Proposition 11.2   Error events $\epsilon(D)$ such that

$$\min_{x \in C} \delta^2(\epsilon, x) = \delta_{\min}^2 = d_{\min}^2(1 - 4\alpha)^2 = 4(1 - 4\alpha)^2$$

take the following form:

$$\epsilon(D) = \sum_{i=0}^{l-1} (-1)^i D^i, \quad l \geq 5$$

For all other error sequences for which $d^2(\epsilon) = 4$, we have $\min_{x \in C} \delta^2(\epsilon, x) = 4(1 - 3\alpha)^2$.

Therefore, an improvement in error-probability performance of this channel can be accomplished by limiting the length of strings of alternating symbols in code sequences to four. For the NRZI type of recording, this can be achieved by a code that limits the runs of successive ones to three. Note that the set of minimum distance error events is smaller than in the case with no ITI. Thus performance improvement can be accomplished by higher rate codes which would not provide any gain on the ideal channel.

Channel equalization to the EPR4 target introduces cross-correlation among noise samples for a range of current linear recording densities (see [27] and references therein). The following result was obtained in [27]:

Proposition 11.3   Error events $\epsilon(D)$ such that

$$\Delta^2(\epsilon) = \Delta_{\min}^2$$

take the following form:

$$\epsilon(D) = \sum_{i=0}^{l-1} (-1)^i D^i, \quad l \geq 3, \quad l \text{ odd}$$

Again, the set of minimum distance error events is smaller than in the ideal case (white noise), and performance improvement can be provided by codes which would not give any gain on the ideal channel. For example, since all minimum distance error events have odd parity, a single parity check code can be used.

11.6 Future Directions

11.6.1 Soft-Output Decoding of Modulation Codes

Detection and decoding in magnetic recording systems is organized as a concatenation of a channel detector, an inner decoder, and an outer decoder, and as such should benefit from techniques known as erasure and list decoding. To declare erasures or generate lists, the inner decoder (or channel detector) needs to assess symbol/sequence reliabilities. Although the information required for this is the same one necessary for producing a single estimate, some additional complexity is usually required. So far, the predicted gains for erasure and list decoding of magnetic recording channels with additive white Gaussian noise were not sufficient to justify the increasing complexity of the channel detector and inner and outer decoder. However, this is not the case for systems employing new magneto-resistive reading heads, for which an important noise source, thermal asperities, is to be handled by passing erasure flags from the inner to the outer decoder.

In recent years, one more reason for developing simple soft-output channel detectors has surfaced. The success of turbo-like coding schemes on memoryless channels has sparked the interest in using them as modulation codes for ISI channels. Several recent results show that the improvements in performance turbo codes offer when applied to magnetic recording channels at moderate linear densities are even more dramatic than in the memoryless case [12, 29]. The decoders for turbo and low density parity check codes (LDPC) either require or perform much better with soft input information which has to be supplied by the channel detector as its soft output. The decoders provide soft outputs which can then be utilized by the outer Reed-Solomon (RS) decoder [22]. A general soft-output sequence detection was introduced in [11], and it is possible to get information on symbol reliabilities by extending those techniques [21, 31].

11.6.2 Reversed Concatenation

Typically, the modulation encoder is the inner encoder, that is, it is placed downstream of an error-correction encoder (ECC) such as an RS encoder; this configuration is known as standard concatenation (Figure 11.5). This is natural since otherwise the ECC encoder might well destroy the modulation properties before passing across the channel. However, this scheme has the disadvantage that the modulation decoder, which must come before the ECC decoder, may propagate channel errors before they can be corrected. This is particularly problematic for modulation encoders of very high rate, based on very long block size. For this reason, a good deal of attention has recently focused on a reversed concatenation scheme, where the encoders are concatenated in the reversed order (Figure 11.6). Special arrangements must be made

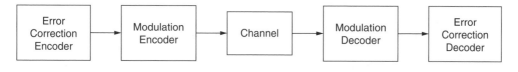

FIGURE 11.5   Standard concatenation.

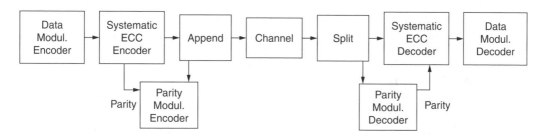

FIGURE 11.6   Reversed concatenation.

to ensure that the output of the ECC encoder satisfies the modulation constraints. Typically, this is done by insisting that this encoder be systematic and then reencoding the parity information using a second modulation encoder (the "parity modulation encoder"), whose corresponding decoder is designed to limit error propagation; the encoded parity is then appended to the modulation-encoded data stream (typically a few merging bits may need to be inserted in between the two streams in order to ensure that the entire stream satisfies the constraint). In this scheme, after passing through the channel the modulation-encoded data stream is split from the modulation-encoded parity stream, and the latter is then decoded via the parity modulation decoder before being passed on to the ECC decoder. In this way, many channel errors can be corrected before the data modulation decoder, thereby mitigating the problem of error propagation. Moreover, if the data modulation encoder has high rate, then the overall scheme will also have high rate because the parity stream is relatively small.

Reversed concatenation was introduced in [3] and later in [23]. Recent interest in the subject has been spurred on by the introduction of a lossless compression scheme, which improves the efficiency of reversed concatenation [15], and an analysis demonstrating the benefits in terms of reduced levels of interleaving [8]; see also [9]. Research on fitting soft decision detection into reversed concatenation can be found in [7, 33].

References

[1] R. Adler, D. Coppersmith, and M. Hassner, Algorithms for sliding-block codes, *IEEE Trans. Inform. Theory,* 29, no. 1, 5–22, January 1983.

[2] J. Ashley and B. Marcus, Time-varying encoders for constrained systems: an approach to limiting error propagation, *IEEE Trans. Inform. Theory,* 46, 1038–1043, 2000.

[3] W. G. Bliss, Circuitry for performing error correction calculations on baseband encoded data to eliminate error propagation, *IBM Tech. Discl. Bull.,* 23, 4633–4634, 1981.

[4] W. G. Bliss, An 8/9 rate time-varying trellis code for high density magnetic recording, *IEEE Trans. Magn.,* 33, no. 5, 2746–2748, September 1997.

[5] T. Conway, A new target response with parity coding for high density magnetic recording, *IEEE Trans. Magnetics,* 34, 2382–2386, 1998.

[6] T. Cover, Enumerative source encoding, *IEEE Trans. Inform. Theory,* 73–77, January 1973.

[7] J. Fan, Constrained coding and soft iterative decoding for storage, Ph.D. Dissertation, Stanford University, 1999.

[8] J. Fan and R. Calderbank, A modified concatenated coding scheme, with applications to magnetic data storage, *IEEE Trans. Inform. Theory,* 44, 1565–1574, 1998.

[9] J. Fan, B. Marcus, and R. Roth, Lossless sliding-block compression of constrained systems, *IEEE Trans. Inform. Theory,* 46, 624–633, 2000.

[10] K. Knudson Fitzpatrick and C. S. Modlin, Time-varying MTR codes for high density magnetic recording, *Proc. 1997 IEEE Global Telecommun. Conf. (GLOBECOM '97),* Phoenix, AZ, 1250–1253, November 1997.

[11] J. Hagenauer and P. Hoeher, A Viterbi algorithm with soft–decision outputs and its applications, *Proc. 1989 IEEE Global Telecommun. Conf. (GLOBECOM '89),* Dallas, TX, 1680–16867, November 1989.

[12] C. Heegard, Turbo coding for magnetic recording, *Proc. 1998 Inform. Theory Workshop,* San Diego, CA, 18–19, February 8–11, 1998.

[13] C. D. Heegard, B. H. Marcus, and P. H. Siegel, Variable-length state splitting with applications to average runlength-constrained (ARC) codes, *IEEE Trans. Inform. Theory,* 37, 759–777, 1991.

[14] H. D. L. Hollmann, On the construction of bounded-delay encodable codes for constrained systems, *IEEE Trans. Inform. Theory,* 41, 1354–1378, 1995.

[15] K. A. Schouhamer Immink, A practical method for approaching the channel capacity of constrained channels, *IEEE Trans. Inform. Theory,* 43, 1389–1399, 1997.

[16] K. A. Schouhamer Immink, P. H. Siegel, and J. K. Wolf, Codes for digital recorders, *IEEE Trans. Inform. Theory,* 44, 2260–2299, October 1998.

[17] K. A. Schouhamer Immink, *Codes for Mass Data Storage,* Shannon Foundation Publishers, The Netherlands, 1999.

[18] R. Karabed and P. H. Siegel, Matched spectral null codes for partial response channels, *IEEE Trans. Inform. Theory,* 37, 818–855, 1991.

[19] R. Karabed and P. H. Siegel, Coding for higher order partial response channels, *Proc. 1995 SPIE Int. Symp. on Voice, Video, and Data Communications,* Philadelphia, PA, 2605, 115–126, October 1995.

[20] R. Karabed, P. H. Siegel, and E. Soljanin, Constrained coding for binary channels with high inter-symbol interference, *IEEE Trans. Inform. Theory,* 45, 1777–1797, September 1999.

[21] K. J. Knudson, J. K. Wolf, and L. B. Milstein, Producing soft–decision information on the output of a class IV partial response Viterbi detector, *Proc. 1991 IEEE Int. Conf. Commn. (ICC '91),* Denver, CO, 26.5.1.–26.5.5, June 1991.

[22] R. Koetter and A. Vardy, preprint 2000.

[23] M. Mansuripur, Enumerative modulation coding with arbitrary constraints and post-modulation error correction coding and data storage systems, *Proc. SPIE,* 1499, 72–86, 1991.

[24] B. Marcus, R. Roth, and P. Siegel, Constrained systems and coding for recording channels, *Handbook of Coding Theory,* V. Pless, C. Huffman, Eds., Elsevier, Amsterdam, 1998, Chap. 20.

[25] D. Modha and B. Marcus, Art of constructing low complexity encoders/decoders for constrained block codes, *IEEE J. Sel. Areas Comm.,* 2001, to appear.

[26] B. E. Moision, A. Orlitsky, and P. H. Siegel, On codes that avoid specified differences, *IEEE Trans. Inform. Theory,* 47, 433–441, January 2001.

[27] B. E. Moision, P. H. Siegel, and E. Soljanin, Distance enhancing codes for high-density magnetic recording channel, *IEEE Trans. Magn.,* January 2001, submitted.

[28] J. Moon and B. Brickner, Maximum transition run codes for data storage systems, *IEEE Trans. Magn.,* 32, 3992–3994, September 1996.

[29] W. Ryan, L. McPheters, and S.W. McLaughlin, Combined turbo coding and turbo equalization for PR4-equalized Lorentzian channels, *Proc. 22nd Annual Conf. Inform. Sciences and Systems,* Princeton, NJ, March 1998.

[30] E. Soljanin, On–track and off–track distance properties of Class 4 partial response channels, *Proc. 1995 SPIE Int. Symp. on Voice, Video, and Data Communications,* Philadelphia, PA, 2605, 92–102, October 1995.

[31] E. Soljanin, Simple soft-output detection for magnetic recording channels, *1998 IEEE Int. Symp. Inform. Theory (ISIT'00),* Sorrento, Italy, June 2000.

[32] A. J. van Wijngaarden and K. A. Schouhamer Immink, Combinatorial construction of high rate runlength-limited codes, *Proc. 1996 IEEE Global Telecommun. Conf. (GLOBECOM '96),* London, U.K., 343–347, November 1996.

[33] A. J. van Wijngaarden and K. A. Schouhamer Immink, Maximum run-length limited codes with error control properties, *IEEE J. Select. Areas Commn.,* 19, April 2001.

[34] A. J. van Wijngaarden and E. Soljanin, A combinatorial technique for constructing high rate MTR–RLL codes, *IEEE J. Select. Areas Commn.,* 19, April 2001.

12

Information Theory of Magnetic Recording Channels

Zheng Zhang

Tolga M. Duman
Arizona State University
Tempe, AZ

Erozan M. Kurtas
Seagate Technology
Pittsburgh, PA

12.1   Introduction **12**-2
12.2   Channel Capacity and Information Rates for
        Magnetic Recording Channels **12**-3
        Capacity of ISI Channels with AWGN • Achievable
        Information Rates for ISI Channels with AWGN
12.3   Achievable Information Rates for Realistic Magnetic
        Recording Channels with Media Noise **12**-12
        Media Noise Model • Estimation of Achievable Information
        Rates
12.4   Conclusions **12**-18

Zheng Zhang: received the B.E. degree with honors from Nanjing University of Aeronautics and Astronautics, Nanjing, China, in 1997, and the M.S. degree from Tsinghua University, Beijing, China, in 2000, both in electronic engineering. Currently, he is working toward the Ph.D. degree in electrical engineering at Arizona State University, Tempe, AZ.

His current research interests are digital communications, wireless/mobile communications, magnetic recording channels, information theory, channel coding, turbo codes and iterative decoding, MIMO systems and multi-user systems.

Tolga M. Duman: received the B.S. degree from Bilkent University in 1993, M.S. and Ph.D. degrees from Northeastern University, Boston, in 1995 and 1998, respectively, all in electrical engineering. Since August 1998, he has been with the Electrical Engineering Department of Arizona State University first as an assistant professor (1998–2004), and currently as an associate professor. Dr. Duman's current research interests are in digital communications, wireless and mobile communications, channel coding, turbo codes, coding for recording channels, and coding for wireless communications.

Dr. Duman is the recipient of the National Science Foundation CAREER Award, IEEE Third Millennium medal, and IEEE Benelux Joint Chapter best paper award (1999). He is a senior member of IEEE, and an editor for IEEE Transactions on Wireless Communications.

Erozan M. Kurtas: received the B.Sc. degree from Bilkent University, Ankara, Turkey, in 1991 and M.Sc. and Ph.D., degrees from Northeastern University, Boston, MA, in 1993 and 1997, respectively.

His research interests cover the general field of digital communication and information theory with special emphasis on coding and detection for inter-symbol interference channels. He has published over 75 book chapters, journal and conference papers on the general fields of information theory, digital communications and data storage. Dr. Kurtas is the co-editor of the book Coding and Signal Processing for Magnetic Recording System (to be published by CRC press in 2004). He has seven pending patent applications. Dr. Kurtas is currently the Research Director of the Channels Department at the research division of Seagate Technology.

Abstract

In this chapter, we review the information theoretical results for the magnetic recording channels. Such results are very important in that they provide theoretical limits on the performance of practical coding/decoding schemes. Since recording channels can be modeled as *intersymbol interference* (ISI) channels with binary inputs, we first present the capacity of the general ISI channels with additive white Gaussian noise, which is achieved by correlated Gaussian inputs. When taking the constraint of binary inputs into account, no closed-form solutions exist, however, several upper and lower bounds are derived. More recently, Monte-Carlo simulation techniques have been used to estimate the achievable information rates of ISI channels, which are also described. The simulation-based techniques can be extended further to compute the achievable information rates of the magnetic recording channels with media noise present in high density recording systems. The results are illustrated by a comprehensive set of examples.

12.1   Introduction

Channel capacity is defined as the highest rate in bits per channel use at which information can be transmitted with an arbitrarily low probability of error [1]. The computation of the channel capacity is desirable in that it determines exactly how far the performance achieved by a specific channel coding scheme is from the ultimate theoretical limit. The capacities for some channels are known, such as the *additive white Gaussian noise* (AWGN) channel under the input energy constraints [1]. However, for more complicated channel models and with specific input constraints, there are still many unsolved information theoretical problems.

This chapter will focus on the magnetic recording channels and summarize the available information theoretical results. Since the magnetic recording channel can be modeled as an *intersymbol interference* (ISI) channel, we are in fact considering the channel capacity and achievable information rates for ISI channels.

The capacity of the ISI channels with AWGN is derived in [2,3] and it is shown that the channel capacity is achieved when the channel inputs are correlated Gaussian random variables. However, in most practical cases, the channel inputs cannot be selected arbitrarily. For instance, in magnetic recording, the inputs are restricted to be binary. Although closed-form solutions for the channel capacity do not yet exist for the case of binary-input ISI channels with AWGN, upper and lower bounds on the achievable information rates are provided in [4–6]. More recently, Arnold and Loeliger have derived a simulation-based method to compute the information rates of ISI channels with AWGN under the constraint that the inputs are chosen from a finite alphabet with a certain distribution [7]. Meanwhile, other authors independently propose similar algorithms to estimate the achievable information rates by Monte-Carlo simulations [8,9]. Furthermore, Kavčić presents an iterative algorithm to maximize the achievable information rates over Markov inputs with a certain memory for these channels [10].

In high-density magnetic recording channels, signal-dependent noise, called *media noise* [11–13], exists, with which the magnetic recording channels can be appropriately modeled as ISI channels with signal-dependent correlated noise [13,14]. The achievable information rates of such channels with signal-dependent correlated Gaussian noise are computed in [15,16] by extending the techniques in [7].

This chapter is organized as follows. In Section 12.2, we present the channel capacity for the ISI channels with AWGN, as well as the achievable information rates over such channels with inputs chosen from a finite alphabet, for example, binary inputs for the case of magnetic recording channels. The upper and lower bounds on the achievable information rates are first given since there are no closed-form solutions, then the simulation-based techniques are described to estimate the information rates. We use both the ideal *partial response* (PR) channels and the realistic longitudinal and perpendicular magnetic recording channels as examples. The magnetic recording channels with media noise are considered in Section 12.3, where the achievable information rates are computed based on the extension of the techniques originally presented for the ISI channels with AWGN. We conclude the chapter in Section 12.4.

12.2 Channel Capacity and Information Rates for Magnetic Recording Channels

The block diagram of a communication system over a band-limited channel with AWGN is shown in Figure 12.1, where the symbols $\{x_k\}$ are the (binary) inputs with symbol energy constraint E_s, $h(t)$ denotes the channel pulse response and $n(t)$ represents the AWGN with two sided power spectral density of $N_0/2$. The received signal $r(t)$ can be represented by

$$r(t) = \sum_{k=-\infty}^{\infty} x_k h(t - kT) + n(t) \tag{12.1}$$

where T is the symbol duration.

For magnetic recording channels, the pulse response $h(t)$ can be expressed by $h(t) = p(t) - p(t - T)$ where $p(t)$ is the response of the channel to each signal transition. For longitudinal recording channels, we can model $p(t)$ by the Lorentzian pulse [17]

$$p(t) = \frac{1}{1 + (2t/D_n)^2} \tag{12.2}$$

where D_n is the normalized density of the recording channel. Similarly, the pulse response of the perpendicular recording channels is given by [18]

$$p(t) = \frac{1}{2}\text{erf}\left(2\sqrt{\log 2}\,\frac{t}{D_n}\right) \tag{12.3}$$

where $\text{erf}(\cdot)$ denotes the error function defined as

$$\text{erf}(t) = \frac{2}{\sqrt{\pi}} \int_0^t e^{-x^2}\, dx \tag{12.4}$$

Here, the symbol duration is set as $T = 1$ without loss of generality. Define $b_k = x_k - x_{k-1}$ as the transition symbol, then, equivalent to Equation 12.1, we have

$$r(t) = \sum_{k=-\infty}^{\infty} b_k p(t - kT) + n(t) \tag{12.5}$$

It is well known that, for a band-limited channel with AWGN, the optimum receiver is the matched filter followed by a sampler operating at the symbol rate, and the *maximum likelihood sequence detector* (MLSD) [19,20]. The noise samples at the output of the matched filter are colored, therefore, to obtain an equivalent intersymbol interference channel with AWGN, a noise-whitening filter is necessary. The output of the noise-whitening filter, y_k, can be considered as the overall output, as shown in Figure 12.1, and can be used in the MLSD.

Let $\{g_k\}$ represent the samples of the autocorrelation function of $h(t)$, that is,

$$g_k = \int_{-\infty}^{\infty} h^*(t) h(t + kT)\, dt \tag{12.6}$$

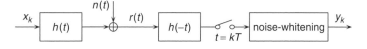

FIGURE 12.1 Block diagram for ISI channels with AWGN.

where we assume that $g_k = 0$ for $|k| > M$. Then, its z transform, $G(z)$, is given by

$$G(z) = \sum_{k=-M}^{M} g_k z^{-k} \tag{12.7}$$

and the output of the equivalent discrete-time channel can be expressed as

$$y_k = \sum_{m=0}^{M} f_m x_{k-m} + z_k \tag{12.8}$$

where $\{z_k\}$ is the AWGN with variance $\sigma_z^2 = N_0/2$ and $\{f_k\}$ is the set of tap coefficients of the equivalent channel with memory M, whose z transform satisfies

$$F(z)F^*(z^{-1}) = G(z) \tag{12.9}$$

We observe from Equation 12.8 that, the channel output consists of the desired symbol scaled by f_0, the intersymbol interference and the noise term. Clearly, this channel model includes the memoryless or ISI-free AWGN channel as a special case with $M = 0$.

We note that the channel coefficients $\{f_k\}$ for both the longitudinal and perpendicular recording channels can be computed easily.

Although the optimal receiver for ISI channels is easy to describe, the complexity may become an issue. This is true, in particular, when the normalized density of the recording channel is increased, since the number of taps of the channel is increased as well. More precisely, the complexity for the maximum likelihood sequence estimation is exponential to the channel memory M. To alleviate this problem, in practice, the channel output is usually equalized to an appropriate partial response target using a linear equalizer before being input to the detector. For example, a class of partial response targets suitable for the longitudinal recording channels is expressed by $(1 - D)(1 + D)^n$ where D denotes the delay operation and n is a nonnegative number. By setting n as 0, 1, 2, and 3, we obtain the dicode, PR4, EPR4, and E^{2}PR4 channels, respectively [21]. As a very simple recording channel model, we can consider the ideal PR channels (e.g., dicode, PR4, EPR4, or E^{2}PR4) with AWGN.

We define the *signal-to-noise ratio* (SNR) as

$$\text{SNR} = \frac{E_{nf} \cdot E_s}{N_0} \tag{12.10}$$

where E_{nf} is the normalization factor. For the realistic magnetic recording channels, we set E_{nf} as $\int_{-\infty}^{\infty} h^2(t)\, dt$. While for the ideal PR channels, $E_{nf} = \sum_{m=0}^{M} f_m^2$.

Before the discussion, let us describe our notation. We use $\mathbf{y}_a^b$ to represent the sequence $(y_a, y_{a+1}, \ldots, y_b)$, where $b > a$, and for simplicity, $\mathbf{y}^b$ is used when $a = 1$. In addition, $I(x; y)$ is used to denote the mutual information between two random variables, and $I(\mathbf{x}^b; \mathbf{y}^b)$ is for the mutual information between two random sequences.

12.2.1 Capacity of ISI Channels with AWGN

The capacity of the discrete-time ISI channels with AWGN is derived in [2] for inputs with the block energy constraint. In [3], Hirt and Massey proved that this capacity also equals to the one for the ISI channels with the symbol energy constraint (which is more strict).

The capacity of the ISI channels is defined as [22]

$$C = \lim_{n \to \infty} \frac{1}{n} \max_{p(\mathbf{x}^n)} I(\mathbf{x}^n; \mathbf{y}^n) \tag{12.11}$$

where $\mathbf{x}^n$ and $\mathbf{y}^n$ denote the input and output sequences, respectively, and the maximum of the mutual information is taken over all input *probability density functions* (PDF) $p(\mathbf{x}^n)$ that satisfy the symbol energy constraint given by

$$E\left[x_k^2\right] \le E_s \tag{12.12}$$

with $E[\cdot]$ denoting the expectation.

In [3], the authors first set up a hypothetical channel model and derive the capacity of this channel by using *discrete Fourier transform* (DFT) and water-filling in the transform domain. Then, it is proved that this capacity is equal to the one of the discrete-time ISI channel with AWGN, and it is given by

$$C = \frac{1}{2\pi} \int_0^\pi \log_2[\max(\Theta|F(\lambda)|^2, 1)] \, d\lambda \tag{12.13}$$

where $F(\lambda)$ is the channel transfer function

$$F(\lambda) = \sum_{m=0}^M f_m e^{-jm\lambda}, \quad j = \sqrt{-1} \tag{12.14}$$

and the parameter Θ is the solution of the equation

$$\int_{\substack{0 \\ F(\lambda)\ne 0}}^\pi \max(\Theta - |F(\lambda)|^{-2}, 0) \, d\lambda = 2\pi \frac{E_s}{N_0} \tag{12.15}$$

The capacity is achieved when the inputs are correlated Gaussian random variables with zero mean and covariance given by

$$E\left[x_{k+n}x_k\right] = \frac{1}{\pi} \int_0^\pi S_x(\lambda) \cos(n\lambda) \, d\lambda \tag{12.16}$$

where $S_x(\lambda)$ satisfies

$$S_x(\lambda) = \begin{cases} \frac{N_0}{2}(\Theta - |F(\lambda)|^{-2}), & \Theta|F(\lambda)|^2 > 1, |\lambda| \le \pi \\ 0, & \text{otherwise} \end{cases} \tag{12.17}$$

and Θ was defined in Equation 12.15.

In Figure 12.2, the capacities, in bits per channel use, of the PR4, EPR4, and E^2PR4 channels with AWGN are shown, where the SNR is defined as in Equation 12.10. Also shown is the capacity of the ISI-free AWGN channel, which is expressed by $\frac{1}{2}\log_2(1 + 2E_s/N_0)$. From the figure, we observe that the intersymbol interference decreases the channel capacity in the high SNR region, which is expected. However, in the low SNR region, with the use of water-filling at the transmitter, the capacity of the ISI channels can be higher than that of the ISI-free channel. This is explained as follows. When the average channel powers for the ISI channel and ISI-free channel are taken the same (which is the case with our SNR definition), the frequency response of the ISI channel exhibits a peak larger than unity (i.e., the gain for the ISI-free case) for some short range of frequencies. Then, with water-filling, one can exploit this to increase the achievable information rates beyond those offered by the ISI-free channel, in particular, when the noise level is high. In [23], the authors propose to use a different SNR definition (i.e., normalization) to compare the capacities of the ISI channels and the ISI-free channel. They argue that the peak of the channel transfer function should be normalized (as opposed to the average power) for a fair comparison. In this case, the ISI-free channel is shown to have a larger capacity for the entire SNR range.

12.2.2 Achievable Information Rates for ISI Channels with AWGN

As stated above, the capacity-achieving inputs are correlated Gaussian distributed. But in many applications, the channel inputs have specific constraints, for example, they are chosen from a finite alphabet.

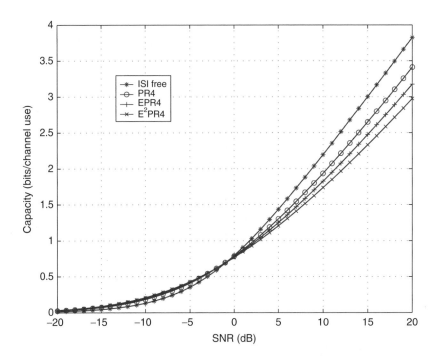

FIGURE 12.2   Capacities of several ISI channels.

The magnetic recording channel is such an example where the inputs are restricted to be binary. Under the input constraint, the capacity shown above cannot be achieved in general, especially in the high SNR region. Therefore, it is necessary to compute the achievable information rates with the specific input constraint, which can be defined by Equation 12.11 with the maximization taken over the inputs satisfying the specific constraint. In some cases, we may be interested in the achievable information rates under specific input constraints with a fixed distribution, such as the commonly used constraint of *independent identically distributed* (i.i.d.) symmetric inputs. In this case, there is no maximization involved, and the information rate achievable over the ISI channel is given by

$$I(X;Y) = \lim_{n \to \infty} \frac{1}{n} I(\mathbf{x^n}; \mathbf{y^n}) \tag{12.18}$$

where $\mathbf{x^n}$ is the input sequence with the specific constraints and $\mathbf{y^n}$ is the corresponding channel output sequence.

For a memoryless AWGN channel, we can easily obtain this information rate under the i.i.d. binary symmetric (or, equiprobable) input constraint as follows [6],

$$I^b(E_s/N_0) = 1 - \int_{-\infty}^{+\infty} \frac{e^{-\tau^2/2}}{\sqrt{2\pi}} \log_2[1 + e^{-2\tau\sqrt{2E_s/N_0} - 4E_s/N_0}] \, d\tau \tag{12.19}$$

which is in fact the constrained capacity for such channels with binary inputs.

Unfortunately, for the ISI channels, there is no closed-form solution to simplify the expression in Equation 12.18. Therefore, various bounds on the information rates are derived, and several simulation-based algorithms are developed.

12.2.2.1 Bounds on Achievable Information Rates

A lower bound on the information rates of ISI channels with i.i.d. inputs chosen from a finite alphabet, denoted by I_L, is derived in [4]. Using the ISI channel model defined by Equation 12.8, I_L can be expressed by the achievable information rate of a memoryless channel with a degradation factor ρ as

$$I_L = I(x; \rho x + z) \tag{12.20}$$

where x is the i.i.d. constrained input with the same distribution as x_k, z is the white Gaussian noise with the same variance as z_k and $\rho = |f_0|$ for a causal ISI channel. This lower bound is explained in [4] as the information rate of the channel with ideal decision feedback equalizer and error-free past decisions. For the channel with i.i.d. binary symmetric inputs, we replace E_s in Equation 12.19 by $\rho^2 E_s$ to obtain the lower bound I_L^b as

$$I_L^b = I^b(\rho^2 E_s / N_0) \tag{12.21}$$

Since I_L is determined only by the first channel coefficient, it is a loose lower bound especially when the memory of the ISI channel is large and the interfering coefficients are not negligible.

In [5], a conjectured lower bound on the information rates for channels with i.i.d. inputs is proposed, which is found to be a much tighter one though it is not proved formally. The bound is given by

$$I_{CL} = I(x; x + v) \tag{12.22}$$

where v is a zero-mean Gaussian random variable with variance σ_v^2. It is shown in [5] that

$$E_s / \sigma_v^2 = 2^{2I_G} - 1 \tag{12.23}$$

where I_G is the information rate of the ISI channel with i.i.d. Gaussian inputs, given by [3]

$$I_G = \frac{1}{2\pi} \int_0^\pi \log_2 \left[1 + 2 \frac{E_s}{N_0} |F(\lambda)|^2 \right] d\lambda \tag{12.24}$$

Therefore, with i.i.d. symmetric binary inputs, the conjectured lower bound can be computed using Equation 12.19 as

$$I_{CL}^b = I^b((2^{2I_G} - 1)/2) \tag{12.25}$$

Clearly, the information rate given in Equation 12.24 is in fact an upper bound for the ISI channels with i.i.d. inputs [4], and we call it Gaussian upper bound. Another upper bound, called matched filter bound, is derived in [4]. This bound states that

$$I_U = I(x; \| f \| x + z) \tag{12.26}$$

where $\| f \|$ is the enhancement factor given by $\| f \| = \sqrt{\sum_{m=0}^M f_m^2}$. This upper bound corresponds to the transmission of one single symbol without interference from others. Similarly, for binary inputs, we can compute the upper bound I_U^b by

$$I_U^b = I^b(\| f \|^2 E_s / N_0) \tag{12.27}$$

Several other bounds, not necessarily for i.i.d. inputs, are given in [4,5]. Furthermore, in [6], Monte-Carlo simulation is used to estimate bounds for channels with i.i.d. inputs, which may be tight, but at the cost of increased computational complexity.

In order to demonstrate the tightness of the bounds, we use the ideal PR4 channel as an example and show the different bounds on the achievable information rates with i.i.d. symmetric binary inputs in Figure 12.3, together with the capacity of the PR4 channel, which can be viewed as a loose upper bound. In addition, the exact information rates with i.i.d. symmetric binary inputs are shown as a reference, which are obtained by the simulation-based technique described in the next section. We observe that, the conjectured lower bound and the Gaussian upper bound can provide a very accurate estimation for the

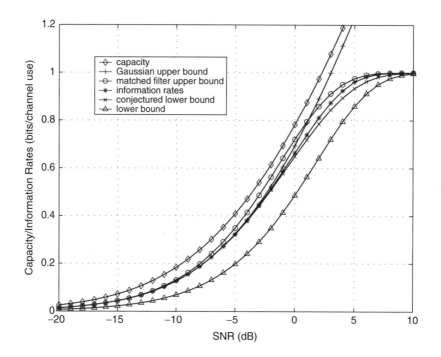

FIGURE 12.3   Bounds on the information rates of the PR4 channel with i.i.d. inputs.

achievable information rates in the low-to-medium SNR region. In the high SNR region, we can use the conjectured lower bound and the matched filter upper bound to estimate the information rates. Other bounds are relatively loose and do not provide accurate estimates. Although not shown here, the bounds become looser for the channels with more severe ISI, such as the EPR4 and E^2PR4 channels. We can also clearly observe the difference between the achievable information rates and the capacity. For instance, to achieve a rate of 0.9 bits per channel use, there is a gap of about 2.3 dB for the required SNR for the case of Gaussian inputs versus binary i.i.d. inputs.

12.2.2.2   Estimation of Achievable Information Rates

From above discussion, we know that the bounds on the achievable information rates may not be tight for the ISI channels, especially in the high SNR region. Thus, it is desirable to have a technique to compute the exact achievable information rates. Recently, Arnold and Loeliger, and other authors have independently derived a simulation-based method to compute the information rates of ISI channels with AWGN under the constraint that the inputs are chosen from a finite alphabet with a certain distribution [7–9].

We rewrite the information rate defined in Equation 12.18 as

$$I(X;Y) = \lim_{n \to \infty} \frac{1}{n} I(\mathbf{x^n}; \mathbf{y^n})$$
$$= h(Y) - h(Z \mid X) \tag{12.28}$$

where $h(\cdot)$ denotes the differential entropy, and

$$h(Y) = \lim_{n \to \infty} \frac{1}{n} h(\mathbf{y^n}) \tag{12.29}$$

$$h(Z \mid X) = \lim_{n \to \infty} \frac{1}{n} h(\mathbf{z^n} \mid \mathbf{x^n}) \tag{12.30}$$

with $\mathbf{z^n}$ denoting the noise sequence. When the noise is independent of the input, $h(Z \mid X)$ can also be expressed as $h(Z)$, which is given by

$$h(Z) = \lim_{n \to \infty} \frac{1}{n} h(\mathbf{z^n}) \qquad (12.31)$$

We can easily compute $h(Z)$ for AWGN provided that its variance is known, that is, $h(Z) = \frac{1}{2} \log(2\pi e \sigma_z^2)$ with the variance $\sigma_z^2 = N_0/2$. Therefore, the problem reduces to the computation of $h(Y)$, which can be further expressed as

$$h(Y) = -\lim_{n \to \infty} \frac{1}{n} E\left[\log(p(\mathbf{y^n}))\right] \qquad (12.32)$$

where $p(\mathbf{y^n})$ is the probability density function of the output $\mathbf{y^n}$. So if we can estimate $p(\mathbf{y^n})$ by simulation, the problem can be solved by using Monte Carlo techniques. In [7], the forward recursion of the BCJR algorithm [24] is used to compute $p(\mathbf{y^n})$ for a given ISI channel and its output with length n. Let us now briefly describe the procedure.

At first, we consider the case with i.i.d. inputs. It is well known that ISI channels with inputs chosen from a finite alphabet can be represented by a trellis. Assume that the size of the alphabet is K and the memory of the ISI channel is M, then there are K^M states in the trellis. In addition, there are K valid transitions, or outgoing paths, from every trellis state and K incoming paths to every state.

Suppose that we have the channel output sequence $\mathbf{y^n}$ with a large block length n excited by the constrained inputs. By denoting the trellis state at time instance k as S_k $(0 \le S_k \le K^M - 1)$, we define

$$\alpha_k(j) = p(\mathbf{y^k}, S_k = j) \qquad (12.33)$$

It is clear that we can compute $p(\mathbf{y^n})$ as

$$p(\mathbf{y^n}) = \sum_j \alpha_n(j) \qquad (12.34)$$

To compute $\alpha_n(j)$, we define

$$\begin{aligned} \gamma_k(i, j) &= p(S_k = j, y_k \mid S_{k-1} = i) \\ &= p(y_k \mid S_{k-1} = i, S_k = j) \cdot p(S_k = j \mid S_{k-1} = i) \end{aligned} \qquad (12.35)$$

As shown in [24], we have

$$\alpha_k(j) = \sum_i \gamma_k(i, j) \cdot \alpha_{k-1}(i) \qquad (12.36)$$

which holds because the transition probability does not depend on the previous output samples and y_k is independent of the previous output samples conditioned on the state transition when the noise is white.

It is easy to compute $\gamma_k(i, j)$ in Equation 12.35. We know that the transition probability is determined by the distribution of the inputs. For example, when the inputs are i.i.d. equiprobable binary, $p(S_k = j \mid S_{k-1} = i) = \frac{1}{2}$ for valid transitions and $p(S_k = j \mid S_{k-1} = i) = 0$ for the others, and given the trellis transition, y_k is Gaussian distributed.

We need to initialize $\alpha_k(j)$ for $k = 0$ according to the stationary distribution of the inputs. For i.i.d. binary inputs with equal probability, we may set $\alpha_0(j) = 1/2^M (j = 0, 1, \ldots, 2^M - 1)$. Thus we can compute $\alpha_k(j)$ by the forward recursion given in Equation 12.36.

In order to estimate the expectation of $\log(p(\mathbf{y^n}))$, we can perform the simulation N times and compute the average of $\log(p(\mathbf{y^n}))$. Let ρ_l be the computed $p(\mathbf{y^n})$ for the lth simulation. Then we have

$$-\frac{1}{n} E\left[\log(p(\mathbf{y^n}))\right] = -\lim_{N \to \infty} \frac{1}{N \cdot n} \sum_{l=1}^{N} \log(\rho_l) \qquad (12.37)$$

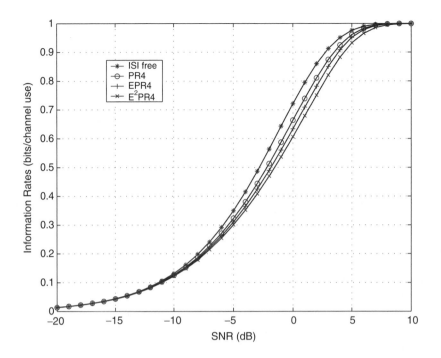

FIGURE 12.4   Information rates of several channels with i.i.d. binary inputs.

Since Y is a stationary ergodic hidden-Markov process, the Shannon-McMillan-Breiman theorem holds [25,26] and we can use one single simulation with long enough block length instead of N simulations. Therefore, we compute $h(Y)$ and $I(X;Y)$ as:

$$h(Y) = -\lim_{n \to \infty} \frac{1}{n} \log(p(\mathbf{y^n}))$$ (12.38)

and

$$I(X;Y) = h(Y) - \frac{1}{2} \log\left(2\pi e \sigma_z^2\right)$$ (12.39)

Figure 12.4 shows the information rates of the ISI-free channel, PR4, EPR4 and E^2PR4 channels with i.i.d. equiprobable binary inputs. It is obvious that the intersymbol interference decreases the achievable information rates in the entire SNR range since no water-filling is employed (i.e., the inputs are i.i.d.). For example, to achieve a rate of 0.9 bits per channel use, there is an SNR loss of about 1.6 dB for the E^2PR4 channel compared to the ISI-free channel.

We now give several examples for the realistic recording channels. The achievable information rates for the longitudinal and perpendicular recording channels with different normalized densities are shown in Figure 12.5 and Figure 12.6, where i.i.d. binary symmetric inputs are employed. We observe that, with higher normalized densities, the achievable information rates are reduced, especially in the high SNR region.

12.2.2.3   Information Rates with Correlated Inputs

The simulation-based technique can also be used to compute the achievable information rates for the channels with correlated inputs, such as the the constrained coded inputs that are commonly employed in the magnetic recording channels. For example, the (d, k) *run-length limited* (RLL) codes are used in order to reduce the ISI and provide easier synchronization [19,27]. With the restriction defined by the constrained codes, the maximum rate achieved over the noiseless channel is decreased. This is referred as

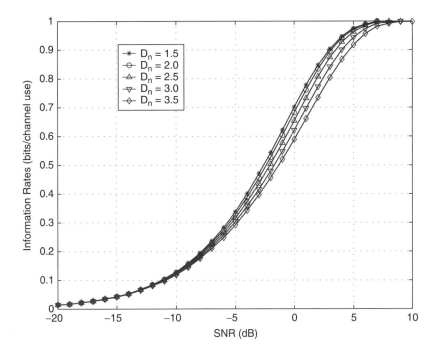

FIGURE 12.5 Information rates of longitudinal recording channels with i.i.d. inputs.

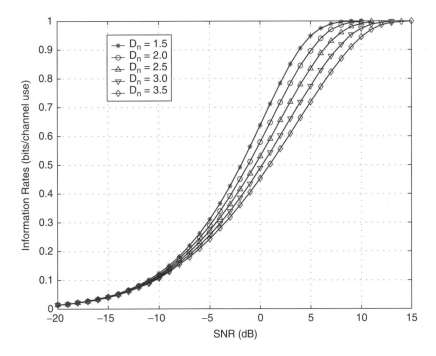

FIGURE 12.6 Information rates of perpendicular recording channels with i.i.d. inputs.

the capacity of the constrained codes, which can be computed using the labeled graph of the constrained codes [19,27]. However, for the achievable information rates as a function of the SNR, we may resort to the simulation-based algorithm. To accomplish this, we need to use a joint trellis that incorporates both the constrained code and the ISI channel [28–30].

Furthermore, we can use the water-filling technique for the ISI channels with inputs chosen from a finite alphabet. By taking the correlated inputs into consideration, we may achieve a higher information rate than the one with i.i.d. inputs, especially in the low SNR region. In [7], the maximized information rates of the dicode channel are computed for Markov inputs with memory 1 and 2. Instead of using the exhaustive search for the optimized transition probabilities, Kavčić presents an algorithm to maximize the achievable information rates over the Markov inputs with a certain memory for the ISI channels with AWGN [10]. When the memory goes to infinity, we achieve the highest information rates under the constraint of the inputs chosen from the specific finite alphabet. This approach updates the transition probabilities iteratively in order to maximize the information rates which can be estimated by the BCJR algorithm.

It is shown in [7,10,31] that, with the increase of the Markov memory, higher information rates can be obtained, especially in the low-to-medium SNR region. But in the high SNR region or the high rate region, i.i.d. symmetric input (equivalent to the case with Markov memory 0) is almost optimum. As a limiting case, we know that i.i.d. inputs achieve the highest entropy rate, that is, $\log_2(K)$ bits per channel use when SNR goes to infinity. In addition, we observe that the increase becomes very small when the Markov memory is large. However, even with a very large memory, one cannot make sure that we have obtained the highest achievable information rate, or the capacity under the input constraint. Thus the obtained information rate can be viewed as a tight lower bound on the real capacity and some upper bounds are proposed in [32,33]. For the details, the reader is referred to [31].

12.3 Achievable Information Rates for Realistic Magnetic Recording Channels with Media Noise

In high-density magnetic recording channels, in addition to the electronic noise, a kind of signal-dependent noise, called media noise [34], exists, which poses a significant challenge to the computation of the channel capacity and the design of the detectors. In [35,36], the upper and lower bounds on the channel capacity are studied where the media noise is assumed to be signal independent. Similar channel model is used in [37], however, the binary-input constraint is taken into account and upper and lower bounds on the achievable information rates are considered. In [38], the media noise is modeled as signal-dependent Gaussian noise and upper and lower bounds are computed for two simplified channel models. A conjectured lower bound on the achievable information rate is presented for the magnetic recording channels with media noise in [39,40] based on the conjectured lower bound reported in [5] for the ISI channels with AWGN, which is claimed to be close to the symmetric information rates achieved by the i.i.d. equiprobable binary inputs. In [15], the authors extend the simulation techniques proposed first for the ISI channels with AWGN [7–9] to the ISI channels with signal-dependent correlated Gaussian noise, which is used to model the media noise in the recording channels. A similar simulation-based algorithm is proposed in [16] for channels with signal-dependent autoregressive noise, where the noise is assumed to be Gauss-Markov. In this section, we follow the channel model and approach in [15] and compute the achievable information rates for the ISI channels with signal-dependent correlated Gaussian noise.

12.3.1 Media Noise Model

One of the main sources of the media noise is the fluctuation of the position of the transition pulse, which is also called *jitter noise* [41,42]. Taking the jitter noise into account, the received signal is given by

$$r(t) = \sum_{k=-\infty}^{\infty} b_k \, p(t + j_k - kT) + n(t) \qquad (12.40)$$

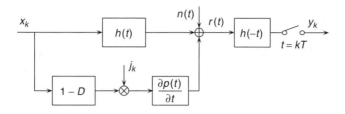

FIGURE 12.7 Magnetic recording channel with jitter noise.

where j_k is usually modeled as an independent zero-mean Gaussian noise term that reflects the amount of position jitter. In the examples given in this chapter, we only consider the position jitter as the media noise. In [15,41,43], width variations are also considered.

Using the Taylor series expansion and keeping the first two terms, $p(t + j_k) \approx p(t) + j_k \partial p(t)/\partial t$, we can obtain a first-order approximation to the jitter noise, which is accurate for the channels with low media noise levels [14]. The block diagram of the first-order channel model is shown in Figure 12.7, where the symbol D represents the delay operator. The equivalent discrete-time channel model can be obtained as shown in Figure 12.8, where the channel coefficients $\{g_k\}$ are defined as in Equation 12.6 and p_k^j can be computed by using

$$p_k^j = \int_{-\infty}^{\infty} \frac{\partial p(t)}{\partial t} h(t - kT) \, dt \tag{12.41}$$

and the covariance of the colored Gaussian noise v_k is

$$E\left[v_k v_{n+k}\right] = \sigma_n^2 \cdot g_k \tag{12.42}$$

where $\sigma_n^2 = N_0/2$.

From Figure 12.8, we can see that the overall noise consists of two different types of correlated Gaussian noise terms, one of which (i.e., media noise) also depends on the transmitted signal. Therefore, the overall noise can also be viewed as signal-dependent correlated Gaussian noise.

The SNR is defined as in Equation 12.10, and we define the standard deviation of j_k as σ_j. By this definition, we can study the effects of the media noise on the magnetic recording channels, which cannot be mitigated or eliminated by increasing the SNR. On the other hand, we can also define the SNR as the ratio of the signal energy to the energy of the total received noise, composed of the electronic noise and the media noise. In this way, we can compare the relative influences of the two kinds of noise as in [15], which shows that the media noise may be preferable in terms of the achievable information rates, compared to the electronic noise.

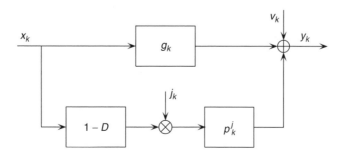

FIGURE 12.8 Equivalent discrete-time magnetic recording channel with jitter noise.

12.3.2   Estimation of Achievable Information Rates

As described in Section 12.2.2, we can generate a long simulation of the channel and estimate the achievable information rates by computing the joint probability of the output samples. In this case, since the noise is colored, the output sample of the channel is related to previous samples for a given trellis state transition. Therefore, we modify the forward recursion of the BCJR algorithm [24] by defining

$$\gamma'_k(i, j) = p(S_k = j, y_k \mid \mathbf{y}^{k-1}, S_{k-1} = i)$$

$$= p(y_k \mid \mathbf{y}^{k-1}, S_{k-1} = i, S_k = j) \cdot p(S_k = j \mid S_{k-1} = i) \quad (12.43)$$

Thus, the forward recursion becomes

$$\alpha'_k(j) = p(\mathbf{y}^k, S_k = j) \quad (12.44)$$

$$= \sum_i p(\mathbf{y}^k, S_{k-1} = i, S_k = j)$$

$$= \sum_i p(y_k \mid \mathbf{y}^{k-1}, S_{k-1} = i, S_k = j) \cdot p(S_k = j \mid S_{k-1} = i) \cdot p(\mathbf{y}^{k-1}, S_{k-1} = i)$$

$$= \sum_i \gamma'_k(i, j) \cdot \alpha'_{k-1}(i) \quad (12.45)$$

Finally, $p(\mathbf{y^n})$ can be computed by

$$p(\mathbf{y^n}) = \sum_j \alpha'_n(j) \quad (12.46)$$

We notice that $p(y_k \mid \mathbf{y}^{k-1}, S_{k-1} = i, S_k = j)$ is impractical to compute as k increases. But we know that, in reality, the correlation of two noise samples conditioned on other samples between them should decrease if they are further apart. This is also true for output samples given the trellis state transition. Based on this, we set a parameter, L, called the noise correlation delay. If L is chosen to be large enough, it is reasonable to use the following approximation

$$p(y_k \mid \mathbf{y}^{k-1}, S_{k-1} = i, S_k = j) \approx p\left(y_k \mid \mathbf{y}_{k-L}^{k-1}, S_{k-1} = i, S_k = j\right) \quad (12.47)$$

where

$$p\left(y_k \mid \mathbf{y}_{k-L}^{k-1}, S_{k-1} = i, S_k = j\right) = \frac{p\left(\mathbf{y}_{k-L}^{k}, S_{k-1} = i, S_k = j\right)}{p\left(\mathbf{y}_{k-L}^{k-1}, S_{k-1} = i, S_k = j\right)} \quad (12.48)$$

Both the denominator and the numerator can be computed given the covariance matrix of the colored Gaussian noise.

We can also compute the conditional probability using another method. Assume that $\mathbf{y} = [y_{k-L}, \ldots, y_k]^T$, $\boldsymbol{\mu} = E[\mathbf{y} \mid S_{k-1} = i, S_k = j] = [\mu_{k-L}, \ldots, \mu_k]^T$ and $\Sigma = E[(\mathbf{y} - \boldsymbol{\mu})(\mathbf{y} - \boldsymbol{\mu})^T]$, where ^{T} represents the transpose operator. We further denote $\mathbf{y}'_k = [y_{k-L}, \ldots, y_{k-1}]^T$ and $\boldsymbol{\mu}'_k = [\mu_{k-L}, \ldots, \mu_{k-1}]^T$, and partition $\boldsymbol{\mu}$ and Σ into

$$\boldsymbol{\mu} = \begin{pmatrix} \boldsymbol{\mu}'_k \\ \mu_k \end{pmatrix} \quad (12.49)$$

and

$$\Sigma = \begin{bmatrix} \Sigma_{11} & \Sigma_{12} \\ \Sigma_{21} & \Sigma_{22} \end{bmatrix} \quad (12.50)$$

where $\Sigma_{11} = E[(\mathbf{y}'_k - \boldsymbol{\mu}'_k)(\mathbf{y}'_k - \boldsymbol{\mu}'_k)^T]$, $\Sigma_{22} = E[(y_k - \mu_k)^2]$, $\Sigma_{12} = E[(\mathbf{y}'_k - \boldsymbol{\mu}'_k)(y_k - \mu_k)]$ and $\Sigma_{21} = (\Sigma_{12})^T$. Then, the conditional distribution of y_k given the previous L samples and the trellis state transition is normal with mean μ^* and variance Σ^*, given by [44]

$$\mu^* = \mu_k + \Sigma_{21}\Sigma_{11}^{-1}(\mathbf{y}_k' - \boldsymbol{\mu}_k') \tag{12.51}$$

and

$$\Sigma^* = \Sigma_{22} - \Sigma_{21}\Sigma_{11}^{-1}\Sigma_{12} \tag{12.52}$$

Therefore, we have

$$p\left(y_k \mid \mathbf{y}_{\mathbf{k-L}}^{\mathbf{k-1}}, S_{k-1} = i, S_k = j\right) = \frac{1}{\sqrt{2\pi\Sigma^*}}\exp\left\{-\frac{(y_k - \mu^*)^2}{2\Sigma^*}\right\} \tag{12.53}$$

Compared to Equation 12.48, its complexity is lower since we only need the variance Σ^* and the vector $\boldsymbol{\xi} = \Sigma_{21}\Sigma_{11}^{-1}$ in the computation.

By making the approximation in Equation 12.47, we indeed compute an upper bound on $h(Y)$. If $p(\mathbf{y^n})$ is the real PDF of the output sequence $\mathbf{y^n}$ and $q(\mathbf{y^n})$ is the PDF after the approximation (or any other PDF), then the relative entropy $D(p \mid q) = -E\left[\log(q(\mathbf{y^n}))\right] - h(\mathbf{y^n}) \geq 0$ [25], where $-E\left[\log(q(\mathbf{y^n}))\right]$ is in fact what we compute to estimate $h(\mathbf{y^n})$. Therefore, the estimation is an upper bound that will match the real entropy asymptotically with the increase of L. Since the complexity is increased exponentially with L, we choose L carefully based on the property of the correlated noise. One way to determine whether L is large enough for accurate estimation is to compute $h(Y)$ for increasing L until no further significant change occurs.

Here we also note that, if the noise is assumed to be Gauss-Markov as discussed in [13,45], the left and the right-hand sides of Equation 12.47 are exactly equal to each other when L is selected to be the Markov memory length of the noise.

Suppose that the memory of the ISI channel is M, we should set up a trellis with 2^{M+L} states. In addition, since the noise is colored, $h(Z)$ should be computed based on its covariance matrix. Assuming that the covariance matrix of the colored Gaussian noise is K_z and $K_z = \sigma_z^2 K_z'$, where σ_z^2 is the noise variance, then the differential entropy of the noise can be computed as

$$h(Z) = \frac{1}{2}\log\left(2\pi e\sigma_z^2\right) + \lim_{n\to\infty}\frac{1}{2n}\log(|K_z'|) \tag{12.54}$$

The second term is a constant independent of the noise power or the SNR value and can be estimated. However, for accurate results, a large n is necessary. A better way to solve this problem is to use the fact that K_z' is a Hermitian Toeplitz matrix [46]. If we define $t_{i-j} = K_z'(i, j)$, we can write

$$\lim_{n\to\infty}\frac{1}{n}\log(|K_z'|) = \frac{1}{2\pi}\int_0^{2\pi}\log f(\lambda)\, d\lambda \tag{12.55}$$

where

$$f(\lambda) = \sum_{k=-\infty}^{\infty} t_k e^{\sqrt{-1}k\lambda} \tag{12.56}$$

Since $t_k = t_{-k}$, $f(\lambda)$ can also be expressed as

$$f(\lambda) = t_0 + \sum_{k=1}^{\infty} 2t_k \cos(k\lambda) \tag{12.57}$$

As a result, only a single numerical integration is needed to obtain the differential entropy of the colored Gaussian noise.

To further extend the above approach to the ISI channels with signal-dependent correlated noise, we note that, at each stage of the trellis processing, the covariance of the output sequence is directly related to the transmitted signals and the trellis state. Therefore, to every branch in the trellis, we need to assign

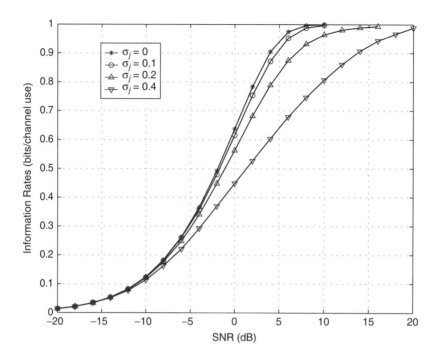

FIGURE 12.9   Information rates of the perpendicular recording channels with $D_n = 1.5$.

specific covariance matrices, which will be stored and used in the computation of the joint probabilities in Equation 12.48, or we need specific Σ^* and ξ for different signal patterns as needed in Equation 12.53. The number of different signal patterns is determined by the memory of the filters in the channel model [15]. In addition, since the noise is dependent on the input signal, we should compute $h(Z \mid X)$, instead of $h(Z)$. We can use a similar approach to estimate $h(Z \mid X)$ as the one used to estimate $h(Y)$, or employ another simulation-based method as described in [15].

In Figure 12.9, we show the information rates achievable over the perpendicular recording channel with i.i.d. symmetric binary inputs when the normalized recording density $D_n = 1.5$. We observe that, in the high SNR region, the jitter noise degrades the information rates much, while in the low SNR region, the jitter noise has less effects since the electronic noise dominates over the jitter noise.

Finally, we extend the optimization algorithm proposed in [10] to the channels with media noise to maximize the achievable information rates. The idea is to apply the BCJR algorithm (including both the forward recursion and backward recursion) for the ISI channels with signal-dependent correlated Gaussian noise. In addition to the definitions of $\alpha'_k(j)$ and $\gamma'_k(i,j)$ as in Equation 12.44 and Equation 12.43, we define

$$\beta'_k(j) = p\left(\mathbf{y}^{\mathbf{n}}_{\mathbf{k+1}} \mid \mathbf{y}^{\mathbf{k}}, S_k = j\right) \tag{12.58}$$

In this way, $\beta'_{k-1}(i)$ can be computed by the backward recursion as

$$\beta'_{k-1}(i) = \sum_j p\left(\mathbf{y}^{\mathbf{n}}_{\mathbf{k}}, S_k = j \mid \mathbf{y}^{\mathbf{k-1}}, S_{k-1} = i\right)$$

$$= \sum_j p(y_k, S_k = j \mid \mathbf{y}^{\mathbf{k-1}}, S_{k-1} = i) \cdot p\left(\mathbf{y}^{\mathbf{n}}_{\mathbf{k+1}} \mid \mathbf{y}^{\mathbf{k}}, S_{k-1} = i, S_k = j\right)$$

$$= \sum_j \gamma'_k(i,j) \cdot \beta'_k(j) \tag{12.59}$$

We compute $\gamma'_k(i,j)$ based on Equation 12.47 with signal-dependent covariance matrix of the noise.

To compute $p(S_{k-1} = i, S_k = j, \mathbf{y^n})$, we have

$$
\begin{aligned}
p(S_{k-1} = i, S_k = j, \mathbf{y^n}) &= p(S_{k-1} = i, \mathbf{y^{k-1}}) \cdot p\left(S_k = j, y_k, \mathbf{y^n_{k+1}} \mid S_{k-1} = i, \mathbf{y^{k-1}}\right) \\
&= p(S_{k-1} = i, \mathbf{y^{k-1}}) \cdot p(S_k = j, y_k \mid \mathbf{y^{k-1}}, S_{k-1} = i) \cdot p\left(\mathbf{y^n_{k+1}} \mid \mathbf{y^k}, S_k = j\right) \\
&= \alpha'_{k-1}(i) \cdot \gamma'_k(i, j) \cdot \beta'_k(j).
\end{aligned}
\tag{12.60}
$$

Thus, we have

$$
p(S_{k-1} = i, S_k = j \mid \mathbf{y^n}) = \alpha'_{k-1}(i) \cdot \gamma'_k(i, j) \cdot \beta'_k(j) / p(\mathbf{y^n})
\tag{12.61}
$$

where $p(\mathbf{y^n})$ can be computed by Equation 12.46. Then, $p(S_{k-1} = i \mid \mathbf{y^n})$ can be obtained by

$$
p(S_{k-1} = i \mid \mathbf{y^n}) = \sum_j p(S_{k-1} = i, S_k = j \mid \mathbf{y^n})
\tag{12.62}
$$

or,

$$
p(S_{k-1} = i \mid \mathbf{y^n}) = \alpha'_{k-1}(i)\beta'_{k-1}(i) / p(\mathbf{y^n})
\tag{12.63}
$$

Based on the above computations, we can run the maximization algorithm as described in [10]. We emphasize that we need a large enough L to achieve an accurate result.

We present an example for the perpendicular recording channels with normalized density of 1.5 when $\sigma_j = 0$ and 0.4, respectively. In Figure 12.10, we show both the symmetric information rates achieved by the i.i.d. symmetric inputs and the maximized information rates achieved by the optimized Markov inputs with a large memory. We see that in the low-to-medium SNR region, the information rates with optimized Markov inputs are significantly larger than the symmetric ones, while in the high SNR region (or, the high rate region), there is almost no observable difference.

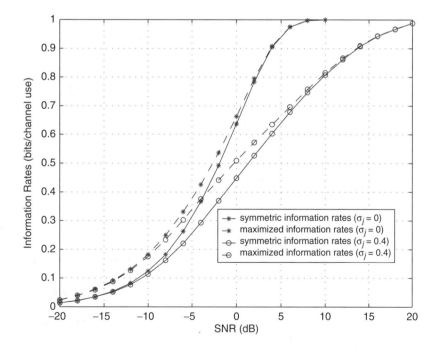

FIGURE 12.10 Maximized information rates of the perpendicular recording channels with $D_n = 1.5$.

12.4   Conclusions

In this chapter, we presented a survey of the information theoretical results for magnetic recording channels, which can be modeled as ISI channels. The capacity of such channels with AWGN, and the achievable information rates under the binary input constraints with or without the media noise are reviewed. Specifically, we discussed several lower and upper bounds, and described the simulation-based algorithms to estimate the achievable information rates with accuracy even when the first-order media noise is present. Both the cases with i.i.d. inputs and with (optimized) Markov inputs were discussed.

The simulation-based algorithm can be extended further to the more general case where the signal-dependent correlated noise is not restricted to be Gaussian distributed. This is implemented by using the canonical techniques of multivariate density estimation as considered in [47]. This approach may result in more accurate information rate estimates for realistic recording channels, in particular, when the media noise level is high.

References

[1] C.E. Shannon, A mathematical theory of communication, *Bell Sys. Tech. J.*, 27, 379–423, 623–656, July, October 1948.

[2] B.S. Tsybakov, Capacity of a discrete-time Gaussian channel with a filter, *Problemy Peredachi Informatsii*, 6, 3, 78–82, July–September 1970.

[3] W. Hirt and J.L. Massey, Capacity of the discrete-time Gaussian channel with intersymbol interference, *IEEE Trans. Inform. Theory*, 34, 3, 380–388, May 1988.

[4] S. Shamai, L.H. Ozarow, and A.D. Wyner, Information rates for a discrete-time Gaussian channel with intersymbol interference and stationary inputs, *IEEE Trans. Inform. Theory*, 37, 6, 1527–1539, November 1991.

[5] S. Shamai and R. Laroia, The intersymbol interference channel: Lower bounds on capacity and channel precoding loss, *IEEE Trans. Inform. Theory*, 42, 5, 1388–1404, September 1996.

[6] W. Hirt, Capacity and information rates of discrete-time channels with memory, Doctoral Dissert. (Diss. ETH No. 8671), Swiss Federal Inst. of Technol. (ETH), Zurich, Switzerland, 1988.

[7] D. Arnold and H.-A. Loeliger, On the information rate of binary-input channels with memory, in *Proceedings of IEEE International Conference on Communications (ICC)*, vol. 9, June 2001, pp. 2692–2695.

[8] H.D. Pfister, J.B. Soriaga, and P.H. Siegel, On the achievable information rates of finite state ISI channels, in *Proceedings of IEEE Global Communications Conference (GLOBECOM)*, vol. 5, November 2001, pp. 2992–2996.

[9] V. Sharma and S.K. Singh, Entropy and channel capacity in the regenerative setup with applications to Markov channels, in *Proceedings of IEEE International Symposium on Information Theory (ISIT)*, June 2001, p. 283.

[10] A. Kavčić, On the capacity of Markov sources over noisy channels, in *Proceedings of IEEE Global Communications Conference (GLOBECOM)*, vol. 5, November 2001, pp. 2997–3001.

[11] J.G. Zhu and H. Wang, Noise characteristics of interacting transitions in longitudinal thin film media, *IEEE Trans. Magn.*, 31, 2, 1065–1070, March 1995.

[12] J. Caroselli and J.K. Wolf, Applications of a new simulation model for media noise limited magnetic recording channels, *IEEE Trans. Magn.*, 32, 5, 3917–3919, September 1996.

[13] A. Kavčić and A. Patapoutian, A signal-dependent autoregressive channel model, *IEEE Trans. Magn.*, 35, 5, 2316–2318, September 1999.

[14] T.R. Oenning and J. Moon, Modeling the Lorentzian magnetic recording channel with transition noise, *IEEE Trans. Magn.*, 37, 1, 583–591, January 2001.

[15] Z. Zhang, T.M. Duman, and E.M. Kurtas, Information rates of binary-input intersymbol interference channels with signal-dependent media noise, *IEEE Trans. Magn.*, 39, 1, 599–607, January 2003.

[16] S. Yang and A. Kavčić, On the capacity of data-dependent autoregressive noise channels, in *Proceedings of Allerton Conference on Communications, Control and Computing*, Urbana, IL, October 2002.

[17] J.G. Proakis, Equalization techniques for high-density magnetic recording, *IEEE Signal Processing Mag.*, 15, 4, 73–82, July 1998.

[18] M.F. Erden, I. Ozgunes, E.M. Kurtas, and W. Eppler, General transform filters in perpendicular recording architectures, *IEEE Trans. Magn.*, 38, 5, 2334–2336, September 2002.

[19] J.G. Proakis, *Digital Communications*, McGraw-Hill, Inc., New York, NY, 2001.

[20] G.D. Forney, Maximum-likelihood sequence estimation of digital sequences in the presence of intersymbol interference, *IEEE Trans. Inform. Theory*, 18, 3, 363–378, May 1972.

[21] H.K. Thapar and A.M. Patel, A class of partial response systems for increasing storage density in magnetic recording, *IEEE Trans. Magn.*, 23, 5, 3666–3668, September 1987.

[22] R.G. Gallager, *Information Theory and Reliable Communication*, John Wiley & Sons, New York, NY, 1968.

[23] W. Xiang and S.S. Pietrobon, On the capacity and normalization of ISI channels, *IEEE Trans. Inform. Theory*, 49, 9, 2263–2268, September 2003.

[24] L.R. Bahl, J. Cocke, F. Jelinek, and J. Raviv, Optimal decoding of linear codes for minimizing symbol error rate, *IEEE Trans. Inform. Theory*, 20, 284–287, March 1974.

[25] T.M. Cover and J.A. Thomas, *Elements of Information Theory*, John Wiley & Sons, New York, NY, 1991.

[26] B.G. Leroux, Maximum-likelihood estimation for hidden Markov models, *Stochastic Processes and Their Applications*, 40, 127–143, 1992.

[27] B.H. Marcus, R.M. Roth, and P.H. Siegel, An introduction to coding for constrained systems, [online]. Available: http://www.stanford.edu/class/ee392p.

[28] J.L. Fan, T.L. Poo, and B.H. Marcus, Constraint gain, in *Proceedings of IEEE International Symposium on Information Theory (ISIT)*, June-July 2002, p. 326.

[29] Z. Zhang, T.M. Duman, and E.M. Kurtas, Achievable information rates for magnetic recording channels with constrained codes, in *Proceedings of 41st Allerton Conference on Communications, Control and Computing*, October 2003, pp. 853–862.

[30] Z. Zhang, T.M. Duman, and E.M. Kurtas, Quantifying information rate increase by constrained codes for recording channels with media noise, *IEE Electron. Lett.*, 39, 11, 852–854, May 2003.

[31] S. Yang and A. Kavčić, Capacity of partial response channels, in *Coding and Recording Systems Handbook*, B. Vasic and E. M. Kurtas, Eds., CRC Press, Boca Raton 2003.

[32] P.O. Vontobel and D.M. Arnold, An upper bound on the capacity of channels with memory and constraint input, in *Proceedings of IEEE Information Theory Workshop (ITW)*, September 2001, pp. 147–149.

[33] S. Yang, A. Kavčić, and S. Tatikonda, Delayed feedback capacity of finite-state machine channels: Upper bounds on the feedforward capacity, in *Proceedings of IEEE International Symposium on Information Theory (ISIT)*, June-July 2003, p. 290.

[34] H.N. Bertram, *Theory of Magnetic Recording*, Cambridge University Press, Cambridge, U.K. 1994.

[35] C.A. French and J.K. Wolf, Bounds on the capacity of a peak power constrained Gaussian channel, *IEEE Trans. Magn.*, 24, 5, 2247–2262, September 1988.

[36] S.W. McLaughlin and D.L. Neuhoff, Upper bounds on the capacity of the digital magnetic recording channel, *IEEE Trans. Magn.*, 29, 1, 59–66, January 1993.

[37] Z.-N. Wu, S. Lin, and J.M. Cioffi, Numerical results on capacity bounds for magnetic recording channels, in *Proceedings of IEEE Global Communications Conference (GLOBECOM)*, vol. 6, November 1998, 3385–3390.

[38] T.R. Oenning and J. Moon, The effect of jitter noise on binary input intersymbol interference channel capacity, in *Proceedings of IEEE International Conference on Communications (ICC)*, vol. 8, June 2001, pp. 2416–2420.

[39] D. Arnold and E. Eleftheriou, On the information-theoretic capacity of magnetic recording systems in the presence of medium noise, *IEEE Trans. Magn.*, 38, 5, 2319–2321, September 2002.

[40] D. Arnold and E. Eleftheriou, Computing information rates of magnetic recording channels in the presence of medium noise, in *Proceedings of IEEE Global Communications Conference (GLOBECOM)*, vol. 2, November 2002, pp. 1344–1348.

[41] J. Moon, Discrete-time modeling of transition-noise-dominant channels and study of detection performance, *IEEE Trans. Magn.*, 27, 6, 4573–4578, November 1991.

[42] N.R. Belk, P.K. George, and G.S. Mowry, Noise in high performance thin-film longitudinal magnetic recording media, *IEEE Trans. Magn.*, 21, 5, 1350–1355, September 1985.

[43] J. Moon and L.R. Carley, Detection performance in the presence of transition noise, *IEEE Trans. Magn.*, 26, 5, 2172–2174, September 1990.

[44] Y.L. Tong, *The Multivariate Normal Distribution,* Springer-Verlag, New York, NY, 1990.

[45] A. Kavčić and J.M.F. Moura, The Viterbi algorithm and Markov noise memory, *IEEE Trans. Inform. Theory,* 46, 1, 291–301, January 2000.

[46] R.M. Gray, Toeplitz and circulant matrices: A review, [online]. Available: http://ee-www.stanford.edu/~gray/toeplitz.pdf.

[47] Z. Zhang, T.M. Duman, and E.M. Kurtas, Calculating the capacity of recording channels with multivariate density estimation, in *Digests of 7th Perpendicular Magnetic Recording Conference (PMRC)*, May-June 2004, pp. 239–240.

13

Capacity of Partial Response Channels

13.1 Introduction **13**-1
13.2 Channel Model **13**-2
13.3 Information Rate and Channel Capacity **13**-3
13.4 Computing the Information Rate for a
Markov Source.................................. **13**-4
A Stationary Markov Source of Order M_s • Reformulating
the Expression of the Mutual Information Rate • A Monte
Carlo Method to Compute the Mutual Information Rate
13.5 Tight Channel Capacity Lower Bounds — Maximal
Information Rates for Markov Sources **13**-7
Reformulating the Expression of the Mutual Information Rate
• Iterative Markov Source Optimization Algorithm
13.6 A Tight Channel Capacity Upper Bound — Delayed
Feedback Capacity **13**-11
Markov Sources Achieve the Feedback Capacity C^{fb}
• A Stochastic Control Formulation for Feedback Capacity
Computation • Feedback Capacity Computation
• Delayed Feedback Capacity Computation
13.7 Vontobel-Arnold Upper Bound **13**-15
13.8 Conclusion and Extensions Beyond Partial Response
Channels **13**-17
Media Noise • Long Impulse Responses

Shaohua Yang

Aleksandar Kavčić
Harvard University
Boston, MA

13.1 Introduction

The partial response (PR) channel is the simplest model of a magnetic recording channel. Despite its simplicity, the computation of the capacity of a PR channel has long remained an open problem [1]. Only very recently, algorithmic techniques to very tightly bound (from below and above) the capacity of a PR channel have emerged. This monograph presents a summary of these techniques for bounding the capacity of a PR channel.

Most algorithms that we will present here rest on the algorithm independently proposed by Arnold and Loeliger [2], and by Pfister et al. [3] (see also [4]), which uses a Monte Carlo simulation to compute the information rate of a PR channel when the channel input is a Markov process. Subsequently, an algorithm was proposed by Kavčić [5] to optimize the Markov source and thus obtain (tight) lower bounds on the capacity. Two methods are known for computing (tight) upper bounds. The first, proposed by Yang et al. [6], is obtained by computing the feedback capacity (and the delayed feedback capacity) of a PR channel. The second is a method by Vontobel and Arnold [**?**], which rests on an information-theoretic inequality for the channel output sequence obtained by a (close to) optimal input Markov sequence.

Of course, to compute the capacity of a real magnetic recording channel, we cannot consider only simple PR channel models. Real magnetic recording channels have infinitely long impulse responses, and are corrupted by signal-dependent media noise. We will only give a superficial survey of the information rate computation methods for such realistic channels at the end of the monograph, but the reader is referred to [7], where these methods are more thoroughly presented.

In Section 13.2, we introduce the PR channel. In Section 13.3, we define information rates and capacities of PR channels. The computation of the information rate of a PR channel for a Markov channel input process is given in Section 13.4. In Section 13.5, we present a method for optimizing Markov sources to produce tight lower bounds on the channel capacity. Two different upper-bounding methods, one based on computing the feedback capacity and the other based on an information-theoretic inequality, are presented in Section 13.6 and Section 13.7, respectively. In Section 13.8 we conclude the monograph and give a survey of the methods for information rate and capacity computations beyond simple PR channels.

13.2   Channel Model

We consider the partial response (PR) channel [8], that is, the binary-input linear intersymbol interference (ISI) channel with a finite-length input response. Let the channel input at time t be X_t and its realization be $x_t \in \mathcal{X} = \{+1, -1\}$. Let the channel output be Y_t and its realization be $y_t \in \mathbb{R}$. A partial response (PR) channel (X, Y) with a memory length M_c is then defined by the following channel law

$$Y_t = \sum_{k=0}^{M_c} a_k X_{t-k} + N_t \tag{13.1}$$

Here, N_t is additive white Gaussian noise (AWGN) with variance σ^2, whose realization is denoted by n_t, and the channel coefficients $a_k \in \mathbb{R}$, for $k = 0, 1, \ldots, M_c$, characterize the intersymbol interference (ISI) in the channel. The channel coefficients a_k may also be specified by their D transform $a(D) \triangleq \sum_{k=0}^{M_c} a_k D^k$. We assume that the channel coefficients are normalized such that $\sum_{k=0}^{M_c} (a_k)^2 = 1$. Then the signal to noise ratio (SNR) is SNR $\triangleq 1/\sigma^2$. For such a channel law, the channel output is statistically dependent on $M_c + 1$ channel input symbols only, that is, the conditional probability density function (pdf) of the channel output Y_t satisfies

$$f_{Y_t | X_{-\infty}^t, Y_{-\infty}^{t-1}}\left(y_t \mid x_{-\infty}^t, y_{-\infty}^{t-1}\right) = f_{Y_t | X_{t-M_c}^t}\left(y_t \mid x_{t-M_c}^t\right) \tag{13.2}$$

Example 13.1 (The dicode partial response channel)

The PR channel with channel coefficients specified by $a(D) = (1 - D)/\sqrt{2}$ is also called as the dicode channel. For $X_t \in \mathcal{X} = \{+1, -1\}$, the channel law for the dicode PR channel is

$$Y_t = \sqrt{\frac{1}{2}}(X_t - X_{t-1}) + N_t \tag{13.3}$$

where N_t is additive white Gaussian noise with variance σ^2, denoted by $N_t \sim \mathcal{N}(0, \sigma^2)$. The conditional density function of the channel output equals

$$f_{Y_t | X_{t-1}, X_t}(y_t \mid x_{t-1}, x_t) = \frac{1}{\sqrt{2\pi\sigma^2}} e^{-\frac{|y_t - (x_t - x_{t-1})/\sqrt{2}|^2}{2\sigma^2}} \tag{13.4}$$

13.3   Information Rate and Channel Capacity

For the PR channel, we define the information rate between the channel input (source) and the channel output as

$$\mathcal{I}(X;Y) \overset{\triangle}{=} \lim_{n\to\infty} \frac{1}{n} I\left(X_1^n; Y_1^n \,\middle|\, X_{-\infty}^0\right) \tag{13.5}$$

where $I(X_1^n; Y_1^n \mid X_{-\infty}^0)$ is the conditional mutual information between the channel inputs X_1^n and outputs Y_1^n conditioned on the initial channel status $X_{-\infty}^0$. The conditional mutual information [9] between two random variables (or vectors) U and V conditioned on W is defined as the following expectation

$$I(U;V \mid W) \overset{\triangle}{=} \mathrm{E}\left[\log\left(\frac{P(U,V \mid W)}{P(U \mid W)P(V \mid W)}\right)\right] \tag{13.6}$$

where the notation $P(\cdot \mid \cdot)$ represents either the conditional probability mass function of a discrete random variable (or vector), or the conditional probability density function (pdf) of a continuous random variable (or vector). We use the base-2 logarithm throughout, thus the unit of information is measured in *bits*. Obviously, the value of the information rate $\mathcal{I}(X;Y)$ is dependent on the source distribution, that is, the distribution of the channel input sequence $[X_1, X_2, \ldots]$

$$P\left(X_t \,\middle|\, X_{-\infty}^{t-1}\right) = \Pr\left(X_t \,\middle|\, X_{-\infty}^{t-1}\right), \quad \text{for } t = 1, 2, \ldots \tag{13.7}$$

We define the channel capacity C as the supremum of the mutual information rate

$$C \overset{\triangle}{=} \sup \mathcal{I}(X;Y) \tag{13.8}$$

where the supremum is taken over all the possible source distributions (Equation 13.7). Shown by Shannon [10], the channel capacity C equals the supremum of achievable code rates, and can be equivalently defined as follows:

1. An (n, M, ϵ) code is a code whose block length is n, whose number of codewords is M, and whose average probability of error is not larger than ϵ. The rate of the code is $\frac{\log M}{n}$.
2. A rate $R \geq 0$ is called an ϵ-achievable rate if, for every $\delta > 0$, and any sufficiently large n, there exists an (n, M, ϵ) code with rate

$$\frac{\log M}{n} > R - \delta \tag{13.9}$$

3. The maximum ϵ-achievable rate is called the ϵ-capacity C_ϵ.
4. The capacity C is defined as the maximal rate that is ϵ-achievable for all $0 < \epsilon < 1$.

We note that the PR channel generally has a nonzero input memory length M_c. To achieve the capacity C of a channel with memory, we typically need to utilize nonlinear codes [?, 11]. Random linear codes, such as the low density parity check (LDPC) code [12] or the turbo code [13], typically cannot approach the capacity C. For the PR channel, the code-rate limit of random linear codes equals the so called i.u.d. channel capacity C_{iud}, which is the information rate induced by an independent and uniformly distributed (i.u.d.) channel input sequence, that is

$$C_{\text{iud}} \overset{\triangle}{=} \mathcal{I}(X;Y) \tag{13.10}$$

where

$$P\left(X_t \,\middle|\, X_{-\infty}^{t-1}\right) = P(X_t) = \frac{1}{2}, \quad \text{for any } t > 0 \tag{13.11}$$

The i.u.d. source typically does not achieve the supremum in Equation 13.8, and we have

$$C_{\text{iud}} \leq C \tag{13.12}$$

where the inequality is usually strict for PR channels.

13.4 Computing the Information Rate for a Markov Source

In this section, we show a Monte Carlo algorithm [2–4] to compute the information rate for a stationary Markov source distribution, which is the basic tool we will use throughout this chapter.

13.4.1 A Stationary Markov Source of Order M_s

Definition 13.1 (Stationary Markov source distribution)   A stationary Markov source satisfies

$$\Pr\left(X_t = x_t \mid X_{-\infty}^{t-1} = x_{-\infty}^{t-1}\right) = \Pr\left(X_t = x_t \mid X_{t-M_s}^{t-1} = x_{t-M_s}^{t-1}\right) = P\left(x_t \mid x_{t-M_s}^{t-1}\right) \tag{13.13}$$

where the source distribution function $P(\cdot \mid \cdot)$ is time-invariant and the integer M_s is called the *order* (or memory length) of the stationary Markov source.

Definition 13.2 (Joint source/channel state and memory length)   For a PR channel with memory length M_c, if a stationary Markov source of order M_s is used, then we define the joint source/channel memory length as

$$M_{sc} \triangleq \max(M_c, M_s) \tag{13.14}$$

and define the joint source/channel state as

$$S_t \triangleq X_{t-M_{sc}+1}^t = [X_{t-M_{sc}+1}, X_{t-M_{sc}+2}, \ldots, X_t], \quad \text{for } t = 1, 2, \ldots \tag{13.15}$$

The realization of the source/channel state S_t is denoted by s_t.

By Definition 13.2, the set of all possible source/channel states is $\mathcal{S} \triangleq \mathcal{X}^{M_{sc}} = \{+1, -1\}^{M_{sc}}$. It is convenient to index the elements of the set $\mathcal{S}$ and refer to each element $s_t \in \mathcal{S}$ by its index. Here, we index the state s_t by the integer value corresponding to the binary string $[x_{t-M_{sc}+1}, x_{t-M_{sc}+2}, \ldots, x_t]$, that is, the index of state s_t is computed as $s_t = \sum_{k=0}^{M_{sc}-1}(x_{t-k} + 1)2^{k-1}$. Thus, we have $\mathcal{S} = \{0, 1, 2, \ldots, 2^{M_{sc}} - 1\}$.

The source/channel state S_t forms a stationary irreducible Markov process. We note that not all state pairs (i, j), where $i, j \in \mathcal{S}$, are valid, for example, the state pair $(s_{t-1}, s_t) = (0, 2^{M_{sc}-1})$ is not valid, this is because $s_{t-1} = 0$ requires $x_{t-M_{sc}+1} = -1$ and $s_t = 2^{M_{sc}-1}$ requires $x_{t-M_{sc}+1} = +1$, which cause a contradiction. A state pair (i, j) is *valid* if and only if $(i, j) \in \mathcal{T} \triangleq \mathcal{X}^{M_{sc}+1}$, where by definition $\mathcal{T}$ is the set of valid state transitions ($\mathcal{T}$ is also known as a trellis section). In the following, we only consider valid state transitions $(i, j) \in \mathcal{T}$. The Markov source (Equation 13.13) can be equivalently represented by the set of state transition probabilities

$$P_{ij} \triangleq \Pr(S_t = j \mid S_{t-1} = i), \quad \text{for any } t \geq 1, \quad \text{and} \quad \text{any } (i, j) \in \mathcal{T} \tag{13.16}$$

For the PR channel, the initial source/channel state s_0 is usually assumed to be known. Given the initial source/channel state $s_0 = x_{-M_{sc}+1}^0 = [x_{-M_{sc}+1}, x_{-M_{sc}+2}, \ldots, x_0]$, the state sequence s_1^t and the input sequence x_1^t determine each other uniquely. Thus, we may interchangeably use the state symbol S_t for the input symbol X_t. By Equation 13.2, the channel output Y_t is statistically dependent only on the state transition $s_{t-1}^t \in \mathcal{T}$, that is,

$$f_{Y_t \mid S_{-\infty}^t, Y_{-\infty}^{t-1}}\left(y_t \mid s_{-\infty}^t, y_{-\infty}^{t-1}\right) = f_{Y_t \mid S_{t-1}^t}\left(y_t \mid s_{t-1}^t\right) \tag{13.17}$$

s_t	$[x_{t-2}, x_{t-1}, x_t]$
0	$[-1, -1, -1]$
1	$[-1, -1, +1]$
2	$[-1, +1, -1]$
3	$[-1, +1, +1]$
4	$[+1, -1, -1]$
5	$[+1, -1, +1]$
6	$[+1, +1, -1]$
7	$[+1, +1, +1]$

$$\mathcal{T} = \begin{cases} (0,0), (0,1), \\ (1,2), (1,3), \\ (2,4), (2,5), \\ (3,6), (3,7), \\ (4,0), (4,1), \\ (5,2), (5,3), \\ (6,4), (6,5), \\ (7,6), (7,7) \end{cases}$$

FIGURE 13.1 Source/channel states for the dicode PR channel and a second-order Markov source.

Definition 13.3 (Stationary source/channel state distribution) Let μ_i be the steady state probability of state $i \in \mathcal{S}$, that is,

$$\mu_i \triangleq \lim_{t \to \infty} \Pr(S_t = i) \tag{13.18}$$

For every possible value $j \in \mathcal{S}$, we have $\mu_j = \sum_{i:(i,j) \in \mathcal{T}} \mu_i P_{ij}$ and $\mu_j \geq 0$.

Example 13.2 (The dicode partial response channel with a second-order Markov source)

For the dicode channel in Example 13.1, assuming that a second-order Markov source is used, we have $M_c = 1$, $M_s = 2$ and $M_{sc} = \max(M_c, M_s) = 2$. The states and the set $\mathcal{T}$ of valid state transitions are depicted in Figure 13.1. The total number of states is $2^3 = 8$, and there are 16 valid state transitions in $\mathcal{T}$.

13.4.2 Reformulating the Expression of the Mutual Information Rate

For the PR channel, if a Markov source (Equation 13.16) is used to generate the channel inputs (equivalently, generate the channel states), the information rate (Equation 13.5) becomes

$$\mathcal{I}(X;Y) = \lim_{n \to \infty} \frac{1}{n} I\left(X_1^n; Y_1^n \mid X_{-M_{sc}+1}^0\right) \tag{13.19}$$

$$= \lim_{n \to \infty} \frac{1}{n} I\left(S_1^n; Y_1^n \mid S_0\right) \tag{13.20}$$

$$= \lim_{n \to \infty} \frac{1}{n} \sum_{t=1}^{n} \left[h\left(Y_t \mid S_0, Y_1^{t-1}\right) - h\left(Y_t \mid S_0^n, Y_1^{t-1}\right)\right] \tag{13.21}$$

where $h(U \mid V)$ denotes the conditional differential entropy of a continuous random variable (or random vector) U given the random variable (or random vector) V, which is defined as the following expectation

$$h(U \mid V) \triangleq \mathrm{E}[-\log(f_{U|V}(U \mid V))] \tag{13.22}$$

Since the channel output Y_t is dependent only on the state transition (S_{t-1}, S_t) and the channel noise is AWGN with variance σ^2, we have [9]

$$h\left(Y_t \mid S_0^n, Y_1^{t-1}\right) = h\left(Y_t \mid S_{t-1}^t\right) = \frac{1}{2} \log(2\pi e \sigma^2) \tag{13.23}$$

By substituting Equation 13.23 into Equation 13.21, we have

$$\mathcal{I}(X;Y) = \lim_{n \to \infty} \frac{1}{n} \sum_{t=1}^{n} \left[h\left(Y_t \mid S_0, Y_1^{t-1}\right) \right] - \frac{1}{2} \log(2\pi e \sigma^2) \tag{13.24}$$

$$= \lim_{n \to \infty} \frac{1}{n} \mathrm{E} \left[-\sum_{t=1}^{n} \log\left(f_{Y_t \mid S_0, Y_1^{t-1}}\left(Y_t \mid S_0, Y_1^{t-1}\right) \right) \right] - \frac{1}{2} \log(2\pi e \sigma^2) \tag{13.25}$$

where the expectation is over the random variables Y_1^n and S_0. As shown in [2–4], the following equality holds for the stationary Markov source (Equation 13.26)

$$\mathcal{I}(X;Y) \stackrel{\text{w.p.}1}{=} \lim_{n \to \infty} \frac{1}{n} \left[-\sum_{t=1}^{n} \log\left(f_{Y_t \mid S_0, Y_1^{t-1}}\left(y_t \mid s_0, y_1^{t-1}\right) \right) \right] - \frac{1}{2} \log(2\pi e \sigma^2) \tag{13.26}$$

where $\stackrel{\text{w.p.}1}{=}$ represents equality with probability 1 (or almost surely).

13.4.3 A Monte Carlo Method to Compute the Mutual Information Rate

In order to evaluate the conditional pdf's $f_{Y_t \mid S_0, Y_1^{t-1}}(y_t \mid s_0, y_1^{t-1})$ in the mutual information rate formula (Equation 13.26), we first review the forward sum-product recursion of the BCJR (or Baum-Welch) algorithm [14–17].

Definition 13.4 (Posterior state distribution)   Define the time-dependent function $\alpha_t(s)$, for $s \in \mathcal{S}$ and $t \geq 0$, as the posterior source/channel state probability mass function, that is

$$\alpha_t(s) \stackrel{\triangle}{=} \Pr\left(S_t = s \mid S_0 = s_0, Y_1^t = y_1^t \right) \tag{13.27}$$

Obviously, the function $\alpha_t(s)$ must satisfy $\sum_{s \in \mathcal{S}} \alpha_t(s) = 1$ and $\alpha_t(s) \geq 0$.

Algorithm 13.1

The Forward Recursion of the BCJR (or Baum-Welch) Algorithm.

Input: Obtain the values $\alpha_0(s)$ for all $s \in \mathcal{S}$, and the realization of y_0^n.
Recursions: For $t = 1, 2, \ldots, n$, recursively compute $f_{Y_t \mid S_0, Y_1^{t-1}}(y_t \mid s_0, y_1^{t-1})$ and $\alpha_t(s)$ as

$$f_{Y_t \mid S_0, Y_1^{t-1}}\left(y_t \mid s_0, y_1^{t-1}\right) = \sum_{i,j:(i,j)\in\mathcal{T}} \alpha_{t-1}(i) P_{ij} f_{Y_t \mid S_{t-1}, S_t}(y_t \mid i, j), \tag{13.28}$$

$$\alpha_t(s) = \frac{\displaystyle\sum_{i:(i,s)\in\mathcal{T}} \alpha_{t-1}(i) P_{is} f_{Y_t \mid S_{t-1}, S_t}(y_t \mid i, s)}{\displaystyle\sum_{i,j:(i,j)\in\mathcal{T}} \alpha_{t-1}(i) P_{ij} f_{Y_t \mid S_{t-1}, S_t}(y_t \mid i, j)}, \quad \text{for } s \in \mathcal{S}. \tag{13.29}$$

Output: Return the values of $f_{Y_t \mid S_0, Y_1^{t-1}}(y_t \mid s_0, y_1^{t-1})$ and $\alpha_t(s)$ for $1 \leq t \leq n$ and all $s \in \mathcal{S}$.

With the help of the forward sum-product recursion of the BCJR (or Baum-Welch) algorithm (Algorithm 13.1), we can compute the information rate $\mathcal{I}(X;Y)$ for a stationary Markov source distribution P_{ij}, where $(i,j) \in \mathcal{T}$, using the following algorithm [2–4].

Algorithm 13.2

Computing Information Rate for a Markov Source. (Arnold-Loeliger, Pfister-Soriaga-Siegel, Sharma-Singh)

Step 1: Select a large integer n, e.g., $n = 10^7$.

Step 2: Let $s_0 = 0$, generate s_1^n using the Markov source distribution $\{P_{ij} : (i, j) \in \mathcal{T}\}$ and pass them through the noisy PR channel to get y_1^n.

Step 3: Let $\alpha_0(s) = \begin{cases} 1 \text{ if } s = 0, \\ 0 \text{ otherwise.} \end{cases}$ and apply Algorithm 1 to compute the values of $f_{Y_t|S_0, Y_1^{t-1}}(y_t \mid s_0, y_1^{t-1})$

for $t = 1, 2, \dots, n$.

Step 4: Compute the information rate $\mathcal{I}(X; Y)$ as

$$\mathcal{I}(X; Y) \approx \frac{1}{n} \left[-\sum_{t=1}^n \log\left(f_{Y_t|S_0, Y_1^{t-1}}\left(y_t \mid s_0, y_1^{t-1}\right) \right) \right] - \frac{1}{2} \log(2\pi e \sigma^2) \tag{13.30}$$

We note that the i.u.d. source is a stationary Markov source of zero order, that is, $M_s = 0$. Thus Algorithm 13.2 is also used to compute the i.u.d. channel capacity C_{iud}.

Example 13.3 (The i.u.d. channel capacity for the dicode partial response channel)

The dicode PR channel was defined in Example 13.1. For the i.u.d. source distribution (Equation 13.11), we have

$$M_{sc} = \max(M_s, M_c) = \max(0, 1) = 1 \tag{13.31}$$

$$S_t \overset{\triangle}{=} X_t \tag{13.32}$$

We index the state as $s_t = \sum_{k=0}^{M_{sc}-1}(X_k + 1)2^{k-1} = (X_k + 1)/2$, that is, $s_t = 0$ corresponds to $x_t = -1$ and $s_t = 1$ corresponds to $x_t = +1$. The set of valid state transitions $\{(s_{t-1}, s_t)\}$ is

$$\mathcal{T} = \{(0, 0), (0, 1), (1, 0), (1, 1)\} \tag{13.33}$$

The set of state transition probabilities $\{P_{ij} \overset{\triangle}{=} \Pr(S_t = j \mid S_{t-1} = i)\}$ is

$$\{P_{00} = 0.5, P_{01} = 0.5, P_{10} = 0.5, P_{11} = 0.5\} \tag{13.34}$$

The i.u.d. channel capacity C_{iud} computed by Algorithm 13.2 is depicted in Figure 13.2 at the end of this monograph.

13.5 Tight Channel Capacity Lower Bounds — Maximal Information Rates for Markov Sources

In this section, we compute the maximal information rate for a stationary Markov source of order M_s, which serves as a lower bound on the channel capacity C.

Definition 13.5 (Maximal information rate for a Markov source) We denote C_{M_s} as the maximal information rate for a Markov source of order M_s, that is,

$$C_{M_s} \overset{\triangle}{=} \max \mathcal{I}(X; Y) \tag{13.35}$$

where the maximization is over the transition probabilities of all Markov sources whose order (or memory length) is $M_s > 0$.

In general, for the maximal information rate C_{M_s} achieved by a Markov source of order M_s, we have

$$C \geq \cdots \geq C_2 \geq C_1 \geq C_{\text{iud}} \tag{13.36}$$

$$C = \lim_{M_s \to \infty} C_{M_s} \tag{13.37}$$

Suggested by Equation 13.36 and Equation 13.37, we can use C_{M_s} to lower bound the channel capacity C, and this lower bound can be made arbitrarily close to C by increasing the source memory M_s.

In this section, we apply Kavčić's iterative optimization algorithm [5] to optimize the Markov source distribution and thus compute (or estimate) C_{M_s}. For the purpose of bounding the channel capacity C, it suffices to only consider the case when $M_{sc} = M_s \geq M_c$. Thus, we consider the following optimization problem

$$C_{M_s} = \max_{\{P_{ij}:(i,j)\in T\}} \mathcal{I}(X;Y) \tag{13.38}$$

where the maximization is over the Markov transition probabilities $\{P_{ij} : (i,j) \in T\}$. Our goal is to find the transition probabilities P_{ij} that achieve the maximization in Equation 13.38.

13.5.1 Reformulating the Expression of the Mutual Information Rate

Using the chain rule, the Markov property and stationarity, we rewrite the mutual information rate (Equation 13.5) as

$$
\begin{aligned}
\mathcal{I}(X;Y) &= \lim_{n\to} \frac{1}{n} \sum_{t=1}^{n} I\left(S_t; Y_1^n \mid S_0^{t-1}\right) \\
&= \lim_{n\to} \frac{1}{n} \sum_{t=1}^{n} I\left(S_t; Y_1^n \mid S_{t-1}\right) \\
&= \underbrace{\lim_{n\to\infty} \frac{1}{n} \sum_{t=1}^{n} H(S_t \mid S_{t-1})}_{\sum_{i,j:(i,j)\in T} \mu_i P_{ij} \log \frac{1}{P_{ij}}} - \lim_{n\to\infty} \frac{1}{n} \sum_{t=1}^{n} H\left(S_t \mid S_{t-1}, Y_1^n\right)
\end{aligned} \tag{13.39}
$$

where the function $H(U \mid V)$ denotes the entropy of the discrete random variable (or vector) U given the random variable (or random vector) V and is defined as

$$H(U \mid V) \triangleq \mathrm{E}[-\log(f_{U|V}(U \mid V))] \tag{13.40}$$

We express the conditional entropy in the second term in Equation 13.39 as

$$
\begin{aligned}
-H\left(S_t \mid S_{t-1}, Y_1^n\right) &= \mathrm{E}\left[\log \Pr\left(S_t \mid S_{t-1}, Y_1^n\right)\right] \\
&= \mathrm{E}\left[\log \Pr\left(S_{t-1}, S_t \mid Y_1^n\right)\right] - \mathrm{E}\left[\log \Pr\left(S_{t-1} \mid Y_1^n\right)\right] \\
&= \sum_{i,j:(i,j)\in T} \mu_i P_{ij} \mathrm{E}_{Y_1^n|i,j}\left[\log \Pr\left(S_{t-1} = i, S_t = j \mid Y_1^n\right)\right] \\
&\quad - \sum_{i:i\in S} \mu_i \mathrm{E}_{Y_1^n|i}\left[\log \Pr\left(S_{t-1} = i \mid Y_1^n\right)\right] \\
&= \sum_{i,j:(i,j)\in T} \mu_i P_{ij} \mathrm{E}_{Y_1^n|i,j}\left[\log \Pr\left(S_{t-1} = i, S_t = j \mid Y_1^n\right)\right] \\
&\quad - \sum_{i,j:(i,j)\in T} \mu_i P_{ij} \mathrm{E}_{Y_1^n|i,j}\left[\log \Pr\left(S_{t-1} = i \mid Y_1^n\right)\right]
\end{aligned} \tag{13.41}
$$

Here, $\mathrm{E}_{Y_1^n|i,j}$ denotes the conditional expectation taken over the variable Y_1^n when the pair (S_{t-1}, S_t) equals (i, j). Similarly, $\mathrm{E}_{Y_1^n|i}$ is the conditional expectation taken over Y_1^n when $S_{t-1} = i$. Define the expectation T_{ij} as

$$T_{ij} \triangleq \lim_{n \to \infty} \frac{1}{n} \sum_{t=1}^{n} \left\{ E_{Y_1^n | i, j} \left[\log \Pr \left(S_{t-1} = i, S_t = j \mid Y_1^n \right) \right] - E_{Y_1^n | i, j} \left[\log \Pr \left(S_{t-1} = i \mid Y_1^n \right) \right] \right\} \quad (13.42)$$

Using the Bayes rule, the expectation T_{ij} may be alternatively expressed as

$$T_{ij} = \lim_{n \to \infty} \frac{1}{n} \sum_{t=1}^{n} E \left[\log \frac{P_t \left(i, j \mid Y_1^n \right)^{\frac{P_t(i,j|Y_1^n)}{\mu_i P_{ij}}}}{P_t \left(i \mid Y_1^n \right)^{\frac{P_t(i|Y_1^n)}{\mu_i}}} \right] \quad (13.43)$$

where $P_t(i, j \mid Y_1^n) \triangleq \Pr(S_{t-1} = i, S_t = j \mid Y_1^n)$ and $P_t(i \mid Y_1^n) \triangleq \Pr(S_{t-1} = i \mid Y_1^n)$. The expression in Equation 13.43 is advantageous for numerical evaluations because it does not involve the conditional expectation as does Equation 13.42.

Combining Equation 13.39, Equation 13.41 and Equation 13.42, we may express the mutual information rate as

$$\mathcal{I}(X; Y) = \sum_{i, j : (i,j) \in \mathcal{T}} \mu_i P_{ij} \left[\log \frac{1}{P_{ij}} + T_{ij} \right] \quad (13.44)$$

13.5.2 Iterative Markov Source Optimization Algorithm

We note that the values of $\{T_{ij} : (i, j) \in \mathcal{T}\}$ are dependent on the Markov source transition probabilities $\{P_{ij} : (i, j) \in \mathcal{T}\}$, and this dependence cannot even be described explicitly by a closed-form function. Thus, the following maximization problem is extremely hard, if not impossible, to be solved directly

$$C_{M_s} = \max_{\{P_{ij} : (i,j) \in \mathcal{T}\}} \mathcal{I}(X; Y) = \max_{\{P_{ij} : (i,j) \in \mathcal{T}\}} \sum_{i, j : (i,j) \in \mathcal{T}} \mu_i P_{ij} \left[\log \frac{1}{P_{ij}} + T_{ij} \right] \quad (13.45)$$

To solve the maximization in Equation 13.45, we can use the iterative maximization procedure proposed in [5].

Algorithm 13.3

Iterative Procedures to Optimize the Markov Source Distribution. (Kavčić)

Initialization: Pick an arbitrary source distribution $\{P_{ij} : (i, j) \in \mathcal{T}\}$ that satisfies
1. $0 < P_{ij} < 1$
2. for any $i \in \mathcal{S}$, $\sum_{j : (i,j) \in \mathcal{T}} P_{ij} = 1$

Iterations: Run the following two steps until convergence.

Step 1: For the Markov source $\{P_{ij} : (i, j) \in \mathcal{T}\}$, compute the expectations T_{ij}.

Step 2: While keeping all T_{ij} fixed, find all P_{ij} (and the corresponding values $\mu_j = \sum_i \mu_i P_{ij}$) to achieve the following maximization

$$\{P_{ij} : (i, j) \in \mathcal{T}\} = \arg \max_{\{P_{ij} : (i,j) \in \mathcal{T}\}} \sum_{i, j : (i,j) \in \mathcal{T}} \mu_i P_{ij} \left[\log \frac{1}{P_{ij}} + T_{ij} \right] \quad (13.46)$$

In order to implement the iterations in Algorithm 13.3, we need to compute the expectations T_{ij} for a given Markov source distribution $\{P_{ij} : (i, j) \in \mathcal{T}\}$, and also solve the optimization problem Equation 13.46 for fixed values of T_{ij}.

For a given source distribution $\{P_{ij} : (i,j) \in T\}$, the values of T_{ij} can be conveniently evaluated with a Monte Carlo method. We note that in Equation 13.42 and Equation 13.43, the probabilities $P_t(i,j \mid Y_1^n) = \Pr(S_{t-1} = i, S_t = j \mid Y_1^n)$ and $P_{t-1}(i \mid Y_1^n) = \Pr(S_{t-1} = i \mid Y_1^n)$ are exactly the outputs of the forward-backward BCJR (or Baum-Welch) algorithm [14–16]. For n large, generate a realization s_0^n of the state sequence S_0^n according to the transition probabilities P_{ij}. Pass the realization s_0^n of the state sequence through the noisy channel to get a realization y_1^n of the output sequence Y_1^n. Now run the forward-backward BCJR (or Baum-Welch) algorithm [14–16] and compute the outputs $P_t(i,j \mid y_1^n) = \Pr(S_{t-1} = i, S_t = j \mid y_1^n)$ and $P_t(i \mid y_1^n) = \Pr(S_{t-1} = i \mid y_1^n)$ for all $1 \le t \le n$, all $i \in S$ and all pairs $(i,j) \in T$. Next for $(i,j) \in T$ estimate Equation 13.43 as the empirical expectation

$$\hat{T}_{ij} = \frac{1}{n} \sum_{t=1}^{n} \left[\log \frac{P_t(i,j \mid y_1^n)^{\frac{P_t(i,j\mid y_1^n)}{\mu_i P_{ij}}}}{P_t(i \mid y_1^n)^{\frac{P_t(i\mid y_1^n)}{\mu_i}}} \right] \tag{13.47}$$

By the ergodicity assumption, invoking the law of large numbers, we have (with probability 1) $\lim_{n\to\infty} \hat{T}_{ij} = T_{ij}$.

For fixed values of T_{ij}, the solution to Equation 13.46 is given by the following generalization of Shannon's result for the maximal achievable (noise-free) entropy rate of a Markov process [10]. Form a *noisy adjacency matrix* $\mathbf{A}$ whose elements are defined as

$$A_{ij} = \begin{cases} 2^{T_{ij}} & \text{if } (i,j) \in T \\ 0 & \text{otherwise} \end{cases} \tag{13.48}$$

Let $W_{\max}$ be the maximal real eigenvalue of $\mathbf{A}$, and let $b = [b_1, b_2, \ldots, b_M]^T$ be the corresponding eigenvector. Then the maximization in Equation 13.46 is achieved by

$$P_{ij} = \frac{b_j}{b_i} \cdot \frac{A_{ij}}{W_{\max}} \tag{13.49}$$

and the maximal value of $\sum_{i,j:(i,j)\in T} \mu_i P_{ij} [\log \frac{1}{P_{ij}} + T_{ij}]$ is

$$C(T_{ij}) = \log W_{\max} \tag{13.50}$$

Thus, the iterative Algorithm 13.3 to optimize the Markov source can be implemented [5] by the Monte Carlo method in the following Algorithm 13.4.

Algorithm 13.4

Optimizing the Markov Source Distribution. (Kavčić)

Initialization: Pick an arbitrary source distribution $\{P_{ij} : (i,j) \in T\}$ that satisfies
1. $0 < P_{ij} < 1$
2. for any $i \in S$, $\sum_{j:(i,j)\in T} P_{ij} = 1$

Iterations: Run the following four steps until convergence.

Step 1: For n large, generate s_0^n according to the transition probabilities P_{ij} and pass them through the noisy channel to get y_1^n.

Step 2: Run the forward-backward sum-product BCJR (or Baum-Welch) algorithm [14–16] and compute $\hat{T}_{ij}$ according to Equation 13.47.

Step 3: Estimate the noisy adjacency matrix as

$$\hat{A}_{ij} = \begin{cases} 2^{\hat{T}_{ij}} & \text{if } (i,j) \in T \\ 0 & \text{otherwise} \end{cases} \tag{13.51}$$

and find its maximal eigenvalue $\hat{W}_{\max}$ and the corresponding eigenvector $\hat{b} = [\hat{b}_1, \hat{b}_2, \cdots, \hat{b}_M]^\mathsf{T}$.

Step 4: Compute the new transition probabilities for all $(i, j) \in \mathcal{T}$ as $P_{ij} = \frac{\hat{b}_j}{\hat{b}_i} \cdot \frac{\hat{A}_{ij}}{\hat{W}_{\max}}$.

At the end of the execution of Algorithm 13.4, the optimized information rate C_{M_s} can be evaluated using Algorithm 13.2, or using the information rate expression (Equation 13.44), or simply using (Equation 13.50). Numerical evaluations of Algorithm 13.4 on low-dimensional Markov sources show that Algorithm 13.3 achieves the same Markov transition probabilities as a brute-force information rate maximization. The same numerical evaluations have shown that at the end of the execution of Algorithm 13.4, the optimized information rate can be just as accurately evaluated by $\hat{C}_{M_s} = \log \hat{W}_{\max}$. This supports the conjecture that Algorithm 13.3 converges to the capacity-achieving distribution. Rigorous convergence results are given in [18].

13.6 A Tight Channel Capacity Upper Bound — Delayed Feedback Capacity

In this section, we derive a tight upper bound on the channel capacity C. The upper bound is computed as the feedback capacity or the delayed feedback capacity of the same PR channel. That is, we assume that the transmitter, before sending out the symbol X_t, knows without error all the values of previous channel outputs $Y_1^{t-1-\nu} = y_1^{t-1-\nu}$, where $\nu \geq 0$ is the feedback delay. When the feedback is instantaneous, that is, $\nu = 0$, the channel capacity is called the *feedback capacity* and is denoted by C^{fb}. When the feedback is delayed, that is, $\nu > 0$, the channel capacity is called *the delayed feedback capacity* and is denoted by C_ν^{fb}. For channels with memory, we generally have

$$C \leq \cdots \leq C_2^{\mathrm{fb}} \leq C_1^{\mathrm{fb}} \leq C_0^{\mathrm{fb}} = C^{\mathrm{fb}} \tag{13.52}$$

$$C = \lim_{\nu \to \infty} C_\nu^{\mathrm{fb}} \tag{13.53}$$

Thus, we may bound the channel capacity C of a PR channel from above by using the feedback capacity C^{fb} or the delayed feedback capacity C_ν^{fb}, and this bound can be made arbitrarily tight by selecting a large enough feedback delay ν.

13.6.1 Markov Sources Achieve the Feedback Capacity C^{fb}

For the instantaneous feedback ($\nu = 0$), shown by Tatikonda [19], the feedback capacity equals the supremum of the directed information rate. The directed information *rate* is defined by Massey [20] as

$$\mathcal{I}(X \to Y) \triangleq \lim_{n \to \infty} \frac{1}{n} \sum_{t=1}^{n} I\left(X_1^t; Y_t \mid X_{-\infty}^0, Y_1^{t-1}\right) \tag{13.54}$$

The feedback capacity may be expressed as

$$C^{\mathrm{fb}} = \sup_{\mathcal{P}} \mathcal{I}(X \to Y) \tag{13.55}$$

where the supremum is taken over the set of all feedback-dependent source distributions

$$\mathcal{P} = \left\{ \Pr\left(X_t = x_t \mid X_{-\infty}^{t-1} = x_{-\infty}^{t-1}, Y_1^{t-1} = y_1^{t-1}\right), \quad t = 1, 2, \ldots \right\} \tag{13.56}$$

The source distribution (Equation 13.56) has an unbounded number of parameters when $t \to \infty$, so it is not feasible to solve the maximization in Equation 13.55 directly. Fortunately, for the PR channel, we have the following two theorems [21, 22] which simplify the feedback capacity computation problem dramatically.

Theorem 13.1   *The feedback capacity of a PR channel is achieved by a feedback-dependent Markov source, whose memory length equals the channel memory length.*

Theorem 13.2   *The posterior channel state distribution is a sufficient statistic of the channel output feedback.*

By Theorem 13.1 and Theorem 13.2, the feedback capacity of a partial response channel is

$$C^{\text{fb}} = \sup_{\mathcal{P}_{\alpha}^{\text{Markov}}} \mathcal{I}(X \to Y) \tag{13.57}$$

where the supremum is over the feedback-dependent Markov source $(M_s = M_c = M_{sc})$

$$\mathcal{P}_{\alpha}^{\text{Markov}} = \left\{ P\left(j \mid i, \underline{\alpha}_{t-1}\right) \stackrel{\triangle}{=} \Pr(S_t = j \mid S_{t-1} = i, \underline{\alpha}_{t-1}) : (i,j) \in \mathcal{T} \right\} \tag{13.58}$$

and $\underline{\alpha}_{t-1}$ denotes the set of posterior probabilities of all channel states $s \in \mathcal{S}$

$$\underline{\alpha}_{t-1} \stackrel{\triangle}{=} \left\{ \alpha_{t-1}(s) = \Pr\left(S_{t-1} = s \mid Y_1^{t-1} = y_1^{t-1}\right) : s \in \mathcal{S} \right\} \tag{13.59}$$

13.6.2   A Stochastic Control Formulation for Feedback Capacity Computation

For the feedback-dependent Markov source distribution (Equation 13.58), we can use the chain rule, Markovianity and the channel property to reformulate the directed information rate (Equation 13.54) as

$$\mathcal{I}(X \to Y) = \lim_{n \to \infty} \frac{1}{n} \sum_{t=1}^{n} I\left(S_1^t; Y_t \mid S_0^{t-1}, Y_1^{t-1}\right) \tag{13.60}$$

$$= \lim_{n \to \infty} \frac{1}{n} \sum_{t=1}^{n} \left[h\left(Y_t \mid S_0, Y_1^{t-1}\right) - h\left(Y_t \mid S_0^t, Y_1^{t-1}\right) \right] \tag{13.61}$$

$$= \lim_{n \to \infty} \frac{1}{n} \sum_{t=1}^{n} \left[h\left(Y_t \mid \underline{\alpha}_{t-1}\right) - h\left(Y_t \mid S_{t-1}^t, \underline{\alpha}_{t-1}\right) \right] \tag{13.62}$$

$$= \lim_{n \to \infty} \frac{1}{n} \sum_{t=1}^{n} I\left(S_{t-1}^t; Y_t \mid \underline{\alpha}_{t-1}\right). \tag{13.63}$$

The problem of maximizing the directed information rate (Equation 13.63) over the Markov source (Equation 13.58) can be formulated as a stochastic control problem. We state this stochastic control problem here.

Consider the dynamic system whose *state* is the causal posterior channel state distribution $\underline{\alpha}_{t-1}$. The *control* or *policy* is the set of Markov transition probabilities $\{P(j \mid i, \underline{\alpha}_{t-1})\}$, which is dependent on the dynamic system state $\underline{\alpha}_{t-1}$. The system *disturbance* is Y_t which has the following conditional distribution

$$f_{Y_t \mid \underline{\alpha}_{t-1}}(y_t \mid \underline{\alpha}_{t-1}) = f_{Y_t \mid S_0, Y_1^{t-1}}\left(y_t \mid s_0, y_1^{t-1}\right) \tag{13.64}$$

$$= \sum_{i,j:(i,j) \in \mathcal{T}} \alpha_{t-1}(i) P(j \mid i, \underline{\alpha}_{t-1}) f_{Y_t \mid S_{t-1}, S_t}(y_t \mid i, j) \tag{13.65}$$

The system equation of this dynamic system is the forward sum-product recursion of the BCJR (or Baum-Welch) algorithm [14–17]

$$\underline{\alpha}_t = F_{\text{BCJR}}(\underline{\alpha}_{t-1}, \{P(j \mid i, \underline{\alpha}_{t-1})\} y_t) \tag{13.66}$$

To be explicit, the entries $\alpha_t(s)$ of the state $\underline{\alpha}_t$ evolve as follows (cf. Algorithm 13.1)

$$\alpha_t(s) = \frac{\displaystyle\sum_{i:(i,s)\in\mathcal{T}} \alpha_{t-1}(i)P(s \mid i,\underline{\alpha}_{t-1})f_{Y_t|S_{t-1},S_t}(y_t \mid i,s)}{\displaystyle\sum_{i,j:(i,j)\in\mathcal{T}} \alpha_{t-1}(i)P(j \mid i,\underline{\alpha}_{t-1})f_{Y_t|S_{t-1},S_t}(y_t \mid i,j)} \tag{13.67}$$

In this dynamic system, let the *reward* for each stage t be

$$\phi(\underline{\alpha}_{t-1},\{P(j \mid i,\underline{\alpha}_{t-1})\}) \triangleq I\left(S_{t-1}^t; Y_t \mid \underline{\alpha}_{t-1}\right) = I\left(S_{t-1}^t; Y_t \mid s_0, y_1^{t-1}\right) \tag{13.68}$$

where the distribution of S_{t-1}^t, Y_t for given $\underline{\alpha}_{t-1}$ (or $Y_1^{t-1} = y_1^{t-1}$) is determined as

$$f_{S_{t-1},S_t,Y_t|\underline{\alpha}_{t-1}}(i,j,y_t \mid \underline{\alpha}_{t-1}) = f_{S_{t-1},S_t,Y_t|S_0,Y_1^{t-1}}\left(i,j,y_t \mid s_0, y_1^{t-1}\right) \tag{13.69}$$

$$= \sum_{i:(i,j)\in\mathcal{T}} \alpha_{t-1}(i)P(j \mid i,\underline{\alpha}_{t-1})f_{Y_t|S_{t-1},S_t}(y_t \mid i,j) \tag{13.70}$$

According to Equation 13.63, the expectation of the average reward per stage equals the directed information rate (Equation 13.54), that is

$$\mathcal{I}(X \to Y) = \lim_{n\to\infty} \frac{1}{n}\mathrm{E}\left[\sum_{t=1}^n \phi(\underline{\alpha}_{t-1},\{P(j \mid i,\underline{\alpha}_{t-1})\})\right] \tag{13.71}$$

Thus, finding the optimal source distribution that maximizes the directed information rate is an *average-reward-per-stage* stochastic control problem [23].

13.6.3 Feedback Capacity Computation

We now describe a dynamic programming value iteration method used to optimize the feedback-dependent Markov source distribution (Equation 13.58). Let $\underline{a}$ represent a possible realization of $\underline{\alpha}_t$. Define $J_0(\underline{a}) = 0$ to be the *terminal* reward function [23], and recursively generate the optimal k-stage reward-to-go functions $J_k(\underline{a})$ and sources $\mathcal{P}_{\alpha,k}^{\mathrm{Markov}}$, for $k = 1, 2, \ldots$

$$J_k(\underline{a}) = \max_{\{P(j|i,\underline{a})\}} \{\phi(\underline{a},\{P(j \mid i,\underline{a})\}) + \mathrm{E}_Y[J_{k-1}(F_{\mathrm{BCJR}}(\underline{a},\{P(j \mid i,\underline{a})\},Y))]\}, \tag{13.72}$$

$$\mathcal{P}_{\alpha,k}^{\mathrm{Markov}} = \arg\max_{\{P(j|i,\underline{a})\}} \{\phi(\underline{a},\{P(j \mid i,\underline{a})\}) + \mathrm{E}_Y[J_{k-1}(F_{\mathrm{BCJR}}(\underline{a},\{P(j \mid i,\underline{a})\},Y))]\}, \tag{13.73}$$

where the reward-per-stage $\phi(\underline{a},\{P(j \mid i,\underline{a})\})$ is defined in Equation 13.68, the maximization is over the stationary Markov source $\{P(j \mid i,\underline{a})\}$, and the expectation $\mathrm{E}_Y[\cdot]$ is over the random variable Y which has the following distribution

$$f_{Y|\underline{\alpha}}(y \mid \underline{a}) = \sum_{i,j} a(i)P(j \mid i,\underline{a})f_{Y_t|S_{t-1},S_t}(y \mid i,j) \tag{13.74}$$

The directed information rate $\mathcal{I}(X \to Y)$ induced by the source $\mathcal{P}_{\alpha,k}^{\mathrm{Markov}}$ converges to the feedback capacity C^{fb} when $k \to \infty$ (see [23], p. 390). Thus, the source distribution determined by Equation 13.73 as $k \to \infty$ is an optimal source distribution.

In general, it is hard to find the optimal source distribution $\{P(j \mid i,\underline{a})\}$ in closed form by applying the above value iteration procedures. The following is the quantization-based *numerical approximation* of the value iteration method [21].

Algorithm 13.5

Optimizing Feedback-Dependent Markov Sources Distributions. (Yang-Kavčić)

Initialization:

1. Choose a finite-level quantizer $\underline{\hat{a}} = Q(\underline{a})$.
2. Initialize the terminal reward function as $J_0(\underline{\hat{a}}) = 0$.
3. Choose a large positive integer K.

Recursions: For $1 \leq k \leq K$, numerically compute the k-stage reward-to-go function as

$$J_k(\underline{\hat{a}}) = \max_{\{P(j|i,\underline{\hat{a}})\}} \{\phi(\underline{\hat{a}}, \{P(j \mid i, \underline{\hat{a}})\}) + \mathrm{E}_Y[J_{k-1}(Q(F_{\mathrm{BCJR}}(\underline{\hat{a}}, \{P(j \mid i, \underline{\hat{a}})\}, Y)))]\} \tag{13.75}$$

Optimized source: For $\underline{\hat{a}} = Q(\underline{a})$, the optimized source $\mathcal{P}_\alpha^{\mathrm{Markov}}$ is taken as

$$\{P(j \mid i, \underline{a})\} = \arg \max_{\{P(j|i,\underline{\hat{a}})\}} \{\phi(\underline{\hat{a}}, \{P(j \mid i, \underline{\hat{a}})\}) + \mathrm{E}_Y[J_K(Q(F_{\mathrm{BCJR}}(\underline{\hat{a}}, \{P(j \mid i, \underline{\hat{a}})\}, Y)))]\} \tag{13.76}$$

End.

In Equation 13.75 and Equation 13.76 of Algorithm 13.5, the maximization is taken by exhaustively searching the Markov source distribution $\{P(j \mid i, \underline{\hat{a}})\}$ where finite-precision numerical approximations are necessary due to limited computing resources. The accuracy of the resulting optimal source distribution Equation 13.76 is thus affected by the resolution of the quantizer $Q(\cdot)$, by the finite value iteration number K and by the finite-precision approximation of $\{P(j \mid i, \underline{\hat{a}})\}$. So, strictly speaking, the directed information rate $\mathcal{I}(X \to Y)$ induced by the source $\mathcal{P}_\alpha^{\mathrm{Markov}}$ optimized in Equation 13.76 is only a lower bound on the feedback capacity. The information rate induced by $\mathcal{P}_\alpha^{\mathrm{Markov}}$ equals the feedback capacity only if Algorithm 13.5 is ideally executed with an infinitely fine quantizer $Q(\cdot)$ for $K \to \infty$.

After the optimized feedback-dependent Markov source Equation 13.76 is obtained, we can numerically compute the feedback capacity as

$$\hat{C}^{\mathrm{fb}} = \frac{1}{n} \left[\sum_{t=1}^{n} \phi(\underline{\alpha}_{t-1}, \{P(j \mid i, \underline{\alpha}_{t-1})\}) \right] \tag{13.77}$$

where n is a large enough integer, for example, $n = 10^7$. We note that by substituting P_{ij} with $P(j \mid i, \underline{\alpha}_{t-1})$, we can use Algorithm 13.2 to compute the directed information rate.

13.6.4 Delayed Feedback Capacity Computation

Let the PR channel be

$$Y_t = \sum_{k=0}^{M_c} a_k X_{t-k} + N_t \tag{13.78}$$

and let the feedback delay be $v > 0$. We next show that, the PR channel (X, Y) is equivalent to the following PR channel $(X, \tilde{Y})$ of memory length $M_c + v$ whose channel output is

$$\tilde{Y}_t = Y_{t-v} = \sum_{k=0}^{v-1} 0 \cdot X_{t-k} + \sum_{k=v}^{M_c+v} a_{k-v} X_{t-k} + N_{t-v} \tag{13.79}$$

$$= \sum_{k=0}^{M_c+v} \tilde{a}_k X_{t-k} + \tilde{N}_t \tag{13.80}$$

where $\tilde{a}_k = 0$ for $k = 0, 1, \ldots, v-1$ and $\tilde{a}_k = a_{k-v}$ for $k = v, v+1, \ldots, M_c + v$, and $\tilde{N}_t$ is an additive white Gaussian noise with a variance σ^2. Thus, the channel coefficient polynomial of the channel $(X, \tilde{Y})$ is $\tilde{a}(D) = D^v a(D)$. Further, the v-time delayed feedback of the original channel (X, Y) is equivalent to the instantaneous feedback of the augmented channel $(X, \tilde{Y})$. Thus, the delayed feedback capacity C_v^{fb} of the original channel (X, Y) equals the feedback capacity $\tilde{C}^{\mathrm{fb}}$ of the augmented channel $(X, \tilde{Y})$, that is

$$C_v^{\mathrm{fb}} = \tilde{C}^{\mathrm{fb}} \tag{13.81}$$

and the value of $\tilde{C}^{\mathrm{fb}}$ can be computed by applying Algorithm 13.5 over the channel $(X, \tilde{Y})$.

Example 13.4 (The dicode PR channel with 1-time delayed feedback)

Let $v = 1$, which corresponds to a 1-tap delayed channel output feedback. The delayed feedback capacity C_1^{fb} of the dicode channel equals the instantaneous feedback capacity $\tilde{C}^{\text{fb}}$ of the following PR channel

$$\tilde{Y}_t = 0 \cdot X_t + \sqrt{\frac{1}{2}}(X_{t-1} - X_{t-2}) + \tilde{N}_t = \sqrt{\frac{1}{2}}(X_{t-1} - X_{t-2}) + \tilde{N}_t. \tag{13.82}$$

By Theorem 13.1 and Theorem 13.2, we only need to consider the feedback-dependent Markov source of order $M_s = M_c = 2$. We index the source/channel state $s_t \in \mathcal{S} \stackrel{\triangle}{=} \mathcal{X}^2$ by $s_t = \sum_{k=0}^{1}(x_t + 1)2^{k-1}$. Thus, the source/channel state space is $\mathcal{S} = \{0, 1, 2, 3\}$, and we have the following state and input mapping

$$
\begin{aligned}
s_t = 0 &\quad \leftrightarrow \quad [x_{t-1}, x_t] = [-1, -1], \\
s_t = 1 &\quad \leftrightarrow \quad [x_{t-1}, x_t] = [-1, +1], \\
s_t = 2 &\quad \leftrightarrow \quad [x_{t-1}, x_t] = [+1, -1], \\
s_t = 3 &\quad \leftrightarrow \quad [x_{t-1}, x_t] = [+1, +1].
\end{aligned}
$$

The set of valid state transitions is $\mathcal{T} = \{(0,0), (0,1), (1,2), (1,3), (2,0), (2,1), (3,2), (3,3)\}$. Let $\underline{\alpha}_{t-1}$ represent the posterior source/channel state distribution, that is, the set of conditional probabilities

$$\underline{\alpha}_{t-1} \stackrel{\triangle}{=} \{\alpha_{t-1}(s) = \Pr\left(S_{t-1} = s \mid Y_1^{t-1} = y_1^{t-1}\right) : s \in \mathcal{S}\} \tag{13.83}$$

We can apply Algorithm 13.5 to optimize the Markov source

$$\mathcal{P}_\alpha^{\text{Markov}} = \{P(j \mid i, \underline{\alpha}_{t-1}) \stackrel{\triangle}{=} \Pr(S_t = j \mid S_{t-1} = i, \underline{\alpha}_{t-1}) : (i, j) \in \mathcal{T}\} \tag{13.84}$$

and then compute the corresponding feedback capacity $\tilde{C}^{\text{fb}}$ by Equation 13.77 for a large n, which equals the 1-time delayed feedback capacity C_1^{fb} of the dicode PR channel.

13.7 Vontobel-Arnold Upper Bound

Let $P(X_1^n, Y_1^n \mid x_{-\infty}^0)$ and $P(Y_1^n \mid x_{-\infty}^n)$ be the distribution functions for the channel inputs X_1^n and outputs Y_1^n determined by the channel law (Equation 13.1). Let $Q(Y_1^n)$ be any valid probability density function of the sequence $Y_1^n \in \mathbb{R}^n$. Vontobel and Arnold [?] observed that

$$C = \lim_{n \to \infty} \frac{1}{n} \max_{P(X_1^n)} I\left(X_1^n; Y_1^n \mid x_{-\infty}^0\right) \tag{13.85}$$

$$= \lim_{n \to \infty} \frac{1}{n} \max_{P(X_1^n)} E_{X_1^n, Y_1^n \mid x_{-\infty}^0}\left[\log\left(\frac{P\left(Y_1^n \mid X_1^n, x_{-\infty}^0\right)}{P\left(Y_1^n \mid x_{-\infty}^0\right)}\right)\right] \tag{13.86}$$

$$\stackrel{(a)}{\leq} \lim_{n \to \infty} \frac{1}{n} \max_{P(X_1^n)} E_{X_1^n, Y_1^n \mid x_{-\infty}^0}\left[\log\left(\frac{P\left(Y_1^n \mid X_1^n, x_{-\infty}^0\right)}{Q\left(Y_1^n\right)}\right)\right] \tag{13.87}$$

$$\leq \lim_{n \to \infty} \frac{1}{n} \max_{x_1^n} E_{Y_1^n \mid x_{-\infty}^n}\left[\log\left(\frac{P\left(Y_1^n \mid x_{-\infty}^n\right)}{Q\left(Y_1^n\right)}\right)\right] \tag{13.88}$$

$$= \lim_{n \to \infty} \frac{1}{n} \max_{x_1^n} E_{Y_1^n \mid x_{-\infty}^n}\left[\sum_{t=1}^{n} \log\left(\frac{P\left(Y_t \mid x_{t-M_c}^t\right)}{Q\left(Y_t \mid Y_1^{t-1}\right)}\right)\right] \tag{13.89}$$

where the notation $E_{X_1^n, Y_1^n \mid x_{-\infty}^0}$ represents the expectation over X_1^n, Y_1^n whose distribution function is $P(x_1^n, y_1^n \mid x_{-\infty}^0)$, similarly the notation $E_{Y_1^n \mid x_{-\infty}^n}$ represents the expectation over Y_1^n whose distribution is $P(y_1^n \mid x_{-\infty}^n)$, and the inequality (a) is a generalization of Lemma 13.8.1 in [9] which reduces to equality if and only if $Q(y_1^n) \equiv P(y_1^n \mid x_{-\infty}^0)$.

To make the upper bound Equation 13.89 computable, let the process defined by the distribution $Q(\cdot)$ be a stationary Markov process of order L, that is,

$$Q\left(Y_t \mid Y_{-\infty}^{t-1}\right) = Q\left(Y_t \mid Y_{t-L}^{t-1}\right) \tag{13.90}$$

Further, to make the bound tight, let $Q(Y_{t-L}^t)$ equal the marginal density function of the PR channel outputs Y_{t-L}^t in the steady state when an optimal Markov source (e.g., computed by Algorithm 13.4) of order $M_s = M_{sc} \geq M_c$ is used. Thus, for any $t > L$, we have

$$Q\left(Y_{t-L}^t = y_{t-L}^t\right) \triangleq \lim_{\tau \to \infty} P\left(Y_{t-L}^\tau = y_{t-L}^t \mid x_{-M_{sc}+1}^0\right) \tag{13.91}$$

With such a definition for the function $Q(\cdot)$, the upper bound Equation 13.89 becomes

$$C \leq \lim_{n \to \infty} \frac{1}{n} \max_{x_1^n} \sum_{t=1}^{n} \mathrm{E}_{Y_{t-L}^t \mid x_{t-L-M_{sc}}^t} \left[\log\left(\frac{P\left(Y_t \mid x_{t-M_c}^t\right)}{Q\left(Y_t \mid Y_{t-L}^{t-1}\right)} \right) \right] \tag{13.92}$$

$$= \lim_{n \to \infty} \frac{1}{n} \max_{x_1^n} \sum_{t=1}^{n} \Omega\left(x_{t-L-M_{sc}}^t\right) \tag{13.93}$$

where the function $\Omega(x_{t-L-M_{sc}}^t)$ is defined as

$$\Omega\left(x_{t-L-M_{sc}}^t\right) \triangleq \mathrm{E}_{Y_{t-L}^t \mid x_{t-L-M_{sc}}^t} \left[\log\left(\frac{P\left(Y_t \mid x_{t-M_c}^t\right)}{Q\left(Y_t \mid Y_{t-L}^{t-1}\right)} \right) \right] \tag{13.94}$$

$$= \mathrm{E}_{Y_t \mid x_{t-M_c}^t} \left[\log\left(P\left(Y_t \mid x_{t-M_c}^t\right) \right) \right] - \mathrm{E}_{Y_{t-L}^t \mid x_{t-L-M_{sc}}^t} \left[\log\left(Q\left(Y_t \mid Y_{t-L}^{t-1}\right) \right) \right] \tag{13.95}$$

$$= -\frac{1}{2} \log\left(2\pi e \sigma^2 \right) - \mathrm{E}_{Y_{t-L}^t \mid x_{t-L-M_{sc}}^t} \left[\log\left(Q\left(Y_t \mid Y_{t-L}^{t-1}\right) \right) \right] \tag{13.96}$$

Here, the function $\Omega(x_{t-L-M_{sc}}^t)$ can not be expressed in a closed form, but we can conveniently estimate it by a Monte Carlo method.

Algorithm 13.6

A Monte Carlo Method to Compute the Function $\Omega(x_{t-L-M_{sc}}^t)$. (Vontobel-Arnold)

Initialization:
    1. Choose a large positive integer n, e.g., $n = 10^4$.
    2. Find an optimal Markov source $\mathcal{P} = \{P_{ij}\}$ of order M_{sc} using Algorithm 13.4.
Simulation: For $k = 1, 2, \ldots, n$ obtain $\Omega^{(k)}(x_{t-L-M_{sc}}^t)$ as follows
Setup: Pass $x_{t-L-M_{sc}}^t$ through the noisy PR channel and generate channel outputs y_{t-L}^t. Let $\alpha_{t-L-1}(s) = \mu_s$, where μ_s is the stationary probability of state s induced by the Markov source $\mathcal{P}$.
Recursions: For $\tau = t - L, \ldots, t$, recursively compute $Q(y_\tau \mid y_{t-L}^{\tau-1})$ and $\alpha_\tau(s)$ by

$$Q\left(y_\tau \mid y_{t-L}^{\tau-1}\right) = \sum_{i,j:(i,j)\in\mathcal{T}} \alpha_{\tau-1}(i) P_{ij} f_{Y_\tau \mid S_{\tau-1}, S_\tau}(y_\tau \mid i, j) \tag{13.97}$$

$$\alpha_\tau(s) = \frac{\displaystyle\sum_{i:(i,s)\in\mathcal{T}} \alpha_{\tau-1}(i) P_{is} f_{Y_\tau \mid S_{\tau-1}, S_\tau}(y_\tau \mid i, s)}{\displaystyle\sum_{i,j:(i,j)\in\mathcal{T}} \alpha_{\tau-1}(i) P_{ij} f_{Y_\tau \mid S_{\tau-1}, S_\tau}(y_\tau \mid i, j)}, \quad \text{for } s \in \mathcal{S} \tag{13.98}$$

Evaluation: $\Omega^{(k)}(x_{t-L-M_{sc}}^t) = -\frac{1}{2}\log(2\pi e\sigma^2) - \log(Q(y_t \mid y_{t-L}^{t-1}))$.

Averaging: Approximate $\Omega(x_{t-L-M_{sc}}^t)$ by $\hat{\Omega}(x_{t-L-M_{sc}}^t) = \frac{1}{n}\sum_{k=1}^n \Omega^{(k)}(x_{t-L-M_{sc}}^t)$.

Definition 13.6 (Auxiliary trellis to compute the Vontobel-Arnold upper bound) Let the nodes of the trellis be $u_t \stackrel{\triangle}{=} x_{t-L-M_{sc}+1}^t$. Then, the total number of trellis nodes is $2^{L+M_{sc}}$, and the total number of branches (u_{t-1}, u_t) in one trellis section is $2^{L+M_{sc}+1}$. Let the metric for each branch connecting the two nodes $u_{t-1} = x_{t-L-M_{sc}}^{t-1}$ and $u_t = x_{t-L-M_{sc}+1}^t$ be the value of $\Omega(u_{t-1}, u_t) \stackrel{\triangle}{=} \Omega(x_{t-L-M_{sc}}^t)$.

Obviously, the upper bound Equation 13.93 equals the maximum average metric of the auxiliary trellis. Since the trellis is stationary and the metric values are deterministic, there exists at least one maximum-average-weight-achieving (or worst) periodic sequence u_t whose period is no greater than $2^{M_{sc}+L}$. This worst periodic sequence u_t can be found by using the Viterbi Algorithm [24, 25].

Generally, the Vontobel-Arnold upper bound can be made tight by choosing large L and M_{sc}. However, for some PR channels, this might be very difficult if not impossible, for example, this upper bound is fairly loose for the dicode PR channel at high SNR's for any reasonable values of M_{sc} and L, as reported in [?].

13.8 Conclusion and Extensions Beyond Partial Response Channels

We have presented numerical methods to accurately estimate the capacities of partial response (PR) channels. These include lower and upper bounds for tightly sandwiching the capacity. In practice, perhaps the most useful tool is the computation of the information rate for any Markov source (Algorithm 13.2) and the lower-bounding technique (Algorithm 13.4) which not only delivers tight lower bounds, but also gives a source that achieves the bound. The claim that the lower bounds are tight is supported by computing (tight) upper bounds (Algorithms 13.5 and 13.6), see Figure 13.2. In practice, when a quick evaluation of the capacity is needed, upper bounds are typically too cumbersome to evaluate. However, given that the

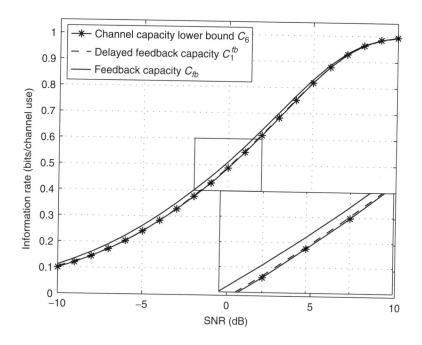

FIGURE 13.2 Capacity bounds for the dicode channel.

upper bounds are tight, one can resort to computing lower bounds only for an accurate estimate of the capacity.

A drawback of the methods presented in this monograph is that they apply only to PR channels with short memory lengths (up to four or five symbol periods) because the algorithms otherwise become too computationally complex. However, in practice, many magnetic recording channels have much longer (even infinitely long) impulse responses. A further complication is that magnetic recording channels are typically corrupted by signal-dependent media noise. For coverage of detections in media noise, see [?], while for capacity estimation methods in media noise, see [7]. Here we only briefly summarize the capacity estimation methods for channels that reach beyond simple PR channels.

13.8.1 Media Noise

For channels with media noise, the capacity may be estimated with algorithms similar to Algorithm 13.2, see [7] for details. A method for estimating the the information rate of a channel with media noise using the Shamai-Laroia conjecture has also been proposed in [26]. If the noise is assumed to be autoregressive, an appropriate modification of Algorithm 13.4 may be applied [27]. Consequently, after proper media noise model extractions, the capacities of one-dimensional [28] and two-dimensional [29] magnetic recording channels can be numerically computed. Upper bounds are still unknown. Note that the accuracy of these methods is strongly dependent on the accuracy of the noise model that is used to describe the signal-dependent media noise.

13.8.2 Long Impulse Responses

When the channel impulse response is very long (or even infinitely long), only estimates of information rates (but not capacities) are known. In [30], a method for computing the upper bound on the information rate for a known source is given, and a lower bound is conjectured. In [31], further methods for computing upper and lower bounds of these information rates are presented. These methods are based on reduced-state trellis methods and on channel approximations by truncation or quantization.

In short, the computation (or accurate estimation) of the capacity of a magnetic recording channel with a very (or infinitely) long impulse response corrupted by media noise, remains an open problem.

References

[1] K. A. S. Immink, P. H. Siegel, and J. K. Wolf, "Codes for digital recorders," *IEEE Trans. Inform. Theory*, vol. 44, pp. 2260–2299, October 1998.

[2] D. Arnold and H.-A. Loeliger, "On the information rate of binary-input channels with memory," in *Proc. IEEE ICC 2001*, Helsinki, Finland, pp. 2692–2695, June 2001.

[3] H. D. Pfister, J. B. Soriaga, and P. H. Siegel, "On the achievable information rates of finite state ISI channels," in *Proc. IEEE GLOBECOM 2001*, San Antonio, Texas, pp. 2992–2996, November 2001.

[4] V. Sharma and S. K. Singh, "Entropy and channel capacity in the regenerative setup with applications to Markov channels," in *Proc. IEEE ISIT 2001*, Washington, DC, p. 283, June 2001.

[5] A. Kavčić, "On the capacity of Markov sources over noisy channels," in *Proc. IEEE GLOBECOM 2001*, San Antonio, Texas, pp. 2997–3001, November 2001.

[6] S. Yang, A. Kavčić, and S. Tatikonda, "Delayed feedback capacity of finite-state machine channels: Upper bounds on the feedforward capacity," in *IEEE ISIT 2003*, Yokohoma, Japan, July 2003.

[7] Z. Zhang and T. Duman, "Information theory of magnetic recording channels." another monograph in this book.

[8] J. G. Proakis, *Digital Communications*. McGraw-Hill, New York, 4th ed., 2000.

[9] T. M. Cover and J. A. Thomas, *Elements of Information Theory*. John Wiley & Sons, New York, 1991.

[10] C. E. Shannon, "A mathematical theory of communications," *Bell Syst. Tech. J.*, vol. 27, pp. 379–423 (part I) and 623–656 (part II), 1948.

[11] X. Ma, A. Kavčić, and N. Varnica, "Matched information rate codes for partial response channels." submitted to *IEEE Transactions on Information Theory*, June 2002.

[12] R. G. Gallager, *Low-Density Parity-Check Codes*. MIT Press, Cambridge, MA, 1962.

[13] C. Berrou, A. Glavieux, and P. Thitimajshima, "Near Shannon limit error-correcting coding and decoding: Turbo-codes," in *Proc. IEEE Int. Conf. on Communications*, Geneva, Switzerland, pp. 1064–1070, May 1993.

[14] R. W. Chang and J. C. Hancock, "On receiver structures for channels having memory," *IEEE. Trans. Inform. Theory*, vol. 12, pp. 463–468, October 1966.

[15] L. E. Baum and T. Petrie, "Statistical interference for probabilistic functions of finite state markov chains," *Ann. Math. Statist.*, vol. 37, pp. 1559–1563, 1966.

[16] L. R. Bahl, J. Cocke, F. Jelinek, and J. Raviv, "Optimal decoding of linear codes for minimizing symbol error rate," *IEEE Trans. Inform. Theory*, vol. 20, pp. 284–287, September 1974.

[17] F. R. Kschischang, B. J. Frey, and H.-A. Loeliger, "Factor graphs and the sum-product algorithm," *IEEE Trans. Info. Theory*, vol. 47, pp. 498–519, February 2001.

[18] P. O. Vontobel, "A generalized blahut-arimoto algorithm," in *Proc. IEEE ISIT 2003*, Yokohama, Japan, July 2003.

[19] S. C. Tatikonda, *Control Under Communications Constraints*. Ph.D. thesis, Massachusetts Institute of Technology, Cambridge, MA, September 2000. available at http://pantheon.yale.edu/~sct29/thesis.html.

[20] J. L. Massey, "Causality, feedback and directed information," in *Proc. 1990 Intl. Symp. on Info. Th. and its Appli.*, pp. 27–30, November 1990.

[21] S. Yang and A. Kavčić, "Sum-product algorithm and feedback capacity," in *Proc. Int. Symp. Math. Theory Networks Sys.*, South Bend, Indiana, August 2002. available at http://hrl.harvard.edu/~kavcic/publications.html.

[22] S. Yang, A. Kavčić, and S. Tatikonda, "The feedback capacity of finite-state machine channels." Submitted to *IEEE Trans. Inform. Theory*, December 2002.

[23] D. P. Bertsekas, *Dynamic Programming and Optimal Control* 2nd ed., vol. 1, Athena Scientific, Belmont, MA, 2001.

[24] A. D. Viterbi, "Error bounds for convolutional codes and an asymptotically optimum decoding algorithm," *IEEE Trans. Info. Theory*, vol. 13, pp. 260–269, April 1967.

[25] G. D. Forney Jr., "Maximum-likelihood sequence estimation of digital sequences in the presence of intersymbol interference," *IEEE Trans. Info. Theory*, vol. 18, pp. 363–378, March 1972.

[26] D. Arnold and E. Eleftheriou, "On the information-theoretic capacity of magnetic recording systems in the presence of medium noise," *IEEE Trans. Magn.*, vol. 38, pp. 2319–2321, September 2002.

[27] S. Yang and A. Kavčić, "On the capacity of data-dependent autoregressive noise channels," in *Proc. Allerton Conf. Commn. Control*, 2002.

[28] W. Ryan, "Optimal code rates for Lorentzian channel models" in *Proc. IEEE ICC 2003*, (Anchorage, AK), May 2003.

[29] S. Yang, A. Kavčić, and W. Ryan, "Optimizing the Bit Aspect Ratio of a recording system using an Information-theoretic criterion," in *IEEE INTERMAG 2003*, Boston, Massachusetts, April 2003.

[30] D. Arnold and H.-A. Loeliger, "On finite-state information rates from channel simulations," in *Proc. IEEE ISIT 2002*, Lausanne, Switzerland, p. 164, July 2002.

[31] D. Arnold, A. Kavčić, H.-A. Loeliger, P. O. Vontobel, and W. Zeng, "Simulation based computation of information rates: Upper and lower bounds," Yokohama, Japan, July 2003.

III

Introduction to Read Channels

14 Recording Physics and Organization of Data on a Disk *Bane Vasic, Miroslav Despotović and Vojin Šenk* **14**-1
Magnetic Recording Basics • Physical Organization of Data on Disk • Logical Organization of Data on a Disk • Increasing Recording Density • Physical Limits on Recording Density • Future

15 Read Channels for Hard Drives *Bane Vasic, Pervez M. Aziz, and Necip Sayiner* ... **15**-1
Analog Front End • Partial Response Signaling with Maximum Likelihood Sequence Estimation • Adaptive Equalization • Viterbi Detection • Timing Recovery • Read Channel Servo Information Detection • Precompensation • The Effect of Thermal Asperites • Postprocessor • Modulation Coding • Error Control Coding • Error Performance Measures

16 An Overview of Hard Drive Controller Functionality *Bruce Buch* **16**-1
Introduction • Hardware Subsystems of a Hard Disk Controller • Global Controller Strategies

14

Recording Physics and Organization of Data on a Disk

Bane Vasic
University of Arizona
Tucson, AZ

Miroslav Despotović

Vojin Šenk
Technical University of Novi Sad
Novi Sad, Yugoslavia

14.1 Magnetic Recording Basics **14**-1
14.2 Physical Organization of Data on Disk **14**-6
14.3 Logical Organization of Data on a Disk **14**-7
14.4 Increasing Recording Density **14**-8
14.5 Physical Limits on Recording Density **14**-8
14.6 Future **14**-9

14.1 Magnetic Recording Basics

The basic elements of a magnetic recording system are read/write head, which is an electromagnet with a carefully shaped ferrous core, and a rotating disk with a ferromagnetic surface. Since the core of the electromagnet is ferrous, the magnetic flux preferentially travels through the core. The core is deliberately broken at an air gap. In the air gap, the flux creates a fringing field that extends some distance from the core.

To record data on a surface of a disk, the modulated signal current, typically bipolar, is passed through the electromagnet coils thus generating a fringing magnetic field. The fringing magnetic field creates a remanent magnetization on the ferromagnetic surface, that is, the ferromagnetic surface becomes permanently magnetic. The magnetic domains in the surface act like tiny magnets themselves and create their own fringing magnetic field above the ferromagnetic surface. The data are recorded in concentric tracks as a sequence of small magnetic domains with two senses of magnetization depending on a sign of writing current. In this, so called *saturation recording*, the amplitude of two writing current signal levels are chosen sufficiently large so as to magnetize to saturation the magnetic medium in one of two directions. In this way, the nonlinear hysteresis effect does not affect domains recorded over previously recorded ones. (Figure 14.1 illustrates the writting process for longitudinal and perpendicular recording [7] explained in Ch. 2.).

In a simple reading scenario the reading head flies over the disk-spinning surface (at head-to-medium velocity — v) and passes through the fringing magnetic fields above the magnetic domains Figure 14.2. Depending on a head type, the output voltage induced in the electromagnet is proportional to the spatial derivative of the magnetic field created by the permanent magnetization in the material in the case of inductive heads, or is proportional to the fringing magnetic field in the case of magneto-resistive heads. Today's hard drives use magneto-resistive heads for reading, because of their higher sensitivity. Pulses sensed by a head in response to transition on the medium are amplified and then detected to retrieve back the recorded data. For both types of heads, it is arranged that the head readback signal responds primarily

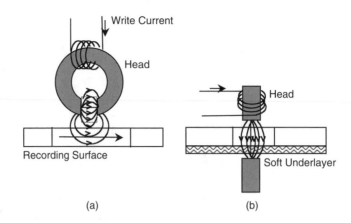

FIGURE 14.1   (a) Longitudinal recording, (b) Perpendicular recording.

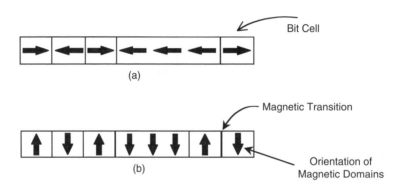

FIGURE 14.2   Magnetic Domains Representing Bits.

to transitions of the magnetization pattern. The simplest, single parameter model for an isolated magnetic *transition response* is the so-called *Lorenzian pulse*

$$g(t) = \frac{1}{1 + \left(\frac{2t}{t_{50}}\right)^2}$$

where t_{50} is a parameter representing the pulse width at 50% of the maximum amplitude. Simplicity and relatively good approximation of the channel response are the main reasons for attractiveness of this model. The family of $g(t)$ curves for different t_{50} values is depicted in Figure 14.3. The width at half amplitude defines the recording process *resolution*, that is, PW_{50},[1] as a spatial, while t_{50}, as a temporal measure, is alternatively in use ($PW_{50} = v \cdot t_{50}$).

Ideal conditions for readback process would be to have a head that is sensing the medium in an infinitely narrow strip in front of the head. However, head resolution is limited, so that the head output depends on "past" and "future" bit cell magnetization patterns. Such dependence causes superposition of isolated transition responses partly canceling each other. This phenomenon is known as *intersymbol interference* (ISI). The interference is largest for transitions at minimum spacing, that is a spacing of a single bit cell T.

---

[1]This is not so strict because, contrary to this, some authors use PW_{50} designating temporal resolution.

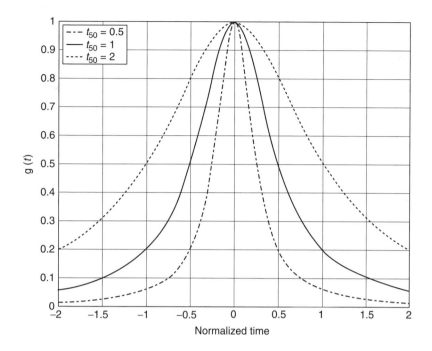

FIGURE 14.3 Transition response $g(t)$ — mathematical model.

The response to two physically adjacent transitions is designated *dibit response* or *symbol response*, that is, $h(t) = g(t) - g(t - T)$. Typical readback waveform illustrating these types of responses is depicted in Figure 14.4.

Mathematically, we can express noiseless input-output relationship as

$$y(t) = \sum_{i=-\infty}^{\infty} x_i h(t - iT) = \sum_{i=-\infty}^{\infty} (x_i - x_{i-1})g(t - iT)$$

where $\mathbf{y}$ and $\mathbf{x} \in \{-1, +1\}$ are readback and recorded sequences respectively. Notice that every transition between adjacent bit cells yields a response $\pm 2g(t)$, while no transition in recorded sequence produces zero output.

Normalized measure of the *information density*[2] is defined as the ratio $D = t_{50}/T$ showing how many channel bits are packed "under" the dispersed pulse of duration t_{50}. Case in which we are increasing density $(D > 2)$ is accompanied by an increase of duration of $h(t)$ expressed in units of T, as well as rapid decrease of the amplitude of dibit response, which is equivalent to lowering of signal-to-noise ratio in the channel. As a consequence, any given bit will interfere with successively more preceding and subsequent bits producing more severe ISI. At low normalized information densities peaks of the transition responses are clearly separated, so it is possible to read recorded data in simple manner by detecting these peaks, that is, *peak detectors*. Contrary to this, high density detectors have to implement more sophisticated detection methods in order to resolve combination of these effects. One of the most important techniques to combat intersymbol interference in magnetic recording channels is *partial-response* (PR) signaling with *maximum-likelihood* (ML) sequence detection, that is, PRML detection, (See Ch. 8 and Ch. 32.) The applicability

---

[2]When channel coding is introduced, this density is greater than the user information density, because of added redundancy in the recorded channel sequence.

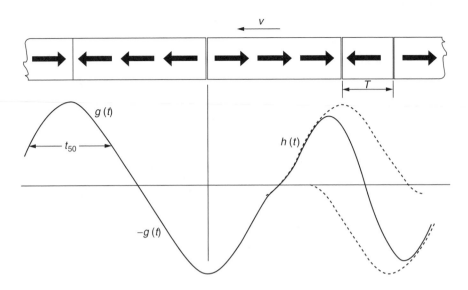

FIGURE 14.4   Sketch of a typical readback waveform in magnetic recording.

of this scheme in magnetic recording channels was suggested over 30 years ago [4], but the advance in technology enabled first disk detectors of this type at the beginning of nineties [2].

Basic idea of partial response system is that certain, controlled amount of ISI, at the channel equalizer output, is left for a detector to combat with. The nature of controlled ISI is defined by a partial response. This method avoids full channel equalization and intolerable noise enhancement induced by it in situation when amplitude distortions, as a result of increased density, are severe. In magnetic recording systems the PR detector reconstructs recorded sequence from samples of a suitable equalized readback signal at time instants $t = iT$, $i \geq 0$. The equalization result is designed in a manner that produces just a finite number of *nonzero* $h(t)$ samples $h_0 = h(0)$, $h_1 = h(T)$, $h_2 = h(2T), \ldots, h_K = h(KT)$. This is usually represented in a compact *"partial-response polynomial"* notation $h(D) = h_0 + h_1 D + h_2 D^2 + \cdots + h_K D^K$, where the dummy variable D^i signifies a delay of i time units T. Then the "sampled" input-output relationship is of the form

$$y(jT) = \sum_{i=j-K}^{j} x_i h(jT - iT)$$

For channel bit densities around $D \approx 2$, the target partial response channels is usually the class-4 partial-response (PR4), described by $h(D) = 1 - D^2 = (1 - D)(1 + D)$. At higher recording densities Thapar and Patel [6] introduced a general class of partial-response models with partial-response polynomial in the form $h_n(D) = (1 - D)(1 + D)^n$, $n \geq 1$ that is a better match to the actual channel discrete-time symbol response. Notice that the PR4 model corresponds to the $n = 1$ case. The channel models with $n \geq 2$ are usually referred to as "Extended class-4" models, and denoted by $E^{n-1}PR4$ (EPR4, E^2PR4). Recently, the *modified* E^2PR4 (ME^2PR4) channel, $h(D) = (1 - D^2)(5 + 4D + 2D^2)$, was suggested due to its robustness in high-density recordings. Notice that as the degree of partial-response polynomials gets higher, the transition response, $g(t)$, becomes wider and wider in terms of channel bit intervals, T (EPR4 response extends over 3 bit periods, E^2PR4 over 4), that is the remaining ISI is more severe.

The transfer characteristics of the Lorentzian model of the PR4 saturation recording channel (at densities $D \approx 2$), is close to transition response given by

$$g(t) = \frac{\sin\left(\pi \frac{t}{T}\right)}{\pi \frac{t}{T}} + \frac{\sin\left(\pi \frac{t-T}{T}\right)}{\pi \frac{t-T}{T}}$$

generating the output waveform described by

$$y(t) = \sum_{i=-\infty}^{\infty} (x_i - x_{i-1}) g(t - iT) = \sum_{i=-\infty}^{\infty} (x_i - x_{i-2}) \frac{\sin\left(\pi \frac{t-iT}{T}\right)}{\pi \frac{t-iT}{T}}$$

Note that $g(t) = 1$ at consecutive sample times $t = 0$ *and* $t = T$, while at all other discrete time instants, its value is 0. Such transition response results in known ISI at sample times, leading to output sample values that, in the absence of noise, take values from the ternary alphabet $\{0, \pm 2\}$. In order to decode the readback partial-response sequence it is useful to describe the channel using the *trellis state diagram*. This is diagram similar to any other graph describing a finite-state machine, where states indicate the content of the channel memory and branches between states are labeled with output symbols as a response to the certain input (the usual convention is that for the upper branch leaving a state we associate input -1 and $+1$ for the lower). The EPR4 channel has memory length 3, its trellis has $2^3 = 8$ states, and any input sequence is tracing the path through adjacent trellis segments. An example of the trellis diagram is given in Figure 14.5. for the EPR4 channel. However, notice that in partial-response trellis there are also distinct states for which there exist mutually identical paths (output sequences) that start from those states, so that we can never distinguish between them (e.g. the all-zero paths emerging from the top and bottom states of the EPR4 trellis). Obviously, such a behavior can easily lead to great problems in detection in situations when noise can confuse us in resolving the current trellis state (e.g., the bottom one for the upper in the running example). Such a trellis is so-called *quasi-catastrophic* trellis and further details on this subject could be found in [3].

A common approach to studying the partial-response channel characteristics is to analyze its frequency spectra. Basically, when the recording density is low ($D \approx 0.5$) and readback pulses are narrow compared to the distance between transitions, such a signal contains a high frequency component (highest frequency components correspond to the fastest edge of the signal). With the growth of density, the spectral energy distribution move towards lower frequency range. This means that for the system with $D = 2$, the signal spectrum is concentrated below half of the channel bit rate given by $1/T$. The power of the highest spectral components outside this half-bandwidth range is negligible. This means that for high-density recording we can limit the channel bandwidth to $1/2T$ without loss of information and filtering the high frequencies containing noise only.

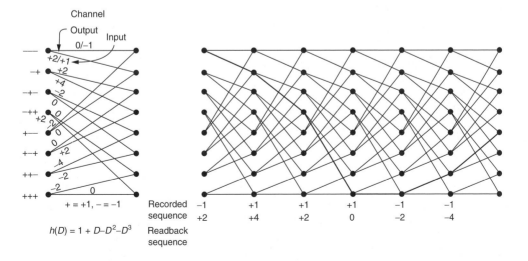

FIGURE 14.5   Trellis representation of EPR4 channel outputs.

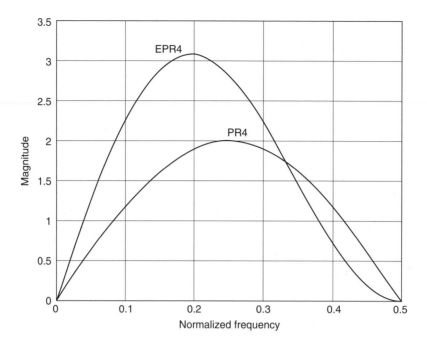

FIGURE 14.6   Frequency response of PR4 and EPR4 channel.

Finding the Fourier transform of the dibit response we obtain the frequency response for the PR4 channel given by

$$H(w) = 1 - e^{i2wT}; \quad |H(w)| = 2\sin(wT) \quad 0 \le w \le \frac{\pi}{T}$$

For higher order partial-response channels we have different transition responses and accordingly different frequency responses calculated in a similar fashion as for the PR4 channel. The frequency response for these channels is shown in Figure 14.6.

These lower frequency spectrum distributions of PR channels are closer to a typical frequency content of raw non-equalized pulses. Hence, equalization for extended partial-response channels can become less critical and requires less high frequency boost that may improve signal to noise ratio, (see Ch. 8 and Ch. 19)

14.2   Physical Organization of Data on Disk

In most designs the head is mounted on a slider, a small sled-like structure with rails. Sliders are designed to form an air bearing that gives the lift force to keep the slider-mounted head flying at the small and closely controlled height (the so-called Winchester technology). A small flying height is desirable because it amplifies the readback amplitude and reduces the amount of field from neighboring magnetic domains picked by the head, thus enabling sharper transitions in the readback signal and recording more data on a disk. However, the surface imperfections and dust particles can cause the head to "crash." Controlling the head-medium spacing is of critical importance to ensure high readback signal, and stable signal range. It is also important during reading to keep the head center above the track being read to reduce magnetization picked up from neighboring tracks. The signal induced in the head as a result of magnetic transitions in a neighboring track is known as a cross-talk or intertrack interference. In order to position the head, special, periodic, wedge-like areas, the so-called *servo wedges*, are reserved on a disk surface for radial position information. They typically consume 5–10% of the disk surface available. An arch of a track laying in a servo wedge is called a *servo sector*. The area between servo wedges is used to record data, and a portion

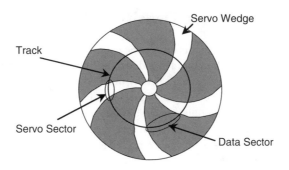

FIGURE 14.7   Data and servo sectors.

of a track between two servo sectors is referred to as a *data sector* or *data field* (see Figure 14.7). In other words, the data and servo fields are time multiplexed, or using disk drive terminology, the servo field is *embedded* in the data stream. To estimate radial position a periodic waveform in a servo sectors is detected, and the radial position error signal is calculated based on the current estimated position of a head and the actual position of the track to be followed, and then used in a head positioning servo system.

14.3   Logical Organization of Data on a Disk

On a disk data are organized in *sectors*. For a computer system, the sector is a sequence of (eight-bit) bytes within each addressable portion of data on a disk drive. The sector size is typically 512 bytes. For an error control system, the sector is a sequence of error control codewords or blocks. For interleaved error control systems, each sector contains as many blocks as there are interleave cycles. The block elements are symbols that are not necessarily eight bit long. In the most general terms, the symbols are used in the error control coding (ECC) system to calculate and describe error locations and values.

A read channel sees a sector as a sequence of modulation codewords together with synchronization bits. Synchronization field is partitioned into a sector address mark, or sync mark, typically of length around 20 bits, and *phase lock loop*, PLL, field, a periodic pattern whose length is about 100 bits used for PLL synchronization. In addition to this, a secondary sync mark is placed within a data field and used for increased reliability. A zero-phase start pattern of length 8–16 bits is used for initial estimation of phase in the phase lock loop. Figure 14.8 illustrates the format of user data on a disk.

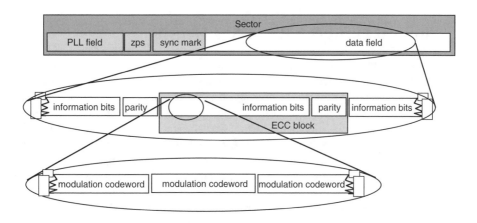

FIGURE 14.8   Format of data on a disk.

14.4   Increasing Recording Density

Increasing areal density of the data stored on a disk can be achieved by reducing lengths of magnetic domains along the track (increasing linear density), and by reducing a track width and track pitch (increasing radial density.) While the radial density is mostly limited by the mechanical design of the drive and ability to accurately follow the track, the linear density is a function of properties of magnetic materials and head designs, and ability to detect and demodulate recorded data.

As linear density increases, the magnetic domains on a surface become smaller and thus thermally unstable, which means that lower energy of an external field is sufficient to demagnetize them. This effect is known as a superparamagnetic effect [1]. Another physical effect is the finite sensitivity of the read head, that is, at extremely high densities, since the domains are too small; the signal energy becomes so small as to be comparable with the ambient thermal noise energy.

The orientation of magnetization on a disk can be longitudinal, which is typical for today's systems, or perpendicular. The choice of the media influences the way the magnetization is recorded on the disk. Media with needle shaped particles oriented longitudinally tend to have a much higher remanent magnetization in the longitudinal direction, and favor longitudinal recording. The head design must support the favorable orientation of magnetization. Longitudinal orientation requires head shapes that promote longitudinal fields such as ring heads. Similarly, some media are composed of crystallites oriented perpendicularly to the field. Such media have a much higher remanent magnetization in the perpendicular direction, and favor perpendicular recording. If a head design promotes perpendicular fields, such as single pole heads, the result is perpendicularly recorded magnetization.

Some recent experiments have shown that media that favor perpendicular recording have better thermal stability. This is why lately the perpendicular recording is attracting a considerable attention in magnetic recording community. Typically in perpendicular recording a recording surface is made of a hard ferromagnetic material, that is, material requiring large applied fields to permanently magnetize it. Once magnetized, the domains remain very stable, that is, large fields are required to reverse the magnetization. The recording layer is made thick so that, since each magnetic domain contains a large number of magnetic particles, larger energy is required for demagnetization. The low remanence, low coercivity, materials (the so-called *soft materials*) are placed beneath hard ferromagnetic surface (soft underlayer) and used to conduct magnetic field back to another electromagnet pole. A pole-head geometry is used, so that the medium can effectively travel through the head gap, and be exposed to stronger magnetic field. A pole-head/soft-underlayer configuration can produce about twice the field that a ring head produces. In this way sharp transitions can be supported on relatively thick perpendicular media, and high frequencies (that get attenuated during readback) are written firmly. However, effects of demagnetizing fields are much more pronounced in perpendicular recording systems, because in longitudinal media the transitions are not that sharp.

14.5   Physical Limits on Recording Density

At extremely high areal densities each bit of information is written on a very small area. The track width is small and magnetic domains contain relatively small numbers of magnetic particles. Since the particles have random positions and sizes, there are large statistical fluctuations or noise on the recovered signal. The signal to noise ratio is proportional to the track width, and is inversely proportional to the mean size of the particle and the standard deviation of the particle size. Therefore, increasing the track size, increasing the number of particles by increasing media thickness, and decreasing the particle size will improve the signal to noise ratio. Uniaxial orientation of magnetic particles also gives higher signal to noise ratio. However the requirement for thermal stability over periods of years dictates a lower limit to the size of magnetic particles in a magnetic domain because ambient thermal energy causes the magnetic signals to decay. Achieving both small particle size and thermal stability over time can be done by using magnetic materials with higher coercivity. However, there is a strong practical upper limit to the coercivity that can be written, and it is determined by the saturation magnetization of the head material.

In addition to the basic physics, there are a number of practical engineering factors that must be considered at extremely high densities. In particular, these factors include the ability to manufacture accurately the desired head geometries and control media thickness, the ability to closely follow the written tracks, to control head flying height, and the ability to maintain a very small, stable magnetic separation.

14.6 Future

The hard drive areal densities have grown at an annual rate approaching 100%. Recently a 20Gbit/sqinch has been demonstrated [5], and some theoretical indications of feasibility of extremely high densities approaching 1 Tbit/sqinch have been given [8, 9].[3] Although the consideration related to user needs including higher capacity, speed, error performance, reliability, environment condition tolerances etc. are important, the factors affecting cost tend to dominate read channel architecture and design considerations. Thus achieving highest recording density with lowest component costs at high manufacturing yields is the ultimate goal.

With areal densities growing at an annual rate approaching 100%, real concern continues to be expressed that we may be approaching a limit to conventional magnetic recording technology. However, as long as the read channel is concerned, there are large opportunities to improve on the existing signal processing, both with detectors better matched to the channel and by applying more advanced detection, modulation and coding schemes.

Acknowledgement

This work is supported by the NSF under grant CCR-0208597.

References

[1] S. H. Charrap, P. L. Lu, and Y. He, "Thermal stability of recorded information at high densities," *IEEE Trans. Magn.*, part 2, vol. 33, no. 1, pp. 978–983, January 1997.

[2] J. D. Coker, et al., Implementation of PRML in a rigid disk drive, *Digest of Magnetic Recording Conf.* 1991, paper D3, June 1991.

[3] G. D. Forney and A. R. Calderbank, Coset codes for partial response channels; or, coset codes with spectral nulls, *IEEE Trans. Info. Theory*, vol. IT-35, no. 5, pp. 925–943, September 1989.

[4] H. Kobayashi and D. T. Tang, Application of partial response channel coding to magnetic recording systems, *IBM J. Res. Dev.*, vol. 14, pp. 368–375, July 1970.

[5] M. Madison et al., "20 Gb/sq in: Using a merged notched head on advanced low noise media," in MMM Conference, November 1999.

[6] H. Thapar and A. Patel, A class of partial-response systems for increasing storage density in magnetic recording, *IEEE Trans. Magn.*, vol. MAG-23, pp. 3666–3668, September 1987.

[7] H. Osawa, Y. Kurihara, Y. Okamoto, H. Saito, H. Muraoka, and Y. Nakamura, PRML systems for perpendicular magnetic recording, *J. Magn. Soc. Japan*, vol. 21, no. S2, 1997.

[8] R. Wood, Detection and capacity limits in magnetic media noise, *IEEE Trans. Magn.*, vol. MAG-34, no. 4, pp. 1848–1850, July 1998.

[9] R. Wood, The feasibility of magnetic recording at 1 Terabit per square inch, 36 *IEEE Trans. Magn.*, vol. 36, no. 1, pp. 36–42, January 2000

---

[3]Many new results have occured after the book was submitted to the publisher.

15

Read Channels for Hard Drives

Bane Vasic

University of Arizona
Tucson, AZ

Pervez M. Aziz

Agere Systems, Dallas
Dallas, TX

Necip Sayiner

Agere Systems
Allentown, PA

15.1   Analog Front End **15**-3
15.2   Partial Response Signaling with Maximum
        Likelihood Sequence Estimation **15**-3
15.3   Adaptive Equalization **15**-4
15.4   Viterbi Detection **15**-4
15.5   Timing Recovery **15**-5
15.6   Read Channel Servo Information Detection **15**-6
15.7   Precompensation **15**-7
15.8   The Effect of Thermal Asperites **15**-7
15.9   Postprocessor **15**-7
15.10   Modulation Coding **15**-8
15.11   Error Control Coding **15**-9
15.12   Error Performance Measures **15**-9

The read channel is a device situated between the drive's controller and the recording head's preamplifier (Figure 15.1). The read channel provides an interface between the controller and the analog recording head, so that digital data can be recorded and read back from the disk. Furthermore it reads back the head positioning information from a disk and presents it to the head positioning servo system that resides in the controller. A typical read channel architecture is shown in Figure 15.2. During a read operation, the head generates a pulse in response to magnetic transitions on the media. Pulses are then amplified by the preamplifier that resides in the arm electronics module, and fed to the read channel. In the read channel the readback signal is additionally amplified and filtered to remove noise and to shape the waveform, and then the data sequence is detected (Figure 15.2). The data to be written on a disk are sent from a read channel to a write driver that converts them into a bipolar current that is passed through the electromagnet coils. Prior to sending to read channel, user data coming from computer (or from a network in the network attached storage devices) are encoded by an error control system. Redundant bits are added in such a way to enable a recovery from random errors that may occur during reading data from a disk. The errors occur due to a number of reasons including: demagnetization effects, magnetic field fluctuations, noise in electronic components, dust and other contaminants, thermal effects etc. Traditionally, the read channel and drive controller have been separate chips. The latest architectures have integrated them into so called "super-chips."

0-8493-1524-7/05/$0.00+$1.50
© 2005 by CRC Press, LLC

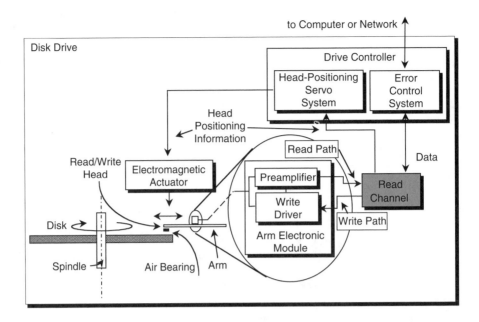

FIGURE 15.1   The block diagram of a disk drive.

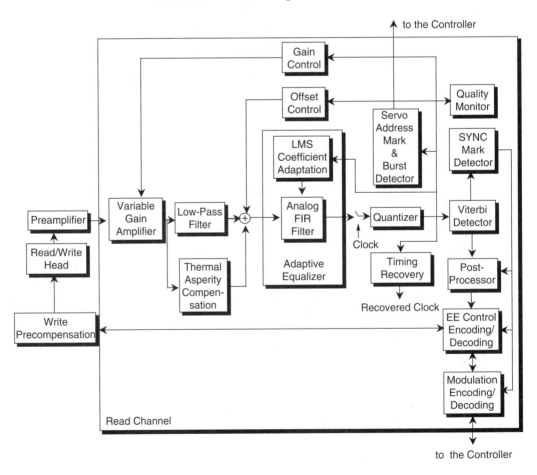

FIGURE 15.2   A typical read channel architecture.

15.1 Analog Front End

As a first step, the read signal is normalized with respect to gain and offset so that it falls into an expected signal range. Variation of gain and offset is a result of variations in the head media spacing, variations in magnetic and mechanical and electronic components in the drive, preamplifier and read channel. The front end also contains a thermal asperity (TA) circuit compensation. A thermal asperity occurs when the head hits a dust particle or some other imperfection on a disk surface. At the moment of impact, the temperature of the head rises, and a large signal at the head's output is generated. During a TA a useful readback signal appears as riding on the back of a low frequency signal of much higher energy. The beginning of this "background" signal can be easily predicted and the TA signal itself suppressed by a relatively simple filter.

High frequency noise is then removed with a continuous-time low pass filter to permit a sampling of the signal without aliasing of high frequency noise back into the signal spectrum. The filter frequently includes programmable cut-off frequency which can be used to shape the signal to optimize data detection. A programmable cut-off frequency is essential since the disk rotates with constant angular velocity, and data rate varies by approximately a factor of 2 from the inner to outer radius of the disk. It is also important for the analog filter bandwidth to be switched to allow for low cut-off frequencies when processing servo sector information.

15.2 Partial Response Signaling with Maximum Likelihood Sequence Estimation

After sampling with a rate $1/T$, the read signal is passed trough an analog or digital front end filter and detected using a maximum likelihood sequence detector. The partial response signaling with maximum likelihood sequence estimation PRML was proposed for use in magnetic recording by Kobayashi 30 years ago [15, 16]. In 1990 IBM produced the first disk drives employing partial-response signaling with maximum-likelihood detection. Today's all read channels are based on some version of the PRML. Cidecyan et al. [3] described a complete PRML system including equalization, gain and timing control, and Viterbi detector. All basic functions of a PRML system have remained practically unchanged, until the introduction of a postprocessor that performs a special type of soft error correction after maximum likelihood sequence detection. Also, significant improvements in all the subsystems have been made during last 10 years. The term "partial response" comes from the fact that the sample of the equalized signal at, say, time nT (T is a signaling interval), contains information not only on data bits at time nT, but also on neighboring bits, that is magnetic transitions. The number of adjacent bits that determine the sample at nT is referred to as *channel memory*. The channel memory is a parameter that can be selected in the process of defining a read channel architecture. The channel memory and the details of the partial response selection are made based on an attempt to have the partial response be as close a match to the channel as possible. Since the complexity of a maximum likelihood detector is an exponential function of a memory, it is desirable to keep the memory low, but, the equalization required to achieve this might boost the high frequency noise, which result in decrease of signal to noise ratio, called *equalization loss*. The typical value of channel memory in today's read channels is four. The value of an equalized sample at time nT, y_n can be written as

$$y_n = \sum_{k=0}^{L_h} h_k \cdot x_{n-k}$$

where x_n is a user-data bit recorded at time $n (x_n \in \{-1, +1\})$, and L_h is a channel memory. The coefficients h_k form, $h(D) = \sum_{k=0}^{L_h} h_k \cdot D^k$, a *partial response polynomial* or *partial response target* (D is a formal, time-delay variable). The main idea in partial response equalization is to equalize the channel to a known and short target that is matched to the channel spectrum so that noise enhancement is minimum. Therefore, the deliberate intersymbol interference is introduced, but since the target is known, the data can be recovered, as explained in the previous article.

15.3   Adaptive Equalization

To properly detect the user-data it is of essential importance to maintain the partial response target during the detection. This implies that channel gain, finite-impulse response (FIR) filter coefficients, and sampling phase must be adaptively controlled in real-time. Continuous automatic adaptation allows the read channel to compensate for signal variations and changes that occur when drive operation is affected by changes in temperature and when the input signals is altered by component aging. Comparing the equalizer output samples with the expected partial response samples generates an error signal, which is used to produce adaptive control signals for each of the adaptive loops. For filter coefficients control, a least-mean square (LMS) algorithm is used [4]. LMS operates in the time domain to find filter coefficients that minimize the mean-squared error between the samples and the desired response. Initial setting of the filter coefficients is accomplished by training the filter with an on-board training sequence, and the adaptation is continued while the chip is reading data. Adaptation can be in principle performed on all coefficients simultaneously at the lower clock rate or on one coefficient at a time. Since disk channel variations are slow relative to the data rate, the time-shared coefficient adaptation achieves the same optimum filter response while consuming less power and taking up less chip area. Sometimes, to achieve better loop stability, not all filter coefficients are adapted during reading data. Also, before writing, data are scrambled to whiten the power spectral density and ensure proper adaptation filter adaptation.

The FIR filter also compensates for the large amount of group-delay variation that may be caused by a low-pass filter with a nonlinear phase characteristic. Filters with nonlinear characteristics, such as a Butterworth filter, are preferred over, say, an equi-ripple design of the same circuit complexity, because they have much better roll-off characteristics. The number of FIR filter coefficients in practical read channels has been as low as 3 and as high as 10 with various tradeoffs associated with the different choices which can be made.

15.4   Viterbi Detection

In many communications systems a symbol by symbol detector is used to convert individual received samples at the output of the channel to corresponding detected bits. In today's PRML channels, a Viterbi detector is a maximum likelihood detector which converts an entire *sequence* of received equalized samples to a corresponding sequence of detected bits. Let $y = (y_n)$ be the sequence of received equalized samples corresponding to transmitted bit sequence $x = (x_n)$. Maximum likelihood sequence estimation maximizes the probability density $p(y \mid x)$ across all choices of transmitted sequence x [7]. In the absence of noise and mis-equalization, the relationship between the noiseless equalized samples, z_n and the corresponding transmitted bits is known by the Viterbi detector and is given by

$$z_n = \sum_{k=0}^{L} h_k \cdot x_{n-k} \tag{15.1}$$

In the presence of noise and mis-equalization the received samples will deviate from the noiseless values. The Viterbi detector considers various bit sequences and efficiently compares the corresponding expected PR channel output values with those actually received. For Gaussian noise at the output of the equalizer and equally probable input bits, maximizing $p(y \mid x)$ is equivalent to choosing as the correct bit sequence the one closest in a (squared) Euclidean distance sense to the received samples. Therefore, we wish to minimize

$$\min_{x_k} \left(\sum_{n=0}^{P-1} \left[y_n - \sum_{k=0}^{L} h_k \cdot x_{n-k} \right]^2 \right) \tag{15.2}$$

The various components of Equation 15.3 are also known as branch metrics.

The Viterbi detector accomplishes the minimization in an efficient manner using a trellis based search rather than an exhaustive search. The search is effectively performed over a finite window known as the decision delay or path memory length of the Viterbi detector. Increasing the window length beyond a certain value leads to only insignificant improvements of the bit detection reliability or bit error rate (BER).

Despite the efficient nature of the Viterbi algorithm the complexity of a Viterbi detector increases exponentially with the channel memory of the PR target. A target with channel memory of $L-1$ requires for example a 2^{L-1} state Viterbi detector trellis. For a fully parallel Viterbi implementation, each Viterbi state contains an add-compare-select (ACS) computational unit which is used to sum up the branch metrics of Equation 15.2 and keep the minimum metric paths for different bit sequences. Also required for the hardware is a $2^{L-1} \cdot P$ bit memory to keep a history of potential bit sequences considered across the finite decision delay window.

15.5 Timing Recovery

A phase-locked loop (PLL) is used to regenerate a synchronous clock from the data stream. The PRML detector uses decision directed timing recovery typically with a digital loop filter. The digital loop filter parameters can be easily controlled using programmable registers and changed when a read channel switches from acquisition to tracking mode. Because significant pipelining is necessary in the loop logic to operate at high speeds, the digital loop filter architecture exhibits a relatively large amount of latency. It can affect considerably the acquisition time when the timing loop must acquire significant phase and frequency offsets. To ensure that only small frequency offsets are present, the synchronizer VCO is phase-locked to the synthesizer during nonread times. For fast initial adjustment of the sampling phase, a known preamble is recorded prior to user data. The time adjustment scheme is obtained by applying the stochastic gradient technique to minimize the mean squared difference between equalized samples and data signal

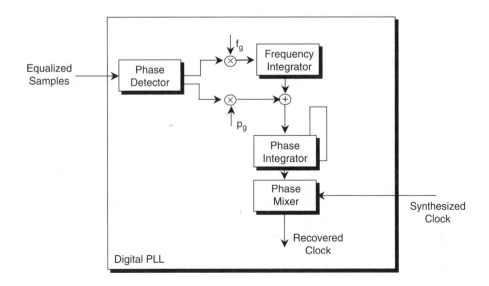

estimates. To compensate for offset between the rate of the signal received and the frequency of the local timing source the loop filter design allows for a factor ΔT_n to be introduced, so that the sample at discrete time n is taken $T + \Delta T_n$ seconds after the sample at discrete time $n - 1$. In acquisition mode, in order to quickly adjust the timing phase, large values for loop gains are chosen. In tracking mode, the loop gains are lowered to reduce loop bandwidth.

15.6 Read Channel Servo Information Detection

In an *embedded* servo system (introduced in the previous article), the radial position of the read head is estimated from two sequences recorded on servo wedges: *track addresses* and *servo-bursts*. The track address provides a unique number for every track on a disk, while a servo-burst pattern is repeated on each track or on a group of tracks. Determining the head position using only the track number is not sufficient because the head has to be centered exactly on a given track. Therefore, the servo-burst waveform is used in conjunction with the track address to determine the head position. Using the servo-burst pattern, it is possible to determine the off-track position of a head with respect to a given track with a high resolution. While positioning the head over a surface, the disk drive can be in either *seeking* or *tracking* operation mode. In seeking mode, the head moves over multiple tracks, trying to reach the track with a desired address as quickly as possible, while in tracking mode, the head tries to maintain its position over a track. The track addresses are therefore used mostly in the seeking mode, while servo-burst information is usually used in the tracking mode [25, 30].

In read channels periodic servo-burst waveforms are detected and used to estimate radial position. The radial position error signal is calculated based on the current estimated position and the position of the track to be followed, and then used in an external head positioning servo system. Generally two types of position estimators are in use: maximum likelihood estimators that are based on a matched filtering, and suboptimal estimators based on averaging the area or the peaks of the incoming periodic servo-burst waveform. A variety of techniques have been used to demodulate servo bursts including amplitude, phase and null servo detectors. Today, most read channels use an amplitude modulation with either peak or area detection demodulators.

Older generation channels generally implemented the servo functions in analog circuitry. The analog circuitry of these servo channels partially duplicates functions present in the digital data channel. Now, several generations of read-channel chips have switched from analog to digital circuits and digital signal processing [8, 34]. These channels reduce duplication of circuits used for servo and data and provide a greater degree of flexibility and programmability in the servo functions.

Typically a single analog to digital converter (ADC) or quantizer is used for both data detection and servo position error signal estimation [8, 20, 27, 34]. However, quantizer requirements are different in data and servo fields. Compared to position error signal estimators, data detectors require a quantizer with higher sampling clock rate. On the other hand position error signal estimators require a quantizer with finer resolution. A typical disk drive has a data resolution requirement of around six bits, and a servo resolution requirement of around seven or eight bits. Furthermore, servo bursts are periodic waveforms as opposed to data streams. In principle, both the lower sampling clock rate requirement in the servo field and the periodicity property of servo-burst signals can be exploited to increase the detector quantization resolution for position error signal estimation. The servo field is oversampled asynchronously to increase the effective quantizer resolution.

Track densities in today's hard drives are higher than 25000 tracks per inch and , and the design of a tracking servo system is far from trivial. Some of the recent results include [2, 24, 25, 26, 31]. Increasing the drive level servo control loop bandwidth is extremely important. Typical bandwidth of a servo system is about 1.5 kHz, and is mainly limited by the parameters that are out of reach of a read channel designer, such as mechanical resonances of voice coil motor, arm holding a magnetic head, suspension, and other mechanical parameters.

Another type of disturbance with mechanical origins, that has to be also detected and controlled in a read channel is *repeatable runout* (RRO) in the position of the head with respect to the track center.

These periodic disturbances are inherent in any rotating machine, and can be as a result of an eccentricity of the track, offset of the track center with respect to the spindle center, bearing geometry and wear and motor geometry. The frequencies of the periodic disturbances are integer multiplies of the frequency of rotation of the disk, and if not compensated they can be a considerable source of tracking error. In essence the control system possesses an adaptive method to learn on-line the true geometry of the track being followed, and a mechanism of continuous-time adaptive runout cancellation [31].

15.7   Precompensation

Nonlinear bit shift in magnetic recording is the shift in position of a written transition due to the demagnetizing field from adjacent transitions. In a PRML channel, the readback waverofm is synchronously sampled at regular intervals, and the sample values depend on the position of written transitions. Therefore nonlinear bit shift leads to error in sample values which, in turn, degrades the channel performance. The write precompensation is employed to counteract the nonlinear bit shift. However, determining the nonlinear bit shift is not simple and straightforward especially when one tries to fine tune each drive for its optimum precompensation. The precompensation circuit generates the write clock signal whose individual transition timing is delayed from its nominal position by the required precompensation amount. The amount of precompensation and the bit patterns requiring precompensation can be found using the extracted pulse shape [10, 18]. Another approach is a frequency-domain technique that offers simple measurement procedure and a possible hardware implementation using a band-pass filter [32] or using PRML sample values [33].

15.8   The Effect of Thermal Asperites

As explained earlier, if a head hits a dust particle, a long thermal asperity will occur, producing a severe transient noise burst, loss of timing synchronization, or even off-track perturbation. Error events caused by thermal asperities (*TA*) are much less frequent than random error events, but they exist and must be taken into account during read channel design. If there were no TA protection in the read channel, a loss of lock in timing recovery system would occur, causing massive numbers of data errors well beyond the error correction capability of any reasonable ECC system. Despite TA protection, the residual error cannot be completely eliminated, and many bits will be detected incorrectly. However, the read channel should be designed to enable proper functioning of timing recovery in the presence of bogus samples. Typically the read channel estimates the beginning and length of TA and sends this information to the ECC system, which may be able to improve its correction capability using so-called erasure information. However, since the TA starting location is not known precisely, and the probability of random error in the same sector is not negligible, the ECC system can misscorrect, which is more dangerous than not to detect the error.

15.9   Postprocessor

Due to the channel memory and noise coloration, maximum likelihood sequence detector (Viterbi detector) produces some error patterns more often than others. They are referred to as *dominant error sequences*, or *error events*, and can be obtained analytically or through experiments and/or simulation. Relative frequencies of error events strongly depend on a recording density.

Parity check processors combine syndrome decoding and soft-decision decoding. Error event likelihoods needed for soft decoding can be computed from a channel sequence by some kind of soft-output Viterbi algorithm. By using a syndrome calculated for a received codeword, a list is created of all possible *positions* where error events can occur, and then error event likelihoods are used to select the most likely position and most likely type of the error event. Decoding is completed by finding the error event position and type. The decoder can make two type of errors: it fails to correct if the syndrome is zero, or it makes a wrong correction if the syndrome is nonzero but the most likely error event or combination of error events do not produce right syndrome.

A code must be able to detect a single error from the list of dominant error events, and should minimize the probability of producing zero syndrome when more than one error event occur in a codeword.

Consider a linear code given by an $(n - k) \times n$ parity check matrix H. We are interested in H capable of correcting or detecting dominant errors. If all errors from a list were contiguous and shorter than m, then a cyclic $n - k = m$ parity bit code could be used to correct a single error event However, in reality, the error sequences are more complex, and occurrence probabilities of error events of lengths 6, 7, 8 or more are not negligible. Furthermore, practical reasons (such as decoding delay, thermal asperities etc.) dictate using short codes, and consequently, in order to keep code rate high, only a relatively small number of parity bits is allowed, making the design of error event detection codes nontrivial.

The detection is based on the fact that we can calculate the likelihoods of each of dominant error sequences at each point in time. The parity bits serve to detect the errors, and to provide some localization in error type and time. The likelihoods are then used to choose the most likely error events (type and location) for corrections. The likelihoods are calculated as the difference in the squared Euclidean distances between the signal and the convolution of maximum likelihood sequence estimate and the channel partial response, versus that between the signal and the convolution of an alternative data pattern and the channel partial response. During each clock cycle, the lowest M are chosen, and the syndromes for these error events are calculated. Throughout the processing of each block, a list is maintained of the N most likely error events, along with their associated error types, positions and syndromes. At the end of the block, when the list of syndromes are calculated for each of six combinations of M candidate error events which are possible. After disqualifying those sets of candidates (with cardinality less then M) which overlap in the time domain, and those sets of candidates which produced a syndrome which does not match the actual syndrome, the set of candidates which remains.

15.10   Modulation Coding

Modulation of constrained coding is used to translate an arbitrary sequence of input data to a channel sequence with special properties required by the physics of the medium [21]. Two large important classes of channel constraints are run-length and spectral constraints. The run-length constraints [12] bound the minimal and/or maximal lengths of certain types of channel subsequences, while the spectral constraints include dc-free [11] and higher order spectral-null constraints [6]. The spectral constraints also include codes that produce spectral zeros at rational submultiples of symbol frequency as well as constraints that give rise to spectral lines. The most important class of runlength constraints is a (d, k) constraint, where $d + 1$ and $k + 1$ represent minimum and maximum number of consecutive like symbols or space between the adjacent transitions. Bounding minimal length consecutive like symbols controls ISI in the excess bandwidth systems and reduces transition noise. Bounding the upper limits of the mark lengths improves timing recovery and automatic gain control. In order to keep the code rate high, today's read channels employ only k constrained codes. Typical code rates are: $16/17, 24/25, 32/34, 48/49$. Modulation decodes can be either block-by-block or sliding-window. Block decoders determine data word by using a single codeword, while sliding-window decoders require so called look-ahead, which means that the output data word is a function of several consecutive codewords. Due to inherent non-linearity, a modulation decoder may produce multiple errors as a result of a single erroneous input bit. When a sliding-window decoding is used an error can affect several consecutive data blocks. This effect is known as an *error propagation*. The block codes are favored because they do not propagate errors.

A mathematically rigorous code design approach based on symbolic dynamics was developed by Marcus and Siegel et al. [19, 22]. The algorithm is know as "state splitting algorithm" or Adler, Coppersmith, and Hassner (ACH) algorithm [1]. Another constrained coding approach, championed by Immink [14] emphasizes the low complexity encoding and decoding algorithms [13]. Despite of this nice mathematical theory, a design of constrained codes remains too difficult to be fully automated, and in the art of designing efficient codes, a human intervention and skill are still necessary.

15.11 Error Control Coding

In a conventional hard disk drives the ECC system does not reside in a read channel, however the ECC performance is linked to the performance of a detection algorithm, error propagation in a modulation code, and it is natural to try to expand the read channel functionality to error control as well. A new trend in industry is aimed toward designing an integrated circuit, so called *super chip* with a such expanded functionality.

In the most general terms, the purpose of ECC is to protect user data, and this is achieved by including redundant, so called *parity* bits along with the data bits. The codes used in hard drives belong to a large class of ECC schemes, called block codes. A block code is a set of codewords (or blocks) of a fixed length n. The length is expressed in number of symbols, and a symbol is a binary word of length m. Other parameters of a block code are k-number of data symbols in the block, and t-number of symbols correctable by the ECC system [18, 35].

Reed–Solomon (RS) codes [28] have been the class of codes most often used in the magnetic disk storage in last 15 years. The reason is their excellent performance in presence of error events that exhibit burstiness, which is typical for magnetic recording channels, and lend themselves to high speed encoding/decoding algorithms required for high speed disk drives [5, 9, 36]. Very often RS codes are interleaved to reduce effect of long error burst, and to reduce the implementation cost by eliminating conversion of bytes to possibly longer code symbols used in encoding and decoding. The parameters of RS codes satisfy the following relations: $n \leq 2^m - 1$, number of parity symbols $n - k \geq 2t$, and code rate of the RS code $r = k/n$.

In today's hard drives typically a part of ECC decoding is performed in hardware with a throughput equal to the data rate, and the other part is performed in firmware with much lower speed. In some cases, such as thermal asperities, no error control is sufficient to recover the data. In this case it is necessary to retry reading the same sector. A choice between hardware of firmware correction depends on the application, the data transfer protocol and the bus transfer rate. In applications such as single-user work-stations, short data transfers dominate, but streaming data transfers occasionally occurs (during boot, large file transfers, etc.). Additionally, data read from the disk drive can be transmitted to the host computer in a physically sequential or in any convenient order. If the bus transfer rate is higher than the ECC hardware throughput, and if sufficiently long ECC firmware buffer is available to store all the sectors, or if sectors are transmitted to the host computer in any convenient order all firmware error recovery can be performed in parallel with disk reads without interrupting streaming read operations. In the case of short packet transfers it is better to perform read retry in parallel with firmware error correction. Retries in conjunction with hardware correction typically consume less time than firmware error correction. On the other hand, for long streaming transfers, correcting errors in firmware in parallel with reading the sector is better strategy-provided that the firmware ECC throughput is high enough to prevent buffer overflow. A detailed treatment of an error control system design considerations can be found in [21].

15.12 Error Performance Measures

A commonly used measure of ECC performance is a bit error rate (BER), which is defined as a ratio of unrecoverable error events and total user data bits. An unrecoverable error event is a block that contains more erroneous symbols than the ECC system can correct, and it may contain as many as exist in a single data block protected by the ECC system. Various applications require different BER's, but they are typically in the range of $10^{-12} - 10^{-15}$. Another ECC performance measure is undetected bit error rate (UBER), which is a number of undetected error events per total number of user bits. In some cases the ECC system detect that the sector contain errors, but is not able to correct them. Then a controller asks a read channel to retry reading the same sector. The *retry rate per bit* is a useful measure of a data throughput . The hard drive standards of a retry rate is 10^{-14}. The performance measure used depends on the application. For example UBER is much more important for bank transactions than for multimedia applications ran on PC. All performance measures depend on a symbol length, number of correctable errors and symbol error statistics. On the other hand symbol error statistics depend on the read channel error event distribution.

Acknowledgment

This work is supported in part by the NSF under grant CCR-0208597.

References

[1] R. L. Adler, D. Coppersmith, and M. Hassner, Algorithms for sliding block codes: An application of symbolic dynamics to information theory, *IEEE Trans. Inform. Theory*, vol. IT-29, pp. *5–22*, January 1983.

[2] D. Cahalan and K. Chopra, Effects of MR head track profile characteristics on servo performance, *IEEE Trans. Magn.*, vol. 30, no. 6, November 1994.

[3] R. D. Cideciyan, F. Dolivo, R. Hermann, W. Hirt, and W. Schott, A PRML system for digital magnetic recording, *IEEE J. Sel. Areas Commn.*, vol. 10, no. 1, pp. 38–56, January 1992.

[4] J. M. Cioffi, W. L. Abbott, H. K. Thapar, C. M. Melas, and K. D. Fisher, Adaptive equalization in magnetic-disk storage channels, *IEEE Comm. Magazine*, pp. 14–20, February 1990.

[5] E. T. Cohen, On the implementation of Reed-Solomon decoders, Ph.D. dissertation, University of California, Berkeley, 1983.

[6] E. Eleftheriou and R. Cideciyan, On codes satisfying M-th order running digital sum constraints, *IEEE Trans. Inform. Theory.* vol. 37, pp. 1294–1313, September 1991.

[7] G. D. Forney, Maximum-likelihood sequence estimation of digital sequences in the presence of intersymbol interference, *IEEE Trans. Inform. Thoery*, vol. 18., no. 3, pp. 363–378, May 1972.

[8] L. Fredrickson et al., Digital servo processing in the Venus PRML read/write channel, *IEEE Trans. Magn.*, vol. 33, pp. 2616–2619, September 1997.

[9] M. Hassner, U. Schwiegelshohn, and S. Winograd, On-the-fly error correction in data storage channels, *IEEE Trans. Magn.*, vol. 31, pp. 1149–1154, March 1995.

[10] R. Hermann, Volterra model of digital magnetic saturation recording channels, *IEEE Trans. Magn.*, vol. MAG-26, no. 5, 2125–2127, September 1990.

[11] K. A. S. Immink, Spectral null codes, *IEEE Trans. Magn.*, vol. 26, pp. 1130–1135, March 1990.

[12] K. A. S. Immink, Runlength-limited sequences, *Proc. IEEE*, vol. 78, pp. 1745–1759, November 1990.

[13] K. A. S. Immink and L. Patrovics, Performance assessment of DC-free multimode codes, *IEEE Trans. Commn.*, vol. 45, pp. 293–299, March 1997.

[14] K. A. S. Immink, *Codes for Mass Data Storage Systems*, Shannon Foundation Publishers, Essen, Germany, 1999.

[15] H. Kobayashi and D. T. Tang, Application of partial-response channel coding to magnetic recording systems, *Bell J. Res. Develop.*, July 1970.

[16] H. Kobayashi, Correlative level coding and maximum-likelihood decoding, *IEEE Trans. Inform. Theory*, vol. IT-17, pp. 586–594, September 1971.

[17] S. Lin, *An Introduction to Error-Correcting Codes*, Prentice-Hall, Englewood Cliffs, NJ, 1970.

[18] Y. Lin and R. Wood, An estimation technique for accurately modeling the magnetic recording channel including nonlinearities, *IEEE Trans. Magn.*, vol. MAG-25, no. 5, pp. 4058–4060, September 1989.

[19] R. Karabed and B. H. Marcus, Sliding-block coding for input-restricted channels, *IEEE Trans. Inform. Theory*, vol. 34, pp. 2–26, January 1988.

[20] H. Kimura, T. Nishiya, T. Nara, and T. Komotsu, A digital servo architecture with 8.8 bit resolution of position error signal for disk drives, in *IEEE Globecom 97*, Phoenix, AZ, 1997, pp. 1268–1271.

[21] B. Marcus, P. Siegel, and J. K. Wolf, Finite-state modulation codes for data storage, *IEEE J. Select. Areas Commn.*, vol. 10, no. 1, pp. 5–37, January 1992.

[22] B. H. Marcus, Sofic systems and encoding data, *IEEE Trans. Inform. Theory*, vol. IT-31, pp. 366–377, May 1985.

[23] C. Monti, and G. Pierobon, Codes with multiple spectral null at zero frequency, *IEEE Trans. Inform. Theory*, vol. 35, no. 2, pp. 463–472, March 1989.

[24] A. Patapoutian, Optimal burst frequency derivation for head positioning, *IEEE Trans. Magn.*, vol. 32, no. 5, part. 1, pp. 3899–3901, September 1996.

[25] A. Patapoutian, Signal space analysis of head positioning formats, *IEEE Trans. on Magn.*, vol. 33, no. 3, pp. 2412–2418, May 1997.

[26] A. Patapoutian, Analog-to-digital converter algorithms for position error signal estimators, *IEEE Trans. Magn.*, vol. 36, no. 1, part. 2, pp. 345–400, January 2000.

[27] D. E. Reed, W. G. Bliss, L. Du, and M. Karsanbhai, Digital servo demodulation in a digital read channel, *IEEE Trans. Magn.*, vol. 34, pp. 13–16, January 1998.

[28] I. S. Reed and G. Solomon, Polynomial codes over certain finite fields, *J. Soc. Indust. Appl. Math.*, vol. 8, pp. 300–304, 1960.

[29] C. M. Riggle, and S. G McCarthy, Design of error correction systems for disk drives, *IEEE Trans. Magn.*, vol. 34 , 4 Part: 2, pp. 2362–2371, July 1998.

[30] A. H. Sacks, Position signal generation in magnetic disk drives, Ph.D. dissertation, Carnegie-Mellon University, Pittsburgh, PA, 1995.

[31] A. H. Sacks, M. Bodson, and W. Messner, Advanced methods for repeatable runout compensation [disc drives], *IEEE Trans. Magn.*, vol. 31, no. 2 , pp. 1031–1036, March 1995.

[32] Y. Tang and C. Tsang, A technique for measuring nonlinear bit shift, *IEEE Trans. Magn.*, vol. 27, no. 6, pp. 5326–5318, November 1991.

[33] Y. Tang, R. L. Galbraith, J. D. Coker, P. C. Arnett, and R. W. Wood, Precompesation value determination in a PRML channel, *IEEE Trans. Magn.*, vol. 32, no. 3, pp. 2013–1014, May 1996.

[34] G. T. Tuttle et al., A 130 Mb/s PRML read/write channel with digital-servo detection, in *Proc. IEEE Int. Solid State Circuits Conf. 1996*, San Francisco, CA, February 8–10, 1996, pp. 64–65.

[35] S. B. Wicker, *Error Control Systems for Digital Communication and Storage*, Prentice-Hall, Englewood Cliffs, NJ, 1995.

[36] D. L. Whiting, Bit-serial Reed-Solomon decoders in VLSI, Ph.D. dissertation, California Inst. Tech., Pasadena, 1984.

16

An Overview of Hard Drive Controller Functionality

16.1   Introduction **16**-1
Hard Drive Component Overview • Role of the Controller
• A Historical Perspective

16.2   Hardware Subsystems of a Hard Disk Controller **16**-6
Block Diagram • Disk Format Control Subsystem • Error
Correction Code (ECC) Subsystem • Host Interface
Subsystem • The Processor Subsystem • Buffer Memory
Subsystem • The Servo Subsystem

16.3   Global Controller Strategies **16**-24
Data Integrity Strategies • Drive Caching

Bruce Buch

Maxtor Corporation
Shrewsbury, MA

16.1   Introduction

A hard drive controller choreographs the actions of a diverse set of electronic and mechanical functions to provide the user with a data storage system that implements the high-level behavior described by the drive's user interface (e.g., ATA or SCSI) and the product's specification. Host interface standardization has been very successful in homogenizing the set of signals and protocols exposed at the host interface to a uniform model implemented by all manufacturers. In contrast, what goes on behind the interface is not constrained by standards, and manufacturers are free to implement whatever homegrown strategies they can concoct to outdo the competition in performance and cost.

In years past, the use of catalog parts for hard drive electronics established some de facto conventions for controller designs. The industry has since consolidated to a handful of survivors, most of which use captive ASICs with proprietary designs to implement their controllers. (This may explain the dearth of texts on controller architecture in the literature.) Hence, there is no one standard design for a hard drive controller to draw upon for a summary description. This author has been involved in controller designs for three drive manufacturers and has often seen the same basic problem attacked with three different approaches. Thus, the reader should be aware that what is presented herein is one engineer's attempt at describing some generic workings of a hard drive controller, but that there is no pretense that this is a comprehensive description of all the ways that engineers have devised to implement these functions.

16.1.1   Hard Drive Component Overview

Figure 16.1 shows a hard drive controller ASIC in a representative configuration and introduces some terms used in this chapter.

0-8493-1524-7/05/$0.00+$1.50
© 2005 by CRC Press, LLC

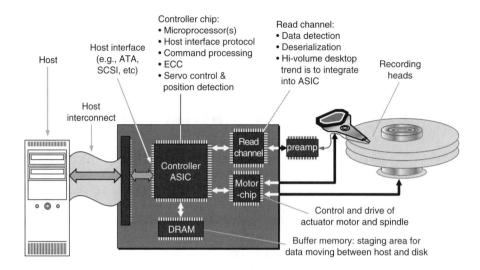

FIGURE 16.1   Location of a hard drive controller in a system.

In rough order from left to right, these elements are described as follows:

- A "host" is an entity for which the hard drive provides a data storage service, and is the source of I/O commands issued to the drive. Some host interfaces allow the connection of multiple hosts to the same drive. A typical host is a PC, or the processor system in a storage server.

- The "host interconnect" is the data connection between the host and the hard drive. Some physical manifestations are the big ribbon cable going to the drive of your PC, or the SCSI backplane of a multidrive array.

- Although the "host interface" label in Figure 16.1 points to where the hard drive connects to the host interconnect, this term usually does not refer to the physical connection per se, but to the protocol for command and information exchange collaboratively implemented by the parties connected by the host interconnect (e.g., the host and the drive), such as ATA (IDE), SCSI, or Fibrechannel.

- The "buffer" is a memory primarily used as a staging point for data moving between the host and the disk. This is sometimes called *the cache*, which is a bit of a misnomer, since a good part of this memory is often used for other collateral functions of the drive beyond data caching.

- The "read channel" is a special signal processing circuit that creates a bridge between the raw analog domain of a drive's magnetic transducers and the digital logic of the controller. Although read signal detection accounts most of the silicon area, it is more precisely a "read/write channel," since the channel also provides essential functionality required for writing. Shown as a separate chip in Figure 16.1, many cost-focused drives use controllers with integrated channel cores.

- The "preamp," also called a *read-write head controller*, is a first stage of amplification from the drive's read transducers, and a final amplifying stage for the write transducers. To minimize the electrical distance from the R/W transducers, the preamp is typically mounted on the actuator instead of on the main printed circuit board with the other electronics. It also provides functionality for switching between transducers of different surfaces and for managing collateral support functions of the transducers under the auspices of the controller.

- The "heads" are the transducers that create a bridge between the magnetic domains on a disk and the electrical signals of the components listed above. There are separate transducers for reading and writing that are different in structure, but colloquially, they are collectively called *the heads*.

- The "VCM" is a motor used to rotate the actuator on which the heads are mounted to position them to a desired track on the media. The name stands for "voice coil motor," and is an artifact of the days when heads were positioned using a structure similar to the voice coil of a loudspeaker. VCMs are now more like part of a rotary DC motor that only rotates through a fraction of a revolution, but the term VCM has stuck.

- The "motorchip" or "powerchip" is a device that provides the power stages, or control of external power devices, that drive the VCM and spindle motor.
- And, in the middle of it all, is the "controller." This usually takes the form of a complex digital or mixed-signal application specific integrated circuit (ASIC) that interprets the host interface protocol and controls the other elements listed above so that their collective action implements the storage system abstraction defined by the host interface.

16.1.2   Role of the Controller

As shown in Figure 16.2, the device's presentation of itself to the host interface is that of a linear and contiguous array of logical data blocks which are read from and stored in a defect-free memory space, as depicted in the "Host's View." A logical block is block of data, usually 512 bytes or longer, that is physically manifested as a tiny stripe containing a sequence of alternating magnetic transitions whose spacing is a code of the data stored (see "physical manifestation" in Figure 16.2). Each stripe is a few mils long and a few microinches wide, and is one in a sea of several million others in a three-dimensional cylindrical array on imperfect surfaces of a stack of magnetic storage media. The flux coupled into the read elements as these transitions move under the heads creates the analog signal presented to the channel ("channel input" in Figure 16.2). To read (retrieve) a block of data from the media, the channel, under direction of the controller, recovers the data coded by this waveform and passes it to the controller as a sequential stream of digital multi-bit symbols. To write (store) a data block, the channel, again under direction of the controller, turns the stream of digital data from the controller into a sequence of alternating current transitions driven to the writing element of a selected head. The head converts this to a sequence of alternating magnetizing fields that coerce a stripe of the media into a coded sequence of magnetic transitions as the media passes under the head. This is the representation of the data ultimately stored by the drive.

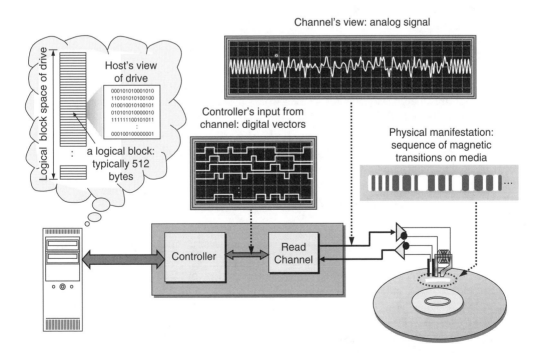

FIGURE 16.2   Manifestation of a data block at various points in a system.

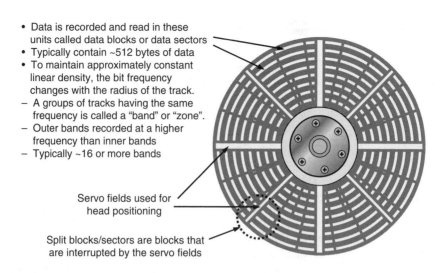

- Data is recorded and read in these units called data blocks or data sectors
- Typically contain ~512 bytes of data
- To maintain approximately constant linear density, the bit frequency changes with the radius of the track.
  - A groups of tracks having the same frequency is called a "band" or "zone".
  - Outer bands recorded at a higher frequency than inner bands
  - Typically ~16 or more bands

Servo fields used for head positioning

Split blocks/sectors are blocks that are interrupted by the servo fields

FIGURE 16.3   Surface format.

The arrangement of data blocks on the media is shown in Figure 16.3. This surface format has eight tracks each containing about a dozen data sectors, which are the media containers for data blocks. A typical drive surface would have tens of thousands of tracks, with roughly a thousand or so data sectors in each track. The data sectors in this example are interrupted by eight equiangular spaced "spokes" (these typically number in the low hundreds). These are "servo" fields used for determining the position of the heads relative to the media. The controller anticipates the precise time of arrival of these fields, and suspends and resumes reading and writing of data blocks when a servo spoke passes under the heads. Although servo functions are sometimes delegated to a separate chip, most modern controllers field the servo information received from the spokes as well as the information from the data sectors.

This part of the introduction concludes with a high-level summary of the steps for reading and writing data as per a host I/O command.

- The controller interprets commands issued by the host to read/write data in a abstract linear memory space of logical blocks.
- The controller determines the physical location on the media associated with the block's location in the abstract linear space, and uses the electromechanical subsystems of the drive to position the heads to that physical location.
- The controller directs the channel to read or write magnetic transitions on the media at the identified physical location, passing data from the channel to the host for a read operation, and data from the host to the channel for a write operation.

16.1.3   A Historical Perspective

The descriptions stated earlier apply to today's notion of a controller, which takes the form of a highly integrated ASIC designed to control a single hard drive, and is indeed an integral part of the drive, residing on a printed circuit board screwed to the casting of the mechanical assembly of the drive.

This was not always the case. Before the era of today's integrated-controller drives, the "controller" was a standalone PC board or cabinet that connected to the I/O bus of a computer, or to a host adapter plugged into the I/O bus. A low-level "drive interface" connected "dumb" drives with minimal electronics and intelligence to the controller, as shown in Figure 16.4. Two widely used drive interfaces were ST506 and Enhanced System Device Interface (ESDI).

This topology looks like a nice modular strategy for connecting simple, and therefore cheap, drives to a controller that gets leveraged across drive generations. But the standardization of the low-level drive

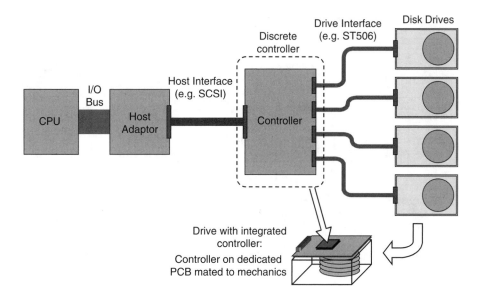

FIGURE 16.4 Discrete controller of past generations.

interfaces that enabled the modularity of this strategy was also a weakness: this architecture required the controller to manage each connected drive at a very low level, to keep track of where the heads on each drive were, to understand the physical format of each drive, and to transfer data only in the form and rate allowed by the drive interface. Hence, improvements in drive functionality were constrained to what the controller could support via the interface, and perhaps more limiting, to presumptions of the generation in which the controller was designed.

With today's integrated controllers, the controller can be designed to implement highly specific functionality for the particular drive generation the controller is to support. This gives designers more rein in the tricks they can use to enhance the performance, increase the capacity, or reduce the cost of a drive. Drives are still constrained to the externally-visible functionality allowed by their host interface, but contemporary host interfaces represent such a high-level abstraction of a drive that tremendous latitude is left for controller/drive implementation within the bounds of what the interfaces support.

Host interfaces have also evolved since the days of the discrete controllers in Figure 16.4. In generations past, host computers had to be aware of the physical geometry of drives, and communicated I/O requests in terms of physical coordinates known as CHS addressing, for cylinder, head, and sector. A "cylinder" is the set of tracks with the same radius in a stack of disk platters. The "head" number referenced a physical head for a specific surface of a platter. The "sector" number referenced one of a fixed number of equiangular arcs along a track containing a data block. CHS addressing was at least cumbersome in supporting drives with different physical geometries, but the restrictive notion of CHS addressing was miserably inapplicable for constant-density recording and drive-controlled defect management, where the number of sectors on a track varies and defective blocks leave discontinuities in the physical space of useable blocks.

CHS addressing has long given way to addressing by logical block address (LBA). An LBA is a simple linear address of one of the blocks in the virtual space of blocks depicted in the "Host View" of a drive in Figure 16.2. With LBA addressing, the drive's controller is responsible for translating an LBA into the coordinates of the corresponding block's physical location, which is now largely irrelevant to the host. This task is greatly simplified for a controller dedicated to a specific drive type, since it doesn't need to cover the general case of logical-to-physical translation. And of course, the task is moot for the host, since it does not need to do translation at all.

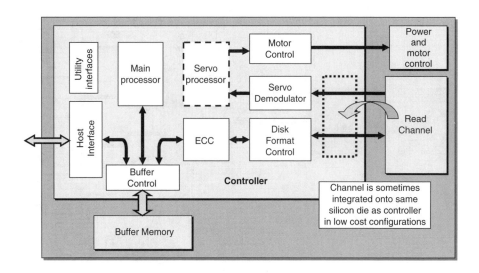

FIGURE 16.5   Block diagram of controller components.

16.2   Hardware Subsystems of a Hard Disk Controller

16.2.1   Block Diagram

Figure 16.5 shows a block diagram representative of the hardware for a typical hard drive controller. Although many cost-focused implementations integrate a read channel core on the same silicon die as the controller, the read channel is still considered a function distinct from the controller. This section describes the operation of the key subsystems listed in Table 16.1.

16.2.2   Disk Format Control Subsystem

The macro view of data sectors and servo spokes is shown in Figure 16.3. Before examining disk format control details, some detail about the fields within data sectors and servo spokes is in order.

16.2.2.1   Data Sector Format Fields

The format of data sectors can be separated into channel fields and controller fields, as shown in Figure 16.6. The channel fields give the channel what it needs to do timing and gain acquisition so it can detect the

TABLE 16.1   Subsystems Detailed in this Chapter

Subsystem	Section	Comments on Subsystem
Disk format control	2.2	Also called "the formatter," this subsystem understands the track format and controls the read channel to read and write the fields of the format.
Error checking and correction	2.3	Simply called "the ECC" subsystem, this is a hardware implementation of error detection and correction algorithms applied to the disk data.
Host interface	2.4	Physical and link layer components and protocol engines for the particular host interface implemented by the controller.
Processor	2.5	One or more microprocessor cores that execute firmware to supervise the controller and implement behavior specified by the host interface.
Buffer memory controller	2.6	Logic for multiplexing data transfers from other subsystems to a single buffer memory, and for arbitrating memory usage by multiple clients.
Servo demodulator	2.7	Logic that uses the position information demodulated by the channel to control the position of the heads as commanded by firmware.

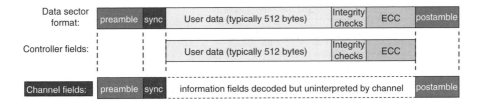

FIGURE 16.6   Data sector fields.

information that follows in the rest of the sector. The controller fields are fields containing the user data and other fields of interest only to the controller. The functions of these fields are discussed later in this chapter. The channel detects the information in these fields, but does not interpret it and makes no distinction between these fields.

The channel knows how to sequence through the channel fields within a data sector, but has no concept of the format beyond the bounds of a sector. Put simply, the channel knows *how* to read data sectors, but doesn't know *when* to read them, nor does it have any context to distinguish the identity of one sector from another. This is the function of the disk format control, specifically,

- To maintain context about where the heads are relative to the track format
- To figure out where the data sector for a particular block occurs on the track
- To tell the channel precisely when to do a read or write operation
- To send the data associated with a sector to the channel when that sector is written
- To receive the data from the channel and direct it to the appropriate destination in buffer memory when a sector is read

16.2.2.2   Servo Spoke Format Fields

Virtually all drives today use an ID-less embedded servo format where the only feedback as to the position of the heads comes from the surface used for data itself, in particular, from reading the servo spokes. Unlike data sectors, which may be written, read, or ignored, the read-only servo spokes are usually read all the time while the drive is up and running. Specific fields within the servo spokes are used to determine the radial position of the heads relative to the array of concentric tracks, and angular position, i.e., where the heads are on the circumferential track relative to some index position.

The format for a servo spoke is shown in Figure 16.7. This is not a "typical" format, since there is a great deal of variation in spoke formats used across the industry. All of the spoke fields shown are uniquely decoded by the channel. In some implementations, the controller may demarcate the digital information into sub-fields containing different values for different attributes of position.

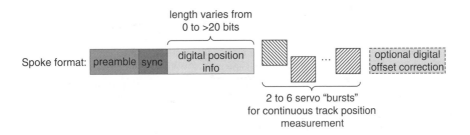

FIGURE 16.7   Servo spoke fields.

TABLE 16.2   Format Attributes of Data Sectors versus spokes

Format Attribute	Data Sectors	Servo Spokes
Track spacing	Written at a pitch that leaves some daylight between the domains of adjacent tracks.	Written with no gap between adjacent tracks.
Adjacent track phase	No phase relationship assumed between data sectors of adjacent tracks.	Radially coherent, i.e., transitions of adjacent servo spokes are aligned.
Linear density	Linear density held nearly constant across tracks, resulting in data frequencies that vary with track radius.	Written at constant data frequency, resulting in linear density that varies with track radius.

16.2.2.3   Format Differences between Data Sectors and Servo Spokes

There are some fundamental differences between the layout of data sectors and servo spokes. These differences are rooted in the requirements for radial head position when reading the two. Data sectors are only assumed to be readable within the narrow bounds of a specific track center position, whereas servo spokes need to be readable at a continuum of radial positions. These differences are summarized in Table 16.2, and illustrated in Figure 16.8.

16.2.2.4   Channel Control Signals

The flowcharts and functional descriptions that follow reference three signals that have become a standard in the controller-channel interface called *read-gate* (RDGATE), *write-gate* (WRGATE), and *servo-gate* (SVOGATE). These signals are shown in Figure 16.9.

16.2.2.5   Spoke Sync Detection: the Master Timing Reference

The detection of a sync mark ("finding sync") in the servo format serves as the main timing reference for the Disk Format Control subsystem. Each servo sync detection is used as a reference for timing subsequent control events up to the next servo spoke. Although the spokes are written at a constant angular interval, under real operating conditions the time interval between spoke syncs can slowly vary due to mechanical eccentricity and torque disturbances that cause spin speed variation. The Disk Format

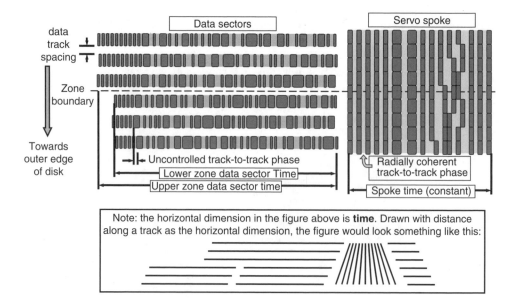

FIGURE 16.8   Data versus servo spoke track format differences.

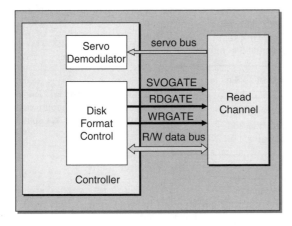

Channel control signal	Channel action upon assertion of signal
SVOGATE	read servo spoke, and output demodulated servo information over servo bus
RDGATE	read data sector and output detected data over R/W data bus
WRGATE	input data over R/W data bus and output to preamp in representation that goes on the media

FIGURE 16.9   Channel control signals.

Controller subsystem tracks these variations as it detects each servo sync. The channel is instructed when to look for servo sync mark within the bounds of a confined detection window, and the controller in turn uses the time of sync detection with the window to adjust the timing of the detection window for the next expected servo sync (Figure 16.10). The reason for narrowly confining the sync detection window is discussed in Section 16.2.2.10, "sync detection windowing."

16.2.2.6   Servo Spoke Read Sequence

While there's a fair degree of similarity across drive manufacturers in the format of data sectors and the procedures for reading and writing them, there is a great deal of variation in the servo spoke formats used to provide position information to the controller. Hence, the flow discussed here and shown in Figure 16.11 is kept at a level sufficiently high above the levels where commonality breaks down, and to avoid biasing the reader's perception of a "typical" servo spoke format.

1. The controller runs a timer referenced to the last spoke sync detection that indicates when to expect the next servo spoke. Prior to this time, data read and write operations are enabled. Other events within this period may be triggered from this timer.
2. The controller asserts the SVOGATE control signal, which instructs the channel to synchronize its signal detection mechanisms to the preamble and sync fields at the start of the servo spoke. Detection of sync provides the framing the channel needs to properly demodulate the fields that

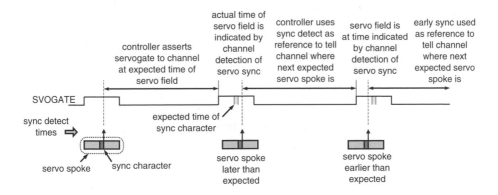

FIGURE 16.10   Running adjustment of spoke sync detection window.

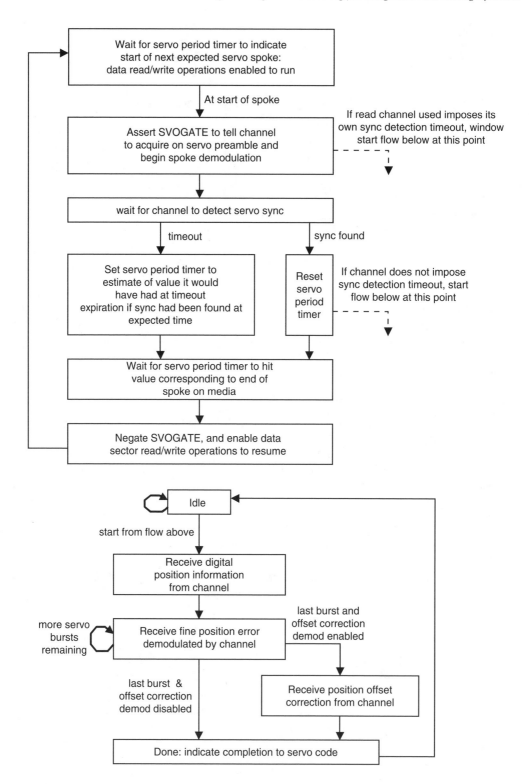

FIGURE 16.11   Servo spoke read sequence.

follow. If sync detection timeouts are imposed and reported by the channel, the controller process (lower flow in Figure 16.11) to receive the servo information expected from the channel is started.

3. The controller waits for the detection of a servo sync character within the bounds of a confined detection window. See Section 16.2.2.10 on Sync Detection Windowing. Implementations vary depending on whether this detection window is imposed by the controller or the channel.

4. Spoke sync detection is the main timing reference for all format events up to the next spoke, and, accordingly, this event resets the servo period timer. Sync detection also provides the framing needed to demodulate the servo fields that follow. If detection timeouts are imposed by the controller for channels that otherwise search for servo sync as long as SVOGATE is asserted, the lower flow of Figure 16.11 is started at this point to receive the servo information expected from the channel

5. When the servo period timer indicates that the heads are at the end of the servo spoke, the controller negates SVOGATE. Because of servo demodulation pipeline delay, the channel may continue to output servo information to the controller after SVOGATE has negated.

6. After SVOGATE is negated, data sector read/write operations are enabled to resume, and the servo spoke flow returns to idle to await the arrival of the next spoke.

The detailed flow for receiving the information demodulated from the servo spoke (lower flow of Figure 16.11) depends on the specific servo format used, but the process is basically one to just receive a sequence of digital values outputs from the channel over a serial or parallel bus. The significance of each value is implied by convention of the specific format by its order in the sequence. The steps below are representative of at least one flow for servo spoke information reception:

1. This flow is started by the flow in the upper part of Figure 16.11 at the point where servo spoke information is to be expected from the channel.

2. The first field(s) following the sync character are typically digital fields that indicate coarse position on the surface. The fields here usually at least include a Gray coded track number or some subfield of the track number. Gray coding is used so that a coherent value can be extracted when the heads are positioned between two servo tracks, and therefore reading the digital value between two adjacent tracks. Some implementations also include a spoke number field, some include an indication of surface/head number. To optimize format efficiency, engineers have strived to identify the minimal amount of information needed for unambiguous position identification, and have used an array of tricks to pack it into as few bits of spoke format as possible. In this step, the controller receives the information for field in the form of one or more digital values output by the channel over a serial or parallel bus.

3. The next fields are usually the position error signal (PES) fields which the channel demodulates to yield an indication of fine track position. There are a number of schemes currently in the field in use for this function (see Section 16.2.7).

4. Some drives use an additional field after the bursts which serves as a digital offset correction for position error written into the bursts.

16.2.2.7 Timing Data Sector R/W Operations

The disk format control logic needs to precisely control the time at which data read/write operations start so that the operation takes place on the designated place on the track. There are usually several data sectors between spokes.

The Format Control logic dead-reckons the start of data sectors in a spoke-to-spoke interval using a calculated delay from the servo sync detection preceding the interval. This delay must take into account the time between spokes, the time allocated for data sectors for the current band, and how the interval starts, that is, whether it starts with a new data sector or with some fraction of a split sector that started in the preceding interval. Another complication to this calculation is that split data sectors are slightly longer than nonsplit sectors since each fragment of a split sector carries the channel fields (e.g., preamble, sync) of a single nonsplit sector. Some methods used for generating the data sector start times include:

- Precalculated lookup tables which are indexed by data sector number from index to get values that tell the disk format hardware where data sectors start.
- Hardware "calculators," where a hardware engine does modulo arithmetic using format parameters of the current track to generate the data sector start times.
- Counter logic: The data sector start times can be generated in real-time by a system of counters that mark out data sector lengths and suspend during servo [1].

16.2.2.8   Data Sector Read/Write Sequences

The disk format control subsystem directs the channel to read and write data sectors as per the condensed flow charts in Figure 16.12 and Figure 16.13. These flows are typically implemented by a programmable sequencer or a set of hardware state machines.

The sequence for reading a data sector shown is shown in Figure 16.12. The list that follows walks through this flow.

1. The disk format controller enters its read flow when it determines that the heads are at the start of a sector to be read.
2. The controller asserts the RDGATE signal to tell the channel to start the read operation. To mitigate uncertainty in where the written preamble starts for the sector, the RDGATE assertion may be slightly delayed to ensure that the read operation starts over valid preamble. Upon the assertion of

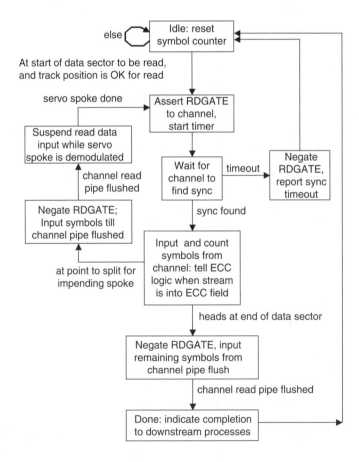

FIGURE 16.12   Data sector read sequence.

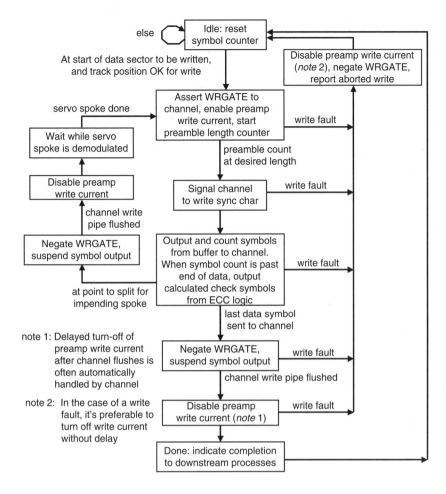

FIGURE 16.13   Data sector write sequence.

RDGATE, the channel trains on the preamble to synchronize its internal clock to the sector, and looks for the data sync character.

3. After asserting RDGATE, the controller waits for the channel to signal that sync was found within a timeout deadline imposed by the controller (see Section 16.2.2.10).

4. Detection of the sync character provides the channel with the reference needed to frame the decoding of the data that follows. Sync detection is typically signaled to the controller by an assertion on the data bus to the controller and the channel immediately begins outputting symbols, usually 8 or 10 bits wide, detected from the read signal from the preamp. A counter in the controller held reset before the operation counts these symbols as they are received.

5. If the sector being read is a split sector, that is, a data sector interrupted by a servo spoke, the controller determines when to pause the read operation for the impending servo spoke.

6. At such a point, the controller negates RDGATE and inputs the symbols remaining in the channel's read pipeline until it is flushed.

7. After the pipeline is flushed, the controller suspends the data read operation until the servo spoke has passed. From the channel's point of view, the read operation has ended.

8. After the spoke, the controller again asserts RDGATE to tell the channel to start another read operation by synchronizing on the preamble and sync preceding the second part of the data sector. From the channel's point of view, this is a new read operation.

9. The controller again waits for the channel to find sync.

10. If the sync character is found within the deadline, the input of data symbols from the channel resumes with the symbol count starting from the state that was saved during the split.

11. The controller knows from its symbol counter when the end of the data sector is imminent. At the end of the sector, the controller negates RDGATE, and inputs the remaining symbols flushed by the channel, after which point the read operation is done.

12. Referring back to the states of waiting for sync detection, if the channel fails to find sync within the controller-imposed deadline, the controller negates RDGATE, terminating the read, and reports the error to higher layers of firmware.

The sequence for writing a data sector is shown in Figure 16.13

1. The disk format controller enters its write flow when the heads arrive at the start of a sector to be written.

2. The controller asserts the WRGATE signal to the channel and puts the preamp in write mode to enables write current to start the write operation. In some configurations, the preamp mode is controlled indirectly through the channel. Upon the assertion of WRGATE, the channel immediately starts writing preamble. The controller waits for a delay corresponding to the format's target preamble length.

3. When the desired length of preamble has been written, the controller signals the channel to write the sync character. This signaling is typically an assertion on the controller-channel data bus which is held deasserted while preamble is written.

4. The controller starts outputting data symbols to the channel, which expects the data immediately after being signaled to write the sync character. The channel usually encodes the data symbols, serializes them, and outputs a single-bit stream to the preamp.

5. If the sector being written is a split sector, that is, a data sector interrupted by a servo spoke, the controller determines when to pause for the servo spoke.

6. At such a point, the controller negates WRGATE, which initiates a run-down of the write operation from the channel's point of view. The preamp is held in write mode by the controller, or by the channel on behalf of the controller, while the channel flushes its write pipeline.

7. After the pipeline is flushed, preamp write mode is de-asserted, and the controller suspends the write operation holding context of the operation until the servo spoke has passed.

8. After the spoke has passed, the controller reasserts WRGATE to the channel and puts the preamp in write mode to write the preamble that precedes the second part of the data sector.

9. The controller again tells the channel to write the sync character after the desired length of preamble has been written.

10. The controller resumes data output and symbol counting from where it left off from the first part of the split sector.

11. When the symbol count indicates that all the symbols for the current sector have been sent to the channel, the controller negates WRGATE to begin running down the write operation. The preamp is held in write mode while the channel flushes its write pipeline.

12. After the flush is done, the preamp is taken out of write mode, and the operation is complete.

13. Note the "write fault" exit conditions from various states. During a write, a number of external conditions are monitored to validate the proper execution of the write operation. In general, a fault on any of these conditions immediately terminates the write operation. See Section 16.2.2.9 on Read/Write Fault Checking.

16.2.2.9 Read/Write Fault Checking

The fault conditions monitored during disk read/write operations include various methods for detecting track misregistration, timing errors, electrical faults, and failure to detect the sync character. Track misregistration is a condition in which the heads are not positioned close enough to the center of the data track for proper operation. This is usually specified as a two-sided limit within some design percentage of track pitch to the center of the track. An example of a timing error would be a read or write operation that fails to complete in an expected time interval. Electrical faults include a plethora of conditions that the preamp monitors about itself and the heads. Some of the preamp conditions monitored are marginal supply voltage and die temperature. Some of the head conditions include shorted or open elements. A fault in anyone of the conditions monitored by the preamp is usually signaled to the controller via a pin either directly from the preamp or indirectly through the channel.

A conspicuous asymmetry in fault handling can be observed from the read/write state flows of Figure 16.12 and Figure 16.13. Some of these are artifacts of the opposing data transfer directions, for example, failure to find the sync character obviously applies to reading only, since it is written during write operations. However, some of the fault handling asymmetry reflects an underlying asymmetry in the consequences of a fault hitting a disk read versus a disk write operation.

Two important aspects that differentiate writes from reads are:

- Writing results in an irreversible change to the state of the media
- There is no explicit feedback from the write operation itself as to the success of the operation

The first aspect makes detecting track misregistration critical when writing. A write operation that proceeds when the heads are off track can overwrite and destroy data stored on an adjacent track. Even if there is no active data on the adjacent track, data written when a disturbance knocks the heads off track center may be difficult to recover with a subsequent read operation. In contrast, the consequences of read that fails due to off-track event is just some performance lost in the extra time it takes to re-read the sector, and there's no risk of destroying data as with a write. Accordingly, the operating limits for track position are usually more generous for reads than for writes.

Another reason for paranoid fault-monitoring during writes is that disk write operations are inherently open loop. The controller is oblivious to any corruption of data due to a fault in the path between the controller output and the media during a write. Applications with critical data integrity requirements sometimes read data after it is written to verify the success of the write, but the performance loss of the extra revolution makes this prohibitive for general application. In contrast, the effect of a fault that occurs during a read operation is usually conspicuous by the failure of the data integrity checks applied to read data as a matter of course. But for writes, the fault mechanisms themselves, rather than their effect on the data, are monitored to verify the health of a write operation.

16.2.2.10 Sync Detection Windowing

The sync detection "window" is the period during which a sync detection is considered to be valid. Failure to detect an expected sync within the detection window is called a *sync timeout*. Reference was made in preceding sections to confining valid sync detection to a narrow window, but the timing uncertainty of an anticipated sync detection would seem to suggest the opposite, that is, the use of a wide detection window, which would seem to be a "bigger net" in which to catch a sync. This is true only for detecting a sync character with a unique encoding that precludes its existence in the fields surrounding the sync character.

Sync characters are usually preceded by constant frequency preamble, and it's easy to choose a pattern that is readily distinguished from preamble. Sync characters are usually followed by fields of arbitrary information, and while it's possible to encode the sync character and the information field such that no span of the information field resembles the sync character, such exclusionary encoding of adjacent fields is hard to do efficiently. Generally, the sync and information coding is such that a serendipitous copy of the sync pattern could indeed exist as a substring within an information field following the "real" sync. The rationale for being not too generous with the sync detection window is to avoid mistaking one of these coincidental sync patterns within the information field for the real sync when the real sync itself is not detected.

Although the consequences of a sync timeout can be the loss of a format timing update or failure to extract information following sync, the consequences of detecting sync in the wrong place is worse. This is the rationale for being stingy with the detection window, that is, to force sync timeouts to occur rather than risk receiving incorrect data that was detected from a bad framing reference.

16.2.3   Error Correction Code (ECC) Subsystem

Although we all depend on hard drives to store the information that keeps the world running, the error rates of magnetic media and of the process of retrieving data from it are so poor in their raw form that without error correcting codes (ECC) hard drives would not be a viable storage technology. With each bit of information occupying a few trillionths of a square inch, it does not take much in the way of contamination or surface defects to create errors, nor much in the way of electronic noise to corrupt the pathetically weak electrical signal produced by the read process. ECC subsystems are an essential part of all disk controllers for converting the error rates of the raw storage medium to the nearly error-free storage device abstraction seen by the users. A busy drive may correct millions of bits in errors in a minute without the user ever knowing about it.

This chapter intends to discuss ECC only as it relates to the surrounding disk data transfer functions of a hard drive controller. A reader interested in the theory of error control coding can refer to Chapters 12, but a brief overview will be given here for sake of context.

The general idea behind ECC is to use the data to be protected in a calculation that generates some number of redundant bits that get appended to the data to create an ECC "codeword." A block of user data might differ from another by only a single bit, which corresponds to a Hamming distance of one. The ECC calculations create a set of codewords from the user data with a much greater Hamming distance. If the code used has a Hamming distance of N, changing a single bit of user data results in a codeword that differs from any other valid codeword in at least N bit locations. A codeword read back from some medium that differs from a valid codeword by fewer than N bits is presumed to have been corrupted by errors so that it is between valid codewords. The valid codeword closest in Hamming distance to the corrupted codeword is most likely to have been the original codeword.

Most hard drives today use Reed-Solomon ECC (RS-ECC) codes. RS-ECC breaks up the user data into fixed-width data symbols, typically 8 or 10 bits wide, and calculates a number of redundant symbols that get appended to the user data when the data is written to the disk. When the data sector is read, potentially with errors, a similar calculation is performed on the data and the redundant symbols to generate a set of symbols called *syndromes*. The syndromes depend only on the errors introduced in the data, and not on the data itself. If the syndromes are all 0s, the data read was free of errors. Otherwise, the nonzero syndromes can be used to generate polynomials whose roots identify the symbol locations and symbol values of the error pattern. The error locations and values can be used to restore the raw data read from the disk to its original error-free form.

Although the calculations are pretty much self-contained function of the ECC subsystem, the disk data path of the controller must accommodate the delays in the data path imposed by the calculations for encoding for writes and decoding and correcting for reads.

16.2.3.1   ECC Encoding

Data fetched for a write operation to the media must be fetched sufficiently far in advance to compensate for delay in the redundant symbol generation. One common technique, shown in Figure 16.14, is to use a shift register or FIFO in the path between the source of write data (buffer memory) and the channel to compensate for encoder delay. In this configuration, the last symbol of user data enters the ECC encoder early with respect to the data stream being output to the channel. This gives the encoder a time equivalent to the depth of the FIFO to finish the calculation of the redundant symbols after the last user data symbol is received. After the last data symbol is output to the channel, the mux is switched to send the redundant symbols to the channel on the heels of the user data.

16.2.3.2   ECC Decoding and Data Correction

The ECC decoding and correction process for reads is more intrusive to the data path than the encoding process. The ECC syndromes can be calculated as data symbols are input from the channel during disk reads, similar to the method for encoding redundant symbols from the write data stream. The time required for the remaining steps of generating the error locator and error vector polynomials, extracting their roots,

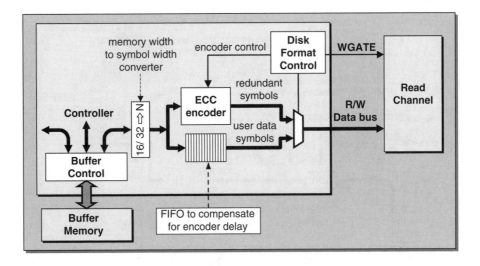

FIGURE 16.14 ECC encoding data path.

and making the indicated corrections can approach or exceed an additional data sector time. The onus is on the controller to:

- Provide a means for the ECC logic to apply the calculated error corrections on the raw data.
- Prevent any potential consumer of the data read from the disk (e.g., the host interface) from using the data read until data is validated or needed corrections have been applied; until such time the data is essentially invalid.

Two common methods have evolved for providing the functions above to the ECC logic: in-buffer correction, and in-flight correction. In-buffer correction is shown in Figure 16.15 and in-flight correction is shown in Figure 16.16.

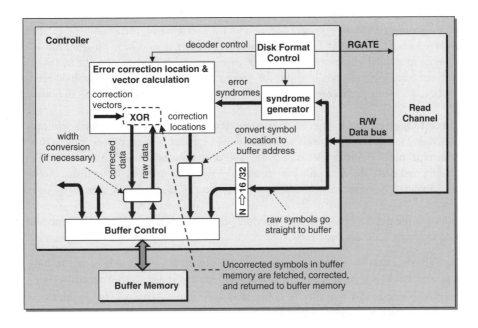

FIGURE 16.15 In-buffer ECC correction.

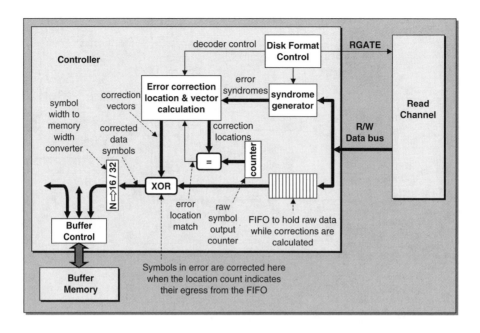

FIGURE 16.16   In-flight ECC correction.

For the case of in-buffer correction, the raw data is written to the external buffer memory as a data sector is read, and resides in its raw form until the error corrections are calculated. The error corrections are then applied later by reading the buffer locations containing the errors, XOR'ing them with the error vectors, and writing the corrected data back to the buffer locations.

With in-flight correction, the raw data on the way to external buffer memory is staged in a FIFO or memory internal to the controller while the corrections are being calculated. The error corrections are applied to the raw data in a local memory, or as the corrupted symbols emerge from a FIFO. Only corrected data is output to the external buffer memory.

In-buffer correction eliminates the area required on the controller for memories to stage the raw data awaiting correction. However, for ECC symbol widths of other than 8 bits, modification of symbols is cumbersome in standard byte-oriented external memories. This method also consumes some extra buffer memory bandwidth, especially if the corrections are applied using small-grained read-modify-write operations which are particularly inefficient for the burst-optimized devices typically used for buffer memory.

In-flight correction requires enough on-chip memory to buffer the raw data internally while corrections are calculated, but for ECC symbol widths other than 8 bits, the correction process can be simplified by the use of internal memories with widths naturally aligned to the symbol width. External buffer memory usage is minimized since the data only makes one trip to buffer memory in its final corrected form, which also simplifies the application of additional data integrity checks on the data on its way to buffer memory.

In addition to the normal encoding and decoding procedures described above, the controller usually supports collateral ECC modes and functions:

- The ECC logic must recognize when the syndromes indicate that the codeword read from the disk is uncorrectable, and must report the inability to recover the data to the requesting user.
- For media testing and diagnostics, ECC-bypass modes are usually provided to disable the appending of redundant symbols when writing and to disable data correction when reading.

- A normal read operation does not send the redundant symbols to the host, and thus not to buffer memory either. But provision must be made to allow the entire codeword to be transferred to the host to support "Read Long" operations defined by some host interfaces.
- Similarly, the controller needs to support the transfer of data from the host with externally appended redundant symbols that are used instead of the calculated redundant symbols to support "Write Long" operations.

16.2.4 Host Interface Subsystem

Disk controllers exist with embedded interfaces for a number of different host interfaces. The most common interfaces at this writing are ATA, SCSI, and Fibrechannel. ATA dominates in desktop PCs, with higher performance SCSI and Fibrechannel drives fulfilling the requirements of the server market. Drives with new serial versions of ATA and SCSI are expected to supplant their parallel counterparts over time.

The differences between these interfaces are too extensive for this chapter to cover the specifics of each in any detail. The reader interested in such detail is referred to the ANSI Standards that define these interfaces [2]. Keeping in the spirit of a generic controller functionality summary, this chapter will instead focus on a high-level model common to all interfaces, and how it shapes the architecture of the host interface logic in a controller.

Most information exchanges between the host and disk can be put into two categories: data and messages. Data exchanges carry the data being written to or read from the drive. The message category includes exchanges *about* the data exchanges: the commands that describe and evoke the read or write operation, the status message confirms the successful or failed completion of the command, and messages that manage the data exchange during the execution of the command.

One distinction between these two categories that is reflected in the host interface hardware is the destination of the information. The end node for host interface data at the drive end is ultimately the media, and thus the buffer as the intermediate staging point. The intended target for messages from the host, and the source for messages to the host, is the entity supervising the transfer, which in general is the processor. Thus, one aspect common to the logic for all host interfaces is a means to steer message exchanges to/from the processor and data exchanges to from the buffer, as shown in Figure 16.17. The means for

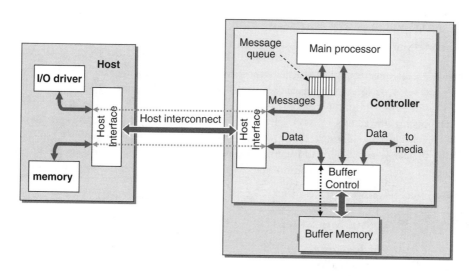

FIGURE 16.17   Bifurcation of host message and data transfers.

sifting between these two categories is fairly explicit in ATA, which uses specific registers in the controller for command/message-type exchanges, and in legacy versions of SCSI, which use specific bus phases for command and messages. The means is more subtle for serial interfaces such as Fibrechannel, where headers within packets must be parsed to identify the type of exchange.

A generic host read operation proceeds as follows:

- The host establishes a connection to the drive (moot step for single-host interfaces like ATA).
- The host sends a read command to the drive.
- The controller directs the read command to the processor's attention.
- For multi-drop host interfaces that support disconnection, the controller may disconnect at this point to free the bus for use by other devices.
- The processor configures the controller hardware to fetch the data from the media into buffer memory.
- The processor configures the data transfer hardware of the host interface to transfer the read data from buffer memory to the host.
- The controller's host interface reconnects to the host (if it disconnected above).
- The controller's host interface sends the read data to the host.
- The processor sends a status message to the host to indicate the completion of the read command.

A generic host write operation proceeds as follows:

- The host establishes a connection to the drive (this step is moot for some interfaces).
- The host sends a write command to the drive.
- The controller directs the write command to the processor's attention.
- For multi-drop host interfaces that support disconnection, the controller may disconnect at this point to free the bus for use by other devices.
- The processor configures the data transfer hardware of the host interface to transfer the pending write data from the host interface to buffer memory.
- The controller's host interface reconnects to the host (if it disconnected above).
- The controller's host interface sends the write data to the host.
- The processor configures the controller hardware to write the data from buffer memory to the media.
- The processor sends a status message to the host to indicate the completion of the write command.

Host interfaces frequently employ special purpose hardware or dedicated sequencers to automatically execute a small set of performance-critical commands. This expedites command execution by obviating the need for intervention from the main processor, which is often busy with other controller functions at the time a host command arrives. This is an effective mix of functionality. Host interfaces may have a lot of commands defined in their repertoire, but only a very small number of them really matter when attempting to maximize the performance of the drive. The special purpose hardware can focus on recognizing and efficiently executing a few particular commands, while leaving the larger set of nonperformance-critical commands to be handled by the generality of the main processor.

16.2.5   The Processor Subsystem

The processor subsystem, or more precisely, the firmware it executes, acts as the supervisor of the operations executed by the controller. The processor subsystem for a processor used to be a standard discrete microcontroller external to the controller, but virtually all controllers now instead use one or more integrated processor cores. The processor's connection to the controller functionality is via hundreds of registers in the controller hardware that are mapped into the address space of the processor.

The division of labor between the controller hardware and the processor firmware varies from drive to drive, but in general

- The controller hardware executes data transfers, real-time control sequencing, expediting of performance-critical operations, and special operations that a general-purpose processor is ill-equipped to do, such as manipulating polynomials of Galois fields for ECC calculations.
- Processor firmware does things that involve decisions, are algorithmically complex, and, to put it plainly, everything left over that the controller doesn't do. Most importantly, the firmware is ultimately responsible for implementing the behaviors and all of the commands specified by the standards of the drive's particular host interface.

Some examples of algorithmically-intense decision-making tasks are:

- Deciding which of a set of queued commands should be activated to maximize performance.
- How to allocate buffer resources, cache replacement policies.
- Computing the physical coordinates of a data sector from its corresponding logical block address (LBA). The LBA to physical-address calculation is not a closed form, involving many intermediate steps which also don't have closed forms:
  - Since the number of data sectors per track depends on band number
  - The number of tracks per band varies, and may even vary with surface number
  - There are scattered discontinuities in the physical array of blocks created by places left unused because of surface defects

Probably the single most demanding task in terms of total processor usage is controlling the head positioning mechanism. This is implemented as a sampled data control system (see Section 16.2.7) where the processor responsible for servo control executes a position control algorithm upon the receipt of position information demodulated from each spoke. Servo sample rates are typically in the tens of kHz.

It is economically desirable to use a single processor in the controller that implements servo control in an interrupt service routine (ISR) evoked by a interrupt generated as each servo spoke arrives under the heads. The remaining controller tasks (e.g., host I/O command processing) are executed as background tasks in the remaining processor cycles. However, this uniprocessor strategy is problematic for drives with aggressive servo sample rates and/or computationally demanding control algorithms. As depicted in Figure 16.18, as servo sample rates increase or as the servo ISR time increases, the time left over for background tasks can become vanishingly small. Moreover, the drives with the most demanding servo requirements are generally high-performance drives with aggressive I/O command processing requirements that need the left-over background cycles the most. Hence, high-performance drives typically relegate servo processing to a dedicated processor, thus allowing 100% of the main processor to be used for the nonservo tasks.

16.2.6 Buffer Memory Subsystem

Virtually all controllers use buffered data transfers for moving data between the host and the disk, wherein data moving between the host interface and the disk is first staged in an electronic memory "buffer." Buffered transfers are analogous to the workings of an airport terminal: passengers amble into and out of the terminal at different rates that vary over the course of a day; they are "buffered" into the holding areas we call "gates" until the planes to which they are assigned show up for departure at expected intervals, at which point they are discharged from the terminal.

The buffer for a hard drive controller usually takes the form of a standard dynamic random access memory (DRAM) external to, and managed by the controller. Controllers have been deployed that used small external SRAMs (static RAM) or used memories integrated onto the controller silicon for their buffers. However, use of external Synchronous Dynamic RAM and Dual-Data Rate SDRAM (SDRAM and DDR SDRAM) is almost universal across the industry.

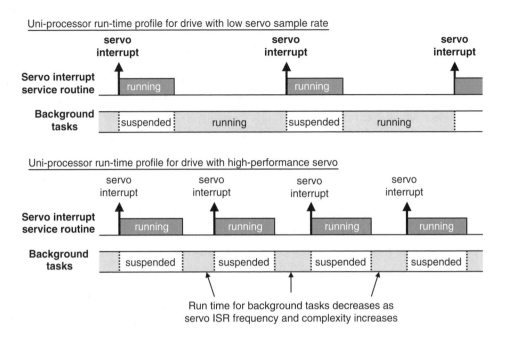

FIGURE 16.18   Effect of servo ISR on background task processing.

There are several reasons why buffered transfer is used:

1. It decouples the time and rate at which data is moved to/from the media from the time and rate it is moved to/from the host interface.
2. It provides a means for data caching, that is, honoring an I/O request from the memory buffer rather than from the disk itself to avoid the disk access time incurred by mechanical movement.
3. It provides a methods for reordering data sent to the disk or host interface relative to the order in which it was received.

The exact time at which data can be transferred to/from a host computer is subject to other activity which can affect the expediency of the host responding to the drive and the availability of shared resources (e.g., buses) between the drive and the host. This means that the controller has to be flexible in moving data to/from a host system: transfers can stall, restart, slow down, and speed up somewhat unpredictably. In contrast, the time at which data can be transferred to/from a specific media location is very predictable and unforgivingly inflexible: data can be transferred when that location passes under the heads, and at that time only. If the opportunity is missed, the controller has to wait for a full rotation of the disk (8.3 mS for a 7200 rpm drive is considered a long time to wait) for the next chance. Controllers complete the data path between the host and disk ports by first moving data into the buffer at a time of the data-producing port's choosing, and then moving data from the buffer to the data-consuming port at a time of that port's choosing.

Buffering is also required for accommodating the disparity in average transfer rates between host interfaces and disks. The data rate from the host is usually constrained to some maximum rate by the particular host interface (e.g., 320 MB/sec for SCSI 320 m), but the actual rate can vary widely. On the other hand, the data rate from a location on the media is rigidly determined by a predetermined linear density (bits per inch) for that location and the rotational speed of the drive, which typically varies less than 0.1% from its nominal rate. The buffer acts as an elastic buffer between the ports: if a fast data-producing port gets ahead of the consumer, the buffer accumulates the excess data; if the data producer is slow, the buffer provides a place where a quorum of data can be accumulated before transfer begins to a faster data consumer.

If a host-disk data transfer was first written in its entirety to the buffer from the data producer, and then read from the buffer by the consumer, the data bandwidth required of the buffer would simply be the sustained transfer bandwidth of the faster of the two ports.

However, this would reduce the maximum end-to-end transfer bandwidth of the drive to the reciprocal sum of the two ports ($1/B_{DRIVE} = 1/B_{HOST} + 1/B_{DISK}$). Instead, buffer systems are usually configured to have at least the sum of the two data port bandwidths so that these two stages of overall end-to-end transfers can run concurrently. By overlapping these two hops, the effective bandwidth of the drive approaches the full bandwidth of the slower port ($B_{DRIVE} = \min(B_{HOST}, B_{DISK})$).

Although the host and disk subsystems account for most of the data transferred to/from buffer memory, other subsystems within the controller also use buffer memory:

- The processor system uses some part of buffer memory for program and data memory
- Some ECC systems read and write buffer memory to implement error correction
- Hardware machines to coordinate the flow of data through the controller may parse and update data structures that reside in buffer memory

The commodity single-port devices used for buffer memory can only read or write to/from a single client at one time, and are managed as a shared resource for multiple clients by a subsystem within the controller called the *buffer controller*.

The buffer controller subsystem implements an apparent multiport transfer concurrency by time-sharing the external memory buffer to service pending transfers from multiple memory clients. A FIFO or some other form of intermediate storage in a client port's datapath to memory can be used to sustain data flow to/from the client for short periods while the memory buffer is servicing other clients.

It would be a simple matter to satisfy the data transfer requirements of each client with a shared memory if there were no bounds on the data transfer bandwidth of the shared memory. However, practical limitations of pins and cost usually compel the controller designer to find a balance in a number of interactive memory system attributes that does the job with the minimum bandwidth possible. Some of the attributes that must be considered are

- The arbitration policy for selecting which of multiple pending requests to service.
- The sizes of FIFOs for intermediate storage between the shared memory and each client.
- The "dwell time" for each client, that is, how long the shared memory is tied up servicing a particular client.

Two simple arbitration policies are fixed priority arbitration and round-robin. Fixed priority assigns a numeric priority to each client, and in the event of multiple pending requests, the client with the highest number is chosen to be serviced. Round-robin priority rotates the priority among the set of clients at each arbitration event. Neither of these simple schemes is an ideal solution to an application involving a set of clients with a complex mix of transfer requirements. For example:

- The host interface usually has the highest bandwidth capability of any single client, thus can consume a lot bandwidth, but transfers can be throttled if needed, and normal transfer lengths make memory latencies largely inconsequential.
- Processors, often with some amount of local memory, may use very little external buffer memory bandwidth, but are very sensitive to latency (an instruction that reads buffer memory stalls the processor until the read is serviced).
- The predictability of disk transfers make them similar to host transfers in bandwidth vs. latency sensitivity, but since they must keep pace with the linear velocity of the media, disk transfers cannot be throttled like host transfers.

In the absence of excess bandwidth, straight priority schemes can lock out the lower-priority requestors under conditions of heavy load. Round-robin guarantees access to each requestor, but can stall critical accesses that must wait while non-critical accesses that could have waited are serviced. Hence, many

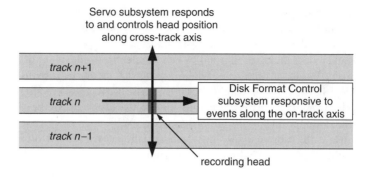

FIGURE 16.19   Servo and disk formatter orthogonality.

controllers have used more elaborate schemes for arbitration tuned to the particular sensitivities of the buffer memory client set.

16.2.7   The Servo Subsystem

While format events occurring along a track are the responsibility of the disk format control subsystem, the orthogonal dimension, that is, controlling head position across tracks, is the domain of the servo subsystem (see Figure 16.19).

The heart of most servo subsystems today is the digital control loop implemented by firmware that executes in either the servo ISR (Interrupt Service Routine) of a uniprocessor controller or in the servo processor of multiprocessor configurations.

The servo control loop gets its input from the servo spoke often in a form consisting of coarse and fine track information. The coarse information comes from a Gray-coded digital field containing an integer value related to track position. The fine information, that is, fractional track position, comes from constant-frequency servo "bursts" of various formats. The fine position comes from comparing the relative magnitude of a sequence of bursts written at fractional track positions, or the phase of bursts written with a phase proportional to track position, or the net amplitude of adjacent bursts written 180° out of phase. The channel typically demodulates the burst amplitudes and/or phases, and outputs the information to the controller as a digital value which is captured and read by the servo processor.

The servo processor inputs the position information demodulated by the channel from the servo spokes, processes the raw information to discern the track position of the heads, and uses the error from the desired position to compute an update to the torque signal that drives the VCM. This control loop is shown in Figure 16.20.

16.3   Global Controller Strategies

The strategies discussed here are in this section because the hardware that supports them is not isolated to any particular section of the controller, and instead are more global policies implemented across subsystem boundaries.

16.3.1   Data Integrity Strategies

ECC strategies have become quite powerful in providing reliable recovery of data from a mechanism plagued with media and electronic noise. However, there are some error scenarios beyond the domain of what ECC can cover:

- If a read operation inadvertently reads a data sector from an adjacent track, and the data it reads is correctable, the data would be correct from the ECC subsystem's point of view, but would still be in error because it was not the data requested by the host.

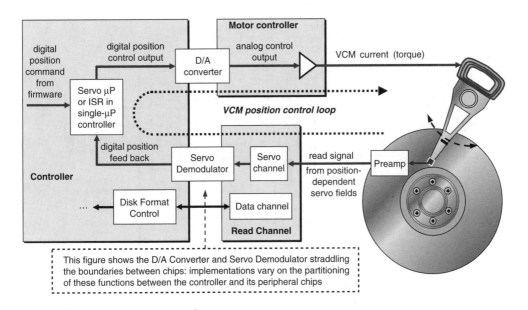

FIGURE 16.20 Servo subsystem control loop.

- If write data received from the host on the way to the disk is corrupted by a soft error while being staged in the buffer memory, then the redundant symbols would be calculated on erroneous data. On readback, the ECC would "correct" the data to the version with the soft error from which the redundant symbols were calculated.

To protect against these scenarios, some controllers append a data integrity check field (Figure 16.6) to the user data. This check field consists of an LBA tag and a cyclic redundancy check (CRC) appended to data as soon as it is received from the host interface.. The LBA tag is either the LBA or some hash thereof of the LBA associated with the block. The CRC is an error detecting code calculated on the user data and the LBA tag. Figure 16.21 shows these check fields "following" a data block as it moves through the controller to buffer, to disk, to buffer, and finally back to the host.

The handling of the data integrity fields at the seven points indicated above is detailed below:

1. Incoming data blocks are received with no integrity checks from the interface, or perhaps with host-interface-specific checks already discarded.
2. As the buffer is received, it is appended with a "tag" of the LBA associated with the block, and with a CRC calculated over the data.
3. The data block is written to buffer memory with the integrity check field appended.
4. When the block is read from buffer memory to be sent to the disk for a write, the LBA tag and CRC field are checked against their expected values. The step checks the following:
   a. The CRC check detects if a soft error corrupted the block while it was in buffer memory.
   b. The LBA tag check ensures that the block read is the one that was intended to be read.
5. If the LBA tag and CRC are written to the media and carried back from the media, the data block is protected by a true end-to-end check, that can guarantee that a block returned to a host is the same as the block once written.
6. The check at this point (after ECC correction) serves two purposes:
   a. If the LBA tag checks here, we can be assured that the data read was from the intended data sector, and not an adjacent track.
   b. The CRC check here provides assurance that the ECC subsystem did not miscorrect severely corrupted data, or that a framing error did not create a cyclically-shifted codeword that would appear valid to the ECC subsystem, but would be the wrong data.

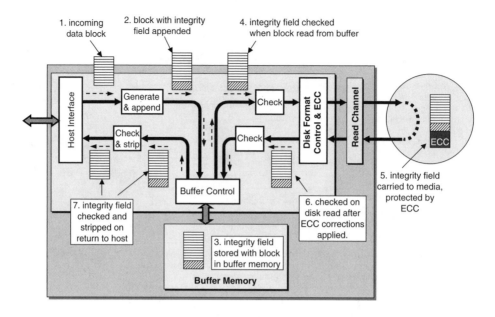

FIGURE 16.21   LBA tagging and end-to-end data check.

7. The check here ensures two thing:
    a. The LBA tag check provides assurance that the data being returned to the host is indeed the LBA that was requested, and not otherwise due to a firmware bug.
    b. The CRC check checks against soft errors that may have occurred in buffer memory while the block awaited transmission to the host.

16.3.2   Drive Caching

The time to read a random disk sector from the disk is slow by electronic standards, even on a "fast" drive. The time to move the heads to the desired track is measured in milliseconds, as is the time it takes for the disk to rotate so that the desired sector is under the heads. In contrast, a data block already in buffer memory can be returned to the host with electronic delays which are orders of magnitude less.

This is the motivation for caching. When a host sends a read command, drives usually try to make the long trip to the destination track worth the trouble by reading the requested data plus some amount of extra data from the same track. The extra data is held in buffer memory in anticipation of a cache "hit," that is, a subsequent host command requesting data already waiting in buffer memory. Locality of reference is common enough in workloads to make this gamble a pretty good bet, and methods to manage buffer memory to maximize its effectiveness as a cache is a subject of much, albeit largely proprietary, research and development work.

References

[1] Buch, Bruce D., On-the-fly splitting of disk data blocks using timed sampling of a data position indicator, U.S. Patent 5,274,509, 1993.
[2] Readers seeking information about the host interfaces used by hard drives can obtain the Standards Documents for these interfaces from the American National Standards Institute (ANSI) at the address below:
    American National Standards Institute (ANSI)
    25 West 43rd Street, 4th Floor
    New York, NY 10036
    1-212-642-4900

or can go to the "ANSI Electronics Standards Store" for on-line documentation, by linking from the ANSI website at www.ansi.org.

Additional information on interface standards can be found at the website for the International Committee for Information Technology Standards at www.incits.org. This is the body that serves as an information technology advisory group for ANSI. The T10 committee is responsible for SCSI, T11 for Fibrechannel, and T13 for ATA. Links to these interfaces appear in the Technical Committees section as "T10 SCSI storage Interfaces", "T11 Fibre Channel Interfaces", and "T13 ATA Storage Interface".

Good sources of information for the serial versions of SCSI and ATA are the websites of The SCSI Trade Association at www.scsita.org, and of The Serial ATA Working Group at www.serialata.org, respectively.

IV

Coding for Read Channels

17 Runlength Limited Sequences   *Kees A. Schouhamer Immink* **17**-1
Introduction • Asymptotic Information Rate • Other Constraints • Codes for the
Noiseless Channel

18 Maximum Transition Run Coding   *Barrett J. Brickner* **18**-1
Introduction • Error Event Characterization • Maximum Transition Run
Codes • Detector Design for MTR Constraints • Simulation Results
• Summary

19 Spectrum Shaping Codes   *Stojan Denic and Bane Vasic* **19**-1
Introduction • Recording System and Spectrum Shaping Codes • Dc-free Codes
• Codes with Higher Order Spectral Zeros • Composite Constrained and Combined
Encoding • Conclusion

20 Introduction to Constrained Binary Codes with Error Correction Capability
Hendrik C. Ferreira and Willem A. Clarke **20**-1
Introduction • Bounds • Example: A Trellis Code Construction • An Overview
of Some Other Code Constructions • Post Combined Coding System
Architectures • Conclusion

21 Constrained Coding and Error-Control Coding   *John L. Fan* **21**-1
Introduction • Configurations • Reverse Concatenation and Soft Iterative Decoding

22 Convolutional Codes for Partial-Response Channels   *Bartolomeu*
F. Uchôa-Filho, Mark A. Herro, Miroslav Despotović, and Vojin Šenk **22**-1
Introduction • Encoding System Description and Preliminaries • Trellis Codes for
Partial-Response Channels Based Upon the Hamming Metric • Trellis-Matched Codes
for Partial-Response Channels • Run-Length Limited Trellis-Matched Codes • Avoiding
Flawed Codewords • The Distance Spectrum Criterion for Trellis Codes • Good
Trellis-Matched Codes for the Partial-Response Channels Based on the Distance
Spectrum Criterion

23 Capacity-Approaching Codes for Partial Response Channels   *Nedeljko Varnica,*
Xiao Ma, and Aleksandar Kavčić **23**-1
Introduction • The Channel Model and Capacity Definitions • Trellis Codes,
Superchannels and Their Information Rates • Matched Information Rate (MIR) Trellis
Codes • Outer LDPC Codes • Optimization Results • Conclusion

24 Coding and Detection for Multitrack Systems   *Bane Vasic and Olgica Milenkovic* **24**-1

Introduction • The Current State of Research in Multitrack Codes • Multitrack Channel Model • Multitrack Constrained Codes • Multitrack Soft Error-Event Correcting Scheme

25 Two-Dimensional Data Detection and Error Control   *Brian M. King and Mark A. Neifeld* **25**-1

Introduction • Two-Dimensional Precompensation • Two-Dimensional Equalization • Two-Dimensional Quasi-Viterbi Methods • Two-Dimensional Joint Detection and Decoding

17

Runlength Limited Sequences

Kees A. Schouhamer Immink
Turing Machines Inc.
Rotterdam, Netherlands
and
University of Essen
Essen, Germany

17.1 Introduction **17**-1
17.2 Asymptotic Information Rate **17**-2
     Counting of Sequences • Capacity
17.3 Other Constraints **17**-4
     MTR Constraints • $(O, G/I)$ Sequences • Weakly Constrained
     Sequences • Two-dimensional RLL Constraints
17.4 Codes for the Noiseless Channel **17**-6

17.1 Introduction

Codes based on runlength-limited sequences have been the state of the art corner stone of current disc recorders whether their nature is magnetic or optical. This chapter provides a detailed description of various properties of runlength-limited sequences and the next section gives a comprehensive review of the code construction methods, ad hoc as well as systematic, that are available.

The length of time usually expressed in channel bits between consecutive transitions is known as the *runlength*. For instance, the runlengths in the word

$$0111100111000000$$

are of length 1, 4, 2, 3, and 6. Runlength-limited (RLL) sequences are characterized by two parameters, $(d+1)$ and $(k+1)$, which stipulate the minimum (with the exception of the very first and last runlength) and maximum runlength, respectively, that may occur in the sequence. The parameter d controls the highest transition frequency and thus has a bearing on intersymbol interference when the sequence is transmitted over a bandwidth-limited channel. In the transmission of binary data it is generally desirable that the received signal is self-synchronizing or self-clocking. Timing is commonly recovered with a phase-locked loop. The maximum runlength parameter k ensures adequate frequency of transitions for synchronization of the read clock.

Recording codes that are based on RLL sequences have found almost universal application in disc recording practice. In consumer electronics, we have the EFM code (rate $= 8/17, d = 2, k = 10$), which is employed in the Compact Disc (CD), and the EFMPlus code (rate $= 8/16, d = 2, k = 10$) used in the DVD.

A dk-limited binary sequence, in short, (dk) sequence, satisfies simultaneously the following two conditions:

1. d constraint — two logical 1s are separated by a run of consecutive 0s of length at least d.
2. k constraint — any run of consecutive 0s is of length at most k.

If only proviso (1.) is satisfied, the sequence is said to be d-limited (with $k = \infty$), and will be termed (d) sequence. In general, a (dk) sequence is not employed in optical or magnetic recording without a simple coding step. A (dk) sequence is converted to a runlength-limited channel sequence in the following way. Let the channel signals be represented by a bipolar sequence $\{y_i\}$, $y_i \in \{-1, 1\}$. The channel signals represent the positive or negative magnetization of the recording medium, or pits or lands when dealing with optical recording. The logical 1s in the (dk) sequence indicate the positions of a transition $1 \to -1$ or $-1 \to 1$ of the corresponding RLL sequence. The (dk) sequence

$$0\ 1\ 0\ 0\ 0\ 1\ 0\ 0\ 1\ 0\ 0\ 0\ 1\ 1\ 0\ 1\ ...$$

would be converted to the RLL channel sequence

$$1\ -1\ -1\ -1\ -1\ 1\ 1\ 1\ -1\ -1\ -1\ -1\ 1\ -1\ -1\ 1\ ....$$

Waveforms that are transmitted without such an intermediate coding step are referred to as non-return-to-zero (NRZ). It can readily be verified that the minimum and maximum distance between consecutive transitions of the RLL sequence derived from a (dk) sequence is $d + 1$ and $k + 1$ symbols, respectively, or in other words, the RLL sequence has the virtue that at least $d + 1$ and at most $k + 1$ consecutive like symbols (runs) occur.

The outline of this chapter is as follows. We start with a discussion of the maximum rate of RLL sequences given the parameters d and k. Thereafter we will present various methods for constructing codes for generating RLL sequences.

17.2 Asymptotic Information Rate

17.2.1 Counting of Sequences

This section addresses the problem of counting the number of sequences of a certain length which comply with given dk constraints. We start for the sake of clerical convenience with the enumeration of (d) sequences. Let $N_d(n)$ denote the number of distinct (d) sequences of length n and define

$$N_d(n) = 0, \quad n < 0$$
$$N_d(0) = 1 \tag{17.1}$$

The number of (d) sequences of length $n > 0$ is found with the recursive relations [1]

$$(i) \quad N_d(n) = n + 1, \ 1 \le n \le d + 1$$
$$(ii) \quad N_d(n) = N_d(n - 1) + N_d(n - d - 1), \ n > d + 1 \tag{17.2}$$

The proof of Equation 17.2, taken from [1], is straightforward.

1. If $n \le d + 1$, a (d) sequence can contain only a single 1 (and there are exactly n such sequences), or the sequence must be the all 0 sequence (and there is only one such sequence).
2. If $n > d + 1$, a (d) sequence can be built by one of the following procedures:
    i. To build any (d) sequence of length n starting with a 0, take the concatenation of a 0 and any (d) sequence of length $n - 1$. There are $N_d(n - 1)$ of such.
    ii. Any (d) sequence of length n starting with a 1 can be constructed by the concatenation of a 1 and d 0s followed by any (d) sequence of length $n - d - 1$. There are $N_d(n - d - 1)$ of such.

Table 17.1 lists the number of distinct (d) sequences as a function of the sequence length n with the minimum runlength d as a parameter.

When $d = 0$, we simply find that $N_0(n) = 2N_0(n - 1)$, or in other words, when there is no restriction at all, the number of combinations doubles when a bit is added, which is, of course, a well-known result.

TABLE 17.1   Number of Distinct (d) Sequences as a Function of the Sequence Length n and the Minimum Runlength d as a Parameter.

$d \setminus n$	2	3	4	5	6	7	8	9	10	11	12	13	14
1	3	5	8	13	21	34	55	89	144	233	377	610	987
2	3	4	6	9	13	19	28	41	60	88	129	189	277
3	3	4	5	7	10	14	19	26	36	50	69	95	131
4	3	4	5	6	8	11	15	20	26	34	45	60	80
5	3	4	5	6	7	9	12	16	21	27	34	43	55

The numbers $N_1(n)$ are

$$1, 2, 3, 5, 8, 13, \ldots,$$

where each number is the sum of its two predecessors. These numbers are called *Fibonacci numbers*.

The number of (dk) sequences of length n can be found in a similar fashion. Let $N(n)$ denote the number of (dk) sequences of length n. (For the sake of simplicity in notation no subscript is used in this case.) Define

$$N(n) = 0, \quad n < 0$$
$$N(0) = 1 \tag{17.3}$$

The number of (dk) sequences of length n is given by

$$N(n) = n + 1, \ 1 \leq n \leq d + 1$$
$$N(n) = N(n-1) + N(n-d-1), \quad d+1 \leq n \leq k$$
$$N(n) = d + k + 1 - n + \sum_{i=d}^{k} N(n-i-1), \quad k < n \leq d+k \tag{17.4}$$
$$N(n) = \sum_{i=d}^{k} N(n-i-1), \quad n > d+k$$

The proof of the above recursion relations is not interesting and therefore omitted (see [1]).

17.2.2   Capacity

An encoder translates arbitrary user (or source) information into, in this particular instance, a sequence that satisfies given dk constraints. On the average, m source symbols are translated into n channel symbols. What is the maximum value of $R = m/n$ that can be attained for some specified values of the minimum and maximum runlength d and k?

The maximum value of the rate, R, that can be achieved by any code is called the *capacity* of a (dk) code. The capacity, or asymptotic information rate, of (dk) sequences, denoted by $C(d,k)$, defined as the number of information bits per channel bit that can maximally be carried by the (dk) sequences, on average, is governed by the specified constraints and is given by

$$C(d,k) = \lim_{n \to \infty} \frac{1}{n} \log_2 N_{dk}(n) \tag{17.5}$$

We simply find

$$C(d,k) = \log_2 \lambda_{dk} \tag{17.6}$$

where λ_{dk} is the largest real root of the characteristic equation

$$z^{k+2} - z^{k+1} - z^{k-d+1} + 1 = 0 \tag{17.7}$$

Table 17.2 lists the capacity $C(d,k)$ versus the parameters d and k.

TABLE 17.2 Capacity $C(d,k)$ versus Runlength Parameters d and k

k	$d = 0$	$d = 1$	$d = 2$	$d = 3$	$d = 4$
1	0.6942				
2	0.8791	0.4057			
3	0.9468	0.5515	0.2878		
4	0.9752	0.6174	0.4057	0.2232	
5	0.9881	0.6509	0.4650	0.3218	0.1823
6	0.9942	0.6690	0.4979	0.3746	0.2669
∞	1.000	0.6942	0.5515	0.4650	0.4057

17.3 Other Constraints

Besides sequences with simple runlength constraints as discussed above, there are a variety of channel constraints that have been reported in the literature.

17.3.1 MTR Constraints

The Maximum transition run (MTR) codes, introduced by Moon and Brickner [2], $d = 0$, have different constraints on the maximum runs of 0s and 1s. The maximum 0 runlength constraint, k_0, is imposed, as in standard RLL constraints, for clock recovery, while the maximum runlength constraint on 1s, denote by k_1, is imposed to bound the maximum number of consecutive transitions (i.e., consecutive 1s). It has been shown by Moon and Brickner [2] that removing said vexatious sequences leads to improved robustness against additive noise. MTR (d,k) constraints, $d > 0$, have been advocated as they are said to improve the detection quality. The MTR constraint limits the number of consecutive strings of the form $0^d 1$, that is, repetitive occurrence of the minimum runlength are limited. In wireless infrared communications applications, the MTR constraint is imposed as otherwise catastrophic receiver failure under near-field may be induced [3, 4]. Implementations of these codes usually have $d = 1$ and rate equal to 2/3.

17.3.2 $(O, G/I)$ Sequences

Partial response signaling in conjunction with maximum likelihood detection [5–8] is a data detection technique commonly used in magnetic recording. Special runlength constraints are needed to avoid vexatious sequences which could foil the detection circuitry. These constraints are characterized by two parameters G and I. The parameter G stipulates the maximum number of allowed 0s between consecutive 1s, while the parameter I stipulates the maximum number of 0s between 1s in both the even and odd numbered positions of the sequence. The G constraint, as the k constraint in dk sequences, is imposed to improve the timing. The I constraint is used to limit the hardware requirements of the detection circuitry. Marcus et al. [9] showed that it is possible to represent $(O, G/I)$ constraints by state-transition diagrams.

To that end, we define three parameters. The quantity g denotes the number of 0s since the last 1, and a and b denote the number of 0s since the last 1 in the even and odd subsequence. It is immediate that

$$g(a,b) = \begin{cases} 2a + 1 & \text{if } a < b \\ 2b & \text{if } a \geq b \end{cases}$$

Each state in the state-transition diagram is labeled with 2-tuples (a, b), where by definition $0 \leq a, b \leq I$ and $g(a, b) \leq G$. A transition between the states numbered by (a, b) to $(b, a + 1)$ (emitting a 0) and (a, b) to $(b, 0)$ (emitting a 1) are easily attached.

By computing the maximum eigenvalue of the above state-transition matrix, we obtain the capacity of the $(O, G/I)$ sequences. Results of computations are listed in Table 17.3.

Examples of implementation of $(O, G/I)$ constrained codes were given by Marcus, Siegel and Patel [10], Eggenberger and Patel [11] and Fitzpatrick and Knudson [12].

TABLE 17.3 Capacity for Selected Values of G and I [9]

G	I	Capacity
4	4	0.9614
4	3	0.9395
3	6	0.9445
3	5	0.9415
3	4	0.9342
3	3	0.9157

17.3.3 Weakly Constrained Sequences

Weakly constrained codes do not follow the letter of the law, as they produce sequences that violate the channel constraints with probability p. It is argued that if the channel is not free of errors, it is pointless to feed the channel with perfectly constrained sequences. In the case of a dk-constrained channel, violation of the d-constraint will very often lead to errors at the receiving site, but a violation of the k-constraint is usually harmless. Clearly, the extra freedom offered by weak constraints will result in an increase of the channel capacity. An analytic expression between the capacity and violation probability of the k-constraint has been derived by Janssen and Immink [13]. Worked examples of weakly constrained codes have been given by Immink [14] and Jin et al. [15].

17.3.4 Two-dimensional RLL Constraints

In conventional recording systems, information is organized along tracks. Interaction between neighboring tracks during writing and reading of the information cannot be neglected. During reading, in particular when tracking, either dynamic or static, is not optimal, both the information track itself plus part of the neighboring tracks are read, and a noisy phenomenon, called *crosstalk,* or *inter-track interference* (ITI) may disturb the reading process. Crosstalk is usually modeled as additive noise, and thus, essentially, the recording process is considered to be one-dimensional. Advanced coding systems that take into account inter-track interference, were developed by Soljanin and Georghiades [16].

It is expected that future mass data systems will show more of their two-dimensional character: the track pitch will become smaller and smaller relative to the reading-head dimensions, and, as a result, the recording process has to be modeled as a two-dimensional process. An example of a type of code, where the two-dimensional character of the medium is exploited to increase the code rate was introduced by Marcellin and Weber [17]. They introduced *multi-track (d,k)-constrained binary codes.* Such n-track codes are extensions of regular (d,k) codes for use in multi-track systems. In an n-track (d,k)-constrained binary code, the d constraint is required to be satisfied on each track, but the k constraint is required to be satisfied only by the bit-wise logical "or" of n consecutive tracks. For example, assume two parallel tracks, where the following sequences might be produced by a 2-track (d,k) code:

$$\begin{array}{ll} \text{track 1} & 000010100010100 \\ \text{track 2} & 010000010000001 \end{array}$$

Note that the $d = 1$ constraint is satisfied in each track, but that the $k = 2$ constraint is satisfied only in a joint manner — there are never more that two consecutive occurrences of 0 on both tracks simultaneously. Although n-track codes can provide significant capacity increase over regular (d,k) codes, they suffer from the fact that a single faulty track (as caused by media defects, for example) may cause loss of synchronization and hence loss of the data on all tracks. To overcome this flaw Swanson and Wolf [18] introduced a class of codes, where a first track satisfies the regular (d,k) constraint, while the k-constraint of the second track is satisfied in the "joint" manner. Orcutt and Marcellin [19, 20] computed the capacity of *redundant* multi-track (d,k)-constrained binary codes, which allow only r tracks to be faulty at every time instant. Vasic computed capacity bounds and spectral properties [21–23]. Further improvements of n-track systems with faulty tracks were given by Ke and Marcellin [24].

In holographic recording, data is stored using optical means in the form of two-dimensional binary patterns. In order to safeguard the reliability of these patterns, certain channel constraints have been proposed. More information on holographic memories and channel constraints can be found in [25, 26].

Codes that take into account the two-dimensional character have been investigated by several authors. Talyansky, Etzion and Roth [27] studied efficient coding algorithms for two types of constraints on two-dimensional binary arrays. The first constraint considered is that of the t-conservative arrays, where each row and column of the array has at least t transitions. Blaum, Siegel, Sincerbox and Vardy [28–30] disclosed a code which eliminates long periodic stretches of contiguous light or dark regions in any of the dimensions of the holographic medium such that interference between adjacent images recorded in the same volume is effectively minimized.

Kato and Zeger [31] considered two-dimensional RLL constraints. A two-dimensional binary pattern of 1s and 0s arranged in an $m \times n$ rectangle is said to satisfy a two-dimensional (d, k) constraint if it satisfies a one-dimensional (d, k)-constraint both horizontally and vertically. In contrast to the one-dimensional capacity, there is little known about the two-dimensional capacity. It was shown by Calkin and Wilf that $C(d, k)$ is bounded as $0.587891 \leq C(d, k) \leq 0.588339$ [32]. Bounds on $C(d, k)$ have been derived by Kato and Zeger [31] and Siegel and Wolf [33].

17.4 Codes for the Noiseless Channel

In the present section, we take a look at the techniques that are available to produce constrained sequences in a practical manner. Encoders have the task of translating arbitrary source information onto a constrained sequence. It is most important that this be done as efficiently as possible within some practical considerations. Efficiency is usually measured in terms of the ratio of code rate R and capacity C of the constrained channel. A good encoder algorithm realizes a code rate close to the capacity of the constrained sequences, uses a simple implementation, and avoids the propagation of errors in the process of decoding.

In coding practice, the source sequence is partitioned into blocks of length p, and under the code rules such blocks are mapped onto words of q channel symbols. The rate of such an encoder is $R = p/q \leq C$. A code may be state dependent, in which case the codeword used to represent a given source block is a function of the channel or encoder state, or the code may be state independent. State independence implies that codewords can be freely concatenated without violating the sequence constraints. When the encoder is state dependent, it typically takes the form of a synchronous finite-state machine.

A decoder is preferably state independent. Due to errors made during transmission, a state-dependent decoder could easily lose track of the encoder state, and as a result the decoder could possibly make error after error with no guarantee of recovery. In order to avoid error propagation, a decoder should preferably use a finite observation interval of channel bits for decoding, thus limiting the span in which errors may occur. Such a decoder is called a *sliding block decoder*. A sliding block decoder makes a decision on a received word on the basis of the q-bit word itself, as well as a number of m preceding q-bit words and a upcoming q-bit words. Essentially, the decoder comprises a register of length $(m + a + 1)$ and a logic function $f(.)$ that translates the contents of the register into the retrieved q-bit source word. Since the constants m and a are finite, an error in the retrieved sequence can propagate in the decoded sequence only for a finite distance, at most the decoder window length $(m + a + 1)$. An important subclass of the sliding-block decoder is the *block decoder*, which uses only a single codeword for reproducing the source word, that is, $m = a = 0$ [34]. The above parameters define the playing field of the code designer. Early players are Tang and Bahl [1], Franaszek [35–38], and Cattermole [39]. In addition, important contributions were made by Jacoby [40, 41], Lempel [42], Patel [43], Cohen [44], and many others. Tutorial expositions can be found in [9, 45, 46].

In its simplest form, the set of encoder states, called *principal states*, is a subset of the channel states used to describe the constraints. From each of the principal states there are at least 2^p constrained words beginning at such a state and ending in a principal state. The set of principal states can be found by invoking Franaszek's procedure [35]. Flawless concatenation of the words is implied by the structure of the finite-state machine describing the constraints.

Concatenation of codewords can also be established by using *merging* bits between constrained words [40, 47, 48]. Merging bits are used, for example, in the EFM code employed in the Compact Disc [49]. Each source word has a unique q'-bit channel representation. We require one look-up table for translating source words into constrained words of length q' plus some logic circuitry for determining the $q - q'$ merging bits. Decoding is extremely simple: discard the merging bits and translate the q'-bit word into the p-bit source word. For (dk) codes the relation between Franaszek's principal state and the merging bit procedures was found by Gu and Fuja [50]. Immink [51] gave a constructive proof that (dk) codes with merging bits can be made for which $C - R < 1/(2q)$. As a result, (dk) codes with a rate only 0.1% less than Shannon's capacity can be constructed with codewords of length $q \approx 500$. The number of codewords grows exponentially with the codeword length, and the key obstacle to practically approaching capacity is the massive hardware required for the translation. The massiveness problem can be solved by using a technique called *enumeration* [52], which makes it possible to translate source words into codewords and vice versa by invoking an algorithmic procedure rather than performing the translation with a look-up table. Single channel bit errors could corrupt the entire data in the decoded word, and, of course, the longer the codeword the greater the number of data symbols affected. This difficulty can be solved by a special configuration of the error correcting code and the recording code [51, 53].

A breakthrough in code design occurred in the 1980s with the elegant construction method presented by Adler, Coppersmith, and Hassner (ACH) [54]. A generalized procedure was published by Ashley and Marcus [55]. The ACH algorithm, also called *state-splitting algorithm*, gives a step-by-step approach for designing constrained codes. The guarantee of a sliding-block decoder and the explicit bound on the decoder window length are the key strengths of the ACH algorithm. Roughly speaking, the state-splitting algorithm proceeds by iteratively modifying the FSTD. At each round of iteration, the maximum weight (greater than unity) is reduced, so that we eventually reach an FSTD whose approximate eigenvector has binary components. Complexity issues related to the number of encoder states and window length are an active field of research, which is exemplified by, for example, [56–58].

The *sequence replacement technique* [59] converts source words of length p into $(0, k)$-constrained words of length $q = p + 1$. The control bit is set to 1 and appended at the beginning of the p-bit source word. If this $(p + 1)$-bit sequence satisfies the prescribed constraint it is transmitted. If, on the other hand, the constraint is violated, that is, a runlength of at least $k + 1$ 0s occur, we remove the trespassing $k + 1$ 0s. The position where the start of the violation was found is encoded in $k + 1$ bits, which are appended at the beginning of the $p + 1$-bit word. Such a modification is signaled to the receiver by setting the control bit to 0s. The codeword remains of length $p + 1$. The above procedure is repeated until all forbidden subsequences have been removed. The receiver can reconstruct the source word as the position information is stored at a predefined position in the codeword. In certain situations the entire source word has to be modified which makes the procedure prone to error propagation. The class of rate $(q - 1)/q$, $(0, k)$-constrained codes, $k = 1 + \lfloor q/3 \rfloor$, $q \geq 9$, was constructed to minimize error propagation [60]. Error propagation is confined to one decoded byte irrespective of the codeword length q.

Recently, the publications by Fair et al. [61] and Immink & Patrovics [62] on *guided scrambling* brought new insights into high-rate code design. Guided scrambling is a member of a larger class of related coding schemes called *multi-mode* codes. In multi-mode codes, the p-bit source word is mapped into $(m + p)$-bit codewords. Each source word $\mathbf{x}$ can be represented by a member of a *selection set* consisting of $L = 2^m$ codewords. Examples of such mappings are the guided scrambling algorithm presented by Fair et al. [61], and the scrambling using a Reed-Solomon code by Kunisa et al. [63].

The encoder opts for transmitting that codeword that minimizes, according to a prescribed criterion, for example, the low-frequency spectral contents of the encoded sequence. There are two key elements which need to be chosen judiciously: (a) the mapping between the source words and their corresponding selection sets, and (b) the criterion used to select the "best" word. Provided that 2^m is large enough and the selection set contains sufficiently different codewords, multi-mode codes can also be used to satisfy almost any channel constraint with a suitably chosen selection method. A clear disadvantage is that the encoder needs to generate all 2^m possible codewords, compute the criterion, and make the decision.

References

[1] D.T. Tang and L.R. Bahl, Block codes for a class of constrained noiseless channels, *Info. Control*, vol. 17, pp. 436–461, 1970.

[2] J. Moon and B. Brickner, Design of a rate 6/7 maximum transition run code, *IEEE Trans. Magn.*, vol. 33, pp. 2749–2751, September 1997.

[3] M.A. Hassner, N. Heise, W. Hirt, B.M. Trager, Method and Means for Invertibly Mapping Binary Sequences into Rate 2/3, $(1, k)$ Run-Length-Limited Coded Sequences with Maximum Transition Density Constraints, U.S. Patent 6,195,025, February 2001.

[4] W. Hirt, M. Hassner, and N. Heise, IrDA-VFlr (16 Mb/s): Modulation code and system design, *IEEE Personal Commn.*, pp. 58–71, February 2001.

[5] H. Kobayashi, A Survey of coding schemes for transmission or recording of digital data, *IEEE Trans. Commn.*, vol. COM-19, pp. 1087–1099, December 1971.

[6] R. Cideciyan, F. Dolivo, R. Hermann, W. Hirt, and W. Schott, A PRML system for digital magnetic recording, *IEEE J. Selected Areas in Commn.*, vol. 10, pp. 38–56, January 1992.

[7] H. Kobayashi and D.T. Tang, Application of partial response channel coding to magnetic recording systems, *IBM J. Res. Develop.*, vol. 14, pp. 368–375, July 1970.

[8] R.W. Wood and D.A. Petersen, Viterbi detection of class IV partial response on a magnetic recording channel, *IEEE Trans. Commn.*, vol. COM-34, pp. 454–461, May 1986.

[9] B.H. Marcus, P.H. Siegel, and J.K. Wolf, Finite-state modulation codes for data storage, *IEEE J. Selected Areas in Commn.*, vol. 10, no. 1, pp. 5–37, January 1992.

[10] B.H. Marcus, A.M. Patel, and P.H. Siegel, Method and Apparatus for Implementing a PRML code, U.S. Patent 4,786,890, November 1988.

[11] J.S. Eggenberger and A.M. Patel, Method and Apparatus for Implementing Optimum PRML Codes, U.S. Patent 4,707,681, November 17, 1987.

[12] J. Fitzpatrick and K.J. Knudson, Rate 16/17, $(d = 0, G = 6/I = 7)$ modulation code for a magnetic recording channel, U.S. Patent 5,635,933, June 1997.

[13] A.J.E.M. Janssen and K.A.S. Immink, An entropy theorem for computing the capacity of weakly (d, k)-constrained sequences, *IEEE Trans. Inform. Theory*, vol. IT-46, no. 5, pp. 1034–1038, May 2000.

[14] K.A.S. Immink, Weakly constrained codes, *Electronics Letters*, vol. 33, no. 23, pp. 1943–1944, November 1997.

[15] Ming Jin, K.A.S. Immink, and B. Farhang-Boroujeny, Design techniques for weakly constrained codes, *Trans. Commn.*, vol. 51, no. 5, pp. 709–714, May 2003.

[16] E. Soljanin and C.N. Georghiades, Coding for two-head recording systems, *IEEE Trans. Inform. Theory*, vol. IT-41, no. 3, pp. 794–755, May 1995.

[17] M.W. Marcellin and H.J. Weber, Two-dimensional modulation codes, *IEEE J. Selected Areas Commn.*, vol. 10, no. 1, pp. 254–266, January 1992.

[18] R.D. Swanson and J.K. Wolf, A new Class of Two-dimensional RLL recording codes, *IEEE Trans. Magn.*, vol. 28, pp. 3407–3416, November 1992.

[19] E.K. Orcutt and M.W. Marcellin, Enumerable multi-track (d, k) block codes, *IEEE Trans. Inform. Theory*, vol. IT-39, pp. 1738–1743, September 1993.

[20] E.K. Orcutt and M.W. Marcellin, Redundant multi-track (d, k) codes, *IEEE Trans. Inform. Theory*, vol. IT-39, pp. 1744–1750, September 1993.

[21] B.V. Vasic, Capacity of channels with redundant multi-track (d, k) constraints: The $k < d$ Case, *IEEE Trans. Inform. Theory*, vol. IT-42, no. 5, pp. 1546–1548, September 1996.

[22] B.V. Vasic, Shannon capacity of M-ary redundant multi-track runlength limited codes, *IEEE Trans. Inform. Theory*, vol. IT-44, no. 2, pp. 766–774, March 1998.

[23] B.V. Vasic, Spectral analysis of maximum entropy multi-track modulation codes, *IEEE Trans. Inform. Theory*, vol. IT-44, no. 4, pp. 1574–1587, July 1998.

[24] L. Ke and M.W. Marcellin, A new construction for n-track (d, k) codes with redundancy, *IEEE Trans. Inform. Theory*, vol. IT-41, no. 4, pp. 1107–1115, July 1995.

[25] J.F. Heanue, M.C. Bashaw, and L. Hesselink, Volume holographic storage and retrieval of digital data, *Science*, pp. 749–752, 1994.

[26] J.F. Heanue, M.C. Bashaw, and L. Hesselink, Channel codes for digital holographic data storage, *J. Opt. Soc. Am.*, vol. 12, 1995.

[27] R. Talyansky, T. Etzion, and R.M. Roth, Efficient code construction for certain two-dimensional constraints, *IEEE Trans. Inform. Theory*, vol. IT-45, no. 2, pp. 794–799, March 1999.

[28] M. Blaum, P.H. Siegel, G.T. Sincerbox, and A. Vardy, Method and apparatus for modulation of multi-dimensional data in holographic storage, U.S. Patent 5,510,912, April 1996.

[29] M. Blaum, P.H. Siegel, G.T. Sincerbox, and A. Vardy, Method and apparatus for modulation of multi-dimensional data in holographic storage, U.S. Patent 5,727,226, March 1998.

[30] A. Vardy, M. Blaum, P.H. Siegel, and G.T. Sincerbox, Conservative arrays: multi-dimensional modulation codes for holographic recording, *IEEE Trans. Inform. Theory*, vol. IT-42, no. 1, pp. 227–230, January 1996.

[31] A. Kato and K. Zeger, On the capacity of two-dimensional run-length constrained channels, *IEEE Trans. Inform. Theory*, vol. IT-45, no. 5, pp. 1527–1540, July 1999.

[32] N.J. Calkin and H.S. Wilf, The number of independent sets in a grid graph, *SIAM J. Discr. Math.*, vol. 11, pp. 54–60, February 1998.

[33] P.H. Siegel and J.K. Wolf, Bit-stuffing bounds on the capacity of two-dimensional constrained arrays, *Proc. 1998 IEEE Int. Symp. Inform. Theory*, pp. 323, 1998.

[34] P. Chaichanavong and B. Marcus, Optimal block-type-decodable encoders for constrained systems, *IEEE Trans. Inform. Theory*, vol. IT-49, no. 5, pp. 1231–1250, May 2003.

[35] P.A. Franaszek, Sequence-state encoding for digital transmission, *Bell Syst. Tech. J.*, vol. 47, pp. 143–157, January 1968.

[36] P.A. Franaszek, Sequence-state methods for run-length-limited coding, *IBM J. Res. Develop.*, vol. 14, pp. 376–383, July 1970.

[37] P.A. Franaszek, Run-length-limited variable length coding with error propagation limitation, U.S. Patent 3,689,899, September 1972.

[38] P.A. Franaszek, On future-dependent block coding for input-restricted channels, *IBM J. Res. Develop.*, vol. 23, pp. 75–81, 1979.

[39] K.W. Cattermole, *Principles of Pulse Code Modulation*, Iliffe Books Ltd, London, 1969.

[40] G.V. Jacoby, A new look-ahead code for increasing data density, *IEEE Trans. Magn.*, vol. MAG-13, no. 5, pp. 1202–1204, September 1977.

[41] G.V. Jacoby and R. Kost, Binary two-thirds rate code with full word look-ahead, *IEEE Trans. Magn.*, vol. MAG-20, no. 5, pp. 709–714, September 1984.

[42] A. Lempel and M. Cohn, Look-ahead coding for input-restricted channels, *IEEE Trans. Inform. Theory*, vol. IT-28, no. 6, pp. 933–937, November 1982.

[43] A.M. Patel, Zero-modulation encoding in magnetic recording, *IBM J. Res. Develop.*, vol. 19, pp. 366–378, July 1975.

[44] M. Cohn and G.V. Jacoby, Run-length reduction of 3PM code via look-ahead technique, *IEEE Trans. Magn.*, vol. MAG-18, pp. 1253–1255, November 1982.

[45] B.H. Marcus, R.M. Roth, and P.H. Siegel, Constrained systems and coding for recording channels, in *Handbook of Coding Theory*, Brualdi R., Huffman C., and Pless V., Eds., Amsterdam, The Netherlands, Elsevier Press, 1996.

[46] K.A.S. Immink, Runlength-limited sequences, *Proc. IEEE*, vol. 78, no. 11, pp. 1745–1759, November 1990.

[47] G.F.M. Beenker and K.A.S. Immink, A generalized method for encoding and decoding runlength-limited binary sequences, *IEEE Trans. Inform. Theory*, vol. IT-29, no. 5, pp. 751–754, September 1983.

[48] K.A.S. Immink, Constructions of almost block-decodable runlength-limited codes, *IEEE Trans. Inform. Theory,* vol. IT-41, no. 1, pp. 284–287, January 1995.

[49] J.P.J. Heemskerk and K.A.S. Immink, Compact disc: system aspects and modulation, *Philips Techn. Review,* vol. 40, no. 6, pp. 157–164, 1982.

[50] J. Gu and T. Fuja, A new approach to constructing optimal block codes for runlength-limited channels, *IEEE Trans. Inform. Theory,* vol IT-40, no. 3, pp. 774–785, 1994.

[51] K.A.S. Immink, A practical method for approaching the channel capacity of constrained channels, *IEEE Trans. Inform. Theory,* vol. IT-43, no. 5, pp. 1389–1399, September 1997.

[52] T.M. Cover, Enumerative source coding, *IEEE Trans. Inform. Theory,* vol. IT-19, no. 1, pp. 73–77, January 1973.

[53] J.L. Fan and A.R. Calderbank, A modified concatenated coding scheme with applications to magnetic recording, *IEEE Trans. Inform. Theory,* vol. IT-44, pp. 1565–1574, July 1998.

[54] R.L. Adler, D. Coppersmith, and M. Hassner, Algorithms for sliding block codes: an application of symbolic dynamics to information theory, *IEEE Trans. Inform. Theory,* vol. IT-29, no. 1, pp. 5–22, January 1983.

[55] J.J. Ashley and B.H. Marcus, A generalized state-splitting algorithm, *IEEE Trans. Inform. Theory,* vol. IT-43, no. 4, pp. 1326–1338, July 1997.

[56] J.J. Ashley, R. Karabed, and P.H. Siegel, Complexity and sliding-block decodability, *IEEE Trans. Inform. Theory,* vol. IT-42, pp. 1925–1947, 1996.

[57] J.J. Ashley and B.H. Marcus, Canonical encoders for sliding-block decoders, *SIAM J. Discrete Math.,* vol. 8, pp. 555–605, 1995.

[58] B.H. Marcus and R.M. Roth, Bounds on the number of states in encoder graphs for input-constrained channels, *IEEE Trans. Inform. Theory,* vol. IT-37, no. 3, part 2, pp. 742–758, May 1991.

[59] A.J. de Lind van Wijngaarden and K.A.S. Immink, Construction of constrained codes using sequence replacement techniques, Submitted *IEEE Trans. Inform. Theory,* 1997.

[60] K.A.S. Immink and A.J. de Lind van Wijngaarden, Simple high-rate constrained codes, *Electronics Letters,* vol. 32, no. 20, pp. 1877, September 1996

[61] I.J. Fair, W.D. Gover, W.A. Krzymien, and R.I. MacDonald, Guided scrambling: a new line coding technique for high bit rate fiber optic transmission systems, *IEEE Trans. Commn.,* vol. COM-39, no. 2, pp. 289–297, February 1991.

[62] K.A.S. Immink and L. Patrovics, Performance assessment of DC-free multimode codes, *IEEE Trans. Commn.,* vol. COM-45, no. 3, March 1997.

[63] A. Kunisa, S. Takahashi, and N. Itoh, Digital modulation method for recordable digital video disc, *IEEE Trans. Consumer Electr.,* vol. 42, pp. 820–825, August 1996.

18

Maximum Transition Run Coding

18.1 Introduction **18**-1
18.2 Error Event Characterization **18**-2
18.3 Maximum Transition Run Codes **18**-4
18.4 Detector Design for MTR Constraints **18**-10
18.5 Simulation Results **18**-11
18.6 Summary **18**-12

Barrett J. Brickner

Bermai, Inc.
Minnetonka, MN

18.1 Introduction

The written data and corresponding readback waveform in any recording system are subject to noise and distortions that limit reliability of the system. An error correction code allows a certain amount of data corruption to be corrected in the decoder by providing specific redundancy in the recorded data. If the system is more susceptible to errors in specific data patterns, a more direct approach is to employ a channel coding constraints that prevents these troublesome patterns so that the error simply does not occur.

In the text that follows, a specific class of codes designed to improve the performance of recording systems is explored. The discussion follows a methodology that is generally applicable to the development of code constraints to improve minimum distance properties of communications or recording systems. Initially, the system is characterized in terms of types of error events and the probability with which they may occur, a value expressed as a distance in a geometric context. Having identified the error types that are most likely to corrupt the system, pairs of code bit sequences that produce these errors are determined. An error occurs when noise, combined with intersymbol interference (ISI) causes the received signal produced by one code bit sequence to resemble that produced by another. This ambiguity is resolved by simply enforcing a constraint in the encoder, which prevents one or both of the code bit sequences. When only one of the two error-producing code bit sequences is removed, the detector and/or decoder must be modified to choose in favor of the valid sequence. The resulting encoder and detector/decoder work in concert to improve the system performance. However, the addition of a code constraint reduces the code rate and the amount of information conveyed in a sequence of code bits. An analysis or simulation is then used to verify that a net gain results from removing error events at the expense of a lower code rate.

A properly chosen maximum transition run (MTR) constraint, which limits the number of consecutive transitions, is shown to prevent minimum-distance errors for a variety of channel responses applicable to recording systems. This idea of using a code constraint to prevent problematic bit sequences is not new to recording. For years, RLL(d, k) codes which specify a minimum, d, and maximum, k, number of nontransitions between transitions have been used with $d > 0$ to help peak detector base read circuits by reducing the effects of ISI in adjacent transitions. A $d = 1$ code has also been used to improve the distance properties of a high order partial-response maximum likelihood (PRML) channel.[1] Moreover, while an MTR constraint is sufficient to remove certain, minimum-distance error events, it is not a unique solution to this problem. Other coding schemes based on forbidding one of the pairs of error generating code sequences have been shown to give similar distance gains.[2,3] For recording systems, RLL $d > 0$ and MTR constraints are of particular interest because many of the disturbances that fall outside the simple linear ISI and additive white Gaussian noise (AWGN) model occur when transitions are brought in close proximity to one another.

18.2   Error Event Characterization

For convenience and ease of analysis, the channel is assumed to be linear with additive white Gaussian noise such that the received signal is written as

$$r_k = s_k + n_k = \sum_{i=0}^{L-1} h_k a_{k-i} + n_k$$

where the discrete-time channel is represented as $h(D) = \sum_{k=0}^{L} h_k D^k$, and the input data are taken from the binary alphabet $\{0, 1\}$. Typically, $h(D)$ is formed by equalizing the received signal with a noise-whitened matched filter such that the noise statistics are preserved, and $h(D)$ is the combined response of the channel and equalizer. The noise is assumed to be additive white Gaussian noise with variance σ_n^2. The maximum likelihood sequence detection (MLSD) estimates an N-sample input sequence $\mathbf{a}_k = [a_k, a_{k-1}, \ldots, a_{k-N+1}]$ using

$$\hat{\mathbf{a}}_k = \arg \left\{ \min_{\mathbf{a}_k} \sum_{j=0}^{N+L-2} \left(r_{k-j} - \sum_{i=0}^{L-1} h_i a_{k-i-j} \right)^2 \right\}$$

where $a_{k-i} = 0$ for $i \geq N$. An error is produced whenever the estimated sequence does not match the input sequence in one or more locations, that is, $\hat{\mathbf{a}}_k \neq \mathbf{a}_k$ such that $e_k = \left\{ \begin{smallmatrix} a_k - \hat{a}_k, & k=0...N-1 \\ 0, & \text{elsewhere} \end{smallmatrix} \right.$. Using Marcum's Q function

$$Q(x) = \Pr[X > x] = \frac{1}{\sqrt{2\pi}} \int_x^\infty e^{-z^2/2} \, dz$$

the probability of a particular error is $\Pr(\hat{\mathbf{a}} \neq \mathbf{a} \mid \mathbf{a}) \approx Q(d_{\hat{\mathbf{a}},\mathbf{a}}/2\sigma_n)$ where

$$d_{\hat{\mathbf{a}},\mathbf{a}} = \sqrt{\sum_{i=0}^{N+L-2} \left(\sum_{j=0}^{L-1} h_j e_{i-j} \right)^2}$$

This value is the Euclidean distance between two points whose coordinates correspond to the noiseless received signals generated by two valid input sequences. For low error rate situations, a small change in the distance d results an exponential change in the error probability. Therefore, the performance of the system is dominated by those error events that produce the minimum distance, $d_{\min}$.

To compute the minimum distance as shown above, the corresponding error event must be known. In the general case, the length of the input sequence and, therefore, the possible error event sequences are unbounded. However, by considering all error events up to a particular length, the minimum distance can be bounded by

$$\min_{\mathbf{e}_k} \sum_{k=0}^{N-1} \left(\sum_{j=0}^{L-1} f_j e_{k-j} \right)^2 \leq d_{\min}^2 \leq \min_{\mathbf{e}_k} \sum_{k=0}^{N+L-2} \left(\sum_{j=0}^{L-1} f_j e_{k-j} \right)^2$$

where N is the length of the error event $\mathbf{e}_k$.[4] The lower-bound has particular relevance to the fixed delay tree search with decision feedback (FDTS/DF) detector as it gives the exact minimum distance when $N - 1$ is equal to the depth of the tree search.[5] It is often useful to compare the result of this calculation to the matched filter bound

$$d_{MF} = \sqrt{\sum_{k=0}^{L-1} f_k^2}$$

which indicates the distance for the case where a single bit is transmitted. For channels with little ISI, it is common for, $d_{\min} = d_{MF}$; however, for many of the responses seen in data storage channels, the ISI structure is more severe, and $d_{\min} < d_{MF}$. In any case, the minimum distance for the uncoded channel is never larger than d_{MF}.

The error events of interest here are closed error events, or, in other words, those that are finite in duration. Open, or quasi-catastrophic, error events can extend indefinitely and are generally handled by additional code constraints in practical MLSD implementations. In this discussion, error events are shown to start and end with nonzero terms e_k. To be an independent error event, these sequences are both preceded and followed by $L - 1$ zero terms. Often, the errors of interest are short, or have a predictable form, so that they can be identified a simple exhaustive search of all events up to a particular length N. However, to guarantee that all events have been identified, a more rigorous approach is followed. Altekar et al. describe a suitable method for finding error events in partial response channels up to a certain distance.[6]

With low-order partial response channels, such as PR1, PR4, and EPR4 characterized by the response polynomials $(1 + D)$, $(1 - D)(1 + D)$, and $(1 - D)(1 + D)^2$, respectively, $d_{\min} = d_{MF}$. However, as shown in Table 18.1, error events other than a single bit error are dominant for channels with a greater high frequency roll-off; that is, those with higher order $(1 + D)$ factors. In particular, these errors consist of groups of terms for which the signs of consecutive error bits alternate. An error event written as $\pm\{\mathbf{e}_1, \langle\mathbf{e}_2\rangle\}$ indicates that the same distance is obtained by concatenating error sequence $\mathbf{e}_1$ with a nonnegative integer number $(0, 1, 2, \ldots)$ of repetitions of sequence $\mathbf{e}_2$. Because these errors correspond to a signal difference with a significant high-frequency component, it is not too surprising that they are prevalent in channels that are inherently low-pass in nature.

The basic pattern that emerges in these error sequences for the high-order partial response channels is that they are defined by, or contain, the sequences $\pm\{+1, -1\}$ and $\pm\{+1, -1, +1\}$. Two particular cases of interest are PR2 with a response $(1 + D)^2$ and E2PR4 with a response $(1 - D)(1 + D)^3$. Error events with distances up through d_{MF} are given for these two channels in Table 18.2 and Table 18.3, respectively. In longitudinal magnetic disc recording, a Lorentzian pulse is often used to model the transition response

TABLE 18.1 Distance to Matched Filter Bound for Partial Response Channels

Response	Polynomial	$d_{\min}^2$	d_{MF}^2	$d_{\min}/d_{MF}$ (dB)	Minimum Distance Errors
PR1	$(1 + D)$	2	2	0	$\pm\{+1, \langle-1, +1\rangle\}$
					$\pm\{+1, -1, \langle+1, -1\rangle\}$
PR2	$(1 + D)^2$	4	6	-1.76	$\pm\{+1, -1, \langle+1, -1\rangle\}$
					$\pm\{+1, -1, +1, \langle-1, +1\rangle\}$
EPR2	$(1 + D)^3$	10	20	-3.01	$\pm\{+1, -1\}$
Dicode	$(1 - D)$	2	2	0	$\pm\{+1, \langle+1\rangle\}$
PR4	$(1 - D)(1 + D)$	2	2	0	$\pm\{+1, \langle0, +1\rangle\}$
EPR4	$(1 - D)(1 + D)^2$	4	4	0	$\pm\{+1, \langle0, +1\rangle\}$
					$\pm\{+1, -1, +1, \langle-1, +1\rangle\}$
					$\pm\{+1, -1, +1, -1, \langle+1, -1\rangle\}$
E2PR4	$(1 - D)(1 + D)^3$	6	10	-2.22	$\pm\{+1, -1, +1\}$
E3PR4	$(1 - D)(1 + D)^4$	12	26	-3.36	$\pm\{+1, -1, +1\}$

TABLE 18.2 Dominant Closed Error Events for a PR2 $(1 + D)^2$ Channel

d^2	d^2/d_{MF}^2 (dB)	Error Events
4	−1.76	$\pm\{+1, -1, \langle+1, -1\rangle\}$
		$\pm\{+1, -1, +1, \langle-1, +1\rangle\}$
6	0.00	± 1
		$\pm\{+1, -1, 0, +1, -1, +1, \langle-1, +1\rangle\}$
		$\pm\{+1, -1, \langle+1, -1\rangle, 0, +1, -1, \langle+1, -1\rangle\}$
		$\pm\{+1, -1, +1, \langle-1, +1\rangle, 0, -1, +1, \langle-1, +1\rangle\}$

TABLE 18.3 Dominant Closed Error Events for an E2PR4 $(1 - D)(1 + D)^3$ Channel

d^2	d^2/d_{MF}^2 (dB)	Error Events
6	−2.22	$\pm\{+1, -1 +1, \langle-1, +1\rangle\}$
8	−0.97	$\pm\{+1, -1 +1, -1, \langle+1, -1\rangle\}$
		$\pm\{+1, -1, +1, -1, +1, \langle-1, +1\rangle\}$
		$\pm\{+1, -1, +1, 0, 0, +1, -1, +1\}$
10	0.00	± 1
		$\pm\{+1, 0, 0, +1, -1, +1\}$
		$\pm\{+1, -1, +1, 0, -1, +1, -1\}$
		$\pm\{+1, -1, +1, 0, 0, +1\}$
		$\pm\{+1, -1, +1, 0, 0, +1, -1, +1, -1, \langle+1, -1\rangle\}$
		$\pm\{+1, -1, +1, 0, 0, +1, -1, +1, -1, +1, \langle-1, +1\rangle\}$
		$\pm\{+1, -1, +1, 0, 0, 0 + 1, -1, +1\}$
		$\pm\{+1, -1, +1, 0, 0, +1, 0, 0, +1, -1, +1\}$
		$\pm\{+1, -1, +1, 0, 0, +1, -1, +1, 0, 0, +1, -1, +1\}$
		$\pm\{+1, -1, +1, -1, \langle+1, -1\rangle, 0, 0, -1, +1, -1\}$
		$\pm\{+1, -1, +1, -1, +1, \langle-1, +1\rangle, 0, 0, +1, -1, +1\}$

of the magnetic channel. This transition response is defined by

$$p(t) = \frac{A}{1 + \left(\frac{2t}{PW_{50}}\right)^2}$$

where PW_{50} parameterizes the width of the pulse. For a symbol clock period of T, the symbol density, defined as $D_s = PW_{50}/T$, gives a measure of the severity of the channel intersymbol interference (ISI). Figure 18.1 shows the relevant distance between several key error events as a function of density. As symbol density increases, the high-frequency roll-off of the channel response also increases, and just as with the partial response polynomials, the error events with alternating signs become dominant. Although the Lorentzian response with AWGN provides a reasonable approximation for the recording channel, it is not purported to be an exact model. As such, there will always be a place for analysis of errors in specific recording systems.[7]

18.3 Maximum Transition Run Codes

For channels where the minimum distance is not produced by a single bit error, a code can be constructed to prevent the error from occurring by prohibiting one or both of the pairs of sequences whose difference produces the error. Because the constraints used to prevent errors will require a greater density of code bits to maintain a particular data density, this approach does not always provide a net gain. Fortunately, there are a number of cases where the distance gain does exceed the code rate penalty.

First, consider the high density Lorentzian channels ($D_s \sim 3$) and E2PR4 response. For these, the dominant error events have the form $\pm\{+1, -1, +1\}$. The pairs of input sequences, which generate these

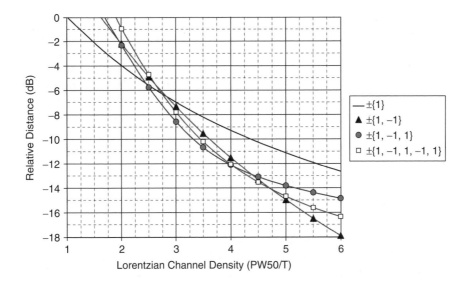

FIGURE 18.1 Euclidean distance for key error events in a Lorentzian channel.

errors, contain at least one pattern with three or more consecutive transitions as shown in Figure 18.2. If all input sequences that contain three or more consecutive transitions are eliminated, a detector can be constructed such that the original minimum distance error events are suppressed. A maximum transition run code with constraint parameter j is one that limits the number of consecutive transitions to j. In the case of the E2PR4 channel, a code with $j = 2$ is sufficient to prevent the $\pm\{+1, -1, +1\}$ error event. More generally, this constraint will prevent error events of the form $\pm\{+1, -1, +1, (-1, +1)\}$. Specifically, this constraint will allow the suppression of an error event where one or both of the sequences involved in the error contain $j + 1$ or more transitions. Because an MTR code usually includes a runlength limited k constraint to assist with timing recovery, the code parameters are encapsulated as $\mathrm{MTR}(j; k)$.

The MTR code with $j = 1$ will also prevent any error events of the form $\pm\{+1, -1, \langle+1, -1\rangle\}$. This constraint is implied by $\mathrm{RLL}(d, k)$ codes with $d > 0$. Specifically, the MTR $j = 1$ constraint is the same as the more common RLL $d = 1$. Because $d = 1$ codes automatically prevent transition runs, they have the same distance-enhancing properties as MTR $j = 2$ codes, albeit at a lower code rate. This distance enhancing property of the $d = 1$ constraint has been exploited for both E2PR4 decoders and FDTS/DF.[1,5]

It is convenient to describe the MTR constraints in terms of an NRZI format where a 1 and 0 represent, respectively, the presence and absence of a transition. In this form, j indicates the maximum number of consecutive 1s, and k is the maximum number of consecutive 0s. A precoder of the form $P(D) = 1/(1 \oplus D)$ is used to generate the mapping from NRZI bits, denoted x_k, to the NRZ channel input symbols a_k. Thus, the precoder implements $a_k = a_{k-1} \oplus x_k$.

A simple rate 4/5 MTR(2; 8) code can be constructed by removing from the list of 32 5-bit codewords, the all 0s codeword, all codewords with three or more consecutive transitions, and all codewords with more

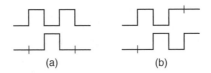

(a)   (b)

FIGURE 18.2 Pairs of write sequences which produce a $\pm\{+1, -1, +1\}$ error event.

than a single transition at the beginning or end. The set of 16 valid codewords is given by {00001, 00010, 00100, 00101, 00110, 01000, 01001, 01010, 01100, 01101, 10000, 10001, 10010, 10100, 10101, 10110}.

The MTR constraint is useful when the specific objective of the code designer is to limit the length of transition runs. With partial response channels such as PR4, an INRZI precoder given by $P(D) = 1/(1 \oplus D)^2$ is often used to convert $(0, G/I)$ codes into bit streams that avoid certain quasi-catastrophic error events (those errors that can occur in a Viterbi decoder truncated to have a finite decision delay). If an RLL$(0, k)$ coded sequence intended to be NRZI precoded is, instead, INRZI precoded, the output satisfies NRZI precoded MTR$(k + 1; k + 1)$ constraints. The MTR$(j; k)$ constraints remove all E2PR4 quasi-catastrophic error events except those that contain repetitions of $\pm\{+1, 0\}$. This particular event can be prevented by eliminating repetitions of the NRZI sequence $\{1, 1, 0, 0\}$ from the encoder output.[8] Cideciyan et al. define a formal constraint t to limit the number of consecutive $\{0, 0\}$ and $\{1, 1\}$ pairs.[9]

The use of the MTR constraint here is justified by the minimum distance gain it provides; however, an MTR constraint can provide additional benefits in the magnetic recording channel. In the presence of intertrack interference, the use of an MTR $j = 3$ constraint has been suggested as a means to reduce the $\pm(+1, -1, +1, -1, +1, \langle-1, +1\rangle)$ error event in EPR4 channels.[10] When transitions in a longitudinal recording system are written with a narrow spacing, the zigzag nature of the transitions allows portions of pairs of transitions to overlap and partially erase the transition pair. With an MTR $j = 2$ coded system, the dibit transition pairs can be recorded with the leading transition written early and the trailing transition written late to mitigate the effects of partial erasure.[11] However, if the increased separation is too large, it will be difficult to resolve the position of adjacent dibits as the two transitions separated by a nontransition are moved closer to one another. To avoid this difficulty, the code can be further constrained to require a minimum of two nontransitions between successive dibit patterns.[9]

Although the MTR $j = 2$ code will provide the desired distance gain, it is not a unique solution. In Figure 18.2, the error event produced by sequence pair (a) can be eliminated with MTR $j = 3$. Pair (b) is generated by a shifted tribit (three consecutive transitions). If these tribits are prevented from starting at either even or odd clock periods, then the decoder can uniquely resolve the correct sequence, and the error event will be prevented.[12,13] Although the time-varying nature of the constraints yields a more complicated detector, the available code rate is higher. These constraints are written as TMTR$(j - 1/j; k)$ to indicate that the MTR constraint alternates between $j - 1$ and j for sequences starting at either an even or odd time index. In fact, this code is also an MTR$(j; k)$ code, but the added constraint of $j - 1$ for transition runs starting on every other code bit period provides the same distance gain as the MTR$(j - 1; k)$ code.

For the strictly low-pass channels characterized by $(1 + D)^n$, the high-order response polynomials are also sensitive to error events with an even number of consecutive alternating signs. In particular, for PR2$(n = 2)$, minimum distance error event is $\{+1, -1\}$. Although there are several pairs of sequences which can produce this error, if an MTR $j = 2$ constraint is employed, this type of error is only produced by a shifted dibit (two consecutive transitions). Just as a time-varying TMTR$(2/3; k)$ constraint can be used to prevent a shifted tribit, so too can a TMTR$(1/2; k)$ constraint be used to prevent a shifted dibit error, thus yielding a 1.76 dB distance gain.[14] A set of codewords suitable for implementing a rate 3/4 TMTR$(1/2; 6)$ code is given by {0001, 0010, 0100, 0101, 0110, 1000, 1001, 1010}. Because the MTR constraint is time-varying for even and odd time indices, choosing codewords with an even length simplifies the design by making the constraints uniquely position dependent within the codeword. Although the use of parity codes is beyond the scope of this discussion, it is worth noting that a parity code combined with an MTR$(1/2; k)$ code can prevent single occurrences of error events with distances up to 3.98 dB from d_{min}.[15]

TMTR codes prevent errors due to shifted transition runs by allowing these patterns to begin only on alternating sample indices. A sufficient condition for avoiding shifted transition run errors is to enforce a constraint that uniquely specifies a valid starting position for the transition run. The even/odd timing requirements of the TMTR code provide this constraint. Another approach, proposed as an alternative to the TMTR$(1/2; k)$ code, is to combine an MTR$(2; k)$ code with a constraint that forces dibits to be preceded by one of either an even or odd number of nontransitions.[16] Herein, this type of code is denoted

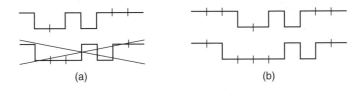

FIGURE 18.3 Pairs of write sequences which produce an error containing a shifted tribit.

MTR($j; k$, odd) if the number of nontransitions preceding a run of j transitions is odd. The same idea can be applied to shifted tribits in an E2PR4 channel with an MTR(3; k, odd) code. An error event with a shifted tribit must be preceded by a single bit error during the period of nontransitions for the error to occur. This condition is illustrated in Figure 18.3; the bottom sequence for the pair labeled (a) is disallowed because it contains an even number of nontransitions before the tribit. In the same figure, the sequence pair labeled (b) shows two valid sequences that can produce an error containing a shifted tribit. Note that the squared distance for this error, $\pm\{+1, 0, 0, +1, -1, +1\}$, in an E2PR4 channel is 10, which is the same as d_{MF}. For isolated transitions and dibits (runs of transitions up to $j - 1$), no constraint is placed on the number of 0s preceding the transition run. A set of codewords which give a rate 3/4 MTR(3; 7, odd) code are given by {00010, 00100, 00101, 00110, 01000, 01010, 01100, 01101, 10000, 10001, 10010, 10100, 10101, 10110, 11000, 11010}.

All the MTR variants discussed to this point will serve to eliminate the targeted error sequences. What varies from one to another is the detector/decoder complexity and available code capacity. The capacity for a set of constraints is an upper bound on the achievable code rate, R. Recording channels, which suffer from ISI are particularly sensitive to code rate because the SNR loss is greater than the $10 \cdot \log_{10} R$ that would be expected if the only penalty were increased noise bandwidth. Assuming that the rate loss is incurred to remove error events with distances less than d_{MF}, the penalty can be computed from Figure 18.1 as a function of density. For convenience, define user density as the number of data bits per PW_{50}, that is, the symbol density is $D_s = D_u/R$. The rate loss penalty for a code with rate $R = 1/2$ increases from 4 dB at user density $Du = 1$ to 5.31 dB at $Du = 2$, and 5.67 dB at $Du = 3$. Ultimately, the penalty will be higher in a real system; the computation here is for a linear ISI channel, and additional rate dependent loss mechanisms such as partial erasure and transition noise are neglected.

To compute capacity, a finite state transition diagram (FSTD) representing the code constraints is constructed. All valid coded bit streams can be obtained by traversing states in the FSTD and concatenating the corresponding edge labels. In addition to computing capacity, the FSTD can be used as the basis for certain code construction techniques.[17] The FSTD for a MTR($j; k$) code is shown in Figure 18.4 where an NRZI format is assumed for the code bits. Note that the dotted lines indicate additional states, like those previous, may be added as necessary to give the proper number of states.

For an FSTD with N states, an adjacency matrix describing the edges is constructed. This is an N-by-N matrix **A**, where each entry a_{ij} is the number of edges from state i to state j. As an example, consider an

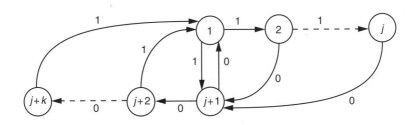

FIGURE 18.4 Finite state transition diagram for an MTR($j; k$) code.

TABLE 18.4   Capacities for MTR($j; k$) Codes

	$j = 1$	$j = 2$	$j = 3$	$j = 4$	$j = 5$	$j = 6$
$k = 1$	0.0000	0.4057	0.5515	0.6174	0.6509	0.6690
$k = 2$	0.4057	0.6942	0.7947	0.8376	0.8579	0.8680
$k = 3$	0.5515	0.7947	0.8791	0.9146	0.9309	0.9388
$k = 4$	0.6174	0.8376	0.9146	0.9468	0.9614	0.9684
$k = 5$	0.6509	0.8579	0.9309	0.9614	0.9752	0.9818
$k = 6$	0.6690	0.8680	0.9388	0.9684	0.9818	0.9881
$k = 7$	0.6793	0.8732	0.9427	0.9718	0.9850	0.9912
$k = 8$	0.6853	0.8760	0.9447	0.9735	0.9865	0.9927
$k = 9$	0.6888	0.8774	0.9457	0.9744	0.9873	0.9934
$k = 10$	0.6909	0.8782	0.9462	0.9748	0.9877	0.9938
$k = 11$	0.6922	0.8786	0.9465	0.9750	0.9879	0.9940
$k = 12$	0.6930	0.8789	0.9466	0.9751	0.9880	0.9941
$k = \infty$	0.6942	0.8791	0.9468	0.9752	0.9881	0.9942

MTR(2; 4) constraint. The adjacency matrix is then

$$
\mathbf{A} =
\begin{bmatrix}
0 & 1 & 1 & 0 & 0 & 0 \\
0 & 0 & 1 & 0 & 0 & 0 \\
1 & 0 & 0 & 1 & 0 & 0 \\
1 & 0 & 0 & 0 & 1 & 0 \\
1 & 0 & 0 & 0 & 0 & 1 \\
1 & 0 & 0 & 0 & 0 & 0
\end{bmatrix}
$$

Given $\mathbf{A}$, capacity is then computed as

$$
C = \log_2 \lambda(\mathbf{A})
$$

where $\lambda(\mathbf{A})$ is the largest real eigenvalue of the matrix $\mathbf{A}$.[18] For the MTR(2; 4) example, the capacity is $C = \log_2(1.7871) = 0.8376$. Capacities for different values of j and k are provided in Table 18.4. For MTR constraints of $j = 4$, the impact to capacity is minimal, and codes with rates of 24/25 can be constructed. Typically, MTR codes are specified with $j < k$, but because j and k represent the maximum number of consecutive 1s and 0s, respectively, the capacity for MTR($j; k$) is the same as for MTR($k; j$).

The TMTR codes were proposed as a means of obtaining the same distance gain, but at a higher capacity. The FSTD for a TMTR($j - 1/j; k$) code is shown in Figure 18.5. In this figure, the states are shown with

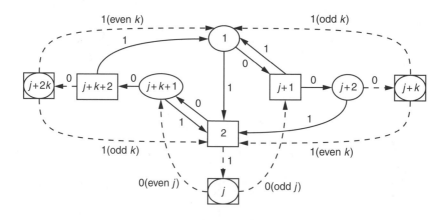

FIGURE 18.5   Finite state transition diagram for a TMTR($j - 1/j; k$) code.

TABLE 18.5　Capacities for MTR($j - 1/j; k$) Codes

	$j = 1/2$	$j = 2/3$	$j = 3/4$	$j = 4/5$
$k = 1$	0.0000	0.5000	0.5840	0.6358
$k = 2$	0.5706	0.7507	0.8170	0.8482
$k = 3$	0.6804	0.8423	0.8974	0.9231
$k = 4$	0.7381	0.8802	0.9312	0.9543
$k = 5$	0.7619	0.8983	0.9466	0.9685
$k = 6$	0.7764	0.9070	0.9540	0.9753
$k = 7$	0.7831	0.9115	0.9576	0.9786
$k = 8$	0.7874	0.9137	0.9594	0.9802
$k = 9$	0.7894	0.9149	0.9604	0.9810
$k = 10$	0.7908	0.9156	0.9608	0.9814
$k = 11$	0.7915	0.9159	0.9611	0.9816
$k = 12$	0.7919	0.9161	0.9612	0.9817
$k = \infty$	0.7925	0.9163	0.9613	0.9818

circles or squares to indicate alternating clock periods, that is, the squares represent an odd time index and the circles an even time index, or vice versa. The end states for runs of 0s and ones are determined by whether k and j are even or odd. To be valid, edges can only connect from squares to circles and from circles to squares. Table 18.5 lists the capacities for different values of j and k. Of interest to E2PR4 channels are the TMTR($2/3; k$) codes. For large values of k, the capacity of these codes approaches 0.9163, which is significantly better than the 0.8791 attained by MTR($2; k$) codes. For PR2 channels, the MTR $j = 1$ (or RLL $d = 1$) codes are limited to a capacity of 0.6942, while TMTR($1/2; k$) codes are available with capacities up to 0.7925.

Even greater capacities are available from the MTR($j; k$, odd) codes where an odd number of 0s must precede a run of j 1s. The state diagram for this code is shown in Figure 18.6. If the RLL k constraint is an

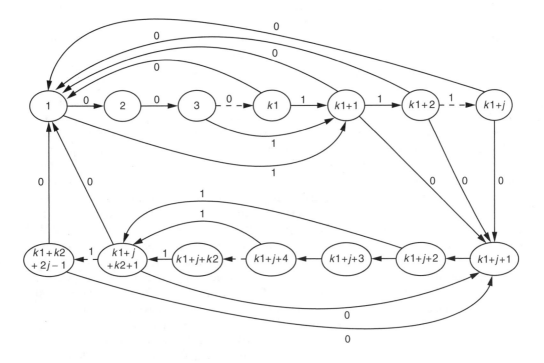

FIGURE 18.6　Finite state transition diagram for an MTR($j; k$, odd) code.

TABLE 18.6   Capacities for MTR($j; k$, odd) Codes

	$j = 2$	$j = 3$	$j = 4$	$j = 5$
$k = 1$	0.0000	0.0000	0.0000	0.0000
$k = 2$	0.6054	0.7618	0.8232	0.8510
$k = 3$	0.7366	0.8569	0.9049	0.9263
$k = 4$	0.7720	0.8903	0.9363	0.9566
$k = 5$	0.7996	0.9089	0.9519	0.9709
$k = 6$	0.8086	0.9165	0.9588	0.9774
$k = 7$	0.8161	0.9211	0.9625	0.9807
$k = 8$	0.8187	0.9231	0.9642	0.9823
$k = 9$	0.8210	0.9243	0.9651	0.9831
$k = 10$	0.8218	0.9248	0.9656	0.9835
$k = 11$	0.8225	0.9252	0.9658	0.9837
$k = 12$	0.8227	0.9253	0.9659	0.9838
$k = \infty$	0.8232	0.9255	0.9661	0.9839

odd number, then $k_1 = k$, and $k_2 = k - 1$; otherwise, $k_1 = k - 1$, and $k_2 = k$. In any case, $k_1 + k_2 = 2k - 1$, and the total number of states is $2(k + j - 1)$. Capacities for different parameter values are listed in Table 18.6. For E2PR4 channels, an MTR(3; k, odd) code provides capacities up to 0.9255 compared with 0.9163 for the TMTR(2/3; k) code. The increase in capacity is even more dramatic for PR2 channels where an MTR(2; k, odd) code has capacities up to 0.8232 compared with 0.7925 for TMTR(1/2; k).

18.4   Detector Design for MTR Constraints

The code constraints considered to this point provide a distance gain by allowing sequences that contain, at most, one type of bit sequence involved in generating an error event. However, it is the detector which must determine the correct sequence from the receive signal waveform. In order to resolve the ambiguity about which pair of error-generating sequences was actually written, the detector must be modified to prevent detection of the disallowed sequences.

A common detector choice is MLSD, implemented with a Viterbi detector.[19] If the channel response has length L time-samples (e.g., $L = 5$ for E2PR4), each detector state may correspond to a sequence with up to $L - 2$ transitions. Therefore, if $L > j + 2$, at least two states in the detector will correspond to forbidden sequences. Considering edges from previous states, one additional transition can be implied, so for $L > j + 1$, at least two edges in the trellis will correspond to illegal sequences. In the case of a TMTR code, these illegal states and edges will also be time-varying to match the code constraints.

As an example, consider the $N = 2^{L-1} = 16$ state trellis section shown in Figure 18.7. For the TMTR(2/3; k) code, the states with the NRZ sequence labels 0101 and 1010, which correspond to three consecutive transitions, are illegal during time periods for which only sequences of two transitions are allowed. At all time periods, transitions between 0101 and 1010 are removed because they correspond to four consecutive transitions. By removing these states and transitions, the Viterbi path memory can only contain sequences that are permitted by the TMTR constraints. For a static MTR(2; k) constraint, the states 0101 and 1010 are always illegal. The additional states and edges removed by $j = 2$ are shown with dotted lines in the figure. The MTR($j; k$, odd) codes require a slightly different approach. Accommodating the constraint within the trellis structure by pruning states and edges would require a 2^{j+k+1} state trellis such that a sequence with a transition, followed by k nontransitions and then j transitions is represented. Clearly, this is unreasonable for large values of k. Alternatively, for states corresponding to j consecutive transitions, the detector can look back through the path memory associated with the edges leading into the state in question. If the number of nontransitions preceding the j-transition run is even, the edge is disallowed.

Fixed delay tree search with decision feedback (FDTS/DF) is another detector structure that has been suggested for use with MTR coded system. This detector uses a decision feedback equalizer (DFE) to remove ISI due to symbols beyond a chosen truncation point in the response. A distance metric, such as is

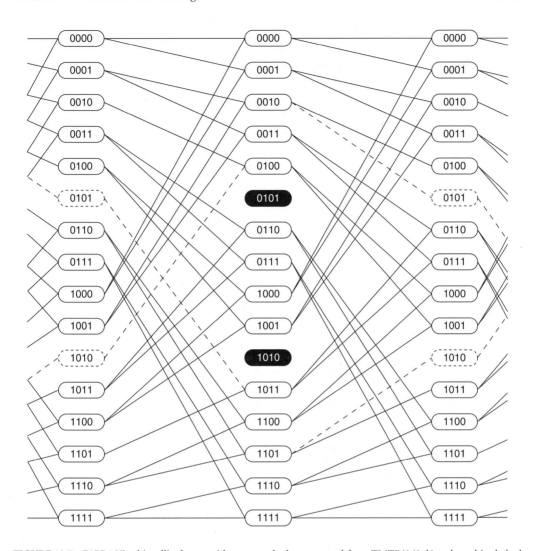

FIGURE 18.7 E2PR4 Viterbi trellis shown with states and edges removed for a TMTR$(2/3; k)$ code and in dashed lines for an MTR$(2; k)$ code.

used with MLSD, is computed for possible sequences constructed from the remaining ISI terms. Decisions are made at a fixed time-delay determined by the tree depth or number of ISI terms considered. For an MTR $j = 2$ code and FDTS/DF with $t = 2$ (sequence estimation using the $f_0 + f_1 D + f_2 D$ portion of the channel response), Figure 18.8 shows the relevant sequences in the tree search eliminated by the code. As shown, the previous decision is used to dynamically deselect one of two detector paths, thereby preventing the illegal tribit patterns.

18.5 Simulation Results

The net SNR gain for a particular MTR coding scheme can be estimated by subtracting the rate loss penalty from the distance gain. Because this value is only an approximation, simulations are used to provide a more accurate result. Here, the coding gain on a Lorentzian channel equalized to an E2PR4 target is examined. Although it is the difference in SNR that is of interest here, the definition used in the plot is the squared peak of the isolated transition response to the integrated noise power in the $1/PW_{50}$ frequency band.

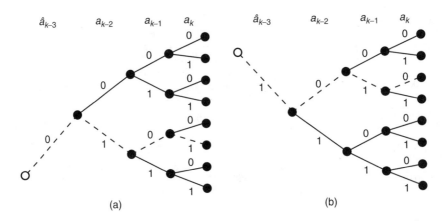

FIGURE 18.8   Fixed delay tree search detector shown with sequences disallowed for an MTR $j = 2$ code.

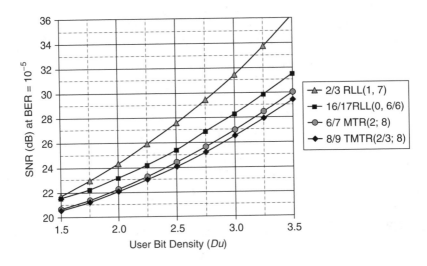

FIGURE 18.9   Simulation results for an E2PR4 equalized Lorentzian channel.

For several types of codes, the SNR required to obtain a bit error rate (BER) of 10^{-5} is shown as a function of user density in Figure 18.9. To isolate the distance gain of the code, ideal clock synchronization and gain control are assumed. The receive signal is equalized to the E2PR4 response with a relatively long FIR filter designed to minimize the mean squared error at the equalizer output.

A rate 16/17 RLL(0,6/6) code is used in lieu of an uncoded reference. From the figure, it is clear that although the RLL(1,7) code provides a significant distance gain, its code rate loss exceeds the distance gain, giving a net loss. Of course, such a code is often chosen to mitigate transition noise and nonlinearities, which are not reflected in this model. At a user density of $D_u = 2.5$, the rate 6/7 MTR codes provides a net gain of about 1 dB while, as expected, the rate 8/9 TMTR provides a slightly larger gain.

18.6   Summary

Performance in recording channels, subject to severe intersymbol interference, is limited by error event sequences that produce a minimum distance less than the matched filter bound. High density Lorentzian, PR2, and E2PR4 channels are all subject to error events that can be avoided by properly constraining the maximum number of consecutive transitions. A variety of code constraints with varying capacities can be formed around this basic idea. For a simple linear channel with additive Gaussian noise, these codes provide a net performance gain, despite the fact that the MTR constraint reduces the code rate.

References

[1] Behrens, R. and Armstrong, A., An advanced read/write channel for magnetic disk storage, *Proc. 26th Asilomar Conf. on Signals, Systems, and Computers*, 956, 1992.

[2] Karabed, R. and Siegel, P. H., Coding for higher order partial response channels, *Proc. SPIE Int. Symp. on Voice, Video, and Data Commn.*, 2605, 115, 1995.

[3] Karabed, R., Siegel, P. H., and Soljanin, E., Constrained coding for binary channels with high intersymbol interference, *IEEE Trans. Inform. Th.*, 45, 1777, 1999.

[4] Messerschmitt, D. G., A geometric theory of intersymbol interference, Part II: Performance of the maximum likelihood detector, *Bell Syst. J.*, 52, 1973.

[5] Moon, J. and Carley, L. R., Efficient sequence detection for intersymbol interference channels with run-length constraint, *IEEE Trans. Commn.*, 42, 2654, 1994.

[6] Altekar, S. A. et al., Error-event characterization on partial-response channels, *IEEE Trans. Info. Th.*, 45, 241, 1999.

[7] Xu, C. and Keirn, Z., Error event analysis of EPR4 and ME2PR4 channels: Captured head signals versus Lorentzian signal models, *IEEE Trans. Magn.*, 36, 2200, 2000.

[8] Brickner, B. *Maximum Transition Run Coding and Pragmatic Signal Space Detection for Digital Magnetic Recording*, Ph.D. Thesis, University of Minnesota, 1998.

[9] Cideciyan, R. D., et al., Maximum transition run codes for generalized partial response channels, *IEEE J. Sel. Areas Commn.*, 19, 619, 2001.

[10] Soljanin, E., On-track and off-track distance properties of class 4 partial response channels, *Proc. SPIE Int. Symp. on Voice, Video, and Data Commn.*, 2605, 92, 1995.

[11] Brickner, B. and Moon, J., Combatting partial erasure and transition jitter in magnetic recording, *IEEE Trans. Magn.*, 36, 532, 2000.

[12] Bliss, W. G., An 8/9 rate time-varying trellis code for high density magnetic recording, *IEEE Trans. Magn.*, 33, 2746, 1997.

[13] Knudson Fitzpatrick, K. and Modlin, C. S., Time-varying MTR codes for high density magnetic recording, *Proc. IEEE Global Telecom. Conf.*, 1997.

[14] Moision, B. E., Siegel, P. H., and Soljanin, E., Distance-enhancing codes for digital recording, *IEEE Trans. Magn.*, 34, 69, 1998.

[15] Brickner, B. and Padukone, P., Partial Response Channel Having Combined MTR and Parity Constraints, U.S. Patent 6,388,587, 2002.

[16] Liu, Pi-Hai, Distance-Enhancing Coding Method, U. S. Patent 6,538,585, 2003.

[17] Marcus, B., Siegel, P. H., and Wolf, J. K., Finite-state modulation codes for data storage, *IEEE J. Sel. Areas Commun.*, 10, 5, 1992.

[18] Shannon, C. E., A mathematical theory of communication, *Bell Syst. Tech. J.*, 27, 379, 1948.

[19] Forney, D. G., Maximum-likelihood sequence estimation of digital sequences in the presence of intersymbol interference, *IEEE Trans. Inform. Th.*, 18, 363, 1972.

19

Spectrum Shaping Codes

19.1 Introduction **19**-1
19.2 Recording System and Spectrum Shaping Codes..... **19**-2
19.3 Dc-free Codes.................................... **19**-2
Introduction • Dc-free Constraint Sequences • Capacity of dc-free Constraint • Spectral Characteristics of dc-free Constraint • Encoding and Decoding of dc-free Constraints
19.4 Codes with Higher Order Spectral Zeros **19**-9
Introduction • Sequences with Higher Order Zeros at $f = 0$ • K-RDS$_f$ Sequences • K-RDS$_f$ Sequences on Partial Response Channels
19.5 Composite Constrained and Combined Encoding ... **19**-15
19.6 Conclusion **19**-18

Stojan Denic
University of Ottawa
Ottawa, ON

Bane Vasic
University of Arizona
Tucson, AZ

19.1 Introduction

The first application of spectrum shaping codes was related to digital communication systems that used transformers to connect two communication lines. Because transformers do not convey dc-component, and suppress low frequency components, direct transmission of source signals whose power spectral densities contain these frequency components were not possible without significant distortion. That is why dc-free or dc-balanced codes were devised [1–3, 10–13]. Their role is to transform a source sequence into a channel sequence whose spectral characteristic corresponds to spectral characteristic of communication channel. At the end of communication line, the sequence is received by the decoder that generates original sequence without errors in the case of noiseless channel. In recording systems, that can be modeled as any communication system, this kind of codes have been widely used. For instance in digital audio tape systems, they prevent write signal distortion that can occur due to transformer-coupling in write electronics [2]. In optical recording systems they are used to circumvent the interference between recorded signal, and servo tracking system. Further development of spectrum shaping codes for recording systems was driven by requirements for better codes in the sense of larger rejection of low frequency components. The codes providing this feature are codes with higher order spectral zero at $f = 0$, and can be found in [14, 16, 17]. Although the width of suppressed frequencies of these codes is smaller than in the case of dc-balanced codes, the rejection in the vicinity of $f = 0$ is significantly larger. Another class of spectrum shaping codes were invented in order to support the use of frequency multiplexing technique for track following [1], and partial response technique for high density data storage [15]. Both techniques require the spectral nulls of the recorded signal at frequencies that can be different than $f = 0$ in order to enable reliable data storage. The typical example of such codes are codes that have spectral zeros at submultiple of channel symbol frequency.

0-8493-1524-7/05/$0.00+$1.50
© 2005 by CRC Press, LLC

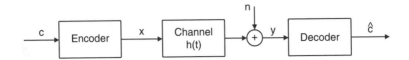

FIGURE 19.1   Recording system.

For this type of codes see, for example, [18–20]. The fourth characteristic group of spectrum shaping codes are those that give rise to spectral lines. Their purpose is to give the reference information to the head positioning servo system that positions and maintains the head accurately over a track in digital recorders [1].

Besides spectral constraint, recorded sequences have to comply with certain time constraints. That is why, it is interesting to say something about compound codes, which generate sequences that in the same time satisfy more than one constraint. Typical representatives are RLL (runlength limited) dc free codes. RLL dc-free sequences have confined minimal, and maximal consecutive like channel symbols, and in the same time, their spectrum has zero at $f = 0$, [21, 22]. Important classes of spectrum shaping codes are dc-free error correcting codes. It was mentioned earlier that the decoder of dc-free code will decode channel sequence without errors if the channel is noiseless. Because a recording channel is not noiseless one, a lot of effort was put into design of dc-free error correcting codes, both block and convolutional codes. The examples of these codes can be found in [27–30].

The goal of this chapter is to give a survey of spectrum shaping codes for digital recording systems from the theoretical, and practical point of view. The organization of the article is as follows. Section 19.2 contains short description of recording system with respect to the role of spectrum shaping codes. In Section 19.3, dc-free codes are considered. Some theoretical basis for studying of dc-free codes are given, which will be important for studying of all types of spectrum shaping codes [4–8]. Section 19.4 discusses codes with higher order spectral zeros at $f = 0$, and codes with zeros at frequencies that are submultiple of channel symbol frequency. In Section 19.5, certain compound constraint codes are mentioned. The specific example of RLL-dc code is given. All codes are considered from the point of view of maxentropic sequences. The channel capacity of maxentropic sequences are computed, and corresponding power spectral densities are given. Also, the basic encoding techniques are described.

19.2   Recording System and Spectrum Shaping Codes

The encoder of spectrum shaping code is the last in the chain of encoders preceding the recording channel, and the decoder of spectrum shaping code is the first in the chain of the decoders that follows the recording channel. As shown in Figure 19.1, the encoder receives a symbol stream $c = \{c_k\}_{k=0}^{\infty}$ from an error correcting code encoder, and transforms it into the stream of channel symbols $x = \{x_k\}_{k=0}^{\infty}$ that matches the spectral characteristics of the recording channel with impulse response $h(t)$. $n(t)$ is additive noise. The decoder accepts data stream from the channel $y = \{y_k\}_{k=0}^{\infty}$, and transforms it to the symbol stream $\hat{c} = \{\hat{c}_k\}_{k=0}^{\infty}$ that is the noisy version of c, and fed to the error correcting code decoder. From the point of view of error correcting code encoder, and decoder, the spectrum shaping code encoder, and decoder are merely part of the recording channel. One can also define the code rate R of the spectrum shaping codes. In the case of block codes, the input stream c is divided into the sourcewords of length k, which are encoded in codewords of length n forming the output stream x. The coding rate is defined as the ration of the word length at the input k, and the codeword length n, $R = k/n$.

19.3   Dc-free Codes

19.3.1   Introduction

Dc-free codes belong to a class of spectrum shaping codes that transform the spectrum of the input sequence into an encoded sequence whose spectrum has zero at the zero frequency. Dc-free codes emerged first in digital communications. The intrinsic part of communication lines were coupling devices whose characteristic as well as the characteristic of digital recording channels is that they suppress the low frequency components of the transmitted signal or recorded data respectively. This implies the necessity of a signal processing techniques, which will reshape the spectrum of the original source sequence, and in that way match the characteristics of the communication or recording channels. Not only that dc-free codes give zero at dc, but also the low frequency components of the encoded sequence are suppressed depending on the choice of code parameters. But this is not the only reason for employing dc-free codes. The other reason for using dc-free codes is that the optical data storage systems use high-pass filters to diminish the effect of dirt on system performance caused for example by fingerprints on an optical medium. Another reason is to prevent the mutual interference between recorded data, and servomechanism for track tracking that operates at a low frequency [1]. All those arguments indicate the importance of dc-free codes for data recording systems.

Dc-free codes attracted considerable attention of the data storage community, and a number of papers, and patents have been published till now, and a large number of references can be found in [2].

This part of the chapter is organized in the following manner. In the Section 19.3.2, theoretical background of dc-free constraint sequences is given. Section 19.3.3 considers the finite state transition diagram (FSTD) description of dc-free constraints and method for computing the noiseless channel capacity of the constrained channel. Section 19.3.4 presents power spectral density (PSD) characteristics of dc-fee constraints, and discusses the most important parameters that determine PSD, and channel capacity, and their mutual dependence. Section 19.3.5 contains survey of simple and well-known coding techniques for generating dc-free sequences.

19.3.2   Dc-free Constraint Sequences

The starting point for analysis and design of dc-free codes is the result by Pierobon [3]. Before stating the result of Pierobon the notion of running digital sum is defined. The running digital sum (RDS) at moment n, z_n, of sequence of symbols $x = \{x_k\}_{k=0}^{\infty}$ is defined as

$$z_n = \sum_{k=0}^{n} x_k \tag{19.1}$$

Pierobon proved that the power spectral density $S_x(f)$ of a sequence x vanishes at zero frequency, regardless of the sequence distribution, if and only if the running digital sum $z = \{z_k\}_{k=0}^{\infty}$ of the encoded sequence is bounded, that is, $|z_n| \leq N$, for each n, when $N < \infty$. This condition describes the constraint posed on the sequences that are eligible for recording on a recording medium. In other words, any sequence that violates the condition on RDS will not be permitted in the recording channel, and sequences that satisfy this constraint are called dc-free sequences. It follows that recording channels are channels with the constraint that is called dc-free constraint. Further, the effect of RDS is twofold; first, it affects the shape of power spectral density of constrained sequences, and second the upper bound N on RDS determines the capacity of a constrained recording channel [1]. In general case, the capacity of noiseless constrained channels was defined by Shannon [4] as

$$C = \lim_{T \to \infty} \frac{\log_2 n(T)}{T} \tag{19.2}$$

$n(T)$ is total number of admissible sequences, and T is sequence duration. The channel capacity determines the maximal theoretical code rate R of an constrained code. It will be explained later that the upper bound of RDS, N, has contradictory effects on the channel capacity, and desired power spectral density of recording sequences so that code design is a compromise between two opposite requirements.

19.3.3   Capacity of dc-free Constraint

In order to compute the channel capacity of dc-free constrained channels, the notion of digital sum variation (DSV) is introduced. If N_{max} is the maximum value of z, and N_{min} is its minimum value then digital sum variation is defined as $N = N_{max} - N_{min} + 1$. Actually, DSV is equal to the number of different values that z can take. Intuitively, it can be seen that the total number of admissible sequences $n(T)$ depends on DSV, that is, the larger N, the larger $n(T)$, and the larger channel capacity C.

The common tool for description of constrained sequences is a finite state transition diagram (FSTD). It was shown by Shannon [4] that the FSTD description of a constrained channel can be used for computation of its channel capacity. In this section that result is used to determine the channel capacity of M-ary dc-free constraint, where M is the number of levels that are permitted in the recording channel [5]. One example of FSTD representing dc-free constraint is given in Figure 19.2. It depicts M-ary dc-free constraint when the channel symbol alphabet is $\{-(M-1)/2, \ldots, -1, 0, 1, \ldots, (M-1)/2\}$, M odd. If M is even then the channel symbol alphabet can be chosen as $\{-(M-1), -(M-3) \ldots, -1, 0, 1, \ldots, (M-3), (M-1)\}$. The state of the FSTD represents the value of RDS, z_k, at time instant k. An edge between two states represents the transition between states, while a label above the edge denotes the input symbol generated during the transition between states. The number of states is equal to DSV, N. To compute the channel capacity, the connection matrix of FSTD is introduced. The connection matrix D of FSTD is a square matrix of dimension N, where the entry d_{ij} of matrix D represents number of edges emanating from state i, and ending at state j. The connection matrix for FSTD shown in Figure 19.2 is

$$D = \begin{bmatrix} 1 & 1 & 0 \\ 1 & 1 & 1 \\ 0 & 1 & 1 \end{bmatrix} \tag{19.3}$$

On the other hand, because the transition from one state to another depends only on the current state, each FSTD may be assigned corresponding Markov chain $s = \{s_k\}_{k=0}^{\infty}$, where each state of FSTD is related to a value that Markov chain can take. The finite number of values that Markov chain takes is equal to the number of states of FSTD. Each edge is assigned a transition probability p_{ij}, from state i to state j, where $i, j \in \sum$, where $\sum$ is the set of all states of Markov chain. The measure of uncertainty or measure of information generated by a Markov chain is equal to the entropy of Markov chain. It is calculated as

$$H(X) = \sum_{i=1}^{N} p_i H_i \tag{19.4}$$

where p_i is probability of state $i (i = 1, \ldots, N)$, and H_i is the entropy of the state i, that is, uncertainty of being in state $i (i = 1, \ldots, N)$ at any time instant. It is defined as the entropy of the transition probabilities p_{ij}, $H_i = \sum_{j=1}^{N} p_{ij} \log \frac{1}{p_{ij}}$.

It was proven by Shannon [4] that channel capacity of constrained noiseless channel is

$$C = \max H(X) = \log_2 \lambda_{max} \tag{19.5}$$

λ_{max} is the maximum eigenvalue of connection matrix D of FSTD representing channel constraint. The Equation (19.5) connects the channel capacity of the constrained channel to maximum entropy notion.

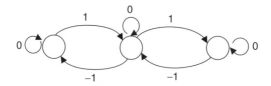

FIGURE 19.2   FSTD for $M = 3$, $N = 3$ dc-free constraint.

TABLE 19.1 Capacity of dc-free Constraint

N	M=2	M=3	M=4	M=5	M=6	M=7	M=9
3	0.5	1.2716	—	1.5850	—	—	—
4	0.6492	1.3885	1	1.8325	—	2	—
5	0.7925	1.4500	1.2925	1.9765	—	2.2159	2.3219
6	0.8495	1.4864	1.4500	2.0642	1.5850	2.3559	2.5114
7	0.8858	1.5098	1.5665	2.1223	1.7952	2.4529	2.6423
8	0.9103	1.5258	1.6508	2.1626	1.9227	2.5211	2.7388
9	0.9276	1.5371	1.7111	2.1919	2.0283	2.5713	2.8117
10	0.9403	1.5455	1.7573	2.2137	2.1085	2.6094	2.8672

It shows that maximum entropy Markov chain generates maximal number of sequences of certain length. The constrained sequences generated by a Markov chain having maximum entropy are called maxentropic sequences. Table 19.1 contains the channel capacities for some M-ary dc-free constraints. Table 19.1 shows that channel capacity increases when two degrees of freedom DSV, and M increase. For $M = 2$, the closed form expression for channel capacity was derived [1]. The derivation is based on the recurrent relation that exists between characteristic polynomials of connection matrices D_N, D_{N-1}, D_{N-2}, where the subscripts denote the DSV of the corresponding channels described by connection matrices. The formula for channel capacity is

$$C(N) = \log_2 2 \cos \frac{\pi}{N+1} \tag{19.6}$$

The results obtained by Equation 19.6 agree with those found in Table 19.1.

19.3.4 Spectral Characteristics of dc-free Constraint

In Section 19.3.2, a necessary and sufficient condition for the null of a power spectral density at $f = 0$ is introduced. In this section the importance of variance of RDS will be considered, and the power spectral densities of maxentropic dc-free sequences will be shown.

In [6], Justesen derived interesting relation between the sum variance (variance of RDS) $E[z_k^2] = \sigma_z^2(N)$ where N is a digital sum variation, and so-called cut-off frequency ω_0

$$2\sigma_z^2(N)\omega_0 \approx 1 \tag{19.7}$$

The cut off frequency is defined as the value of the frequency $\omega_0 = 2\pi f_0$, such that $S_x(f_0) = 0.5$. The cut off frequency determines the bandwidth, called notch width, from $f = 0$ to $f = f_0$, within which the power spectral density of recorded sequence $S_x(f)$ is low. The performance is better if the cut-off frequency f_0 is larger giving the larger notch width. On the other hand, from Equation 19.7, it is clear that the larger cut-off frequency f_0, the smaller sum variance $\sigma_z^2(N)$. Having in mind that the sum variance $\sigma_z^2(N)$, and the channel capacity $C(N)$ both decrease as N decreases, it can be assumed that the capacity of the constrained channel $C(N)$ is smaller as cut-off frequency f_0 grows, implying that the redundancy $1 - C(N)$ is bigger [1]. The previous discussion also points out the equal importance of redundancy $1 - C(N)$, and sum variance $\sigma_z^2(N)$ as performance measures for spectrum shaping codes. That is why in [1], the new measure of performance for spectrum shaping codes was introduced as the product of redundancy, and sum variance $(1 - C(N))\sigma_z^2(N)$. It was shown that this product is tightly bounded, from below, and above for $N > 9$ in the case of maxentropic dc-free sequences. This shows the significance of the relation Equation 19.7, that reveals the conflicting demands posed on the constrained sequences. If one wants to get better suppression of low frequency components of power spectral density, one must pay with bigger redundancy, that is, code inefficiency. Although the previous result is derived for binary recording channels, the conclusions are true for $M > 2$.

In what follows, the power spectral densities for some maxentropic M-ary dc-free sequences will be computed. In order to compute the power spectral density of the constrained channel, FSTD of Moore type, and corresponding Markov chain representation of the constrained channel are used. For example,

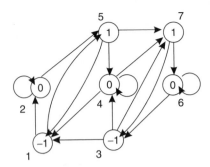

FIGURE 19.3   Moore type FSTD for $M = 3$, $N = 3$ dc-free constraint.

in Figure 19.3, the FSTD of Moore type for $M = 3$, $N = 3$, dc-free constrained channel is depicted. The characteristic of the Moore type FSTD, as opposed to the FSTD found in Figure 19.2, is that all edges entering the same state have the same label, that is, the same channel symbol is generated by visiting particular state. In this example the alphabet of channel symbols is $\{-1, 0, 1\}$. If the initial state is $s_0 = 4$, and $z_0 = 0$, then $z_k \in \{-1, 0, 1\}$ meaning $N = 3$. It can be seen that states 1, and 2 correspond to $z_k = -1$, states 3, 4, and 5 to $z_k = 0$, and 6, and 7 to $z_k = 1$.

Associated with FSTD, there exists a Markov chain generating maxentropic M-ary dc-free sequences with transition probabilities given by [4]

$$p_{ij} = \frac{1}{\lambda_{\max}} d_{ij} \frac{v_j}{v_i} \tag{19.8}$$

The quantities v_i, and v_j are the entries of the right eigenvector corresponding to the maximal eigenvalue $\lambda_{\max}$ of the connection matrix D. The methods for computing power spectral density $S_x(f)$ of Markov sources were proposed in [7, 8], and are used here in order to get the spectra of maxentropic M-ry dc-free sequences. The equation for continuous part of power spectral density of Markov source is given as

$$S_x(f) = C_x(0) + 2 \sum_{k=1}^{\infty} C_x(k) \cos 2\pi k f \tag{19.9}$$

where $C_x(k), k = 0, 1, 2, \ldots$, is autocovariance function of generated sequence x, and can be written as

$$C_x(k) = \xi^T \Pi (P^{|k|} - P_\infty)\xi \tag{19.10}$$

In Equation 19.10, each entree of vector ξ, which has as many entrees as there are states of Markov chain, represents symbol emitted when particular state of Markov chain is visited. Further, matrix Π is a diagonal matrix whose diagonal elements are steady state probabilities of Markov chain, P is a transition probability matrix, and each row of P_∞ is equal to the steady state probability vector of Markov chain. The importance of maxentropic sequences is that their power spectral density corresponds to the power spectral density of the codes whose code rate R is near the channel capacity C, and consequently can be used as a good approximation of the spectrum of those codes. The power spectral density of maxentropic sequences does not depend on the specific encoder and decoder realization, and therefore it is easier to compute than the spectrum of the specific code.

Figure 19.4 shows PSD of maxentropic sequences for $M = 3$, and $N = 3, 4, 5$ versus normalized frequency f where normalization is done by $f_s = 1/T_s$, where T_s is a signaling interval. It is interesting to check the validity of the formula Equation 19.7 for previously computed PSD. For instance, the normalized cut-off frequency for $M = 3, N = 4$, is $f_0 = 0.1$, $\omega_0 = 0.6283$, and by formula Equation 19.7, approximate sum variance is 0.7958. The formula for exact value of the sum variance is given in [6], according to which $\sigma_z^2 = -\sum_{k=1}^{\infty} k R_x(k) = 0.8028$, meaning that Equation 19.7 provides good approximation for relation between sum variance and cut-off frequency.

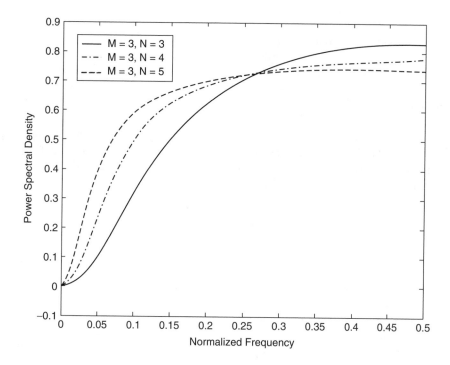

FIGURE 19.4 Continuous part of the power spectral density of $M = 3, N = 3, 4, 5$ maxentropic sequences.

19.3.5 Encoding and Decoding of dc-free Constraints

Finite state transition diagrams of dc-free constraints, and their maxentropic Markov chains are of great importance. They are analytical tools that provide a number of parameters, such as channel capacity, cut-off frequency, sum variance, representing theoretical bounds of dc-free constraints. Those parameters have a practical meaning for the designers of encoders and decoders of dc-free codes. But these analytical tools do not give a recipe how to build practical encoders, and decoders. This section considers the standard solutions for realization of encoders and decoders of dc-free constraints. To determine the quality of particular encoder for constraint channels, two parameters are introduced. The first, called code rate efficiency, is defined as the ratio of code rate R, and channel capacity C, $\eta = R/C$. While the first one is commonly used for any channel coding scheme, the second was specially defined for dc-free codes [1]. The encoder efficiency E is

$$E = \frac{(1 - C(N))\sigma_z^2(N)}{(1 - R)s_z^2} \tag{19.11}$$

The encoder efficiency compares the product of redundancy, and sum variance of maxentropic sequence for particular channel, and product of redundancy, and sum variance of specific encoder for the same channel. The maxentropic sequence was taken as a reference because it achieves the channel capacity of dc-free constraint channel. The importance of sum variance was explained in the previous section.

In general, encoders, and decoders can be divided into two groups, state independent and state dependent encoders and decoders. State independent encoders are usually realized by a look-up tables where there exists one-to-one correspondence between sourcewords and codewords. As opposed to state independent encoders, state dependent encoders are designed as a synchronous finite state machines [9] where the next codeword is the function of a current internal state of the encoder, and a current sourceword. The advantage of state dependent encoding is that sometimes more efficient codes can be constructed (in terms of code rate) [1] with shorter codewords.

Similarly, the output of state dependent decoder depends on the decoder current state, the current input codeword, as well as finitely many upcoming codewords [9]. The weakness of this type of decoders is that if an error occurs due to the noisy channel the decoder could lose track of states. In turn, this can lead to the series of errors, that is, error propagation. In order to prevent this kind of event state independent encoders are introduced. The decoders output depends only on the certain number of preceding codewords, current codeword, and certain number of upcoming codewords. The name for such type of decoders is sliding window decoders [9].

Here, both types of encoders will be presented. It is assumed that recording channel is binary channel with channel alphabet $\{-1, 1\}$. The main idea behind dc-free encoder realization is the fact that the RDS of channel sequence has to be bounded. Having that in mind, the notion of codeword disparity is introduced. The disparity d of codeword $\mathbf{x} = (x_1 x_2 \ldots x_n)$ of length n is defined as

$$d = \sum_{i=1}^{n} x_i \tag{19.12}$$

The simplest idea is to use only codewords that have zero disparity. Each sourceword is assigned a unique codeword of zero disparity. It means that RDS at the end of each codeword will be equal to zero, if $z_0 = 0$. This type of encoding is called zero disparity encoding, and it is obviously state independent. It is simple, but the main shortcoming is inefficiency in terms of code rate R for a fixed codeword length n, because the number of possible codewords with zero disparity of length n is finite, $\binom{n}{n/2}$. The larger efficiency η can be achieved by increasing the codeword length n, which in the same time makes the look-up table realization of encoder more complex.

Another way for achieving better efficiency is to use so-called low disparity codes. The drawback of this technique is the increase of the power of low frequency components as compared to zero disparity coding. The codes that belong to this class of encoding can be found in [10]. Again the goal is to keep the value of RDS within some prescribed bounds. In addition to zero disparity codewords, it is allowed to use the codewords with low disparity. As explained in [1], let S_+ denote all codewords with positive disparity, and let S_- denote all codewords with negative disparity. S_+ is the union, $S_+ = \cup_{j=0}^{K} S_j$, $K \leq n/2$, where S_j is the set of all codewords with disparity $d = 2j$. The set S_- is obtained by inverting codewords of set S_+. $K = 0$ represents the case of zero disparity encoding. If $K = 1$, each sourceword is assigned a pair of codewords of opposite disparity or a codeword of zero disparity. In process of encoding, the value of RDS is tracked. Because the encoder has a choice of two codewords to send to the channel, it chooses one whose disparity minimizes absolute value of RDS. Actually, the encoder has two codebooks, and chooses the codeword from one codebook such that an instantaneous value of RDS is within some bounds.

The polarity bit coding is yet another simple method that generates dc-free sequences [11, 12]. One extra bit, called polarity bit, is added to $n - 1$ source bits comprising codeword of length n. The polarity bit is set to 1. If the disparity of the codeword has the same sign as RDS at the moment of sending the codeword, the inverted codeword is recorded. Otherwise original codeword is recorded. The decoder, based on the polarity bit, recognizes if the codeword was inverted or not.

In [1], the encoder efficiency E versus codeword length n of above coding techniques is shown. The conclusion can be drawn that low disparity codes outperform zero disparity codes, and zero disparity codes outperform polarity bit codes. Also, for small codeword lengths, low, and zero disparity codes have unit efficiency, while the efficiency drops as the codeword length increases.

It is worth to mention two more coding techniques. Their realizations do not rely on look-up tables so that they are convenient if better efficiency is necessary. One is based on enumerating the set of sequences denoted by $T = T(z_n, n, N, z_0)$ [1], where n is a sequence length, N is the maximal value of RDS within the sequence, z_n and z_0 are final and initial value of RDS. It is understood that the minimal value of RDS is 1. The enumeration algorithm establishes one to one correspondence between set T and the set of integers $0, 1, \ldots, |T| - 1$. The enumeration represents the foundation of decoding algorithm. Also algorithm that performs reverse operation was derived, that enables mapping from the set of integers to the set of constrained sequences. The second technique uses the fact that each k-bit binary sourceword can be divided in

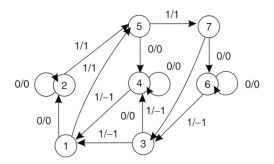

FIGURE 19.5 Encoder for $M = 3$, $N = 3$ dc-free constraint.

TABLE 19.2 Decoding Table
for M=3, N=3, dc-free Code

Window	Decoded Bit
1	1
0	0
−1	1

two segments, each having equal disparity [13]. The zero disparity codeword is obtained by inverting one of two segments. Additional l bits are used to mark the position splitting these two segments. Those additional l bits are usually coded with zero disparity word. It follows that code rate is $R = k/(k + l)$. The encoding is possible because there always exists one to one correspondence between sourcewords, and codewords.

19.3.5.1 State Dependent Encoding

Here, the example of state dependent encoder, and sliding window decoder is presented, for the following parameters $M = 3, N = 3$. It is assumed that the source generates binary sequences with alphabet $\{0, 1\}$. The set of channel symbols is $\{-1, 0, 1\}$. The capacity of this constraint is $C = 1.2716$ bits/symbol. The code rate of $R = 1/1$ bits/symbol is chosen. It means that both sourceword length, and codeword length is 1. The code rate efficiency is $\eta = 0.77$. The starting point for encoder construction is FSTD presented at Figure 19.3, that represents Moore type FSTD of $M = 3, N = 3$ dc-free constraint. This type of FSTD belongs to almost finite type according to Proposition 7 in [9]. According to Theorem 3 in [9], there exists a noncatastrophic finite state encoder accompanied with sliding window decoder. Approximate eigenvector is $v = [1\ 1\ 1\ 1\ 1\ 1\ 1]$ that guarantees the existence of enough number of edges from each state, and there is no need for state splitting. The encoder is shown in Figure 19.5. Every label above an edge consists of pair, sourceword/codeword. From every state emanate as many edges as there are sourcewords, and in this case, it is $2^1 = 2$. The input bits are assigned in such a way to minimize the size of the decoder window. Table 19.2 gives sliding window decoder table. It can be seen that only one symbol is enough to decode a bit.

19.4 Codes with Higher Order Spectral Zeros

19.4.1 Introduction

The first representatives of spectrum shaping codes were dc-free codes discussed in the previous section. Further development of spectrum shaping codes went into two directions. One direction is the improvement in suppressing of low frequencies, and the other is the introduction of spectral zeros at frequencies different than zero frequency.

As it was pointed out at the very beginning, the role of spectrum shaping codes is to match the spectral characteristics of source sequences to the channel characteristics, and in the case of dc-free codes to cancel out dc, and to reduce the low frequency components of the recorded signal. The goodness of some code is measured by notch width determined by cut-off frequency. This parameter determines the frequency bandwidth within which the power spectral density is less than some value, but it does not tell anything how well these spectral components are rejected. Further improvement in terms of better suppression of low frequency components is achieved by generalizing of basic concepts given in previous sections. The better reduction of low frequency components is obtained by putting constraint on so-called kth order RDS [14]. That results in higher order zeros of power spectrum density.

The importance of spectrum shaping codes was rediscovered with introducing of partial response signaling techniques in high density recording devices [15]. The spectrum of such kind of recording channels can have zero not only at $f = 0$ but also at Nyquist frequency $f = f_s/2$, where f_s is a channel symbol frequency. The spectrum of codes that are designed to match partial response recording channels, have to have zeros not only at $f = 0$, but at some other frequencies as well, as required by specific channel. It was revealed that spectrum shaping codes on partial response channels have an additional virtue. It turned out that they exhibit enhanced error correcting capabilities on partial response channels.

In this section codes that give rise to higher order spectral nulls at frequencies $f = 0$, and $f = pf_s/q$, where p, and q are relatively prime, are presented. The results discussed here are due to Eleftheriou, and Cideciyan [14], and Immink [1]. Among the authors that considered these problems are Immink [16], Immink, and Beenker [17], Marcus, and Siegel [18], and Karabed, and Siegel [19].

19.4.2   Sequences with Higher Order Zeros at $f = 0$

First time the notion of sequences whose power spectral densities have higher order spectral nulls was introduced by Immink in [16]. He defined running digital sum sum (RDSS) of sequence $\{x_k\}_{k=0}^n$ as

$$y_n = \sum_{i=0}^n z_i = \sum_{i=0}^n \sum_{j=0}^i x_j \tag{19.13}$$

where z_k is the value of RDS at moment k. It can be proven that power spectral density of sequence $\{x_k\}_{k=0}^\infty$ has second order zero at $f = 0$, that is, $S_x(0) = 0$, and $S_x^{(2)}(0) = 0$, if and only if RDSS is bounded. The sequences whose RDSS is bounded are called dc^2-constrained sequences, and corresponding codes are dc^2-constrained codes. The odd derivatives of power spectral density $S_x(f)$ are zero because $S_x(f)$ is an even function of frequency f.

As in the case of dc-free constraint, similar encoding techniques are applied for higher order constraints, for example, zero disparity encoding, enumerative encoding, state dependent encoding. Zero disparity encoding employs the codewords that have zero disparity with respect to both RDS, and RDSS. The enumerative encoding counts the number of codewords of fixed length n, $x = (x_1 x_2 \ldots x_n)$, denoted by $A_n(d_z, d_y)$, that satisfy RDS disparity $d_x = \sum_{i=1}^n x_i$, and RDSS disparity $d_y = \sum_{j=1}^n \sum_{i=1}^j x_i$. The state dependent encoding are organized in the following way. All codewords are of zero d_z disparity, and are divided in two groups; one consists of codewords having zero, and positive d_y disparity, and the other consists of codewords having zero, and negative d_y disparity. The encoder chooses the next codeword such that at the end of that codeword the RDSS is close to zero.

The larger rejection of low frequencies is accomplished, if the higher derivatives of power spectral density $S_x^{(k)}(f)$, $k > 2$, are zero at $f = 0$. The condition for vanishing of higher order derivatives is defined in [17], and it uses the concept of codeword moment. The kth moment of codeword x is defined as

$$\mu_k^0(x) = \sum_{i=1}^n i^k x_i \tag{19.14}$$

where $k \in \{0, 1, 2, \ldots\}$. The superscript 0 means that the spectral null is at $f = 0$. The first $2K + 1$ derivatives of power spectral density $S_x(f)$ vanish at $f = 0$, if the first $K + 1$ codeword moments $\mu_k^0(x)$

are zero. The concept of kth codeword moment is very useful in computing the error correcting capabilities of codes with higher order spectral zeros, and it is related to kth order running digital sum that is introduced in the following section [14].

19.4.3 K-RDS$_f$ Sequences

In [14], more general approach was taken. The concepts of kth order running digital sum at frequency f, and K-RSD$_f$ FSTD are introduced. They are used to describe necessary, and sufficient conditions that an FSTD has to satisfy in order to be able to generate sequences whose power spectral densities have desired characteristics. Two main theorems of [14], Theorem 2, and Theorem 4, completely describe FSTDs that generate sequences with spectral nulls of order K at $f = 0$, and $f = pf_s/q$ respectively. The frequency f_s is a channel symbol frequency, and p, and q are relatively prime numbers. In this section just Theorem 4 will be presented.

The kth order RDS$_f$ at frequency $f = pf_s/q$ of sequence $\boldsymbol{x} = \{x_k\}_{k=0}^n$ is defined as

$$\sigma_k^f(x) = \sum_{i_1=0}^{n} \sum_{i_2=0}^{i_1} \cdots \sum_{i_k=0}^{i_{k-1}} w^{i_k} x_{i_k} \tag{19.15}$$

where $w = e^{-j2\pi p/q}$. In order to be able to state the main result, another notion has to be defined. For an FSTD is said that it is K-RDS$_f$ FSTD if there is a mapping ψ from the set of states $\sum$ onto a finite set of complex numbers ς such that $x(D) = wz(D)(1 - w^{-1}D)^K$, where $\boldsymbol{x} = \{x_k\}_{k=0}^n$ is a channel sequence, $z_n = \psi(s_{n+1})$, $\boldsymbol{s} = \{s_k\}_{k=0}^n$ is a state Markov process ($s_k \in \sum$), and $a(D) = \sum_{k=0}^{\infty} a_k$ defines D transform of sequence $\boldsymbol{a} = \{a_k\}_{k=0}^{\infty}$. The spectrum of sequence $\boldsymbol{x}$ generated by state process $\boldsymbol{s}$ assigned to an FSTD has a spectral null of order K at $f = pf_s/q$, if and only if FSTD is an K-RDS$_f$ FSTD. The main result of [14], Theorem 4, that completely describes K-RDS$_f$ FSTD, says that the following statements are equivalent

1. There are K functions, $\psi_k, 1 \le k \le K$, that map the set of states $\sum$ onto a finite set of complex numbers ς such that

    $$x_n = w\psi_1(s_{n+1}) - \psi_1(s_n)$$
    $$\psi_{k-1}(s_n) = \psi_k(s_n) - w^{-1}\psi_k(s_{n-1}), \quad 2 \le k \le K \tag{19.16}$$

2. There are K functions, $\psi_k, 1 \le k \le K$, that map the set of states $\sum$ onto a finite set of complex numbers ς such that the kth order RDS$_f$, $1 \le k \le K$, of any channel sequence is given by

    $$\sigma_1^f(x) = w^{n+1}\psi_1(s_{n+1}) - \psi_1(s_0)$$
    $$\sigma_k^f(x) = w^{n+1}\psi_k(s_{n+1}) - \psi_k(s_0) - \sum_{i=1}^{k-1} \psi_{k-i}(s_0)\binom{n+i}{i}, \quad 2 \le k \le K \tag{19.17}$$

3. There are $K - 1$ functions, $\psi_k, 1 \le k \le K - 1$, that map the set of states Σ onto a finite set of complex numbers ς such that for every cycle of states $s_0, s_1, \ldots, s_{n+1} = s_0$ of length that is multiple of q, the following equations are satisfied

    $$\sigma_1^f(x) = 0$$
    $$\sigma_k^f(x) = -\sum_{i=1}^{k-1} \psi_{k-i}(s_0)\binom{n+i}{i}, \quad 2 \le k \le K \tag{19.18}$$

4. K-RDS$_f$ FSTD that goes through the state sequence $\boldsymbol{s} = \{s_k\}_{k=0}^n$ generates channel sequence $\boldsymbol{x} = \{x_k\}_{k=0}^n$, that has a spectral null of order K at $f = pf_s/q$.

It should be noted that spectral null of order K guarantees that first $2K - 1$ derivatives of power spectral density $S_x(f)$ vanish at frequency f. If, in formula (Equation 19.15), and expressions in Theorem 4, w is replaced by 1, and superscript f with 0, Theorem 4 becomes Theorem 2 from [14] that completely characterizes FSTD that generates sequences with K-th order spectral null at $f = 0$. A usefulness of previous results can be seen from manipulation of two formulas in (Equation 19.16), which gives

$$\sigma_{n+1}^f = A\sigma_n^f + w^n 1 x_n \qquad (19.19)$$

where $\sigma_n^f = w^n \psi_n$ is K-dimensional column vector, A is a lower triangular all one matrix, and 1 is K-dimensional all one vector. Again if $w = 1$ then this formula is valid for the case of spectral null at $f = 0$. If $\sigma_0^f = 0$, the kth element of vector σ_n^f represents kth-order RDS$_f$ of sequence x at moment n. The Equation (19.19) describes dynamics of K-RDS$_f$ FSTD with respect to time. Also, the Equation (19.19) is related to the concept of canonical state transition diagrams [18], denoted by D_K^f. These diagrams have a countably infinite number of states, and there exists a finite state subdiagram of D_K that generates sequences with Kth order spectral null at frequency f. One such finite state subdiagram, D_2^0, is shown in Figure 19.6, for the following values of parameters: cardinality of channel symbol alphabet $M = 2$, 1-RDS$_0$, and 2-RDS$_0$ assume values from $\{-1, 0, 1\}$, and $K = 2$. One can notice that 1-RDS$_0$ is obtained by setting $w = 1$, and $k = 1$ in (Equation 19.15), which is classical RDS as defined by (Equation 19.1), while 2-RDS$_0$ is obtained by setting $w = 1$, and $k = 2$ in (Equation 19.15), which is classical RDSS as defined by (Equation 19.15). In this case vector σ_n^f is two-dimensional, where first entree is z_n, and the second is y_n, if the notation from previous section is used. The graph can be drawn such that abscissa represents z_n, and ordinate represents y_n. If V_k denotes the number of different values that kth-order RDS$_f$ can assume then for $K = 2$, the following relation holds between V_1, the number of values that 1-RDS$_0$ can assume, V_2, the number of values that 2-RDS$_0$ can assume

$$V_1 = 2 \lfloor \sqrt{V_2 - 1} \rfloor + 1 \qquad (19.20)$$

where $\lfloor x \rfloor$ defines the largest integer smaller than x. In general, it is proven that finite bound on V_k implies the finite bound on all V_j, $j < k$. These FSTD enables the computation of capacities, and power spectral densities of the constraints, and encoder/decoder design. Table 19.3 gives capacities for some values of constraint parameters. In Figure 19.7, the power spectral densities of two maxentropic sequences, (1) with first order spectral null, and parameters $M = 2$, $N = 3$, and (2) with second-order spectral null, and parameters $M = 2$, $V_1 = V_2 = 3$, 2-RDS$_0$, at frequency $f = 0$ are shown. It can be seen that sequences with second-order spectral null has better rejection of low frequency components although the notch width is wider in the case of sequences with first-order spectral null. The relation between cut-off frequency, and sum variance, Equation 19.7, is no longer valid. Figure 19.8 depicts, power spectrum of two binary memoryless codes with spectral zero at $f = 0.5 f_s$, whose codeword length are (1) $n = 4$, and (2) $n = 8$.

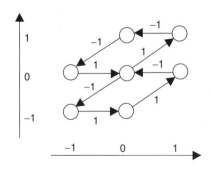

FIGURE 19.6   Canonical state diagram for $M = 2$, $V_1 = V_2 = 3$, 2-RDS$_0$ constraint.

TABLE 19.3 Capacity of Higher Order Zero Constraint for M=2, f=0

K	V_k	C
2	$V_2 = 3$	0.2500
2	$V_2 = 4$	0.3471
2	$V_2 = 5$	0.4428
2	$V_2 = 6$	0.5155
2	$V_2 = 7$	0.5615
3	$V_3 = 5$	0.1250
3	$V_3 = 6$	0.1250
3	$V_3 = 7$	0.1735
3	$V_3 = 8$	0.1946
3	$V_3 = 9$	0.2500

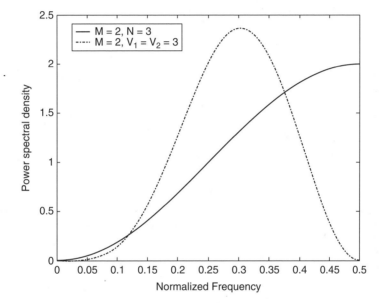

FIGURE 19.7 Power spectral density: (i) $M = 2$, $N = 3$ dc-free constraint; (ii) $M = 2$, $V_1 = V_2 = 3$, 2-RDS$_0$ constraint.

These codes satisfy condition that $\sigma_1^{f_s/2}(x) = 0$, and closed form expression for their power spectral densities is derived in [1]. The larger codeword length, the smaller notch width, saying that better rejection of low frequency components has to be paid by redundancy.

A number of constructions of codes with higher order spectral zeros, based on FSTDs, can be found in [14].

At the end, we give a useful relation between kth codeword moment at frequency f, $\mu_k^f(x)$, and kth-order RDS$_f$. Lemma 1 from [14] says that the following two statements are equivalent

$$\sigma_f^k(x) = 0, \quad 1 \le k \le K$$

$$\mu_f^k(x) = \sum_{i=0}^{n} i^k w^i x_i = 0, \quad 0 \le k \le K - 1$$

for codeword $x = (x_0 x_1 \ldots x_n)$.

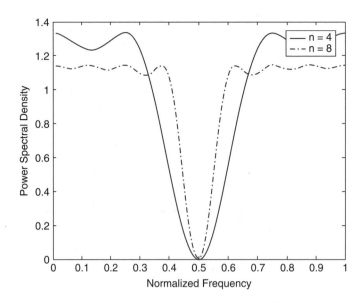

FIGURE 19.8   Power spectral density of codes with spectral zero at $f = f_s/2$: (i) Codeword length $n = 4$; (ii) Codeword length $n = 8$.

19.4.4 K-RDS$_f$ Sequences on Partial Response Channels

As it was mentioned in introduction, codes that generates K-RDS$_f$ constrained sequences improve the performance of recording systems utilizing partial response channels. The most interesting cases are of channels whose transfer functions have $(1 - D)^P$, $(1 + D)^P$ or both as factors. Here, two results will be given that concerns partial response channels. One is related to Hamming distance of binary sequences having one Kth order spectral null [17], and the other considers Euclidean distance of encoded sequences that was conveyed through partial response channel [14]. It is assumed that used codes have spectral zeros identical to spectral zeros of the channels. This notion of matched spectral zeros of codes, and partial response channels was introduced in [20].

Hamming distance. According to [17], the lower bound of minimum Hamming distance $d_{\min}^H$ between two binary sequences that have Kth order spectral null at $f = 0$ or $f = f_s/2$ is given as

$$d_{\min}^H \geq 2K \tag{19.21}$$

Euclidean distance. Before stating the result, the Euclidean distance for this kind of codes and channels will be discussed. Let's consider two sequences of channel symbols $x = \{x_i\}_{i=0}^n$, and $\hat{x} = \{\hat{x}_i\}_{i=0}^n$, generated by the following sequences of states, $\psi = \{\psi_i\}_{i=0}^{n+1}$, and $\hat{\psi} = \{\hat{\psi}_i\}_{i=0}^{n+1}$ respectively, such that $\psi_0 = \hat{\psi}_0$, $\psi_{n+1} = \hat{\psi}_{n+1}$. The error sequence at the input of the channel is defined as $e_i = x_i - \hat{x}_i$, $0 \leq i \leq n$. If the error sequence at the output of the channel with memory H is denoted as $\varepsilon = \{\varepsilon_i\}_{i=0}^{n+H}$, then Euclidean distance is defined as

$$d_{\min}^2 = \min_{\varepsilon} \sum_{i=0}^{n+H} |\varepsilon_i|^2 \tag{19.22}$$

where minimum is taken over all allowable output error sequences and all n.

The lower bound on Euclidean distance $d_{\min}^2$ for sequences with Kth order spectral null at $f = 0$ or $f = f_s/2$ at the output of a partial response channel with a spectral null of order P at $f = 0$ or $f = f_s/2$ is given by

$$d_{\min}^2 \geq 8d^2(K + P) \tag{19.23}$$

where $2d$ is the minimum distance between two amplitude levels. The lower bound in Equation 19.23 can be reached if $M + P \leq 10$.

19.5 Composite Constrained and Combined Encoding

Previous two sections considered the codes that main purpose is to shape the spectrum of the channel stream in order to match spectral characteristics of the channel, and in that way make recording reliable. This type of encoding represents just one of many types of encoding used in digital recording systems. Like any communication system, a recording system employs different encoding techniques such as source encoding, channel encoding, and modulation encoding. It means that a channel sequence has to satisfy different kinds of constraints to be reliable recorded. That's why there is a need for codes that generate sequences satisfying composite constraints. One example of such codes are codes that are in the same time RLL (run length limited), and dc-free codes [21, 22]. RLL codes are widely used in digital recording systems. They confine minimal, and maximal number of consecutive like symbols in a recording channel to fight intersymbol interference, and to enable clock recovery. In wider sense, to this group of codes belongs the combination of dc-free codes and error correcting codes that improves the performance of dc-free codes on noisy recording channels. For instance those codes can be found in [27–30].

The construction of composite constrained codes can be based on so-called composite graphs generating composite constraint. A composite graph represents the composition of two or more graphs that generate different constraints, and the sequence satisfying composite constraint is the one that in the same time satisfies the constraints of composition constituents. Here, the formal approach of graph composition will be given. The basis for spectral analysis of composite constraints can be found in [23–25].

A sofic or constrained system S is the set of all biinfinite sequences generated by walks on a directed graph $G = G(S)$ whose edges are labeled by symbols in a finite alphabet A. The graph $G = (V, E, \pi)$ is given by a finite set of vertices (or states) V, a finite set of directed edges E, and a labeling $\pi : E \to A$. So, for a given sequence of edges $\{e^{(k)}\}(e^{(k)} \in E)$, we have the output sequence $\{a^{(k)} = \pi(e^{(k)})\}$. A graph G is strongly connected if for every two vertices $u, v \in V$ there exists a path (sequence of edges) from u to v. A graph, G, is deterministic if for each state $v \in V$, the outgoing edges from v, $E(v)$, are distinctly labeled.

The connection matrix (or vertex transition matrix) $\mathbf{D}(G) = \mathbf{D} = [D(u, v)]_{u,v \in V}$ of graph G is $|V| \times |V|$ matrix where entry $D(u, v)$ is the number of edges from vertex u to vertex v, and $|V|$ is the number of vertices of the graph G.

One type of composition of two graphs is given by Kronecker's product. The Kronecker's product of the graphs $G_0 = (V_0, E_0, \pi_0)$ and $G_1 = (V_1, E_1, \pi_1)$, $G = G_0 \otimes G_1$, is the graph $G = (V, E, \pi)$ for which $V = V_0 \times V_1$ ($\times$ denotes Cartesian product of the sets) and for every edge e_0 from u_0 to v_0 in G_0 and every edge e_1 from u_1 to v_1 in G_1, there exists an edge $e(e = (e_0, e_1))$ in G, emanating from vertex $u = (u_0, u_1) \in V$ and terminating at $v = (v_0, v_1) \in V$ with the vector label $\pi(e) = \pi(e_0, e_1) = [\pi_0(e_0)\pi_1(e_1)]$.

So, the graph $G = G_0 \otimes G_1$ generates vector sequences $\{a^{(k)}\} = \{[a_0^{(k)} a_1^{(k)}]\}$. If connection matrices of the graphs G_0 and G_1 are $\mathbf{D}_0$ and $\mathbf{D}_1$, then the adjacency matrix of the graph $G = G_0 \otimes G_1$, $\mathbf{D}$, is $\mathbf{D} = \mathbf{D}_0 \otimes \mathbf{D}_1$, wherein now $\otimes$ denotes the Kronecker's product of matrices [26]

$$
\begin{aligned}
\mathbf{D} = [D(u, v)]_{u,v \in V} &= [D(u = (u_0, u_1), v = (v_0, v_1))]_{u_0, v_0 \in V_0; u_1, v_1 \in V_1} \\
&= \mathbf{D}_0(u_0, u_1) \otimes \mathbf{D}_1(u_1, v_1) = [D_0(u_0, v_0) \cdot D_1(u_1, v_1)]_{u_0, v_0 \in V_0; u_1, v_1 \in V}
\end{aligned}
\tag{19.24}
$$

As an example, Figure 19.9 shows two deterministic graphs G_0 and G_1 and their Kronecker's product, G.

The Kronecker's product of the graphs represents the vector constrained system in which the component subsequences of the constrained systems, are generated independently. So the vector sequence carries an average amount of information equal to the sum of average amounts of information of the constituent sequences. Consequently the capacity of the Kronecker's product $G = \otimes_{0 \leq i \leq N-1} G_i = G_0 \otimes G_1 \otimes \cdots \otimes G_{N-1}$ is $C(G) = \sum_{0 \leq i \leq N-1} C(G_i)$. The proof of this statement follows directly from the fact that the set of eigenvalues of $\mathbf{D}(G)$ is the set of all products of eigenvalues of the factor graph connection matrices [26].

As an example we show the composition of two graphs, G_1 representing $M = 3$, $(d = 1, k = 2)$ RLL constraint (Figure 19.10), and G_2 representing $M = 3$, $N = 3$ dc-free constraint (Figure 19.3). Resulting

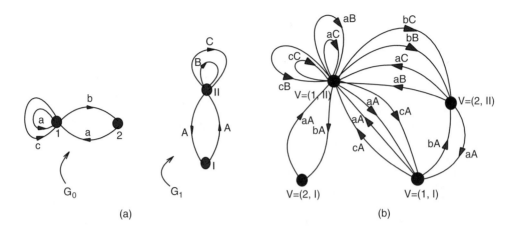

FIGURE 19.9   (a) An example of two graphs G_0 and G_1 ($V_0 = \{1, 2\}$, $V_1 = \{I, II\}$, $A_0 = \{a, b, c\}$, $A_1 = \{A, B, C\}$), and (b) Their Kronecker's product.

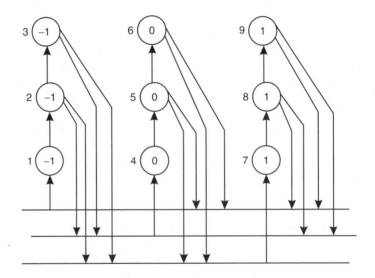

FIGURE 19.10   FSTD of M-ary RLL constrait, $M = 3$, $(d = 1, k = 2)$.

graph G_3 generates sequences that are in the same time RLL, and dc-free (Figure 19.11), with parameters $M = 3$, $(d = 1, k = 2)$, $N = 3$. Both constrained systems use the same alphabet $A = \{-1, 0, 1\}$. This composition is slightly modified as compared to previously defined Kronecker's product. Namely, there exists transition between the states in composite graph, $u = (u_0, u_1) \in V$, and $v = (v_0, v_1) \in V$, if $\pi_0(e_0) = \pi_1(e_1)$. Then, the label of corresponding edge will be $\pi(e) = \pi_0(e_0) = \pi_1(e_1)$. It should be noted that resulting graph is not always strongly connected. That's why, the irreducibel component of G_3 should be found that has the same Shannon capacity C as reducible graph. The final form of the composite graph is obtained by finding the Shannon cover of G_3 [9]. The graph in Figure 19.11 has ten states and it is Moore's type FSTD. The Shannon capacity is $C = 0.4650$ bits/sym, and the chosen code rate is $R = 1/3$ bits/sym, which is less than C. Table with capacities of this constraint is given in [22]. In order to get the encoder, the algorithm from [9] was applied, and it is shown in Figure 19.12. The sliding window decoder for this code can be found in [22].

At the end, in Figure 19.13, the spectrum of M-ary RLL dc-free maxentropic sequences are show, for fixed DSV, $N = 7$, and different values of parameter M.

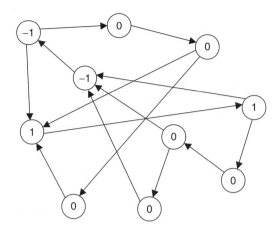

FIGURE 19.11 FSTD of M-ary RLL dc-free constraint, $M = 3$, $(d = 1, k = 2)$, $N = 3$.

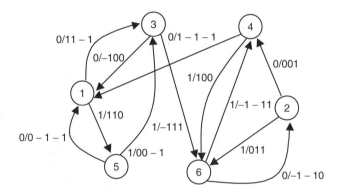

FIGURE 19.12 Encoder of $M = 3$, $(d = 1, k = 2)$, $N = 3$, RLL dc-free codes.

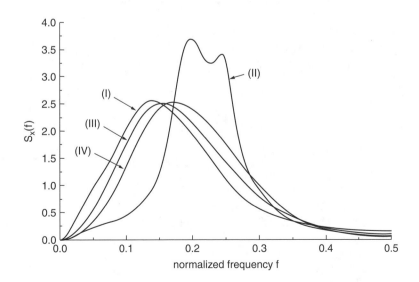

FIGURE 19.13 Spectrum of M-ary RLL dc-free maxentropic sequences: (i) $M = 3$, $(d = 1, k = 3)$, $N = 7$; (ii) $M = 4$, $(d = 1, k = 3)$, $N = 7$; (iii) $M = 5$, $(d = 1, k = 3)$, $N = 7$; (iv) $M = 7$, $(d = 1, k = 3)$, $N = 7$.

19.6   Conclusion

This chapter gives a survey of basic concepts and ideas of spectrum shaping codes for digital recording systems. We considered theoretical, and practical aspects of four groups of spectrum shaping codes, dc-free codes, codes with higher order spectral null at $f = 0$, codes with spectral nulls at the submultiples of channel symbol frequency, and codes with coposite constraints. We provided constrained channel capacities, power spectral densities, and some practical solutions for encoding and decoding schemes.

Acknowledgment

This work is supported in part by the NSF under grant CCR-0208597.

References

[1] Immink, K.A.S., *Coding Techniques for Digital Recorders*, 1st ed., Prentice Hall International, New York, 1991.

[2] Immink, K.A.S., Siegel, P.H, and Wolf, J.K., Codes for digital recorders, *IEEE Trans. Inform. Theory*, 44, 2260, 1998.

[3] Pierobon, G.L., Codes for zero spectral density at zero frequency, *IEEE Trans. Inform. Theory*, 30, 435, 1984.

[4] Shannon, C.E., A mathematical theory of communications, *Bell Syst. Tech. J.*, 27, 379, 1948.

[5] Spielman, S. et al., Using pit-depth modulation to increase the capacity and data transfer rate in optical discs, invited paper, *1997 Optical Data Storage Society Annual Meeting*.

[6] Justesen, J., Information rate and power spectra of digital codes, *IEEE Trans. Inform. Theory*, 28, 457, 1982.

[7] Cariolaro, G.L. and Tronca, G.P., Spectra of block coded digital signals, *IEEE Trans. Commun.*, 22, 1555, 1974.

[8] Bilardi, G., Padovani, R., and Pierobon, G.L., Spectral analysis of functions of Markov chains with applications, *IEEE Trans. Commun.*, 31, 853, 1983.

[9] Marcus, B.H., Siegel, P.H., and Wolf, J.K., Finite state modulation codes for data storage, *IEEE J. Selec. Area Commun.*, 10, 5, 1992.

[10] Cattermole, K.W., Principles of digital line coding, *Int. J. Electron.*, 55, 3, 1983.

[11] Bowers, F.K., U.S. Patent 2957947, 1960.

[12] Carter, R.O., Low disparity binary coding system, *Elect. Lett.*, 1, 65, 1965.

[13] Knuth, D.E., Efficient balanced codes, *IEEE Trans. Inform. Theory*, 32, 51, 1986.

[14] Eleftheriou, E. and Cideciyan, R.D., On codes satisfying Mth order running digital sum constraint, *IEEE Trans. Inform. Theory*, 37, 1294, 1991.

[15] Kobayashi, H., Correlative level coding and maximum likelihood decoding, *IEEE Trans. Inform. Theory*, 17, 586, 1971.

[16] Immink, K.A.S., Spectrum shaping with binary dc^2-constrained channel codes, *Philips J. Res.*, 40, 40, 1985.

[17] Immink, K.A.S. and Beenker, G.F.M, Binary transmission codes with higher order spectral zeros at zero frequency, *IEEE Trans. Inform. Theory*, 33, 452, 1987.

[18] Marcus, B.H. and Siegel, P.H., On codes with spectral nulls at rational submultiples of the symbol frequency, *IEEE Trans. Inform. Theory*, 33, 557, 1987.

[19] Karabed, R. and Siegel, P.H., Matched spectral-null codes for partial-response channels, *IBM Res. Rep.* RJ 7092, April 1990.

[20] Karabed, R. and Siegel, P.H., Matched spectral-null trellis codes for partial-response channels, part II: High rate codes with simplified Viterbi detectors, *IEEE Int. Symp. Inform. Theory*, Kobe, Japan, p. 143, 1988.

[21] Norris, K. and Bloomberg, D.S., Channel capacity of charge-constrained run-length limited codes, *IEEE Trans. Magn.*, 17, 3452, 1981.

[22] Denic, S.Z., Vasic, B., and Stefanovic, M.C., M-ary RLL dc-free codes for optical recording channels, *Elect. Lett.*, 36, 1214, 2000.

[23] Vasic, B., "Spectral analysis of multitrack codes," *IEEE Transactions on Information Theory*, Vol. 44, no. 4, pp. 1574–1587, July 1998.

[24] Vasic, B. and Stefanovic, M., Spectral analysis of coded digital signals by means of difference equation systems, *Elect. Lett.*, 27, 2272, 1991.

[25] Vasic, B., Spectral analysis of codes for magnetic and optical recording, Ph.D. dissertation, University of Nis, May 1993 (in Serbian).

[26] Cvetkovic, D., *Combinatorial Theory of Matrices,* Naucna knjiga, Belgrade, 1985 (in Serbian).

[27] Popplewell, A. and O'Reilly, J., A simple strategy for constructing a class of DC-free error-correcting codes with minimum distance 4, *IEEE Trans. Inform. Theory*, 41, 1134, 1995.

[28] Deng, R.H., Li, Y.X., Herro, and M.A., DC-free error-correcting convolutional codes, *Elect. Lett.*, 29, 1910, 1993.

[29] Lee, S.I., DC- and Nyquist-free error correcting convolutional codes, *Elect. Lett.*, 32, 2196, 1996.

[30] Etzion, T., Constructions of error-correcting DC-free block codes, *IEEE Trans. Inform. Theory*, 36, 899, 1990.

[31] Immink, K.A.S. and Patrovics, L., Performance assessment of dc-free multimode codes, *IEEE Trans. Commun.*, 45, 293, 1997.

20

Introduction to Constrained Binary Codes with Error Correction Capability

20.1 Introduction **20**-1
20.2 Bounds **20**-2
20.3 Example: A Trellis Code Construction **20**-4
20.4 An Overview of Some Other Code Constructions ... **20**-7
    Channel Models and Error Types • Input Constraints • Block and Trellis Code Constructions • Combined Codes Directly Derived from Linear Error Correcting Codes • Constrained Codes Carefully Matched to Error Correcting Codes • Constructions Employing Ideas from Contemporary Developments in Coding Techniques • Restrictions on the (d, k) Sequence • Multilevel Constructions • Constructions using the Lee or Levenshtein Metrics • Spectral Shaping Constraints and Error Correction • Maximum Likelihood Decoding of Standard Constrained Codes
20.5 Post Combined Coding System Architectures **20**-11
20.6 Conclusion **20**-12

Hendrik C. Ferreira

Willem A. Clarke

Rand Afrikaans University
Auckland Park, South Africa

20.1 Introduction

Constrained codes (alternatively called modulation codes, line codes or transmission codes) impose runlength or disparity constraints on the coded sequences, in order to either comply with the input restrictions of some communications channels, as determined by intersymbol interference or bandwidth limitations, or to aid in receiver synchronization and detection processes. Usually, these codes are not designed for error correction although they sometimes have limited error detection capabilities.

Binary runlength constrained codes, that is, (d, k) or (d, k, C) codes, find application on digital magnetic and optical recorders [54, 56]. Here d is the minimum number and k the maximum number of code zeros between consecutive code ones in the nonreturn to zero inverse (NRZI) representation and C the upper bound on the running disparity between ones and zeros in the nonreturn to zero (NRZ) representation.

Binary dc free codes have a bounded running disparity of ones and zeros in the coded sequences. These codes have been employed in early disc systems and later on in tape drives, and they also find widespread

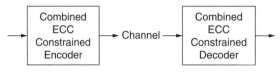

(a) Traditional approach to coding for error control and channel with input constraints.

(b) Combined coding scheme

FIGURE 20.1   Traditional and combined coding schemes.

application on metallic and optical cable systems. Since the minimum Hamming distance d_{min} is at least 2, dc free codes can detect at least one error. Balanced binary codes are dc free codes with equal numbers of ones and zeros in every codeword. This class of codes can also be considered as a subset of the class of constant weight codes.

In Figure 20.1(a), we show the traditional concatenated coding scheme, used to achieve both the goals of conforming to the channel input constraints and of providing error correction. To date, this approach is still used in many recording standards and products. One disadvantage of this scheme is the error propagation at the output of the constraint code's decoder: a single channel error may trigger multiple decoding errors. In general, the closer the coding rate R of the constrained code approaches the capacity C of the input restricted channel, the higher the complexity of such a constrained coding scheme, and the more the errors propagated. The propagated errors furthermore tend to be bursty in nature, which poses an additional load on the error-correcting scheme, and hence even more redundancy may be required. Consequently, since the early 1980s, several researchers have investigated the coding scheme in Figure 20.1(b). In the literature, codes for this scheme have been called combined codes, and on occasion also combi-codes or transcontrol codes.

During the late 1970s and early 1980s, the question was also posed whether some soft decision coding gain could be obtained from the constrained codes employed at that time, by using Viterbi decoding — see for example, [95]. This was partly inspired by the benefits of trellis-coded modulation for bandlimited channels, which was introduced at that time and which furthermore contributed to the impetus to develop combined codes.

It should also be noted that during the period under review here, several combined code constructions were furthermore aimed at partial response channels and later also on channels with two dimensional runlength constraints (e.g., [21]), as well as on channels with multilevel (M-ary instead of binary) symbols (e.g., [76]). However, a complete overview of code constructions for all these channels, is beyond the scope of this chapter.

In this chapter, we thus emphasize codes for the channel with one dimensional (d, k) constraints. The development of these codes went hand in hand with the development of error correcting dc free codes during the same time period, often by the same researchers and often using the same construction techniques, hence we also include some results and references on the latter class of codes.

20.2   Bounds

At first, the construction of error-correcting constrained codes, that is, constrained codes with minimum Hamming distance $d_{min} \geq 3$, appeared to be an elusive goal. Consequently, to find an existence proof, some early lower bounds on the Hamming distance achievable with runlength constrained block codes or balanced block codes were set up, using Gilbert type arguments, and published in [30]. To find an existence proof, the freedom of having infinitely long codewords was assumed.

Briefly, a constrained code with Hamming distance $d_{\min}$ can be formed by selecting a word from the set of all constrained words of length n bits as first code word, and by purging all other words at $d < d_{\min}$ from the set. Subsequently, a second word at $d = d_{\min}$ from the first word can be selected, and all words remaining at $d < d_{\min}$ from this word can also be purged. This purging process can be continued until only the desired set of code words remains.

Thus, since $\binom{n}{i}$ is the maximum number of binary words which may be at distance $i < d_{\min}$ from a code word, we can arrive at the following lower bound on the minimum Hamming distance achievable with an (n, k) constrained block code with $n \to \infty$:

$$2^{nC} \geq 2^k \sum_{i=0}^{d_{\min}-1} \binom{n}{i} \tag{20.1}$$

where C is the capacity of the noiseless input restricted channel. Stated differently, if $C > k/n$, we can obtain an (n, k) block code with desirable $d_{\min}$ by making n large enough.

This bound is also a rather loose lower bound on the minimum k achievable with a specified n and $d_{\min}$, since firstly not all binary words at distance i from a retained codeword satisfy the constraints, and secondly the number of words which has to be purged as the purging process continues, may grow smaller, as some of these words have been purged earlier.

As shown in [30], the bound can be tightened for balanced dc free codes, for which n will be even. Note that the distance between two words with the same weight and hence also the parameter i in Equation 20.1, can only be even for words of the same weight. Setting $i = 2a$, and making use of the balance between ones and zeros in both the constrained words retained and those purged, we can form a bound for balanced (d, k, C) sequences as

$$2^{nC} \geq 2^k \sum_{a=0}^{(d_{\min}-2)/2} \binom{n/2}{a}^2 \tag{20.2}$$

which simplifies for a balanced code with constant weight $w = n/2$ (and minimum runlength $d = 0$) to

$$\binom{n}{n/2} \geq 2^k \sum^{(d_{\min}-2)/2} \binom{n/2}{a}^2 \tag{20.3}$$

For further work and improvements on the topics of both lower and upper bounds, refer to [2, 43, 60, 73, 94, 97, 98]. A few results from these references will be briefly discussed next.

Ytrehus [98] derived recursive upper bounds for several choices of (d, k). Kolesnik and Krachkovsky [62] generalised the Gilbert-Varshamov bound to constrained systems by estimating the average volume of such constrained spheres. They then used a generating function for the distribution of pair wise distances of words in the constrained system, together with the Perron-Frobenius theorem, in order to obtain asymptotic existence results relating the attainable code rate R to a prescribed relative minimum distance $d_{\min}$ when $n \to \infty$. Gu and Fuja [43] improved on this Gilbert-Varshamov bound even further, still using average volume sphere arguments. Marcus [73], use labelled graphs to improve on the bound of Kolesnik and Krachkovsky.

Abdel-Ghaffar and Weber in [2] derive explicit sharp lower and upper bounds on the size of optimal codes that avoid computer search techniques. Using sphere packing arguments, they derive general upper bounds on the sizes of error-correcting codes and apply these bounds to bit-shift error correcting (d, k) codes. These bounds improve on the bounds given by Ytrehus [98].

In [43], Gu and Fuja provide a generalised Gilbert-Varshamov bound, derived via analysis of a code-search algorithm. This bound is applicable to block codes whose codewords must be drawn from irregular sets. It is demonstrated that the average volume of a sphere of a given radius approaches the maximum such volume and so a bound previously expressed in terms of the maximum volume can in fact be expressed in terms of the average volume. This bound is then applied specifically to error-correcting (d, k)-constrained codes.

20.3 Example: A Trellis Code Construction

Some early work on the actual construction of error-correcting constrained codes, can be traced back to [29–34], and interestingly, as later discovered, also to [4]. Subsequently, many widely different and sometimes ad hoc approaches to constructing such codes, evolved. We next present as simple example, suitable for an introductory and tutorial presentation, an approach from the earlier work on trellis codes. In the following section, we shall give an overview and classification of other block and trellis code constructions in the literature.

Trellis code constructions have the advantage that a general decoding algorithm, namely the Viterbi algorithm is immediately available. Constrained codes are usually nonlinear and suitable decoding algorithms may thus be difficult to find or complex to implement. Furthermore, soft decisions may be utilized when doing Viterbi decoding, in order to obtain additional coding gain.

The construction procedure in this section can be used to obtain trellis codes having various coding rates, constraint lengths and free distances, and with complexity commensurate with that of the class of linear binary convolutional codes widely used in practice. It uses the distance preserving mapping technique first described in [34, 35]. For related work, refer to [14, 38, 40, 41, 90].

Refer to Figure 20.2. The mapping table in Figure 20.2 maps the output binary n-tuple code symbols from a $R = k/n$ convolutional code (henceforth called the base code) into constrained binary m-tuples, which in this application are code words from a (d, k) code. The key idea is to find an ordered subset of 2^n m-tuples, out of the set of all possible constrained m-tuples with cardinality $N(m)$, such that the Hamming distance between any two constrained m-tuples is at least as large as the distance between the corresponding convolutional code's output n-tuples which are mapped onto them. This property may be called *distance preserving*, since the Hamming distance of the base code will be at least be conserved, and may sometimes even be increased in the resulting trellis code.

To illustrate this idea, we first present an example. We use the simple, "generic text book example" four state, binary $R = 1/2, v = 2, d_{\text{free}} = 5$ convolutional code in Figure 20.3(a) as base code. At the output of the encoder, we can map the set of binary 2-tuple code symbols, $\{00, 01, 10, 11\}$ onto constrained 4-tuples with $d = 1$, specifically using the set $\{0100, 0010, 1000, 1010\}$. Note that the last bit of each constrained 4-tuple used here is a merging bit, initially set to 0, to allow concatenation of the constrained symbols, without violating the d constraint. The state system of the resulting code appears in Figure 20.3(b). It should be stressed that the intention is to decode this resulting code in one step with the Viterbi algorithm.

In general, the property of *distance preserving* can be verified by setting up the matrices $D = [d_{ij}]$ and $E = [e_{ij}]$. Briefly, let d_{ij} be the Hamming distance between the binary code symbols i and j, where

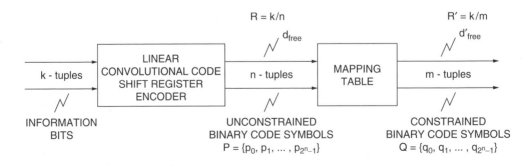

FIGURE 20.2   Distance preserving trellis code: encoding process.

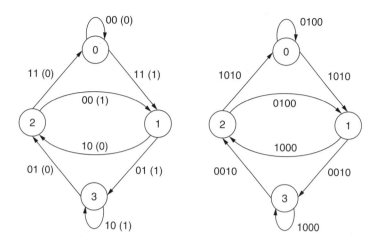

FIGURE 20.3 State systems: (a) Convolutional base code, (b) Constrained trellis code.

$0 \leq i, j \leq 2^n - 1$. The key to this code construction technique is thus to find an ordered subset of 2^n constrained m-tuples such that $e_{ij} \geq d_{ij}$, for all i and j, and where e_{ij} is the Hamming distance between the i'th and j'th m-tuples in the subset. Thus $\boldsymbol{D}$ and $\boldsymbol{E}$ can be set up to verify the mapping for our example code in Figure 20.3:

$$D = \begin{bmatrix} 0 & 1 & 1 & 2 \\ 1 & 0 & 2 & 1 \\ 1 & 2 & 0 & 1 \\ 2 & 1 & 1 & 0 \end{bmatrix} \quad \text{and} \quad E = \begin{bmatrix} 0 & 2 & 2 & 3 \\ 2 & 0 & 2 & 1 \\ 2 & 2 & 0 & 1 \\ 3 & 1 & 1 & 0 \end{bmatrix} \tag{20.4}$$

In this example the base code has $d_{\text{free}} = 5$, consequently the resulting $(d = 1, k)$ constrained code will also have $d_{\text{free}} \geq 5$. It can be shown by inspection that the maximum runlength k can be reduced to $k = 3$, by inverting the merging bit whenever possible.

Furthermore, note that the same mapping can now be applied to any $R = 1/2$ convolutional base code. In this way, we can thus easily construct more powerful trellis codes achieving larger free distances.

The procedure in the above example can be formalized as follows. Let the unconstrained binary n-tuples form a set $U = \{u_i\}$, with cardinality 2^n. The n-bit code symbols of a linear $R = k/n$ convolutional code will always be contained in U. The set of all constrained binary m-tuples may be represented by set $C = \{c_i\}$, with cardinality $N(m)$. The m-bit code symbols of the desired $R' = k/m$ constrained trellis code will be a subset of C.

In general, we want to transform a linear rate $R = k/n$ convolutional code with free distance d_{free} into rate $R' = k/m$, d'_{free} trellis code with $d'_{\text{free}} \geq d_{\text{free}}$. Usually, $m > n$, and to maximize the rate R' of the constrained trellis code derived from a given $R = k/n$ base code, we thus choose to only investigate mappings such that

$$n = \lfloor \log_2 N(m) \rfloor \tag{20.5}$$

The transformation will be per invariable mapping table as shown in Figure 20.3, that is, the same unconstrained binary code symbol will always be mapped onto the same constrained code symbol, irrespective of the base code.

To formalize the above: we want to map the ordered subset containing all unconstrained binary n-tuples, that is, $P = \{p_0, p_1, \ldots, p_{2^n-1}\}$, with elements $p_i \in U$, $p_i < p_{i+1}$, onto an ordered subset of constrained m-tuples, that is, $Q = \{q_0, q_1, \ldots, q_{2^n-1}\}$, with $Q \subset C$. To devise the mapping table, we first compute the Hamming distance matrix $\boldsymbol{D}$ with elements d_{ij}, being the Hamming distance between unconstrained binary n-tuples p_i and p_j, or,

$$d_{ij} = w(p_i + p_j) \quad 0 \le i, j \le 2^n - 1 \tag{20.6}$$

where $w(x)$ denotes the Hamming weight of x and addition is modulo 2. The task is now to find a suitable ordered subset Q, with 2^n elements q_i from the set C, such that the Hamming distance matrix E with elements

$$e_{ij} = w(q_i + q_j) \quad 0 \le i, j \le 2^n - 1 \tag{20.7}$$

has

$$e_{ij} \ge d_{ij}, \quad \text{for all } i, j \tag{20.8}$$

Note that the matrices D and E are square symmetric matrices with all-zero diagonal elements. Furthermore, the total number of permutations of the $N(m)$ constrained m-tuples, taken 2^n at a time, grows very rapidly with the tuple length m. For this reason in [35], we modeled the search as a tree search, similar to the Fano algorithm, but with the number of branches per node in the tree decreasing by one at each new depth. Note also that the main diagonal divides D into an upper and a lower triangular array with equal valued entries, and due to the symmetry only one of these need to be used in the search. The search may furthermore be speed up by reordering P such that the maximum valued elements in D are encountered earlier. See [35] for more details.

A few tests can also be performed before the tree search to establish the nonexistence of a suitable subset of constrained m-tuples or to prove that a specific m-tuple cannot be a member of the ordered subset of m-tuples — refer to [35].

Using prefix constructions, a mapping for some m may be used as kernel and extended to find a mapping for $m + 1$. The principle can be explained as follows. The set of binary $(n + 1)$-tuples, can be ordered following normal lexicography, that is, setting up the standard table of $(n + 1)$-bit binary numbers. It is easy to see that this set is partitioned into two subsets each containing 2^n elements. The first subset of $(n + 1)$-bit binary numbers are obtained by prefixing the set of n-bit binary numbers with a most significant bit 0, and the second subset of $(n + 1)$-bit binary numbers by prefixing the set of binary n-bit binary numbers with a most significant bit 1. Within each subset, the intradistance between elements is determined by the $n \times n$ D matrix, and stays the same. However, the binary prefixes of 0 and 1 account for an additional one unit of distance between two elements from the two different subsets. In a similar way, the ordered subset Q, containing 2^n constrained binary m-tuples can be extended to an ordered subset containing 2^{n+1} constrained $(m + 2)$-tuples by using two prefixes with Hamming distance at least one unit. For the $d = 1$ constraint, we can use the prefixes 00 and 10 and still satisfy the minimum runlength requirement. When using balanced symbols to construct dc free codes, we can use the prefixes 01 and 10.

Explicit mappings for m-tuples with $d = 1$, where it was attempted to maximize the combined code's rate $R' = k/m$, and to minimize k, are reported in [35], as well as mappings using balanced m-tuples from which dc free codes can be obtained. The highest achievable code rates with the mappings published, were $R' = 4/9$ and $R' = 3/6$ respectively. The maximum achievable free distances, will be determined by the underlying $R = 4/5$ and $R = 3/4$ base codes — in the literature on convolutional codes there are many codes of these rates available with different free distances and constraint lengths. Song and Shwedyk [90] later investigated a graph theoretic procedure to enumerate all the distance preserving mappings as in [35].

Finally, as discussed in [38], note that we can expand and generalize the concept of a distance preserving mapping to include a distance conserving mapping (DCM), as well as a distance increasing mapping (DIM) and a controlled distance reducing mapping (DRM).

In this section, we have thus presented a code construction procedure, capable of constructing powerful constrained trellis codes. By using this construction procedure, advantage can be taken of the many results on, and vast literature covering good convolutional codes. Furthermore an important reason for presenting this procedure is that it can also be applied to other constraints and channels as in [38, 14].

20.4   An Overview of Some Other Code Constructions

A literature search, which was not exhaustive, revealed more than one hundred papers on the topic of combining error correction with constrained codes. We attempted to include a representative selection of papers in our bibliography. As can be seen, several disjoint, and sometimes ad hoc procedures evolved. We next present a short overview, attempting to indicate some of the most important trends and directions.

20.4.1   Channel Models and Error Types

Much of the work on error correcting constrained codes focused on the binary symmetric channel, since these codes dominate the theory of linear error correcting codes, and also exhibit certain robustness. Refer to [13, 18, 67] for a few examples of constructions aimed at correcting additive or reversal errors on the binary symmetric channel. It should however be noted that these codes cannot be directly interleaved to correct burst errors — usually the channel's input constraints will be violated.

Some experimental work (see e.g., [51]) showed that peak shift errors, that is, errors represented in NRZI as $010 \rightarrow 001$ or $010 \rightarrow 100$, dominate on many recording channels, and this inspired a body of work on suitable code constructions. In one test of the IBM 3380 disk, 85% of the observed errors were shift errors [47]. For examples of such code constructions, refer to for example [11, 47, 63, 64, 70, 87].

During the late 1980s, as recording densities increased, it was observed that the electronic circuits for bit synchronization might fail more often; hence the topic of correcting bit insertion/deletion errors also received some attention. However, these errors, although they may have very destructive consequences, have a much lower probability of occurrence than the above error types and hence not many papers were published on this topic — see [11, 37, 46, 59] for a few examples of code constructions.

20.4.2   Input Constraints

Simultaneously with the interest in combined codes for magnetic recording, several researchers investigated combined codes for cable systems with somewhat different input constraints, such as a dc free power spectral density, or maximum runlength constraint, with no restriction on the minimum runlength — see for example, [5, 60–61, 79, 82–86].

In this regard, it is interesting to note that the balanced codes, a subset of the family of constant weight codes, are dc free, and bounds on the cardinalities of balanced codes as a function of d_{min}, have been tabulated in a few text books on coding for error correction and papers in the literature even before the new interest in combined codes during the 1980s.

In terms of (d, k) constraints, the $(1, 7)$ and $(2, 7)$ constraints dominated the magnetic recording industry for a long time, hence most researchers of combined codes attempted to conform with these constraints, or at least with $d = 1$ or 2 — see for example, [1, 6].

So far, the success of investigations into the construction of new combined codes appears to be proportional to the capacity of the input constrained channel. For (d, k, C) parameters of practical interest, the channel capacity increases in the same order if the input constraint is relaxed from (d, k, C) to (d, k) to $(0, k, C)$ to $(0, k)$. Consequently, very few results have been published on error correcting (d, k, C) codes. On the other hand, when it was proposed to relax the d constraint for magnetic recording systems, results on $(0, k)$ combined codes followed readily — for example, [93].

20.4.3   Block and Trellis Code Constructions

Most constructions in the vast field of linear error correcting codes can be classified as either a block code or a trellis code, and the same holds for combined codes. For a few results on block codes, refer to [2, 30, 67, 72, 87, 91], and for some results on trellis codes (or convolutional codes), refer to [15, 19, 45, 48–50, 100].

In the highly competitive recording industry, an increase in storage density of a few percentage points can be an important advantage. Although trellis codes have the advantages of Viterbi decoding and soft decision coding gain, some of the combined code constructions using block codes achieve higher code rates and hence the best exploitation of the Shannon capacity of the input-constrained noiseless channel. Furthermore, interleaving of the sequences of combined trellis codes in order to correct burst errors, can violate the (d,k) constraints. On the other hand, block-coding schemes are sometimes based on linear error correcting codes, burst error correcting in nature, such as Reed-Solomon codes.

Several authors (see e.g., [10, 36, 55]) have considered a block code construction which is systematic over a (d,k) constrained sequence. In its simplest form, a finite state machine or lookup table encoder may firstly map the bits from the information source onto a (d,k) sequence of length k bits. Next the parity bits of an (n,k) error correcting codeword can be computed. Finally, these parity bits are appended to the information (d,k) sequence in such a way that the parity sequence also complies with the channel's (d,k) constraints. This constrained parity sequence can be obtained by using means such as a lookup table, buffer bits between parity bits, etc.

Some authors have also parsed the (d,k) sequence into substrings, starting with an 1 and followed by between d and k zeros, or alternatively in reverse order. In [36, 17] the authors went a step further and showed that a unique integer composition can be associated with each (d,k) sequence. By imposing compositional restrictions on the (d,k) sequences, some error detection becomes possible. Error correction can be done by further appending parity bits. An advantage of applying the theory of integer compositions here, is that generating functions and channel capacities followed naturally. However, due to the compositional restrictions, code rates were too low for practical implementations, except when $d > 4$, while historically recording systems employed codes with $d = 1$ or 2, as dictated by physical factors such as detection window width. The parity bits in [36] typically keep track of the number of parts in the composition, or of the sum of the indexes of the positions in which a part occurs, both expressed modulo a small integer, and hence the number of parity bits could stay fixed, irrespective of the codeword length.

20.4.4   Combined Codes Directly Derived from Linear Error Correcting Codes

A natural question posed early in the development of combined codes, was whether a subcode of a linear error correcting code, or a coset code, having the required constraints, might be used as a combined code. This may have the advantage of making the powerful theory of linear codes applicable to the input constrained channel and perhaps using off-the-shelf decoders.

Examples considering subcodes of linear block codes can be found for example in [80], while subcodes of linear convolutional codes can be found in [78].

Pataputian and Kumar in [80] presented a (d,k) subcode of a linear block code. The modulation code is treated as a subcode of the error correcting code (a Hamming code), and they find a subcode of an error correcting code satisfying the additional (d,k) constraints required. They do this by selecting the coset in the standard array of Hamming codes which has the maximum number of (d,k) constrained sequences. This approach requires modulation codes that have very large block sizes when compared to conventional modulation codes in practice. One advantage of this scheme is that off-the-shelf decoders can be used. Systematic (but suboptimum) subcodes are also presented. In a similar approach, Liu and Lin [72] describes a class of (d,k) block codes with minimum Hamming distance of 3 based on $dklr$-sequences (the l and r represent the maximum number of consecutive zeros at the beginning and end of a (d,k)-sequence respectively) and cyclic Hamming codes. A codeword in the constructed code is formed by two subblocks satisfying the $dklr$ constraint. One is the message subblock and the other is the parity check subblock.

Similarly, employing cosets in new code constructions, has been considered in for example [18, 48–50]. Hole in [48] presented cosets of convolutional codes with short maximum zero-run lengths, that is, the k parameter. He achieved this by using cosets of (n,k) convolutional codes to generate the channel inputs. For $k \leq n - 2$ it is shown that there exist cosets with short maximum zero-run length for any constraint length. Any coset of an $(n, n-1)$ code with high rate and/or large constraint length is shown to have a large maximum zero-run length. A systematic procedure for obtaining cosets with short maximum zero-run length (n, k) codes is also given.

20.4.5 Constrained Codes Carefully Matched to Error Correcting Codes

Another natural approach was to carefully match the constrained codewords to the symbols of the error correcting code. In this way error correcting performance may be optimized. See for example [6]. The distance preserving mappings in [35] also fall into this category.

20.4.6 Constructions Employing Ideas from Contemporary Developments in Coding Techniques

One important contemporary idea was the principle of set partitioning applied to the channel signals, borrowed from the field of trellis coded modulation (TCM) where it was applied very successfully to amplitude/phase signals to develop combined coding and modulation schemes for the bandlimited channel. Attempts to apply it to the binary input constrained channel, met with limited success, due to the signal set lacking the same degree of symmetry.

Another important development was the ACH or state splitting algorithm for constructing finite state machine modulation codes for the input restricted channel — see for example, [21, 77] for application to combined codes. Application, again, does not follow directly. Nasiri-Kenari and Rushforth [77] show how the state-splitting and merging procedure can be adapted and applied to the problem of finding efficient (d, k) codes with guaranteed minimum Hamming distance. A second procedure in [77] partitions the encoder for a state-dependent (d, k) code into two subsections, one with memory and one without, and then combine the subsystem having memory with a matched convolutional code.

20.4.7 Restrictions on the (d, k) Sequence

By imposing compositional restrictions on (d, k) sequences, or by alternating between sets of (d, k) sequences with either only odd or even parity, error detecting and ultimately error correcting codes could be constructed — see, [36, 67].

Lee and Wolf [67] first derived a codebook Q consisting of all codewords adhering to the (d, k) constraint. However, concatenation of codewords from this codebook usually violates the (d, k) constraints. Therefore, the codebook Q is divided into maximal concatenatable subsets. Using a finite state code construction algorithm, they then generate a code with minimum free distance of three. Ferreira and Lin [36] presented combinatorial and algebraic techniques for systematically constructing different (d, k) block codes capable of detecting and correcting single bit errors, single peak-shift errors, double adjacent errors and multiple adjacent erasures. Their constructions are based on representing constrained sequences using integer compositions. Codes are obtained by imposing restrictions on such compositions and by appending parity bits to the constrained sequences.

Lee and Madisetti [69] proposes a general construction scheme for error correcting (d, k) codes having a minimum Hamming distance of 4. The proposed method uses a codeword set with a minimum Hamming distance of 2, instead of using two sets obtained by partitioning a concatenatable codeword set as in Lee and Wolf [67]. This means the code rate is always higher than that of Lee and Wolf. This proposed coding scheme is especially beneficial when the code lengths are short.

20.4.8 Multilevel Constructions

Several researchers have considered algorithms with multilevel partitioning of the symbol set, which is especially suitable for the constructing of dc free codes — see for example [13, 57].

On the other hand, M-ary multilevel (d,k) codes with error correction abilities have also received attention [76]. M-ary (d,k) codes are used for recording media that support unsaturated M-ary ($M \geq 3$) signalling. This is different than the normal binary case ($M = 2$) where the media is saturated. In [76], McLaughlin presented codes that achieve high coding densities, with improved minimum distance over an ordinary Adler-Coppersmith-Hassner code designed with the state-splitting algorithm. It is also shown that these codes have comparable minimum distance to Ungerboeck style amplitude modulation trellis codes.

20.4.9   Constructions using the Lee or Levenshtein Metrics

Most of the work under review, employed the Hamming distance metric, since it is widely known and the best understood. Consequently most constructions are directed towards the correction of additive (reversal) errors. After it was realized that codes over the Lee metric could be used to correct shift errors and to some extent insertions/deletion errors, some publications followed [11, 47, 63, 87].

Bours [11] suggested the use of the Lee distance when considering peak-shift errors. Roth and Siegel [87] showed that some of the Lee-metric BCH codes could be used to provide efficient protections against bit-shift or synchronisation errors. For bit-shift correction, these codes possess a smaller redundancy than the codes in Hamming metric. In [63], Krachkovsky et al. proposed another class of fixed length, t-error-correcting codes in the Lee metric. In their codes, the Galois field characteristics may be chosen independently of t and metric parameter q (where q is the alphabet size).

Similarly, a few papers employed the Levenshtein metric to correct insertions/deletions – see for example, [7, 37]. These errors cause catastrophic failures due to loss of synchronisation. Although these errors are not as common as the other types of errors, it is important to have a method to detect them and differentiate them from regular errors. Insertions/deletions, also known as synchronisation errors, are different from peak-shift or bit-shift errors in that all the 1s are shifted after an insertion/deletion. In a peak-shift error, only one 1 is shifted. Often, the Levenshtein distance is used in this environment rather than the Hamming distance.

Blaum et al. [7] used the $(1,7)$ modulation code to present two methods to recover from insertions or deletions. The first method (based on variable length codes) allows for the identification of up to 3 insertions and/or deletions in a given block, permitting quick synchronisation recovery. The second method, based on block codes, allows for the detection of large numbers of insertions and deletions. Kuznetsov and Vinck [64] presented codes for the correction of one of the following: a peak-shift of type $(k - d)/2$ or less, a deletion of $(k - d)/2$ or less zeros between adjacent ones and an insertion of $(k - d)/2$ or less zeros between adjacent ones.

Klove [59] presented codes correcting a single insertion/deletion of a zero or a single peak-shift. Particularly, he considers variable length (d, k) codes of constant Hamming weight.

20.4.10   Spectral Shaping Constraints and Error Correction

Codes with higher order spectral nulls [20, 52] were initially constructed for shaping the spectrum in the frequency domain, without consideration for error correction. Later, it was realized that these codes may also have good additive error correction properties, and in fact also insertion/deletion error correction properties [37]. Thus there seems to be a link to be further investigated.

20.4.11   Maximum Likelihood Decoding of Standard Constrained Codes

Some researchers revisited the approach as in [95], and investigated maximum likelihood or Viterbi decoding of standard constrained codes as employed in products on the market, reporting some gain for the digital compact cassette [12] and the DVD [44].

20.5   Post Combined Coding System Architectures

The movement to construct combined codes, or error correcting constrained codes as defined in Figure 20.1(b), experienced a peak during the 1990s, although a few new papers still appear every year. The emphasis of later work in order to achieve the same goals, has shifted to some extent to reversed concatenation (or post modulation coding) [10, 24, 42, 55] and to iterative decoding techniques – see for example, [25–27].

With these alternative approaches, higher coding rates and hence storage densities may be achieved. In the traditional concatenated coding scheme, the efficiency (R/C) of the constrained code, was more of a limiting factor than the efficiency of the error correcting code, and this influenced the post combined coding architectures later proposed. Many of the proposed combined coding schemes described in this chapter, also suffered the limitation of being too narrowly focused on one type of error.

Immink describes "a practical method for approaching the channel capacity of constrained channels" in [55], expanding a scheme previously proposed by Bliss [10]. Immink's scheme employs very long constrained code words while avoiding the possibly massive error propagation which may be triggered by a single channel error. The technique can be used in conjunction with any constrained code and it reverses the normal hierarchy of the error correction and channel codes. Block diagrams of Immink's scheme are shown in Figure 20.4.

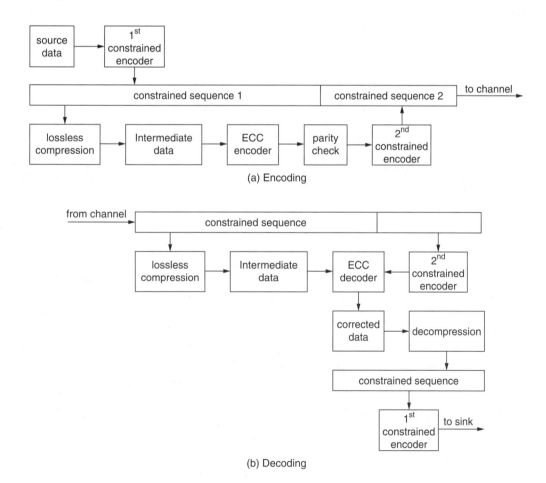

FIGURE 20.4   Immink's post modulation scheme.

Essential to Immink's scheme is the lossless compression step with limited error propagation, used to create the intermediate coding layer. A symbol error-correcting code, such as a byte orientated Reed-Solomon code is used to encode the intermediate layer. Look up tables, which are carefully matched, are used between the layers. Finally, the two sequences generated by the first and second constrained codes are cascaded and transmitted.

At the receiving end, the received sequence is decoded by firstly retrieving the parity symbols under the decoding rules of the second channel decoder. Next the constrained sequence is blocked into q-tuples and by using a lookup table are translated into the symbols of the intermediate sequence. Transmission errors in the intermediate sequence or parity symbols are corrected with the aid of the ECC code. The corrected intermediate sequence is decompressed and a constrained sequence which is essentially error-free, is obtained. This sequence is decoded and the original data retrieved.

In [55] the systematic design of the long block-decodable (d, k) constrained block codes, essential for the new coding method, is also considered. To this effect, Immink employs enumerative encoding and concatenatable (d, k, l, r) sequences, that is, (d, k) sequences with at most l consecutive leading zeros preceding the first one, and at most r consecutive trailing zeros succeeding the last one.

Examples of explicit results in [55] include a rate $R = 256/371 \, (d, k) = (1, 12)$ code which achieves 99.6% of the capacity of the $(1, 12)$ constrained channel, and a $R = 256/466 \, (d, k) = (2, 15)$ code achieving 99.86% of the capacity of the $(2, 15)$ constrained channel.

In conclusion, Immink's scheme made possible the use of long constrained codewords while avoiding error propagation, in order to approach the capacity of the input constrained channel. Important, it also offers the capability of correcting random and burst channel errors with powerful state of the art error correction codes such as Reed-Solomon codes.

20.6 Conclusion

In this chapter, we have attempted to give the reader an introduction to, and overview of the topic of combined codes. It currently appears that, in magnetic recording applications, this field and approach may have been overtaken by other systems architectures. However, it has stimulated much research and debate and has thus contributed to the development of newer architectures. Also, combined codes have potentially other applications in digital communications and the transmission of information. Hopefully, this presentation will thus help to stimulate further research.

References

[1] Abdel-Ghaffar, K.A.S., Blaum M., and Weber J.H., Analysis of coding schemes for modulation and error control, *IEEE Transactions on Information Theory*, vol. 41, no. 6, pp. 1955–1968, November 1995.

[2] Abdel-Ghaffar Khaled A.S. and Weber J.H., Bounds and constructions for runlength-limited error–control block codes, *IEEE Transactions on Information Theory*, vol. 37, no. 3, pp. 789–800, May 1991.

[3] Barg A.M. and Litsyn S.N., DC-constrained codes from Hadamard matrices, *IEEE Transactions on Information Theory*, vol. 37, no. 3, pp. 801–807, May 1991.

[4] Bassalygo A., Correcting codes with an additional property, *Problems of Information Transmission*, vol. 4, no1, pp. 1–5, Spring 1968.

[5] Bergmann E.E., Odlyzko A.M., and Sangani S.H., Half weight block codes for optical communications, *AT&T Technical Journal*, vol 65, no. 3, pp. 85–93, May–June 1986.

[6] Blaum M., Combining ECC with modulation: performance comparisons, *Proceedings of the IEEE Globecom '90 Conference*, San Diego, California, USA, vol. 3, pp. 1778–1781, December 2–5, 1990.

[7] Blaum M., Bruck J., Melas C.M., and van Tilborg H.C.A., Resynchronising (d, k)-constrained sequences in the presence of insertions and deletions, *Proceedings IEEE International Symposium on Information Theory*, p. 126, 17–22 January 1993.

[8] Blaum M., Litsyn S., Buskens V., and van Tilborg H.C.A., Error-correcting codes with bounded running digital sum, *IEEE Transactions on Information Theory*, vol. 39, no. 1, pp. 216–227, January 1993.

[9] Blaum M., Bruck J., Melas C.M., and van Tilborg H.C.A., Methods for Synchronising (d, k)-constrained sequences, *Proceedings of the 1994 IEEE International Conference on Communications*, vol.3, pp. 1800–1808, 1–5 May 1994.

[10] Bliss W.G., Circuitry for performing error correction calculations on baseband encoded data to eliminate error propagation, *IBM Technological Disclosure Bulletin*. vol. 23, pp. 4633–4634, 1981.

[11] Bours P.A.H., Construction of fixed-length insertion/deletion correcting runlength-limited codes, *IEEE Transactions on Information Theory*, vol. 40, no. 6, pp. 1841–1856, November 1994.

[12] Braun V., Schouhamer Immink K.A., Ribeiro M.A., and van den Enden G.J., On the application of sequence estimation algorithms in the digital compact cassette (DCC), *IEEE Transactions on Consumer Electronics*, vol. 40, no. 4, pp. 992–998, November 1994.

[13] Calderbank A.R., Herro M.A., and Telang V., A multilevel approach to the design of dc-free line codes, *IEEE Transactions on Information Theory*, vol. 35, no. 3, pp. 579–583, May 1989.

[14] Chang J.C., Chen R.J., Klove T, and Tsai S.C., Distance preserving mappings from binary vectors to permutations, *IEEE Transactions on Information Theory*, vol. 49, no. 4, pp. 1054–1059, April 2003.

[15] Chiu M.C., DC-free error-correcting codes based on convolutional codes, *IEEE Transactions on Communications*, vol. 49, no. 4, pp. 609–619, April 2001.

[16] Coene W., Pozidis H., and Bergmans J., Run-length limited parity-check coding for transition-shift errors in optical recording, *Proceedings IEEE GLOBECOM, 2001*, pp. 2982–2986, November 2001.

[17] Coetzee C.S., Ferreira H.C., and van Rooyen P.G.W., On the performance and implementation of a class of error and erasure control (d, k) block codes, *IEEE Transactions on Magnetics*, vol. MAG-26, no. 5, pp. 2312–2314, September 1990.

[18] Deng R.H. and Herro M.A., DC-free coset codes, *IEEE Transactions on Information Theory*, vol. 34, no. 4, pp. 786–792, July 1988.

[19] Deng R.H., Li Y.X., and Herro M.A., DC-free error-correcting convolutional codes, *Electronics Letters*, vol. 29, no. 22, pp. 1910–1911, 28th October 1993.

[20] Eleftheriou E. and Cideciyan R.D., On codes satisfying Mth-order running digital sum constraints, *IEEE Transactions on Information Theory*, vol. 37, no. 5, pp. 1294–1313, September 1991.

[21] Erxleben W.H. and Marcellin M.W., Error-correcting two-dimensional modulation codes, *IEEE Transactions on Information Theory*, vol. 41, no. 4, pp. 1116–1126, July 1995.

[22] Etzion T., Cascading methods for runlength-limited arrays, *IEEE Transactions on Information Theory* vol. 43, no. 1, pp. 319–324, January 1997.

[23] Fair I.J. and Xin Y., A method of integrating error control and constrained sequence codes, *2000 Canadian Conference on Electrical and Computer Engineering*, vol. 1, pp. 63–67, 7–10 March 2000.

[24] Fan J.L., Calderbank A.R., A modified concatenated coding scheme with applications to magnetic data storage, *IEEE Trans. Inform. Theory*, vol. 44, pp. 1565–1574, July 1998.

[25] Fan J.L., Constrained coding and soft iterative decoding, *Proceedings IEEE Information Theory Workshop*, 2001, pp. 18–20, 2–7 September 2001.

[26] Farkas P., Pusch W., Taferner M., and Weinrichter H., Turbo-codes with run length constraints, *International Journal of Electronics and Communications*, vol. 53, no. 3, pp. 161–166, January 1999.

[27] Farkas P., Turbo-codes with RLL properties, IEE Colloquium on Turbo Codes in Digital Broadcasting — Could It Double Capacity? (Ref. No. 1999/165), pp. 13/1–13/6, 22 November 1999.

[28] Fernandez E.M.G. and Baldini F.R., A method to find runlength limited block error control codes, *Proceedings of the 1997 IEEE International Symposium on Information Theory*, p. 220, 29 June–4 July 1997.

[29] Ferreira H.C., On dc free magnetic recording codes generated by finite state machines, *IEEE Transactions on Magnetics*, vol. 19, no. 6, pp. 2691–2693, November 1983.

[30] Ferreira H.C., Lower bounds on the minimum Hamming distance achievable with runlength constrained or dc free block codes and the synthesis of a (16, 8) dmin = 4 dc free block code, *IEEE Transactions on Magnetics*, vol. 20, no. 5, pp. 881–883, September 1984.

[31] Ferreira H.C., The synthesis of magnetic recording trellis codes with good Hamming distance properties, *IEEE Transactions on Magnetics*, vol. 21, no. 5, pp. 1356–1358, September 1985.

[32] Ferreira H.C., Hope J.F., and Nel A.L., Binary rate four eighths, runlength constrained, error correcting magnetic recording modulation code, *IEEE Transactions on Magnetics*, vol. 22, no. 5, pp. 1197–1199, September 1986.

[33] Ferreira H.C., Hope J.F., Nel A.L., and van Wyk M.A., Viterbi decoding and the power spectral densities of some rate one half binary dc free modulation codes, *IEEE Transactions on Magnetics*, vol. 23, no. 3, pp. 1928–1934, May 1987.

[34] Ferreira H.C., Wright D.A., and Nel A.L., On generalized error correcting trellis codes with balanced binary symbols, *Proceedings of the 25th Annual Allerton Conference on Communication, Control and Computing*, Monticello, Illinois, USA, pp. 596–597, September 30–October 2, 1987.

[35] Ferreira H.C., Wright D.A., and Nel A.L., Hamming distance preserving mappings and trellis codes with constrained binary symbols, *IEEE Transactions on Information Theory*, vol. 35, no. 5, pp. 1098–1101, September 1989.

[36] Ferreira H.C. and Lin S., Error and erasure control (d, k) block codes, *IEEE Transactions on Information Theory*, vol. 37, no. 5, pp. 1399–1408, September 1991.

[37] Ferreira H.C., Clarke W.A., Helberg A.S.J., Abdel-Ghaffar K.A.S., and Vinck A.J., Insertion/deletion correction with spectral nulls, *IEEE Transactions on Information Theory*, vol. 43, no. 2, pp. 722–732, March 1997.

[38] Ferreira H.C., and Vinck A.J., Interference cancellation with permutation trellis codes, *Proceedings of the IEEE Vehicular Technology Conference Fall 2000*, Boston, MA, USA, pp. 2401–2407, September 24–28, 2000.

[39] Fredrickson L.J. and Wolf J.K., Error detecting multiple block (d, k) codes, *IEEE Transactions on Magnetics*, vol. 25, no. 5, pp. 4096–4098, September 1989.

[40] French C.A., Distance preserving run-length limited codes, *IEEE Transactions on Magnetics*, vol. 25, no. 5, pp. 4093–4095, September 1989.

[41] French C.A. and Lin Y., Performance comparison of combined ECC/RLL codes, *Proceedings of the 1990 IEEE International Conference on Communications*, Atlanta, USA, pp. 1717–1722, April 1990.

[42] Fitingof B. and Mansuripur M., Method and apparatus for implementing post-modulation error correction coding scheme, U.S. Patent 5,311,521, May 1994.

[43] Gu J. and Fuja T., A generalized Gilbert-Varshamov bound derived via analysis of a code-search algorithm, *IEEE Transactions on Information Theory*, vol. 39, no. 3, pp. 1089–1093, May 1993.

[44] Hayashi H., Kobayashi H., Umezawa M., Hosaka S., and Hirano H., DVD players using a Viterbi decoding circuit, *IEEE Transactions on Consumer Electronics*, vol. 44, no. 2, pp. 268–272, May 1998.

[45] Helberg A.S.J. and Ferreira H.C., Some new runlength constrained binary modulation codes with error-correcting capabilities, *Electronics Letters*, vol. 28, no. 2, pp. 137–139, 16th January 1992.

[46] Helberg A.S.J., Clarke W.A., Ferreira H.C., and Vinck A.J.H., A class of dc free, synchronization error correcting codes, *IEEE Transactions on Magnetics*, vol. 29, no. 6, pp. 4048–4049, November 1993.

[47] Hilden H.M., Howa D.G., and Weldon E.J. Jr., Shift error correcting modulation codes, *IEEE Transactions on Magnetics*, vol. 27, no. 6, pp. 4600–4605, November 1991.

[48] Hole K.J., Cosets of convolutional codes with short maximum zero-run lengths, *IEEE Transactions on Information Theory*, vol. 41, no. 4, pp. 1145–1150, July 1995.

[49] Hole K.J. and Ytrehus O., Further results on cosets of convolutional codes with short maximum zero-run lengths, *Proceedings IEEE International Symposium on Information Theory*, pp.146, 17–22 September 1995.

[50] Hole K.J. and Ytrehus O., Cosets of convolutional codes with least possible maximum zero- and one-run lengths, *IEEE Transactions on Information Theory*, vol. 44, no. 1, pp. 423–431, January 1998

[51] Howell T.D., Analysis of correctable errors in the IBM 3380 disk file, *IBM Journal of Research and Development*, vol. 28, no. 2, pp. 206–211, March 1984.

[52] Immink K.A.S. and Beenker G.F.M., Binary transmission codes with higher order spectral zeros at zero frequency, *IEEE Transactions on Information Theory*, vol. 33, no. 3, pp. 452–454, May 1987.

[53] Immink K.A.S., Coding techniques for the noisy magnetic recording channel: a state-of-the art report, *IEEE Transactions on Communications*, vol. COM-37, no. 5, pp. 413–419, May 1989.

[54] Immink K.A.S., Coding Techniques for Digital Recorders, Prentice-Hall, Englewood Cliffs, NJ, 1991.

[55] Immink K.A.S., A practical method for approaching the channel capacity of constrained channels, *IEEE Transactions on Information Theory*, vol. 43, no. 5, pp. 1389–1399, September 1997.

[56] Immink K.A.S., Coding for mass data storage systems, Shannon Foundation, The Netherlands, 1999.

[57] Jeong C.K. and Joo E.K., Generalized algorithm for design of DC-free codes based on multilevel partition chain, *IEEE Communications Letters*, vol. 2, pp. 232–234, August 1998.

[58] Kamabe H., Combinations of finite state line codes and error correcting codes, *Proceedings of the 1999 IEEE Information Theory and Communications Workshop*, p 126, 20–25 June 1999.

[59] Klove T., Codes correcting a single insertion/deletion of a zero or a single peak-shift, *IEEE Transactions on Information Theory*, vol. 41, no. 1, pp. 279–283, January 1995.

[60] Kokkos A., Popplewell A., and O'Reilly J.J., A power efficient coding scheme for low frequency spectral suppression, *IEEE Transactions on Communications*, vol. 41, no.11, pp. 1598–1601, November 1993.

[61] Kokkos A., O'Reilly J.J., Popplewell A., and Williams S., Evaluation of class of error control line codes: an error performance perspective, *IEE Proceedings-I*, vol. 139, no. 3, pp. 128–132, April 1992.

[62] Kolesnik V.D. and Krachkovsky V.Y., Generating functions and lower bounds on rates for limited error-correcting codes, *IEEE Transactions on Information Theory*, vol. 37, no. 3, pp. 778–788, May 1991.

[63] Krachkovsky V.Y., Yuan Xing Lee and Davydov V.A. A new class of codes in Lee metric and their application to error-correcting modulation codes, *IEEE Transactions on Magnetics*, vol. 32, no. 5, pp. 3935–3937, September 1996.

[64] Kuznetsov A. and Vinck A.J.H., A coding scheme for single peak-shift correction in (d, k)-constrained channels, *IEEE Transactions on Information Theory*, vol. 39, no. 4, pp. 1444–1449, July 1993.

[65] Laih S. and Yang C.N., Design of efficient balanced codes with minimum distance 4, *IEE Proceedings-I*, vol. 143, no. 4, pp. 177–181, August 1996.

[66] Lee J. and Lee J., Error correcting RLL codes using high rate RSC or turbo code, *Electronics Letters*, vol. 37, no. 17, pp. 1074–1075, 16th August 2001.

[67] Lee P., and Wolf J.K., A general error-correcting code construction for run-length limited binary channels, *IEEE Transactions on Information Theory*, vol. 35, no. 6, pp. 1330–1335, November 1989.

[68] Lee J. and Madisetti V.K., Error correcting run-length limited codes for magnetic recording, *IEEE Transactions on Magnetics*, vol. 31, no. 6, pp. 3084–3086, November 1995.

[69] Lee J. and Madisetti V.K., Combined modulation and error correction codes for storage channels, *IEEE Transactions on Magnetics*, vol. 32, no. 2, pp. 509–514, March 1996.

[70] Levenshtein V.I. and Han Vinck A.J., Perfect (d,k)-codes capable of correcting single peak-shifts, *IEEE Transactions on Information Theory*, vol. 39, no. 2, pp. 656–662, May 1993.

[71] Lin Y. and Wolf J.K., Combined ECC/RLL codes, *IEEE Transactions on Magnetics*, vol. 24, no. 6, pp. 2527–2529, November 1988.

[72] Pi-Hai Liu and Yinyi Lin, A class of (d, k) block codes with single error correcting capability, *IEEE Transactions on Magnetics*, vol. 33, no. 5, pp. 2758–2760, September 1997.

[73] Marcus B.H. and Roth R.M., Improved Gilbert-Varshamov bound for constrained systems, *IEEE Transactions on Information Theory*, vol. 38, no. 4, pp. 1213–1221, July 1992.

[74] Markarian G. and Honary B., Trellis decoding technique for block RLL/ECC, *IEE Proceedings-I*, vol. 141, no. 5, pp. 297–302, October 1994.

[75] Markarian G., Honary B., and Blaum M., Maximum-likelihood trellis decoding technique for balanced codes, *Electronics Letters*, vol. 31, no. 6, pp. 447–448, 23rd March 1995.

[76] McLaughlin S.W., Improved distance M-ary (d, k) codes for high density recording, *IEEE Transactions on Magnetics*, vol. 31, no. 2, pp. 1155–1160, March 1995.

[77] Nasiri-Kenari M. and Rushforth C.K., Some construction methods for error-correcting (d, k) codes, *IEEE Transactions on Communications*, vol. 42, no. 2/3/4, pp. 958–965, February/March/April 1994.

[78] Nasiri-Kenari M. and Rushforth C.K., A class of DC-free subcodes of convolutional codes, *IEEE Transactions on Communications*, vol. 44, no. 11, pp. 1389–1391, November 1996.

[79] O'Reilly J.J. and Popplewell A., Class of disparity reducing transmission codes with embedded error protection, *IEE Proceedings-I*, vol. 137, no. 2, pp. 73–77, April 1990.

[80] Patapoutian A. and Kumar P.V., The (d, k) subcode of a linear block code, *IEEE Transactions on Information Theory*, vol. 38, no. 4, pp. 1375–1382, July 1992.

[81] Perry P.N., Runlength-limited codes for single error detection in the magnetic recording channel, *IEEE Transactions on Information Theory*, vol. 41, no. 3, pp. 809–815, May 1995.

[82] Popplewell A. and O'Reilly J.J., Spectral characterisation and performance evaluation for a new class of error control line codes, *IEE Proceedings-I*, vol. 137, no. 4, pp. 242–246, August 1990.

[83] Popplewell A. and O'Reilly J.J., Runlength limited codes for random and burst error correction, *Electronics Letters*, vol. 28, no. 10, pp. 970–971, 7th May 1992.

[84] Popplewell A. and O'Reilly J.J., Runlength limited binary error control codes, *IEE Proceedings-I*, vol. 139, no. 3, pp. 349–355, June 1992.

[85] Popplewell A. and O'Reilly J.J., Manchester-like coding with single error correction and double error detection, *Electronics Letters*, vol. 29, no. 6, pp. 524–525, 18th March 1993.

[86] Popplewell A. and O'Reilly J.J., A simple strategy for constructing a class of DC-free error-correcting codes with minimum distance 4, *IEEE Transactions on Information Theory*, vol. 41, no. 4, pp. 1134–1137, July 1995.

[87] Roth R.M. and Siegel P.H., Lee-metric BCH codes and their application to constrained and partial-response channels, *IEEE Transactions on Information Theory*, vol. 40, no. 4, pp. 1083–1096, July 1994.

[88] Saeki K. and Keirn Z., Optimal combination of detection and error correction coding for magnetic recording, *IEEE Transactions on Magnetics*, vol. 37, no.2, pp. 708–713, March 2001.

[89] Sechny M. and Farkas P., Some new runlength-limited convolutional codes, *IEEE Transactions Communications*, vol. 47 no. 7, pp. 962–966, July 1999.

[90] Song S. and Shwedyk E., Graph theoretic approach for constrained error control codes, *Proceedings of the 1990 International Symposium on Information Theory and Its Applications*, Honolulu, HI, USA, pp. 17–18, November 27–30, 1990.

[91] van Tilborg H. and Blaum M., On error-correcting balanced codes, *IEEE Transactions on Information Theory*, vol. 35, no. 5, pp. 1091–1095, September 1989.

[92] van Wijngaarden A.J. and Soljanin E., A combinatorial technique for constructing high-rate MTR-RLL codes, *IEEE Journal Selected Areas in Communication*, vol. 19, pp. 582–588, April 2001.

[93] van Wijngaarden A.J. and Immink K.A.S., Maximum runlength-limited codes with error control capabilities, *IEEE Journal Selected Areas in Communication*, vol. 19, pp. 602–611, April 2001.

[94] Waldman H. and Nisenbaum E., Upper bounds and Hamming spheres under the DC constraint, *IEEE Transactions on Information Theory*, vol. 41, no. 4, pp. 1138–1145, July 1995.

[95] Wood R.W., Viterbi reception of Miller-squared code on a tape channel, *Proceedings of the IERE International Conference on Video and Data Recording*, Southampton, England, pp. 333–344, 26–23 April 1982.

[96] Wood R.W., Further comments on the characteristics of the Hedeman H-1, H-2 and H-3 codes, *IEEE Transactions on Communications*, vol. 31, no. 1, pp. 105–110, January 1983.

[97] Yang S.H. and Winick K.A., Asymptotic bounds on the size of error-correcting recording codes, *IEE Proceedings-I*, vol. 141, no. 6, pp. 365–370, December 1994.

[98] Ytrehus O., Upper bounds on error-correcting runlength-limited block codes, *IEEE Transactions on Information Theory*, vol. 37, no. 3, pp. 941–945, May 1991.

[99] Ytrehus O., Runlength-limited codes for mixed-error channels, *IEEE Transactions on Information Theory*, vol. 37 no. 6, pp. 1577–1585, November 1991.

[100] Ytrehus O. and Hole K., Convolutional codes and magnetic recording, *In Proceedings of the URSI International Symposium on Signals, Systems, and Electronics*, Paris, pp. 803–804, September 1992.

$n_{ECC} = B \cdot n_B$ bits. With Reed-Solomon codes, if any bit in a symbol is incorrect, the entire symbol is considered incorrect. The decoder for the Reed-Solomon code has the ability to correct a limited number of symbol errors (e.g., it can correct $\lfloor \frac{n_B - k_B}{2} \rfloor$ symbol errors).

In many data storage systems, it is necessary to satisfy simultaneously the needs of constrained coding and error-control coding. A conceptually simple solution to this problem is to choose the intersection of the constraint set S_C with an error-control code S_{ECC}. Assuming that the two sets are of the same length ($n_C = n_{ECC}$), then the words in the intersection set $S_C \cap S_{ECC}$ would be suitable for decoding by both the ECC decoder and constraint decoder since they are codewords in both codes. An encoding function f for this situation could be defined by an indexing function for an exhaustive list of the words in $S_C \cap S_{ECC}$. Suppose A is a binary alphabet. Let $k = \lfloor log_2(S_C \cap S_{ECC}) \rfloor$, and index 2^k words of the set $S_C \cap S_{ECC}$ using the numbers $\{0, 1, \ldots, 2^k - 1\}$ in binary representation. An encoder function f can then be defined by mapping this binary index $u \in A^k$ to the corresponding indexed word $w \in S_C \cap S_{ECC} \subset A^n$.

For practical implementations, it is important to choose a scheme that permits the encoding and decoding to be performed within the limits on complexity imposed by existing hardware. In particular, the direct approach of using an encoder function f for the intersection $S_C \cap S_{ECC}$ requires memory that is exponential in the codeword length to store an exhaustive encoder table. In contrast, most constrained codes and error-control codes in use today have associated encoders and decoders which offer efficient implementations; in particular, their complexity and storage requirements do not increase exponentially with the codeword length. This chapter considers a number of approaches to this problem of combined constrained and error-control coding based on configurations that make use of the existing encoders and decoders for the constrained code and the error-control code.

21.2.2 Standard Concatenation

The *standard concatenation* scheme, shown in Figure 21.1, consists of the serial concatenation of an ECC as the outer code and a constrained code as the inner code. In terms of the encoding functions, the user sequence u is mapped $v = f_{ECC}(u)$ and then to $w = f_C(f_{ECC}(u))$. Note that in this case, n_{ECC} is equal to k_C, and in particular, the output word w belongs to S_C, but does not belong to S_{ECC}. The overall rate of standard concatenation is $\frac{k_C}{n_C} \cdot \frac{k_{ECC}}{n_{ECC}}$.

Suppose that the transmitted word is w and the received word is $\hat{w}$. The decoding of the constrained code (also known as *demodulation*) is typically implemented as the inverse of the encoder function: $\hat{v} = f_C^{-1}(\hat{w})$. A problem known as *error-propagation* arises when a small number of errors in the received word becomes magnified by demodulation. Let $d(\cdot, \cdot)$ be a function that measures the Hamming distance between two words, that is, the number of symbols in which the two words differ. Generally speaking, error-propagation is said to occur when $d(w, \hat{w})$ is small but $d(f_C^{-1}(w), f_C^{-1}(\hat{w}))$ is large.

The constrained code is typically implemented in such a way as to reduce the error-propagation caused by the demodulation function f_C^{-1}. The impact of error-propagation depends on the details of the decoder for the error-control code. Common constructions of constrained codes involve block codes or sliding-block codes with short block lengths. As an example, consider a binary block code of length N_C with an encoder function F_C that maps from A^{K_C} to A^{N_C}. (Note that upper-case letters are used to represent parameters of the short block code, while lower-case letters are used to represent parameters of the full code.)

This encoder function F_C for the short block code can be used to implement the encoder function f_C for the entire constrained word. The image of F_C should be chosen such that the concatenation of any

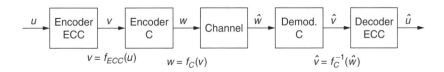

FIGURE 21.1 Standard concatenation scheme.

combination of words yields an overall word of length n_C that belongs to S_C (and the constraint is said to be maintained across blocks). If $u^{(1)}, u^{(2)}, \ldots, u^{(k_C/K_C)}$ are all words of length K_C, then the full constraint encoder f_C can be defined by the block encoder as follows

$$f_C\left(\left\{u^{(1)}, u^{(2)}, \ldots, u^{(k_C/K_C)}\right\}\right) = \left\{F_C\left(u^{(1)}\right), F_C\left(u^{(2)}\right), \ldots, F_C\left(u^{(k_C/K_C)}\right)\right\}$$

For the short block code, the encoding function F_C can be an arbitrary map from A^{K_C} to A^{N_C}. As a result, for the corresponding demodulation map F_C^{-1}, one or more bit errors in a word of N_C bits are assumed to result in an incorrect word upon demodulation, resulting in many bits in error in the demodulated output. The error-propagation effect, however, can be limited through the use of short block codes.

Suppose that a constrained code based on short block codes of rate K_C/N_C is used with an ECC with symbol size B. In the worst case, a single bit error can propagate into $\lceil K_C/B \rceil$ symbol errors, and a short burst of bit errors on the boundary of two blocks can result in $\lceil 2K_C/B \rceil$ symbol errors. The error-propagation effect becomes worse for large K_C, so that it is typical to choose a value of K_C that is matched to the symbol size B, such as $K_C = B$. This need to use short block lengths usually places a significant restriction on the design of the constrained code, making it difficult to use constrained codes whose rate approaches the capacity of the modulation constraint.

21.2.3   Reverse Concatenation

To allow for better combining of the error-control code and the constrained code so as to mitigate the effects of error-propagation, one promising approach is the *reverse concatenation* scheme, which reverses the order of the ECC and constrained code so that the encoding of the constraint takes place before the encoding of the error-control code, as shown in Figure 21.2. This method allows the decoder to perform decoding on most of the received word with the need for demodulation.

This method has the following benefits:

- It allows the use of arbitrary constrained codes
- It reduces error propagation
- It facilitates the use of soft decoding

Reverse concatenation is also known as "modified concatenation," or the "commuted configuration." Its history in the literature goes back to Bliss [4] and Mansuripur [16], and was analyzed in [7] and [11] as a method of preventing error-propagation in magnetic recording systems. It may also be viewed as related to specific schemes presented in [3,10,15,18,19] that combine (d, k) run-length constraints with error-control codes for detecting and correcting bit errors or bit shifts.

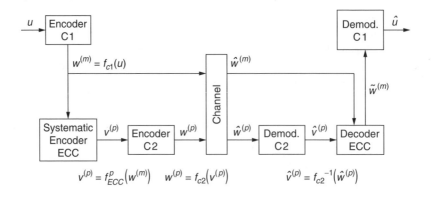

FIGURE 21.2   Reverse concatenation scheme.

Reverse concatenation involves a *main constrained code C*1 and an *auxiliary constrained code C*2, which have rates k_{C1}/n_{C1} and k_{C2}/n_{C2}, respectively. It also assumes the use of a systematic ECC of rate k_{ECC}/n_{ECC}. There are no restrictions on the design of the main constrained code $C1$ since the demodulation of this code takes place after its decoding by the ECC, so that it does not suffer from error-propagation. This allows the usage of constrained codes that have rates arbitrarily close to the capacity of the constraint. The auxiliary constrained code, however, must be designed in such a way as to prevent error propagation during demodulation, much as in the case of the constrained code in standard concatenation.

The encoding for reverse concatenation uses the encoding functions for $C1, C2$ and the ECC as follows: First, the main constrained code maps the user word u to $w^{(m)} = f_{C1}(u)$. This constrained codeword is then encoded by the systematic encoder for the ECC. The parity portion $v^{(p)} = f_{ECC}^p(f_{C1}(u))$ of the ECC codeword is then encoded by the constrained code $C2$ to give $w^{(p)} = f_{C2}(v^{(p)})$. The end result is the word

$$w = \{w^{(m)}, w^{(p)}\} \tag{21.1}$$

$$= \{f_{C1}(u), f_{C2}(f_{ECC}^p(f_{C1}(u)))\} \tag{21.2}$$

The code parameters for reverse concatenation are related as follows: $n_C = n_{C1} + n_{C2}, k_{C2} = n_{ECC} - k_{ECC}$, $k_{ECC} = n_{C1}$, and $k_{C1} = len(u)$. The two constrained codes $C1$ and $C2$ should be chosen such that for any input u and v, placing the two words $f_{C1}(u)$ and $f_{C2}(v)$ together gives a word $w = \{f_{C1}(u), f_{C2}(v)\}$ that belongs to the target constraint set S_C.

Note that the overall rate is

$$\frac{k_{C1}}{n_{C1} + n_{C2}} = \frac{k_{C1}}{n_{C1} + \frac{n_{C2}}{k_{C2}}(n_{ECC} - k_{ECC})} \tag{21.3}$$

$$= \frac{k_{C1}}{n_{C1}} \left(1 + \frac{n_{C2}}{k_{C2}} \left(\frac{n_{ECC}}{k_{ECC}} - 1\right)\right)^{-1} \tag{21.4}$$

If the constrained code $C2$ were not used (i.e., $k_{C2}/n_{C2} = 1$), then the overall rate would be the same as in standard concatenation. The constrained code $C2$ is necessary, however, to make sure that the parity bits also satisfy the constraint.

To reduce error-propagation during demodulation of the auxiliary constrained code $C2$, it is typically implemented using block codes or sliding-window block codes with short block length, as in the case of the constrained code in standard concatenation. For example, for the case of an ECC with symbol size B, the code $C2$ could be implemented by putting together words from a short block code of rate K_{C2}/N_{C2}, where the length K_{C2} is chosen to match B (e.g., it is equal to B, or a multiple of B).

The advantages of reverse concatenation over standard concatenation lie in the decoding process. Error-propagation occurs when demodulation must be performed before decoding takes place. With reverse concatenation, the error-propagation during demodulation is restricted to the parity portion of the ECC codeword. Note that reverse concatenation is most effective when the ECC code rate k_{ECC}/n_{ECC} is high.

The decoding procedure for reverse concatenation is as follows: (For simplicity in exposition in this section, we consider the ECC to be a hard-decision ECC, although this discussion generalizes in a straightforward manner to soft-decoding.) Suppose the channel decoder (e.g., a Viterbi decoder) produces a received sequence $\hat{w}$ that is a possibly erroneous copy of the transmitted word w. This sequence can be divided into a message portion $\hat{w}^{(m)}$ and a parity portion $\hat{w}^{(p)}$, based on the correspondence to $C1$ and $C2$, respectively. The parity portion $\hat{w}^{(p)}$ is first demodulated by the code $C2$, which has limited error-propagation by design (e.g., $K_{C2} = B$), to obtain the word $\hat{v}^{(p)}$. On the other hand, the message portion $\hat{w}^{(m)}$ can go directly to the ECC decoder without any need for demodulation. The ECC decoder performs decoding on the word $\{\hat{w}^{(m)}, \hat{v}^{(p)}\}$. If the ECC decoding is successful, then the output $\bar{w}^{(m)}$ is an error-free version of the constrained message bits. In this case there is no risk of error-propagation during demodulation, and applying demodulation using f_{C1}^{-1} yields a replica $\bar{u}$ of the original user bits u, completing the successful decoding.

To summarize, for reverse concatenation, the encoder's output to the channel satisfies the constraint, while the decoder sees only limited error propagation from demodulation. The main constrained code $C1$ can have arbitrary design, so there is no restriction on the function f_{C1}, and in particular, it is possible to use a constrained code with extremely long block length. This method is most effective when the ECC code rate is high, since the parity portion comprises a small part of the whole codeword. Reverse concatenation gives an effective method to meet a desired modulation constraint using a near-capacity constrained code.

21.2.4 Bit Insertion

For certain classes of constraints, it is possible to use a variation on reverse concatenation is to insert the parity bits into a constrained message sequence in such a way that the resulting sequence does not violate the constraint. This *bit insertion scheme*, shown in Figure 21.3, entails choosing a modulation code $C1$ such that inserting bits into the sequence in some predetermined pattern results in a sequence that still meets the target constraint (e.g., it belongs to the constraint set S_C). In other words, instead of using a constrained code $C2$, the parity bits $v^{(p)}$ from the encoder are inserted directly into pre-specified locations in the sequence $w^{(m)} = f_{C1}(u)$.

The surprising advantage of this approach is that no error-propagation due to demodulation takes place at all. The entire sequence received from the channel decoder can be directly used by the ECC decoder (after possibly some permutation of the order), without the need for demodulation. (This bit insertion method is also beneficial for soft decoding as discussed in Section 21.3, since soft information for all bits is directly usable by the ECC decoder.) As in Section 21.2.3, if the ECC decoding is successful, then the decoder output $\tilde{w}^{(m)}$ is error-free and can be demodulated by f_{C1}^{-1} to yield the original user bits u.

While simple and effective, the bit insertion method is only useful for certain classes of constraints. It has been considered by Anim-Appiah and McLaughlin [1] for using $(0, k)$ modulation codes with Turbo Codes, and has been considered by van Wijngaarden and Immink [20,21] for the $(0, G/I)$-RLL constraint. An extensive analysis of this technique is given by Campello et al. in [5], which considers the "unconstrained positions" in a constrained code, referring to the locations that are suitable for placing parity bits with arbitrary values.

21.2.5 Lossless Compression

One issue with reverse concatenation is that the input to the ECC encoder is increased by a factor of n_{C1}/k_{C1} compared with standard concatenation since f_{ECC} is applied to $f_{C1}(u)$ instead of u directly. For constrained codes with low rate (e.g., $k_{C1}/n_{C1} \approx 0.5$), this poses a problem as the length of the ECC codeword expands proportionally, which leads to increased complexity in the ECC encoder and decoder.

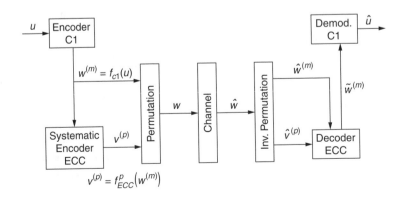

FIGURE 21.3   Bit insertion scheme.

To reduce codeword expansion in reverse concatenation, Immink proposed in [11] the use of a lossless compression code to compress the input to the ECC encoder. This technique is typically used in a scenario with a symbol-oriented hard-decision decoder such as a Reed-Solomon code. In this context, the compression should have the following characteristics:

- The sequence to be compressed satisfies a modulation constraint.
- The compression map should be lossless, so that it is possible to exactly recover the original sequence.
- The compression map should cause little error propagation. (This can be accomplished using short block lengths.)

The lossless compression code can be defined by an encoding function f_L whose image is a *superset* of the constraint set: $Im(f_L) \supset S_C$. The rate is lower-bounded by the constraint capacity: $cap(S_C) \leq k_L/n_L$. (In contrast, with constrained coding, the encoding function f_C has an image that is a *subset* of the constrained set, $Im(f_C) \subset S_C$, and constrained code rate is upper-bounded by the constraint capacity, $k_C/n_C \leq cap(S_C)$.) This map f_L can also be called an *expanding coder*, or "*excoder*" (in analogy to "encoder"). This is an invertible mapping from A^{k_L} to A^{n_L}. The inverse map f_L^{-1} then gives the corresponding *compression map*, which is a bijection from $Im(f_L)$ to A^{k_L} [9].

Applying this idea to reverse concatenation, the excoder map f_L should have an image that is a superset of S_{C1}, where $n_L = n_{C1}$. Then as shown in Figure 21.4, the compression map f_L^{-1} is applied to the constrained word $w^{(m)} = f_{C1}(u)$ to obtain t. Then the ECC encoder produces parity corresponding to t using a systematic ECC encoder to obtain $v^{(p)} = f_{ECC}^p(t)$. Finally, the parity portion is modulated by the constraint $C2$, and the transmitted word has the form:

$$w = \left\{ f_{C1}(u), f_{C2}\left(f_{ECC}^p\left(f_L^{-1}(f_{C1}(u)) \right) \right) \right\}$$

As in reverse concatenation, $C1$ and $C2$ are chosen such that the resulting word w belongs to the constraint set S_C. In this case, $n_L = n_{C1}$, while $k_L = k_{ECC}$.

As for the decoder for this configuration, if the received word is $\hat{w}$, the message portion $\hat{w}^{(m)}$ must be first compressed by f_L^{-1} before it can be used by the ECC decoder. Hence, the compression map f_L^{-1} should be designed to have limited error-propagation. In particular, the goal is to create an excoder function f_L such that the compression map f_L^{-1} has limited error-propagation. In terms of practical implementation, this can be accomplished through short block codes or sliding-block codes.

The rest of the decoding procedure goes as follows, as shown in Figure 21.4. Let $\hat{t} = f_L^{-1}(\hat{w}^{(m)})$ represent the compressed version of the message portion of the received word. Meanwhile, the parity portion $\hat{w}^{(p)}$

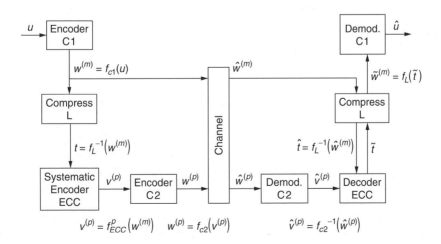

FIGURE 21.4 Lossless compression scheme for reverse concatenation.

must be demodulated by $C2$, so that the input to the ECC decoder is $\{\hat{t}, f_{C2}^{-1}(\hat{w}^{(p)})\}$. Upon successful decoding, a corrected version of the message portion is obtained, which is denoted by $\bar{t}$. Next, the excoder f_L performs decompression on $\bar{t}$ to recover the corrected $\tilde{w}^{(m)} = f_L(\bar{t})$, and finally, demodulation is performed for the constrained code $C1$ to recover the user data $\tilde{u} = f_{C1}^{-1}(\tilde{w}^{(m)})$.

For the design of the lossless compression code, the goal is to decrease the codeword expansion, so that it is desirable to have as low a compression rate K_L/N_L as possible. On the other hand, error propagation must be avoided, which limits the selection of lossless compression codes (e.g., to ones with short block length). At one extreme, the compression could be trivial, so that the output t is exactly the same as the input $w^{(m)}$; this situation corresponds to reverse concatenation (Figure 21.2). At the other extreme, the lossless compression could compress $w^{(m)}$ back to u (demodulating the code $C1$), corresponding to standard concatenation (Figure 21.1). Choosing a compression code in between these two extremes allows for the benefits of reverse concatenation, while minimizing its codeword expansion relative to standard concatenation.

(It should be noted that this lossless compression technique is not generally applicable for soft decoding, since the compression step tends to further obscure the reliability information. In general, the compression code may be an arbitrary assignment of codewords, so that the computation of postcompression reliability information is a difficult task, similar to computing postdemodulation reliability information for an arbitrary constrained code.)

Block codes for lossless compression are given in [7] and [11]. The basic construction is a block codes of rate K_L/N_L and an invertible function F_L that maps from K_L bits to N_L bits. Then the excoder f_L can be constructed using F_L (similarly to how f_C is built from F_C):

$$f_L\left(\left\{v^{(1)}, v^{(2)}, \ldots, v^{(k_L/K_L)}\right\}\right) = \left\{F_L\left(u^{(1)}\right), F_L\left(u^{(2)}\right), \ldots, F_L\left(u^{(k_L/K_L)}\right)\right\}$$

When any sequence from the constraint set S_C is divided into words of length N_L, these words should all lie in the image of F_L and so that they can be mapped via the compression map F_L^{-1} to a word of length K_L.

Sliding-block lossless compression codes are discussed in [9]. The compression map is a sliding-block decoder from sequences of N_L-codewords of the constraint set S_C to unconstrained sequences of length K_L; that is, a N_L-codeword $w^{(i)}$ is compressed into a K_L-frame $t^{(i)}$ as a deterministic function of the current N_L-codeword and perhaps some m preceding and a following N_L-codewords. The *sliding-block window length* is defined by the sum m + a + 1, and the code is said to be *sliding-block compressible*. The excoder, on the other hand, takes the form of a finite-state machine.

The existence of sliding-block compressible codes was shown in [9] for constraints defined as finite-memory constrained systems: if $\mathsf{cap}(S_C) \leq k_L/n_L$, then for this constrained system S_C, there exists a lossless compression code of rate K_L/N_L that is sliding-block compressible. Recall that the state-splitting algorithm for designing encoders for constrained codes (Ref. [17]) is guided by an "approximate eigenvector" that satisfies a certain inequality. For sliding-block lossless compression codes, this reversed inequality is used in a variant of the state-splitting algorithm for constructing finite-state excoders.

21.3   Reverse Concatenation and Soft Iterative Decoding

A very important benefit of reverse concatenation is the ability for the ECC decoder to make use of the full information available from the channel. Many channels provide *soft information*, which is typically a real-valued metric on each bit indicating its reliability, for example, a probability or a log-likelihood ratio (LLR). For many situations, the ability to obtain this soft information on the channel output allows the decoder for the error-control codes to improve its performance.

With standard concatenation, the usual implementations of constrained codes make it difficult to associate soft information to the bits of the demodulated output. In other words, given bit-wise probabilities for the word w, it is not straightforward to obtain bit-wise probabilities for the demodulated word v. With reverse concatenation, however, it is possible to directly use soft information for the message portion $(w^{(m)})$.

This is possible because no demodulation step is necessary for the message portion. (The bit insertion method is even better, allowing the use of soft information on the whole received word! The bit insertion method, however, can only be used on a limited class of constraints.) This ability of reverse concatenation to use soft information directly from the channel is critical for using constrained codes with ECCs such as Turbo codes and low-density parity-check (LDPC) codes whose iterative decoders rely on soft information for their superior performance over other codes. Of the extensive literature on applying these soft iterative ECCs in the context of magnetic and optical storage, a number of papers (e.g., [1,2,8]) also consider their application in the context of a constrained code.

In addition, with reverse concatenation, a soft decoder for the constraint can be used in conjunction with the ECC decoder to obtain additional coding gain from the constraint, as presented in [8]. The basic idea is that the constraint imposes restrictions on the valid codewords, as defined by the constrained set S_C, and the constraint decoder uses knowledge of these restrictions can be used to improve the soft information (e.g., the bit-wise probabilities) for use in the ECC decoder.

While for arbitrary constraint sets S_C, it may be difficult to perform this soft decoding, for certain constraints, it is possible to take advantage of the structure of the constraint to perform a soft-in, soft-out (SISO) decoding to make useful updates to the soft information. For example, as shown in [6,8], for the case of the (d, k)-RLL constraints (and other constraints whose codewords can be represented by traversals on a trellis), it is possible to use the structure of the constraint to define a SISO decoder for the constraint using the BCJR algorithm.

With a SISO decoder for the constraint, it is possible to iterate between the ECC decoder and the constraint decoder as follows: Assume the parity portion has been demodulated according to the auxiliary constrained code C2. Then the ECC decoder receives the message portion and performs a SISO decoding (e.g., for LDPC or Turbo codes) to obtain updated soft output. This is used to create a new estimate of soft information that is passed to the constraint decoder. The SISO decoder for the constraint then uses the structure of the constraint to produce soft output that can then be combined with the channel soft information to create an updated soft input for the ECC decoder. By iterating back and forth between the ECC decoder and the constraint decoder, it is possible to gain additional coding gain by making use of the redundancy that is inherent in the constraint.

Note that the effectiveness of the soft constraint decoder requires the use of reverse concatenation (or the bit insertion method), as opposed to standard concatenation, in which the process of demodulation can distort the soft information. Also, note that the gains from decoding the constraint are more significant for lower rate constraints (e.g., k_C/n_C less than 2/3) since there is more redundancy imposed by the constraint that can be exploited for error-correction purposes. As a result, this technique of decoding the constraint may be more applicable to channels that use lower-rate constraints (e.g., optical storage). Finally, the channel decoder has been largely ignored in this discussion. As intersymbol interference is present in both magnetic and optical recording channels, it is common to use a trellis-based (e.g., Viterbi) detector for the channel. This is an additional issue that needs to be considered in combination with the constraint decoder and the ECC decoder.

References

[1] K. Anim-Appiah and S. McLaughlin, "Turbo codes cascaded with high-rate block codes for $(0, k)$ constrained channels," *IEEE Journal on Selected Areas of Communication* vol. 19, no. 4. pp. 677–685, April 2001.

[2] W. Ryan, S. McLaughlin, K. Anim-Appiah, and M. Yang, "Turbo, LDPC, and RLL codes in magnetic recording," *Proc. 2nd Int. Symp. on Turbo Codes and Related Topics*, September 2000.

[3] A. Bassalygo, Correcting codes with an additional property, *Prob. Inform. Trans.*, vol. 4, no. 1, pp. 1–5, Spring 1968.

[4] W.G. Bliss, Circuitry for performing error correction calculations on baseband encoded data to eliminate error propagation, *IBM Techn. Discl. Bul.*, vol. 23, pp. 4633–4634, 1981.

[5] J. C. de Souza, B.H. Marcus, R. New, and B.A. Wilson, Constrained systems with unconstrained positions, Presented at the DIMACS Workshop on Theoritical Advances in Recording of Information, New Brunswik, NJ, April 2004. Available at http://rutgers.edu

[6] J.L. Fan, *Constrained Coding and Soft Iterative Decoding*, Kluwer Academic Publishers, Boston, 2001.

[7] J.L. Fan and A.R. Calderbank, A modified concatenated coding scheme, with applications to magnetic storage, *IEEE Trans. Inform. Theory*, vol. 44, no. 4, pp. 1565–1574, July 1998.

[8] J.L. Fan and J.M. Cioffi, Constrained coding techniques for soft iterative decoders, *Proc. Globecom* (Rio de Janeiro), 1999.

[9] J.L. Fan, B.H. Marcus, and R.M. Roth, Lossless compression in constrained coding, *Proc. 37th Allerton Conf. on Commun., Control, and Computing*, 1999.

[10] H.M. Hilden, D.G. Howe, and E.J. Weldon, Jr., Shift error correcting modulation codes, *IEEE Trans. Magnetics*, vol. 27, no. 6, pp. 4600–4605, November 1991.

[11] K.A.S. Immink, A practical method for approaching the channel capacity of constrained channels, *IEEE Trans. Inform. Theory*, vol. 43, no. 5, pp. 1389–1399, September 1997.

[12] K.A.S. Immink, *Codes for Mass Data Storage*, Shannon Foundation Press, 1999.

[13] K.A.S. Immink, P.H. Siegel, and J.K. Wolf, Codes for digital recorders, *IEEE Trans. Inform. Theory*, vol. 44, no. 6, pp. 2260–2299, October 1998.

[14] R. Karabed, P.H. Siegel, and E. Soljanin, Constrained coding for binary channels with high intersymbol interference, *IEEE Trans. Inform. Theory*, vol. IT-45, pp. 1777–1797, September 1999.

[15] W.H. Kautz, Fibonacci codes for synchronization control, *IEEE Trans. Inform. Theory*, pp. 284–292, April 1965.

[16] M. Mansuripur, Enumerative modulation coding with arbitrary constraints and post-modulation error correction coding and data storage systems, *Proc. SPIE*, vol. 1499, pp. 72–86, 1991.

[17] B.H. Marcus, R. Roth, and P.H. Siegel, Constrained systems and coding for recording channels, in *Handbook of Coding Theory*, Pless, V.S. and Huffman W.C., Eds., Elsevier, Amsterdam, pp. 1635–1764, 1998.

[18] A. Patapoutian and P.V. Kumar, The (d, k) subcode of a linear block code, *IEEE Trans. Inform. Theory*, vol. 38, no. 4, pp. 1375–1382, July 1992.

[19] P.N. Perry, Runlength-limited codes for single error detection in the magnetic recording channel, *IEEE Trans. Inform. Theory*, vol. 41, no. 3, pp. 809–814, May 1995.

[20] A.J. van Wijngaarden and K.A.S. Immink, Efficient error control schemes for modulation and synchronization codes, *Proc. IEEE ISIT* (Boston), p. 74, 1997.

[21] A.J. van Wijngaarden and K.A.S. Immink, Maximum run-length limited codes with error control capabilities, *IEEE J. Sel. Areas Commn.*, vol. 19, no. 4, April 2001.

22

Convolutional Codes for Partial-Response Channels

Bartolomeu F. Uchôa-Filho
Federal University of Santa Catarina
Florianopolis, Brazil

Mark A. Herro
University College
Dublin, Ireland

Miroslav Despotović

Vojin Šenk
University of Novi Sad
Novi Sad, Yugoslavia

22.1 Introduction **22**-1
22.2 Encoding System Description and Preliminaries **22**-2
22.3 Trellis Codes for Partial-Response Channels
Based Upon the Hamming Metric **22**-4
22.4 Trellis-Matched Codes for Partial-Response
Channels **22**-7
22.5 Run-Length Limited Trellis-Matched Codes **22**-10
Cosets of Convolutional Codes
22.6 Avoiding Flawed Codewords **22**-13
22.7 The Distance Spectrum Criterion for Trellis Codes .. **22**-15
22.8 Good Trellis-Matched Codes for the
Partial-Response Channels Based on
the Distance Spectrum Criterion **22**-15

22.1 Introduction

As described in Chapter 8, partial-response equalization plays an important role in magnetic recording. In particular, the partial-response system described by the polynomial $P_n(D) = (1 - D)(1 + D)^n$, where n is a nonnegative integer and D denotes a symbol delay, represents an interesting and largely adopted model for the high-density magnetic recording channel [2, 3]. The excellent performance of systems based on partial-response models sparked interest in the search for compatible coding techniques. In this chapter, we describe the application of convolutional coding to partial-response channels, a technique that was introduced by Wolf and Ungerboeck [1] for the $(1 \pm D)$ channel in 1986. Many authors have subsequently elaborated on this coded system, such as Zehavi and Wolf [4], Hole [14], Hole and Ytrehus [15], and Siala and Kaleh [16]. The extension of this coding technique to the more general case of the $P_n(D)$ channel has been considered by Uchôa-Filho and Herro [12, 13], and refined by Despotović et al. [19, 20].

The key point in dealing with the convolutionally coded partial-response systems is to view the $P_n(D)$ channel as a finite state machine, as is usually done for convolutional encoders. Assuming binary channel inputs, the trellis associated with the channel has 2^{n+1} states. This channel trellis is then combined with the trellis describing the convolutional code, resulting in a trellis for the overall system that describes all possible coded sequences. Decoding is provided by the Viterbi algorithm operating on this trellis. The properties of this trellis are discussed in detail in this chapter. The combination of a precoder and a convolutional encoder drawn from a restricted set of generator matrices plays a key role in reducing the number of required decoder states. Other issues taken into consideration are the limitation of the length of equal symbol runs, which is addressed through the use of cosets of convolutional codes, and a code search for good codes based upon the (squared Euclidean) distance spectrum criterion.

This chapter is organized as follows. The description of the coded system and some preliminaries are given in Section 22.2. In Section 22.3, we derive a lower bound on the minimum free squared Euclidean distance of the trellis corresponding to the overall system. This bound, which is related to the minimum free Hamming distance of a subcode of the convolutional code, is then shown to be weak (and hence of little significance) when $n > 0$, which suggests the need for a computer-aided search for good codes. In Section 22.4, we point out several structural properties of the coded system and establish the rules which guide the computer search. In this section, we introduce the important theory of trellis matching. Section 22.5 addresses the problem of limiting the length of equal symbol runs for synchronization purposes. Section 22.6 deals with the problem of avoiding flawed codewords. Flawed codewords result whenever two infinitely long paths in the trellis of the overall system have the same labels (symbols from the output of the partial-response channel). In Section 22.7, the distance spectrum criterion for trellis codes is presented, which we adopt for finding good codes. These codes are tabulated for the $(1 - D)(1 + D)^2$ and $(1 - D)(1 + D)^3$ channels in Section 22.8.

22.2   Encoding System Description and Preliminaries

The block diagram of the encoded system that we consider in this chapter is shown in Figure 22.1. In this section we describe the components of this system and introduce some needed definitions. Unless otherwise stated, the signals and transfer functions of the blocks in Figure 22.1 are represented by their D-transforms. Beginning at the left, the input sequence, $\mathbf{U}(D) = [U^1(D), \ldots, U^k(D)]$, is first encoded by the (m, k) convolutional encoder of rate $R_c = k/m$, represented by the generator matrix $\mathbf{G}(D)$. The ith input sequence is represented by the polynomial $U^i(D) = u_0^i + u_1^i D + \cdots$, where $u_t^i \in GF(2)$ for $1 \le i \le k$, and for each time instant $t \ge 0$. The input sequence produces the encoded sequence, $\mathbf{V}(D) = \mathbf{U}(D)\mathbf{G}(D) = [V^1(D), \ldots, V^m(D)]$. The representation of the jth encoded sequence is $V^j(D) = v_0^j + v_1^j D + \cdots$, where $v_t^j \in GF(2)$ for $1 \le j \le m$, and for each time instant $t \ge 0$. The generator matrix $\mathbf{G}(D) \triangleq (G_i^j(D))$, where $G_i^j(D) = \sum_{l=0}^{v} g_{l,i}^j D^l$, and the elements $g_{l,i}^j \in GF(2)$, $1 \le i \le k$, and $1 \le j \le m$, represent the tap connections in the encoder. We define the ith input constraint length of the encoder by $v_i = \max_j \{\deg G_i^j(D)\}$ and the overall constraint length of the encoder by $v = \sum_{i=1}^{k} v_i$. The jth encoded

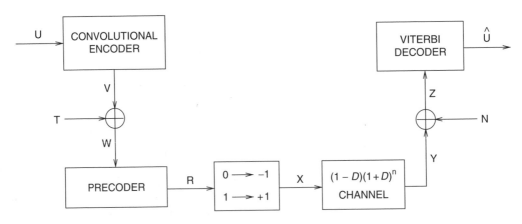

FIGURE 22.1   The convolutionally coded partial-response system. (Copyright ©2001 IEEE. Reproduced with permission.)

sequence, $V^j(D)$, can be written in terms of $\mathbf{U}(D)$ and $\mathbf{G}(D)$ as follows:

$$V^j(D) = \sum_{i=1}^{k} U^i(D)G_i^j(D), \quad 1 \le j \le m \tag{22.1}$$

The m encoded sequences are multiplexed into a single sequence, called the *codeword*, as follows:

$$V(D) = \sum_{j=1}^{m} D^{j-1} V^j(D^m) \tag{22.2}$$

A coset of the code is used to avoid the all zero sequence, since this sequence causes loss of clock synchronization. The coset sequence $\tilde{T}(D)$ is added to the codeword $V(D)$, giving

$$W(D) \equiv V(D) \oplus \tilde{T}(D) \tag{22.3}$$

where $\oplus$ denotes addition of polynomials over $GF(2)$. The sequence $W(D)$ is passed through a precoder. In the communication context, precoding is used as a method of avoiding error propagation in symbol-by-symbol detection of partial-response signals [11]. In this chapter, where maximum-likelihood sequence detection is adopted, the precoder has other purposes, as will become clear in the following sections. The transfer function of the precoder for the $P_n(D) = (1 - D)(1 + D)^n$ channel is chosen to be $(1 \oplus D)^{-n-1}$, which corresponds to $[1/P_n(D)]_{\mathrm{mod}\,2}$. The precoded sequence, $R(D)$, is then given by

$$R(D) \equiv (1 \oplus D)^{-n-1} W(D) \tag{22.4}$$

where we consider the two sequences $R(D)$ and $W(D)$ to have the same length. The polar NRZ modulator (i.e., the map: $0 \to -1$, $1 \to +1$) is introduced to make the system of Figure 22.1 compatible with saturation recording [6]. The output of the modulator is given by

$$X(D) = 2R(D) - 1(D) \tag{22.5}$$

where $1(D)$ is the all one sequence with the same length as $R(D)$. The modulated sequence $X(D)$ is then sent to the $P_n(D) = p_0 + p_1 D + p_2 D^2 + \cdots + p_{n+1} D^{n+1}$ channel, which results in the multilevel channel output sequence

$$Y(D) = P_n(D)X(D) + \sum_{j=1}^{n+1} \left(\sum_{i=j}^{n+1} (-1) p_i \right) D^{j-1} - \left\{ \begin{array}{l} \text{terms in } D^j, \text{ where} \\ j \ge \text{length of } W(D) \end{array} \right\} \tag{22.6}$$

where the second term is the response (independent of $X(D)$) due to the initial content of the unit-delay cells of the $P_n(D)$ channel (namely $-1, -1, \ldots, -1$). We illustrate this with an example.

Example 22.1

Consider the $P_2(D) = (1 - D)(1 + D)^2$ channel and the sequence $W(D) = (10110100) = 1 + D^2 + D^3 + D^5$. Then the precoder output is $R(D) = (11100000) = (1 \oplus D)^{-3} W(D) = 1 + D + D^2$. The modulated sequence is $X(D) = (+ + + - - - - -) = 2R(D) - 1(D) = 1 + D + D^2 - D^3 - D^4 - D^5 - D^6 - D^7$. Finally, the multilevel channel output sequence is $Y(D) = (+2, +4, +2, -2, -4, -2, 0, 0) = X(D)P_2(D) + (1 + 2D + D^2) - (D^8 + 2D^9 + D^{10}) = 2 + 4D + 2D^2 - 2D^3 - 4D^4 - 2D^5$.

A closed form for the number of output levels as a function of n is in general difficult to obtain. But for the most commonly used channels, namely for $n = 0, 1, 2,$ or 3, these numbers are well known. For $n = 0$ and $n = 1$ there are three levels, namely 0 and ± 2. For $n = 2$ there are five levels, namely 0, ± 2, and ± 4. And for $n = 3$ there are seven levels, namely 0, ± 2, ± 4, and ± 6.

After passing through the partial response "channel," zero mean, i.i.d., Gaussian noise $N(D)$ is added to $Y(D)$, producing the noisy sequence $Z(D)$. A Viterbi detector is then used to find the maximum likelihood estimate $\hat{U}(D)$ of $U(D)$ from the noisy channel output $Z(D)$.

22.3 Trellis Codes for Partial-Response Channels Based Upon the Hamming Metric

For the trellis code formed at the output of the $1 - D$ channel, in the encoded system of Figure 22.1 (with $n = 0$), Wolf and Ungerboeck [1] derived a lower bound on the minimum free squared Euclidean distance, which is monotonically related to the minimum free Hamming distance of the convolutional code. From this result, it follows that convolutional codes with large free Hamming distance generate trellis codes with large free squared Euclidean distance. In this section, we present the extension of the result for the $1 - D$ channel to the $P_n(D)$ channel case, and conclude that similar guidelines exist for choosing a convolutional code that leads to a trellis code with large free squared Euclidean distance. We also introduce some concepts which will be useful in the forthcoming sections.

The set of all output sequences $V(D)$, as the input sequence $\mathbf{U}(D)$ ranges over all possible values, is called the *convolutional code*, denoted by C, and generated by the encoder $\mathbf{G}(D)$. Let $\omega_H(V(D))$ be the Hamming weight of the coded sequence $V(D)$, that is, the number of nonzero coefficients in $V(D)$. The minimum free Hamming distance of C, denoted by $d_H(C)$, is defined as the minimum of the set $\{\omega_H(V(D)) \mid V(D) \in C, V(D) \neq 0\}$.

We denote the *channel code* by the set of noiseless output sequences $Y(D)$ from the $P_n(D)$ channel in Figure 22.1. This code is multilevel and nonlinear trellis. The trellis for the channel code is called the *decoder trellis*, which is the trellis for the overall system modelled as the combination of the convolutional encoder, the precoder, and the $P_n(D)$ channel. A *channel codeword* of the channel code consists of the multilevel labels on any path in the decoder trellis that starts in any state and ends in any state (not necessarily the same as the starting state). We say that an *error event* of length l has occurred when two paths in the decoder trellis, say $Y(D)$ and $Y'(D)$, diverge from each other at time t and remerge at time $t + l - 1$. We may assume with no loss of generality that $t = 0$ and that the correct channel codeword $Y(D) = y_0 + y_1 D + \cdots + y_{l-1}D^{l-1}$ and the incorrect channel codeword $Y'(D) = y'_0 + y'_1 D + \cdots + y'_{l-1}D^{l-1}$. A typical error event is shown in Figure 22.2.

The squared Euclidean distance between $Y(D)$ and $Y'(D)$ is defined as

$$d_E^2(Y, Y') \triangleq \sum_{i=0}^{l-1} (y_i - y'_i)^2 \tag{22.7}$$

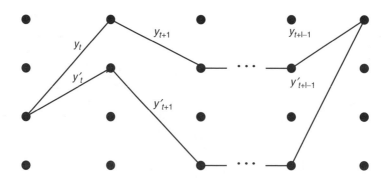

FIGURE 22.2 A typical error event in a decoder trellis. (Modified from Uchôa Filho, B. F. and Herro, M. A., *IEEE Trans. Inform. Theory*, vol. 43, No. 2, pp. 441–453, Mar. 1997. Copyright ©1997 IEEE. With permission.)

The minimum *free* squared Euclidean distance of the channel code is defined as

$$d^2_{\text{free}} \stackrel{\Delta}{=} \min_{Y \neq Y'} \left(d^2_E(Y, Y') \right) \tag{22.8}$$

where the minimization is over all possible error events.

We now present a lower bound on the minimum free squared Euclidean distance, d^2_{free}, of the channel code for the $P_n(D)$ channel, generated by the convolutional code C. This lower bound relates $d_H(C)$ to d^2_{free}. In the derivation we consider $\tilde{T}(D) \equiv 0$. However, it can easily be shown that adding $\tilde{T}(D) \not\equiv 0$ modulo 2 to every codeword of C, as given in Equation 22.3, will not change $d_H(C)$. Thus the bound is also valid for $\tilde{T}(D) \not\equiv 0$.

We define the function $|Y(D)|$ for polynomials $Y(D)$ with even coefficients as

$$|Y(D)| \stackrel{\Delta}{=} \frac{y_0}{2} + \frac{y_1}{2}D + \frac{y_2}{2}D^2 + \cdots \pmod 2 \tag{22.9}$$

where $|Y(D)|$ reduces to a binary sequence because the (mod 2) reduction is carried out on each coefficient of $Y(D)$, after the division by 2. As an example, consider the sequence $Y(D) = 2 + 4D + 2D^2 - 2D^3 - 4D^4 - 2D^5$ given in Example 22.1. We have that $|Y(D)| \equiv 1 + 2D + D^2 - D^3 - 2D^4 - D^5 \pmod 2 \equiv 1 + D^2 + D^3 + D^5$.

It is shown in Appendix A that the correspondence between $Y(D)$ and $W(D)$ is given by

$$W(D) \equiv |Y(D)| \tag{22.10}$$

Note that the binary sequence obtained in the previous paragraph is the sequence $W(D)$ in Example 22.1. It can be seen from Equation 22.10 that the sequence $W(D)$ can be recovered from the noiseless channel output $Y(D)$ in a symbol-by-symbol fashion. Moreover, because of Equation 22.10, whenever a bit w is 0, the corresponding channel output y is divisible by 4, and whenever a bit w is 1, the corresponding channel output y is divisible by 4 with remainder 2. This can be expressed analytically as:

$$w \equiv 0 \pmod 2 \Longleftrightarrow y \in 4\mathbb{Z} \stackrel{\Delta}{=} \{0, \pm 4, \pm 8, \dots\}$$

$$w \equiv 1 \pmod 2 \Longleftrightarrow y \in 4\mathbb{Z} + 2 \stackrel{\Delta}{=} \{2, \pm 6, \pm 10, \dots\} \tag{22.11}$$

where $\mathbb{Z}$ is the set of all integers.

We now define the binary representation of an error event $E(D)$, as done in references [5, 15]. The binary representation of an error event $E(D)$ is the modulo 2 sum of the two binary sequences $W(D)$ and $W'(D)$ that produce, at the output of the channel, the sequences $Y(D)$ and $Y'(D)$, respectively. Using the function defined in Equation 22.9 we may write

$$\begin{aligned} E(D) &\stackrel{\Delta}{=} (w_0 \oplus w'_0) \oplus (w_1 \oplus w'_1)D \oplus \cdots \oplus (w_{l-1} \oplus w'_{l-1})D^{l-1} \\ &\stackrel{\Delta}{=} (|y_0| \oplus |y'_0|) \oplus (|y_1| \oplus |y'_1|)D \oplus \cdots \oplus (|y_{l-1}| \oplus |y'_{l-1}|)D^{l-1} \\ &\stackrel{\Delta}{=} e_0 \oplus e_1 D \oplus \cdots \oplus e_{l-1}D^{l-1} \end{aligned} \tag{22.12}$$

Note that $E(D)$ is independent of $\tilde{T}(D)$ since $E(D) \equiv W(D) \oplus W'(D) \equiv V(D) \oplus \tilde{T}(D) \oplus V'(D) \oplus \tilde{T}(D) \equiv V(D) \oplus V'(D)$. It also follows from these equivalences that $E(D) \equiv V(D) \oplus V'(D)$ is a codeword of C, since $V(D)$ and $V'(D)$ are two codewords of C, and C is a linear code. Zehavi and Wolf [5] observed that since $E(D)$ must be a codeword of the convolutional code, only sequences that are codewords can be potential error events, that is, not every sequence is the binary representation of an error event. In the next lemma we reduce even further the set of such possible binary representations.

Define the subset C_n of the convolutional code C as

$$C_n \triangleq \{V(D) \in C \mid (1 \oplus D)^{n+1} \text{ divides } V(D)\}$$

Clearly C_n is a linear subcode of C and $C_{-1} = C \supseteq C_0 \supseteq C_1 \supseteq C_2 \supseteq \cdots$.

Lemma 22.1   If $E(D)$ is the binary representation of an error event, then $E(D) \in C_n$.

Proof 22.1   Consider the difference sequence $Y(D) - Y'(D)$. From Equation 22.6 we have that

$$Y(D) - Y'(D) = (X(D) - X'(D))P_n(D)$$

since the terms independent of $X(D)$ cancel out. If we apply the polynomial function defined in Equation 22.9 to the previous equation and use the results in Appendix B, we have the following equivalences:

$$|Y(D) - Y'(D)| \equiv |(X(D) - X'(D))P_n(D)|$$
$$\equiv |X(D) - X'(D)|[P_n(D)]_{\text{mod } 2}$$
$$\equiv |X(D) - X'(D)|(1 \oplus D)^{n+1}$$
$$\equiv E(D)$$

where the last equivalence, namely $E(D) \equiv |Y(D) - Y'(D)|$, comes from Equation 22.12. Therefore $(1 \oplus D)^{n+1}$ divides $E(D)$ and consequently $E(D) \in C_n$ as claimed.   □

We are now ready to state the lower bound on d_{free}^2.

Theorem 22.1   The minimum free squared Euclidean distance of the channel code, for the precoded $P_n(D) = (1 - D)(1 + D)^n$ channel, generated by the convolutional code C, according to the encoded system of Figure 22.1, is lower bounded by:

$$d_{\text{free}}^2 \geq 4d_H(C_n)$$

Proof 22.2   First note that because of Equation 22.11 whenever two encoded bits, say w and w', differ from each other, the squared Euclidean distance between their corresponding channel outputs, namely y and y', is lower bounded by the intersubset squared Euclidean distance between the subsets $4\mathbb{Z}$ and $4\mathbb{Z} + 2$, which is 4. The rest of the proof follows from Lemma 22.1. The binary representation of any error event in the decoder trellis is a codeword in C_n, and this has at least $d_H(C_n)$ ones. Therefore d_{free}^2 is lower bounded by $4d_H(C_n)$.   □

Remark 22.1   For $n = 0$ this bound is equivalent to the bound of Wolf and Ungerboeck (Lemma 22.2, [1]), namely $d_{\text{free}}^2 \geq 4d_H^{(e)}$, where $d_H^{(e)}$ is the weight of the lowest even weight codeword in C. To see this equivalence note that, for $n = 0$, C_0 is the set of all codewords of C that have even weight, since a polynomial $V(D)$ is divisible by $(1 \oplus D)$ in $GF(2)$ if and only if $V(D)$ has even weight. Thus, for $n = 0$, $d_H(C_0) = d_H^{(e)}$.

The immediate consequence of Theorem 22.1 is that convolutional codes whose subcode C_n has large minimum free Hamming distance $d_H(C_n)$ are good candidates to generate channel codes with large d_{free}^2. The advantage of this construction is that many convolutional codes designed for the Hamming metric have already been found and are tabulated in references such as [10]. Many of these channel codes for the $1 - D$ channel were found in [1]. For this simpler channel, the bound given in Theorem 22.1 is always tight [1]. However, for the $P_n(D)$ channel $(n > 0)$, the actual d_{free}^2 of the channel code is often much larger than the lower bound. We can see this by means of an example. Assume $n = 2$, so that the output levels are $0, \pm 2$, and ± 4. The bit $w = 0$ may be converted to the output level $y = -4$, and the bit $w' = 1$ may be converted to $y' = +2$. Then $d_E^2(y, y') = d_E^2(-4, +2) = 36$, as opposed to 4 given by the bound in Theorem 22.1. This discrepancy between the actual d_{free}^2 and the bound becomes even more prominent when $\tilde{T}(D) \not\equiv 0$, as we shall verify in Section 22.4. We now give an example of channel codes designed using this method.

Example 22.2

Consider the $(4, 1)$ convolutional code C with constraint length $v = 1$ generated by $\mathbf{G}(D) = [1+D, 1, 0, 1]$. This code has $d_H(C) = d_H(C_0) = d_H(C_2) = 6$. Hence the bound given in Theorem 22.1 becomes $d_{\text{free}}^2 \geq 4 \times 6 = 24$. For $n = 0$, the channel code generated by C has $d_{\text{free}}^2 = 24$. But for $n = 2$, $d_{\text{free}}^2 = 56 > 24$.

Since the bound given in Theorem 22.1 is weak when $n > 0$, basing a search for channel codes with large d_{free}^2 solely on convolutional codes with large d_H appears to be too restrictive and does not exploit the channel memory in an efficient way. A computer-aided search is thus required to find good channel codes. The theoretical elements related to this search are presented next in Section 22.4.

22.4 Trellis-Matched Codes for Partial-Response Channels

In this section we point out some structural properties of channel codes and establish the rules (for choosing $\mathbf{G}(D)$ and $\tilde{T}(D)$) which guided the computer search that led to the channel codes given in Section 22.8.

The first concern is with the state complexity of the decoder trellises for the channel codes generated by the convolutional encoders. As mentioned in the introductory section, under binary inputs the $P_n(D)$ channel has 2^{n+1} states. The precoder does not increase the number of states since it shares the same (up to a one-to-one map — the polar NRZ map) states with the $P_n(D)$ channel. In other words, *once the state of the precoder is known, the state of the channel trellis can be determined*. On the other hand, since the decoder trellis is obtained from the product of the trellis of the convolutional code (which has 2^v states) and the channel trellis, the number of states of the decoder trellis is in general 2^{n+v+1}. However, depending on a structural property of the encoder, the decoder trellis can have fewer states. We use the following classification as we refer to the number of states of the decoder trellis.

We say that a channel code is unmatched (UM) to the $P_n(D)$ channel if the decoder trellis has 2^{n+v+1} states; that it is partially trellis matched (PTM) to the $P_n(D)$ channel if the decoder trellis has $2^{n'+v+1}$ states, where $0 \leq n' < n$; and that it is totally trellis matched (TTM) (or simply trellis matched (TM)) to the $P_n(D)$ channel if the decoder trellis has at most 2^v states. Since reduced-complexity decoder trellises are preferred, this chapter focuses on channel codes that are TTM to the $P_n(D)$ channel.

In Theorem 22.2 we will present a sufficient condition for a convolutional encoder to generate a channel code that is TTM to the $P_n(D)$ channel. To do this we require the following definitions.

Definition 22.1   We define the *input generator vector*, $\mathbf{G_{in}}(D)$, as another representation for an (m, k) convolutional encoder which is equivalent to and derived from $\mathbf{G}(D)$ as given below:

$$\mathbf{G_{in}}(D) = \left[G_{\text{in}}^1(D), \ldots, G_{\text{in}}^k(D) \right]$$

where $G_{\text{in}}^i(D) = \sum_{j=1}^{m} D^{j-1} G_i^j(D^m)$, $1 \leq i \leq k$. Note that $G_{\text{in}}^i(D)$ is the result of multiplexing the polynomials in the ith row of $\mathbf{G}(D)$.

The input generator vector, for a particular encoder, is given in the following example.

Example 22.3

Consider the (5,2) convolutional encoder with constraint length $v = 2$ represented by the following generator matrix:

$$\mathbf{G}(D) = \begin{pmatrix} D & 1+D & 1 & 0 & 0 \\ 1 & 0 & 0 & D & 0 \end{pmatrix}$$

Then $m = 5$, $k = 2$, and $\mathbf{G}_{\text{in}}(D) = [G_{\text{in}}^1(D), G_{\text{in}}^2(D)] = [D + D^2 + D^5 + D^6, 1 + D^8]$.

Definition 22.2   An encoder is said to be *feedback free* if none of the $G_i^j(D)$ are rational polynomials.

Clearly the encoder of Example 22.3 is a feedback free encoder. All encoders considered in this chapter are feedback free encoders. In the following theorem, $\tilde{T}(D) \equiv 0$. The case $\tilde{T}(D) \not\equiv 0$ will be treated later in this section.

Theorem 22.2   Let n' be the largest nonnegative integer (if it exists) such that the input generator vector $\mathbf{G}_{\text{in}}(D)$ for C may be factored as:

$$\mathbf{G}_{\text{in}}(D) = (1 \oplus D)^{n'+1}\, \mathbf{G}'_{\text{in}}(D)$$

Then the channel code generated by C is TTM to the $P_n(D)$ channel if $n' \geq n$.

***Proof* 22.3**   We begin this proof by writing the overall output sequence $V(D)$ in terms of $\mathbf{G}_{\text{in}}(D)$. Substituting Equation 22.1 into Equation 22.2 yields

$$V(D) = \sum_{j=1}^{m} D^{j-1} \left(\sum_{i=1}^{k} U^i(D^m)G_i^j(D^m) \right)$$

$$= \sum_{i=1}^{k} U^i(D^m) \sum_{j=1}^{m} D^{j-1}\, G_i^j(D^m)$$

$$= \sum_{i=1}^{k} U^i(D^m)G_{\text{in}}^i(D)$$

From the assumption of the theorem we may write $V(D)$ as

$$V(D) = (1 \oplus D)^{n'+1} \sum_{i=1}^{k} U^i(D^m)G_{\text{in}}'^i(D) \tag{22.13}$$

Then the precoded output in equivalence (22.4), for $\tilde{T}(D) \equiv 0$, becomes

$$R(D) = (1 \oplus D)^{-n-1}V(D) = (1 \oplus D)^{n'-n} \sum_{i=1}^{k} U^i(D^m)G_{\text{in}}'^i(D)$$

Now we note that if $n' \geq n$ the feedback free modified encoder

$$(1 \oplus D)^{n'-n}\mathbf{G}'_{\text{in}}(D) \tag{22.14}$$

may be used to produce $R(D)$ directly from the input $U(D)$, and it has at most 2^v states. To see this note that since the degree of $G_{\text{in}}^i(D)$ satisfies $mv_i \leq \deg G_{\text{in}}^i(D) \leq mv_i + m - 1$, the input constraint length of the encoder, v_i, can be determined by:

$$v_i = \left\lfloor \frac{\deg G_{\text{in}}^i(D)}{m} \right\rfloor$$

where $[x]$ is the largest integer less than or equal to x. From Equation 22.14, the corresponding degree for the modified encoder is equal to $n' - n + \deg G_{\text{in}}^{\prime i}(D) = n' - n + (\deg G_{\text{in}}^i(D) - n' - 1) = \deg G_{\text{in}}^i(D) - n - 1$. Then the input constraint length of the modified encoder, $v_i(\text{modified})$, can be similarly determined by:

$$\begin{aligned}
v_i(\text{modified}) &= \left\lfloor \frac{\deg G_{\text{in}}^i(D) - n - 1}{m} \right\rfloor \\
&\leq \left\lfloor \frac{mv_i + m - n - 2}{m} \right\rfloor \\
&\leq v_i,
\end{aligned}$$

which shows that the modified encoder has at most 2^v states (since $v = \sum_{i=1}^{k} v_i$).

It follows from these facts that all states of the precoder can be determined by at most 2^v states, the states of the modified encoder. We know, from the second paragraph of this section, that the states of the precoded $P_n(D)$ channel trellis can be determined from the states of the precoder. Therefore the states of the precoded $P_n(D)$ channel trellis can be determined from the states of the modified encoder, and consequently the decoder trellis has at most 2^v states if $n' \geq n$. □

In the following we define a class of convolutional encoders for which an encoder, combined with the precoder for the $P_n(D)$ channel, generates a decoder trellis with exactly 2^v states.

Definition 22.3   There can exist many encoders $\mathbf{G}(D)$ that generate the same code C. An encoder $\mathbf{G}(D)$ of constraint length v (realized with 2^v states) that generates C is said to be a *minimal encoder* for C if no other encoder with constraint length smaller than v (realized with fewer than 2^v states) generates C.

Lemma 22.2   If the convolutional encoder $\mathbf{G}(D)$ is minimal and satisfies the condition in Theorem 22.2, then it generates a decoder trellis with *exactly* 2^v states.

***Proof* 22.4**   Let C be the convolutional code generated by a minimal convolutional encoder $\mathbf{G}(D)$ (or $\mathbf{G}_{\text{in}}(D)$), with constraint length v, satisfying the condition in Theorem 22.2. Let C be the channel code generated by C. Suppose that a decoder trellis for C with fewer than 2^v states exists. If we apply the equivalence in Equation 22.10 to each channel codeword $Y(D)$ of C, the resulting set of sequences is clearly the convolutional code C. Hence, by this procedure, we have found a trellis for C with fewer than 2^v states, which contradicts the assumption on the minimality of $G(D)$. Therefore, from Theorem 22.2, the decoder trellis has exactly 2^v states. □

A criterion for determining whether or not a convolutional encoder is minimal was established by Forney [8]. He showed that an encoder $\mathbf{G}(D)$ is minimal if and only if each k by k subdeterminant of $\mathbf{G}(D)$ has degree not exceeding v, and the greatest common divisor of all such subdeterminants is equal to 1. We treat the case where the encoder is nonminimal later in this section.

In Theorem 22.2, we note that since $n' \geq n$, Equation 22.13 implies that *every codeword $V(D)$ of C is divisible by $(1 \oplus D)^{n+1}$*, which further implies that $C_n = C$. Hence, from Lemma 22.1, *every codeword $V(D)$ of C is the binary representation of some error event*. From this fact, and from Lemma 22.2, it follows that the decoder trellis and the trellis of the convolutional code are the same, except for the labels. We illustrate this by means of an example.

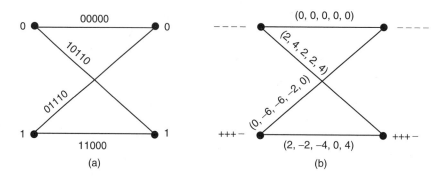

FIGURE 22.3   (a) The trellis of the convolutional code and; (b) the decoder trellis of Example 22.4. (Copyright ©1997 IEEE. Reproduced with permission.)

Example 22.4

Consider the minimal (5,1) convolutional encoder with constraint length $\nu = 1$, represented by the following generator matrix $\mathbf{G}(D) = [1, D, 1 + D, 1 + D, 0]$. The input generator vector can be factored as $\mathbf{G}_{\text{in}}(D) \equiv [1 \oplus D^2 \oplus D^3 \oplus D^6 \oplus D^7 \oplus D^8] \equiv (1 \oplus D)^5[1 \oplus D \oplus D^3]$. The trellis of the convolutional code and the decoder trellis when the $P_3(D) = (1-D)(1+D)^3$ channel is used are shown in Figure 22.3.

Remark 22.2   In this chapter, we label the states of the decoder trellis with the contents of the unit-delay cells of the precoded $(1 - D)(1 + D)^n$ partial-response system when the inputs are the coset branch bits. In the decoder trellis of Figure 22.3(b), for example, the state labels are "$- - - -$" and "$+ + + -$".

In Example 22.4, $n' = 4 > 3 = n$. Note that both trellises are the same. Also note the correspondence between the bits in the trellis of the convolutional code and the multilevel labels in the decoder trellis. It satisfies the equivalence in Equation 22.10. Every codeword $V(D)$ of this convolutional code is divisible by $(1 \oplus D)^5$, which implies that they are also divisible by $(1 \oplus D)^4$, where $4 = n + 1$. Recall that we can also represent a codeword of a convolutional code by the labels on a path in the trellis of the convolutional code that starts and ends in the zero state. For example, consider the codeword "101101100001110," which corresponds to the path in the trellis of Figure 22.3(a) that leaves state zero, goes to state one and remains there for one time instant, and then returns to state zero again. The D-transform of this codeword is $V(D) \equiv 1 \oplus D^2 \oplus D^3 \oplus D^5 \oplus D^6 \oplus D^{11} \oplus D^{12} \oplus D^{13} \equiv (1 \oplus D)^5(1 \oplus D \oplus D^3 \oplus D^5 \oplus D^6 \oplus D^8)$, that is, $V(D)$ is divisible by $(1 \oplus D)^5$.

The convolutional code in Example 22.4 has $d_H(C) = d_H(C_3) = 6$, so that the bound of Theorem 22.1 becomes $d_{\text{free}}^2 \geq 4 \times 6 = 24$. However the corresponding channel code has $d_{\text{free}}^2 = 108$, more than four times larger than the bound. This example also supports the fact that the bound of Theorem 22.1 is rather weak.

22.5   Run-Length Limited Trellis-Matched Codes

22.5.1   Cosets of Convolutional Codes

Now we begin to consider nontrivial cosets of convolutional codes, that is, we start to analyze the case $\tilde{T}(D) \neq 0$. Assume that $\tilde{T}(D)$ is not a codeword of C. Then the set

$$C \oplus \tilde{T} \triangleq \{V(D) \oplus \tilde{T}(D) | V(D) \in C\}$$

is a *nontrivial coset* of the convolutional code C. The trellis for the coset $C \oplus \tilde{T}$ is called the *coset trellis* [15]. The sequence $\tilde{T}(D)$ is called the *coset representative* and is not unique. The binary labels on any path in the coset trellis that starts and ends in the all zero state is called a *coset word*. $\tilde{T}(D)$ is chosen to be a periodic sequence of period m, the number of output lines of the convolutional encoder. Let $T(D)$ be the polynomial of degree at most $m - 1$ which corresponds to a period of $\tilde{T}(D)$. With this choice the binary addition in Equation 22.3, which generates the coset of C, can be implemented by simply inverting the bits on the output lines of the convolutional encoder corresponding to ones in $T(D)$. The periodic sequence $\tilde{T}(D)$ can be written in terms of $T(D)$ as below:

$$\tilde{T}(D) = \sum_{\ell=1}^{n_b} D^{(\ell-1)m} T(D) \tag{22.15}$$

where n_b is the number of branches in the path whose labels form the codeword $V(D)$ in Equation 22.3.

We denote the *maximum zero-run length* of the channel code, that is, the maximum run of zeros between two consecutive nonzero symbols in any path of the decoder trellis, by $L_{\max}$. We can see from Equation 22.11 that the channel output $y = 0$ only if the coset bit $w = 0$. Hence $L_{\max}$ can be upper bounded by the maximum zero-run length of the coset. Focusing on the $(1 - D)$ channel, Hole and Ytrehus [15] have observed that a good choice of $T(D)$ may not only limit $L_{\max}$, but may also yield a trellis code with d_{free}^2 larger than the lower bound given in Theorem 22.1. It was shown in [13] that this assertion remains true for the $P_n(D)$ channel. We verify this by means of an example.

Example 22.5

Consider the convolutional encoder of Example 22.4. The coset trellis and the decoder trellis for the choice $T(D) = [10001] = 1 + D^4$ are given in Figure 22.4.

The channel code given in Example 22.5 has $d_{\text{free}}^2 = 264$. This is 11 times larger than the bound of Theorem 22.1. Note also that the maximum zero-run length of the coset is 3. However, we can see by inspection that the decoder trellis of Figure 22.4(b) has $L_{\max} = 2 < 3$. This is because some of the bits equal to "0" in the coset are converted into either -4 or 4 in the decoder trellis.

It is important to note that the output lines of the convolutional encoder can not be permuted. This would not necessarily imply the corresponding permutation of the multilevel labels in the branches of the decoder trellis. This is due to the fact that $P_n(D)$ channels have memory. The coset representative cannot be changed either. For example, if we choose $T(D) = [01111] = D + D^2 + D^3 + D^4$ instead of the $T(D)$ given in Example 22.5, the channel code generated by this coset can not be represented by a trellis with only two states. Four states are needed instead. To see this consider the four input sequences: $U_1^{(1)} = [0]$,

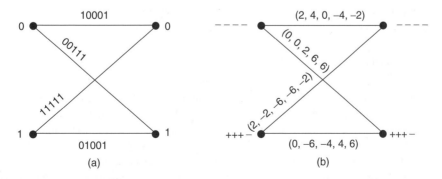

FIGURE 22.4 (a) The coset trellis and; (b) the decoder trellis of Example 22.5. (Copyright ©1997 IEEE. Reproduced with permission.)

$U_1^{(2)} = [1]$, $U_1^{(3)} = [1, 1]$, and $U_1^{(4)} = [1, 1, 0]$. The branch labels in the coset trellis for these inputs are: $R^{(1)} = [01111]$, $R^{(2)} = [11001]$, $R^{(3)} = [11001, 10111]$ and $R^{(4)} = [11001, 10111, 00001]$, respectively. These branch labels in turn drive the channel state from "$----$" to "$----$", "$---+$", "$+++-$" and "$+++$," respectively. Any other sequence will drive the channel state from "$----$" to one of the channel states above. Therefore, the decoder trellis has four states. After these observations we can see that a good choice of the coset $C \oplus \tilde{T}$ is necessary to achieve good results. We devote the remainder of this section to establishing the rules that lead to good choices of $T(D)$.

In the last paragraph, we noted that the choice of $T(D)$ may also have an effect on the structure of the channel code. In particular, while C generates a channel code that is TTM to the $P_n(D)$ channel, the coset $C \oplus \tilde{T}$ may not do so. It is therefore important to know for which polynomials $T(D)$ the channel code generated by the coset $C \oplus \tilde{T}$ is TTM to the $P_n(D)$ channel. This is addressed in the theorem below.

Theorem 22.3 Consider an (m, k) convolutional encoder that generates C and satisfies the condition in Theorem 22.2, that is, C generates a channel code that is TTM to the $P_n(D)$ channel. Let $C \oplus \tilde{T}$ be a coset of C, where $T(D)$ is a polynomial of degree at most $m - 1$, and $\tilde{T}(D) \notin C$. Then the channel code generated by $C \oplus \tilde{T}$ is TTM to the $P_n(D)$ channel if $(1 \oplus D)^{n+1}$ divides $T(D)$, which in turn requires that $m \geq n + 2$.

Proof 22.5 From Equation 22.15, since $(1 \oplus D)^{n+1}$ divides $T(D)$, $(1 \oplus D)^{n+1}$ certainly divides $\tilde{T}(D)$. By assumption, $(1 \oplus D)^{n+1}$ divides $V(D)$. Consequently $(1 \oplus D)^{n+1}$ also divides the coset sequence $W(D) \equiv V(D) \oplus \tilde{T}(D)$ in Equation 22.3. If we replace $V(D)$ by $W(D) \equiv V(D) \oplus \tilde{T}(D)$ in Theorem 22.2, and repeat the proof of that theorem, we have proved the first part of Theorem 22.3. The requirement on the minimum number of encoder output lines m can be seen from the fact that $T(D)$ must have degree at least $n + 1$ for $(1 \oplus D)^{n+1}$ to divide $T(D)$. Since the degree of $T(D)$ is at most $m - 1$, the result follows immediately. □

Note that in Example 22.5, $T(D) \equiv 1 \oplus D^4 \equiv (1 \oplus D)^4$, which satisfies the condition in Theorem 22.3.

The remainder of this section discusses the search for good channel codes performed in [13, 20], whether adding constraints to $G(D)$ and $T(D)$ can lead to good channel codes while reducing search time, or whether these constraints overly restrict the choices of $G(D)$ and $T(D)$, making it difficult to find good channel codes.

Note that the requirement on the minimum number of encoder output lines $m \geq n + 2$, from Theorem 22.3, imposes a constraint on the choice of code rates. For example, run-length limited time-invariant channel codes of rate $R_c = k/3$, TTM to the $(1 - D)(1 + D)^2$ channel, do not exist. Neither do run-length limited time-invariant channel codes of rate $R_c = k/4$, TTM to the $(1 - D)(1 + D)^3$ channel. The choice of $T(D)$ as in Theorem 22.3 always leads to a time-invariant decoder trellis. If $T(D)$ is not so restricted, however, there may still exist TTM channel codes with a time-varying decoder trellis. In this chapter, only the time-invariant case, that is, only the case of cosets that satisfy the condition in Theorem 22.3, is considered.

22.5.1.1 Bit Stuffing

A very simple technique to limit $L_{\max}$ is the so-called *bit stuffing* technique. In this technique we form the (m, k) convolutional code as follows. First a convolutional code with parameters $(m - b, k)$, with $1 \leq b < m - k$, is used. Then b "ones" are "stuffed" into the b bit positions (left uncoded), and this leads to the desired (m, k) code. A simple bound on $L_{\max}$ can be derived when this technique is used. For example, let "1" represent a "stuffed" bit and "x" an encoded bit. Suppose that $m = 8$ and $b = 2$, and we choose "$xx1x1xxx$". Since "x" can be either "0" or "1," the worst scenario would be the following: $\ldots xx1x1000001x1xxx \ldots$. Thus $L_{\max}$ can be bounded by $L_{\max} \leq 5$. Another equivalent representation for the encoder in Figure 22.1 which is particularly helpful to visualize the bit stuffing technique is as follows.

Definition 22.4   *We define the output generator vector as*

$$\mathbf{G}_{\text{out}}(D) = \left[G_{\text{out}}^1(D), \ldots, G_{\text{out}}^m(D) \right]$$

where $G_{\text{out}}^j(D) = \sum_{i=1}^k D^{i-1} G_i^j(D^k)$, $1 \le j \le m$.

The representation given in Definition 22.4, for the encoder in Example 22.3, is $\mathbf{G}_{\text{out}}(D) = [G_{\text{out}}^1(D),$ $G_{\text{out}}^2(D), G_{\text{out}}^3(D), G_{\text{out}}^4(D), G_{\text{out}}^5(D)] = [D + D^2, 1 + D^2, 1, D^3, 0]$. Note that $\mathbf{G}_{\text{out}}(D)$ is the modified 1 by m generator matrix $\mathbf{G_m}(D)$ introduced for the same purpose in [15]. We can see that an encoder $\mathbf{G}(D)$ with b all zero columns, or equivalently, an encoder $\mathbf{G}_{\text{out}}(D)$ with b coordinates equal to zero, along with a choice of $T(D)$ having ones in at least one of the b positions that are equal to zero in $\mathbf{G}_{\text{out}}(D)$, may be used to accomplish the bit stuffing technique. In other words, the encoded system of Figure 22.1 is general enough to realize bit stuffing by an appropriate choice of $\mathbf{G}_{\text{out}}(D)$ and $T(D)$. Despite its simplicity, this technique can generate channel codes with large d_{free}^2 and very low values of $L_{\max}$. However, the bit stuffing technique considerably limits the number of possible choices of encoders. As a result, no good codes, and in some cases no code at all, exist for certain code rates. In a code search, the bit stuffing technique is used whenever it leads to good channel codes. When this is not the case, $L_{\max}$ is determined by inspection.

22.6   Avoiding Flawed Codewords

A channel codeword is said to be a *flawed* channel codeword if the initial state is not uniquely determined from the multilevel labels alone [7, 15]. Such a codeword exists in a channel code if and only if there are two infinitely long paths in the decoder trellis with the same multilevel labels. This is an undesirable situation since it may lead to ambiguity in the decoding process. For the $1 - D$ channel, Hole and Ytrehus [15] showed that minimal encoders generate channel codes that do not contain flawed codewords. Their proof was based on a property of the convolutional codes generated by minimal encoders derived by Forney [9]. It can be similarly shown that the same is true for the $P_n(D)$ channels. However, in this case, some nonminimal encoders also generate channel codes not containing flawed codewords. This is due to the fact that each of the coset bits "0" and "1" can be converted, at the output of the $P_n(D)$ channel, into more than one multilevel symbol, if $n > 0$. Consequently, two infinitely long paths in the trellis of the convolutional code with the same binary labels may be converted by the $P_n(D)$ channel into two paths in the decoder trellis with different labels. In other words, a convolutional code containing flawed codewords may generate a channel code free of flawed channel codewords. In addition, there exist many nonminimal encoders that satisfy the condition in Theorem 22.2. Note that minimality of convolutional encoders by itself does not imply the total match of a channel code, i.e., the minimum number of states for the decoder trellis. In fact minimality is not a condition in Theorem 22.2. Therefore some nonminimal encoders can lead to trellis matched channel codes, whereas some minimal encoders may not. In the next example, we give a nonminimal (4,2) convolutional encoder with constraint length $\nu = 2$ for the $(1 - D)(1 + D)^2$ channel. For the same parameters, no minimal encoder exists that leads to a good channel code TTM to the $(1 - D)(1 + D)^2$ channel.

Example 22.6

Consider the nonminimal (4,2) convolutional encoder with constraint length $\nu = 2$ represented by the following generator matrix:

$$\mathbf{G}(D) = \begin{pmatrix} 0 & 1 + D & 0 & 1 + D \\ 0 & 1 + D & 1 + D & 1 + D \end{pmatrix}$$

The decoder trellis for the channel code TTM to the $(1 - D)(1 + D)^2$ channel is shown in Figure 22.5.

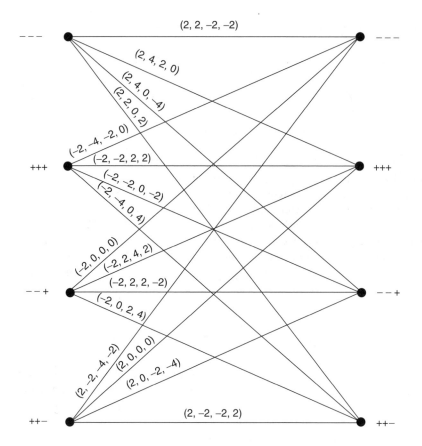

FIGURE 22.5 The decoder trellis of Example 22.6. (Copyright ©1997 IEEE. Reproduced with permission.)

This channel code has $d_{\text{free}}^2 = 40$ and $L_{\max} = 3$. It can be seen from Figure 22.5 that this channel code does not contain any flawed codewords. In a code search, when minimal encoders do not lead to good channel codes, the minimality constraint is relaxed and the best channel code generated by a nonminimal encoder (when this exists) is presented.

We end this section with a summary of the conditions that limit the choices of $\mathbf{G}(D)$ and $T(D)$ according to the results of this section.

Requirements on $\mathbf{G}(D)$ and $\mathbf{T}(D)$:

   (i) $m \geq n + 2$
  (ii) $1 \leq k < m$
 (iii) $\mathbf{G}(D)$ is a feedback free encoder
 (iv) $\mathbf{G}(D)$ is a minimal encoder
  (v) $\mathbf{G}_{\text{in}}(D) = (1 \oplus D)^{n'+1} \mathbf{G}'_{\text{in}}(D),\ n' \geq n$
 (vi) $\mathbf{G}_{\text{out}}(D)$ has at least one coordinate equal to zero, for bit stuffing
 (vii) $(1 \oplus D)^{n+1}$ divides $T(D)$
(viii) $T(D)$ has "1" in at least one of the positions where $\mathbf{G}_{\text{out}}(D)$ is zero, for bit stuffing
 (ix) $1 \leq \nu \leq 5$
  (x) $\nu_i \geq 1$, for $1 \leq i \leq k$, to avoid parallel transitions in the decoder trellis

If the bit stuffing technique is too restrictive, then conditions (vi) and (viii) are eliminated. If minimality is too restrictive, then condition (iv) is eliminated.

22.7 The Distance Spectrum Criterion for Trellis Codes

The *distance spectrum* of a trellis code is defined as the collection of ordered pairs (d_i, M_i), $i = 1, 2, \ldots$, of all distances d_i, where $d_i < d_{i+1}$, together with the average number (multiplicity), M_i, of paths in the trellis at distance d_i from a given (reference) path in the trellis, under the condition that the two paths diverge from the same node at $t = 0$ and remerge only once at some later time. The average is taken over all paths as reference paths. The pair (d_i, M_i) is called the ith *spectral line*.

For the trellis codes considered in this chapter, the distance d_i represents the squared Euclidean distance between sequences and, by convention, $d_{\text{free}}^2 \overset{\Delta}{=} d_1$. It is widely known that, under the assumption of AWGN and high SNR, higher d_{free}^2 implies lower probability of error. At moderate to low SNR, however, this need not be true since the whole distance spectrum influences the upper bound on the probability of error of the code [17], and a code optimized for some specific SNR may not be optimal if the SNR is changed. Nevertheless, it is easily observable that a higher d_{free}^2 code should have better distance spectrum altogether, since codewords at distance d_i from the reference codeword should be mutually separated by at least d_{free}^2. Thus, if d_{free}^2 is smaller, then there is more space available for a larger number of codewords at d_i, potentially increasing M_i. Naturally, counterexamples can be constructed with low d_{free}^2 and good spectrum at d_i, but high d_{free}^2 codes surely cannot be bad in this regard. The same argument may be used for the number M_j of codewords at distance d_j, with regard to distance d_i, $j > i$. Consequently, it is reasonable to assume, and it has been confirmed by all the examples we encountered, that the cumulative distance spectra [18] of two codes, defined as $\sum_{d_i < d_{\text{lim}}} M_i$, almost never intersect as d_{lim} increases. Consequently, the error performance curves of these two codes almost never intersect if all the other parameters of the two codes remain the same [20, 21]

In the next section, we use a simple yet reasonably robust criterion for achieving good codes, called the (squared Euclidean) *distance spectrum* (DS) criterion, defined as follows. In comparing (say) code $\mathcal{C}$ with code $\mathcal{C}'$ with distance spectra (d_i, M_i) and (d_i, M_i'), $i = 1, 2, \ldots$, respectively, the distance spectrum criterion declares that code $\mathcal{C}$ is better than code $\mathcal{C}'$ if there exists a positive integer j such that $M_i = M_i'$ for all $i < j$ and $0 \leq M_j < M_j'$. Note that the code with larger d_{free}^2 is considered better according to this criterion, regardless of the other spectral lines.

22.8 Good Trellis-Matched Codes for the Partial-Response Channels Based on the Distance Spectrum Criterion

Regarding the coded system of Figure 22.1, the best channel codes [13, 20], along with their respective distance spectra (first five spectral lines), are shown in Table 22.1 for the EPR4 channel and in Table 22.2 for the E²PR4 channel. Similar tables for the $1 - D$ channel can be found in [15]. In Figure 22.1, the column $T(D)$ refers to the coset representative (see [6, Equation 22.15 and tables description] for details). The input generator vector, $\mathbf{G_{in}}(D)$, and $T(D)$ are given in the standard octal notation. For example, $\mathbf{G_{in}}(D) = 6^4$ [24, 34] and $T(D) = 74$ denote: $\mathbf{G_{in}}(D) = (110)^4$ [010100, 011100] $= (1+D)^4[D+D^3, D+D^2+D^3]$ and $T(D) = 111100 = 1 + D + D^2 + D^3$. In the $L_{\max}$ column, the maximum identical symbol-run length is listed. Most frequently, this is the zero-run length, but in two cases (shown in Table 22.2), the maximum run-length channel output symbol is nonzero, and is listed in parenthesis following the length of the run. For instance, 4(2) indicates that the run "2222" of length 4 is the longest identical symbol-run. All codes shown in Table 22.1 and Table 22.2 are free of flawed codewords.

Appendix A

In this appendix we show the equivalence in Equation 22.10. First we observe that the coefficients of $P_n(D)$ satisfy $p_i = -p_{n+1-i}$. If we substitute Equation 22.5 into Equation 22.6, and disregard the terms in D^j

TABLE 22.1   Distance Spectra for Channel Codes TTM to the EPR4 Channel

R_c	ν	G_{in}	$T(D)$	L_{max}	Distance spectrum
1/5	1	$6^3[67]$	36	1	(88,0.25)(104,0.25)(120,0.25)(136,0.25)(144,0.125)
1/5	1	$6^3[76]$	42	2	(128,1)(224,0.5)(288,0.5)(320,0.5)(384,0.5)
1/5	2	$6^3[7754]$	42	3	(200,0.25)(216,0.25)(240,0.25)(256,0.75)(264,0.25)
1/5	3	$6^3[55777]$	42	2	(224,0.25)(240,0.5625)(256,0.25)(272,0.25)(288,0.25)
1/5	4	$6^3[5574076]$	42	2	(288,0.375)(304,0.5)(312,0.03125)(320,0.38281)(328,0.09375)
1/5	5	$6^5[63155246]$	42	2	(328,0.15625)(344,0.28125)(360,0.25)(376,0.29688)(392,0.21094)
1/4	1	$6^6[1]$	74	1	(80,1)(144,0.5)(176,0.5)(208,0.25)(240,0.5)
1/4	2	$6^6[26]$	74	1	(104,1)(184,0.5)(192,0.5)(224,0.5)(232,0.5)
1/4	3	$6^6[2134]$	74	2	(168,0.25)(184,0.25)(200,0.5)(216,0.75)(232,0.5)
2/5	2	$6^3[174,514]$	74	4	(48,1.5)(56,0.25)(64,0.75)(72,0.375)(80,1.03125)
2/5	3	$6^3[74,5124]$	42	2	(80,0.75)(88,0.5)(96,1.5)(104,0.5)(112,2)
2/4	2	$6^4[24,34]$	74	3	(40,2)(56,1)(72,1.5)(80,1)(88,3.25)
2/4	3	$6^4[5,436]$	74	8	(48,1.28125)(56,0.63281)(64,1.30469)(72,1.90039)(80,3.67847)
2/4[a]	4	$6^4[202,552]$	74	3	(56,0.875)(64,1.35156)(72,2.18945)(80,3.16211)(88,3.92285)
3/5	3	$6^3[26,15,404]$	74	11	(32,2.0625)(40,2.00391)(48,4.2544)(56,7.86763)(64,12.7282)
3/5	4	$6^3[14,61,4044]$	74	14	(40,2.875)(48,1.75879)(56,5.93607)(64,9.51839)(72,18.3108)

[a] Code found in [20]. All other codes were obtained from [13].

Source: Despotović, M., Šenk, V., and Uchôa-Filho, B. F., *IEEE Trans. Commun.*, vol. 49, no. 7, pp. 1121–1124, July 2001. Copyright ©2001 IEEE. Reproduced with permission.

TABLE 22.2   Distance Spectra for Channel Codes TTM to the E²PR4 Channel

R_c	ν	G_{in}	$T(D)$	L_{max}	Distance spectrum
1/5	1	$6^5[64]$	42	2	(264,1)(512,0.5)(544,0.5)(760,0.25)(792,0.5)
1/5	2	$6^4[7734]$	42	2	(464,0.125)(480,0.125)(496,0.25)(504,0.125)(512,0.25)
1/5	3	$6^8[7416]$	42	2	(536,0.25)(552,0.25)(616,0.25)(632,0.25)(656,0.0625)
1/5	4	$6^5[5253524]$	42	3	(664,0.0625)(680,0.0625)(688,0.01563)(720,0.01563)(728,0.0625)
1/5	5	$6^5[42116304]$	42	3	(792,0.01563)(800,0.03125)(808,0.125)(824,0.125)(848,0.03516)
2/5[a]	2	$6^5[12,16]$	42	3	(112,2)(184,1)(192,0.5)(216,0.5)(224,1)
2/5[a]	2	$6^4[7,47]$	42	4	(120,0.25)(152,1.5)(160,0.5)(200,0.28125)(224,0.1875)
2/5[a]	3	$6^4[3004,74]$	42	3	(152,2)(200,0.375)(216,0.0625)(224,1.375)(232,0.51563)
2/5[a]	3	$6^4[37,7074]$	42	11	(160,0.46875)(192,1.5)(200,0.30469)(232,0.03125)(240,0.66406)
2/5[a]	4	$6^4[74,30601]$	42	3	(216,0.5)(224,2)(232,0.5)(240,1)(256,1)
3/5	3	$6^4[3,24,62]$	42	4	(48,1.5)(64,1.25)(72,0.875)(80,0.34375)(88,1.25781)
3/5	3	$6^4[54,42,21]$	42	8	(72,1)(80,1.25)(88,1.5625)(96,0.35156)(104,0.76563)
3/5	4	$6^5[2,3,43]$	42	4	(48,1.5)(64,0.125)(72,0.85156)(80,0.49414)(88,0.70313)
3/5	4	$6^4[7,26,201]$	42	4(2)	(64,0.0625)(72,0.28125)(80,1.28125)(88,1.41016)(96,1.04297)
3/5	4	$6^4[3,43,4604]$	42	11	(80,1.6875)(88,0.6875)(96,0.625)(104,0.57813)(112,1.55469)
4/5	4	$6^4[2,1,02,05]$	42	4(-2)	(40,3.0625)(48,2.75)(56,1.75)(64,1.47656)(72,2.51758)

[a] Code found in [20]. All other codes were obtained from [13].

Source: Despotović, M., Šenk, V., and Uchôa-Filho, B. F., *IEEE Trans. Commun.*, vol. 49, no. 7, pp. 1121–1124, July 2001. Copyright ©2001 IEEE. Reproduced with permission.

where $j \geq$ length of $W(D)$, the channel output sequence becomes:

$$Y(D) = 2P_n(D)R(D) - \underbrace{\mathbf{1}(D)P_n(D)}_{a} + \underbrace{\sum_{j=1}^{n+1}\left(\sum_{i=j}^{n+1}(-1)p_i\right)D^{j-1}}_{b}$$

Expanding a and b we have:

$$a = p_0 + (p_1 + p_0)D + (p_2 + p_1 + p_0)D^2 + \cdots + (p_{n+1} + \cdots + p_1 + p_0)D^{n+1}$$
$$+ (p_{n+1} + \cdots + p_1 + p_0)D^{n+2} + (p_{n+1} + \cdots + p_1 + p_0)D^{n+3} + \cdots$$

and

$$b = (-p_1 - p_2 - \cdots - p_{n+1}) + (-p_2 - p_3 - \cdots - p_{n+1})D + \cdots + (-p_{n+1})D^n$$

Thus,

$$-a + b = (-p_{n+1} - \cdots - p_1 - p_0) \sum_i D^i = 0$$

since $p_i = -p_{n+1-i}$. Then the channel output sequence becomes:

$$Y(D) = 2P_n(D)R(D) = 2 \underbrace{P_n(D)(1 \oplus D)^{-n-1}}_{\equiv 1 \,(\text{mod}\, 2)} W(D)$$

and we find that

$$W(D) \equiv \frac{Y(D)}{2} \,(\text{mod}\, 2)$$

Appendix B

In this appendix we show the equivalence $|(X(D) - X'(D))P_n(D)| \equiv |X(D) - X'(D)|(1 \oplus D)^{n+1}$, used in the proof of Lemma 22.1. Note that $x_i - x_i' \in \{0, \pm 2\}$ and $p_i \in \mathbb{Z}$. Then the division by 2 in the function defined in Equation 22.9 can be carried out in the polynomial $(X(D) - X'(D))$. Thus, we can have

$$
\begin{aligned}
|(X(D) - X'(D))P_n(D)| &\equiv \frac{(X(D) - X'(D))P_n(D)}{2} \,(\text{mod}\, 2) \\
&\equiv \frac{(X(D) - X'(D))}{2} P_n(D) \,(\text{mod}\, 2) \\
&\equiv |(X(D) - X'(D))| [P_n(D)]_{\text{mod}\, 2} \\
&\equiv |(X(D) - X'(D))|(1 \oplus D)^{n+1}
\end{aligned}
$$

References

[1] J. K. Wolf and G. Ungerboeck, Trellis coding for partial-response channels, *IEEE Trans. Commun.*, vol. COM-34, pp. 765–773, August 1986.

[2] H. Kobayashi and D. T. Tang, Application of partial-response channel coding to magnetic recording systems, *IBM J. Des. Dev.*, vol. 14, pp. 368–375, July 1970.

[3] H. K. Thapar and A. M. Patel, A class of partial response systems for increasing storage density in magnetic recording, *IEEE Trans. Magn.*, vol. MAG-25, pp. 3666–3668, September 1987.

[4] E. Zehavi and J. K. Wolf, On saving decoder states for some trellis codes and partial response channels, *IEEE Trans. Commun.*, vol. COM-36, pp. 222–224, February 1988.

[5] E. Zehavi and J. K. Wolf, On the performance evaluation of trellis codes, *IEEE Trans. Inform. Theory*, vol. IT-33, pp. 196–202, March 1988.

[6] P. H. Siegel and J. K. Wolf, Modulation and coding for information storage, *IEEE Commun Mag.*, vol. 29, pp. 68–86, December 1991.

[7] T. A. Lee and C. Heegard, An inversion technique for the design of binary convolutional codes for the $1 - D^N$ Channel, *Proceedings IEEE regional meeting*, Johns Hopkins University, Baltimore, MD, February 1985.

[8] G. D. Forney, Jr., Convolutional codes I: algebraic structure, *IEEE Trans. Inform. Theory*, vol. IT-16, pp. 720–738, November 1970.

[9] G. D. Forney, Jr., Structural analysis of convolutional codes via dual codes, *IEEE Trans. Inform. Theory*, vol. IT-19, pp. 512–518, July 1973.

[10] S. Lin and D. J. Costello, Jr., *Error Control Coding: Fundamentals and Applications*, Prentice Hall, Englewood Cliffs, NJ, 1983.

[11] J. G. Proakis and M. Salehi, *Communication Systems Engineering*, Prentice Hall, Englewood Cliffs, NJ, 1994.

[12] B. F. Uchôa Filho and M. A. Herro, Convolutional Codes for the High Density $(1 - D)(1 + D)^n$ Magnetic Recording Channel," *Proceedings IEEE Int. Symp. Inform. Theory*, pp. 211, Trondheim, Norway, June-July 1994.

[13] B. F. Uchôa Filho and M. A. Herro, Good convolutional codes for the precoded $(1 - D)(1 + D)^n$ partial response channels," *IEEE Trans. Inform. Theory*, vol. 43, no. 2, pp. 441–453, March 1997.

[14] K. J. Hole, Punctured convolutional codes for the $1 - D$ partial-response channel, *IEEE Trans. Inform. Theory*, vol. I-37, pp. 808–817, May 1991.

[15] K. J. Hole and Ø. Ytrehus, Improved coding techniques for precoded partial-response channels, *IEEE Trans. Inform. Theory*, vol. I-40, pp. 482–493, March 1994.

[16] M. Siala and G. K. Kaleh, Block and trellis codes for the binary $(1 - D)$ partial response channel with simple maximum likelihood decoders, *IEEE Trans. on Commun.*, vol. 44, no. 12, pp. 1613–1615, December 1996.

[17] M. Rouanne and D. J. Costello, Jr., An algorithm for computing the distance spectrum of trellis codes," *IEEE J. Select. Areas Commun.*, vol. SAC-7, pp. 929–940, August 1989.

[18] M. Despotović and V. Šenk, Distance spectrum of channels trellis codes on precoded partial-response $1 - D$ channel, FACTA UNIVERSITATIS (NIS), series: Electronics and Energetics vol. 1 (1995), pp. 57–72, http://factaee.elfak.ni.ac.yu/.

[19] M. Despotović, V. Šenk, and B. F. Uchôa-Filho, Convolutional codes with optimized distance spectrum for the EPR4 and EEPR4 channels, *Proceedings the 1999 IEEE Int. Conference Commun.*, vol. 3, pp. 1658–1662, June 1999.

[20] M. Despotović, V. Šenk, and B. F. Uchôa-Filho, Distance spectra of convolutional codes over partial-response channels, *IEEE Trans. Commun.*, vol. 49, no. 7, pp. 1121–1124, July 2001.

23

Capacity-Approaching Codes for Partial Response Channels

Nedeljko Varnica

Xiao Ma

Aleksandar Kavčić
Harvard University
Boston, MA

23.1   Introduction **23**-1
23.2   The Channel Model and Capacity Definitions **23**-2
     The Channel Model • The Channel Capacity • Trellis
     Representations • The Markov Channel Capacity •
     Computing the Markov Channel Capacity
23.3   Trellis Codes, Superchannels and Their Information
     Rates **23**-6
     Coding Theorems for Superchannels
23.4   Matched Information Rate (MIR) Trellis Codes **23**-9
     Choosing the Extended Channel Trellis and the Superchannel
     Code Rate • Choosing the Number of States in the
     Superchannel • Choosing the Branch Type Numbers in the
     Superchannel • Choosing the Branch Connections
23.5   Outer LDPC Codes **23**-13
     Encoding/Decoding System • Choosing the Branch Input-Bit
     Assignment • Determining the Outer Subcode Rates •
     Subcode Optimization
23.6   Optimization Results **23**-16
     Dicode Channel • A Channel without Spectral Nulls
23.7   Conclusion **23**-19

23.1   Introduction

A partial response (PR) channel is an intersymbol interference (ISI) channel with a binary input alphabet and additive white Gaussian noise (AWGN). The capacity of a PR channel is strictly greater than its i.u.d. capacity, which is defined as the information rate when the PR channel inputs are independent and uniformly distributed (i.u.d.) random variables.

The computations of the capacity C and the i.u.d. capacity $C_{i.u.d.}$ of a PR channel have been subjects of research for some time [1–3]. The information rates and the capacities of finite-state machines and Markov chains are closely related to the capacities of PR channels and have been studied in [4–7]. For a summary of capacity computation methods, see [8].

Recently, a Monte Carlo method for computing the information rate of a finite-state machine channel whose inputs are Markov processes was proposed independently by Arnold and Loeliger [9] and Pfister et al. [10]. This method can be used to compute $C_{i.u.d.}$, which is a lower bound on C. In [11], Kavčić proposed a Markov process optimization algorithm to tighten the lower bounds, and in [12] Vontobel and Arnold used this algorithm to compute tight upper bounds. The feedback capacity of PR channels computed in [13] is also an upper bound on C, and is in some cases tighter than the Vontobel-Arnold bound [12]. These methods (summarized in [8]) give bounds that are so tight that, for all practical purposes, we consider the capacities of PR channels to be known.

Error-correction and modulation codes for PR channels are numerous. Here, we only consider error-correcting codes that are aimed at achieving the capacities of PR channels. Before the invention of turbo codes [14] the codes for PR channels were dominated by trellis codes [15–17]. Matched spectral null (MSN) trellis codes whose binary codeword sequences have nulls in specific frequency locations matching the spectral nulls of the channels were proposed in [18].

The advent of turbo [14] and low-density parity-check (LDPC) [19, 20] codes has sparked research in iteratively decodable codes for PR channels. Typically the BCJR algorithm [21] is used to perform the maximum a posteriori symbol detection on the channel, while iterative decoding algorithms are used to provide constituent decoders in the iterative decoding schemes. Early iteratively decodable codes for PR channels were parallel and serially concatenated turbo codes [22–25]. Subsequent schemes concentrated on simplifying turbo equalization [26, 27]. The LDPC codes and turbo product codes over PR channels have been investigated in [28–30]. Luby et al. [31] introduced irregular LDPC codes and proposed a method to analyze the asymptotic performance of these codes over erasure channels. The analysis was soon thereafter expanded to various other memoryless channels by Richardson and Urbanke [32]. They used the *density evolution* analysis tool for the computation of noise thresholds [32], and the optimization of code degree sequences [33] that produced the capacity-achieving codes for memoryless channels [34]. Kavčić et al. [35] modified the density evolution to fit the PR channels, computed noise thresholds of *random* LDPC codes over PR channels and proved that these codes can achieve rates that approach the i.u.d. capacity, but not higher. This was supported by evidence of code constructions that approach $C_{\text{i.u.d.}}$ very closely [36].

Here, we construct codes that achieve rates higher than $C_{\text{i.u.d.}}$. Our approach is to (1) compute a (maximized) channel information rate achievable by an input Markov process of reasonably low complexity, (2) construct an inner trellis code that mimics this Markov process and therefore achieves a comparable information rate over the channel of interest, and (3) construct an outer LDPC code that ensures that the information rate of the inner code is approached. The strategy is therefore to compute an achievable information rate, and then construct a concatenation of two codes whose overall code rate matches the computed information rate. Hence the name *matched information rate* (MIR) code.

This chapter is organized into six sections. Section 23.2 and Section 23.3, are tutorial-like expositions of the channel capacity and trellis code terminologies. In Section 23.4 and Section 23.5, we design inner MIR trellis codes and outer LDPC codes, respectively. In Section 23.6 the success of the methodology is demonstrated with two examples of constructed codes (one for a channel with a spectral null, and the other for a channel without a spectral null).

23.2 The Channel Model and Capacity Definitions

23.2.1 The Channel Model

Let $t \in \mathbb{Z}$ denote a discrete-time variable. Denote the channel input as a sequence of random variables X_t, and the channel output as a sequence of random variables Y_t. The channel law is

$$Y_t = \sum_{k=0}^{J} h_k X_{t-k} + W_t \tag{23.1}$$

where $J > 0$ is the ISI length and $h_0, h_1, \ldots, h_J$ capture the memory in the channel. Often the channel memory is represented by a partial response polynomial $h(D) = \sum_{k=0}^{J} h_k D^k$. For example, the partial response polynomial of the *dicode* channel ($Y_t = X_t - X_{t-1} + W_t$) is $h(D) = 1 - D$. We assume that the noise process W_t is white and Gaussian, with mean 0 and variance σ^2. We also assume that the realizations x_t of the random variables X_t take values in a binary alphabet $\mathcal{X} = \{-1, 1\}$ which is of practical relevance in many binary recording and communications channels. We will interchangeably use the input alphabets $\mathcal{X} = \{-1, 1\}$ and $\mathcal{A} = \{0, 1\}$. Thereby, we will use the standard conversion $X_t = 1 - 2A_t$, where A_t is a discrete-time random process with binary realizations $a_t \in \mathcal{A} = \{0, 1\}$.

The signal-to-noise ratio (in dB) of a PR channel is defined as

$$\text{SNR} = 10 \log_{10} \frac{\sum_{k=0}^{J} h_k^2}{\sigma^2} \tag{23.2}$$

The most comprehensive known model to which our methods apply is the finite-state model with autoregressive observations described in [37, 38]. However, since all methods for the model in Equation 23.1 canonically extend to the channel model in [37, 38], without loss of generality, it suffices to consider the model in Equation 23.1 with binary inputs and AWGN.

23.2.2 The Channel Capacity

The channel capacity of a PR channel is defined as [39]

$$C = \lim_{N \to \infty} \frac{1}{N} \left[\max_{P_{X_1^N}(x_1^N)} I\left(X_1^N; Y_1^N \mid X_{-J+1}^0 = x_{-J+1}^0\right) \right] = \lim_{N \to \infty} \frac{1}{N} \left[\max_{P_{X_1^N}(x_1^N)} I\left(X_1^N; Y_1^N\right) \right] \tag{23.3}$$

where $I(X_1^N; Y_1^N \mid X_{-J+1}^0 = x_{-J+1}^0)$ is the mutual information rate between the sequence of channel inputs $X_1^N = (X_1, X_2, \ldots, X_N)$ and the sequence of channel outputs $Y_1^N = (Y_1, Y_2, \ldots, Y_N)$, given that the symbols transmitted over the channel at time $t = -J + 1$ through $t = 0$ are x_{-J+1}^0. Following [1], we can drop the conditioning on the symbols $X_{-J+1}^0 = x_{-J+1}^0$ because the channel is indecomposable [39].

Another capacity of interest is the i.u.d. capacity, defined as the information rate when the input sequence X_1^N consists of i.u.d. random variables X_t

$$C_{\text{i.u.d.}} = \lim_{N \to \infty} \frac{1}{N} \cdot I\left(X_1^N; Y_1^N\right) \bigg|_{P_{X_1^N}(x_1^N) = 2^{-N}} \tag{23.4}$$

23.2.3 Trellis Representations

We introduce a general notion of a channel trellis. At time t, the channel trellis has 2^L states (where $L \geq J$), indexed by $\mathcal{S} = \{0, 1, \ldots, 2^L - 1\}$. Exactly 2^n branches emanate form each trellis state, where $n \geq 1$. A branch at time t is determined by a 4-tuple $b_t = (s_{t-1}, u_t, v_t, s_t)$, where $s_{t-1} \in \mathcal{S}$ and $s_t \in \mathcal{S}$ are the starting state and the ending state, respectively. The symbol u_t is an n-tuple of input symbols, that is, $u_t \in \mathcal{A}^n = \{0, 1\}^n$ or $u_t \in \mathcal{X}^n = \{-1, 1\}^n$. The symbol $v_t \in \mathbb{R}^n$ is an n-tuple of real-valued *noiseless* channel outputs. Every branch is uniquely determined by its starting state s_{t-1} and its input n-tuple u_t. That is, v_t and s_t are (deterministic) functions of s_{t-1} and u_t. The random 4-tuple $B_t = (S_{t-1}, U_t, V_t, S_t)$ stands for a random branch, whose realizations are the 4-tuples b_t.

To characterize a time-invariant channel trellis, we need only specify one trellis section. We distinguish 3 trellis types determined by the properties of their trellis sections:

- A minimal channel trellis is a channel trellis whose trellis section has the smallest possible number of states (i.e., 2^J states) and corresponds to a single input and a single output (i.e., $n = 1$).
- An original channel trellis is a channel trellis whose trellis section has $2^L \geq 2^J$ states and corresponds to a single input and a single output (i.e., $n = 1$).
- An extended channel trellis is a channel trellis whose trellis section corresponds to n channel inputs and n channel outputs, where $n > 1$. The number of states is $2^L \geq 2^J$.

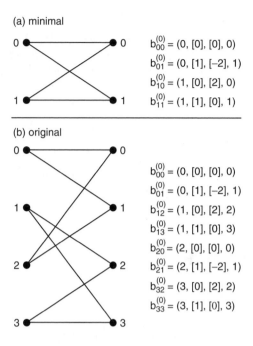

FIGURE 23.1   (a) The minimal trellis for $1 - D$ channel. (b) An original trellis for $1 - D$ channel with $L = 2$ and $n = 1$.

The minimal trellis for the dicode channel with $2^J = 2$ states and $n = 1$ input (and output) symbol per section is shown in Figure 23.1(a). An original trellis for the dicode channel with $2^L = 4$ states and $n = 1$ input (and output) symbol per section is shown in Figure 23.1(b). The 3rd order extension of the minimal trellis is shown in Figure 23.2(a). This trellis is obtained by concatenating $n = 3$ sections of the minimal trellis in Figure 23.1(a). The 2nd order extension of the original trellis from Figure 23.1(b) is shown in Figure 23.2(b).

23.2.4   The Markov Channel Capacity

Let $i \in \mathcal{S}$ and $j \in \mathcal{S}$ denote the starting and the ending state of a branch in the trellis section, respectively. If the trellis is an original channel trellis, then there exists at most one branch connecting state i to state j. If, however, the trellis is an extended channel trellis, then there may be several branches connecting a pair of states i and j. Let $\mathcal{L}$ be the number of distinct branches that connect states i and j. These branches are denoted by $b_{ij}^{(\ell)} = (i, u_{ij}^{(\ell)}, v_{ij}^{(\ell)}, j)$, where $0 \le \ell \le \mathcal{L} - 1$, $u_{ij}^{(\ell)} \in \mathcal{A}^n$ is the binary channel input n-tuple and $v_{ij}^{(\ell)} \in \mathbb{R}^n$ is the real-valued noiseless channel output n-tuple.

Denote by τ the set of all triples (i, ℓ, j) for which a branch $b_{ij}^{(\ell)}$ exists in the trellis section. Since the trellis section uniquely determines the trellis, we say that τ represents a channel trellis. We call a branch $b_{ij}^{(\ell)}$ valid if $(i, \ell, j) \in \tau$. By assigning a probability $P_{ij}^{(\ell)} = \Pr(B_t = b_{ij}^{(\ell)} \mid S_{t-1} = i) = \Pr(S_t = j, B_t = b_{ij}^{(\ell)} \mid S_{t-1} = i)$ to every branch $b_{ij}^{(\ell)}$ of the channel trellis section we define an input Markov process on τ. Denote the probability of state i by $\mu_i = \Pr(S_t = i)$. The Markov process is stationary if for every $j \in \mathcal{S}$,

$$\mu_j = \sum_{i,\ell:(i,\ell,j)\in\tau} \mu_i P_{ij}^{(\ell)} \tag{23.5}$$

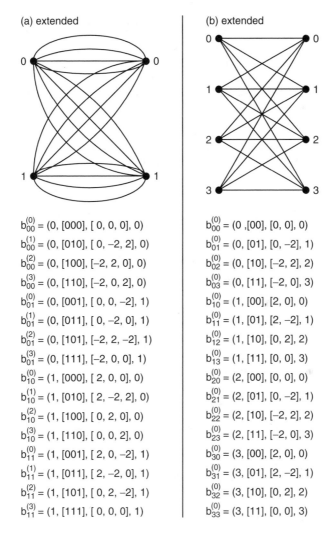

FIGURE 23.2 (a) The 3rd order extension of the minimal trellis. (b) The 2nd order extension of the original trellis in 23.1(b).

Denote by $\mathcal{P}(\tau)$ the collection of Markov transition probabilities $P_{ij}^{(\ell)}$ for all branches $(i, \ell, j) \in \tau$. For this Markov source, the Markov channel information rate is defined as

$$\mathcal{I}_{\mathcal{P}(\tau)} = \lim_{N \to \infty} \frac{1}{nN} I\left(X_1^{nN}; Y_1^{nN}\right) = \lim_{N \to \infty} \frac{1}{nN} I\left(B_1^N; Y_1^{nN}\right) \tag{23.6}$$

where the branch sequence B_1^N is a random sequence defined by the probabilities $\mathcal{P}(\tau)$.

We define the Markov channel capacity for a trellis τ as

$$C_\tau = \max_{\mathcal{P}(\tau)} \mathcal{I}_{\mathcal{P}(\tau)} \tag{23.7}$$

where the maximization in Equation 23.7 is conducted over all Markov processes $\mathcal{P}(\tau)$ defined over τ. Since the set of stationary Markov processes is a subset of the set of all stationary discrete-time processes, it is clear that $C_\tau \leq C$. Let $\tau_o(L)$ be an *original* channel trellis with $2^L \geq 2^J$. Denote the Markov channel

TABLE 23.1   Algorithm 1 Iterative Optimization of Markov Chain Transition Probabilities

Initialization Pick a channel trellis τ and an arbitrary probability mass function $P_{ij}^{(\ell)}$ defined over τ that satisfies: (1) $0 \leq P_{ij}^{(\ell)} \leq 1$ if $(i, \ell, j) \in \tau$; otherwise $P_{ij}^{(\ell)} = 0$ and (2) $\sum_{j,\ell:(i,\ell,j)\in\tau} P_{ij}^{(\ell)} = 1$ for any i.

Repeat until convergence

1. For N large (say, $N > 10^6$), generate a realization of a sequence of N trellis branches b_1^N according to the Markov probabilities $P_{ij}^{(\ell)}$. Determine the channel input sequence x_1^{nN} that corresponds to the branch sequence b_1^N. Pass x_1^{nN} through the PR channel to get a realization of the channel output y_1^{nN}.

2. Run the forward-backward sum-product (BCJR) algorithm and for all $1 \leq t \leq N$ compute the a posteriori probabilities $R_{ij}^{(\ell)}(t, y_1^{nN}) = \Pr(B_t = b_{ij}^{(\ell)} \mid Y_1^{nN} = y_1^{nN})$ and $R_i(t, y_1^{nN}) = \Pr(S_t = i \mid Y_1^{nN} = y_1^{nN})$.

3. Compute the estimate of the expectation term $\hat{T}_{ij}^{(\ell)} = \frac{1}{N} \sum_{t=1}^{N} \log_2 \frac{\frac{R_{ij}^{(\ell)}(t,y_1^{nN})}{\mu_i P_{ij}^{(\ell)}}}{\frac{R_i(t,y_1^{nN})}{\mu_i}}$.

4. Compute the estimate of the noisy adjacency matrix, with the entries

$$\hat{A}_{ij} = \begin{cases} \sum_{\ell:(i,\ell,j)\in\tau} 2^{\hat{T}_{ij}^{(\ell)}} & \text{if states } i \text{ and } j \text{ are connected by at least one branch} \\ 0 & \text{otherwise} \end{cases},$$

   and find its maximal eigenvalue $\hat{W}_{\max}$ and the corresponding right eigenvector $[\hat{e}_1, \hat{e}_2, \ldots, \hat{e}_M]^T$.

5. For $(i, \ell, j) \in \tau$, compute the new transition probabilities as $P_{ij}^{(\ell)} = \frac{\hat{e}_j}{\hat{e}_i} \cdot \frac{2^{\hat{T}_{ij}^{(\ell)}}}{\hat{W}_{\max}}$ and go back to 1.

Terminate the algorithm and set $\hat{C}_\tau = \frac{1}{n} \log_2 \hat{W}_{\max}$. The input distribution $\mathcal{P}(\tau)$ that achieves $\hat{C}_\tau$ is given by the collection of probabilities $P_{ij}^{(\ell)}$.

---

capacity defined on this trellis by $C_{\tau_o(L)}$, or simply C_L. Then an alternative expression for the channel capacity is

$$C = \lim_{L\to\infty} C_{\tau_o(L)} = \lim_{L\to\infty} C_L \tag{23.8}$$

23.2.5 Computing the Markov Channel Capacity

Table 23.1 gives a method to estimate the capacity C_τ. Since the method in Table 23.1 is a Monte Carlo algorithm, we denote its capacity estimate by $\hat{C}_\tau$.

In Figure 23.3, we show the Markov rates $\hat{C}_\tau$ computed by the algorithm in Table 23.1 for the dicode channel trellises given in Figure 23.1 and Figure 23.2. Also shown in Figure 23.3 are the i.u.d. capacity $C_{\text{i.u.d.}}$, the *numerical capacity* $\hat{C}_6$ (which is the tightest known lower bound on the channel capacity C) and the minimum of all known upper bounds on C (the water-filling bound [39, 40], the feedback capacity [13] and the Vontobel-Arnold bound [12]).

23.3   Trellis Codes, Superchannels and Their Information Rates

We consider *binary* time-invariant trellis codes that map k input bits to n output bits. Each trellis section has N_S states, indexed by $\{0, 1, \ldots, N_S - 1\}$. A branch of a trellis code is described by a 4-tuple $b_t = (s_{t-1}, u_t, v_t, s_t)$, where $s_{t-1} \in \{0, 1, \ldots, N_S - 1\}$ is the starting state, $s_t \in \{0, 1, \ldots, N_S - 1\}$ is the ending state, $u_t \in \mathcal{A}^k$ is the input k-tuple, and $v_t \in \mathcal{A}^n$ is the output n-tuple. The code rate is $r = k/n$. Figure 23.4 shows an example of a Wolf-Ungerboeck (W-U) trellis code [15] with rate 2/3.

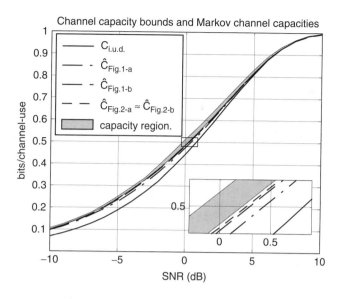

FIGURE 23.3 Markov Capacities for the trellises in Figure 23.1(a,b) and 23.2(a,b). The region between the tightest known lower and upper bound is shaded.

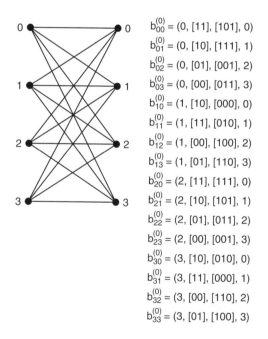

FIGURE 23.4 Wolf-Ungerboeck trellis code of rate $r = 2/3$.

We refer to a superchannel as a concatenation of a trellis code and a PR channel and describe it by a joint code/channel trellis. The number of states in a superchannel trellis may be greater than the number of states in the constituent trellis code. Figure 23.5 gives an example of the superchannel trellis obtained by concatenating the trellis code in Figure 23.4 with the dicode channel in Figure 23.1(a). Note that exactly 2^k branches emanate from each state of the superchannel trellis, that is, one branch for every binary input k-tuple.

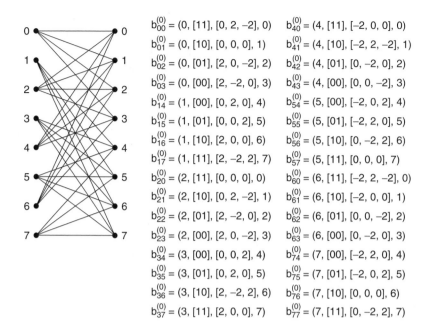

$b_{00}^{(0)} = (0, [11], [0, 2, -2], 0)$ $b_{40}^{(0)} = (4, [11], [-2, 0, 0], 0)$

$b_{01}^{(0)} = (0, [10], [0, 0, 0], 1)$ $b_{41}^{(0)} = (4, [10], [-2, 2, -2], 1)$

$b_{02}^{(0)} = (0, [01], [2, 0, -2], 2)$ $b_{42}^{(0)} = (4, [01], [0, -2, 0], 2)$

$b_{03}^{(0)} = (0, [00], [2, -2, 0], 3)$ $b_{43}^{(0)} = (4, [00], [0, 0, -2], 3)$

$b_{14}^{(0)} = (1, [00], [0, 2, 0], 4)$ $b_{54}^{(0)} = (5, [00], [-2, 0, 2], 4)$

$b_{15}^{(0)} = (1, [01], [0, 0, 2], 5)$ $b_{55}^{(0)} = (5, [01], [-2, 2, 0], 5)$

$b_{16}^{(0)} = (1, [10], [2, 0, 0], 6)$ $b_{56}^{(0)} = (5, [10], [0, -2, 2], 6)$

$b_{17}^{(0)} = (1, [11], [2, -2, 2], 7)$ $b_{57}^{(0)} = (5, [11], [0, 0, 0], 7)$

$b_{20}^{(0)} = (2, [11], [0, 0, 0], 0)$ $b_{60}^{(0)} = (6, [11], [-2, 2, -2], 0)$

$b_{21}^{(0)} = (2, [10], [0, 2, -2], 1)$ $b_{61}^{(0)} = (6, [10], [-2, 0, 0], 1)$

$b_{22}^{(0)} = (2, [01], [2, -2, 0], 2)$ $b_{62}^{(0)} = (6, [01], [0, 0, -2], 2)$

$b_{23}^{(0)} = (2, [00], [2, 0, -2], 3)$ $b_{63}^{(0)} = (6, [00], [0, -2, 0], 3)$

$b_{34}^{(0)} = (3, [00], [0, 0, 2], 4)$ $b_{74}^{(0)} = (7, [00], [-2, 2, 0], 4)$

$b_{35}^{(0)} = (3, [01], [0, 2, 0], 5)$ $b_{75}^{(0)} = (7, [01], [-2, 0, 2], 5)$

$b_{36}^{(0)} = (3, [10], [2, -2, 2], 6)$ $b_{76}^{(0)} = (7, [10], [0, 0, 0], 6)$

$b_{37}^{(0)} = (3, [11], [2, 0, 0], 7)$ $b_{77}^{(0)} = (7, [11], [0, -2, 2], 7)$

FIGURE 23.5   Superchannel obtained by concatenating the trellis code in Figure 23.4 and the dicode channel in Figure 23.1(a).

We assume that the trellis code inputs are i.u.d. symbols. This means that the conditional probability $P_{ij}^{(\ell)} = \Pr(B_t = b_{ij}^{(\ell)} \mid S_t = i)$ of each superchannel trellis branch equals 2^{-k}. Under this i.u.d. assumption, we define the *superchannel information rate* as the information rate between the superchannel input sequence U_1^N and the superchannel output sequence Y_1^{nN}

$$\mathcal{I}_S = \lim_{N \to \infty} \frac{1}{nN} I\left(U_1^N; Y_1^{nN}\right) \bigg|_{P_{U_1^N}(u_1^N) = 2^{-kN}} \tag{23.9}$$

23.3.1 Coding Theorems for Superchannels

Theorem 23.1   *[achievability of $\mathcal{I}_S$ with a linear code] The i.u.d. superchannel information rate $\mathcal{I}_S$ in Equation (23.9) can be achieved by an outer linear (coset) code.*

Proof 23.1   The proof is given in [41].   □

Theorem 23.1 suggests that we could use a concatenation scheme to surpass the i.u.d. capacity $C_{\text{i.u.d.}}$ of the PR channel if we construct a superchannel whose i.u.d. rate is $\mathcal{I}_S > C_{\text{i.u.d.}}$. Consider the block diagram shown in Figure 23.6. The binary vector D_1^M first enters a linear encoder represented by an

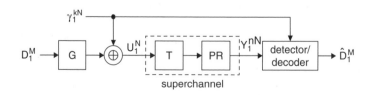

FIGURE 23.6   A concatenation scheme for partial response channels.

$M \times N$ generator matrix G. Before the codeword enters the trellis encoder 'T', a coset-defining binary vector γ_1^N is modulo-2-added on. Different choices of the vector γ_1^N may cause different error rates since the channel has memory. Although we cannot claim that the achievable rate of the best linear (coset) code is upper-bounded by $\mathcal{I}_S$, we have the following "random converse" theorem.

Theorem 23.2 (Random Code Converse) *If $R > \mathcal{I}_S$, there exists $\delta > 0$ such that for any given $M \times N$ generator matrix G (of a linear code) with $M/N > R$ and any given detection/decoding method, the average error rate over all the coset codes is greater than δ.*

Proof 23.2 The proof is given in [41]. □

Theorem 23.2 exposes the difficulties in finding a linear (coset) code that surpasses the superchannel i.u.d. rate $\mathcal{I}_S$. Namely, a randomly chosen linear (coset) code will not achieve a rate higher than $\mathcal{I}_S$. Theorem 23.2 also reveals that without the inner code, the average rate achievable by a linear (coset) code is upper-bounded by the i.u.d. channel capacity $C_{\text{i.u.d.}}$. To surpass $C_{\text{i.u.d.}}$, we separate the complex code construction problem into two smaller problems: designing an inner trellis code, and an outer linear (coset) code. We first construct an inner trellis code whose superchannel i.u.d. rate $\mathcal{I}_S$ satisfies $C \approx \mathcal{I}_S > C_{\text{i.u.d.}}$.

23.4 Matched Information Rate (MIR) Trellis Codes

We design a trellis code for a PR channel, such that the i.u.d. rate $\mathcal{I}_S$ of the resulting superchannel is as close as possible to the (numerical) channel capacity. An interesting feature of our design is that the trellis code is constructed for a specific target code rate r. Our general strategy is to first choose an extended channel trellis τ, and use the algorithm in Table 23.1 to find the optimal Markov probabilities for this trellis at the code rate r. We then construct a superchannel to mimic the optimal Markov process on the trellis τ.

We next specify the design rules to construct MIR superchannels. These rules are derived to formalize the design methodology. We adopt these rules because they satisfy our intuition and deliver good codes. We make no claim regarding their optimality.

23.4.1 Choosing the Extended Channel Trellis and the Superchannel Code Rate

We construct a superchannel trellis with n output symbols per every k binary input symbols. Our first task is to pick k and n. Let r be the target code rate. Pick an integer $n > 0$ and an nth order extended channel trellis τ. For this trellis, run the algorithm in Table 23.1. Denote by $P_{ij}^{(\ell)}$ the optimized probabilities of the trellis τ for which $\hat{C}_\tau = r$.

Rule 1: The rate r_{in} of the inner trellis code should satisfy the constraint

$$r < r_{\text{in}} = \frac{k}{n} \leq \frac{1}{n} \min_{(i,\ell,j) \in \tau} \left[\log_2 \frac{1}{P_{ij}^{(\ell)}} \right] \tag{23.10}$$

The reason for obeying the lower bound $r_{\text{in}} = k/n > r$ is that $r_{\text{in}} = r$ would mean that we would not have the option of using a powerful outer code.

The upper bound $k \leq \min_{(i,\ell,j) \in \tau} [-\log_2 P_{ij}^{(\ell)}]$ is motivated by the unique decodability of the trellis code. To avoid non-unique decodability, we require that all 2^k branches emanating from each super-channel state have distinct noiseless output n-tuples[1]. The assumption that the input to the superchannel is i.u.d. implies that the conditional probability of each of the branches is 2^{-k}. Since the goal is to

---

[1]This is sufficient (but not always necessary) requirement for the construction of a uniquely decodable trellis.

TABLE 23.2   Optimized Transition Probabilities for the Dicode Channel at Rate $r = 1/2$ Using the Extended Channel Trellis in Figure 23.2(a): The Integer Values k, K_i and $n_{ij}^{(\ell)}$ are Determined using Rules 1-3

Branch $b_{ij}^{(\ell)}$	Branch Label $\left(i, u_{ij}^{(\ell)}, v_{ij}^{(\ell)}, j\right)$	Transition Probability	Optimized Integers $k=2, K_0=5, K_1=5$	Integer Probability
$b_{00}^{(0)}$	$(0, [000], [\ 0,\ 0,\ 0], 0)$	$P_{00}^{(0)} = 0.005$	$n_{00}^{(0)} = 0$	$n_{00}^{(0)}/(K_0 \cdot 2^k) = 0.00$
$b_{00}^{(1)}$	$(0, [010], [\ 0, -2,\ 2], 0)$	$P_{00}^{(1)} = 0.146$	$n_{00}^{(1)} = 3$	$n_{00}^{(1)}/(K_0 \cdot 2^k) = 0.15$
$b_{00}^{(2)}$	$(0, [100], [-2,\ 2,\ 0], 0)$	$P_{00}^{(2)} = 0.146$	$n_{00}^{(2)} = 3$	$n_{00}^{(2)}/(K_0 \cdot 2^k) = 0.15$
$b_{00}^{(3)}$	$(0, [110], [-2,\ 0,\ 2], 0)$	$P_{00}^{(3)} = 0.195$	$n_{00}^{(3)} = 4$	$n_{00}^{(3)}/(K_0 \cdot 2^k) = 0.20$
$b_{01}^{(0)}$	$(0, [001], [\ 0,\ 0, -2], 1)$	$P_{01}^{(0)} = 0.066$	$n_{01}^{(0)} = 1$	$n_{01}^{(0)}/(K_0 \cdot 2^k) = 0.05$
$b_{01}^{(1)}$	$(0, [011], [\ 0, -2,\ 0], 1)$	$P_{01}^{(1)} = 0.145$	$n_{01}^{(1)} = 3$	$n_{01}^{(1)}/(K_0 \cdot 2^k) = 0.15$
$b_{01}^{(2)}$	$(0, [101], [-2,\ 2, -2], 1)$	$P_{01}^{(2)} = 0.231$	$n_{01}^{(2)} = 5$	$n_{01}^{(2)}/(K_0 \cdot 2^k) = 0.25$
$b_{01}^{(3)}$	$(0, [111], [-2,\ 0,\ 0], 1)$	$P_{01}^{(3)} = 0.066$	$n_{01}^{(3)} = 1$	$n_{01}^{(3)}/(K_0 \cdot 2^k) = 0.05$
$b_{10}^{(0)}$	$(1, [000], [\ 2,\ 0,\ 0], 0)$	$P_{10}^{(0)} = 0.066$	$n_{10}^{(0)} = 1$	$n_{10}^{(0)}/(K_1 \cdot 2^k) = 0.05$
$b_{10}^{(1)}$	$(1, [010], [\ 2, -2,\ 2], 0)$	$P_{10}^{(1)} = 0.231$	$n_{10}^{(1)} = 5$	$n_{10}^{(1)}/(K_1 \cdot 2^k) = 0.25$
$b_{10}^{(2)}$	$(1, [100], [\ 0,\ 2,\ 0], 0)$	$P_{10}^{(2)} = 0.145$	$n_{10}^{(2)} = 3$	$n_{10}^{(2)}/(K_1 \cdot 2^k) = 0.15$
$b_{10}^{(3)}$	$(1, [110], [\ 0,\ 0,\ 2], 0)$	$P_{10}^{(3)} = 0.066$	$n_{10}^{(3)} = 1$	$n_{10}^{(3)}/(K_1 \cdot 2^k) = 0.05$
$b_{11}^{(0)}$	$(1, [001], [\ 2,\ 0, -2], 1)$	$P_{11}^{(0)} = 0.195$	$n_{11}^{(0)} = 4$	$n_{11}^{(0)}/(K_1 \cdot 2^k) = 0.20$
$b_{11}^{(1)}$	$(1, [011], [\ 2, -2,\ 0], 1)$	$P_{11}^{(1)} = 0.146$	$n_{11}^{(1)} = 3$	$n_{11}^{(1)}/(K_1 \cdot 2^k) = 0.15$
$b_{11}^{(2)}$	$(1, [101], [\ 0,\ 2, -2], 1)$	$P_{11}^{(2)} = 0.146$	$n_{11}^{(2)} = 3$	$n_{11}^{(2)}/(K_1 \cdot 2^k) = 0.15$
$b_{11}^{(3)}$	$(1, [111], [\ 0,\ 0,\ 0], 1)$	$P_{11}^{(3)} = 0.005$	$n_{11}^{(3)} = 0$	$n_{11}^{(3)}/(K_1 \cdot 2^k) = 0.00$

create a superchannel trellis that mimics the optimal Markov process on the extended trellis τ, the occurrence probabilities of the superchannel output n-tuples should match the occurrence probabilities of the noiseless output n-tuples of the extended trellis τ. However, if $P_{ij}^{(\ell)} > 2^{-k}$, this would not be possible.

Our task now becomes finding the smallest possible positive integers k and n such that an n-th order extended trellis τ satisfies Equation 23.10. The search for k and n can be conducted systematically starting with $n = 1$ and increasing n until such a trellis is found. This procedure delivers k, n, the extended trellis τ, the optimized branching probabilities $P_{ij}^{(\ell)}$, and the signal-to-noise ratio $\text{SNR}_{(\tau, r)}$ for which $\hat{C}_\tau = r$.

Consider the dicode channel in Figure 23.1(a). Suppose our target code rate is $r = 1/2$. The simplest extended trellis of this channel for which Rule 1 holds is the extended trellis in Figure 23.2(a) with $n = 3$. Hence, we pick the inner trellis code rate $r_{\text{in}} = k/n = 2/3$ which satisfies Equation 23.10. The corresponding branching probabilities are given in Table 23.2. We shall use this example throughout the chapter to illustrate the design method.

23.4.2 Choosing the Number of States in the Superchannel

We now want to design a superchannel with rate $r_{\text{in}} = k/n$. Let $P_{ij}^{(\ell)}$ be the branching probabilities of the extended trellis τ evaluated at $\text{SNR}_{(\tau, r)}$. Let $\mu_i = \text{Pr}(S_t = i)$ denote the stationary probability of each state $0 \le i \le N_S - 1$ in the extended channel trellis τ.

Let K denote the number of states in the superchannel trellis ($K \le K_{\max}$, where $K_{\max}$ is a predefined maximal number of states in the superchannel trellis). Our strategy is to split each state i of the channel trellis τ into K_i states of the superchannel trellis. We say that these K_i states are of type t_i. Obviously, $K = \sum_{i=0}^{N_S-1} K_i \le K_{\max}$. Our goal is to find integers K_i such that the state types t_i in the superchannel trellis occur with the same probabilities as the state i in the extended channel trellis τ, that is, we desire $\mu_i \approx K_i/K$. Define a probability mass function (pmf) $\kappa = (\kappa_0, \kappa_1, \dots, \kappa_{N_S-1})$, where $\kappa_i = K_i/K$. Denote by μ the pmf $\mu = (\mu_0, \mu_1, \dots, \mu_{N_S-1})$.

Rule 2: Pick the pmf κ such that $D(\kappa \| \mu)$ is minimized under the constraint $N_S \leq K = \sum_{i=0}^{N_S-1} K_i \leq K_{\max}$, where $D(\cdot \| \cdot)$ denotes the Kullback-Leibler distance [40].

Consider again the dicode channel example with $r = 1/2$. Rule 1 gave $k = 2$ and $n = 3$, and the probabilities $P_{ij}^{(\ell)}$ in Table 23.2. Solving Equation 23.5, we get $\mu_0 = \mu_1 = 0.5$. With $K_{\max} = 12$, we get six solutions that satisfy Rule 2. They are $1 \leq K_0 = K_1 \leq 6$. This illustrates that there may not be a unique solution to the optimization problem in Rule 2. If this happens, then further refinement using Rule 3 (presented next) is needed.

23.4.3 Choosing the Branch Type Numbers in the Superchannel

We say that a branch of the superchannel trellis is of type $t_{ij}^{(\ell)}$ if

1. Its starting state is of type t_i and its ending state is of type t_j
2. Its noiseless output n-tuple matches the noiseless output n-tuple of the branch $b_{ij}^{(\ell)}$ in τ

Denote by $n_{ij}^{(\ell)}$ the number of branches in the superchannel trellis of type $t_{ij}^{(\ell)}$. For a fixed $i \in \{0, 1, \ldots, N_S - 1\}$, denote by ν_i the pmf whose individual probabilities are $n_{ij}^{(\ell)}/(K_i \cdot 2^k)$, where i is fixed and ℓ and j are varied under the constraint $(i, \ell, j) \in \tau$. Obviously

$$\sum_{i:(i,\ell,j)\in\tau} \frac{n_{ij}^{(\ell)}}{K_i \cdot 2^k} = 1 \tag{23.11}$$

Similarly, for a fixed $i \in \{0, 1, \ldots, N_S - 1\}$, denote by π_i the pmf whose individual probabilities are $P_{ij}^{(\ell)}$, where i is fixed and ℓ and j are varied under the constraint $(i, \ell, j) \in \tau$.

Rule 3: Determine $0 \leq n_{ij}^{(\ell)} \in \mathbb{Z}$ such that $\sum_{i=0}^{N_S-1} \kappa_i D(\nu_i \| \pi_i)$ is minimized under the constraints $\sum_{j,\ell:(i,\ell,j)\in\tau} n_{ij}^{(\ell)} = K_i \cdot 2^k$ and $\sum_{i,\ell:(i,\ell,j)\in\tau} n_{ij}^{(\ell)} = K_j \cdot 2^k$.

We have established that Rule 2 may not deliver a unique solution. In this case, among all solution candidates for Rule 2, we pick the solution that minimizes the objective function in Rule 3. In the dicode channel example with the target rate $r = 1/2$, Rule 2 delivered a set of solutions $K_0 = K_1 \leq K_{\max}/2 = 6$. Applying Rule 3, we get the integers $n_{ij}^{(\ell)}$ in Table 23.2 and $K_0 = K_1 = 5$, i.e., a superchannel trellis with $K = K_0 + K_1 = 10$ states.

23.4.4 Choosing the Branch Connections

Rules 1-3 guarantee that the marginal probability that the superchannel branch is of type $t_{ij}^{(\ell)}$ is very close to the value $\mu_i P_{ij}^{(\ell)}$. However, this does not guarantee that the resulting output process of the superchannel will mimic the output hidden Markov process of the channel trellis τ. Therefore, we need to choose a branch connection assignment with the following three requirements in mind:

1. Exactly $n_{ij}^{(\ell)}$ branches should be of type $t_{ij}^{(\ell)}$.
2. A branch of type $t_{ij}^{(\ell)}$ must start at a state of type t_i and end at a state of type t_j.
3. Branches emanating from a given state must have distinct types[2], that is, there cannot be two (or more) branches of the same type $t_{ij}^{(\ell)}$ emanating from a given state of type t_i.

Rule 4: Pick the superchannel branch connections that satisfy requirements (1)-(3) and deliver a superchannel with the maximal information rate $\mathcal{I}_S$ evaluated at $\mathrm{SNR}_{(\tau,r)}$.

If the integers K_i and $n_{ij}^{(\ell)}$ are very small, Rule 4 can be satisfied by an exhaustive search. Very often the exhaustive search procedure is too complex and we soften our goal by finding a "good enough"

---

[2]This requirement can be removed if we choose not to obey Rule 1.

TABLE 23.3   Rate $r_{in} = 2/3$ Superchannel Trellis for the $1 - D$ Channel and Design Rate $r = 1/2$

Start State	Super- channel Input k-tuple (Rule 5)	Trellis- code Output n-tuple	Noiseless Superchannel Output n-tuple (Rule 4)			End State
0	0,0	1,0,1	−2,	2,	−2	9
0	0,1	1,0,0	−2,	2,	0	0
0	1,0	0,1,1	0,	−2,	0	5
0	1,1	1,1,0	−2,	0,	2	1
1	0,0	1,1,1	−2,	0,	0	9
1	0,1	1,0,1	−2,	2,	−2	6
1	1,0	0,1,1	0,	−2,	0	8
1	1,1	1,1,0	−2,	0,	2	2
2	0,0	0,1,1	0,	−2,	0	6
2	0,1	1,0,1	−2,	2,	−2	8
2	1,0	0,1,0	0,	−2,	2	3
2	1,1	1,1,0	−2,	0,	2	0
3	0,0	1,0,1	−2,	2,	−2	7
3	0,1	1,0,0	−2,	2,	0	2
3	1,0	0,1,0	0,	−2,	2	4
3	1,1	1,1,0	−2,	0,	2	4
4	0,0	0,0,1	0,	0,	−2	7
4	0,1	1,0,1	−2,	2,	−2	5
4	1,0	0,1,0	0,	−2,	2	1
4	1,1	1,0,0	−2,	2,	0	3
5	0,0	0,1,1	2,	−2,	0	6
5	0,1	1,0,1	0,	2,	−2	8
5	1,0	0,1,0	2,	−2,	2	4
5	1,1	1,1,0	0,	0,	2	2
6	0,0	0,0,1	2,	0,	−2	5
6	0,1	1,0,1	0,	2,	−2	5
6	1,0	0,1,1	2,	−2,	0	7
6	1,1	0,1,0	2,	−2,	2	2
7	0,0	0,0,1	2,	0,	−2	9
7	0,1	1,0,1	0,	2,	−2	6
7	1,0	0,1,0	2,	−2,	2	1
7	1,1	1,0,0	0,	2,	0	3
8	0,0	0,0,1	2,	0,	−2	7
8	0,1	1,0,0	0,	2,	0	1
8	1,0	0,1,0	2,	−2,	2	3
8	1,1	0,0,0	2,	0,	0	0
9	0,0	0,0,1	2,	0,	−2	8
9	0,1	1,0,0	0,	2,	0	4
9	1,0	0,1,1	2,	−2,	0	9
9	1,1	0,1,0	2,	−2,	2	0

superchannel using the following *ordinal optimization* [42] randomized search procedure. We randomly pick (say) 2000 superchannels that satisfy Rules 1-3. For each of these superchannels, we *coarsely* estimate their i.u.d. rates $\mathcal{I}_S$ (say by using a trellis length of 10^4 trellis sections in the Monte Carlo method of [9, 10]). We keep (say) only 10 superchannels that have the 10 highest (coarsely estimated) i.u.d. rates. For these 10 superchannels, we now make *fine* estimates (with long trellises, say 10^8 trellis stages) of the i.u.d. rates $\mathcal{I}_S$, and pick the superchannel with the highest i.u.d. rate $\mathcal{I}_S$.

Applying Rule 4 to the dicode channel example with the target rate $r = 1/2$ delivered the branch connections assignment presented in Table 23.3. The information rate of the constructed MIR super-channel (shown in Figure 23.7) is only 0.1dB away from the Markov capacity $\hat{C}_\tau$ of the extended channel trellis in Figure 23.2(a) at the target rate $r = 1/2$. At $r = 1/2$, the designed MIR super-trellis has a higher

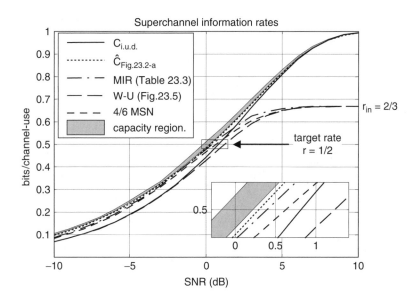

FIGURE 23.7 Information rates of the MIR (Table 23.3), W-U (Figure 23.5) and MSN superchannels with $r_{in} = 2/3$ (target rate $r = 1/2$). $\hat{C}_{Fig.\,23.2-a}$ is the numerical Markov capacity of the trellis in Figure 23.2(a), whose optimal Markov probabilities were used to construct the MIR superchannel. The region between the tightest known lower and upper bound is shaded.

super-trellis information rate than other known trellis codes (the rate-2/3 W-U code [15] in Figure 23.5 and the rate-4/6 MSN code [18]). Much more importantly, we now have a *method* to construct superchannel trellises for *any* channel, even if the channel does not have spectral nulls.

23.5 Outer LDPC Codes

For the dicode channel, Figure 23.7 reveals that the i.u.d. capacity $C_{i.u.d.}$ equals the design rate $r = 1/2$ at SNR $= 0.82$ dB. The i.u.d. rate of the superchannel trellis shown in Table 23.3 equals $\mathcal{I}_S = 1/2$ at SNR $= 0.40$ dB (see Figure 23.7). Theorem 23.1 asserts that there exists at least one linear (coset) code such that, if we apply this code to the superchannel and utilize maximum likelihood (ML) decoding, we may gain 0.42 dB over $C_{i.u.d.}$. Since it is impractical to perform the ML decoding, we construct iteratively decodable outer codes.

23.5.1 Encoding/Decoding System

The normal graph [43] of the entire concatenated coding system is shown in Figure 23.8. We utilize k separate subcodes as the outer code (we explain the reason below), where k is the number of input bits at each stage of the MIR trellis code. Let the rate of the ith subcode be $r^{(i)} = K^{(i)}/N$, where $0 \le i \le k - 1$. Let $\underline{D}$ be an i.u.d. binary sequence of length $K^{(total)} = \sum_{i=0}^{k-1} K^{(i)}$ to be transmitted over the channel. The encoding algorithm can be described by the following three steps (see also Figure 23.8):

1. The sequence $\underline{D}$ is separated into k subsequences $\underline{D} = [\underline{D}^{(0)}, \underline{D}^{(1)}, \ldots, \underline{D}^{(k-1)}]$, where the ith subsequence $\underline{D}^{(i)}$ is of length $K^{(i)}$.
2. The ith subsequence $\underline{D}^{(i)}$ enters the ith LDPC encoder whose code-rate is $r^{(i)} = K^{(i)}/N$. The output sequence from the ith encoder is denoted by $[U^{(i)}]_1^N = [U_1^{(i)}, \ldots, U_N^{(i)}]$.
3. The whole sequence $U_1^N = [U_1, \ldots, U_t, \ldots, U_N]$ enters the MIR trellis encoder, where $U_t = (U_t^{(0)}, \ldots, U_t^{(k-1)})$. The sequence Y_1^{nN} is observed at the channel output.

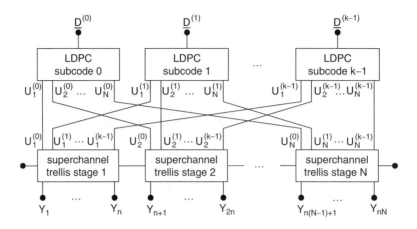

FIGURE 23.8   The normal graph of an outer LDPC code (consisting of k subcodes) and the inner superchannel trellis.

For the purpose of systematically designing good outer codes, we specify a serial multistage decoding schedule [44] depicted in Figure 23.9. To guarantee that the decoding algorithm of Figure 23.9 works well, we have to solve the following three problems:

- A trellis section can be viewed as a collection of branches $\{b_t = (s_{t-1}, u_t, v_t, s_t)\}$. In terms of optimizing the superchannel i.u.d. information rate $\mathcal{I}_S$, it is irrelevant how the input vector u_t of each branch is chosen. However, to construct a good and practically decodable LDPC code we need to choose the branch input-bit assignment judiciously. We develop the input-bit assignment design rule (Rule 5) in Section 23.5.2.

- Consider the soft-output random variables $L_t^{(i)} \overset{\text{def}}{=} \Pr(U_t^{(i)} = 0 \mid Y_1^{nN})$. For a general rate-$k/n$ inner trellis code, when $N \to \infty$, the statistical properties of $L_{t_1}^{(i)}$ and $L_{t_2}^{(j)}$ are the same if and only if $i = j$. In other words, different superchannel bit positions have different soft-output statistics, which is why we utilize k different subcodes. Now the question is: how to determine the rates $r^{(i)}$ of the constituent subcodes? Section 23.5.3 gives the answer.

- For large N, we need to optimize each of the k subcodes. In Section 23.5.4, we develop an optimization method along the lines of [33, 36].

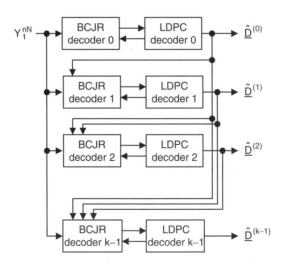

FIGURE 23.9   The iterative decoder. The k codes are successively decoded, each iteratively.

23.5.2 Choosing the Branch Input-Bit Assignment

We need to specify the input k-tuple $u_t = [u_t^{(0)}, u_t^{(1)}, \ldots, u_t^{(k-1)}]$ (where $u_t^{(i)} \in \mathcal{A}$) for every branch $b_t = (s_{t-1}, u_t, v_t, s_t)$ of the superchannel trellis. Generally, if the bit assignments for $u_t^{(0)}, \ldots, u_t^{(i-1)}$ are given, then only the bit assignment for $u_t^{(i)}$ determines the performance of Decoder i in Figure 23.9, irrespective of the bit assignment for $u_t^{(i+1)}, \ldots, u_t^{(k-1)}$. Therefore we use the following rule to determine the bit assignment.

Rule 5: (Greedy bit assignment algorithm)

For $0 \leq i \leq k - 1$

Assume that Decoders 0 through $i - 1$ decode without errors. Find the bit assignment for the ith location $u_t^{(i)}$ such that the ith BCJR decoder in Figure 23.9 delivers the lowest probability of bit error *in the first iteration*.

This algorithm guarantees a good decoding start which is typically sufficient to ensure good decoding properties for the concatenated LDPC decoder [35].

By applying Rule 5, we have selected one good input-bit assignment for the trellis code shown in Table 23.3. The input bit assignment is shown in the third column of Table 23.3.

23.5.3 Determining the Outer Subcode Rates

Ideally, the superchannel i.u.d. rate $\mathcal{I}_S$ should equal the target rate r

$$r = r_{\text{in}} \cdot r_{\text{out}} = \frac{k}{n} \cdot r_{\text{out}} = \lim_{N \to \infty} \frac{1}{nN} \cdot I\left(U_1^N; Y_1^{nN}\right)\Bigg|_{P_{U_1^N}(u_1^N) = 2^{-kN}} = \mathcal{I}_S$$

From the chain rule [40], we have

$$I\left(U_1^N; Y_1^{nN}\right) = I\left([U^{(0)}]_1^N; Y_1^{nN}\right) + \sum_{i=1}^{k-1} I\left([U^{(i)}]_1^N; Y_1^{nN} \mid [U^{(0)}]_1^N, \ldots, [U^{(i-1)}]_1^N\right)$$

where $[U^{(i)}]_1^N = [U_1^{(i)}, U_2^{(i)}, \ldots, U_N^{(i)}]$. Hence, the assumption that Decoders 0 through $i - 1$ perform error-less decoding before Decoder i starts the decoding process is valid only if

$$r^{(i)} \leq \lim_{N \to \infty} \frac{1}{N} \cdot I\left([U^{(i)}]_1^N; Y_1^{nN} \mid [U^{(0)}]_1^N, \ldots, [U^{(i-1)}]_1^N\right) \tag{23.12}$$

Therefore, a reasonable rate-assignment is

$$r^{(i)} = \lim_{N \to \infty} \frac{1}{N} \cdot I\left([U^{(i)}]_1^N; Y_1^N \mid [U^{(0)}]_1^N, \ldots, [U^{(i-1)}]_1^N\right) \tag{23.13}$$

where the sequences $[U^{(i)}]_1^N$ are i.u.d. for all $0 \leq i \leq k - 1$. The rates in Equation 23.13 can be computed by Monte Carlo simulations [9, 10]. Consequently, we get $r_{\text{out}} = \frac{1}{k} \sum_{i=0}^{k-1} r^{(i)}$.

To summarize, first we choose two integers $(K^{(total)}, N)$, where N is large enough, such that $K^{(total)}/(kN)$ is not greater than (and is as close as possible to) $r_{\text{out}} = r/r_{\text{in}}$. Then we choose $K^{(i)}$, for $0 \leq i \leq k - 1$ such that $\sum_{i=0}^{k-1} K^{(i)} = K^{(total)}$ and $K^{(i)}/N \approx r^{(i)}$, where $r^{(i)}$ is computed by Equation 23.13.

23.5.4 Subcode Optimization

23.5.4.1 Outer Code Optimization

To optimize the outer code for a given superchannel we generalize the *single* code optimization method for a PR channel (with no inner code) [33, 34, 36] to a *joint* optimization of k different subcodes. This generalization is summarized as follows:

1. Given a superchannel trellis of rate $r_{\text{in}} = k/n$, we optimize k different constituent subcodes shown, in Figure 23.8 according to the decoding scenario shown in Figure 23.9. That is, while optimizing the degree sequences of the LDPC Subcode i, we assume that the symbols $[U^{(0)}]_1^N$, $[U^{(1)}]_1^N$, ..., $[U^{(i-1)}]_1^N$ are decoded without errors and that no prior information on the symbols $[U^{(i+1)}]_1^N$, ..., $[U^{(k-1)}]_1^N$ is available.

2. The ith constituent subcode of rate $r^{(i)}$ is optimized using the optimization from [36]. We make a change in the density evolution [32, 35] to reflect the fact that Decoders 0 through $i-1$ have decoded symbols $[U^{(0)}]_1^N$ through $[U^{(i-1)}]_1^N$ without errors. This is done by generating i.u.d. realizations of symbols $[u^{(0)}]_1^N$ through $[u^{(i-1)}]_1^N$, and running the BCJR algorithm (in the density evolution step) with prior knowledge of these symbols.

3. We obtain k pairs $(\underline{\lambda}^{(i)}, \underline{\rho}^{(i)})$, $0 \le i \le k-1$, of the optimized edge degree sequences [31], each pair corresponding to one constituent subcode. The noise threshold [32] of the entire code is given by $\sigma^* = \min\{\sigma_0^*, \sigma_1^*, \ldots, \sigma_{k-1}^*\}$, where σ_i^* is the noise threshold for Subcode i.

23.5.4.2 LDPC Decoding

Figure 23.9 seems to suggest that we need k different BCJR decoders and k different LDPC decoders to decode our code. In fact we only need *one* BCJR detector and *one* LDPC decoder. The *single* LDPC code is constructed by interleaving the constituent subcodes.

The parity check matrix of Subcode i is constructed to be a low-density parity-check matrix with degree sequences $\underline{\lambda}^{(i)}$ and $\underline{\rho}^{(i)}$. Denote the parity-check matrix of Subcode i by

$$\mathbf{H}^{(i)} = \left[\underline{h}_1^{(i)} \ \underline{h}_2^{(i)}, \ldots, \underline{h}_N^{(i)} \right]$$

where $\underline{h}_j^{(i)}$ represents the jth column of the matrix $\mathbf{H}^{(i)}$. The parity check matrix $\mathbf{H}$ of the entire code is obtained by interleaving the columns of the subcode parity check matrices

$$\mathbf{H} = \begin{bmatrix} \underline{h}_1^{(0)} & 0 & \cdots & 0 & \underline{h}_2^{(0)} & 0 & \cdots & 0 & & \underline{h}_N^{(0)} & 0 & \cdots & 0 \\ 0 & \underline{h}_1^{(1)} & \cdots & 0 & 0 & \underline{h}_2^{(1)} & \cdots & 0 & & 0 & \underline{h}_N^{(1)} & \cdots & 0 \\ \vdots & \vdots & \ddots & \vdots & \vdots & \vdots & \ddots & \vdots & \cdots & \vdots & \vdots & \ddots & \vdots \\ 0 & 0 & \cdots & \underline{h}_1^{(k-1)} & 0 & 0 & \cdots & \underline{h}_2^{(k-1)} & & 0 & 0 & \cdots & \underline{h}_N^{(k-1)} \end{bmatrix}. \quad (23.14)$$

The size of the matrix $\mathbf{H}$ is $(kN - K^{(total)}) \times (kN)$, where $K^{(total)} = \sum_{i=0}^{k-1} K^{(i)}$.

23.6 Optimization Results

We perform the optimization on the dicode $(1-D)$ channel which has a spectral null at frequency $\omega = 0$, and on the $1 + 3D + D^2$ channel which does not have a spectral null.

23.6.1 Dicode Channel

The inner trellis code with code rate $r_{\text{in}} = k/n = 2/3$ for the dicode channel is given in Table 23.3. Since $k = 2$, we have 2 outer LDPC subcodes. Using Equation 23.13, we found their rates to be $r^{(0)} = 0.66$ and $r^{(1)} = 0.84$, respectively. The rate of the outer code is thus $r_{\text{out}} = \frac{1}{2} \cdot (r^{(0)} + r^{(1)}) = 3/4$. The resulting overall code rate is then $r = r_{\text{in}} \cdot r_{\text{out}} = \frac{2}{3} \cdot \frac{3}{4} = 1/2$, which is exactly our target code rate.

The optimized degree sequences together with their respective thresholds are given in Table 23.4. We constructed an outer LDPC code by interleaving the parity check matrices obtained from the optimized degree sequences, see Equation 23.14. The code block length was set to 10^6 binary symbols. The bit error rate (BER) simulation curve is shown in Figure 23.10. For comparison, Figure 23.10 also shows a tight lower bound $\hat{C}_6$ and an upper bound C_U on the capacity, the superchannel i.u.d. rate $\mathcal{I}_s$, the i.u.d. capacity

TABLE 23.4 Good Degree Sequences and the Noise Thresholds for Separately Coded Even and Odd Bits of the Outer LDPC Code on the Superchannel from Table 23.3

	Subcode 0 (even bits)			Subcode 1 (odd bits)	
	$r^{(0)} = 0.660$			$r^{(1)} = 0.840$	
x	$\lambda_x^{(0)}$	$\rho_x^{(0)}$	x	$\lambda_x^{(1)}$	$\rho_x^{(1)}$
2	0.2225		2	0.2191	
3	0.1611		3	0.1526	
5	0.1627		4	0.1057	
6	0.0164		15	0.0017	
10	0.0024		16	0.0003	
11		0.3325	17		0.0914
13		0.1406	23		0.3855
16		0.3929	49	0.1205	
24		0.1340	50	0.4001	0.1321
48	0.0035		51		0.0698
49	0.4314		59		0.3212
	threshold - Subcode 0 $\sigma_0^* = 1.322$			threshold - Subcode 1 $\sigma_1^* = 1.326$	
		threshold $\sigma^* = 1.322$			
		SNR* $= 10\log_{10} \frac{2}{(\sigma^*)^2} = 0.59\,\text{dB}$			

$C_{\text{i.u.d.}}$ and the noise tolerance threshold SNR* $= 10\log_{10} \frac{\sum_j h_j^2}{(\sigma^*)^2}$ of the code in Table 23.4. We see that the threshold SNR* surpasses the i.u.d. channel capacity $C_{\text{i.u.d.}}$ by 0.23 dB and is 0.19 dB away from the superchannel i.u.d. rate $\mathcal{I}_S$. The code simulation shows that a BER of 10^{-6} is achieved at an SNR that surpasses $C_{\text{i.u.d.}}$ by 0.14 dB.

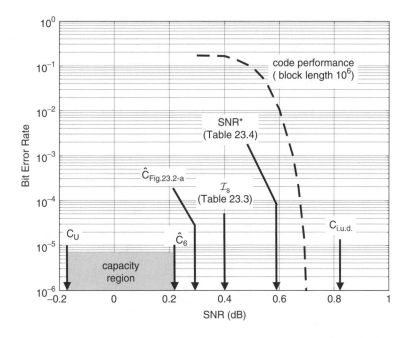

FIGURE 23.10 BER versus SNR for the Dicode channel. Code rate $r = 1/2$; code length 10^6 bits.

TABLE 23.5   Rate $r_{in} = 1/2$ Superchannel Trellis for
the $1 + 3D + D^2$ Channel and Design Rate $r = 1/3$

Start State	Super- Channel Input k-tuple	Trellis- code Output n-tuple	Noiseless Superchannel Output n-tuple	End State
0	0	0,0	5, 5	1
0	1	0,1	5, 3	3
1	0	0,0	5, 5	2
1	1	1,1	3,−3	6
2	0	0,0	5, 5	0
2	1	1,1	3,−3	5
3	0	1,0	−3,−3	4
3	1	1,1	−3,−5	7
4	0	0,0	3, 5	0
4	1	0,1	3, 3	3
5	0	0,0	−3, 3	2
5	1	1,1	−5,−5	7
6	0	0,0	−3, 3	1
6	1	1,1	−5,−5	5
7	0	1,0	−5,−3	4
7	1	1,1	−5,−5	6

23.6.2   A Channel without Spectral Nulls

For the $1 + 3D + D^2$ channel we chose the design rate $r = 1/3$. Using Rules 1-5, we constructed an inner trellis code of rate $r_{in} = k/n = 1/2$ for the $1 + 3D + D^2$ channel; the superchannel trellis is given in Table 23.5. Since $k = 1$, we have only one outer LDPC subcode of rate $r_{out} = 2/3$, which gives the desired code rate $r = \frac{1}{2} \cdot \frac{2}{3} = 1/3$. The optimization of the outer LDPC code delivered the degree sequences and the threshold shown in Table 23.6. Based on these degree sequences we constructed an LDPC code of block length 10^6. The code's BER performance curve is shown in Figure 23.11. We observe from Figure 23.11

TABLE 23.6   Good Degree Sequences and the
Noise Threshold for the Outer LDPC Code
Designed for the Superchannel in Table 23.5

	$r_{out} = r^{(0)} = 0.6667$	
i	λ_i	ρ_i
2	0.2031	
3	0.2195	
4	0.0022	
6	0.1553	
10		0.2974
13		0.3064
15		0.1906
27	0.1616	
28	0.1399	
38		0.2056
50	0.1184	

threshold
$$\sigma_0^* = \sigma^* = 4.475$$

$$\text{SNR}^* = 10\log_{10} \frac{11}{(\sigma^*)^2}$$
$$= -2.60\text{dB}$$

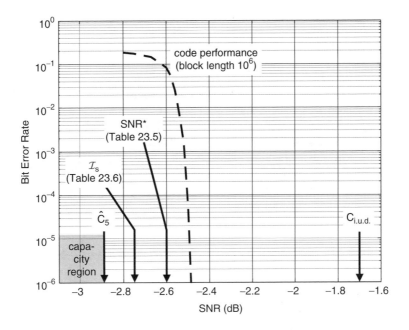

FIGURE 23.11   BER versus SNR for the $1 + 3D + D^2$ channel. Code rate $r = 1/3$; code length 10^6 bits.

that the superchannel i.u.d. rate $\mathcal{I}_S$ is 0.14 dB away from the lower bound on the channel capacity $\hat{C}_5$. The noise threshold SNR* surpasses the i.u.d. channel capacity $C_{\text{i.u.d.}}$ by 0.90 dB and is 0.15 dB away from $\mathcal{I}_S$. The code simulation reveals that our design method yields a code that achieves a BER of 10^{-6} at an SNR that surpasses $C_{\text{i.u.d.}}$ by 0.77 dB.

23.7 Conclusion

We developed a methodology for designing capacity-approaching codes for partial response (PR) channels. Since the PR channel is a channel with memory, its capacity C is greater than its i.u.d. capacity $C_{\text{i.u.d.}}$. We showed that rates above $C_{\text{i.u.d.}}$ cannot be achieved by random linear codes. Given that our goal was to construct a code that surpasses $C_{\text{i.u.d.}}$, we chose a concatenated coding strategy where the inner code is a (generally nonlinear) trellis code, and the outer code is a low-density parity-check (LDPC) code.

The key step in the design of the inner code is to identify a Markov input process that achieves a high (capacity-approaching) information rate over the PR channel of interest. Then, we construct a trellis code that mimics the identified Markov process. Hence, we name it a *matched information rate* (MIR) trellis code. We choose an LDPC code as an outer code to show that the concatenation of the two codes approaches the computed information rate of the identified Markov process. The outer code is optimized by modified density evolution methods to fit our specific inner code and the PR channel.

MIR trellis code constructions are different from any previously known trellis code construction methods in that we do *not* base the code construction on an algebraic criterion. Instead, the code construction is purely *probabilistic*. We provided a set of rules to construct the inner MIR trellis codes. These rules apply to *any* PR channel. Using our design rules and the outer code optimization, we constructed examples of capacity-approaching codes for both channels with and without spectral nulls.

Acknowledgment

The authors would like to thank Michael Mitzenmacher for very helpful discussions over the course of this research.

References

[1] W. Hirt, *Capacity and Information Rates of Discrete-Time Channels with Memory*. Ph.D. thesis, Swiss Federal Institute of Technology (ETH), Zurich, Switzerland, 1988.

[2] S. Shamai (Shitz), L.H. Ozarow, and A.D. Wyner, Information rates for a discrete-time Gaussian channel with intersymbol interference and stationary inputs, *IEEE Trans. Inform. Theory*, vol. 37, pp. 1527–1539, 1991.

[3] S. Shamai (Shitz) and S. Verdú, Worst-case power-constrained noise for binary-input channels, *IEEE Trans. Inform. Theory*, vol. 38, pp. 1494–1511, 1992.

[4] C.E. Shannon, A mathematical theory of communications, *Bell Syst. Tech. J.*, vol. 27, pp. 379–423 (part I) and 623–656 (part II), 1948.

[5] A.S. Khayrallah and D.L. Neuhoff, Coding for channels with cost constraints, *IEEE Trans. Inform. Theory*, vol. 42, pp. 854–867, May 1996.

[6] E. Zehavi and J.K. Wolf, On runlength codes, *IEEE Trans. Inform. Theory*, vol. 34, pp. 45–54, January 1988.

[7] B. Marcus, K. Petersen, and S. Williams, Transmission rates and factors of Markov chains, *Contemporary Mathematics*, vol. 26, pp. 279–293, 1984.

[8] S. Yang and A. Kavčić, Capacity of partial response channels, *Chapter 13*.

[9] D. Arnold and H.-A. Loeliger, On the information rate of binary-input channels with memory, in *Proceedings' IEEE International Conference on Communications 2001*, Helsinki, Finland, June 2001.

[10] H.D. Pfister, J.B. Soriaga, and P.H. Siegel, On the achievable information rates of finite state ISI channels, in *Proceedings IEEE Globecom 2001*, San Antonio, Texas, pp. 2992–2996, November 2001.

[11] A. Kavčić, On the capacity of Markov sources over noisy channels, in *Proceedings IEEE Global Communications Conference 2001*, San Antonio, Texas, pp. 2997–3001, November 2001.

[12] P. Vontobel and D.M. Arnold, An upper bound on the capacity of channels with memory and constraint input, in *IEEE Information Theory Workshop*, Cairns, Australia, September 2001.

[13] S. Yang and A. Kavčić, Markov sources achieve the feedback capacity of finite-state machine channels, in *Proceedings of IEEE International Symposium on Inform. Theory*, Lausanne, Switzerland, July 2002.

[14] C. Berrou, A. Glavieux, and P. Thitimajshima, Near Shannon limit error-correcting coding and decoding: Turbo-codes, in *Proceedings IEEE International Conference on Communications*, pp. 1064–1070, May 1993.

[15] J.K. Wolf and G. Ungerboeck, Trellis coding for partial-response channels," *IEEE Trans. Commun.*, vol. 34, pp. 744–765, March 1986.

[16] A.R. Calderbank, C. Heegard, and T.A. Lee, Binary convolutional codes with application to magnetic recording, *IEEE J. Select. Areas Commun.*, vol. 32, pp. 797–815, November 1986.

[17] K.A.S. Immink, Coding techniques for the noisy magnetic recording channel: A state-of-the-art report, *IEEE Trans. Commun.*, vol. 35, pp. 413–419, May 1987.

[18] R. Karabed and P.H. Siegel, Matched spectral-null codes for partial response channels, *IEEE Trans. Inform. Theory*, vol. 37, pp. 818–855, May 1991.

[19] R.G. Gallager, *Low-Density Parity-Check Codes*, MIT Press, Cambridge, MA, 1962.

[20] D.J.C. MacKay and R.M. Neal, Near Shannon limit performance of low-density parity-check codes, *Electron. Lett.*, vol. 32, pp. 1645–1646, 1996.

[21] L.R. Bahl, J. Cocke, F. Jelinek, and J. Raviv, Optimal decoding of linear codes for minimizing symbol error rate, *IEEE Trans. Inform. Theory*, vol. 20, pp. 284–287, September 1974.

[22] W. Ryan, Performance of high-rate turbo codes on PR4-equalized magnetic recording channels, in *Proceedings of IEEE International Conference on Communications*, Atlanta, GA, pp. 947–951, June 1998.

[23] M.C. Reed and C.B. Schlegel, An iterative receiver for partial response channels, in *Proceedings IEEE International Symposiums Information Theory*, Cambridge, MA, p. 63, August 1998.

[24] T. Souvignier, A. Friedmann, M. Öberg, P. Siegel, R.E. Swanson, and J.K. Wolf, Turbo codes for PR4: Parallel versus serial concatenation, in *Proceedings of IEEE International Conference on Comm.*, pp. 1638–1642, June 1999.

[25] L.L. McPheters, S.W. McLaughlin, and K.R. Narayanan, Precoded PRML, serial concatenation, and iterative (turbo) decoding for digital magnetic recording, *IEEE Trans. Magn.*, vol. 35, September 1999.

[26] M. Tüchler, R. Kötter, and A. Singer, Iterative correction of ISI via equalization and decoding with priors, in *Proceedings of IEEE International Symposiums Information Theory*, Sorrento, Italy, p. 100, June 2000.

[27] J. Park and J. Moon, A new soft-output detection method for magnetic recording channels, in *Proceedings IEEE Global Communications Conference 2001*, San Antonio, Texas, pp. 3002–3006, November 2001.

[28] J. Fan, A. Friedmann, E. Kurtas, and S. McLaughlin, Low density parity check codes for magnetic recording, in *Proceedings Allerton Conference on Communications and Control*, 1999.

[29] M. Oberg and P.H. Siegel, Parity check codes for partial response channels, in *Proceedings IEEE Global Telecommunications Conference*, vol. 1, Rio de Janeiro, pp. 717–722, December 1999.

[30] J. Li, K.R. Narayanan, E. Kurtas, and C.N. Georghiades, On the performance of high rate turbo product codes and low density parity check codes for partial response channels, *IEEE Trans. Commun.*, vol. 50, pp. 723–734, May 2002.

[31] M. Luby, M. Mitzenmacher, M. A. Shokrollahi, and D. Spielman, Improved low-density parity-check codes using irregular graphs, *IEEE Trans. Inform. Theory*, vol. 47, pp. 585–598, February 2001.

[32] T. Richardson and R. Urbanke, The capacity of low-density parity check codes under message-passing decoding, *IEEE Trans. Inform. Theory*, vol. 47, pp. 599–618, February 2001.

[33] T. Richardson, A. Shokrollahi, and R. Urbanke, Design of capacity-approaching low-density parity-check codes, *IEEE Trans. Inform. Theory*, vol. 47, pp. 619–637, February 2001.

[34] S.-Y. Chung, G.D. Forney, T. Richardson, and R. Urbanke, On the design of low-density parity-check codes within 0.0045 dB of the Shannon limit, *IEEE Commun. Lett.*, February 2001.

[35] A. Kavčić, X. Ma, and M. Mitzenmacher, Binary intersymbol interference channels: Gallager codes, density evolution and code performance bounds, *IEEE Trans. Inform. Theory*, pp. 1636–1652, July 2003.

[36] N. Varnica and A. Kavčić, Optimized low-density parity-check codes for partial response channels, *IEEE Commun. Lett.*, vol. 7, pp. 168–170, April 2003.

[37] A. Kavčić and A. Patapoutian, A signal-dependent autoregressive channel model, *IEEE Trans. Magn.*, vol. 35, pp. 2316–2318, September 1999.

[38] E. Kurtas, J. Park, X. Yang, W. Radich, and A. Kavčić, Detection methods for data-dependent noise in storage channels, *Chapter 33*.

[39] R.G. Gallager, *Information Theory and Reliable Communication*, John Wiley and Sons, New York, 1968.

[40] T.M. Cover and J.A. Thomas, *Elements of Information Theory*, John Wiley and Sons, New York, 1991.

[41] A. Kavčić, X. Ma, and N. Varnica, Matched information rate codes for partial response channels, *submitted for publication in IEEE Trans. Inform. Theory*, July 2002.

[42] Y.-C. Ho, An explanation of ordinal optimization: Soft computing for hard problems, *Inform. Sci.*, vol. 113, pp. 169–192, February 1999.

[43] G.D. Forney Jr., Codes on graphs: Normal realizations, *IEEE Trans. Inform. Theory*, vol. 47, pp. 520–548, February 2001.

[44] H. Imai and S. Hirakawa, A new multilevel coding method using error correcting codes, *IEEE Trans. Inform. Theory*, vol. 23, pp. 371–377, May 1977.

24

Coding and Detection for Multitrack Systems

Bane Vasic
University of Arizona
Tucson, AZ

Olgica Milenkovic
University of Colorado
Boulder, CO

24.1 Introduction **24**-1
24.2 The Current State of Research in
      Multitrack Codes **24**-2
24.3 Multitrack Channel Model **24**-2
24.4 Multitrack Constrained Codes **24**-3
      ITI Reducing Codes for PR Channels • Constrained Coding
      for Improved Synchronization • Low-Complexity Encoder
      and Decoder Implementations
24.5 Multitrack Soft Error-Event Correcting Scheme **24**-5

24.1 Introduction

In traditional magnetic disk drive systems user data is recorded on concentric tracks as a sequence of changes of the magnetization of small domains. The direction of the magnetization of a domain depends on the polarity of a data-driven write current. The areal density of data stored on a disk can be increased by either reducing the size of the magnetic domains along tracks (i.e., by increasing the linear density) and/or by reducing the track pitch (i.e., by increasing the track, or radial, density). The linear density is limited by several factors, including the finite sensitivity of the read head, properties of magnetic materials, head design [47, 51], as well as the ability to detect and decode recorded data in the presence of intersymbol interference (ISI) and noise. Most research in disk drive systems has focused on increasing linear density. Extremely high densities, for example, 10 Gbits/in^2, and data rates approaching 1 Gbits/s have been already demonstrated in commercially available systems. Recent progress in heads and media design has opened the possibility of reaching densities of 100 Gbits/in^2 and perhaps even 1 Terabit/in^2. However, the rate of increase of the linear density in future magnetic recording systems is not likely to be as high as in the past. This is due to the fact that as the linear density increases, the magnetic domains on the disk surface become smaller and increasingly thermally unstable. The so-called super-paramagnetic effect [6] represents a fundamental limiting factor for linear density.

Alternative approaches for increasing areal density are therefore required in order to meet the constant demand for increases in data rate and capacity of storage devices, largely driven by the Internet. Since the current linear densities are approaching the super-paramagnetic limit, the obvious alternative to an increase in linear density is an increase in radial density. In modern systems, the radial density is mostly limited by the mechanical design of the drive and the ability to accurately follow a track whose width is of the order of 1 μm. In order to further increase radial density, multiple-head arrays have been developed

[27, 36]. A head array is an arrangement of closely spaced heads that can read and write data simultaneously on multiple tracks. Such heads can potentially provide both high density and high speed, but they suffer from cross-talk or intertrack interference (ITI) [4]. This ITI is the consequence of a signal induced in the heads due to the superposition of magnetic transitions in neighboring tracks. Today's recording systems have a large track pitch, and therefore ITI has a negligible effect on their performance. However, significant advances in coding and signal processing for multitrack recording channels are required before potential large radial densities together with head arrays may be practically applied.

24.2 The Current State of Research in Multitrack Codes

A number of multitrack coding and detection schemes have been proposed in the last decade. They can be categorized as follows:

1. The first category is a class of multitrack codes which exploit the idea that the achievable areal density can be increased *indirectly*. This is done through relaxing the per-track maximum runlength constraint (the so-called k-constraint) in the recording codes and by imposing a constraint across multiple tracks. Such codes have been studied by Marcellin et al. [20–23], Swanson and Wolf [29], and Vasic et al. [39–43].

2. The second category is a class of techniques involving multitrack detection combined with partial-response (PR) equalization [12] and *maximum likelihood* (multiple) sequence detection (MLSD). Such techniques have computational complexity exponential in NM, where N is the number of tracks and M is the memory of the PR channel. Multitrack codes and reduced-complexity detectors for such systems have been studied by Soljanin et al. [31–33] and Kurtas et al. [12, 19]. More recently, a combination of equalization and *maximum a posteriori* (MAP) decoding, reducing both ISI and ITI, was considered by Wu et al. [52] (see also Chugg et al. [7] who first applied these ideas to page oriented holographic storage). The method presented in [12] combines a multitrack version of the BCJR algorithm [1] with iterative Wiener filtering, and performs very well in conjunction with low-density parity check (LDPC) codes. It's major drawback is very high complexity.

3. The third class of techniques uses the idea of imposing a constraint on a recorded bit sequence in such a way that ITI on each PR channel is either completely eliminated or reduced. Recently, Ahmed et al. [1] constructed a two-track runlength-limited code for a Class 4 partial response (PR4) channel. The code forbids any transitions of opposite polarity on adjacent tracks, and results in up to a 23% gain in areal density over an uncoded system. Similar two track schemes, but for a different multitrack constraint, have been proposed by Davey et al. [7, 11] and by Lee and Madisetti [20]. Due to their high complexity, these schemes can be used only for a small number of tracks and low order PR polynomials.

In order to achieve high linear densities, equalization with respect to higher-order PR polynomials is necessary. However, the complexity of read-channel chips increases exponentially with the order of the PR polynomial. Furthermore, the largest contribution to the complexity comes from MLSD, which is already the most complicated subsystem in the "read-channel" electronics and is a primary impediment to high data throughput. Thus, increasing the complexity of a detector by another factor (N) in the exponent, which is required for MLSD detection over N tracks, is not feasible.

24.3 Multitrack Channel Model

The magnetic recording channel is modeled by a discrete-time linear filter with a partial response polynomial typically of the form $h(D) = (1 - D)(1 + D)^M$, or $h(D) = (1 + D)^M, M \geq 1$, depending on whether longitudinal or perpendicular recording is employed [46]. User data is encoded by N separate error control encoders, and are organized in two-dimensional blocks of size $N \times n$ written on N adjacent tracks (see Figure 24.1). The sequence recorded in the kth track is denoted by $\{a_m^{(k)}\}_{m \in Z}$. Adjacent tracks

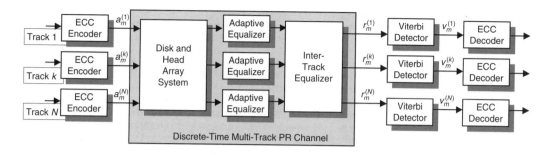

FIGURE 24.1 Discrete-time longitudinal multitrack recording channel model.

are read by an array of N heads. According to the channel model proposed by Vea and Moura [47], the signal read by the kth head is equalized to some partial response target and is given by

$$r^{(k)}(t) = \sum_{1 \le l \le N} \alpha_{|l-k|} \sum_{m \in Z} a_m^{(l)} \cdot h(t - mT) + n(t), \quad 1 \le k \le N$$

where $h(t)$ is the impulse response of the PR target, $n(t)$ is the colored noise process obtained by filtering additive white Gaussian noise (AWGN) by the equalizer; the real numbers $\alpha_d, 0 \le d \le N - 1$ specify the cross-talk between two tracks separated by d other tracks.

For longitudinal recording systems, we assume that the read head response to an isolated (positive-going) transition written on a disk is the Lorentzian pulse, $l(t) = (1 + (2t/PW_{50})^2)^{-1}$. For perpendicular recording, we assume that the transition response is the error function, $g(t) = (2/\sqrt{\pi}) \cdot \int_0^{S \cdot t} \exp(-x^2)\, dx = erf(St)$, where $S = 2\sqrt{\ln 2} \cdot T/PW_{50}$. The *dibit response* is defined as $(g(t + T/2) - g(t - T/2))$. The channel density is defined as $S_c = PW_{50}/T$, where PW_{50} represents the width of the channel impulse response at half of its peak value, and T is the channel bit interval. This models ignores the effects of timing jitter, that is, it assumes that the samples are taken at time instants $t = mT$, where the channel bit interval T is known and fixed. It also ignores the effects of track-following error, that is, it is assumed that the kth head is perfectly aligned with the kth track.

Another equalization scheme was recently proposed by Vasic and Venkateswaran [46] and shown in Figure 24.1. The intertrack interference block is described by the matrix $A = [\alpha_{|i-j|}]_{1 \le i,j \le N}$ [47]. The in-track equalizer is an adaptive least mean-square error equalizer, while the inter-track equalizer is a short zero-forcing equalizer. The equalized data are fed to a bank of N Viterbi detectors, and then to ECC decoders.

24.4 Multitrack Constrained Codes

24.4.1 ITI Reducing Codes for PR Channels

Consider an N-track system in which each track is equalized to a PR channel with polynomial $p(D) = h_0 + h_1 D + \cdots + h_L D^L$. The N-track constrained system is defined as an oriented, strongly connected graph $G = (V, E)$ with vertex set V and edge set E. The vertices are labeled by binary arrays Ψ of dimension $N \times L$, $\Psi = (\Psi_j)_{1 \le j \le L}$, where each Ψ_j is a column vector of length N. The edges are labeled by binary column vectors x of length N (more details on vector constrained systems can be found in [39] and [43].) The response of the multitrack channel to an array of input symbols (Ψ, x) is

$$y = h_0 \cdot x + \sum_{1 \le j \le L} h_j \cdot \Psi_{L+1-j}$$

TABLE 24.1　Shannon Capacities of ITI-Reducing Constraints

N	PR2 'sign'	PR2 'zero'	PR4 'sign'	PR4 'zero'	EPR4 'sign'	EPR4 'zero'
1	$C = 0.91625$	$C = 0.79248$	$C = 0.91096$	$C = 0.79248$	$C = 0.87472$	$C = 0.66540$
2	$C = 0.92466$	$C = 0.77398$	$C = 0.92190$	$C = 0.77398$	$C = 0.87703$	$C = 0.68491$
3	$C = 0.92668$	$C = 0.75000$	$C = 0.92444$	$C = 0.75000$	$C = 0.87680$	$C = 0.62652$
4	$C = 0.92910$	$C = 0.74009$	$C = 0.92731$	$C = 0.74009$	$C = 0.87976$	$C = 0.62696$
5	$C = 0.91625$	$C = 0.79248$	$C = 0.91096$	$C = 0.79248$	$C = 0.87472$	$C = 0.66540$

where the value $y_i, 1 \leq i \leq N$ in the vector y corresponds to the ith track. In this section, we will be interested in the following types of N-track ITI-reducing constraints imposed on elements of the vector y.

1. The number of zeros between two nonzero elements in y is at least $d, d > 0$ (the "zero" constraint)
2. Neighboring elements are either of the same sign or at least one of them is zero (the "sign" constraint)

It is not difficult to see that both of these constraints reduce ITI. The first constraint, also referred to as a perpendicular minimum runlength constraint or d-constraint, requires that the signals in d tracks neighboring the ith track all have zero crossings when the signal in track i is nonzero. This requirement completely cancels ITI. The second constraint does not completely eliminate ITI, but rather allows only "constructive" ITI, which in fact improves performance. This method is a generalization of the approaches taken by Ahmed et al. [1] and Davey et al. [11].

The Shannon noiseless capacities of such constraints for various PR targets of interest are given in Table 24.1. It can be seen that the rate penalty for ITI constraints (especially for the "sign" constraint) is not high. We were able to construct a 100% efficient rate 3/4 three-track code with 256 states for the PR2 and PR4 channels for the "zero" constraint.

24.4.2　Constrained Coding for Improved Synchronization

Multitrack codes that improve timing recovery and the immunity of synchronization schemes to media defects have been introduced and extensively studied by Vasic et al. [26, 40–42]. Synchronization immunity to media defects can be improved by allowing the clock recovery circuit to use any group of l tracks on which the k-constraint is satisfied. This new class of codes was named redundant multitrack (d, k) codes, or (d, k, N) codes, with N being the number of tracks. The redundancy $r = N - l$ is the number of bad tracks out of N tracks that can be tolerated while maintaining synchronization. Orcutt and Marcellin [23] considered the (d, k, N, l) constraint assuming $k \geq d$.

In [39], the starting point of the construction was a class of binary multitrack codes with very good clock recovery properties. These are (d, k, N, l) constrained codes with $k < d$. In the same paper, a reduced-state graph model of the constraint was defined, and based on it the Shannon capacities of the constraint were computed. The same approach was then extended for the case of multiamplitude, multitrack runlength limited (d, k) constrained channels with clock redundancy [41] (mainly for applications in optical recording channels). In [20] the Shannon capacities of these channels were computed and some simple, 100% efficient codes were constructed. The vertex labels of the graph of the constraint were modified to insure that they are independent on the number of tracks. This resulted in significant computational savings for the case when the number of tracks is large. In the same paper it was also shown that the increase of the number of tracks written in parallel provides a significant improvement of per-track capacity for a more restrictive clocking constraint case $k < d$.

24.4.3　Low-Complexity Encoder and Decoder Implementations

Substantial progress has been made in the theory of constrained codes using symbolic dynamics [23, 24], as well as in low complexity encoding and decoding algorithms [15]. Despite this progress, the design of constrained codes remains difficult, especially for large constraint graphs such as multitrack constraint graphs.

One class of codes that is not very difficult to implement is the class of multitrack enumerative codes. The idea of enumerative coding originates from the work of Labin and Aigrain [20], Kautz [16], and has been formulated as a general enumeration scheme for constrained sequences by Tang and Bahl [35] (see also Cover [9]). It was also used by Immink et al. [12] as a practical method for the enumeration of (single-track) (d, k) sequences and by Orcutt and Marcellin [29] for multitrack (d, k) block codes. For this type of codes, it is essential to design an encoder/decoder pair that does not require large memory for storing the constraint graph (because this would be prohibitively complex). This design criterion is met by creating only portions of the graph used in different stages of enumerative encoding/decoding.

24.5 Multitrack Soft Error-Event Correcting Scheme

Multitrack, soft error-event correcting schemes were recently introduced by Vasic and Venkateswaran [46]. This error-correcting scheme supports soft error-event decoding and has complexity slightly higher than the complexity of N MLSD detectors. The idea is to design a multitrack version of the "postprocessor" which has been discussed by Cideciyan et al. [5], Conway [8], and Sonntag and Vasic [34], and is employed in most of today's commercial disk drives. The generalization of the postprocessor concept to multiple tracks is nontrivial because the detector must be designed to mitigate the effect of errors caused both by ISI and ITI, as explained below. For more details, the reader is referred to [46].

As mentioned above, a magnetic recording channel is characterized by a PR polynomial. The appropriate PR polynomial depends on the recording density; this density unavoidably increases when going from the outer sectors of the disk towards the inner sectors. Implementation of hard drive subsystems such as tracking servo, timing recovery, and automatic gain control would be unacceptably costly if the PR polynomials were allowed to vary with recording density. Thus, practical systems typically use only two partial response polynomials, one for high-density regions and another for low-density regions.

As a consequence, the employed PR response polynomial is closely, but not completely, matched to the discrete-time channel response, and the noise samples are not independent. Moreover, the noise samples are neither Gaussian nor stationary because of media noise. The detector complexity, however, dictates the use of an MLSD detector with a squared Euclidean distance metric as opposed to a more complex detector with optimal metric. The metric inaccuracy is in practice compensated by a so-called *postprocessor* [34]. The idea of postprocessing can be generalized so as to apply for the multitrack scenario.

The Viterbi detector produces some error patterns more often than others and the most frequently occurring error patterns are a function of the PR polynomial and the noise coloration. The most frequent patterns, called the *dominant error sequences* or *error events* $E = \{e_i\}_{1 \le i \le I}$, and their probabilities can be obtained experimentally [34] or analytically [2]. The index i referrs to an *error type*. Note that the relative frequencies of error events strongly depend on the recording density.

The block diagram of a multitrack soft error-event correcting system is shown in Figure 24.2. User data is encoded by a high-rate error control code (ECC) as shown in Figure 24.1. The decoding algorithm combines syndrome decoding and soft-decision error correction. The error-event likelihoods needed for soft decoding are computed from the channel sequence by using an algorithm proposed by Conway [8] (see also [34].) By using the syndrome calculated for a received codeword, a list of all possible *positions* where error events might have occurred is created. Then the error-event likelihoods are used to select the most likely position and most likely type of the error event. Decoding is completed by finding the error event position and type [12].

Error detection is based on the fact that one can calculate the likelihoods of each of the dominant error sequences at each point in time. The parity bits introduced by the ECC serve to detect the errors, and to provide some localization of the error type and the position where the error ends. The likelihoods are then used to choose the most likely error events (type and position) for correction. The likelihoods in the k-th

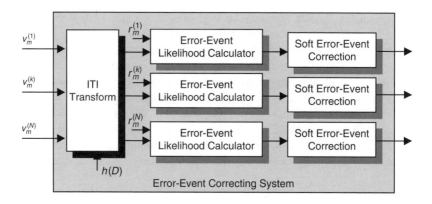

FIGURE 24.2   Block diagram of multitrack soft error-event correction scheme.

track are calculated using the following signals:

a. The read-back signal obtained from the kth head
b. The convolution of the signal obtained by combining the MLSD estimates of all tracks and the channel partial response
c. The convolution of an alternative data pattern (given by a particular error-event) and the channel partial response

The key idea of this approach is that not all candidate data patterns are considered, but only those whose differences with respect to MLSD estimates form the set E.

During each clock cycle, K error-events with largest likelihoods are chosen, and the syndromes for these error events are calculated. Throughout the processing of each block, a list of the K most likely error events, along with their associated error types, positions and syndromes, is maintained. At the end of the block, when the list of candidate error events is finalized, the likelihoods and syndromes are calculated for each of the $\binom{K}{s}$ combinations of s candidate error events that are possible. After disqualifying those s-sets of candidates that overlap in the time domain, and those candidates and s-sets of candidates which produced a syndrome which does not match the actual syndrome, the candidate or s-set which remains and which has the highest likelihood is chosen for correction.

Practical reasons (such as decoding delay and thermal asperities) dictate the use of short codes, and consequently, in order to keep the code rate high, only a relatively small number of parity bits is allowed, making the design of error-event detection codes nontrivial. If all errors from a list E were made up of contiguous bit errors and were shorter than m, then a cyclic code with $n - k = m$ parity bits could be used to detect a single error event [35]. In [44], Vasic introduced a systematic graph-based approach to construct codes for this purpose. This method was further developed in [45].

The performance of multitrack soft error-event decoding will be illustrated on the example of cyclic BCH codes; these codes have fairly large minimum distance, which allows for improvements of the performance of the detector. Single-track state of the art systems use different codes designed to perform well for a given set of dominant error events. However, these codes are not available in the public domain, and therefore will not be discussed here.

Figure 24.3 shows the performance of the multitrack $s = 2$ error-event correction scheme (denoted by 2-EE) based on the (255, 239) BCH code. These results have been obtained for a three-track Lorentzian channel, equalized to the E^2PR4 target, with user bit density of 2.5, and 10% ITI (i.e., $\alpha_1 = 0.1$). The channel bit density due to the [255, 239] BCH code is 2.67. This scheme provides a SNR gain of 3.5 dB at BER $= 10^{-6}$. An extension of the research regarding this multitrack scheme would include the investigation of various PR targets and ECC schemes, the characterization of effects of track misalignment, residual ISI and ITI on the performance of this scheme as well as low-complexity implementations.

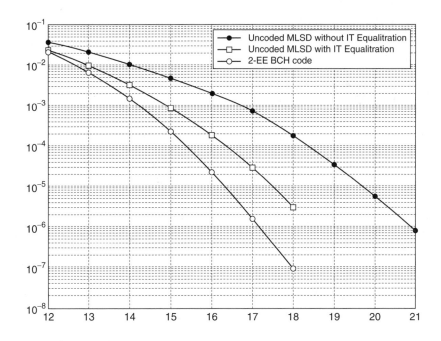

FIGURE 24.3 Comparison of various multitrack schemes.

Acknowledgement

This work is supported by the NSF under grant CCR-0208597.

References

[1] M.Z. Ahmed, T. Donnelly, T., P.J. Davey, and W.W. Clegg, Increased areal density using a two-dimensional approach, *IEEE Trans. Magn.*, vol. 37, no. 4, part: 1, pp. 1896–1898, July 2001.

[2] S.A. Altekar, M. Berggren, B.E. Moison, P.H. Siegel, and J.K. Wolf, Error-event characterization on partial-response channels, *IEEE Trans. Inform. Theory*, vol. 45, no. 1, pp. 241–247, January 1999.

[3] L.R. Bahl, J. Cocke, F. Jelinek, and J. Raviv, Optimal decoding of linear codes for for minimizing symbol error rate, *IEEE Trans. Inform. Theory*, vol. IT-20, no. 2, pp. 284–287, March 1974.

[4] L.C. Barbosa, Simultaneous detection of readback signals from interfering magnetic recording tracks using array heads, *IEEE Trans. Magn.*, vol. 26, pp. 2163–2165, September 1990.

[5] R. Cideciyan, J. Coker, E. Eleftheriou, and R.L. Galbraith, Noise predictive maximum likelihood detection combined with parity-based postprocessing, *IEEE Trans. Magn.*, vol. 37, no. 2 pp. 714–720, March 2001.

[6] S.H. Charrap, P.L. Lu, and Y. He, Thermal Stability of Recorded Information at High Densities, *IEEE Trans. Magn.*, part. 2, vol. 33, no. 1, pp. 978–983, January 1997.

[7] K.M. Chugg, X. Chen, and M.A. Neifeld, Two-dimensional equalization in coherent and incoherent page-oriented optical memory, *J. Opt. Soc. Am.*, vol. 16, no. 3, pp. 549–562, March 1999.

[8] T. Conway, A new target response with parity coding for high density magnetic recording channels, *IEEE Trans. Magn.*, vol. 34, no. 4, pp. 2382–2386, July 1998.

[9] T.M. Cover, Enumerative source coding, *IEEE Trans. Inform. Theory*, vol. IT-19, pp. 73–77, January 1973.

[10] P.J. Davey, T. Donnelly, D.J. Mapps, and N. Darragh, Two-dimensional coding for multi track recording system to combat inter-track interference, *IEEE Trans. Magn.* vol. 34, pp 1949–1951, July 1998.

[11] P.J. Davey, T. Donnelly, and D.J. Mapps, Two-dimensional coding for a multiple-track, maximum-likelihood digital magnetic storage system, *IEEE Trans. Magn.*, vol. 30, no. 6, Part: 1–2, pp. 4212–4214, November 1994.

[12] M. Despotovic and B. Vasic, Hard disk drive recording and data detection, *IEEE International Conference on Telecommunications Cable and Broadcasting Services*, Telsiks, Nis, vol. 2, pp. 555–561, October 8–13, 2001.

[13] I.J. Fair, W.D. Gover, W.A. Krzymien, and R.I. MacDonald, Guided scrambling: A new line coding technique for high bit rate fiber optic transmission systems, *IEEE Trans. Commun.*, vol. 39, pp. 289–297, February 1991.

[14] K.A.S. Immink, A practical method for approaching the channel capacity of constrained channels, *IEEE Trans. Inform. Theory*, vol. 43, no. 5, pp. 1389–1399, September 1997.

[15] K.A.S. Immink, *Coding Techniques for Digital Recorders*, Prentice-Hall Int., Englewood Cliffs, NJ (UK) Ltd., 1991.

[16] W.H. Kautz, Fibonacci codes for synchronization control, *IEEE Trans. Inform. Theory*, vol. IT-11, pp. 284–292, 1965.

[17] H. Kobayashi and D.T. Tang, Application of partial response channel coding to magnetic recording systems, *IBM J. Res. Develop*, vol. 14, pp. 368–375, July 1979.

[18] E. Kurtas, J. Proakis, and M. Salehi, Reduced complexity maximum likelihood sequence estimation for multitrack high density magnetic recording channels, *IEEE. Trans. Magn.* vol. 35, no. 4, pp. 2187–2193, July 1999.

[19] E. Kurtas, J.G. Proakis, and M. Salehi, Coding for multitrack magnetic recording systems, *IEEE Trans. Inform Theory*, vol. 43, no. 6, pp. 2020–2023, November 1997.

[20] E. Labin and P.R. Asgrain, Electric pulse communication system, U.K. Patent 713 614, 1951.

[21] J. Lee and V.K. Madisetti, Combined modulation and error correction codes for storage channels, *IEEE Trans. Magn.*, vol. 32, no. 2, pp. 509–514, March 1996.

[22] M.W. Marcellin and H.J. Weber, Two-dimensional modulation codes, *IEEE J. Select. Areas Commun.*, vol. 10, pp. 254–266, January 1992.

[23] B. Marcus, P. Siegel, and J.K. Wolf, Finite-state modulation codes for data storage, *IEEE J. Select. Areas Commun.*, vol. 10, no. 1, pp. 5–37, January 1992.

[24] B.H. Marcus, Sofic systems and encoding data, *IEEE Trans. Inform. Theory*, vol. IT-31, pp. 366–377, May 1985.

[25] O. Milenkovic and B. Vasic, Power spectral density of multitrack (0, G/I) codes, *Proceedings of the International Symposium on Information Theory*, p. 141, Ulm, Germany 1997.

[26] O. Milenkovic and B. Vasic, Permuation (d, k) codes: efficient enumeration coding and phrase length distribution shaping, *IEEE Trans. Inform. Theory*, vol. 46, no. 7, pp. 2671–2675, November 2000.

[27] H. Muraoka and Y. Nakamura, Multitrack submicron-width recording with a novel integrated single pole head in perpendicular magnetic recording, *IEEE Trans. Magn.*, vol. 30, no. 6, Part: 1–2, pp. 3900–3902, November 1994.

[28] E.K. Orcutt and M.W. Marcelin, Redundant multitrack (d, k) codes, *IEEE Trans. Inform. Theory*, vol. 39., no. 5., pp. 1744–1750, September 1993.

[29] E.K. Orcutt and M.W. Marcellin, Enumerable multitrack (d, k) block codes, *Trans. Inform. Theory*, vol. 39, no. 5, pp. 1738–1744, September 1993.

[30] R.E. Swanson and J.K. Wolf, A new class of two-dimensional RLL recording codes, *IEEE Trans. Magn.*, vol. 28, pp. 3407–3416, November 1992.

[31] E. Soljanin and C.N. Georghiades, Coding for two-head recording systems, *IEEE Trans. Inform. Theory*, vol. 41, no. 3, pp. 747–755, May 1995

[32] E. Soljanin and C.N. Georghiades, A five-head, three-track, magnetic recording channel, *Proceedings, IEEE International Symposium on Information Theory*, p. 244, 1995.

[33] E. Soljanin and C. Georghiades, Multihead detection for multitrack recording channels, *IEEE Trans. Info. Theory*. vol. 44, no. 7, pp. 2988–2997, November 1998.

[34] J.L. Sonntag and B. Vasic, Implementation and bench characterization of a read channel with parity check post processor, *Digest of TMRC 2000*, Santa Clara, CA, August 2000.

[35] D.T. Tang and L.R. Bahl, Block codes for a class of constrained noiseless channels, *Inform. Contr.*, vol. 17, pp. 436–461, 1970.

[36] D.D. Tang, H. Santini, R.E. Lee, K. Ju, and M. Krounbi, A design concept of array heads, *IEEE Trans. Magn.*, vol. 33, no. 3, pp. 2397–2401, May 1997

[37] H. Thapar and A. Patel, A class of partial response systems for increasing storage density in magnetic recording, *IEEE Trans. Magn.*, vol. 23, no. 5, September 1987.

[38] B. Vasic, G. Djordjevic, and M. Tosic, Loose composite constrained codes and their application in DVD, *IEEE J. Select. Areas in Communications*, vol. 19 no. 4, pp. 765 –773, April 2001.

[39] B. Vasic, Capacity of channels with redundant runlength constraints: The $k < d$ case, *IEEE Trans. Inform. Theory*, vol. 42, no. 5, pp. 1567–1569, September 1996.

[40] B. Vasic, O. Milenkovic, and S. McLaughlin, Power spectral density of multitrack (0, G/I) Codes, *IEE Elect. Lett.*, vol. 33, no. 9, pp. 784–786, 1997.

[41] B. Vasic, S. McLaughlin, and O. Milenkovic, Channel capacity of M-ary redundant multitrack runlength limited codes, *IEEE Trans. Inform. Theory*, vol. 44, no. 2, March 1998.

[42] B. Vasic and O. Milenkovic, Cyclic two-dimensional IT reducing codes, *Proceedings of the International Symposium on Information Theory*, p. 414, Ulm, Germany 1997.

[43] B. Vasic, Spectral analysis of multitrack codes, *IEEE Trans. Inform. Theory*, vol. 44, no. 4, pp. 1574–1587, July 1998.

[44] B. Vasic, A graph based construction of high rate soft decodable codes for partial response channels, in *Proceedings of ICC-2001*, vol. 9, pp. 2716–2720, June 11–15, Helsinki, Finland.

[45] B. Vasic, Error event correcting codes for partial response channels, 5th-*IEEE International Conference on Telecommunications Cable and Broadcasting Services*, Telsiks, Nis, vol. 2, pp. 562–566, October 8–13, 2001.

[46] B. Vasic and V. Venkateswaran, Soft error-event decoding for multitrack magnetic recording channels, *IEEE Trans. on Magn.*, Vol. 40, No. 2, pp. 492–497, March 2004.

[47] M.P. Vea and J.M.F. Moura, Magnetic recording channel model with inter-track interference, *IEEE Trans. Magn.*, vol. 27, no. 6, pp. 4834–4836, November 1991.

[48] A. van Wijngaarden and K.A.S. Immink, Construction of constrained codes using sequence replacement techniques, *Proceedings of 1997 IEEE International Symposium on Information Theory*, p. 144, 1997.

[49] J.K. Wolf and D. Chun, The single burst error detection performance of binary cyclic codes, *IEEE Trans. Commun.*, vol. 42, no. 1, pp. 11–13, January 1994.

[50] R. Wood, Detection and Capacity Limits in Magnetic Media Noise, *IEEE Trans. Magn.*, vol. MAG-34, No. 4, pp. 1848–1850, July 1998.

[51] R. Wood, The feasibility of magnetic recording at 1 Terabit per square inch," *IEEE Trans. Magn.*, vol. 36, no. 1, pp. 36–42, January 2000.

[52] Y. Wu, J.A. O'Sullivan, R.S. Indeck, and N. Singla, Iterative detection and decoding for seperable two-dimensional intersymbol interference, *IEEE Trans. on Magn.*, Vol. 39, No. 4, pp. 2115–2120, July 2003.

25

Two-Dimensional Data Detection and Error Control

Brian M. King
Worcester Polytechnic Institute
Worcester, MA

Mark A. Neifeld
University of Arizona
Tucson, AZ

25.1 Introduction **25**-1
25.2 Two-Dimensional Precompensation **25**-3
25.3 Two-Dimensional Equalization **25**-4
    Linear Minimum Mean-Square-Error Equalization
    • Partial Response Equalization
25.4 Two-Dimensional Quasi-Viterbi Methods........... **25**-8
25.5 Two-Dimensional Joint Detection and Decoding **25**-13

25.1 Introduction

Data detection refers to the process by which a continuous-valued signal, appearing at the output of a storage channel is converted into a discrete-valued signal representing an estimate of the original (possibly encoded) stored data. Implicit in this definition is the notion that channel outputs are continuous valued. This will be true even in cases for which the storage medium itself may be highly nonlinear or even bistable. A magnetic storage medium, for example, may have only two stable directions of magnetization; however, the readout process is necessarily imperfect (e.g., fringing fields, low-pass electronics, noise) resulting in a continuous-valued readout signal [1]. This continuous-valued signal serves as the input to the data detection algorithm whose job is to use its prior knowledge of the various corrupting influences of the channel along with its knowledge of the alphabet from which the channel inputs were drawn, to generate an estimate of the stored data. It is important to note that the continuous-valued channel output contains everything we can know about the stored data. The process of converting continuous-valued retrieved signals into discrete-valued codewords or user symbols is, therefore, critical in terms of insuring the highest possible data fidelity.

The storage channel has traditionally been viewed as one-dimensional (1D). If we consider, for example, a magnetic storage system, then data along a single track can be viewed as a 1D signal in space (i.e., on a circle). The rotation of the disc will cause a single read-head situated above this track to produce a 1D readout signal in time. This 1D framework has been very successful, and has enabled the development of powerful data detection methods such as precompensation, linear and nonlinear equalization, maximum-likelihood (i.e., Viterbi) detection, and joint techniques for detection and decoding. The reader is referred to any one of several excellent text books to learn more about these traditional 1D methods of data detection [2–5]. Two phenomena however, have made such a 1D viewpoint less appropriate. The first is the continuing trend toward higher spatial densities in nearly every type of storage system. Consider once

0-8493-1524-7/05/$0.00+$1.50

again the single read-head situated above a single track. As the intertrack spacing decreases, it becomes difficult to insure that signals from adjacent tracks do not corrupt the readout signal. At densities for which this intertrack interference (ITI) becomes unavoidable, the channel can no longer be modeled as 1D. A two-dimensional (2D) model will be needed in order to quantify and mitigate the resulting ITI. The second phenomenon making the 1D channel viewpoint appear antiquated, is the development of novel volumetric storage paradigms (e.g., holographic) in which data is recorded/retrieved as 2D arrays. Such systems exploit the 2D nature of the channel to enhance both capacity and data rate [6–9].

As suggested above, a single platter within a high density magnetic or optical disc drive may be viewed as a 2D storage medium. Time sequential recording/reading along one track provides access to a single dimension of this 2D memory. The use of head arrays (i.e., parallel readout) will make 2D access possible thus increasing both the capacity (via increased track density) and data rate (via multi-track readout) of both optical and magnetic storage systems [10, 11]. Realization of these potential benefits will require data detection techniques that can mitigate both along-track inter-symbol interference (ISI) and across track ITI. Figure 25.1 depicts two types of 2D storage channel corrupted by ISI and ITI. In Figure 25.1(a) a high-density optical tape system is displayed. The binary features are smaller than the focused laser illuminating the material. As a result the reflected beam will be influenced by a number of bits in the track. In addition

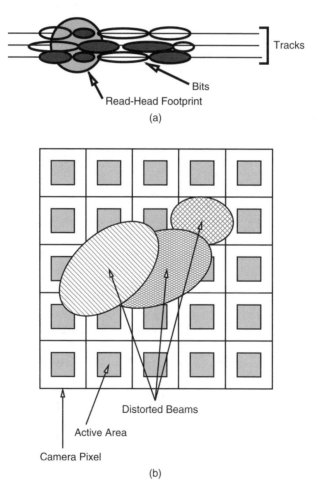

FIGURE 25.1 Types of 2D storage channels. (a) A high-density optical tape or optical disc system. The read-head simultaneously illuminates multiple tracks and multiple bits within a track. (b) A 2D optical memory. The graphic depicts three distorted beams focusing on the 2D array of camera pixels.

as the track density increases, bits in neighboring tracks will directly contribute to the readout signal. Figure 25.2(b) describes a high-density 2D optical storage system, such as a holographic optical memory. Beams corresponding to different bits in the 2D array are focused onto the individual pixels of the camera sensor. However, due to the high-density and small features sizes, the focusing beam associated with each pixel cannot be completely confined to the desired detector which results in cross-talk in both dimensions.

Recent efforts within both the magnetic and optical storage communities have resulted in several important 2D methods of data detection. In this chapter, we will summarize some of this recent work. Section 25.2 describes a technique for 2D precompensation that is analogous to the 1D nonlinear precompensation used in commercial magnetic storage systems. Measurement and correction of channel nonlinearities and alignment errors are the focus of this method. Section 25.3 presents 2D linear equalization techniques for 2D channels. These methods model the channel corruption as a linear shift-invariant filter and utilize an approximate inverse filter to undo the effects of the channel. Section 25.4 will discuss two methods based on the Viterbi algorithm. This algorithm can perform maximum-likelihood sequence detection in 1D by combining accurate channel knowledge with the fact that channel inputs were drawn from a finite alphabet. It does not extend directly to 2D and in this section we describe two quasi-Viterbi methods: one of these is serial and the other is parallel. Section 25.5 will highlight an application in which 2D data detection is performed jointly with the soft-decision decoding of a 2D array code.

25.2 Two-Dimensional Precompensation

In this section we focus on precompensation for 2D channels. Examples of such channels include holographic optical storage and high-density magnetic and optical tape drives. At first glance, magnetic drives appear to be a 1D channel. However, the 2D channel model is applicable because at high-density the neighboring bit domains along and across tracks are spaced close enough to interact directly with the measurement of the desired domain under the head.

Precompensation refers to the technique of adjusting the waveform transmitted into storage channel in order to correct for known channel distortions. These distortions may arise from a variety of sources including amplifier nonlinearities, media irregularities, and variation in read/write channel parameters (disk rotation speed, head fly-height). Precompensation is essentially a form of channel equalization performed *before* the information is transmitted into the channel. As a consequence, the transmitting side of the channel must know precisely the relevant parameters of the channel. This information can be obtained by either directly interrogating the channel or in a bidirectional communication channel by feedback sent from the other end of the channel.

In a general digital communication system, the data begins as discretely-valued symbols (assumed to be binary throughout this chapter) and proceeds through the channel typically as an analog or a continuous-valued function. The receiver measures the signal and converts it back to a discrete form. Even though the internal portion of the system may be analog, we can still model the system end-to-end in terms of the discrete transfer function between the digital input and output. The system consists of a cascaded realization of linear time-invariant systems, hence we can encapsulate the linear behavior of the entire channel in terms of a single discrete-time channel impulse response, $h(i, j)$, where i and j index the discrete taps of the two-dimensional (2D) impulse response. The overall channel response includes the contributions from each of the components in the system. In a magnetic storage system h would be determined by the pulse shape, the impulse response of the writing/reading head, and the magnetic media response. In a 2D optical memory the elements in the discrete impulse response are the spatial light modulator (SLM) pixel shape and period, the optical transfer function, the storage material properties, and the detector response.

Precompensation is necessary when the channel parameters change enough in practical situations to induce an unacceptable drop in the signal-to-noise ratio (SNR) if ignored. The communication system architecture strongly influences this sensitivity and determines if precompensation is necessary. For example, on an optical disc reader (CD, DVD) the same laser and optical head are used to access every bit hence the contributions of those elements to the overall impulse response will be nearly identical. Because of the quality of the disc manufacturing process, the data feature's shape will also be very consistent from bit to bit. At the other extreme, consider a 2D holographic data writer. Data are stored in 2D pages where the whole page is recorded simultaneously. Let the data page be a $N_1 \times N_2$ element array composed of binary pixels. A spatial light modulator (SLM) imprints the 2D data array on the spatial profile of the laser beam. A precisely calibrated optical system images each of the pixels in the 2D data array through the holographic media onto a 2D camera array of matching dimensions. One can think of the 2D data as representing $N_1 N_2$ parallel channels. Each channel is identical in principle but often measurably different in practice. This is primarily because each of the $N_1 N_2$ channels can have different transmitted powers (nonuniform laser illumination of the SLM, varying manufactured SLM pixel transmittances), different optical transfer functions (each pixel uses a different portion of the imaging optics), and separate receivers (each camera pixel's responsivity and electrical noise contributions will vary slightly in manufacturing). For this type of system, precompensation will offer a significant improvement.

Both magnetic and optical systems are also subject to perturbations induced in the channel due to systematic misalignment. For example, the detector may be shifted from its ideal position, resulting in a loss of signal amplitude on the desired pixel and an increase in cross-talk. Calibration or fiducial marks stored or embedded in the media can be used to provide the necessary feedback to estimate the systematic error and adjust the recording parameters in order to improve the SNR or fidelity of the stored data.

As an example of the potential improvement offered by 2D precompensation, we consider the experimental results presented in reference [12] for a 2D holographic optical storage system. The IBM experimental system consists of a 320 × 240 SLM imaged through holographic media (iron-doped lithium niobate, LiNbO$_3$:Fe) onto a charge-coupled device (CCD) camera. The SLM images can be updated at a 60 Hz frame rate [13, 14]. Even though the optical elements were precisely fabricated for the system, residual optical aberrations introduced a narrow but different optical point spread function across the 2D array. In addition to spatial optical nonuniformities, the photosensitive holographic material responds nonlinearly to the incident image from the SLM.

The experiment proceeds by recording a single test hologram in the material. This hologram acts a training pattern that measures both the optical and material distortions to the channel. Based on the measurement, the individual exposure times (equivalent to amplitude modulation in traditional communication systems) of each pixel are adjusted to achieve upon readout of the CCD a uniform 2D array of values. At high-density the experiments demonstrated an improvement of the raw bit-error-rate (BER) from 10^{-4} to 10^{-12}. An uncorrected 2D array had an SNR of 2.82 dB and after applying precompensation the SNR was increased to 10.25 dB for a processing gain of 7.43 dB [13].

25.3 Two-Dimensional Equalization

Digital communication over a band-limited channel is improved by incorporating a linear filter in the receiver to account for the distortion introduced by the channel's band-limited response. A necessary outcome of the limited bandwidth is the spreading of the data pulse shape temporally and spatially, resulting in cross-talk between neighboring bits. For a magnetic storage channel, we typically consider the cross-talk between bits in track as intersymbol interference (ISI) and between bits in neighboring tracks as intertrack interference (ITI). For high-density storage, both contributions will be significant. An equalization filter is designed to correct for the known ISI and ITI at the receiver, but due to the channel bandwidth constraint, the equalizer cannot completely restore the ideal channel response.

As in the previous section, we model the end-to-end digital system in terms of the overall discrete-time channel impulse response, $h(i, j)$, where i indexes the symbols along the track and j indexes the symbols across the tracks. In a 2D optical system, i and j index the row and column of the pixel in the 2D array. The discrete convolution operation determines the output data array value $s(i, j)$ from the input data array $a(i, j)$ according to:

$$s(i, j) = a(i, j) * h(i, j) \tag{25.1}$$

$$= \sum_k \sum_l a(i - k, j - l) h(k, l) \tag{25.2}$$

where the summation indices k and l run over $(-\infty, \infty)$ in principle, but in practice the channel response is negligible beyond some effective channel support length L_1 and L_2, that is,

$$h(k,l) = 0 \, \forall \, |k| > L_1 \quad \text{or} \quad |l| > L_2 \tag{25.3}$$

The channel is modeled to include an additive white Gaussian noise (AWGN) contribution resulting in a measured value,

$$m(i, j) = s(i, j) + n(i, j) \tag{25.4}$$

where $n(i, j)$ is distributed Gaussian with zero-mean and variance σ^2.

The equalizer operates on the measured data array to produce the output

$$y(i, j) = m(i, j) * g(i, j) \tag{25.5}$$

where $g(i, j)$ is the discrete impulse response of the equalizer.

In this section we consider two equalizer designs. First, a minimum mean-square-error design seeks to completely eliminate the ISI by minimizing the variance of the estimator error between $a(i, j)$ (the input data) and $y(i, j)$ (the equalized output). Second, partial response equalization is considered to deterministically control and remove the ISI instead of directly minimizing it.

To compare algorithms and performances we conducted simulations on two incoherent 2D optical channels denoted as channel C and channel D. Channel C represents a diffraction-limited optical imaging system operating at the Sparrow resolution criteria. Channel D represents the channel operating at approximately 25% beyond the Sparrow resolution limit.

25.3.1 Linear Minimum Mean-Square-Error Equalization

Intersymbol interference and Intertrack interference are direct consequences of trying to send data with a bandwidth near or beyond the available channel bandwidth. The overlap or smearing of the bits on the detector can quickly degrade the system SNR and limit operation at higher data rates. The linear minimum-mean-square error (LMMSE) equalizer combats ISI and ITI by designing a receiver filter to minimize the error between the filter output, $y(i, j)$, and the original data array, $a(i, j)$. The reader is directed to references [15–20] for a complete discussion of 2D high-density magnetic and optical equalizer design.

At first appearances, it may seem that choosing the spectrum of the equalization filter to simply invert the channel spectrum, $H(z_1, z_2)$ would be a good choice. In this case $G(z_1, z_2) = 1/H(z_1, z_2)$, producing an overall transfer function of an ideal band-limited channel. However, the inverse filter approach suffers from well-known problems, such as spectral nulls, noise amplification, and an infinite impulse response (IIR) filter implementation [3]. Instead if the design criteria for the filter is to minimize the output BER by thresholding the noisy output of the finite-support filter, we arrive at the LMMSE filter [2]. It is the optimal linear detection filter when the received data is corrupted by additive white Gaussian noise under the conditions of finite support. The difference between the inverse filter and the LMMSE filter is the choice of the filter gain at data frequencies with low magnitude. The inverse filter will boost the gain resulting in an overamplification of the noise, while the LMMSE filter adjusts the gain to minimize the detection BER. In the limit as the noise spectral power density approaches zero, the LMMSE equalizer converges to the inverse filter.

The computation of the LMMSE filter coefficients involves the solution of the Wiener-Hopf (WH) equations. The equations arise from setting the derivative of the expression for the mean-square-error to zero. Because the filter is a linear combination of the input data, there is only one extremum point on the error energy surface and it corresponds to a minimum. Equivalently, the orthogonality principle can be directly applied to derive the WH equations:

$$E\{[a(i, j) - y(i, j)] \, m(i - k, j - l)\} \equiv 0 \quad \forall \, k, l \tag{25.6}$$

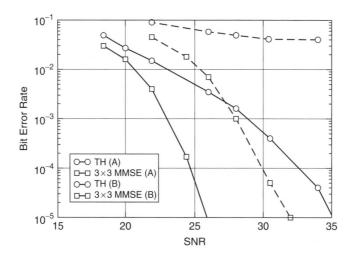

FIGURE 25.2   Results of thresholding and 2D linear-minimum-mean-squared-error equalization. The LMMSE equalizer has a footprint of 3×3. Solid lines refer to channel A and dashed lines refer to channel B.

where $E\{\cdot\}$ is the expectation operator. Expressing $y(i, j)$ as a convolution of $m(i, j)$ and $g(i, j)$ and rearranging the resulting expression yields the Wiener-Hopf equations:

$$\sum_{p} \sum_{q} g(p, q) E\{m(i - k, j - l)m(i - p, j - q)\} = E\{a(i, j)m(i - k, j - l)\} \quad \forall \, k, l \quad (25.7)$$

Since the Wiener-Hopf equations must be satisfied for all values of k and l, Equation 25.7 represents a set of equations that can be represented directly in matrix form:

$$Kg = c \qquad (25.8)$$

K will be a block Toeplitz matrix when the channel impulse response is shift-invariant (for optical systems) or time-invariant (for magnetic drives). g is the desired filter taps rastered from the 2D form into a 1D vector. c is the vector of constraint values that results from evaluating the expectation operator with knowledge of the specific noise statistics. Equation 25.8 can be solved easily by a variety of methods. Note that in Equation 25.7 the channel response, $h(i, j)$, is represented in terms of the noisy received data array values $m(i, j)$.

Figure 25.2 shows the results of applying two detection schemes to data from channels A and B. The threshold approach (TH) does not apply the equalizer filter but instead applies a threshold directly to the output of the channel. The precomputed threshold is chosen to minimize the BER. The second detection approach is LMMSE equalization using a filter of support 3×3. We see that on channel A, LMMSE offers a SNR advantage of 7.9 dB for a system operating at a raw BER of 10^{-4}. On channel B, we see that the cross-talk induced by the channel does not allow low BER operation of a thresholded detector for *any* SNR.

25.3.2 Partial Response Equalization

Instead of minimizing the contribution of ISI, partial response equalization seeks to control the ISI instead of eliminating it. Again, an equalization filter is applied to the output of the channel to reshape the overall channel transfer function. A practical consequence of a system operating without ISI is the requirement of a data symbol rate that is less than the Nyquist rate (twice the bandwidth of the channel). By allowing ISI, symbol rates at exactly the Nyquist rate are possible using physically realizable partial response signals [2, 21].

References [6, 22–28] discuss partial response signaling in more detail for magnetic, magneto-optical, optical disc, and holographic optical channels.

Perhaps the simplest form of partial response (PR) signaling is $1 + D$ PR precoding. In a 1D system, the equalizer is chosen to produce an overall discrete impulse response of one for the first two taps and zero for all other taps. In this case, an input bit is broadened to encompass exactly two bits on the output. Because the interference is deterministic, we can account for it in the receiver or the transmitter. When using antipodal signaling, we send either the $1 + D$ PR pulse shape, $p(t)$ or its negative, $-p(t)$. The receiver determines if a "1" was sent by measuring a transition in the data sequence. If there is no transition than a "0" was transmitted. This is the same decision rule as differential binary phase-shift keying (DPSK) and as a result it suffers from error propagation as well. If a bit decision is incorrect, it will encourage further errors.

Because the data sequence is known perfectly at the transmitter, error propagation can be avoided entirely by preencoding the data such that direct detection of the interference of the two bits produces the ISI-free measurement. Figure 25.3 shows a 1D example of the duobinary pulse shape, $p(t)$, used for the $1 + D$ target (a) as well as two constructively (b) and destructively (c) interfering pulses to produce a

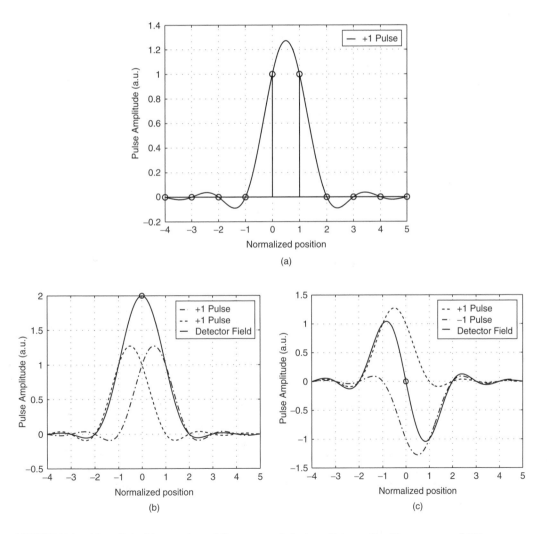

FIGURE 25.3 (a) $1 + D$ duobinary pulse and discrete-sampled values. Only taps 0 and 1 are nonzero. (b) Two consecutive positive pulses constructively add to produce a "1" bit. (c) Two consecutive opposite sign pulses deconstructively add to produce a "0" bit.

detected "1" or "0." The dashed lines show the pulse shape from the respective bits. The solid line denotes the total field impinging on the detector. Duobinary-encoding can be extended to 2D by considering a pulse that is broadened to have significant contributions at exactly two pixels in both dimensions [25]. Again, a simple extension of the 1D procedure for precoding exists and allows efficient precoding of the 2D data array. However, as a consequence of the 2D extension, detection will be multi-valued. Three possible detected values are possible: 0, 1, or 2. Detected values of 0 or 2 correspond to a "0" bit being transmitted and a value of 1 represents transmission of a "1" bit.

One of the advantages of this approach is that it does not need any signal processing or algorithms applied to the detected data. Massively parallel system such as a 2D optical channels typically mandate a low or no complexity detection solution. 2D detection proceeds by applying two thresholds to the detector output. When the value is between the threshold levels, a "1" is output, otherwise a "0" is output.

Figure 25.4 shows example histograms of the detector intensity measurement for a diffraction-limited 2D optical system (a) without partial response signaling and (b) with 2D $1 + D$ precoding. We can see the reduction in the variance of the "1"s distribution, clearly indicating the removal of the interference in both directions (along and across track in optical disks). In Figure 25.4(b), we have excluded the distribution of "0" bit intensity values centered around a normalized intensity of 4. Because the distribution is significantly farther from the "1"s distribution than the "0"s centered around zero, it will not limit the BER.

Figure 25.5 shows the BER performance versus the SNR defined as SNR $= -20 \log(\sigma)$. σ is the standard deviation of the electrical noise on the signal output from 2D camera array. The two curves compare the performance of a 2D optical storage system with and without the use of 2D $1 + D$ partial response precoding. We see at a target BER of 10^{-4} that $1 + D$ precoding offers a 1 dB advantage.

25.4 Two-Dimensional Quasi-Viterbi Methods

The various data detection techniques that have been described in the previous sections are suboptimal: they do not minimize the sequence error rate. This means that none of these techniques is guaranteed to produce the most likely data array (a) that gave rise to the measured array (m). It is possible to formulate an optimal method of mapping m back to a. Consider the case of 1D binary-valued data arrays (sequences) corrupted by a 1D ISI channel and AWGN. The optimal (i.e., maximum-likelihood) estimate $\hat{a}$, is the sequence a that minimizes the Euclidean distance between the measurement m and the candidate measurement $\hat{m} = a * h$, where h is the channel impulse response. Although this algorithm is conceptually straightforward its implementation is computationally impractical, requiring that all possible 2^N sequences be tested. In 1967 an algorithm based on dynamic programming was adapted for use in this domain [29]. The so-called Viterbi algorithm accomplishes the required search for a maximum-likelihood solution with complexity that is only linear in the sequence length N. This trellis-based search technique has become one of the most important methods of data detection for storage and communication channels alike. The reader is referred to references [30, 31] for a description of the conventional 1D Viterbi algorithm.

In more than one dimension the brute-force search method described above remains conceptually valid. In 2D, for example, the $N_1 \times N_2$ element array a, that minimizes the Euclidean distance $d = |m - a * h|$ is the maximum-likelihood solution. Unfortunately, there is no known 2D extension of the Viterbi algorithm that can produce an optimal estimate $\hat{a}$ with complexity that is only linear in the array size $N_1 N_2$. Consider a 2D binary-valued array corrupted by ISI/ITI and AWGN. A straightforward application of the Viterbi algorithm to such an environment can be envisioned in which each column of the array is treated as a symbol in a 2^{N_1} element alphabet. The 2D array can now be viewed as a 1D sequence of 2^{N_1}-ary symbols and the Viterbi algorithm can be applied directly to such a sequence. If the equivalent channel memory is M, then the complexity of the resulting Viterbi search scales as $N_2 2^{M N_1}$ making such an approach feasible only for small $M N_1$.

Our inability to find an efficient (i.e., linear complexity), optimal method of data detection for 2D channels is related to the lack of a natural ordering on the plane. Despite this theoretical bottleneck, two

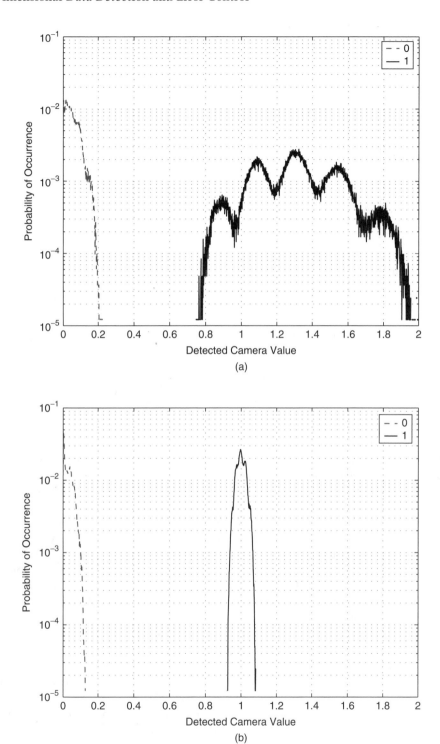

FIGURE 25.4 Example conditional probability density functions (histograms) upon detection for (a) no partial response signaling and (b) using 2D 1 + D PR precoding.

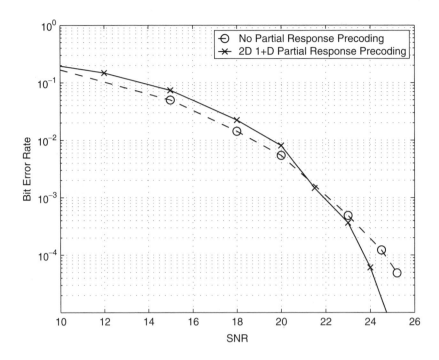

FIGURE 25.5   Bit-error-rate performance versus electrical noise SNR for a 2D optical system with and without the use of 2D 1 + D partial response precoding.

notable adaptations of the 1D Viterbi algorithm have been pursued for use in 2D applications. The first of these was described in [32, 33] and is essentially a sliding-block Viterbi algorithm combined with decision-feedback. It is clear from the previous discussion that any contiguous set of L rows in a 2D data array can be viewed as a nonbinary 2^L-ary sequence. The algorithm described in [32] treats a 2D data array as a set of such sequences. The measured 2D channel output therefore can be viewed as set of output sequences, each of which has been corrupted by a 1D ISI channel. The Viterbi algorithm can be used on each such measured sequence to obtain an estimate of the original stored data. Because $L < N_1$ the resulting Viterbi search will be sub-optimal; however, if L is several times larger than the ITI memory then reasonable performance can be obtained. Performance is improved if the set of output sequences are not treated independently. This algorithm therefore considers the output array to define a set of $N_1 - L$ overlapping L-row strips and for this reason the Viterbi algorithm for each strip only "decides" data values of the first as-yet-undecided row. Figure 25.6(a) is a graphical depiction of the mapping under this algorithm, from a 2D data array to Viterbi states. In this example, we have assumed a channel with ISI memory = 3 and ITI memory = 3 and we have set $L = 5$. The sliding-block approach allows us to perform near-optimal detection on a row-by-row basis. Each row gives rise to a separate trellis and search procedure and the overall algorithm can be realized in a fully row-parallel fashion. Returning to our example we find that the Viterbi search for an arbitrary row of the array will require $2^{15} = 32,768$ states/stage with $2^6 = 64$ paths leaving each state. For many applications this level of complexity remains impractical. In order to further reduce the complexity of this method decision-feedback can be used. Figure 25.6(b) shows how this might be done in our example using two rows of decision feedback. Consider processing the measured array sequentially beginning with the top row. The Viterbi algorithm that operates on the top row can benefit from the knowledge that there was no data above it. Zeros can therefore be assumed to reside in these positions. Upon completion of the trellis search for the top row, we now have estimates of the data from this row that can be used within the Viterbi algorithm for the next row. Because we are exploiting hard-information from two previously decided rows, the number of states/stage of the trellis is reduced from 32,768 to $2^9 = 512$.

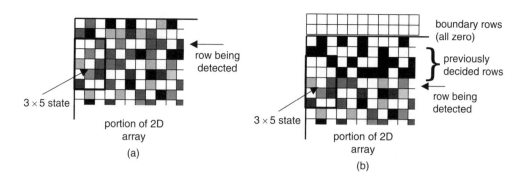

FIGURE 25.6 The application of the Viterbi algorithm to 2D data. (a) Viterbi operates along a strip from left to right. (b) Viterbi complexity can be reduced by employing decision-feedback in which previously decided rows participate in the Viterbi algorithm for subsequent rows.

The algorithm described above is sometimes called the decision-feedback Viterbi (DF-V) algorithm. Its performance has been tested on several important 2D channels and two examples are presented here. These channels represent incoherent optical systems operating near (channel A) and beyond (channel B) the Sparrow resolution limit. Channels such as these might be encountered in a high-density magneto-optic storage environment utilizing parallel (i.e., multiple-head) readout or in a volume holographic storage environment [20]. A 2D Gaussian impulse response is assumed along with AWGN whose variance is σ^2. The discrete channels are truncated to 5×5 support and normalized to unit power. The unique channel taps for channels A and B are given in Table 25.1 and Table 25.2. The BER versus SNR graphs depicting DF-V performance are shown in Figure 25.7. As before we define SNR $= -20\log(\sigma)$. Figure 25.7(a) presents the results for channel A; while, Figure 25.7(b) presents the results for channel B. Although these channels are slightly different than those discussed earlier in the chapter, the performance of an optimal threshold is provided for comparison purposes. We note that for channel A, the DF-V algorithm provides a gain of more than 5 dB at BER $= 10^{-4}$ as compared with simple threshold detection. Channel B represents more severe ISI and ITI and we observe from Figure 25.7(b), that threshold detection fails

TABLE 25.1 2D Discrete Impulse Response for Channel A

	Channel A				
	-2	-1	0	1	2
-2	0.0000	0.0001	0.0003	0.0001	0.0000
-1	0.0001	0.0175	0.0972	0.0175	0.0001
0	0.0003	0.0972	0.5394	0.0972	0.0003
1	0.0001	0.0175	0.0972	0.0175	0.0001
2	0.0000	0.0001	0.0003	0.0001	0.0000

TABLE 25.2 2D Discrete Impulse Response for Channel B

	Channel B				
	-2	-1	0	1	2
-2	0.0001	0.0016	0.0046	0.0016	0.0001
-1	0.0016	0.0411	0.1173	0.0411	0.0016
0	0.0046	0.1173	0.3349	0.1173	0.0046
1	0.0016	0.0411	0.1173	0.0411	0.0016
2	0.0001	0.0016	0.0046	0.0016	0.0001

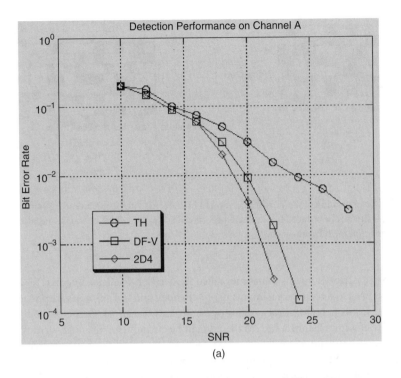

(a)

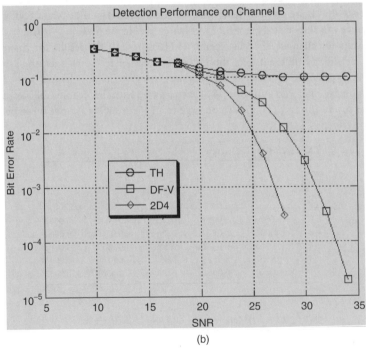

(b)

FIGURE 25.7   2D detection performance using three algorithms: TH = threshold, DF-V = decision-feedback Viterbi and 2D4 = two-dimensional distributed data detection. (a) BER versus SNR for channel A. (b) BER versus SNR for channel B.

for this channel; whereas, the DF-V algorithm is able to achieve the desired BER $= 10^{-4}$ level with SNR $= 44$ dB.

Another adaptation of the Viterbi algorithm to 2D channels was first described in [20]. It was originally motivated by the observation that the DF-V algorithm lacks symmetry. The DF-V algorithm has two preferred directions and it performs data detection differently in these two directions: along track ISI is treated within the trellis search and across track ITI is treated both within the source alphabet and using decision-feedback. A 2D channel in which the ISI and ITI components are nonseparable may not be well suited to such an asymmetric solution. Chen et al. [20] sought a method by which each location in the 2D array would appear identical from the perspective of the data detection algorithm. The result is an algorithm that attempts to maximize the conditional probability $P(m \mid a(i, j))$ at every position in the array. In order to do this exactly, we must average over all data arrays a, consistent with the hypothesized value of $a(i, j)$. Unfortunately this method yields another exponential search. An approximation can be obtained however, by considering only that portion of the array in the neighborhood of $a(i, j)$. The resulting iterative algorithm operates in two phases. First is the likelihood propagation phase in which we compute the update at the *kth* iteration:

$$\Delta^{(k)}[a(i, j)] = \sum_{N \in \Omega} P[m(i, j)|N] \prod_{a(l,m) \in N} L^{(k-1)}[a(l, m)]$$

Note that the sum is taken over all possible binary-valued configurations of a in the neighborhood of $a(i, j)$. We refer to such a configuration as a neighborhood N. For a channel with $L \times L$ pixel support the set Ω is therefore a set of 2^{2L} 2D arrays. The product term of the likelihood update generates the probability of a particular neighborhood $P(N)$ and is based on the notion of "propagating extrinsic information" commonly used within turbo-decoding algorithms [34, 35]. The update term written above therefore represents our instantaneous estimate of the likelihood $L[a(i, j)]$, based on the measured data $m(i, j)$ and our estimate (at the $k - 1$ iteration) of a in the neighborhood of $a(i, j)$. Phase two of the algorithm is a filtering step in which we combine our likelihood estimates from the previous iteration with the update term discussed above. Specifically we have

$$L^{(k)}[a(i, j)] = (1 - \beta)L^{(k-1)}[a(i, j)] + \beta \Delta^{(k)}[a(i, j)]$$

where β is a filtering parameter.

The algorithm outlined above can operate simultaneously at every position in a 2D array. It is a fully 2D parallel algorithm and is sometimes referred to as two-dimensional distributed data detection (2D4). In a single iteration of 2D4 each pixel aggregates information from pixels in its immediate neighborhood. As the algorithm continues the additional iterations allow for aggregation of information from progressively larger portions of the measured array. Because distant pixels exhibit less influence then nearby pixels, very good performance can be obtained in relatively few iterations. Considering once again the A and B channels introduced above we can compare the performance of 2D4 with that of DF-V. The 2D4 results are shown by the diamond-symbol curves in Figure 25.7(a) and Figure 25.7(b). Only five iterations were used to obtain these results. We observe that 2D4 is consistently superior to DF-V on these channels, providing a gain of 5 dB on channel B at BER $= 10^{-4}$. Although the complexity of 2D4 can sometimes be greater than that of the DF-V algorithm, temporary decision-feedback can also be incorporated within the 2D4 algorithm to provide comparable complexity improvements. The reader is referred to reference [20] for more details on 2D4 implementation.

25.5 Two-Dimensional Joint Detection and Decoding

The viewpoint that we have taken in this chapter thus far, is one in which the data detection algorithm generates a binary-valued output array $\hat{a}$ that will later serve as the input to an error decoder. From our discussion of 2D4 and its likelihood-based processing, it is clear that the data detection algorithm may also be capable of providing "reliability" information associated with its decisions. This reliability information can be very important in determining the most likely error pattern and thus the lowest BER user data. Error decoders that can exploit this reliability information are referred to as "soft-decision" decoders and generally provide significant performance advantages over their hard-decision counterparts. It is in fact the existence of soft-decision decoders for both turbo-codes and low-density parity check codes, that is largely responsible for their impressive near-Shannon performance [36].

This discussion leads us to revisit the relationship between detection and decoding. Data detection algorithms are used to undo the correlation introduced by the channel. Knowledge of this correlation (i.e., the ISI and ITI) is used together with the finite-alphabet constraint to determine the most likely stored data. Similarly, decoding algorithms are used to undo the correlation introduced by the encoder. Decoders use knowledge of this correlation (i.e., the encoding algorithm) to determine the most likely user data. In theory, therefore, these two functions could be combined in a single algorithm that had knowledge of both the channel and the codeword structure. In practice however, the correlation structure of error correction codes is often too complex to be treated within a data detection framework. What can be done in this case is to use the exchange of soft-information between data detection and error decoding algorithms so that the correlation structure in each can "inform" the processing of the other. The result is a powerful framework for joint detection and decoding.

The 2D4 algorithm described in the previous section provides a natural mechanism through which soft-information can be exchanged with a suitable error decoder. We will illustrate this idea using a 2D array code [37]. Specifically, we consider a systematic error correcting code obtained by starting with a 2D $(k \times k)$ array of data. A code word is generated by appending four parity columns. The first parity column is obtained by computing one parity bit for each row of the original data array. The next parity column is obtained in an analogous way, by computing parity on the columns of the original data array. The next two parity columns are obtained by computing parity on all possible left and right diagonals (with wraparound). This so-called RCDS code has a rate of $k^2/(k^2+4k)$ and its minimum distance is 4. A soft-decision decoder for this code is described in [38] along with the details of the likelihood exchange process using a 2D4 data

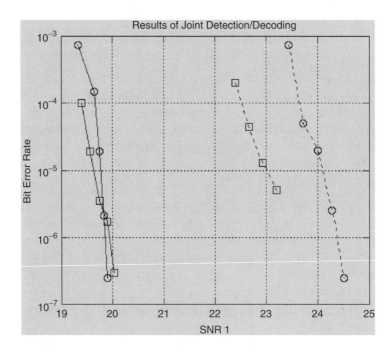

FIGURE 25.8   Results of 2D joint detection/decoding. LMMSE + RS = linear minimum-mean-squared-error equalization combined with a Reed-Solomon n = 127, T = 12 error correction code. 2D4 + RCDS = joint detection/decoding using the 2D4 algorithm and a simple 2D RCDS array code.

detection algorithm. The performance of this combination (2D4 + RCDS) has been compared with that of LMMSE equalization in combination with a sophisticated hard-decision Reed-Solomon error correction code (LMMSE + RS). The results are shown in Figure 25.8 for two different incoherent optical channels. Although these two channels are not the A and B channels underlying the results shown in Figure 25.7, they do represent physical environments at (LPBW = 1) and beyond (LPBW = 0.6) the resolution limit. These results are based on a RS code with length $n = 127$ symbols and error correcting ability $T = 12$ and a RCDS code with $k = 71$. From the results in Figure 25.8, we see that when channel degradation is not severe (LPBW = 1) the LMMSE + RS performance is nearly identical to the 2D4 + RCDS performance. This demonstrates the power of soft-decision decoding to make a simple 2-error-correcting RCDS code whose rate is $r = 0.95$, comparable to a 12-error-correcting RS code with rate $r = 0.81$. Another important observation to be made concerning Figure 25.8 is the benefit of joint detection/decoding in the presence of severe ISI and ITI. The joint approach is seen to offer more than 1 dB advantage over LMMSE + RS at BER = 10^{-4}.

References

[1] H.N. Bertram, *Theory of Magnetic Recording*, Cambridge University Press, 1994.

[2] J.G. Proakis, *Digital Communications*, 3rd ed., McGraw-Hill, 1995.

[3] E.A. Lee and D.G. Messerschmitt, *Digital Communications*, Kluwer Academic Publishers, 1988.

[4] A. Oppenheim and R. Schaffer, *Digital Signal Processing*, Prentice Hall, 1975.

[5] S.S. Haykin, *Communication Systems*, John Wiley & Sons, 1994.

[6] H.J. Coufal, D. Psaltis, and G.T. Sincerbox, Eds., *Holographic Data Storage*, Springer-Verlag, 2000.

[7] J.F. Heanue, M.C. Bashaw, and L. Hesselink, Volume holographic storage and retrieval of digital data, *Science*, vol. 265, pp. 749–752, August 1994.

[8] D. Psaltis and F. Mok, Holographic memories, *Scientific American*, vol. 273, pp. 52–58, November 1995.

[9] G.W. Burr, C.M. Jefferson, H. Coufal, M. Jurich, J.A. Hoffnagle, R.M. Macfarlane, and R.M. Shelby, Volume holographic data storage at an areal density of 250 gigapixels in^2, *Opt. Lett.*, vol. 26, pp. 444–446, 2001.

[10] P.J. Davey and T. Donnelly, Two-dimensional coding for a multi-track recording system to combat inter-track interference, *IEEE Trans. Mag.*, vol. 34, pp. 1949–1951, July 1998.

[11] Y. Shimizu, I. Tagawa, H. Muraoka, and Y. Nakamura, An analysis on multi-track submicron-width recording in perpendicular magnetic recording, *IEEE Trans. Mag.*, vol. 31, pp. 3096–3098, November 1995.

[12] G.W. Burr, H. Coufal, R.K. Grygier, J.A. Hoffnagle, and C.M. Jefferson, Noise reduction of page-oriented data storage by inverse filtering during recording, *Opt. Lett.*, vol. 23, pp. 289–291, February 1998.

[13] J. Ashley, M. Bernal, G.W. Burr, H. Coufal, H. Guenther, J.A. Hoffnagle, C.M. Jefferson, B. Marcus, R.M. Macfarlane, R.M. Shelby, and G.T. Sincerbox, Holographic data storage, *IBM J. Res. Develop.*, vol. 44, pp. 341–368, May 2000.

[14] G.W. Burr, W.-C. Chou, M.A. Neifeld, H. Coufal, J.A. Hoffnagle, and C.M. Jefferson, Experimental evaluation of user capacity in holographic data-storage systems, *Appl. Opt.*, vol. 37, pp. 5431–5443, August 1998.

[15] P.S. Kumar and S. Roy, Two-Dimensional equalization: theory and applications to high density magnetic recording, *IEEE Trans. Comm.*, vol. 42, pp. 386–395, 1994.

[16] K.M. Chugg, X. Chen, and M.A. Neifeld, Two-dimensional equalization in coherent and incoherent page-oriented optical memory, *J. Opt. Soc. Amer. A*, vol. 16, pp. 549–562, March 1999.

[17] M. Keskinoz and B.V.K.V. Kumar, Application of linear minimum mean-squared-error equalization for volume holographic data storage, *Appl. Opt.*, vol. 38, pp. 4387–4393, July 1999.

[18] V. Vadde and B.V.K.V. Kumar, Channel modeling and estimation for intrapage equalization in pixel-matched volume holographic data storage, *Appl. Opt.*, vol. 38, pp. 4374–4386, 1999.

[19] B.M. King and M.A. Neifeld, Parallel detection algorithm for page-oriented optical memories, *Appl. Opt.*, vol. 37, pp. 6275–6298, September 1998.

[20] X. Chen, K.M. Chugg, and M.A. Neifeld, Near-optimal parallel distributed data detection for page-oriented optical memories, *IEEE. J. Sel. Topics in Quantum Elec.*, vol. 4, pp. 866–879, 1997.

[21] P. Kabal and S. Pasupathy, Partial-response signaling, *IEEE Trans. Comm.*, vol. COM-23, pp. 921–934, 1975.

[22] G.L. Silvus and B.V.K.V. Kumar, A comparison of detection methods in the presence of non-linearities, *IEEE Trans. Mag.*, vol. 34, pp. 98–103, January 1998.

[23] W.E. Ryan and B.M. Zafer, A study of class I partial response signaling for magnetic recording, *IEEE Trans. Mag.*, vol. 33, pp. 4543–4550, November 1997.

[24] B.H. Olson and S.C. Esener, Partial response precoding for parallel-readout optical memories, *Opt. Lett.*, vol. 19, pp. 661–663, May 1994.

[25] B.H. Olson and S.C. Esener, Multidimensional partial response for parallel readout optical memories, in *Proc. SPIE*, vol. 2297, pp. 331–337, July 1997.

[26] L. Cheng, M. Mansuripur, and D.G. Howe, Partial-response equalization in magneto-optical disk readout: a theoretical investigation, *Appl. Opt.*, vol. 34, pp. 5153–5166, August 1995.

[27] C. Peng and M. Mansuripur, Evaluation of partial-response maximum-likelihood detection for phase-change optical data storage, *Appl. Opt.*, vol. 38, pp. 4394–4405, July 1999.

[28] C. Peng and M. Mansuripur, Partial-response signalling for phase-change optical data storage without electronic equalization, *Appl. Opt.*, vol. 41, pp. 3479–3486, June 2002.

[29] A.J. Viterbi, Error bounds for convolutional codes and an asymptotically optimum decoding algorithm, *IEEE Trans. Infor. Theory*, vol. IT-13, pp. 260–269, 1967.

[30] J.G. David Forney, The viterbi algorithm, *Proc. IEEE*, vol. 61, pp. 268–278, 1973.

[31] J.W.M. Bergmans, *Digital Baseband Transmission and Recording*, Kluwer Academic Publishers, Boston, August 1996.

[32] J.F. Heanue, K. Gürkan, and L. Hesselink, Signal detection for page-access optical memories with intersymbol interference, *Appl. Opt.*, vol. 35, pp. 2431–2438, May 1996.

[33] M.A. Neifeld, R. Xuan, and M.W. Marcellin, Communication theoretic image restoration for binary-valued imagery, *Appl. Opt.*, vol. 39, pp. 269–276, 2000.

[34] C. Berrou, A. Glavieux, and P. Thitimajshima, Near Shannon limit error correcting coding and decoding: turbo-codes I, in *Proc. IEEE 1989 Intl. Conf. Comm.*, pp. 1064–1070, 1993.

[35] S. Benedetto, D. Divsalar, G. Montorsi, and F. Pollara, Serial concatenation of interleaved codes: performance analysis, design, and iterative decoding, *IEEE Trans. Inform. Theory*, vol. IT-44, pp. 909–926, 1998.

[36] T.J. Richardson and R.L. Urbanke, The capacity of low-density-parity-check codes under message passing decoding, *IEEE Trans. Infor. Theory*, vol. IT-47, pp. 599–618, 2001.

[37] R.J.G. Smith, Easily decodable efficient self-orthogonal block codes, *Elec. Let.*, vol. 13, pp. 2449–2457, 1995.

[38] W.-C. Chou and M.A. Neifeld, Soft-decision array decoding for volume holographic memory systems, *J. Opt. Soc. Amer. A*, vol. 18, pp. 185–194, 2001.

V

Signal Processing for Read Channels

26 **Adaptive Timing Recovery for Partial Response Channels**   *Pervez M. Aziz and Viswanath Annampedu* 26-1
Introduction • Timing Recovery Basics • Symbol Rate Timing Recovery Schemes • Performance Analysis of Symbol Rate Timing Loops • Jitter and BER Simulation Results • Loss of Lock Rate (LOLR) • Conclusions

27 **Interpolated Timing Recovery**   *Piya Kovintavewat, John R. Barry, M. Fatih Erden, and Erozan M. Kurtas* 27-1
Introduction • Conventional Timing Recovery Architecture • Interpolated Timing Recovery Architecture • Implementation Issues • Applications of ITR to Magnetic Recording • Conclusion

28 **Adaptive Equalization Architectures for Partial Response Channels**   *Pervez M. Aziz* 28-1
Introduction • Equalization Architectures and Strategies • CTF Configurations • FIR Filter and The LMS Algorithm • Gain Equalization and Other Issues • Performance Characterization • Actual Equalizer Architectures • Conclusions

29 **Head Position Estimation**   *Ara Patapoutian* 29-1
Introduction • Servo Writing • The Digital Field • The Burst Field

30 **Servo Signal Processing**   *Pervez M. Aziz and Viswanath Annampedu* 30-1
Introduction • Servo Coding Gain • Noise Prediction in Servo Detection • Track Follow Mode Peformance • Seek Mode Performance • Servo Burst Demodulation • Conclusions

31 **Evaluation of Thermal Asperity in Magnetic Recording**   *M. Fatih Erden, Erozan M. Kurtas, and Michael J. Link* 31-1
Introduction • Modeling Thermal Asperities • Effect of TA on System Performance • TA Detection and Cancelation Techniques • Conclusion

32 **Data Detection**   *Miroslav Despotović and Vojin Šenk* 32-1
Introduction • Partial Response Equalization • Decision Feedback Equalization • RAM-Based DFE Detection • Basic Breadth-First Algorithms • Noise Predictive Maximum Likelihood Detectors • Postprocessor • Advanced Algorithms and Algorithms Under Investigation

33 Detection Methods for Data-dependent Noise in Storage Channels
Erozan M. Kurtas, Jongseung Park, Xueshi Yang, and William Radich,
and Aleksandar Kavčić **33**-1
Introduction • Detection Methods in the AWGN Channel • Recording System
Model • Detection Methods in Data-Dependent Noise • Simulations • Conclusions

34 Read/Write Channel Implementation   *Borivoje Nikolić,*
Michael Leung, Engling Yeo, and Kiyoshi Fukahori **34**-1
Introduction • Read/Write Channel Architectures • Partial Response Equalization
with Maximum-Likelihood Detection • Magnetic Disk Drive Read/Write Channel
Integration • Analog Front-Ends • Equalizer Architectures in Read Channels • Viterbi
Detector • Future Detection: Iterative Decoders • Timing Recovery • Write
Path • Overview of Recently Published Disk-Drive Read/Write Channels • Challenges
of Further Integration

26

Adaptive Timing Recovery for Partial Response Channels

26.1 Introduction **26**-1
26.2 Timing Recovery Basics **26**-2
      Symbol Rate VCO versus Interpolative Timing Recovery
      • Timing Loop Modes
26.3 Symbol Rate Timing Recovery Schemes **26**-4
      MMSE Slope Lookup Table (SLT) Timing Recovery
      • Quantized SLT Phase Detector • Mueller and Muller
      (MM) Timing Loop
26.4 Performance Analysis of Symbol
      Rate Timing Loops **26**-9
      Qualitative Loop Filter Description • Noise Jitter
      Analysis of Timing Loop
26.5 Jitter and BER Simulation Results **26**-12
26.6 Loss of Lock Rate (LOLR) **26**-14
      LOLR Detection • LOLR Simulations
26.7 Conclusions **26**-15

Pervez M. Aziz
Agere Systems
Dallas, TX

Viswanath Annampedu
Agere Systems
Allentown, PA

26.1 Introduction

In storage systems such as PRML magnetic recording channels a clock is used to sample the analog waveform to provide discrete samples to symbol-by-symbol (s/s) and sequence (Viterbi) detectors. Improper synchronization of these discrete samples with respect to those expected by the detectors for a given partial response will degrade the eventual bit error rate (BER) of the system. The goal of adaptive timing recovery is to produce samples for the s/s or sequence detector which are at the desired sampling instances for the partial response being used. In this section we will review the basics of timing recovery as well as commonly used algorithms for timing recovery in magnetic recording channels. We will start by presenting two classes of timing recovery algorithms: symbol rate VCO- and interpolation-based algorithms. After a discussion of the tradeoffs between these two types of algorithms we will focus on the traditional symbol rate VCO algorithms for the rest of our discussion. We will derive one of these timing recovery algorithms from first principles. We will then provide an analytical framework for comparing the performance of such algorithms using timing loop noise-induced output jitter as the performance criterion. Finally, we will provide quantitative comparative performance data for some of these algorithms based on the jitter analysis as well as simulations which measure timing loop jitter, BER, and loss of lock rate (LOLR).

0-8493-1524-7/05/$0.00+$1.50
© 2005 by CRC Press, LLC

26.2 Timing Recovery Basics

26.2.1 Symbol Rate VCO versus Interpolative Timing Recovery

Timing recovery schemes which have been considered for magnetic recording channels can be broadly classified into two groups: (i) traditional symbol rate VCO-based schemes (ii) interpolative schemes [1, 2] which sample slightly above the symbol rate. The key difference between the schemes is that the symbol rate VCO scheme adapts or adjusts the phase and frequency of the sampling clock to produce the desired samples whereas interpolative timing recovery (ITR) samples the analog waveform using a uniformly sampled clock to produce samples from which the desired samples are interpolated.

Figure 26.1 shows high level block diagrams of both approaches. Let us describe the VCO-based approach first. For the sake of the discussion the VCO approach is shown with an analog FIR equalizer. Consequently the sampling occurs at the input of the FIR equalizer. The noisy equalized output $y(k)$ must be used to detect the timing error present in these samples. This is done using a phase detector. The phase detector transforms an amplitude error in the samples to $\Delta(k)$ which is related to the desired change in the sampling phase. The phase detector output is also called the timing gradient.

The phase detector may require the use of the noisy equalized samples $y(k)$ or other signals derived from it. The other signals may be the preliminary or tentative decisions $\hat{d}(k)$s, decision directed estimates of $y(k)$ which are $\hat{y}(k)$ or other signals. These auxiliary signals are generated by the block labeled "Signal Generation for Phase Detector." The $y(k)$s are used to generate preliminary (tentative) decisions $\hat{d}(k)$, and an error signal $e(k)$, and a decision directed estimate of the ideal equalized sample value, $\hat{y}(k)$.

The phase detector output is filtered by a loop filter $T(z)$. The loop filter $T(z)$ is usually a second-order digital PLL (DPLL) with an additional delay term z^{-L} which models any latency through the timing loop. Such latency arises from the group delay of the FIR, computations in the DPLL, calculating the signals

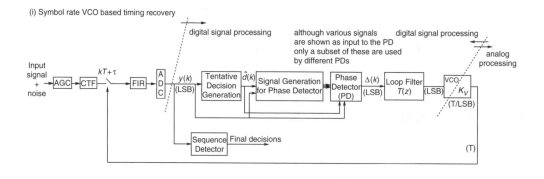

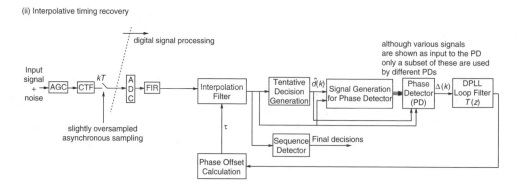

FIGURE 26.1   (i) Symbol rate VCO based timing recovery loop (ii) Interpolative timing recovery loop.

needed by the phase detector, etc. The filtered phase detector output is the input to a VCO which causes the actual sampling phase τ to change. The VCO is an analog circuit under digital control. The analog part can be a phase mixer which is capable of adjusting the timing phase by small fractions of T where T is the channel bit period. In such a case, the VCO acts as an amplitude to time converter and so modeled as a simple gain. To give physical meaning to the system, the units of the signals are noted: the equalized output after quantization by the ADC is in amplitude units of LSBs, the timing gradient $\Delta(k)$ or phase detector output is proportional to an amplitude error and so is also in LSBs. The loop filter provides a frequency dependent gain, so the input of the VCO is LSBs. The VCO has a gain of K_V in units of T/LSB, so the output of the VCO has units of time, T. The VCO gain can also be thought of as a clock update gain. For the specific system we will consider later, the phase mixer can make changes in the sampling phase in steps of $0, \pm 1, \pm 2T/64$ or more. The choice of this factor of 64 is such that the quantization of timing phase adjustment is well below the ADC quantization noise floor.

Let us now describe the ITR loop of Figure 26.1. As noted, with this scheme, an asynchronous clock is used to sample the input to the ADC after which a FIR filter performs the necessary equalization. The asynchronous equalized samples are now used to interpolate samples at the correct sampling instances dictated by the partial response. This is done with the interpolation filter which can be thought of as a filter which delays its input by an amount τ which is a fraction of the channel bit period T [1]. Such an interpolation filter's transfer function is $z^{-\tau}$. The samples $y(k)$ at the output of the interpolation filter drive the phase detector and loop filter as in the VCO-based timing loop. The loop filter output after being processed by the phase offset calculator produces the required sampling phase change. For good operation the loop must be able to produce a large number of fractional delays (such as 32 or 64) and correspondingly would require as many such filters for each of these delays. Figure 26.1 noted that the asynchronous sampling was performed at slightly above the Nyquist rate. The reasons for this is to accomodate a frequency offset between the written signal and the clock used to perform the asynchronous sampling. The magnitude of this frequency offset is usually limited in practical systems to 1% or less and so very little oversampling is required. However, oversampling ratios of up to 5% produce some improvement in performance by reducing the sensitivity of the aliasing with respect to the phase of the asynchronous sampling clock.

The advantages of the ITR-based timing loops are that they are all digital timing loops which are more amenable for design, verification and less susceptible to process variations. Also, for the ITR timing loop, the delays in the equalization filter and ADC do not contribute to the timing loop latency. However, the interpolation filter is not an extremely small piece of hardware and could make the ITR timing loop consume more chip area and power than a VCO-based loop. We have also not discussed practical design issues with the ITR-based system such as adaptation of the equalizer based on asynchronous samples [2] and design of the interpolation filter. From a performance point of view there is no significant difference between the ITR- or VCO-based approaches as indicated by simulation results in [1]. This also seems reasonable based on our observation in chapter 28 where we will note that a read channel system with practical equalization and timing recovery performed within a few tenths of a dB of the corresponding "ideal" system. Therefore, the choice between an all digital ITR-based system or a conventional VCO-based system needs to be based on the relative merits of both systems from an ease of design/verification and area/power standpoint.

26.2.2 Timing Loop Modes

Let us now further describe the operation of the entire timing loop. The entire timing recovery process occurs in several steps: zero phase start (ZPS), acquisition mode (ACQ), and tracking mode (TRK). During the ZPS and ACQ modes the disk controller must guarantee that the read channel is reading a preamble signal known to the timing loop. The preamble signal for almost all magnetic recording channels is a 2T pattern which is the periodic data sequence "... 11001100 ...." The purpose of the ZPS is to obtain a good estimate of the initial phase error between the readback signal and the desired samples for the 2T pattern. Once this estimate is obtained the sampling clock's phase is changed by the calculated amount to approximate the desired sampling phase. The next step is the ACQ process where the sampling phase error

Mode Name	ZPS	Acquisition (ACQ)	Tracking (TRK)
Assumed Data	Preamble	Preamble	Excess preamble/ regular random data
Loop Filter Gains	Zero	High/ medium	Medium/low

FIGURE 26.2   Timing loop operational modes: zero phase start, acquisition, tracking.

is further reduced and the frequency offset between the input signal and the sampling clock is compensated for to produce even more accurately sampled preamble samples. Since the preamble is a known signal pattern, timing recovery is facilitated in that the preliminary decisions can be obtained more reliably with less loop latency. Consequently, high loop filter update gains can be used. Once this initial acquisition is complete, the timing loop transitions into a tracking (TRK) mode which is intended for tracking slow variations in timing. In this mode the signal may contain any excess preamble as well as random data but no apriori assumption about the signal is made. The tentative decisions in the TRK mode are obtained with more loop latency and are not as reliable. The loop filter update gains are correspondingly lower. A summary of the operation described is provided in Figure 26.2. More fine gradations of the loop filter gains (beyond the high/medium/low gains shown in Figure 26.2 can be made across ACQ and TRK to produce improved performance [3]. Of course, there is a tradeoff between improved performance and somewhat enhanced circuit complexity so that one would choose to increase the complexity only until diminishing returns in performance is reached.

26.3   Symbol Rate Timing Recovery Schemes

We now consider in more detail traditional symbol rate VCO-based schemes. A decision directed baud or symbol rate timing recovery algorithm was first proposed by Mueller and Muller [4]. Their technique relied on the concept of a "timing function" $f(\tau)$ which generates the proper amount of timing phase adjustment for a given phase shift, τ, in the signal. The function should be monotonic, and have a zero output for zero sampling phase error. The Mueller and Muller (MM) technique provides a means to derive a timing function from a linear combination of samples of the channel's impulse response. In practice, one can design timing gradients whose expected value equals the suitably defined timing function. The timing gradients can be used to obtain the corresponding phase adjustment signal. In some magnetic recording systems using a PR4 target, a MM timing gradient with a second-order DPLL was used to produce the necessary timing phase updates [5, 6].

One can also derive timing recovery schemes based on other criteria such as the minimum mean square error (MMSE) criterion. MMSE methods seek to minimize the expectation of the square of an error signal $e(k, \tau)$ with respect to the timing phase. The error signal is obtained by subtracting the received equalized samples $y(k, \tau)$ from the corresponding "ideal" samples $\hat{y}(k)$. The minimization is done by adjusting the timing phase in the direction opposite to the derivative of the expected value of the squared error. In practice, one ignores the expected value and minimizes the squared error resulting in a stochastic gradient algorithm. MMSE timing recovery has been proposed in [7] and examined to some degree in [8] for PR magnetic recording channels. Another criterion, the maximum likelihood (ML) criterion, has also been used to derive a phase detector [9].

We first review the derivation of the MMSE gradient and note that the MMSE gradient yields suitable timing functions. We then formulate MMSE timing recovery in the framework of a slope lookup table (SLT) instead of a discrete time filtered version of symbol rate spaced equalized samples $y(k, \tau)$. The SLT approach leads to an efficient implementation with slopes expressed directly in terms of a discrete time filtered version of the data bits $d(k)$ instead of the equalized signal samples.

We describe a methodology for an analytical performance evaluation of the timing loop where the timing loop output noise jitter is the performance criterion. The analysis is described in detail for the SLT-based MMSE timing loop and also applied to the MM timing loop. The quantitative results from this technique are used to compare the SLT and MM timing loops. The ML loop is not considered further here as it has somewhat adverse jitter properties compared with the other two timing loops [10]. Finally, we present simulations results comparing the SLT and MM timing loops in terms of output noise jitter as well as BER performance.

26.3.1 MMSE Slope Lookup Table (SLT) Timing Recovery

Let us review MMSE timing recovery from first principles. The discussion is along the lines of [7, 8]. The expectation of the square of the error, $e(k, \tau) = y(k, \tau) - \hat{y}(k)$ is minimized with respect to the timing or sampling phase. Here,

$$\hat{y}(k) = \sum_{p=0}^{P-1} h(p)\hat{d}(k - p) \tag{26.1}$$

where $h(p)$ are the P coefficients of the partial response channel. In the absence of any channel impairments we would have $\hat{d}(k) = d(k)$ and $\hat{y}(k) = y(k)$. We need to obtain the derivative of the expectation with respect to τ. Ignoring the expectation operator we obtain a stochastic gradient algorithm [7]:

$$\frac{\partial}{\partial \tau}(e^2(k, \tau)) = 2y(k, \tau)\frac{\partial y(k, \tau)}{\partial \tau} - 2\hat{y}(k)\frac{\partial y(k, \tau)}{\partial \tau} = -2e(k, \tau)\frac{\partial y(k, \tau)}{\partial \tau}$$

$$= -2e(k, \tau)\left[\frac{dy(t)}{dt}\right]_{t=kT+\tau} = -2e(k, \tau)s(k, \tau) \tag{26.2}$$

Note that the MM approach was used to generate a timing gradient from a suitably defined timing function $f(\tau)$. Here, we have derived a timing gradient from the MMSE criterion. However, the resulting timing gradient should be a valid timing function, that is, be monotonic, and have a zero-crossing for zero sampling phase error. This has been shown in [8]. An expression for the timing function in terms of the PR channel coefficients is [10]

$$f(\tau) = \sum_{l=-\infty}^{\infty}\left[\underbrace{\sum_{p=0}^{P-1} h(p)\frac{(-1)^{l-p}}{(l-p)T}}_{l\neq p}\right]^2 \tag{26.3}$$

The result of plotting $f(\tau)$ in Equation 26.3 for EPR4 is shown in Figure 26.3. Let us now consider a MMSE based timing gradient or phase detector formulated in terms of a SLT. We model the signal slope in terms of a slope generating filter which when used to filter the data $d(k)$ produces the slopes:

$$s(k) = d(k) * \psi(k) = \sum_{c=-C_1}^{C_2} \psi(c)d(k - c) \tag{26.4}$$

where the negative coefficient index indicates that the slope at time k depends on future bits (accomodated by delaying the slope and adding the delay into the model as additional latency). $C_1 + C_2 + 1$ is the number of nonzero coefficients of the slope filter's impulse response, ψ, and where $*$ denotes the convolution operation. The SLT output $\hat{s}(k)$ approximates $s(k)$ which depends on the data pattern. Such a SLT can be derived for any PR by correlating the data with the actual signal slopes. In practice it is enough to use

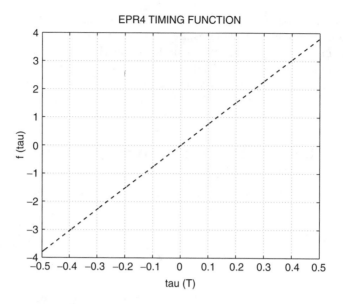

FIGURE 26.3   Timing function for a EPR4 partial response channel.

fewer terms from the filter. Therefore, the simplified SLT output can be represented as

$$\hat{s}(k) = \sum_{b=-B_1}^{B_2} \psi(b)d(k-b) \tag{26.5}$$

where $B = B_1 + B_2 + 1$ is the size of the slope table input that is, the number of data bits used in calculating the slope. The SLT-based gradient is then,

$$\Delta(k) = e(k)\hat{s}(k) \tag{26.6}$$

where the factor of -2 in Equation 26.2 can be absorbed in the lookup table. In our analysis we need the slope generating filter coefficients $\psi(c)$. These coefficients $\psi(c)$s are obtained in the process of numerically generating the signal slopes which are correlated with the data.

26.3.1.1 Phase Detector Properties

Before computing the output noise jitter of the entire timing loop we must first analyze the properties of the phase detector. Quantities important for the performance of the phase detector are its *KPD* and output noise standard deviation ratio ($\frac{KPD}{\sigma_{n_o}}$). The *KPD* is the ratio of the mean phase detector output to a constant sampling phase error, τ. The *KPD* can thus be thought of as the signal gain of the timing loop where the signal is the sampling phase error. The output noise $n_o(k)$ is the equivalent noise at the output of the phase detector for a given input noise $n(k)$ at the phase detector input. The error, $e(k)$, at the equalizer output is a combination of contributions from the sampling phase error, $\tau(k)$ and noise. Let $n(k)$ represent the noise at the equalizer output (intersymbol interference + filtered equalized noise). We then have,

$$e(k) = \tau(k)s(k) + n(k) \tag{26.7}$$

The phase detector output, $\Delta(k)$, is then

$$\Delta(k) = [\tau(k)s(k) + n(k)]\,\hat{s}(k) = \tau(k)s(k)\hat{s}(k) + n_o(k) \tag{26.8}$$

Figure 26.4 shows in detail the timing loop of Figure 26.1 with the details of the SLT phase detector and the composition of the error signal from the sampling phase and noise per Equation 26.7.

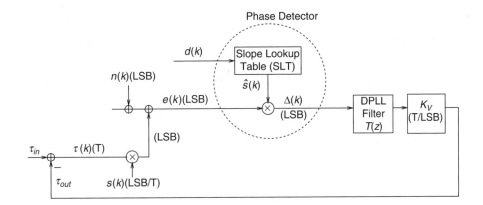

FIGURE 26.4 Timing loop with SLT phase detector.

We now find the statistical properties of *KPD* and n_o using $\mathcal{E}$ as the expectation operator. For a tractable analysis we assume $n(k)$ is AWG. To easily relate σ_n to the error event rate (EER) at the output of the Viterbi detector we assume that channel errors are dominated by a minimum distance error event (with distance $d_{\min}$).

$$\sigma_n = \frac{d_{\min}/2}{Q^{-1}(EER)} \tag{26.9}$$

The EER is the bit error rate (BER) divided by the number of bit errors in the dominant error event. In Equation 26.9 Q refers to the well-known Q function defined by,

$$Q(x) = \frac{1}{2\pi} \int_x^\infty \exp\left(\frac{y^2}{2}\right) dy \tag{26.10}$$

26.3.1.2 Signal Gain (*KPD*) of the Phase Detector

Using the definition of *KPD* we obtain, for a constant sampling phase error τ,

$$KPD = \frac{\mathcal{E}\{\tau\hat{s}(k)s(k) + n(k)\hat{s}(k)\}}{\tau} = \mathcal{E}\{\hat{s}(k)s(k)\} + \frac{\mathcal{E}\{n(k)\hat{s}(k)\}}{\tau} \tag{26.11}$$

Consider $\mathcal{E}\{n(k)\hat{s}(k)\}$, where $\hat{s}(k)$ is a linear function of the data bits which can be realistically assumed to be uncorrelated with the noise $n(k)$. Therefore, this term is zero and as we should expect, the noise does not contribute to the *mean* phase detector output. Thus,

$$KPD = \mathcal{E}\{\hat{s}(k)s(k)\} = \sum_{b=-B_1}^{B_2} \sum_{c=-C_1}^{C_2-1} \psi(b)\psi(c)\mathcal{E}\{d(k-b)d(k-c)\} \tag{26.12}$$

If d is uncoded, hence white, with zero mean, $\mathcal{E}\{d(k-b)d(k-c)\} = \sigma_d^2$, if $b = c$, and is 0, if $b \neq c$. Consequently, the *KPD* is

$$KPD = \sigma_d^2 \sum_{b=-B_1}^{B_2} \psi^2(b) \tag{26.13}$$

where it is assumed that the slope table ouput is based on fewer than $C_1 + C_2 + 1$ terms to reduce the summation to be from $b = -B_1$ to B_2. We note that the *KPD* values obtained here are equivalent to the slopes of the $f(\tau)$ versus τ curve plotted in Figure 26.3.

26.3.1.3 Output Noise of the Phase Detector

Computing the autocorrelation,

$$Rn_o(l) = \mathcal{E}\left\{n_o(k)n_o(k+l)\right\} = \mathcal{E}\left\{n(k)\hat{s}(k)n(k+l)\hat{s}(k+l)\right\}$$

Since $\hat{s}(k)$ are a filtered version of $d(k)$ which are uncorrelated with n, n and $\hat{s}$ are uncorrelated. Therefore,

$$\mathcal{E}\left\{n_o(k)n_o(k+l)\right\} = \mathcal{E}\left\{n(k)n(k+l)\right\}\mathcal{E}\left\{\hat{s}(k)\hat{s}(k+l)\right\}$$

$$= R_n(l)\mathcal{E}\left\{\hat{s}(k)\hat{s}(k+l)\right\}$$

$$\mathcal{E}\left\{\hat{s}(k)\hat{s}(k+l)\right\} = \sum_{b=-B_1}^{B_2}\sum_{b'=-B_1}^{B_2}\psi(b)\psi(b')\mathcal{E}\left\{d(k-b)d(k+l-b')\right\} \qquad (26.14)$$

With d being uncoded (hence white) and zero mean, $\mathcal{E}\{d(k-b)d(k+l-b')\} = \sigma_d^2$, if $b' = b+l$ and 0, if $b' \neq b+l$. Also assuming, $R_n(l) = \sigma_n^2\delta[l]$, that is, n to be white, we need to consider only $l = 0$ in which case we have $b = b'$. In that case,

$$Rn_o(l) = \mathcal{E}\left\{n_o(k)n_o(k+l)\right\} = \sigma_n^2\sigma_d^2\delta[l]\sum_{b=-B_1}^{B_2}\psi^2(b) \qquad (26.15)$$

We observe that the noise at the phase detector output is indeed white with standard deviation,

$$\sigma_{n_o} = \sigma_n\sigma_d\sqrt{\sum_{b=-B_1}^{B_2}\psi^2(b)} \qquad (26.16)$$

26.3.2 Quantized SLT Phase Detector

So far we have not considered any quantization of the inputs to the phase detector. The MM phase detector's inputs are $y(k)$ and $\hat{y}(k)$ (and their delayed versions). The SLT phase detector inputs are $e(k)$ and $\hat{s}(k)$. All of these inputs are multi-bit quantities. One can quantize $\hat{s}(k)$ to three levels of $1, 0, -1$ to obtain a drastic reduction in the implementation complexity by simplifying the $\Delta(k) = e(k)\hat{s}(k)$ multiplication to a sign inversion. We call this scheme the Qnt. (quantized) SLT timing loop.

26.3.3 Mueller and Muller (MM) Timing Loop

We now examine the properties of the Mueller and Muller (MM) timing gradient. This gradient is obtained as

$$\Delta(k) = y(k)\hat{y}(k-1) - y(k-1)\hat{y}(k) \qquad (26.17)$$

in terms of the equalized signal $y(k)$ and its delayed version as well as the corresponding estimates of the "ideal" values $\hat{y}$ for these signals. A block diagram of a MM timing loop using this gradient is shown in Figure 26.5. We need to evaluate this phase detector's *KPD* and noise performance. This is accomplished by writing $y(k)$ as $\hat{y}(k) + e(k)$, expanding $e(k)$ as in Equation 26.7 from which $s(k)$ is further expressed in terms of the slope generating filter based on Equation 26.4. Likewise $\hat{y}(k)$ is expressed in terms of the PR coefficients as per Equation 26.1. The analysis makes the usual assumptions about the data and noise

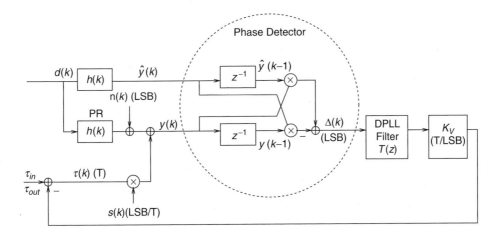

FIGURE 26.5 Timing loop with Mueller/Muller phase detector.

$n(k)$ being white. The details of the analysis can be found in [10] which yields,

$$KPD = \sigma_d^2 \left(\underbrace{\sum_c \sum_m \psi(c)h(m)}_{m-c=-1} - \underbrace{\sum_c \sum_m \psi(c)h(m)}_{m-c=1} \right) \tag{26.18}$$

where the sum over m is from 0 to $P-1$ and that over c is from $-C_1$ to C_2+1.

The autocorrelation, $R_{n_o}(l)$ for the noise at the output of the phase detector, assuming the data to be white, is also computed in [10]. It is shown that even with AWG noise at the phase detector input, that is, noise with autocorelation $R_n(l) = \sigma_n^2 \delta[l]$, noise at the phase detector output is not white. However, it is shown that if $R_n(l) = \sigma_n^2 \delta[l]$ the autocorrelation of $R_{n_o}(l)$ will be limited to only the first delay terms that is, $l = 1$ and -1 so we have,

$$R_{n_o}(0) = \sigma_{no}^2 = 2\sigma_d^2 \sigma_n^2 \sum_{p=0}^{P-1} h(p)^2 \tag{26.19}$$

and

$$R_{n_o}(1) = R_{n_o}(-1) = -\sigma_d^2 \sigma_n^2 \underbrace{\sum_{p=0}^{P-1} \sum_{m=0}^{P-1} h(m)h(p)}_{p-m=2} \tag{26.20}$$

26.4 Performance Analysis of Symbol Rate Timing Loops

We have so far examined the properties of the SLT and MM timing gradients or phase detectors. If the noise at the phase detector output for both systems were white we could directly compare their performance by comparing their respective *KPD* to σ_{no} ratio as a kind of signal to noise ratio of the phase detector. The ratio would measure a signal gain (experienced by sampling phase errors) to noise gain across the entire bandwidth. If the noise had been white for both systems this ratio would scale similarly for both systems when measured over the effective noise bandwidth determined by the loop filter. However for the MM loop we observed that the noise at the phase detector output was not white. Therefore, we must examine the timing loop performance at the output of the loop filter not just at the output of the phase detector.

The overall loop analysis will make use of the results of the previous sections where we computed *KPD* and $Rn_o(l)$ for each phase detector. Before continuing our analysis let us make some qualitative comments about the loop filter.

26.4.1 Qualitative Loop Filter Description

A timing loop is a feedback control loop. Therefore, the stability/loop dynamics are determined by the "gain" (in converting observed amplitude error to a timing update) of the phase detector and the details of the loop filter. If the timing loop were needed to remove the effect of a sampling phase error, a first-order DPLL would be sufficient. However, the timing loop must also recover the proper frequency with which to sample the signal. Therefore, the use of a second-order DPLL loop filter is needed. This allows the timing loop to continually generate small phase updates to produce a clock which not only has the correct sampling phase within a symbol interval *T* but which also has the correct value for the symbol interval that is, the correct clock frequency. DPLL here refers to the portion of the overall loop filter transfer function $T(z)$ without the fixed delay term z^{-L}. In addition, important to the performance of the loop is its noise performance that is, for a given level of input noise, the effect on the jitter in sampling phase updates. The jitter properties are determined by the noise gain of the phase detector as well as the loop filter properties. The loop filters out noise beyond the bandwidth of interest, this bandwidth being determined by how rapidly the loop is designed to react to timing changes. As mentioned earlier the DPLL loop filter is a second-order filter with an additional latency term. Its transfer function is given by:

$$T(z) = z^{-L} \left(\frac{f_g z^{-1}}{1 - z^{-1}} + p_g \right) \left(\frac{z^{-1}}{1 - z^{-1}} \right) \tag{26.21}$$

where f_g and p_g are frequency and phase update gains for the second- and first-order sections, respectively while *L* is the loop latency. A block diagram of $T(z)$ is also shown in Figure 26.6(a).

26.4.2 Noise Jitter Analysis of Timing Loop

Linearized *Z* domain analysis of the DPLL is now performed by replacing the phase detector with its *KPD* (denoted by K_p in the equations for readability). In evaluating the SLT and MM DPLLs we will use three sets or combinations of p_g and f_g: "LOW", "MED", and "HGH" where the LOW gains are relatively low update gains which would be used in tracking mode, MED gains are moderate gains, and HGH gains are high gains which might be used during acquisition. For the SLT phase detector, the effect of the phase detector on the input noise is taken into account by the *KPD* of Fig 26.6(b) as seen from Eqs (26.13) and (26.16) and the total jitter is simply the above σj.

The open loop DPLL transfer function, $G(z)$ incorporating the loop filter $L(z)$ and clock update gain is

$$G(z) = K_V \left(\frac{f_g z^{-1}}{1 - z^{-1}} + p_g \right) z^{-L} \left(\frac{z^{-1}}{1 - z^{-1}} \right)$$

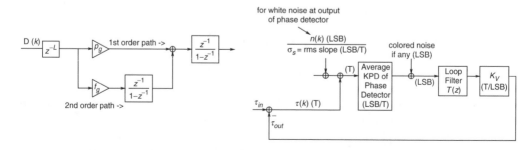

FIGURE 26.6 Linearized model (a) Second-order DPLL loop filter (b) Timing loop with phase detector modeled by its average signal gain.

Referring to the timing loop model of Figure 26.6(b), the *closed* loop transfer function $\left(\frac{T_{out}}{T_{in}}\right) = H(z)$ is

$$H(z) = \frac{K_p G(z)}{1 + K_p G(z)} \tag{26.22}$$

Note that K_p has dimensions of LSB/T, K_V and $G(z)$ have dimensions of T/LSB, and $H(z)$ is a transfer function with respect to two time quantities. The effective noise bandwidth is then,

$$ENB = 2 \int_0^{0.5} |H(f)|^2 \, df$$

An example closed loop transfer function for the SLT DPLL is shown in Figure 26.7 for LOW update gains. To find the effect of AWG noise, $n(k)$, we first convert the σ_n to an effective timing noise by dividing by the rms slope, σ_s, of the signal which is obtained during the numerical generation of the signal slopes and calculation of the slope generating filter coefficients. Now it can be multiplied by the square root of the *ENB* to determine the corresponding noise induced timing jitter σ_j (units of T). Therefore,

$$\sigma_j = \frac{\sigma_n}{\sigma_s} \sqrt{ENB} \tag{26.23}$$

The equivalent model for the above method of analysis is shown in Figure 26.6(b).

For the SLT phase detector, the effect of the phase detector on the input noise is taken into account by the *KPD* of Fig 26.6(b) as seen from Eqs (26.13) and (26.16) and the total jitter is simply the above σ_j. For the MM DPLL the phase detector output noise is colored; however, we know its properties here from Eqs (26.19)–(26.20) and can examine its effect from this point onwards. The only difference is that the closed loop transfer function seen by the MM phase detector output noise is

$$F(z) = \frac{G(z)}{1 + K_p G(z)} \tag{26.24}$$

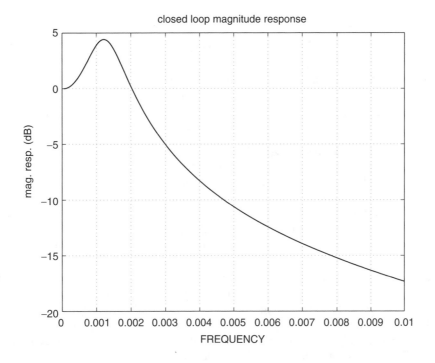

FIGURE 26.7   Closed loop frequency response of SLT DPLL for low p_g and f_g update gains.

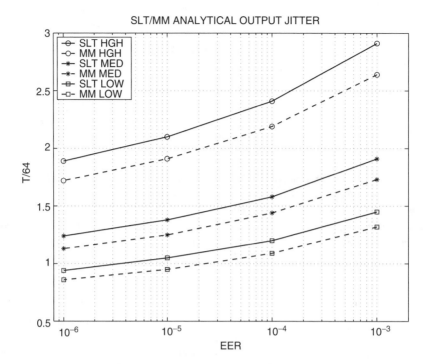

FIGURE 26.8   Analytically calculated output jitter for SLT and MM timing loops.

The noise jitter is then obtained as,

$$\sigma_j = \sqrt{2 \int_0^{0.5} P_n(f)|F(f)|^2 \, df} \tag{26.25}$$

where $P_n(f)$ is the noise p.s.d. at the phase detector output.

Figure 26.8 plots the jitter performance of the SLT- and MM-based DPLLs for three sets of (p_g, f_g): LOW, MED, HGH. Shown are the output noise-induced timing jitter of the loop for four channel error event rates. We observe that the MM timing loop's output noise jitter is almost the same but slightly better than that of the SLT-based timing loop.

26.5   Jitter and BER Simulation Results

Simulations on the SLT-based timing loop and the MM loop are run within the simulator framework described in Figure 26.1. Until now we have performed an analytical comparison of the MM and SLT timing loops based on multi-bit quantities, that is, not coarsely quantizing any of the inputs to the phase detectors. In our simulation based comparisons we also examine the performance of the three level *quantized* (Qnt.) SLT phase detector which is of more practical interest from an implementation perspective. The same DPLL loop filter structure is used for both systems.

The input signal is generated at a channel bit density of 2.8 from an isolated pulse extracted from a realistic signal while the colored input noise has a spectrum same as that of noise extracted from a realistic signal. The SNR definition for the following simulations is isolated pulse peak to r.m.s. noise in the Nyquist band. For the simulations, the same latency and tentative decision detector are used for the timing loops.

For the jitter measurements, simulations are run without noise and SNRs which correspond with channel EERs of 10^{-4} and 10^{-2}. We examine the steady state jitter in the DPLL output phase and the response of the timing loop to a phase step in the data field. Figure 26.9 shows the transient phase response plots of the

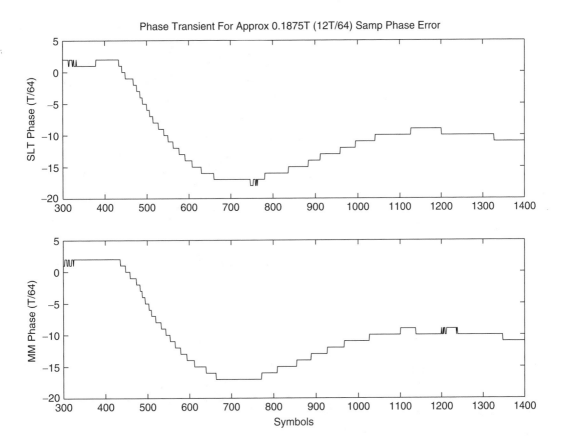

FIGURE 26.9 SLT and MM DPLL reaction to 0.1875T (12T/64) phase step. LOW p_g, f_g gains. No noise in this simulation.

SLT and MM DPLLs responding to a 0.1875 T (12 T/64) phase step in the data field for the same for the same LOW p_g and f_g settings. Note that they have very similar responses. Table 26.1 shows the steady state output jitter of the two timing loops for various combinations of gains and noise levels corresponding to EERs of 10^{-4} and 10^{-2}. The settled DPLL phases show some nonzero jitter without additive noise — from quantization effects. Timing jitter at the DPLL output is measured by measuring the standard deviation of the DPLL phase. We observe that the two timing loops have very similar jitter numbers although the MM timing loop jitter is slightly lower. Even though the analytical performance of the SLT loop was for the unquantized SLT, these simulations comparing the Qnt. SLT loop with the mulit-bit MM loop suggest that there is very little degradation in jitter performance due to the three level quantization of $\hat{s}(k)$ as compared with the performance of the multi-bit MMPD-based timing loop.

TABLE 26.1 Simulation Based Timing Loop Output Jitter σ_{j_t} (Units of T/64) Performance of Qnt. SLT and MM Timing Loops: Final EERs of Zero (noiseless), 10^{-4}, and 10^{-2}

p_g, f_g Gains	SLT			MMPD		
	EER 0	EER 10^{-4}	EER 10^{-2}	EER 0	EER 10^{-4}	EER 10^{-2}
LOW	0.49	1.30	2.18	0.45	1.16	1.86
MED	0.49	1.69	2.99	0.46	1.56	2.51
HGH	0.67	2.67	4.86	0.70	2.67	4.38

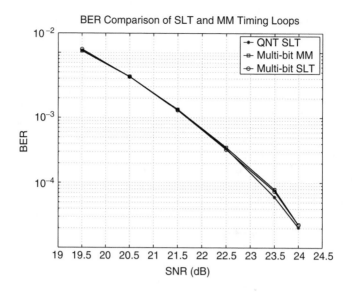

FIGURE 26.10   Simulated BERs of practical read channel using SLT and MM timing loops.

We now examine the Viterbi detector BER performance instead of the timing loop jitter performance for the read channel architecture of Figure 26.1 employing the multi-bit MM and SLT as well the Qnt. SLT timing loops. In addition to additive noise, the signal for any read sector simulated, the initial phase error is chosen from a uniform distribution on $[0, T)$. We observe from Figure 26.10 that the BERs of the three systems are practically indistinguishable.

26.6   Loss of Lock Rate (LOLR)

The steady state performance of the timing loop is measured by its jitter performance, that is, the effect of noise on the jitter of timing or sampling phase updates. The BER of the system is also mostly related to the steady state properties of the timing loop. However, another criterion becomes significant and should be used if channels are operated at lower SNRs with the recent advances in channel detector gains based on approaches such as soft parity codes and iterative decoding schemes. This criterion is the loss of lock rate (LOLR) of the timing loop. Loss of lock occurs when the timing loop phase drifts so far from the desired sampling instances that the detected data is mis-synchronized by more than a channel symbol period causing the sector to be read incorrectly due to the burst errors created with mis-synchronization. Loss of lock occurs usually during changes of phase caused by phase steps or a frequency error between the signal and the recovered clock. Therefore, the LOL performance takes into account the steady state as well has dynamical performance of the loop including any nonlinear behavior.

26.6.1   LOLR Detection

Loss of lock can be detected by a large burst of bit or byte errors. Given the low SNRs being considered the raw byte error rates are very high. Therefore the challenge is to detect actual loss of lock without falsely detecting one as a result of the raw high byte error rate due to random errors produced by noise. We use the following LOL criterion as explained below. We examine byte errors defining a byte to be comprised of S bits, d_i, at the output of the Viterbi detector. We use $S = 8$, that is, the standard byte length. Let d_i be the corresponding correct value of the bits. Let γ_k denote whether the kth byte in an observation window is in error or not. This can be written as

$$
\gamma_k = \begin{cases} 1 & \text{if } \left[\sum_{i=1}^{S} 1 \cdot (\hat{d}_i \neq d_i) \right] > 0 \\ 0 & \text{otherwise} \end{cases}
\tag{26.26}
$$

Let $\overline{\gamma_L} = [\gamma_1, \gamma_2, \ldots, \gamma_L]^T = [\gamma_1 \overline{\gamma} \gamma_L]$ be the vector of binary values of size $(1 \times L)$ with $\overline{\gamma}$ being comprised of $L-2$ bytes $\gamma_k, k \neq 1, L$ as defined above. Note that $\gamma_1 = \gamma_L = 1$, that is, the vector begins and ends with a byte error. We declare a loss of lock when any vector across the read event $\overline{\gamma_L}$ beginning and ending with an error byte does not contain more than Z number of consecutive zeros (good bytes). Thus, loss of lock occurring across one sector can be declared by

$$
\left[\sum_{j=n}^{j=n+1+Z} 1 \cdot (\gamma_j = 0) \right] \leq Z
$$
$$
\forall \, n \text{ such that } \gamma_n = \gamma_{n+1+Z} = 1 \quad \text{and} \quad \gamma_j \epsilon \overline{\gamma_L}
\tag{26.27}
$$

The loss of lock rate, LOLR, is the total number of LOL events observed across the number of sectors divided by the number of sectors. For our simulations we have selected $L = 50$, $Z = 1$ as the parameters for detecting LOL. By using this LOL detector with only random errors due to noise it was verified that the false loss of lock detection rate was negligible compared to the actual loss of lock rate due to timing errors.

26.6.2 LOLR Simulations

In addition to additive noise and initial phase error, the frequency error between the signal and the sampling clock is a significant contributor to loss of lock. For our simulations, this frequency error is chosen from a truncated Gaussian distribution with a σ of 0.03%. The distribution is truncated so that the maximum frequency error possible is 0.4%. The choice of these parameters is consistent with realistic conditions. Note that the frequency error can also be thought of as a linearly increasing/decreasing walk of the input signal's timing phase across the read event.

We now examine the LOL performance of the timing loops. Figure 26.11 presents example signatures of a LOL event showing the phase transient deviating by more than a symbol period and the resulting burst of bit/byte errors. The LOLR results for the three timing loops are plotted in Figure 26.12. For these simulations, the p_g and f_g gains are optimized with respect to LOLR for each phase detector. The optimum gains turned out to be the same for all of them. We observe that the Qnt. SLT timing loop performs about 0.25 dB worse than the multi-bit input SLT timing loop. The multi-bit input MM timing loop performs about 0.3 to 0.4 dB worse. Therefore, despite near equivalent performance of the multi-bit MM and SLT timing loops predicted by the linearized steady state analysis in [10], the large signal dynamical behavior which affects the LOLR is modestly different between the two loops. Consequently, it is important to use LOLR as an important performance criterion even though steady state criteria such as jitter and BER serve as an important guide and provide insight into the performance tradeoffs of timing loops.

26.7 Conclusions

We have provided an overview of timing recovery methods for PRML magnetic recording channels including interpolative and traditional symbol rate VCO-based timing recovery methods. We have reviewed MMSE timing recovery from first principles and its formulation in the framework of a SLT-based timing

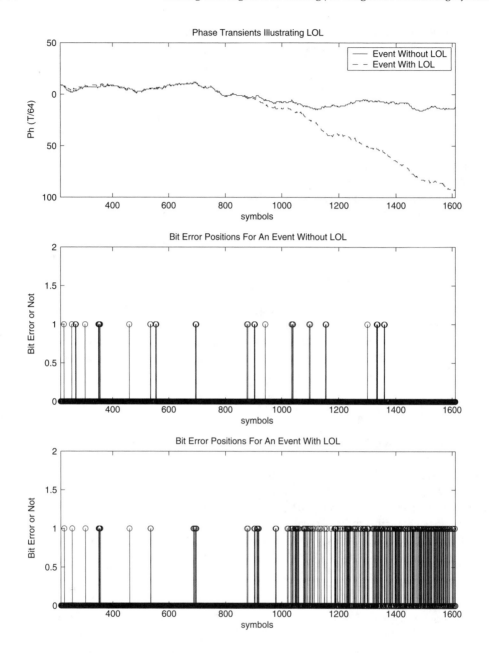

FIGURE 26.11   Example phase transient and BER profile for a loss of lock event.

gradient. We have provided a framework for analyzing the performance of the timing loops in terms of output noise jitter. The jitter calculation is based on obtaining linearized Z domain closed loop transfer functions of the timing loop. We have compared the output timing jitter, due to input noise, of the SLT and Mueller and Muller (MM) timing loops — two commonly used timing loops. The jitter performance of the MM loop is almost the same but slightly better than that obtained with the SLT-based timing loop. However, the Viterbi BER performance of read channel systems employing the two timing loops are practically indistinguishable. We have suggested that LOLR serve as another important criterion with which to compare timing loop performance. The MM timing loop performance is several tenths of a dB worse than the SLT timing loop based on LOLR.

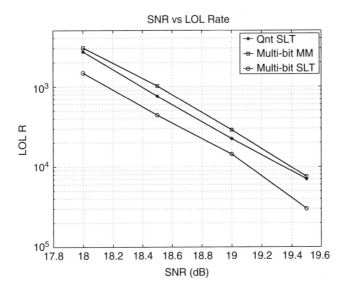

FIGURE 26.12   Loss of lock rates (LOLR) of timing loops.

References

[1] M. Spurbeck and R. Behrens, Interpolated timing recovery for hard disk drive read channels, *Proceedings of IEEE Intl. Conf. on Commn.*, pp. 1618–1624, 1997.

[2] Z. Wu and J. Cioffi, A MMSE interpolated timing recovery scheme for the magnetic recording channel, *Proceedings of IEEE Intl. Conf. on Commn.* pp. 1625–1629, 1997.

[3] A. Patapoutian, On phase-locked loops and kalman filters, *IEEE Trans. Commn.*, pp. 670–672, May 1999.

[4] K. Mueller and M. Muller, Timing recovery in digital synchronous data receivers, *IEEE Trans. Commn.*, pp. 516–531, May 1976.

[5] R. Cideciyan, F. Dolivo, et al., A PRML system for digital magnetic recording, *IEEE J. Select. Areas Commn.*, pp. 38–56, January 1992.

[6] F. Dolivo, W. Schott, G. Ungerbock, Fast timing recovery for partial response signaling systems, *IEEE Conf. on Commn.*, pp. 18.5.1–18.5.4, 1989.

[7] S. Qureshi, Timing recovery for equalized partial-response systems, *IEEE Trans. Commn.*, pp. 1326–1331, December 1976.

[8] H. Shafiee, Timing recovery for sampling detectors in digital magnetic recording, *IEEE Conf. on Commn.*, pp. 577–581, 1996.

[9] J. Bergmans, *Digital Baseband Transmission and Recording*, Kluwer Academic Publishers, Dordrecht, Netherlands, pp. 500–513, 1996.

[10] P. Aziz and S. Surendran, Symbol rate timing recovery for higher order partial response channels, *IEEE J. Select. Areas in Communications*, pp. 635–648, April 2001.

27

Interpolated Timing Recovery

Piya Kovintavewat

John R. Barry
Georgia Institute of Technology
Atlanta, GA

M. Fatih Erden

Erozan M. Kurtas
Seagate Technology
Pittsburgh, PA

27.1   Introduction **27**-1
27.2   Conventional Timing Recovery Architecture **27**-2
27.3   Interpolated Timing Recovery Architecture **27**-3
       Fully Digital Timing Recovery • MMSE Interpolation Filter
       without Channel Knowledge • MMSE Interpolation Filter
       with Channel Knowledge
27.4   Implementation Issues **27**-7
       The Estimate of the Initial Sampling Phase
       • Implementation of the Interpolation Filter
       • Implementation of the Equalizer • Effect of Employing the
       Oversampled System • FIR-ITR Loop Structure
27.5   Applications of ITR to Magnetic Recording **27**-10
       System Model • SSNR Performance • BER Performance
27.6   Conclusion **27**-16

27.1   Introduction

Timing recovery is an essential part of the state-of-the-art read channels employed by high-density data storage systems. It is utilized to adjust the sampling phase offset used to sample the received analog signal so that the sampler output will be synchronized with the transmitted symbol. Conventional timing recovery is performed in the analog domain on a symbol-by-symbol basis with the aid of a second-order phase-locked loop (PLL) [1] and an analog voltage-controlled oscillator (VCO) that can be expensive to implement. As the data rates increase and design cycles decrease, there are both performance and cost benefits to moving the analog parts of the synchronization circuitry to the digital domain.

With a fully digital timing recovery scheme, the received signal is sampled asynchronously by an analog-to-digital (A/D) converter; therefore, neither its frequency nor its phase is synchronized with the transmitted signal. The timing adjustment must be done by digital methods to obtain the synchronized samples. One way to do this is to compute the value of the signal at the desired time instant using interpolation techniques, based on a set of asynchronous samples and the timing information obtained from a digital PLL. Such a system is known as *interpolated timing recovery* (ITR).

The advantage of using asynchronous sampling is that the sampling frequency does not need to be a multiple of the symbol frequency. It only has to be above the Nyquist frequency so as to avoid aliasing. Also, ITR is a fundamental component of iterative timing recovery schemes, which require a whole data sector to be stored before resampling it at the desired time instants.

After briefly describing conventional timing recovery in Section 27.2, we explain how to realize a fully digital timing recovery scheme and describe the design of the interpolation filters based on the minimum mean-squared error (MMSE) approach in Section 27.3. Implementation issues related to ITR are discussed in Section 27.4. Section 27.5 compares the performance of ITR with conventional timing recovery in magnetic recording channels under a variety of operating conditions. Finally, conclusions are given in Section 27.6.

27.2 Conventional Timing Recovery Architecture

Today's magnetic recording read-channel chip architecture uses conventional timing recovery to acquire synchronization. As shown in Figure 27.1, conventional timing recovery is based on a second-order PLL consisting of a timing error detector (TED), a loop filter and a VCO.

A decision-directed TED [1] is used to compute the estimated timing error, $\hat{e}_k$, which is the misalignment between the phase of the received signal and that of the sampling clock. Several TED algorithms have been proposed in the literature [1], depending on how they incorporate the information available at the TED input. In this chapter, we consider the well-known Mueller and Müller TED algorithm [2], which is expressed as

$$\hat{e}_k = y_k \hat{d}_{k-1} - y_{k-1}\hat{d}_k \qquad (27.1)$$

where $y_k = y(kT + \hat{\tau}_k)$ is the kth sampler output, T is the bit period, $\hat{\tau}_k$ is the kth sampling phase offset adjusted by a PLL, and $\hat{d}_k$ is the kth desired sample.

Next, the estimated timing error is filtered by a loop filter to eliminate the noise in the timing error signal. Finally, the next sampling phase offset is updated by a second-order PLL according to

$$\hat{\theta}_k = \hat{\theta}_{k-1} + \beta\hat{e}_k \qquad (27.2)$$

$$\hat{\tau}_{k+1} = \hat{\tau}_k + \alpha\hat{e}_k + \hat{\theta}_k \qquad (27.3)$$

where $\hat{\theta}_k$ represents the frequency error, and α and β are the PLL gain parameters. Note that the PLL gain parameters determine the loop bandwidth and the rate of convergence. The larger the value of PLL gain parameters, the larger the loop bandwidth, the faster the convergence rate, and thus the more the noise allowed to perturb the system. Practically, a known data pattern called a *preamble* (or a training sequence) is usually sent during acquisition mode to help PLL achieve fast synchronization. Since this preamble is known at the receiver, large values of α and β can be used to expedite the convergence rate. However, the values of α and β should be lowered during tracking mode so as to reduce the effect of the noise [3]. Therefore, designers must tradeoff between the loop bandwidth and the convergence rate when designing α and β.

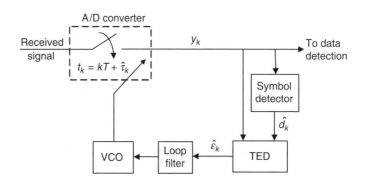

FIGURE 27.1   A conventional VCO-based timing recovery.

27.3 Interpolated Timing Recovery Architecture

This section focuses on how to design the interpolation filters that work well with a minimal amount of oversampling. We first describe how to realize a fully digital timing recovery scheme. Then, the design of the interpolation filters will be explained.

27.3.1 Fully Digital Timing Recovery

All components of the conventional PLL can be implemented digitally except the VCO. To replace the VCO with a fully digital circuit, we need: (1) a fixed sampling clock to sample the received signal asynchronously; (2) a digital accumulator; (3) an interpolation control unit to find the sampling location index; and (4) an interpolation filter to resample the data so as to obtain a synchronized sample. Figure 27.2 shows the fully digital timing recovery scheme known as ITR.

The received signal in Figure 27.2 is practically bandlimited by a low-pass filter whose cutoff frequency is at $1/(2T)$. Then, it is sampled asynchronously by an A/D converter operating at a fixed sampling frequency, $1/T_s$, where T_s is a sampling period (usually $T_s \leq T$ to avoid aliasing). The interpolation filter uses a set of the asynchronous samples (in T_s-domain) and the timing information obtained from the interpolator control unit to output the synchronized samples (in T-domain). The digital accumulator utilizes Eqaution 27.2 and Equation 27.3 to update the next sampling phase offset.

Since $\hat{\tau}_k$ represents the sampling phase offset in T-domain, it cannot be directly used by the interpolation filter, which requires the sampling phase offset in T_s-domain. In doing so, the interpolator control unit needs to know T, T_s and $\hat{\tau}_k$ in order to compute the sampling location index, t_k, which can be written as

$$t_k = kT + \hat{\tau}_k = (m_k + \mu_k)T_s \tag{27.4}$$

$$m_k = \text{int}[(kT + \hat{\tau}_k)/T_s] \tag{27.5}$$

$$\mu_k = (kT + \hat{\tau}_k)/T_s - m_k \tag{27.6}$$

where m_k is an integer-valued *basepoint index* [4], int[x] denotes the largest integer not exceeding x, and $0 \leq \mu_k < 1$ is a *fractional interval* [4] (we will refer to it as a *sampling phase*). Although it is difficult to obtain the exact value of T and T_s, one can use $T_s = E[T]/1.05$ [5], where $E[T]$ is the mean value of T, and 5% oversampling rate is employed.

Many interpolation filters have been proposed in the literature [4–8, 10, 15]. If the received signal is oversampled at a high enough rate, for example, four or eight times, a simple interpolation filter such as a linear interpolation filter can be sufficiently used to recover the desired data. Nonetheless, the read-channel chip used in today's hard disk drives operates at a very high data rate; therefore, such a high oversampling

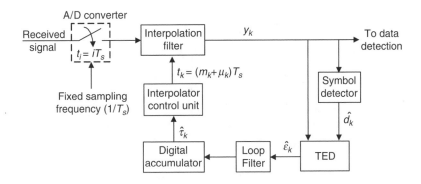

FIGURE 27.2 An interpolated timing recovery.

rate is impractical. This motivates us to look for other interpolation filters that perform well with only a small amount of oversampling (e.g., 5%). This small oversampling rate ensures that the sampling frequency is always above the Nyquist frequency so as to avoid aliasing. In next two sections, we will explain how to design such the interpolation filters based on an MMSE approach [5–8].

27.3.2 MMSE Interpolation Filter without Channel Knowledge

Consider the continuous-time bandlimited signal $x(t)$ with one-sided signal bandwidth $1/(2T)$. The sampling theorem [9] states that the signal $x(t)$ can be perfectly reconstructed from its samples, $\{x(kT_s)\}$, if it was sampled at rate at least twice its highest frequency component (i.e., if $T_s \leq T$) using the ideal low-pass filter of which the cutoff frequency is at $1/(2T_s)$. This ideal low-pass filter has a frequency response [7]

$$F_I(e^{j\omega T_s}, \mu T_s) = \frac{1}{T_s} \sum_{n=-\infty}^{\infty} F_I\left(\omega - \frac{2\pi}{T_s}n, \mu T_s\right) \tag{27.7}$$

where

$$F_I(\omega, \mu T_s) = \begin{cases} T_s \exp(j\omega T_s \mu) & \left|\frac{\omega}{2\pi}\right| < \frac{1}{2T_s} \\ 0, & \text{otherwise} \end{cases}$$

The task of the interpolation filter is to compute the signal value among samples $\{x(kT_s)\}$. For example, the signal value at time $(m + \mu)T_s$ can be obtained by

$$x_\mu(m) = x(t)|_{mT_s + \mu T_s} = \sum_{k=-\infty}^{\infty} x(kT_s)\frac{\sin((mT_s + \mu T_s - kT_s)\pi/T_s)}{(mT_s + \mu T_s - kT_s)\pi/T_s}. \tag{27.8}$$

Equation 27.8 can be viewed as passing the signal samples $\{x(kT_s)\}$ to the filter

$$f_\mu(m) = f(mT_s, \mu T_s) = \frac{\sin((mT_s + \mu T_s)\pi/T_s)}{(mT_s + \mu T_s)\pi/T_s} \quad \text{for } m = \cdots, -1, 0, 1, \ldots \tag{27.9}$$

where $f_\mu(m)$ is the coefficients of the ideal linear interpolation filter. Since this filter has an infinite number of taps, it is impossible to implement in real applications. To solve this problem, we approximate it by a finite-order FIR filter of length $N_1 + N_2 + 1$, that is,

$$\hat{F}(e^{j\omega T_s}, \mu T_s) = \sum_{m=-N_2}^{N_1} f_\mu(m)e^{-j\omega T_s m} \tag{27.10}$$

where N_1 and N_2 are integers. A rule-of-thumb for choosing N_1 and N_2 to obtain the best performance [10] is $N_1 = N - 1$ and $N_2 = N$, where N is an integer number.

The MMSE interpolation filter without channel knowledge can then be obtained by minimizing the quadratic error between the frequency response of the ideal interpolation filter and its approximation [7], that is,

$$e^2(\mu) = \int_{-2\pi B}^{2\pi B} \left| e^{j\omega T_s \mu} - \sum_{m=-N_2}^{N_1} f_\mu(m)e^{-j\omega T_s m} \right|^2 d\omega, \tag{27.11}$$

TABLE 27.1 Coefficients of the 8-tap MMSE Interpolation Filter without Channel Knowledge for some μ's with 5% Oversampling Rate (i.e., $B = 1/(2.1T_s)$)

m/μ	0.1	0.2	0.3	0.4	0.5
-4	-0.0232	-0.0450	-0.0632	-0.0759	-0.0817
-3	0.0327	0.0641	0.0911	0.1107	0.1206
-2	-0.0514	-0.1028	-0.1493	-0.1860	-0.2082
-1	0.1097	0.2343	0.3679	0.5039	0.6353
0	0.9823	0.9335	0.8562	0.7549	0.6353
1	-0.0873	-0.1523	-0.1937	-0.2117	-0.2082
2	0.0439	0.0799	0.1054	0.1191	0.1206
3	-0.0280	-0.0518	-0.0694	-0.0796	-0.0817

where B is one-sided signal bandwidth. It is apparent from Equation 27.11 that

$$f_{\mu=0}(m) = \begin{cases} 1 & m = 0 \\ 0, & \text{otherwise} \end{cases} \qquad (27.12)$$

and

$$f_{\mu=1}(m) = \begin{cases} 1 & m = -1 \\ 0, & \text{otherwise} \end{cases} \qquad (27.13)$$

The constraints in Equation 27.12 and Equation 27.13 ensure that the interpolation filter works properly when $\mu = 0$ and $\mu = 1$. By differentiating Equation 27.11 with respect to $f_\mu(i)$ for $i = -N_2, \ldots, N_1$, and setting the result to zero, one obtains, after some manipulations,

$$\sum_{m=-N_2}^{N_1} f_\mu(m) \frac{\sin(2\pi B T_s(m - i))}{m - i} = \frac{\sin(2\pi B T_s(\mu + i))}{\mu + i} \quad \text{for } i = -N_2, \ldots, N_1 \qquad (27.14)$$

This system of linear equation has $2N$ equations and $2N$ unknowns. Therefore, it is straightforward to solve Equation 27.14 for the coefficients $f_\mu(m)$. It is clear that the resulting interpolation filter is only a function of the sampling phase μ, not the channel or the target response [11]. Note that the coefficient values for $\mu > 0.5$ can be obtained from those for $\mu < 0.5$ according to

$$f_{1-\mu}(m) = f_\mu(-m - 1) \qquad (27.15)$$

Table 27.1 shows the coefficients of the 8-tap MMSE interpolation filter without channel knowledge for some values of μ's when 5% oversampling rate is employed.

27.3.3 MMSE Interpolation Filter with Channel Knowledge

The MMSE interpolation filter obtained in Section 27.3.2 does not exploit the channel knowledge. We now consider another MMSE interpolation filter that exploits the information about the channel impulse response. To do so, we consider a more realistic system model shown in Figure 27.3, where the equalizer output can be written as

$$I(r T_s) = \sum_{k=-\infty}^{\infty} a_k h(r T_s - kT) + w(r T_s) \qquad (27.16)$$

where $a_k \in \{\pm 1\}$ is a binary input sequence with bit period T, $h(t)$ is the target response [11], and $w(t)$ is colored noise because of misequalization with known power spectral density $S_w(\omega)$.

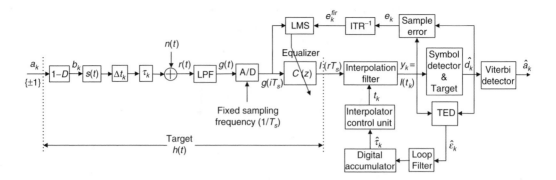

FIGURE 27.3   System model with an ITR.

The receiver needs to obtain the synchronized samples, $I(t_k)$, at the output of the interpolation filter. Unlike Equation 27.4, we first assume that $\hat{\tau}_k = 0$ for all k, and estimate the sampling location index $t_k = kT$ as $(m_k + \mu_k)T_s$, where m_k and μ_k are given in Equation 27.5 and Equation 27.6, respectively. It has been shown in [10] that a finite-order FIR filter is generally preferred to implement the interpolation filter than an IIR filter because of stability. Furthermore, for good performance the coefficients of the interpolation filter should be a function of the sampling phase μ_k. Let $f_{\mu_k}(l)$, for $l = -N_2, \ldots, N_1$, denote the coefficients of the interpolation filter at the sampling phase μ_k. Then, the output of the interpolation filter can be written as

$$I(kT) = I((m_k + \mu_k)T_s) = \sum_{l=-N_2}^{N_1} f_{\mu_k}(l)I((m_k - l)T_s)$$

$$= \sum_{l=-N_2}^{N_1} f_{\mu_k}(l)I(kT - \mu_k T_s - l T_s) \tag{27.17}$$

One way to obtain the coefficients of the interpolation filter is to maximize the signal-to-sampling-noise ratio (SSNR) [5], which is given by

$$\text{SSNR} = \frac{E[\tilde{I}(kT)^2]}{E[(\tilde{I}(kT) - I(kT))^2]} \tag{27.18}$$

where $\tilde{I}(kT)$ denote the noiseless desired (or ideal) samples at the output of the equalizer. Note that Equation 27.18 can also be used as a measure to compare the performance of different interpolation filters. Maximizing Equation 27.18 is equivalent to minimizing its denominator term, which is referred to as the mean-squared error (MSE) between the ideal and the interpolated samples. Therefore, the MMSE interpolation filter with channel knowledge [5, 6] is obtained by minimizing

$$\text{MSE} = E[(\tilde{I}(kT) - I(kT))^2]$$

$$= E\left[\left(\sum_{m=-\infty}^{\infty} a_m h(kT - mT)\right.\right.$$

$$\left.\left. - \sum_{l=-N_2}^{N_1} f_{\mu_k}(l)\left\{\sum_{j=-\infty}^{\infty} a_j h(kT - \mu_k T_s - l T_s - jT) + w(kT - \mu_k T_s - l T_s)\right\}\right)^2\right] \tag{27.19}$$

Let us define some notation as follows:

$$\mathbf{A} = \begin{bmatrix} a_{-\infty} \\ \vdots \\ a_{\infty} \end{bmatrix} \tag{27.20}$$

$$\mathbf{H}_1 = \begin{bmatrix} h(kT - (-\infty)T) \\ \vdots \\ h(kT - \infty T) \end{bmatrix} \tag{27.21}$$

$$\mathbf{H}_2 = \begin{bmatrix} h(kT - \mu_k T_s + N_2 T_s - (-\infty)T) & \cdots & h(kT - \mu_k T_s - N_1 T_s - (-\infty)T) \\ \vdots & \ddots & \vdots \\ h(kT - \mu_k T_s + N_2 T_s - \infty T) & \cdots & h(kT - \mu_k T_s - N_1 T_s - \infty T) \end{bmatrix} \tag{27.22}$$

$$\mathbf{W} = \begin{bmatrix} w(kT - \mu_k T_s + N_2 T_s) \\ \vdots \\ w(kT - \mu_k T_s - N_1 T_s) \end{bmatrix} \tag{27.23}$$

and

$$\mathbf{F}_{\mu_k} = \begin{bmatrix} f_{\mu_k}(-N_2) \\ \vdots \\ f_{\mu_k}(N_1) \end{bmatrix} \tag{27.24}$$

Assuming that the input data sequence and the colored noise sequence are uncorrelated, Equation 27.19 can then be written in a matrix form as [5]

$$\begin{aligned} \mathrm{MSE} &= E\left[(\mathbf{A}^\mathrm{T}\mathbf{H}_1 - \mathbf{A}^\mathrm{T}\mathbf{H}_2\mathbf{F}_{\mu_k} - \mathbf{W}^\mathrm{T}\mathbf{F}_{\mu_k})^2\right] \\ &= \mathbf{H}_1^\mathrm{T}\mathbf{R}_{\mathrm{AA}}\mathbf{H}_1 + \mathbf{F}_{\mu_k}^\mathrm{T}\mathbf{H}_2^\mathrm{T}\mathbf{R}_{\mathrm{AA}}\mathbf{H}_2\mathbf{F}_{\mu_k} - 2\mathbf{H}_1^\mathrm{T}\mathbf{R}_{\mathrm{AA}}\mathbf{H}_2\mathbf{F}_{\mu_k} + \mathbf{F}_{\mu_k}^\mathrm{T}\mathbf{R}_{\mathrm{WW}}\mathbf{F}_{\mu_k} \end{aligned} \tag{27.25}$$

where $[\cdot]^\mathrm{T}$ denotes a transpose matrix, and $\mathbf{R}_{\mathrm{AA}}$ and $\mathbf{R}_{\mathrm{WW}}$ are the autocorrelation matrices of the input data and the colored noise sequences, respectively. Apparently, the matrices $\mathbf{H}_1$ and $\mathbf{H}_2$ consist of the synchronized and the asynchronous samples of the target response, respectively. Since the target response is designed to match as close to the channel impulse response as possible, one can say that the resulting interpolation filter exploits the channel knowledge.

By taking the derivative of Equation 27.25 with respect to $\mathbf{F}_{\mu_k}$ and setting the result to zero, the interpolation filter is found to be [5]

$$\mathbf{F}_{\mu_k} = \left(\mathbf{H}_2^\mathrm{T}\mathbf{R}_{\mathrm{AA}}\mathbf{H}_2 + \mathbf{R}_{\mathrm{WW}}\right)^{-1}\mathbf{H}_2^\mathrm{T}\mathbf{R}_{\mathrm{AA}}\mathbf{H}_1 \tag{27.26}$$

Note that the constraints in Equation 27.12 and Equation 27.13 must also be imposed in Equation 27.26.

27.4 Implementation Issues

There are some implementation issues worth considering when employing the ITR architecture in magnetic recording channels, which can be summarized as follows.

27.4.1 The Estimate of the Initial Sampling Phase

The initial sampling phase (or the initial fractional interval), μ_0, can be estimated by the process called *digital zero phase start* (DZPS) [1, 5] with the aid of the preamble. The aim of DZPS is to help the timing recovery achieve fast synchronization.

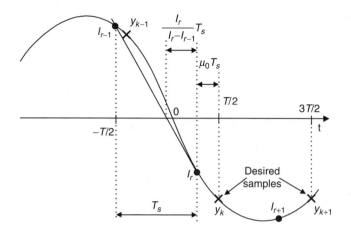

FIGURE 27.4   Digital zero phase start for odd-number target length.

Consider a PR-IV target (i.e., $H(D) = 1 - D^2$) and let the preamble take the form of $\{+1 +1 -1 -1 +1 +1 -1 -1 \ldots\}$, which makes the signal $I(t)$ look like a sinusoidal waveform with a period of $4T$. Denote $I_r = I(rT_s)$ as the rth sample (in T_s-domain) at the input of the interpolation filter. The DZPS uses two adjacent samples of which the current sample, I_r, is negative-valued and the previous sample, I_{r-1}, is positive-valued, to estimate μ_0 using a linear interpolation. If the target length is an odd number (e.g., a PR-IV target), we know that the desired samples must be taken at time $t_n = nT/2$ where n is an odd number. The estimated μ_0, $\hat{\mu}_0$, is obtained by (see Figure 27.4)

$$\hat{\mu}_0 \approx \frac{0.5T}{T_s} - \frac{I_r}{I_r - I_{r-1}} \tag{27.27}$$

However, if the target length is an even number, the desired samples are then located at time $t_n = nT/2$, where n is an even number. The $\hat{\mu}_0$ is thus given by (see Figure 27.5)

$$\hat{\mu}_0 \approx \frac{T}{T_s} - \frac{I_r}{I_r - I_{r-1}} \tag{27.28}$$

In the real system, more than one pair of samples should be used to reduce the effects of the DC offset and the noise on the estimation [7, 8].

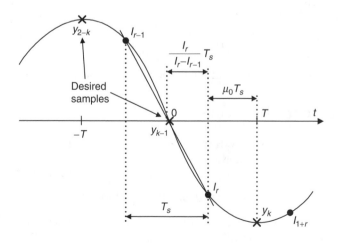

FIGURE 27.5   Digital zero phase start for even-number target length.

27.4.2 Implementation of the Interpolation Filter

To implement the MMSE interpolation filter with channel knowledge, the target response in the continuous-time domain, $h(t)$, is needed to construct the matrices $\mathbf{H}_1$ and $\mathbf{H}_2$. Although only the discrete-time target response is obtained from the target design [11], its continuous-time version can be easily obtained by passing it through the ideal low-pass filter. Additionally, we also approximate the values $\{kT - (-\infty)T\}$ and $\{kT - \infty T\}$ in Equation 27.21 and Equation 27.22 equal to $\{L\,T\}$ and $\{-L\,T\}$, respectively, where L is a large integer number such that the amplitude of $h(kT)$ for $|k| > L$ is insignificant. With this approximation, the matrices $\mathbf{H}_1$ and $\mathbf{H}_2$ can therefore be obtained.

Note that the sampling phase μ is normally time-varying and takes on values over the range $[0, 1)$. Practically, the interpolation filter should not recompute the coefficients of $\mathbf{F}_\mu$ for every clock cycle. Thus, we quantize the sampling phase μ into M values in the range $[0, 1)$. For each quantized phase, storing precalculated coefficients in a look-up table indexed by μ will simplify the implementation. At least five most significant bits [8] are sufficient to determine the interpolation coefficients.

27.4.3 Implementation of the Equalizer

As shown in Figure 27.3, the equalizer operating in T_s-domain is needed (i.e., a T_s-spaced equalizer is required). To obtain such a T_s-spaced equalizer, the following steps are performed. We first use the target design technique in [11] to simultaneously obtain the target and its corresponding equalizer (both in T-domain). Then, we get the desired samples in T_s-domain by interpolating the target output samples in T-domain to those in T_s-domain using a simple interpolation algorithm. Based on the T_s-spaced desired samples and the T_s-spaced equalizer input, the T_s-spaced equalizer can be obtained by using an adaptive algorithm, such as an LMS algorithm [1].

27.4.4 Effect of Employing the Oversampled System

In magnetic recording channels, spin speed variation can lead to undersampling, which results in a fewer clock cycles for processing than bits to be detected. This problem can be solved by slightly oversampling the received signal to ensure that there are always more clock cycles than bits to be detected. However, with 5% oversampling rate, for example, one of every 20 clock cycles (on average) is not needed. To retain synchronization, ITR must stop all processes during the unneeded clock cycles. Additionally, it has been shown in [8] that it is sufficient to apply 3 to 5% oversampling rate in the ITR architecture because beyond 5% an additional performance gain is insignificant.

27.4.5 FIR-ITR Loop Structure

In the conventional partial response maximum-likelihood (PRML) system, the equalizer comes after the sampler, which introduces a large delay in the timing loop. However, the ITR architecture presented here places the equalizer before the sampler. This structure is denoted as the "FIR-ITR" loop structure [5], which has the advantage of reducing the amount of the delay in the timing loop and shaping the overall channel impulse response to the target.

If the adaptive equalizer is used to adjust the equalizer coefficients at each time instant, the inverse interpolation (denoted as ITR^{-1} [5]) block must be employed in order to compute the error sample in T_s-domain, e_k^{fir}, used in an adaptive algorithm (e.g., an LMS algorithm). This is because only the error sample in T-domain, e_k, is generated after the interpolation filter because of the fact that the decisions can only be made on the synchronous samples. Therefore, the detector must perform the inverse interpolation in order to compute the e_k^{fir}.

This ITR^{-1} is actually another interpolation problem where the detector now interpolates among the T-spaced error samples to get the T_s-spaced error samples. Since the detector does not need a very accurate error estimation to run the adaptive algorithm, we can use the linear interpolation for ITR^{-1}. As shown

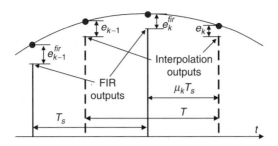

FIGURE 27.6   Inverse ITR.

in Figure 27.6, the error sample in T_s-domain can be expressed as

$$e_k^{fir} \approx \left(1 - \mu_k \frac{T_s}{T}\right) \cdot e_k + \mu_k \frac{T_s}{T} \cdot e_{k-1} \tag{27.29}$$

27.5   Applications of ITR to Magnetic Recording

In this section, we first describe the system model for magnetic recording channels. Then, the performance of different interpolation filters is compared in terms of SSNR. Finally, we compare the BER performance of ITR and conventional timing recovery in both longitudinal and perpendicular systems.

27.5.1   System Model

Referring to Figure 27.3, a binary input sequence $a_k \in \{\pm 1\}$ with bit period T is filtered by an ideal differentiator $(1 - D)$ to form a transition sequence $b_k \in \{-2, 0, 2\}$, where $b_k = \{\pm 2\}$ corresponds to a positive or a negative transition and $b_k = 0$ corresponds to the absence of a transition. The transition response, $s(t)$, for longitudinal recording is $s(t) = 1/(1 + (2t/PW_{50})^2)$ [1], where PW_{50} determines the width of $s(t)$ at half of its peak value, whereas that for perpendicular recording is $s(t) = \text{erf}(2t\sqrt{2}/PW_{50})$ [12], where $\text{erf}(\cdot)$ is an error function and PW_{50} indicates the width of the derivative of $s(t)$ at half its maximum. In the context of magnetic recording, a normalized recording density is defined as $\text{ND} = PW_{50}/T$, which determines how many data bits can be packed within the resolution unit PW_{50}. The media jitter noise, Δt_k, is modeled as a random shift in the *transition position* with a Gaussian probability distribution function with zero mean and variance $|b_k/2|\sigma_j^2$ (i.e., $\Delta t_k \sim \mathcal{N}(0, |b_k/2|\sigma_j^2)$) truncated to $T/2$, where $|x|$ takes the absolute value of x. The clock jitter noise, τ_k, is modeled as a random walk according to $\tau_{k+1} = \tau_k + \mathcal{N}(0, \sigma_w^2)$. A random walk model was chosen because of its simplicity to represent a variety of channels by changing only one parameter.

The readback signal, $r(t)$, can therefore be written as [13]

$$r(t) = \sum_{k=-\infty}^{\infty} a_k \{g(t - kT - \Delta t_k - \tau_k) - g(t - (k+1)T - \Delta t_{k+1} - \tau_k)\} + n(t) \tag{27.30}$$

where $n(t)$ is additive white Gaussian noise (AWGN) with two-sided power spectral density $N_0/2$. The readback signal $r(t)$ is filtered by a seventh-order Butterworth low-pass filter and sampled at time $t_i = iT_s$ to obtain a T_s-spaced sequence, $g(iT_s)$. The sequence $g(iT_s)$ is then equalized by a T_s-spaced equalizer, $C(z)$. The interpolation filter uses a set of the asynchronous samples, $I(rT_s)$, and the sampling location index, t_k, to output the synchronized sample, y_k. This sample y_k is used in the timing update operation and is decoded by a Viterbi detector [14] with a decision delay of $60T$. Note that the symbol detector used in the timing loop is also the Viterbi detector with a decision delay of $4T$.

TABLE 27.2 Coefficients of the 8-tap MMSE Interpolation Filter with Channel Knowledge (of the Perfectly Equalized PR-IV channel) for some μ's with 5% Oversampling Rate

m/μ	0.1	0.2	0.3	0.4	0.5
−4	−0.0137	−0.0261	−0.0367	−0.0442	−0.0480
−3	0.0212	0.0413	0.0591	0.0726	0.0795
−2	−0.0494	−0.0974	−0.1415	−0.1762	−0.1968
−1	0.1066	0.2268	0.3574	0.4908	0.6199
0	0.9756	0.9229	0.8418	0.7387	0.6195
1	−0.0823	−0.1439	−0.1821	−0.1988	−0.1963
2	0.0297	0.0535	0.0701	0.0786	0.0795
3	−0.0164	−0.0301	−0.0404	−0.0460	−0.0478

27.5.2 SSNR Performance

To compare the sampling noise resulted from different interpolation filters, we assume that the channel is perfectly equalized to the PR-IV target, $\hat{t}_k = 0$ for all k, and there is no noise because of misequalization. This experiment will determine how close the output of each interpolation filter is when compared with the ideal sample. The coefficients of the 8-tap MMSE with channel knowledge for this system condition are listed in Table 27.2 for some μ's.

Figure 27.7(a) compares the SSNR performance of different interpolation filters as a function of the sampling phase μ, where a 4-tap cubic interpolation filter is obtained from [15] and an 8-tap spline interpolation filter is obtained from MATLAB. Clearly, both MMSE interpolation filters perform better than the spline and the cubic interpolation filters, and the MMSE interpolation filter with channel knowledge yields better performance than that without channel knowledge.

We also plot the overall SSNR of different interpolation filters as a function of the filter lengths in Figure 27.7(b). Again, the MMSE interpolation filter with channel knowledge provides a better performance than that without channel knowledge for all filter lengths, and both always perform better than the spline interpolation filter. Note that the overall SSNR of a 4-tap cubic interpolation filter is approximately 21.1 dB, which is smaller than that of the 4-tap MMSE interpolation filters (see Figure 27.7(b)). As the filter length increases, the SSNR also increases until a certain filter length because beyond that an additional performance improvement gain is insignificant. Note that the larger the number of filter taps, the higher the implementation cost. Then, one must trade-off between the performance gain and the implementation cost when choosing the number of filter taps.

27.5.3 BER Performance

We consider the system model in Figure 27.3 in moderate condition with ND = 2.5 for both longitudinal and perpendicular recording channels, 5% oversampling rate, $\sigma_j/T = 3\%$ media jitter noise, $\sigma_w/T = 0.5\%$ clock jitter noise, and 0.4% frequency offset. The SNR is defined as

$$\text{SNR} = 10 \log_{10}\left(\frac{E_i}{N_0}\right) \quad (\text{dB}) \tag{27.31}$$

where E_i is the energy of the normalized channel impulse response assumed to be one for convenience. The 5-tap target with the monic constraint [11] and a 21-tap equalizer were designed at the SNR required to achieve BER = 10^{-5}. One data packet consists of 256-bit preamble (4T pattern) and a 4096-bit input data sequence. Each BER point was computed using as many data packets as needed to collect at least 2000 error bits. Eventually, the PLL gain parameters, α and β, were designed to recover phase and frequency changes within 256 samples as the preamble length is 256 bits. We consider the case where the same PLL gain parameters are employed during acquisition and tracking modes.

The ITR architecture using different interpolation filters is compared with conventional timing recovery, where the T-spaced equalizer is used and placed after the sampler (i.e., inside the timing loop). Therefore,

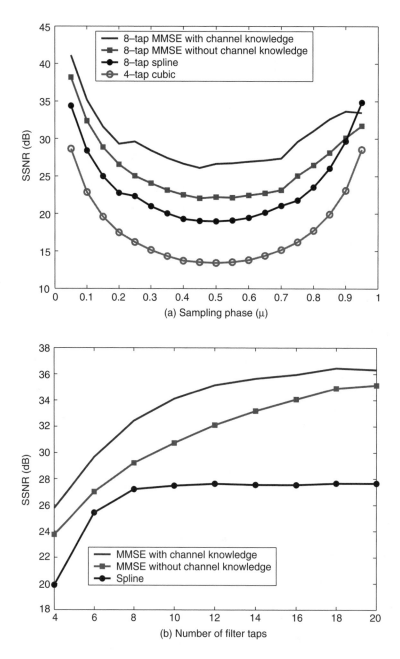

FIGURE 27.7   (a) SSNR performance of different interpolation filters as a function of μs, and (b) the overall SSNR of different interpolation filters as a function of filter lengths for the perfectly equalized PR-IV channel.

it is obvious to show that the total loop delay of ITR is of $4T$, while that of conventional timing recovery is of $14T$ (as the symbol detector and the equalizer introduce delays of $4T$ and $10T$, respectively).

27.5.3.1 Longitudinal Recording

A 5-tap target for longitudinal recording is $H(D) = 1 + 0.613D - 0.478D^2 - 0.626D^3 - 0.291D^4$ [13]. The PLL gain parameters for ITR and conventional timing recovery are shown in Table 27.3.

First, we look at the BER performance of different MMSE interpolation filters as shown in Figure 27.8(a). The MMSE interpolation filter with channel knowledge performs slightly better than that without channel

TABLE 27.3 PLL Gain Parameters for Different System Conditions

Timing Recovery	Longitudinal Recording		Perpendicular Recording	
Architectures	α	β	α	β
ITR ($D = 4T$)	0.0029	3.6e-5	0.0036	4.5e-5
Conventional timing recovery ($D = 14T$)	0.0027	3.1e-5	0.0033	3.8e-5

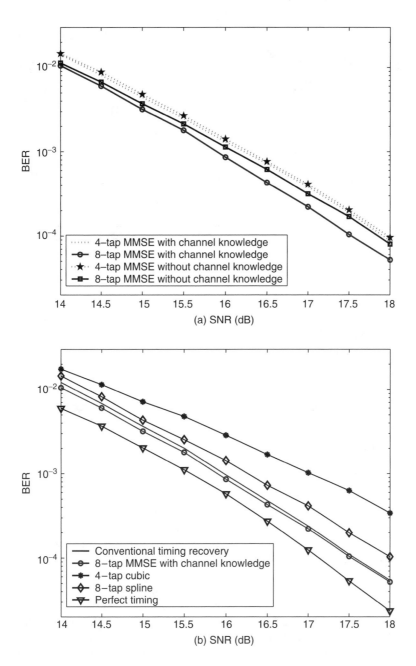

FIGURE 27.8 (a) BER performance of different MMSE interpolation filters, and (b) BER performance of different interpolation filters for longitudinal recording.

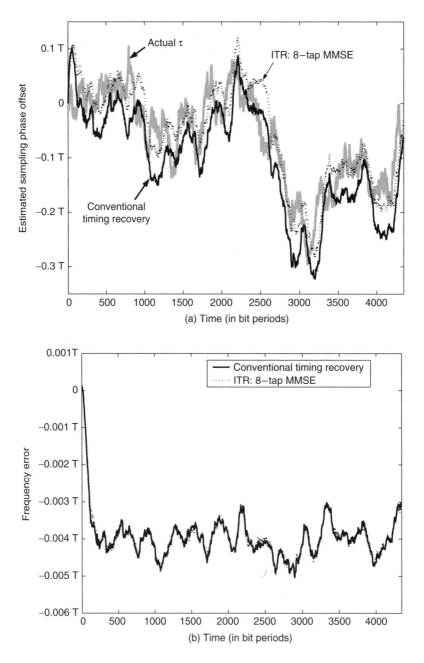

FIGURE 27.9   (a) The sampling phase offset, $\hat{\tau}_k$, and (b) the frequency error, $\hat{\theta}_k$, performances during acquisition and tracking modes in longitudinal recording at SNR = 17 dB.

knowledge. We also compare the performance of different timing recovery schemes in Figure 27.8(b), where the curve labeled "Perfect timing" means the conventional timing recovery uses $\hat{t}_k = \tau_k$ to sample the received signal and the system has no frequency offset. The 8-tap MMSE interpolation filters yields better performance than other interpolation filters and performs as good as the conventional timing recovery, which is approximately 0.45 dB (at BER $= 10^{-4}$) away from the system with perfect timing. We also observed that an oversampling rate beyond 5% gives an insignificant performance improvement (not shown).

Finally, we plot the timing performance of ITR and conventional timing recovery at SNR $= 17$ dB in Figures 27.9(a) and Figures 27.9(b). These two plots indicate that the estimated sampling phase offset $\hat{t}_k$

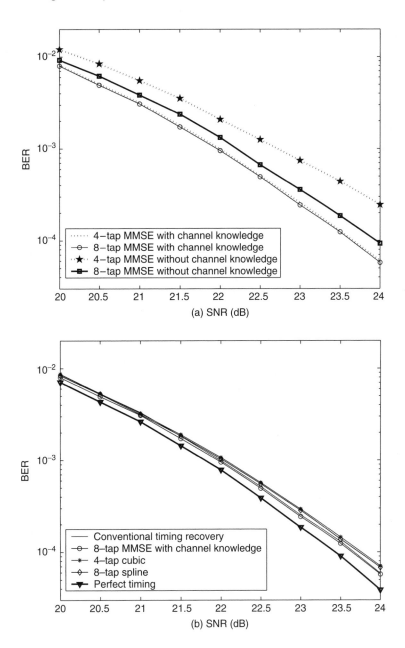

FIGURE 27.10 (a) BER performance of different MMSE interpolation filters, and (b) BER performance of different interpolation filters for perpendicular recording.

of both systems coincides with the actual clock jitter noise, τ_k, and both systems can keep track of the 0.4% frequency offset in the system. Therefore, it can be implied that ITR and conventional timing recovery work properly.

27.5.3.2 Perpendicular Recording

A 5-tap target for perpendicular recording is $H(D) = 1 + 1.429D + 1.097D^2 + 0.465D^3 + 0.099D^4$ [13]. The PLL gain parameters for ITR and conventional timing recovery are also listed in Table 27.3.

First, we look at the BER performance of different MMSE interpolation filters as shown in Figure 27.10(a). Clearly, the MMSE interpolation filter with channel knowledge performs much better than that without channel knowledge, especially when the filter length is small. Also, the 4-tap and 8-tap MMSE interpolation filters with channel knowledge perform alike. Unlike longitudinal recording, the MMSE interpolation filter without channel knowledge does not work well for this channel, especially when the filter length is small. We also compare the performance of different timing recovery schemes in Figure 27.10(b). Similarly, the 8-tap MMSE interpolation filter and the conventional timing recovery perform slightly better than other interpolation filters, and both are about 0.25 dB (at BER = 10^{-4}) away from the system with perfect timing. Like longitudinal recording, we observed that 5% oversampling rate is sufficient to employ for this channel.

27.6   Conclusion

We described the design of timing recovery systems in a fully digital sense, using a technique known as interpolated timing recovery (ITR). The design of the two MMSE interpolation filters with and without channel knowledge was also given.

We observed that the ITR based on the MMSE interpolation filter with channel knowledge performs as good as the conventional timing recovery in both longitudinal and perpendicular recording channels.

References

[1] Bergmans, J.W.M., *Digital Baseband Transmission and Recording*, Kluwer Academic Publishers, Boston, 1996.

[2] Mueller, K.H. and Müller, M., Timing recovery in digital synchronous data receivers, *IEEE Trans. Commn.*, vol. 24, no. 5, pp. 516–531, May 1976.

[3] Shafiee, H., Timing recovery for sampling detectors in digital magnetic recording, *IEEE International Conference on Communications*, vol. 1, pp. 577–581, 1996.

[4] Gardner, F.M., Interpolation in digital modems — Part I: fundamentals, *IEEE Trans. Commn.*, vol. 41, no. 3, pp. 501–507, March 1993.

[5] Wu, Z., Cioffi, J.M., and Fisher, K.D., A MMSE interpolated timing recovery scheme for the magnetic recording channel, *IEEE International Conference on Communications*, vol. 3, pp. 1625–1629, 1997.

[6] Kim, D., Narasimha, M.J., and Cox, D.C., Design of optimal interpolation filter for symbol timing recovery, *IEEE Trans. Commn.*, vol. 45, no. 7, pp. 877–884, July 1997.

[7] Meyr, H., Moeneclaey, M., and Fechtel, S.A., *Digital Communication Receivers: Synchronization, Channel Estimation, and Signal Processing*, John Wiley & Sons, New York, 1998.

[8] Spurbeck, M. and Behrens, R.T., Interpolated timing recovery for hard disk drive read channels, *IEEE International Conference on Communications*, vol. 3, pp. 1618–1624, 1997.

[9] Oppenheim, A.V. and Schafer, R.W., *Discrete-Time Signal Processing*, Prentice Hall, New Jersey, 1989.

[10] Schafer, R.W. and Rabiner, L.R., A digital signal processing approach to interpolation, *Proceedings of the IEEE*, vol. 61, no. 6, pp. 692–702, June 1973.

[11] Moon, J. and Zeng, W., Equalization for maximum likelihood detector, *IEEE Trans. Magn.*, vol. 31, no. 2, pp. 1083–1088, March 1995.

[12] Roscamp, T.A., Boerner, E.D., and Parker, G.J., Three-dimensional modeling of perpendicular recording with soft underlayer, *J. Appl. Phys.*, vol. 91, no. 10, May 2002.

[13] Kovintavewat, P. et al., A new timing recovery architecture for fast convergence, *IEEE International Symposium on Circuits and Systems*, vol. 2, pp. 13–16, May 2003.

[14] Forney, G.D., Maximum-likelihood sequence estimation of digital sequences in the presence of intersymbol interference, *IEEE Trans. Info. Theory*, vol. IT-18, no. 3, pp. 363–378, May 1972.

[15] Erup, L., Gardner, F.M., and Harris, R.A., Interpolation in digital modems — Part II: Implementation and performance, *IEEE Trans. Commn.*, vol. 41, no. 6, pp. 998–1008, June 1993.

28

Adaptive Equalization Architectures for Partial Response Channels

28.1 Introduction **28**-1
28.2 Equalization Architectures and Strategies **28**-2
28.3 CTF Configurations **28**-3
28.4 FIR Filter and The LMS Algorithm................. **28**-5
28.5 Gain Equalization and Other Issues **28**-5
28.6 Performance Characterization **28**-6
    Simulation Environment and Optimization Procedure
    • Results
28.7 Actual Equalizer Architectures **28**-11
28.8 Conclusions **28**-12

Pervez M. Aziz

Agere Systems
Dallas, TX

Adaptive signal processing plays a crucial role in storage systems. Proper detection of the readback signal by a Viterbi detector assumes that the signal has the right gain, is equalized to the partial response, and is sampled at the proper sampling instances. In this chapter we will focus on equalization. We review some of the basic algorithms employed in equalization. We will present various architectures and algorithms which have been used in state of the art read channels. Finally, we will present comparative performance data for some of these architectures.

28.1 Introduction

What is equalization? It is the act of shaping the read back magnetic recording signal to look like a target signal specified by the partial response (PR). The equalized signal is made to look like the target signal in both the time and frequency domain. In this section we first review various equalization architectures and strategies which have been popular historically and still being used in present day read channels. We then provide a quick review of the well-known least mean square (LMS) algorithm used for adaptive equalizers. Finally, we explore the performance implications of selecting several different equalizer architectures. This performance is measured in terms of bit error rate (BER) at the output of the read channel's Viterbi detector.

28.2 Equalization Architectures and Strategies

In PRML channels the read back signal will be sampled at some point in the data path for further digital signal processing. A continuous time filter (CTF) with a low-pass characteristic will be present as an anti-aliasing filter [1] prior to the sampling operation so that high frequency noise is not aliased into the signal band. This same CTF may also play a role in equalizing the read back signal to the target partial response. Various architectures can be used to perform the required equalization. The equalizer architecture can consist of a CTF, a finite impulse response filter (FIR) or both. The CTF parameters may be fixed, programmable, or adaptive. The FIR filter coefficients may be fixed, programmable or adaptive. In addition, the FIR operation may occur in the sampled data analog domain or digital domain. Following equalization the data is detected using a Viterbi detector. Of course, quantization by an analog to digital converter (ADC) occurs at some point before the Viterbi detector.

Figure 28.1 shows some examples of various equalizer architecture configurations. The first architecture (Type 1) consists of a CTF only equalizer. The CTF is comprised of an all-pole low-pass filter section whose purpose is to reject high frequency noise for anti-aliasing. One key parameter in the CTF is its low-pass bandwidth determined by its cutoff or corner frequency, f_c. The type of CTF, f_c, and its order (or the number of poles it contains) will determine its low-pass rolloff characteristic. If the CTF is expected to take part in equalization it must also be able to provide some boost and does so by typically having one or two real zeros at some frequency f_z in its transfer function. These parameters are noted in the figure.

The second architecture (Type 2) is one where both the CTF and an analog FIR are involved in performing equalization. The third architecture (Type 3) is an analog FIR only architecture in that the CTF design does not consist of any zeros that is,t its main role is to perform anti-aliasing and not provide any boost for equalization. Finally, the last architecture (Type 4) is one where a CTF and FIR are both involved in equalization except that the FIR operation is done digitally.

In general there is a clear tradeoff between the degree of flexibility of the equalizer and implementation complexity. The read back signal characteristics change across the disk surface as manifested by somewhat different channel bit densities (*cbd*s) or $\frac{pw_{50}}{T}$. Consequently, programmability of the equalizer parameters is a mimimum requirement for optimum performance. The signal or some of the equalizer parameters themselve may change with chip ageing and temperature variations [2]. Therefore, it is often desirable for some of the equalizer parameters to be continually adaptive to be able to compensate for these effects.

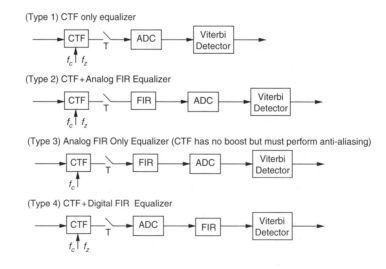

FIGURE 28.1   Various equalizer architectures.

28.3 CTF Configurations

Two common types of CTFs which have been used in read channels are Butterworth filters and equiripple linear phase filters. Butterworth filters have a maximally flat magnitude response but nonlinear phase response. Equiripple linear phase filters as their name implies have linear phase and constant group delay over the passband [3, 4]. For a given order, the Butterworth filters will have sharper rolloff characteristics. One could also consider *mixed* filters whose poles are chosen to lie some percentage of the distance in between the poles of a Butterworth and equiripple linear phase filter. Figure 28.2 shows the normalized pole location on the s plane for a 6th-order Butterworth, a 6th-order equiripple linear phase filter, as well as the poles for various 6th-order mixed filters which are 25, 50, 75, and 90% away from the poles of the equiripple filter. Note that the Butterworth poles lie on the unit circle. Figure 28.3 shows the corresponding magnitude responses for the filters while Figure 28.4 shows the group delay responses. As can be observed, the Butterworth has the sharpest low-pass rolloff and the equiripple filter has the shallowest rolloff but constant group delay over the passband.

The CTF parameters can be programmable or adaptive [5, 6]. However, most CTFs which have been used in actual read channels have had programmable bandwidth and boosts — any adaptivity being left to the FIR. Adaptive CTF systems face some challenging issues as discussed in [7]. We also mention that there has been some work to analytically determine the optimum CTF transfer functions [8, 9].

We will compare the performance of equalizers involving several CTF configurations: 4th-order Butterworth *(b4)*, 6th-order Butterworth *(b6)*, 7th-order equiripple linear phase *(e7)* all with single zeros. We also examine a 7th-order equiripple linear phase CTF with two zeros *(e7tz)*. The linear phase of the all pole section is kept in the *e7tz* filter. Another filter considered is the 4th-order 50% mixed filter with one zero *(em4)*.

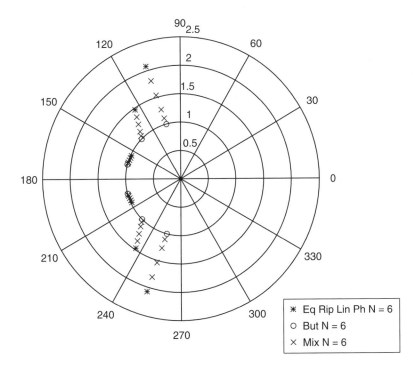

FIGURE 28.2 Normalized pole locations for 6th order Butterworth and equiripple linear phase filters as well as mixed filters.

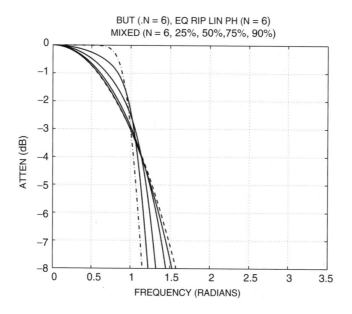

FIGURE 28.3   Magnitude respone for 6th order Butterworth and equiripple linear phase filters as well as mixed filters.

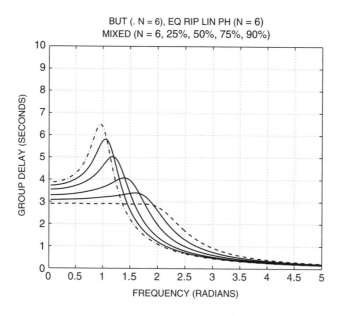

FIGURE 28.4   Group delay respone for 6th order Butterworth and equiripple linear phase filters as well as mixed filters.

28.4 FIR Filter and The LMS Algorithm

We now focus on the FIR filter which is important for equalization. Whether implemented as an analog sampled data filter or a digital filter the FIR filter produces output samples $y(n)$ in terms of input samples $x(n)$ as

$$y(n) = \sum_{k=0}^{L-1} h(k)x(n-k) \tag{28.1}$$

where $h(k)$ are the FIR filter tap weights. As noted, it is very desirable for the FIR to be adaptive. The FIR taps are adapted based on the well-known LMS algorithm [10, 11]. Other adaptive algorithms can also be found in [12, 13].

The basic idea is to minimize the mean squared error with respect to some desired or ideal signal. Let the desired or ideal signal be $\hat{y}(n)$ in which case the error is $e(n) = y(n) - \hat{y}(n)$. This minimization is achieved by adjusting the tap value in the direction opposite to the derivative (with respect to the tap values) of the expected value of the mean squared error. Dispensing with the expected value leads to the LMS or stochastic gradient algorithm. The stochastic gradient for the kth tap weight is

$$\Delta(k,n) = -\frac{\partial}{\partial h(k)}(e^2(n)) = -2e(n)\frac{\partial\hat{e}(n)}{\partial h(k)} = -2e(n)\left[\frac{\partial y(n)}{\partial h(k)} - \hat{y}(k)\frac{\partial\hat{y}(n)}{\partial h(k)}\right]$$

$$= -2e(n)\frac{\partial y(n)}{\partial h(k)} \tag{28.2}$$

where the partial derivative of $\hat{y}(n)$ with respect to $h(k)$ is zero. We can now expand $y(n)$ as in Equation 28.1 to further obtain,

$$\Delta(k,n) = -2e(n)x(n-k) \tag{28.3}$$

The gradient would actually be scaled by some tap weight update gain t_{ug} to give the following tap update equation:

$$h(k, n+1) = h(k, n) - 2t_{ug}e(n)x(n-k) \tag{28.4}$$

The choice of this update gain depends on several factors: (a) it should not be too large so as to cause the tap adaptation loop to become unstable; (b) it should be large enough that the taps converge within a reasonable amount of time; (c) it should be small enough that after convergence the adaptation noise is small and does not degrade the bit error rate performance. In practice, during drive optimization in the factory the adaptation could take place in two steps, initially with higher update gain and then with lower update gain. During the factory optimization different converged taps will be obtained for different radii on the disk surface. Starting from factory optimized values means that the tap weights don't have to adapt extremely fast and so allows the use of lower update gains during drive operation. Also, this means that the tap weights need not all adapt every clock cycle — instead a round robin approach can be taken which allows for sharing of the adaptation hardware across the various taps. A simpler implementation can also be obtained by using the signed LMS algorithm whereby the tap update equation is based on using 2 or 3 level quantized version of $x(n-k)$. For read channel applications, this can be done without hardly any loss in performance.

28.5 Gain Equalization and Other Issues

We now discuss adaptive gain equalization and some other adaptive loop issues. If all the FIR taps were adapted based on the LMS algorithm, the FIR would provide any necessary gain to equalize the signal to the PR target values. However, during a read event the initial gain error can be very large (e.g., ±6 dB or more). Attempting to perform initial gain equalization through the FIR could result in excessive tap convergence times. In practice a preamble field written before the user data is used to aid in initial gain (as well as timing or sampling phase) recovery or equalization. The preamble field is almost always some length of the periodic 2T pattern ...00110011.... Let us now examine an adaptive gain algorithm. Consider a signal $y(n)$ to be produced from a signal $v(n)$ with a gain scale factor of g. If the gain g is not correct, the resulting error signal is $e(n) = gv(n) - \hat{y}(n)$ where $\hat{y}(n)$ is the desired output signal. Minimizing the squared error with respect to the gain we obtain the stochastic gain gradient as,

$$\Delta_g(n) = -2v(n)e(n) \tag{28.5}$$

Similar to the tap adaptation, the gain $g(n)$ can be adapted based on a product of some gain update gain g_{ug} term and the gradient as shown in the following equation:

$$g(n+1) = g(n) - 2g_g v(n)e(n) \tag{28.6}$$

As with tap adaptation, other adaptive gain equalization loops are possible [13] such as an adaptive zero forcing gain loop as used in [16]. The use of the known preamble field aids in a rapid acquisition of gain using adaptive gain algorithms because the error $e(n)$ and other signals used in the algorithms can be computed relatively accurately and with low loop latency.

We now touch upon few other issues concerning the adaptive FIR. After the initial gain and timing acquisition over the preamble field, the system must still perform adaptive gain and timing equalization at the same time the LMS algorithm is used to adapt the FIR filter. The FIR filter equalizes the signal at the same time that the gain and timing loops are operating. It is necessary to minimize the interaction between these adaptive loops. The FIR filter will typically have one tap as a "main" tap which is fixed to minimize its interaction with the gain loop. This has the additional benefit that the dynamic range needed in the signal path can be partitioned between the FIR taps and the separate gain control block. Another tap such as the one preceeding or following the main tap can be fixed (but allowed to be programmable) to minimize interaction with the timing loop [14]. In some situations it may be advantageous to have additional constraints to minimize the interaction with the timing loop [15]. Adaptive timing recovery is considered in detail in Chapter 26.

28.6   Performance Characterization

We now characterize the performance of various equalizer architectures based on bit error rate simulations. The equalizer types (with reference to Figure 28.1) actually simulated are of Types 2 (CTF+analog FIR), and 3 (anti-aliasing CTF + analog FIR). One can consider the case where there are very few taps as an approximation of the Type 1 (CTF only) equalizer. Although many actual read channel architectures do use digital FIRs (Type 4) we do not consider this type for simulations here. Although a digital FIR filter may be cost effective for implementation given a particular technology, it does have two disadvantages compared with the analog FIR. With the analog FIR, quantization noise is added *after* equalization and so is not enhanced through the equalizer whereas for the digital FIR the quantization noise does pass through the equalizer and could be enhanced. Consequently, fewer quantization levels can be used and this results in reduced chip area and power dissipation with the analog FIR. Also, the digital FIR is likely to have more latency in producing its final output and this extra latency may not hurt significantly but is nonetheless not beneficial for the timing loop.

28.6.1   Simulation Environment and Optimization Procedure

We now describe the simulation environment including two system simulation models. We also describe the optimization methodology by which the optimum performance is obtained for each equalizer architecture. Finally, we present bit error rate (BER) results quantifying the performance of the various architectures.

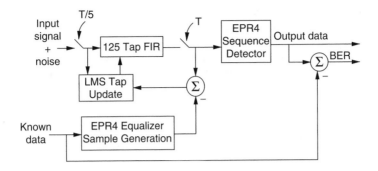

FIGURE 28.5   Block diagram of system with 5x oversampled equalizer.

To obtain a simulation bound for the performance of the best possible equalizer we use the system of Figure 28.5. The signal + noise is fractionally sampled at a rate of $\frac{T}{5}$ and filtered with a fractionally spaced FIR filter equalizer which equalizes the signal to an EPR4 target. The channel bit period is T. The output of the equalizer is then sampled at the channel bit rate of T and these samples are presented to the EPR4 Viterbi. The FIR has 125 taps (spanning $25T$). The FIR tap weights are adapted from zero starting values using the LMS algorithm. There is no quantization, AGC, or timing recovery. Therefore, the performance is solely determined by the noise.

Pseudo random data is 16/17 modulation encoded to generate a signal at various *cbd*s based on a Lorentzian pulse shape. For each *cbd*, the SNR needed by the "ideal" $\frac{T}{5}$ oversampled system of Figure 28.5 to produce a BER of 10^{-5} is determined. SNR is defined as the ratio of the isolated pulse peak to rms noise in the Nyquist band. The (*cbd*, SNR) pairs are used for performing simulations with the practical system of Figure 28.6 which accurately models an actual read channel chip and a version of the $\frac{T}{5}$ system where the equalized samples are quantized before being detected with the Viterbi detector. We examine signals at several *cbd*s or $\frac{PW50}{T}$ values: 1.9, 2.4, and 2.8. The SNRs needed for 1e-5 BER for these densities are 21.66, 22.90, and 24.70 dB respectively.

Let us now describe the simulation model for the actual read channel system. A block diagram of this system is shown in Figure 28.6. The system consists of AGC, CTF, T rate sampled analog FIR equalizer, ADC quantizing the equalizer output, and an EPR4 Viterbi detector. There are three decision directed adaptive loops: LMS tap update loop, AGC loop, and digital PLL (DPLL) timing recovery loop. Note that

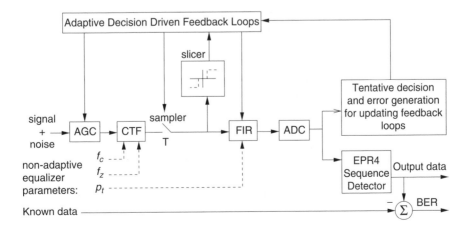

FIGURE 28.6   Block diagram and simulation model of practical symbol rate sampled read channel system.

in this practical system the adaptive feedback loops are updated not based on known data but on tentative or preliminary decisions made by the read channel. The algorithm used for the tap update is the signed LMS algorithm as implied by the three level slicer shown in the figure.

Using the practical system model, BER simulations are performed for the various CTFs mentioned earlier and FIRs of various number of taps. The simulations are performed with this realistic system using the SNRs mentioned earlier. This allows us to calculate the BER degradation of the realistic system with respect to the $\frac{T}{5}$ system for a given *cbd* and SNR.

For each CTF type, *cbd*, and FIR length we simulate the BER of the system across a space of equalizer parameters: f_c, f_z (which determines CTF boost), and the fixed programmable tap of the FIR which is labeled as *pt* in Figure 28.6. The parameters are varied across the following ranges: f_c is varied between 20 and 38% of the channel bit rate, f_z is varied to provide boosts between 2.6 and 8.6 dB, while the programmable tap is varied between −40 and 60% of the main tap value. For CTFs with two zeros, the zeros are adjusted such that the total boost is in the above range. For the 10 tap FIR the 4th tap is chosen to be the fixed main tap while for the 6 and 3 tap filters the 2nd tap is chosen as the main tap. For the other taps, the analog tap range was kept to be relatively large at ±80% of the main tap value. In a real system, one would choose smaller ranges based on tap settings fitting into the smaller range which produced good results. For the equalizer configuration involving the FIR only equalizer, FIRs with 4 to 20 taps are examined. The programmable tap *pt* is reoptimized for each *cbd* and FIR filter length.

28.6.2   Results

Before comparing BER results across different equalizers, we illustrate some results from the equalizer optimization procedure. Figure 28.7 shows a contour plot of the BER obtained with the *b*4 CTF with a 10 tap FIR at a *cbd* of 2.4 and a SNR which gives close to 10^{-5} BER. The horizontal axis is CTF corner frequency (f_c) and the vertical axis is CTF boost in dB. The plot is for one particular value of the programmable FIR tap *pt*. The numbers on the contour plot are 10*log10(BER) so that 10^{-5} BER would correspond to 100. We observe that good BERs result for a large range of boosts and range of f_cs centered in the plot.

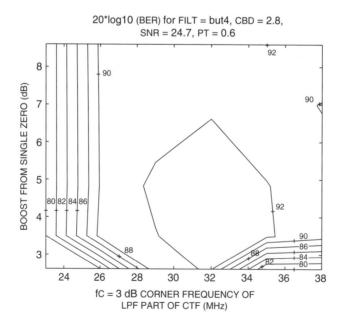

FIGURE 28.7   Boost bandwidth optimization.

Upon examining contour plots for all the CTFs, we concluded that the $b4$ CTF achieves good BERs for f_cs typically in the center of the range explored while the $b6$ CTF the performance is better at somewhat higher f_cs. However, the linear phase CTFs achieve good BERs at very low f_cs. This is because CTFs with worse rolloff characteristics require a smaller f_c to provide enough attenuation at the Nyquist frequency for anti-aliasing. We observe that the BER performance is mostly insensitive to the boost. This is because the adaptive FIR is able to provide any remaining equalization needed. In practice, there will be a tradeoff in how much boost the CTF is able to provide and the corresponding analog tap ranges required by the FIR to provide any necessary remaining equalization — the more equalization the FIR has to provide the larger tap ranges it will require.

We now compare the BER performance of various Type 2 (CTF + analog FIR) equalizers. Figure 28.8 shows the BER performance of different CTFs with a 10 tap FIR. The horizontal axis is cbd and the vertical axis is the BER degradation (in dB) of the optimum BER with respect to the BER of 10^{-5} achieved by the ideal oversampled system using the $\frac{T}{5}$ equalizer. The performance of the CTFs is similar across all cbds— they perform within 0.15 dB of one another. All perform within 0.25 to 0.4 dB of the 10^{-5} BER achieved by the $\frac{T}{5}$ system. The linear phase of the 7th-order CTFs does not necessarily yield superior performance. A final comment is needed about the plot — one should not expect a fixed or monotonic relationship between cbd and the practical system BER in this plot. This is due to the finite resolution of the equalizer optimization search and the fact that BERs are based on observing 100 (versus even larger number) bit errors.

As noted, the above results were with a 10 tap FIR. Further simulations of the various CTFs with a 6 tap or even 3 tap show that the *optimum* BER performance is not very different than that with a 10 tap FIR. These results are presented in Figure 28.9 where the BER degradation (again with respect to the 10^{-5} achieved by the ideal system) of the optimum BER obtained for the various CTFs is plotted vs. the number of FIR taps for $cbd = 2.4$. This initially appears to be a surprising result. However, this not unreasonable when one observes that that with fewer number of taps, a large percentage of the CTF programmings result in poor BER performance. This effect is shown in Figure 28.10 which plots the the percentage of CTF programmings (with respect to the total number of programmings producing convergent tap weights) producing BER worse than 4×10^{-5}. With more taps the percentage of poor programmings decreases. Thus, FIR filters with a few taps, with appropriately optimized CTF settings, can perform as well as a FIR

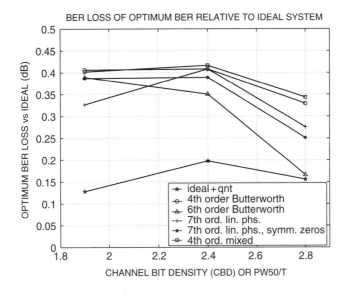

FIGURE 28.8 CTF BER performance degradation with respect to oversampled ideal system vs. *cbd* with 10 tap FIR.

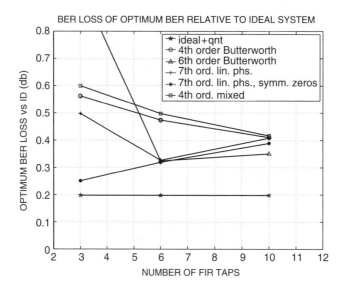

FIGURE 28.9   CTF BER degradation with respect to oversampled ideal system vs. number of taps ($cbd = 2.4$).

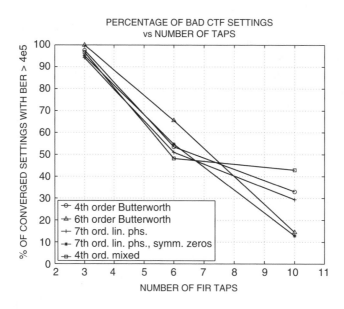

FIGURE 28.10   Percentage of bad CTF settings vs. # of taps ($cbd = 2.4$)

with 10 taps. However, the difficulty in keeping the nonadaptive CTF parameters correct in the presence of realistic device conditions makes such a FIR with few taps impractical to use.

We finally examine the performance of Type 3 equalizers. Here the anti-aliasing CTF is a 7th-order linear phase filter. We examine the performance of this equalizer as a function of the number of FIR taps. The BER performance are shown in Figure 28.11. The vertical axis is again the degradation with respect to the ideal system BER of 10^{-5}. The programmable tap of the FIR is optimized to yield the best performance in each case. The main tap is placed roughly in the center. There is benefit in increasing the number of taps from 4 to 6 to 10. Beyond 10 taps, however, there is more latency in the timing loop as the main tap

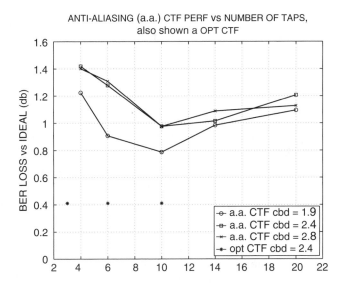

FIGURE 28.11 BER degradation with respect to oversampled ideal system vs. # of taps with anti-aliasing CTF (no boost).

position is more delayed. This causes increased phase errors to enter the timing loop and outweighs the benefit of enhanced equalization obtained with more taps. Although one could increase the number of taps while keeping the main tap location mostly fixed, the FIR will then not be able to cancel the precursor ISI as well with a CTF which is not involved in equalization. Also shown (dashed plot) is the performance of a Type 2 equalizer (CTF with its corner frequency optimized and with an optimized zero included to provide boost). Clearly the Type 2 equalizer outperforms the Type 3 equalizer.

28.7   Actual Equalizer Architectures

We have considered various equalization architectures and examined their performance. Let us now examine what actual architectures read channel vendors have used. Table 28.1 summarizes some of the most commonly used architectures. For example Agere Systems [note: storage products unit of AT&T was spun off to Lucent Technologies in 1996 and again spun off to Agere Systems in 2001) has been using a Type 2 architecture with a 4th order Butterworth CTF and 10 tap analog FIR. The CTF has a programmable corner frequency and zero for providing boost. This architecture is still in place now as this chapter is written. Most other vendors have used Type 4 architectures (digital FIR) but with 7th-order equiripple linear phase filters. The linear phase filters typically have two programmable zeros to provide boost. In the examples of the Cirrus equalizers, the digital FIR does not appear to be adaptive. Some vendors such as

TABLE 28.1   Examples of Equalizers Implemented on Read Channel Chips

| Company | CTF | | | FIR | | | Type | | |
	Type	Order	Zeros	Taps	Adaptive ?	Analog/Digital	(Fig 28.1)	Ref./Yr.	Comments
Agere	But	4th	2	10	Yes	analog	2	[16], 95	8 Adaptive taps
Cirrus	EqRip	7th	2	3	No	digital	2	[17], 95	—
Cirrus	EqRip	7th	2	5	No	digital	2	[18], 96	—
Datapath	EqRip	7th	2	N/A	N/A	N/A	1	[19], 97	No FIR
Marvell	EqRip	7th	?	7	Yes	digital	2	[20], 97	

Cirrus and Marvell seem to have increased the number of FIR taps or the number of adaptive (versus only programmable) taps as the years have gone by. The Datapath equalizer cited is one of the few examples of an all CTF equalizer.

28.8 Conclusions

We have quantified the performance of various continuous time filter (CTF) + adaptive analog FIR (Type 2) equalizers in equalizing a signal to an EPR4 target. It is shown that regardless of the number of taps in the FIR and CTF type, the BER performance of the CTF + FIR equalizers is approximately the same if the optimum fixed equalizer parameters (CTF corner frequency, boost, FIR fixed tap) are chosen.

Therefore, the choice of CTF type should be based on other constraints such as area, power, speed (data rate), as well the benefit of having one less analog block. We have also shown that as the number of taps is increased, the space of CTF parameter programmings producing BERs close to the optimum increases significantly. Therefore, one can tradeoff the cost of the FIR filter versus required accuracy in the CTF setting and the sensitivity of the resulting performance.

We have also examined the performance of Type 3 equalizers consisting of a T spaced FIR filter with only a Nyquist anti-aliasing CTF. We have found that the Type 3 equalizer cannot approach the performance of a system whose CTF is involved in equalization and is optimized. Therefore, to make a valid comparison between FIR and CTF equalizers, one must include a reasonably optimum CTF prior to the FIR.

We have demonstrated that a wide variety of optimized CTF + FIR equalizers can perform within 0.25 dB of the quantized system using the oversampled $\frac{T}{5}$ equalizer. As this 0.25 dB includes performance losses due to AGC and timing recovery, there is very little space left for improved equalization with any other equalizer architecture.

References

[1] A. Oppenheim and R. Schafer, *Discrete Time Signal Processing*, Prentice Hall, New York, 1989.

[2] K. Fisher, W. Abbott, J. Sonntag, and R. Nesin, PRML detection boosts hard-disk drive capacity, *IEEE Spectrum*, pp. 70–76, November, 1996.

[3] M.E. Van Valkenburg, *Analog Filter Design*, Holt Rinehart Winston, 1982.

[4] R. Schaumann, M. Ghausi, K. Laker, *Design of Analog Filters*, Prentice Hall, New York, 1990.

[5] J. Park and L.R. Carley, Analog complex graphic equalizer for EPR4 channels, *IEEE Trans. Magn.*, pp. 2785–2787, September, 1997.

[6] A. Bishop et al., A 300 Mb/s BiCMOS disk drive channel with adaptive analog equalizer, *Digests, Intl. Solid State Circuits Conf.*, pp. 46–47, 1999.

[7] P. Pai, A. Brewster, and A. Abidi, Analog front end architectures for high speed PRML magnetic recording channels, *IEEE Trans. on Magn.*, pp. 1103–1108, March, 1995.

[8] R. Cideciyan, F. Dolivo et al., A PRML system for digital magnetic recording, *IEEE J. Select. Areas Communi.*, pp. 38–56, January, 1992.

[9] G. Mathew et al., Design of analog equalizers for partial response detection in magnetic recording, *IEEE Trans. Magn.*, pp. 2098–2107, July 2000.

[10] S. Qureshi, Adaptive Equalization, *Proceedings of the IEEE*, pp. 1349–1387, September, 1973.

[11] S. Haykin, *Communication Systems*, John Wiley & Sons, New York, pp. 487–497, 1992.

[12] S. Haykin, *Adaptive Filter Theory*, Prentice-Hall, New York, 1996.

[13] J. Bergman, *Digital Baseband Transmission and Recording*, Kluwer Academic Publishers, Dordrecht, Netherlands, 1996.

[14] P. Aziz and J. Sonntag, Equalizer architecture tradeoffs for magnetic recording channels, *IEEE Trans. Magn.*, pp. 2728–2730, September, 1997.

[15] L. Du, M. Spurbeck, and R. Behrens, A linearly constrained adaptive FIR filter for hard disk drive read channels, *Proc., IEEE Intl. Conf. on Communi.*, pp. 1613–1617.

[16] J. Sonntag et al., A high speed low power PRML read channel device, *IEEE Trans. Magn.*, pp. 1186–1189, March, 1995.

[17] D. Welland et al., Implementation of a digital read/write channel with EEPR4 detection, *IEEE Trans. Magn.*, pp. 1180–1185, March, 1995.

[18] G. Tuttle et al., A 130 Mb/s PRML read/write channel with digital-servo detection, *Digests, IEEE Intl. Solid State Circuits Conf.*, 1996.

[19] J. Chern et al., An EPRML digital read/write channel IC, *Digests, IEEE Intl. Solid State Circuits Conf.*, 1997.

[20] N. Nazari, A 500 Mb/s disk drive read channel in 0.25 um CMOS incorporating programmable noise predictive Viterbi detection and Trellis coding, *Digests, Intl. Solid State Circuits Conf.*, pp. 78–79, 2000.

29

Head Position Estimation

	29.1	Introduction	**29**-1
	29.2	Servo Writing.......................................	**29**-3
	29.3	The Digital Field	**29**-4
		Format Efficiency • Offtrack Detection	

Ara Patapoutian

Maxtor Corporation
Shrewsbury, MA

| | 29.4 | The Burst Field .... | **29**-7 |
| | | Impairments • Formatting Strategies • Position Estimators | |

29.1 Introduction

Data in a disk drive is stored in concentric tracks on one or multiple disk surfaces. As the disks spin, a magnetic transducer known as a read/write head, transfers information between magnetic fields on a disk surface and electric currents that are decoded as strings of binary sequences [1]. When a given track needs to be accessed, the head assembly moves the read head to the appropriate radial location. This positioning of the head on top of a given track is achieved by the use of a feedback servo system shown in Figure 29.1. First, a position sensor generates a noisy estimate of the head location. Then a controller generates a signal to adjust the *actuator* by comparing the position estimate to the desired position. To access a particular sector in a track the angular position of the head with respect to the disk surface should also be determined. In this chapter we review methods to estimate the head position which is uniquely determined from the radial and angular position of the head with respect to a disk surface.

The read head, primarily designed to detect recorded user data patterns, can also be used to estimate the head position by sensing dedicated, position specific magnetic marks recorded on a disk surface. The position dependent magnetic marks are read in the form of waveforms and then decoded into a head position estimate using statistical signal-processing and coding techniques. In this chapter we will concentrate on both head position formats that are recorded on the disk and on the decoding of these marks into head position estimates.

In an *embedded servo* scheme a portion of each disk surface, divided into multiple *wedges*, is reserved to provide radial and angular position information for the read head. These reserved wedges are referred to as *servo fields* and are shown in Figure 29.2. The number of wedges per surface varies amongst different products depending on the capability of the drive to detect external shocks and by the servo disturbance bandwidth that the drive is designed to track. A typical servo wedge provides radial estimates in two steps. On a disk surface, each track is assigned a number known as the *track address*. These addresses are included in the servo field, providing complete radial position information with accuracy of up to a single track. In other words, the information provided by a track address is complete but coarse. The *positional error*

0-8493-1524-7/05/$0.00+$1.50

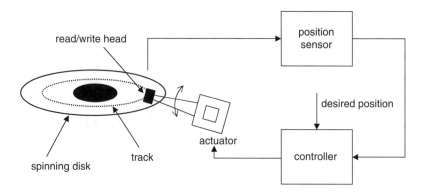

FIGURE 29.1   Position control loop for a disk drive.

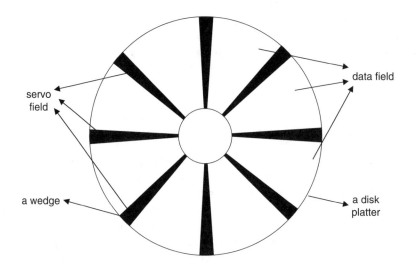

FIGURE 29.2   Data and servo fields on a disk drive.

signal (PES) complements the track address by providing a more accurate estimate within a track. By combining these two estimates, a complete and accurate radial position estimate can be obtained.

The servo field may also contain angular position information by recording the wedge address on all tracks. The user data field, with its own address mark and timing capability, can complement the wedge address by providing finer angular position estimates.

A typical servo field will have multiple subfields as shown in Figure 29.3. A periodic waveform, known as a *preamble*, provides information to calibrate the amplitude of the waveform and if necessary to acquire the

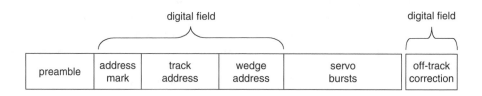

FIGURE 29.3   A generic servo field formats.

timing of the recorded pattern. Frame synchronization, or the start of a wedge, is determined by detecting a special sequence known as the *servo address mark or servo ID* [2]. The track and wedge addresses follow the servo address mark. Next, the *servo burst* that provides information regarding the PES ends the servo field. When a servo field is recorded with a radial offset then it is possible to remove this repeatable offset by appending an off-track correction value field that encodes a radial correction factor of the PES as shown in Figure 29.3.

The subfields in a servo field can be divided into two classes. The address mark, track address, wedge address and off-track correction values are all encoded as binary strings and are referred to as the *digital field* as shown in Figure 29.3. By contrast, ignoring quantization effects of the *read channel*, the periodic servo burst field is decoded to a *real* number representing the analog radial position. Thus, the recorded formats as well as the demodulation techniques for the digital and burst fields are different. The digital field demodulator is known as the *detector* while the servo burst field demodulator is known as the *estimator*.

Despite their differences, the two fields are not typically designed independently of each other. For example, having common sample rates and common front-end hardware simplifies the receiver architecture significantly. Furthermore, it makes sense to use coherent or synchronous detection algorithms with coherent estimation algorithms.

A dedicated servo field comes at the expense of user data capacity. When designing the servo field and the associated demodulators, the primary goal is to minimize the servo field overhead given cost, reliability and performance constraints.

In the rest of this chapter we review position sensing formats and demodulators. The content of this chapter is based on [3]. Here, we have updated recent developments and added the latest references. Since estimation and detection are established subjects presented in multiple textbooks [4, 5], we concentrate on issues that differentiate disk drive position sensors. Furthermore, the statistical signal processing aspects of position sensing rather than the servo control loop design are presented. For a general introduction to disk drive servo control design, the reader is referred to [6], where the design of a disk drive servo is presented as a case study of a control design problem. The literature regarding head position sensing, with the exception of patents, is limited to a relatively few published articles.

When a disk drive is first assembled in a factory, the servo fields need to be recorded on the disk surfaces. Once a drive leaves the factory, these fields will only be read and never re-written. Traditionally, an external device, known as the *servo writer* and introduced in Section 29.2, records the servo fields. In general, the servo writing process constrains and affects the servo field format choices as well as the demodulator performance. In Section 29.3, the digital field format and detection approaches are addressed while in Section 29.4 the servo burst format and PES estimation approaches are introduced.

29.2 Servo Writing

Servo writing is the manufacturing process where head positional information is recorded in the servo fields of a disk surface. A *servo writer* is an external device that assists the disk drive with servo writing. In this section we shortly discuss a servo writer and some alternatives.

A servo writer interacts with a disk drive and helps the disk drive record positional information. A servo writer should be able to estimate both the radial and angular position of a head. Furthermore, a servo writer should be able to move head assembly radially so that the native write-head of the drive can record the servo fields, one track at a time. A common method to estimate the radial head position is through an optical device that emits a laser beam and then measures the reflected beam. The angular position could be sensed from a dedicated external read head that is locked on a recorded periodic waveform on the disk media. Finally, an external mechanical device could be used to move the head assembly.

Servo writers have some disadvantages. They are expensive and require clean room environment not to contaminate the disk drive interior. Furthermore, frequent calibrations may be required. As a result servo writers present a costly solution to perform servo writing.

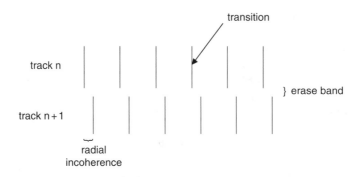

FIGURE 29.4   Servo writer impairments: erase bands and radial incoherence.

From the above discussion, the *servo writing time* per drive is an important metric that disk manufacturers try to minimize. The servo writing time is proportional to the number of tracks per disk surface, to the spin of the disk drive, and to the number of passes needed to record a track. Since the number of tracks per disk is increasing faster than the spin speed, the servo writer time becomes a parameter that needs to be contained. To this end, the disk drive industry has attempted to minimize both servo writer time and the servo writer cost.

Self servo writing is an alternative to external servo writing where the wedges are written by the disk drive itself without the assistance of a servo writer [7, 8]. Since clean room environment and the need of an external device are not needed, the process becomes less costly and the servo writing time can be relaxed.

There are also many hybrid strategies that first record some initial reference marks using an external device and then complete writing the final wedges using the drive itself. The goal of such an approach is to minimize the cost of the external device. One approach is to first record position reference information externally using for example the printed media approach [9, 10] where a master mold 'stamps' each disk with low frequency patterns. Another example is to servo write a reference surface as in [11]. Once the positional marks are recorded and the drive assembled, the drive obtains the position estimates using the reference marks and writes the wedges without any external help. The original reference marks are erased or overwritten by data. Another hybrid approach is given in [12] where the servo writer records a portion of the wedges and the drive completes writing the remaining fields by itself.

The processes of recording the wedges by the native write head has its limitations. The write head is not capable of recording an arbitrary wedge format. For example, it is very difficult for a write head to record a continuous angled transition across the radius since data is recorded naturally track-by-track. Furthermore, a write head may create an *erase band* between tracks [13] where no transition is recorded in a narrow band between two adjacent tracks. Finally, because of uncertainties in both radial and angular positions, two written tracks may not be aligned properly. An angular uncertainty results in *radial incoherence* between adjacent tracks. A radial uncertainty results in squeezing two adjacent tracks unless offtrack correction field is inserted as discussed in the next section. Radial incoherence and erase band impairments are illustrated in Figure 29.4.

29.3 The Digital Field

The digital servo field has similarities to the user data field and to a binary baseband communications system [4, 14]. What differentiates a digital servo field from a data field is its short block length, but more importantly its off-track detection requirement.

A magnetic recording channel is *an intersymbol interference* (ISI) channel. When read back, the information recorded in a given location modifies the read waveform not only at that given location but also in the neighboring locations. Finite length ISI channels can be optimally detected using *sequence detectors*

[15], where at least theoretically all the field samples are observed before detecting them as a string of ones and zeros. For more than a decade now such sequence detectors have been employed in disk drives to detect the user data field.

29.3.1 Format Efficiency

The servo field is very short when compared to the data field. In numbers, present data sector length is around 512 bytes long while a digital servo field is only few bytes long. Therefore, in terms of percent overhead, fixed length fields such as the preamble and address mark have a different impact on long data fields than digital servo fields. Furthermore, coding algorithms to improve reliability are more effective on larger blocks than on shorter blocks. Hence, containing servo field overhead is a difficult but important task. Next we discuss strategies that minimize the servo field format.

The primary goal of the preamble is to estimate the phase or the timing of the recorded waveform. Obtaining the phase estimate is necessary with synchronous detection. However, the preamble format could be reduced if the servo field could be demodulated *asynchronously* [16].

Another approach to minimize the digital field is to record only partial information regarding the track and wedge addresses [17]. Strategies that record partial radial and angular information improve format efficiency at the expense of performance, robustness and complexity. A state machine can predict a wedge address. For example, a single bit of information inserted in the wedge address may signal a reference wedge in a track, and a state machine then could count the wedge numbers starting from that reference point. One disadvantage of such a system is the initial time spent to locate a reference wedge. An improved method is described in [18] where a pseudorandom sequence is stored in the wedges, one bit per wedge. If the pseudorandom sequence is of length L, then after reading $\log(L)$ wedges, it is possible to uniquely identify the wedge address.

Similarly, the track address need not be completely recorded. Given a radial velocity estimate and the current radial position estimate, a controller could predict the upcoming track address within a radial band.

29.3.2 Offtrack Detection

The requirement to detect data reliably at *any* radial position is the primary factor that differentiates digital servo fields from other channels. In contrast, a user data field needs to be read reliably only if the read head is directly above that specific track. As will be discussed shortly, such a constraint influences the choice of both detection as well as the ECC or coding strategies.

We concentrate next on track addresses since for a given wedge the track address varies from one track to the next. Other digital fields such as preamble, address mark and wedge address do not vary with respect to tracks. Let X and Y represent the binary address representations of two adjacent tracks. When the read head is in the middle of these two tracks, the read waveform is the superposition of the waveforms generated from each of the respective track addresses. In general, the resulting waveform will not be decoded reliably to either of the two track addresses. A common solution to this problem is the use of a *Gray code* to encode track addresses, as shown in Figure 29.5, where any two adjacent tracks differ in their binary address representation in only a single position. Hence, if we ignore ISI, when the head is midway

X_G	+	−	+	+	+	−
Y_G	+	−	−	+	+	−

FIGURE 29.5 An example of two Gray coded track addresses. The two addresses are different only in the third location.

between adjacent tracks, the detector will decode the address bits correctly except for the bit location where the two adjacent tracks differ. That is for two track addresses encoded as X_G and Y_G, the decoder will decode the incoming waveform to either the track address X or Y, introducing a radial position error of less than a single track. By designing a radially periodic servo burst field, with period of at least two track widths, track number ambiguity generated by track addresses is resolved. However, as will be discussed next, Gray codes complicate the use of codes that improve reliability, and degrade the sequence detector performance.

A Gray code restricts two adjacent tracks to differ in only a single position, or equivalently forces the *Hamming distance* between two adjacent track addresses to be one. Adding redundancy field to improve data reliability is desirable in the presence of miscellaneous impairments. Depending on how they are decoded, traditional distance enhancing codes maximize the minimum Hamming distance or the minimum Euclidean distance. The former codes, also known as error correcting codes or ECC, use hard decoders that is, operate on binary data. On the other hand soft decoders that operate in Euclidean space outperform hard decoders but are more complicated to implement. In an ISI channel, both ISI detection and code decoding are merged into a single Viterbi detector with a larger trellis. Alternatively, since the Viterbi detector/decoder complexity may become prohibitive, a post-processor could simplify and approximate the Viterbi detector.

Any code that improves reliability is associated with a minimum Hamming distance larger than one. That is, it is not possible to have two adjacent track-addresses be Gray coded and at the same time distance coded to improve reliability.

A second complication of introducing Gray codes results from the interaction of the Gray codes with the ISI channel. First, consider an ISI free channel where the magnetic transitions are written ideally and where the read head is allowed to be anywhere between two adjacent Gray coded track addresses X_G and Y_G. As was discussed earlier the track address reliability, or the probability that the decoded address is neither X nor Y, is independent of the exact read head position. Next we show that for an ISI channel the detector performance depends on the radial position. In particular, consider the simple ISI channel with pulse response 1-D, (the output is the difference between present and previous samples) which approximates a magnetic recording channel. For such a channel, error events of length two are almost as probable as errors of length one (same distance but different number of neighbors). Now, as the head moves from track X to Y, the waveform modification introduced by addresses X_G and Y_G, from that one location where the two tracks differ, can trigger an error event of length two. The detector may decode the received waveform to a different track address Z, which may lie far from addresses X or Y. In other words, in an ISI channel, whenever the head is between two tracks X and Y, the probability that the received waveform is decoded to some other address Z increases.

For readers familiar with signal space representation of codewords, the ISI free and 1-D channels are shown for the three Gray coded addresses X_G, Y_G, and Z_G with their decision regions in Figure 29.6. Let d denote half the minimum Euclidean distance of the code. As shown in Figure 29.6, when the head is midway between tracks corresponding to codewords X_G and Y_G, the shortest distance to cross the decision boundaries of codeword Z_G is reduced by a factor of $\sqrt{3}$ (or 4.77 dB). Therefore, when the head is in the middle of two tracks, addressed as X and Y, the probability that the decoded address is Z increases significantly. For an arbitrary ISI channel this reduction factor in the smallest distance varies, and can be shown to be at most $\sqrt{3}$.

In summary, Gray codes complicate the use of distance enhancing codes and degrade the sequence detection performance. A simple solution is to use no codes and to write address bits far enough from each other to be able to ignore ISI effects. Then a simple symbol-by-symbol detector is sufficient to detect the track addresses without the need for a sequence detector. Actually this was a common approach taken in many disk drive designs. However, dropping distance enhancing code capability affects reliability and relaxing the bit interval requires additional format.

To be able to use distance-enhancing codes, the Gray constraint has to be relaxed or eliminated. For example, the Gray constraint can be eliminated if the track addresses are written on multiple locations at varying radial shifts so that, at any position, the head is always above a track address [19]. Alternatively,

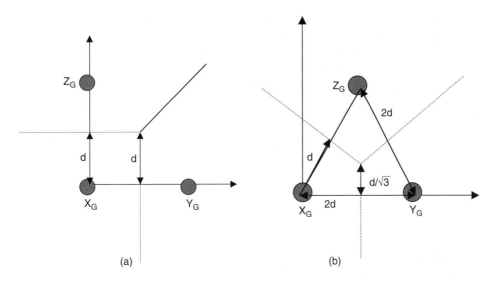

FIGURE 29.6 Signal space representation of three codewords. Configuration (a) ISI free (b) with ISI.

Gray codes can be generalized such that with hard decoding, two adjacent tracks differ in exactly d_h positions and non-adjacent tracks differ in at least d_h positions [20, 21]. Alternatively, with soft decoding, associated waveforms from two adjacent tracks differ in minimum Euclidean distance d_e [22]. In [21] track addresses are constructed such that any adjacent track addresses have a Hamming distance of d_h, and any nonadjacent tracks within a radial band have distance of at least $2d_h$. Since the controller should predict a track address within a band, such generalized Gray code constructions can correct $(d_h - 1)/2$ errors. In [22] short length codes are constructed that maximize the Euclidean distance. Hence, these solutions represent a tradeoff between data reliability and format efficiency.

Radial incoherence and the erase field generate noise statistics that are extremely correlated. If these two phenomena become the dominant impairments in the servo field, the benefits of coding become less obvious. This may suggest the use of modulation schemes that maximize symbol reliability rather than codes that maximize servo blocks. In other words, data reliability is improved by increasing the Euclidean distance of each symbol. An example of such a format is the biphase code [19, 23, 24]. Such codes can be decoded by a Viterbi detector [23], or by a simple threshold circuit since a biphase code produces a positive pulse at the middle of the symbol for a symbol one and a negative pulse for a symbol zero. When the symbols intervals are chosen large enough it can be shown that the ISI related degradations are minimized and the detector performance is improved.

29.4 The Burst Field

In Section 29.3, track addresses provided head position information to within a single track accuracy. However, to be able to read and write user data reliably, it is essential to position the read head over a given track with a higher resolution. For this purpose the track number addresses are complemented with the servo burst field that contain analog radial position information.

The burst field is a periodic pattern both in radial and angular directions. Since a track address provides an absolute radial position, the periodicity of the burst field in the radial direction does not create any position ambiguity, as long as the radial burst period is longer than the radial ambiguity left after decoding a track address. Along a track, a servo burst is recorded as a periodic binary pattern. The read back waveform, at the head output, is periodic and will contain both the fundamental and higher harmonics. The sinusoidal waveform is obtained by retaining only the fundamental harmonic. There are three ways to encode a parameter representing the radial position in a sinusoidal waveform: amplitude, phase and frequency [5]. As will be discussed later, multiple burst formats can be designed that encode the radial positional information in one of these three parameters.

It is possible to maximize the power of the read back waveform by optimizing the fundamental period [25]. If the recorded transitions get too close then ISI reduces most of the signal power. On the other hand, for longitudinal recording, if transitions are far from each other then the read back waveform contain few pulses with small power.

Next, the impairment sources in the burst field are identified. Various servo burst formats and their performances are discussed [26, 27]. Finally, various estimator options and their characteristics are reviewed.

29.4.1 Impairments

Impairments in a servo burst field are classified into three categories: induced by servo-writing, read head and read channel. Not all impairments are present in all servo burst formats.

As was discussed in Section 29.2, while servo writing adjacent servo fields, erase band as well as radial incoherence may be generated between tracks, degrading the performance of some of the servo burst formats. An additional source of servo writing impairment includes duty cycle error of the recorded period that is different from the intended 50%. Finally, write process limitations result in nonideal recorded transitions.

The electronic noise is modeled by additive white Gaussian noise (AWGN), and is generated by the read head element as well as from the preamplifier that amplifies the incoming signal. Also, in many drives the width of the read head element may be shorter than the servo burst radial width as shown in Figure 29.7(a). As will be discussed shortly, for some formats, this creates *saturated* radial regions where the radial estimates are not reliable [13]. Finally, the two dimensional rectangular approximation of the read head element shown in Figure 29.7(a) is not accurate [28, 29]. The head response depends on the total magnetic flux it receives from the surrounding of the head location. A three dimensional *read head profile* more accurately models the read head. When approximated as a two-dimensional profile the shape of the read head may be significantly different than a rectangle.

The read channel, while processing the read waveform, induces a third class of errors. Most present estimators are digitally implemented, and have to cope with *quantization error*. If only the first harmonic of the received waveform is desired then suppressing higher harmonics may leave residues that may interact with the first harmonic inside the estimator. Furthermore, sampling a waveform with higher harmonic residues creates *aliasing effects* where higher harmonics fold into the first harmonic.

Many read channel estimators require that the phase, frequency or both phase and frequency of the incoming waveform are known. Any discrepancy results in estimator performance degradation. Finally,

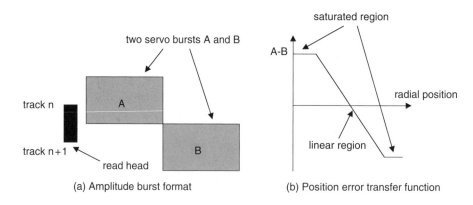

(a) Amplitude burst format   (b) Position error transfer function

FIGURE 29.7 The amplitude burst format and the its position error transfer function as the head moves from center-track *n* to center-track *n* + 1.

estimator complexity constraints result in suboptimal estimators, further degrading the accuracy of the position estimate.

29.4.2 Formatting Strategies

The amplitude servo burst format, shown in Figure 29.7(a), is a simple, robust and popular format. Depending on the radial position of the read head, the overlap between the head and the bursts A and B varies. Through this overlap, or amplitude variation at the estimator input, it is possible to estimate the radial position. First, the waveforms extracted from each burst field A and B is transformed into amplitude estimates. These amplitude estimates are then subtracted from each other and scaled to get a positional estimate. As the head moves from track center n to track center $(n + 1)$, the noiseless positional estimate, known as *position error transfer function*, is plotted in Figure 29.7(b). Since the radial width of the servo burst is larger than the read element, any radial position falls into either the *linear* region, where radial estimate is accurate, or in the *saturated* region, where the radial estimate is not accurate [13]. One solution to withstand saturated regions is to include multiple burst pairs, such that any radial position would fall in the linear region of at least one pair of bursts. The obvious drawback of such a strategy is the additional format loss. The amplitude format just presented does not suffer from radial incoherence since two bursts are not recorded radially adjacent to each other.

Nonrecorded areas in Figure 29.7(a) do not generate any signal. Therefore, only 50% of the servo burst format is utilized. In an effort to improve the position estimate performance, the whole allocated servo area can be recorded. As a result, at least two alternative formats have emerged, both illustrated in Figure 29.8.

In the first improved format, burst A is radially surrounded by an *antipodal* or "opposite polarity" burst A'. For example, if burst A is recorded as $+ + - - + + - - \ldots$ then burst A' is recorded as $- - + + - - + + \ldots$ This format is known as the *antipodal* format or the *null* format since when the head is midway between two tracks no signal is read. For readers familiar with digital communications, the difference between the amplitude and antipodal servo burst formats can be compared to the difference between *on-off* and antipodal servo burst formats. In on-off signaling, a symbol zero or one is transmitted while in antipodal signaling one or minus one is transmitted. Antipodal signaling is 6 dB more efficient than on-off signaling. Similarly, it can be shown that the antipodal servo burst format gives a 6 dB advantage with respect to amplitude servo burst format under the AWGN assumption [27]. To contain radial incoherence is a challenge with the antipodal format since the performance of the estimator is sensitive to such an impairment.

Instead of recording A' to be the opposite polarity of A, another alternative is to record a pattern A' that is orthogonal to A. For example, it is possible to pick up two sinusoids with different frequencies such that the two waveforms are orthogonal over a finite burst length interval. The resulting format is known as the *dual frequency* format [30]. Inside the read channel two independent estimates of the head position are obtained from two estimators, each tuned to one of the two frequencies. In general the first harmonic of the two frequencies will have different amplitudes. If we assume the two first harmonic

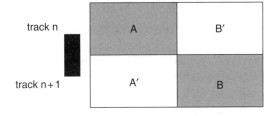

FIGURE 29.8 Alternatively burst formats where A' and B' are either orthogonal to or of opposite polarity of A and B, respectively.

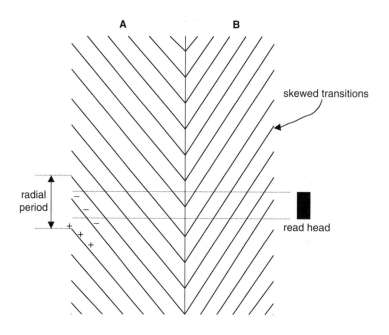

FIGURE 29.9   The phase format.

amplitudes are equal then the final radial estimate is the mean of the two estimates, resulting in a 3 dB improvement when compared with the amplitude format, again under the AWGN assumption. As in the amplitude format, the estimators corresponding to a dual frequency format are not sensitive to radial incoherence, since the orthogonality between the two fields is generated from frequency rather than phase differences.

A different format is presented in Figure 29.9. Here, the transitions are skewed and the periodic pattern gradually shifts in the angular direction as the radius changes. Since the radial information is stored in the phase of the period we call it the phase format. In Figure 29.9 two burst fields A and B are presented where the transition slopes have the same magnitude but opposite polarities. An estimator makes two phase estimates, one from the sinusoid in field A and another one from the sinusoid in field B. By subtracting the second phase estimate from the first and then by scaling the result the radial position estimate can be obtained. Similar to the antipodal format, it can be shown that the phase pattern is 6 dB superior to the amplitude pattern [27] under AWGN. As in the antipodal format, a challenge for the phase format is to successfully record the skewed transitions on a disk surface without radial incoherence.

29.4.3   Position Estimators

Estimating various parameters of a sinusoid is well documented in textbooks [5]. Whereas a decade ago position estimators were mostly implemented by analog circuitry, at present digital implementation is the norm and the one considered in this chapter [31–35]. One way of classifying estimators is to determine whether the phase and/or the frequency of the incoming waveform is known.

Assume the amplitude of a noisy sinusoid needs to be determined. If the phase of this waveform is known then a matched filter can be used to generate the amplitude estimate. This is known as coherent estimation. Such a filter becomes optimal under certain assumptions and performance criteria. When the phase of the waveform is not known, but the frequency is known, then two matched filters can be used, one tuned to a sine waveform while the other filter is tuned to a cosine waveform. The outputs of the two filters are squared and added to give the energy estimate of the waveform. This is known as noncoherent estimation and is equivalent to computing the Fourier transform at the first harmonic. Other ad hoc estimators include the peak estimator and digital area estimators [36], which respectively estimate the averaged peak and the mean value of the unsigned waveform. Neither of these estimators require the phase or the frequency value of the waveform.

For the amplitude format, all the estimators discussed above can be used to get an estimate. For the antipodal format the amplitude as well as the sign of the waveform is desired. The amplitude reflects the head deviation from track center while the sign determines the direction of the deviation. Both the amplitude and the sign can be determined from a coherent estimation using a single matched filter. For dual frequency format, two estimators are needed, each tuned to a given frequency. Since the two waveforms are orthogonal to each other, an estimator tuned to one of the waveforms will not observe the other waveform. Each estimator can utilize a single matched filter for coherent estimation or two matched filters for noncoherent estimation. Finally, for phase estimation, two matched filters are utilized, similar to noncoherent estimation. However, rather than squaring and adding the inphase and quadrature components, the inverse tangent function is performed on the ratio of two filter outputs.

References

[1] Comstock, R.L. and Workman, M.L., Data storage in rigid disks, in *Magnetic Storage Handbook*, 2nd ed., Mee, C.D. and Daniel, E.D., Eds., McGraw-Hill, New York, 1996, chap. 2.

[2] Blaum, M., Hetzler S., and Kabelac, W., An efficicient method for servo-ID synchronization with error-correcting properties, *International Symposium on Communication Theory and Applications*, Charlotte Mason Collledge, UK, July 1999.

[3] Patapoutian, A., *The Computer Engineering Handbook*, Oklobdzija V. Ed., CRC press, 2001.

[4] Proakis, J.G., *Digital Communications*, 4th ed., McGraw-Hill, New York, 2000.

[5] Kay, S.M., *Fundamentals of Statistical Signal Processing: Estimation Theory*, Prentice Hall, Englewood Cliffs, 1993.

[6] Franklin, G.F., Powell, D.J., and Workman, M.L., *Digital Control of Dynamic Systems*, 3rd ed., Addison Wesley, Reading, MA, 1997, chap. 14.

[7] Brown, D.H. et al., Self-Servo Writing File, U.S. patent 06,040,955, 2000.

[8] Liu, B., Hu, S.B., and Chen, Q.S., A novel method for reduction of the cross track profile asymmetry of MR head during self servo-writing, *IEEE Trans. on Mag.*, 34, 1901, 1998.

[9] Bernard, W.R. and Buslik, W.S., Magnetic Pattern Recording, U.S. patent 03,869,711, 1975.

[10] Tanaka, S. et al., Characterization of magnetizing process for pre-embossed servo pattern of plastic hard disks, *IEEE Trans. on Mag.*, 30, 4209, 1994.

[11] Belser, K.A. and Sacks A.H., Advanced Servo Writing Method for Hard Disk Drives, U.S. patent 6,411,459, 2002.

[12] Ehrlich, R.M. et al., Hard Disk Drive having Self Written Servo Burst Patterns, U.S. patent 6,519,107, 2003.

[13] Ho, H.T. and Doan T., Distortion effects on servo position error transfer function, *IEEE Trans. Mag.*, 33, 2569, 1997.

[14] Bergmans, J.W.M., *Digital Baseband Transmission and Recording*, Kluwar, Dordrecht, 1996.

[15] Forney, G.D., Maximum-likelihood sequence estimation of digital sequences in the presence of intersymbol interference, *IEEE Trans. Info. Thy.*, 18, 363, 1972.

[16] Patapoutian, A., McEwen P., Veiga E., and Buch B., Asynchronously sampling wide bi-phase code, U.S. patent application 20020163748.

[17] Chevalier, D., Servo Pattern for Location and Positioning of Information in a Disk Drive, U.S. patent 05,253,131, 1993.

[18] Buch, B., Servo Area Numbering Strategy for Computer Disk Drives, U.S. patent 6,501,608, 2002.

[19] Leis, M.D. et al., Synchronous Detection of Wide Bi-phase Coded Servo Information for Disk Drive, U.S. patent 05,862,005, 1999.

[20] van Zanten, Minimal-change order and separability in linear codes, *IEEE Trans. Info. Theory*, 39, 1988, 1993.

[21] Blaum, M., Kabelac, W.J., and Yu, M.M., Encoded TID with Error Detection and Correction Ability, U.S. patent 6,226,138, 2001.

[22] Aziz, P.M, Rate (M/N) encoder detector and decoder for control data, U.S. paten 6,480,984, 2002.

[23] Makinwa, K.A.A., Bergmans, J.W.M., and Voorman, J.O., Analysis of a biphase-based servo format for hard disk drives, *IEEE Trans. Mag.*, 36, 4019, 2000.

[24] Patapoutian, A., Vea, M.P., and Hung, N.C., Wide Biphase Digital Servo Information Detection, and Estimation for Disk Drive using Servo Viterbi Detector, U.S. patent 05,661,760, 1997.

[25] Patapoutian, A., Optimal burst frequency derivation for head positioning, *IEEE Trans. on Mag.*, 32, 3899, 1996.

[26] Sacks, A.H., Position Signal Generation in Magnetic Disk Drives, Ph.D. dissertation Carnegie Mellon University, Pittsburgh, 1995.

[27] Patapoutian, A., Signal space analysis of head positioning formats, *IEEE Trans. on Mag.*, 33, 2412, 1997.

[28] Cahalan, D. and Chopra, K., Effects of MR head track profile characteristics on servo performance, *IEEE Trans. Mag.*, 30, 4203, 1994.

[29] Sacks, A.H. and Messner, W.C., MR head effects on PES generation: simulation and experiment, *IEEE Trans. Mag.*, 32, 1773, 1996.

[30] Cheung, W.L., Digital demodulation of a complementary two-frequency servo PES pattern, U.S. patent 06,025,970, 2000.

[31] Tuttle, G. T. et al., A 130 Mb/s PRML read/write channel with digital-servo detection, *Proc. IEEE ISSCC '96*, 64, 1996.

[32] Fredrickson, L. et al., Digital servo processing in the Venus PRML read/write channel, *IEEE Trans. Mag.*, 33, 2616, 1997.

[33] Yada, H. and Takeda, T., A coherent maximum-likelihood, head position estimator for PERM disk drives, *IEEE Trans. Mag.*, 32, 1867, 1996.

[34] Kimura, H. et al., A digital servo architecture with 8.8 bit resolution of position error signal for disk drives, *IEEE Globecom '97*, 1268, 1997.

[35] Patapoutian, A., Analog-to-digital converter algorithms for position error signal estimators, *IEEE Trans. Mag.*, 36, 395, 2000.

[36] Reed, D.E. et al., Digital servo demodulation in a digital read channel, *IEEE Trans. Mag.*, 34, 13, 1998.

30

Servo Signal Processing

30.1 Introduction 30-1
30.2 Servo Coding Gain 30-3
Example Codes and Their Minimum Distance • Coding
Gain/Rate Tradeoff for Servo
30.3 Noise Prediction in Servo Detection 30-4
30.4 Track Follow Mode Peformance 30-4
Channel Model and Simulation Environment • Ontrack
Results • 50% Offtrack Results
30.5 Seek Mode Performance.......................... 30-6
Model of Radial Incoherence • Simulation Results
30.6 Servo Burst Demodulation 30-9
Burst Amplitude Calculation • Burst Demodulation SNR
• Some Other Issues
30.7 Conclusions 30-11

Pervez M. Aziz
Agere Systems
Dallas, TX

Viswanath Annampedu
Agere Systems
Allentown, PA

30.1 Introduction

Disk drives read and detect data written on tracks. Each track is comprised of mostly user or "read" data sectors as well as "servo" sectors embedded between read sectors. Information read by the read channel from the servo sectors allows the servo positioning control system to properly position the read/write head at the correct location on the disk. Let us now discuss what information comprises each servo sector. Every servo sector contains a synchronization (sync.) field or preamble pattern allowing the system to perform timing and gain recovery from the servo data. This preamble pattern is commonly some length of the periodic $2T$ pattern ... 00110011 .... The servo sectors are identified by a servo address mark (SAM) which follows the sync. field. The preamble and SAM are the same for all servo sectors. The SAM is followed by servo Gray data which represents the track/cylinder number or coarse positioning information of the servo sector in the disk. As the name Gray data implies the Gray field changes by only 1 bit from one track to the next. The Gray data are followed by a burst demodulation field which represents fine positioning information which allows the servo positioning system to determine whether the head is on track center or off track. The servo sector format described above is shown in Figure 30.1.

When the read head is moving slowly across a track it is said to be in "track follow" mode. In this mode the servo signal is impaired by additive noise and fixed phase, frequency and gain offsets between the written and readback signals — impairments also experienced by the read signal. In the track follow mode, the read head is said to be "ontrack" if it is exactly in the middle of a track — otherwise it is "offtrack."

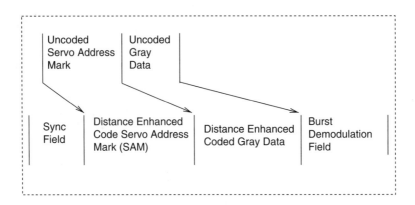

FIGURE 30.1   Format of a servo track.

In the worst case, the head can be 50% offtrack that is, be exactly halfway between two tracks. If the head is offtrack, it will see a superposition of the signals written at adjacent tracks as opposed to the signal from a single track. During the track follow mode it is important to accurately detect the SAM, Gray, and burst demodulation fields whether the head is exactly ontrack or even 50% offtrack. However, in "seek" mode, the head moves fast radially across the disk and must traverse multiple tracks within a short span of time. With ideal servo track writing, the signals on adjacent tracks would be written with the same phase. However, due to servo track writing impairments, the signals at the adjacent tracks are written phase incoherently that is at different phases. This phenomenon is called radial incoherence (RI) and is illustrated in Figure 30.2. Thus, not only can the head be offtrack and see a superposition of the signals written at adjacent tracks — the superposition is based on the tracks being written at different phases. The head will experience phase changes in the signal as it seeks. In seek mode it is typically not as important to read the burst demodulation field accurately.

Techniques such as partial response maximum likelihood (PRML) Viterbi detection [1], distance enhancing coding [2], and noise prediction [3] have been used to improve the detection reliability of read sectors. Here we show that these techniques are also applicable to the detection of the servo SAM and Gray fields. We discuss coding gain for servo and present some example servo codes with good performance. We show that the performance can be further enhanced by incorporating noise prediction into the detection. We then present a model of radial incoherence to characterize seek mode performance. We show that the ontrack coding gain also benefits the servo detection performance in seek mode when RI impairs the detection performance. Finally, we present a brief overview of signal processing issues for burst demodulation performed in the servo channel.

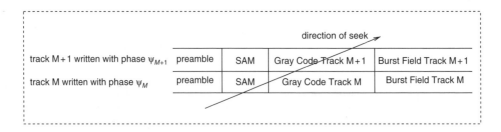

FIGURE 30.2   Phase incoherence in writing a servo track.

30.2 Servo Coding Gain

30.2.1 Example Codes and Their Minimum Distance

Improvements in the bit error rate (BER) achieved by codes can be estimated by the ratio of the minimum distance, d_m, obtained in one system to the d_m of the other system. The ratio in dB is the coding gain: $g = 20 \log(\frac{d_{m1}}{d_{m2}})$. Minimum distance is defined as the minimum Euclidean distance between two error events which are most likely to be mistaken for each other:

$$d_m = \min \forall E \left(d_E = \sqrt{\sum_l [y_{B1}(l) - y_{B2}(l)]^2} \right) \qquad (30.1)$$

where the error event $E = B1 - B2$ results from mistaking one bit pattern $B1$ with another $B2$, and y_{B1} and y_{B2} are equalized samples at the output of the PR channel corresponding to patterns $B1$ and $B2$. The sum is over the error event span. To estimate d_m we take a practical approach — we search distances based on Viterbi allowed codeword pairs which can be confused for each other.

We now consider some servo codes which improve the performance over an uncoded PRML servo channel. One straightforward way of servo encoding is to use a repetition code. For example, one can consider a $2T$ code with rate $1/2$, mapping 0 to 00 and 1 to 11. The d_m event is to mistake a 00 for a 11 or vice versa. Performing the d_m calculations show that the $2T$ code achieves a 3.98 dB coding gain over the single bit error event (mistaking a 1 for a 0 or vice versa) in an uncoded PRML servo channel with an EPR4 target. Different codes with different criteria [4, 5] could be considered. Some codes may perform better than others for different partial responses. Here, we only consider another example code with good performance with the EPR4 target. The code is a rate $1/4$ code which maps 00 to 00001111, 11 to 11110000, 01 to 00111100, and 10 to 11000011. This $2/8$ code provides an additional 3.42 dB of coding gain beyond the $2T$ code.

In order to realize the desired coding gains, the eight state EPR4 Viterbi detector must be modified to enforce code constraints such that the detected Viterbi data satisfies the constraints. To do this we first write out successive codewords to determine which branch transitions are allowed by the code. We must then prune the trellis such that all other paths through the trellis are eventually discarded. For the $2T$ code this results in the trellis of Figure 30.3. The trellis is periodic with 2 channel bits. During the first phase, the trellis is identical to a standard eight state trellis. During the second phase, the trellis is pruned such that each state has one incoming branch transition which is disallowed. One can derive a similar trellis for the $2/8$ code. This trellis will have a periodicity of eight channel bits.

30.2.2 Coding Gain/Rate Tradeoff for Servo

Although the $2T$ and $2/8$ codes provide substantial coding gain over an uncoded EPR4 servo channel, the gain comes at the expense of code rate loss. We consider the impact of this rate loss on the servo system vis a vis normal read detection. In the case of read detection the rate loss offsets the coding gain. Consider read detection with white noise impairing the channel. If two codes have rates r_1 and r_2 the channel bit density (*cbd*) for those codes is ubd/r_1 and ubd/r_2 where ubd is the desired user bit density for the two codes. The code with higher *cbd* requires a higher channel clock frequency resulting in proportionately more noise in the Nyquist band and an effective reduction in the SNR. However, for servo channels, the channel clock frequency is usually fixed so the amount of noise in the Nyquist band is independent of the code rate. There is no direct SNR reduction as in the case of read detection. However, the code with lower rate requires more servo format for the same number of "user" SAM/Gray bits. To translate the format penalty to an effective SNR penalty is not straightforward. However, servo system designers seem to find rates as low as $1/3$ or $1/4$ to be acceptable.

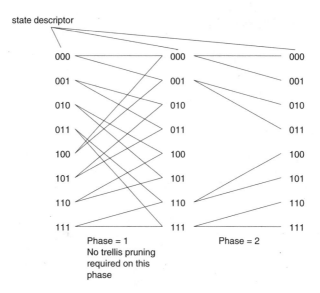

state descriptor

Phase = 1
No trellis pruning
required on this
phase

Phase = 2

FIGURE 30.3 Pruned time varying trellis for the 2T code. Trellis has periodicity of 2 channel bits. No pruning required on one of the phases.

30.3 Noise Prediction in Servo Detection

As done in [3] for read detection, we incorporate noise prediction (NP) for servo detection. We assume a N tap linear predictive model of the equalizer error $e(n)$ such that $e(n) = \sum_{i=1}^{N} p(i)e(n-i)$. The predictor coefficients $p(i)$ can be computed analytically from equalizer error statistics with the Yule-Walker equations or through an adaptive LMS algorithm. With noise prediction, the branch metric $bm(s_j, s_k)$ from state s_j to s_k at time n (in the trellis of Figure 30.3 for the 2T code) becomes

$$bm(s_j, s_k) = \left[y(n) - \sum_{l=0}^{M} h(l)a(n-l) - \sum_{i=1}^{N} p(i)e(n-i) \right]^2 \tag{30.2}$$

The term $h(l)$ is the lth (of M) coefficients of the partial response target and $a(i)$ denote the trellis state dependent data.

30.4 Track Follow Mode Peformance

30.4.1 Channel Model and Simulation Environment

We now examine the ontrack performance of the $2T$ and $2/8$ servo codes using bit error rate (BER) simulations. We use a Lorentzian-Gaussian (LG) pulse with 70 to 30% mixture to produce servo signals in the range of practical *cbd*s from 0.5 to 2.0. To equalize the servo signal to the EPR4 target, we use a fractionally spaced equalizer trained using the LMS algorithm on known servo data. Using this over-sampled equalizer removes the burden of having to optimize a practical equalizer. A Viterbi sequence detector follows the equalizer. The Viterbi trellis is time varying for the coded simulations. A block diagram of this simulation environment is shown in Figure 30.4. Simulations involving NPML are run with eight NPML taps where the branch metrics of the Viterbi detector in Figure 30.4 are modified as per Equation 30.2. We define SNR as isolated pulse peak to r.m.s. noise in the Nyquist band. Simulations are performed with white noise as the input to the channel. Simulations with 85% colored noise are

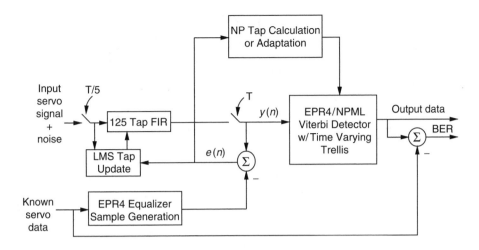

FIGURE 30.4 Servo simulation environment for ontrack results using an oversampled equalizer.

also performed. For the colored case, the colored part of the noise is obtained by filtering white noise with the LG dipulse.

30.4.2 Ontrack Results

Figure 30.5 shows the BER performance simulation results for the servo system with white noise. In white noise the 2T code performs about 4 dB better than the uncoded EPR4 channel at most *cbd*s but 4.5 dB better at the highest *cbd* of 2.0. The 2/8 code performs anywhere between 3.46 and 3.82 dB better than the 2T code. The coding gains correlate closely with the analytical results but of course do not match them exactly since the simple analytical calculation does not take into account the noise color at the equalizer output.

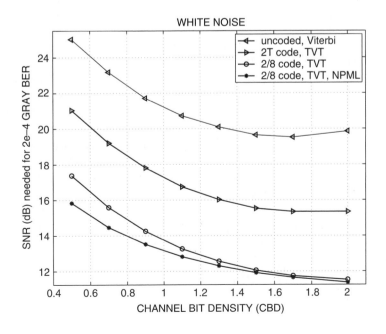

FIGURE 30.5 Ontrack results for white noise, SNR needed for 2e-4 BER.

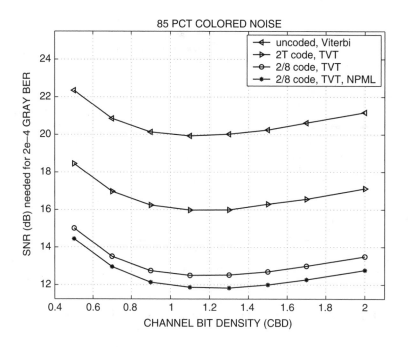

FIGURE 30.6   Ontrack results for colored noise, SNR needed for 2e-4 BER.

The gain from noise prediction with the 2/8 code starts at 1.54 dB at a *cbd* of 0.5, drops to a minimum gain of 0.09 dB at a *cbd* of 1.7 and then apparently starts to increase again with a gain of 0.15 dB at a *cbd* of 2.0. This is not surprising since with white noise input to the channel, EPR4 matches the channel well at the higher servo *cbd*s but does not match it well at very low *cbd*s.

The colored noise results are shown in Figure 30.6. The coding gains are similar. However, with noise prediction, the gain over the 2/8 code ranges from 0.54 to 0.72 dB vs *cbd*.

30.4.3 50% Offtrack Results

We have examined simulation results for an ontrack signals at different *cbd*s. When the head is 50% offtrack the servo channel will see an equally weighted sum of signals from the adjacent tracks. The Gray code fields of these adjacent tracks will differ by 1 Gray code bit. The bit position at which they are different is referred to as the indeterminate bit position. When the Gray code field is detected and decoded, the indeterminate bit will be detected as a 0 or 1 with near equal probability which is to be expected being 50% offtrack. However, it is important that the BER of the nonindeterminate bits do not degrade in this situation due to intersymbol interference from the indeterminate bit. If this is the case, this implies that the Gray code will most of the time decode to either of the adjacent track Gray code values as opposed to some other Gray code value. Having run such 50% offtrack simulations we observe that the BER degradation for the nonindeterminate bits is about 0.25 dB for the 2*T* code relative to the ontrack performance. The corresponding degradation for the 2/8 code is less than 0.1 dB.

30.5   Seek Mode Performance

We now wish to discuss the seek mode detection performance of the servo system. Before doing that we discuss a model of radial incoherence (RI) which can model the impairments the servo signal will experience during seek mode.

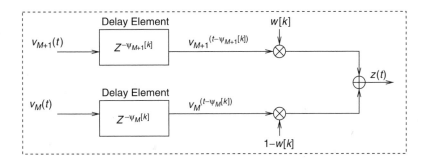

FIGURE 30.7   Seek mode model for radial incoherence.

30.5.1   Model of Radial Incoherence

To model a servo signal impaired with RI, we can compute it as a weighted sum of the phase incoherently written signals from the adjacent tracks [6]. Let $v_{M+1}(t)$ be the signal from track $M+1$ and $v_M(t)$ be the signal from track M where these signals differ only in the Gray code field by 1 user Gray bit. Let the phase or delays for these tracks be $\psi_{M+1}[k]$ and $\psi_M[k]$ (in units of channel bits, T) and the weighting function be $w[k]$ where these parameters vary with time k during a seek. Let the RI impaired signal be $z(t)$. The following equation and Figure 30.7 describe the model for the RI impaired signal:

$$z(t) = v_{M+1}(t - \psi_{M+1}[k])w[k] + v_M(t - \psi_M[k])(1 - w[k]) \tag{30.3}$$

The weighting function and phase change at a rate determined by the track crossing time denoted as C and measured in channel bit times. The smaller C is, the higher is the seek velocity. The fractional phase delays $\psi_{M+1}[k]$ and $\psi_M[k]$ are characterized as Gaussian random variables with standard deviation σ. Since the phase incoherence is due to the phase *difference* of $\psi_{M+1}[k]$ and $\psi_M[k]$, the standard deviation of the RI is characterized by $\sigma_R = \sqrt{2}\sigma$. The $\psi_{M+1}[k]$ and $\psi_M[k]$ change values every C bits. The weighting function $w[k]$, which represents the amplitude weighting from the two tracks is modeled as a triangular function with a period of C bits. The servo is ontrack when $w[k] = 0$ or $w[k] = 1$. It is 50% offtrack when $w[k] = 0.5$. We should note that although the $z(t)$ is modeled in continuous time, the parameters, $\psi_{M+1}[k], \psi_M[k], w[k]$ which control $z(t)$ are modeled as changing in units of discrete time, k. Figure 30.8 shows an example of a sampled preamble $2T$ pattern signal (solid line) as well the corresponding signal impaired with RI (dashed line). The unimpaired signal is sampled at the peaks and zero crossings of the underlying sinusoid. Note that over time the RI impaired signal experiences a change in phase causing the samples to become shoulder sampled and reducing the peak sampled value.

30.5.2   Simulation Results

We now examine simulation results comparing the detection performance of the 2T and 2/8 coded systems with RI impairments. The simulations are conducted with a symbol rate equalizer, practical digital phase locked loop (DPLL) based timing recovery loop, and an adaptive gain loop. A block diagram of the simulation environment is shown in Figure 30.9. Equalization is performed by a combination of a continuous time filter (CTF) and a discrete time analog FIR filter. The FIR is programmable but not adapted during the servo event. The arrows from the SAM detection to the Gray code decoding and burst demodulation blocks reflect the fact that these operations are timed relative based on the detected SAM. Using this simulation environment we measure the BER of the 2T and 2/8 coded systems at a *cbd* of 1.0 at a colored noise SNR of 20 dB. The track crossing time is chosen to be a realistic value, $C = 50$ channel bits. No noise prediction is used with the 2/8 code for these simulations. The BER versus σ_R is measured and plotted in Figure 30.10. We conclude that for a given error rate, the 2/8 code can tolerate more than

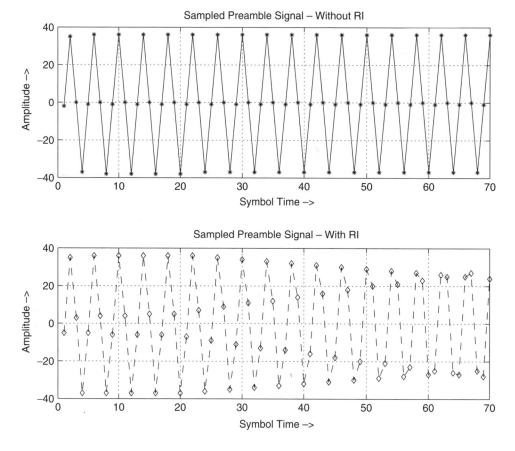

FIGURE 30.8   Preamble signal impaired with radial incoherence.

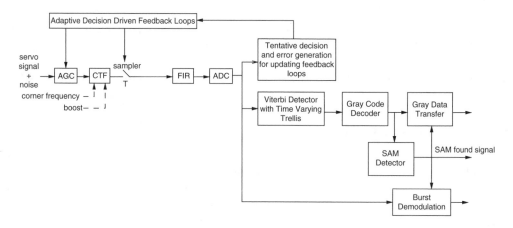

FIGURE 30.9   Servo simulation environment for RI results using a practical symbol rate equalizer and digital PLL (DPLL) based timing recovery.

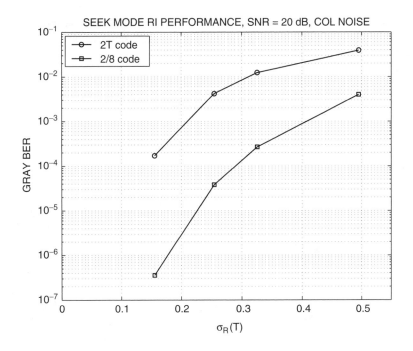

FIGURE 30.10 Seek mode Gray BER performance with RI for 2T and 2/8 codes. SNR $= 20$ dB, *cbd* $= 1.0$.

twice the amount of RI than the $2T$ code. Thus the 2/8 code provides significantly more tolerance to RI during a seek as well as ontrack coding gain.

30.6 Servo Burst Demodulation

In this section we review signal processing for servo burst demodulation. As discussed in [7, 8] there are numerous burst demodulation schemes which are based on different burst formats and different methods in calculating the position error signal (PES). However, most schemes involve writing of a sinusoidal signal in the burst demodulation field. The sinusoid amplitude changes as the head moves offtrack and this dependence of amplitude on position in the burst field is used by the servo control system to calculate the PES which drives the servo control loop to move the head to the correct track center for reading or writing. The sinusoid is typically written at one quarter the servo channel clock frequency that is at frequency $\frac{1}{4T}$ where T is the clock period. Therefore, the sinusoid can be written by writing the periodic $2T$ pattern . . . 11001100 . . . as was the case for the servo preamble field. The function of the servo channel is to detect the amplitude of the written sinusoid and provide that to the servo control system for PES calculation.

30.6.1 Burst Amplitude Calculation

Before discussing the burst field in detail let us quickly review the operations which take place during the entire servo sector. The servo channel will first acquire gain and timing during the preamble field where it may typically sample the preamble sinusoid at its peak and zero crossing samples. Since the burst sinusoid and the preamble signal are the same, the timing and gain information acquired during the preamble also allows the burst sinusoid amplitude to be calculated correctly.

After timing and gain acquisition, the SAM and Gray fields are detected. Note that the number of bits in the SAM and Gray fields is known *apriori*. Therefore, provided the SAM is detected correctly, the beginning and end of the burst field is known *apriori*.

There are a number of methods to calculate the burst sinusoid amplitude such as peak detection, area integration, etc [7]. However, we will focus on one popular method which is Fourier demodulation based on symbol rate (T rate) samples of the burst field. This method computes "sine" and "cosine" components of the sinusoid as follows:

$$z_c = \sum_{k=0}^{4N_p-1} y(k)b(k) \tag{30.4}$$

$$z_s = \sum_{k=0}^{4N_p-1} y(k)b(k-1) \tag{30.5}$$

where $b(k)$ is the periodic pattern $1, 0, -1, 0, \ldots$ with period $4T$. The signal $y(k)$ is the ADC output in Figure 30.9. Ideally $y(k)$ will be a sampled sinusoid that is, $y(k) = A\sin(\frac{2\pi}{4T}(kT - \phi))$ where ϕ is the sampling phase of the sinusoid, $0 < \phi < T$ and where we have assumed there is no residual frequency error between the written and read back signal. The amplitude of the equalized sinusoid is A. The number of periods of the sinusoid written for the burst is N_p, each period producing $4T$ spaced samples. If the timing recovery subsystem of Figure 30.9 is working well, this will be the case. Also, the timing recovery subsystem will typically drive ϕ to be as close to zero as possible during the preamble field. Therefore $y(k)$ will closely represent the peak and zero crossing samples of the sinusoid. In the absence of any noise or other impairments and when the head is ontrack to produce a full amplitude burst, we will have $z_s = 2N_p A$ and $z_c = 0$. The values of z_s and z_c will scale proportionately with offtrack position.

The magnitude of the burst is calculated as,

$$m = \sqrt{z_c^2 + z_s^2} \tag{30.6}$$

Note that m is insensitive to the acquisition phase error ϕ.

30.6.2 Burst Demodulation SNR

The ultimate performance metric for the burst demodulation should be the quality of the PES signal calculated by the servo control system based on the burst amplitude it receives from the servo channel. Nevertheless, it is still useful to characterize the performance at the servo channel output. One common metric is the burst demodulation SNR which is defined as the ratio of the mean of the burst magnitude to its standard deviation.

$$SNR_m = 20\log 10\left(\frac{\text{mean}(m)}{\sigma_m}\right) \tag{30.7}$$

Ideally the mean will be $2N_p A$ as noted earlier. The standard deviation σ_m will depend on the equalized noise on the burst samples $y(k)$ as well as any other system perturbations. For the sake of simple analysis let us assume an additive zero mean white Gaussian noise is the only impairment on the $y(k)$ samples that is, the timing and gain recovery of the burst samples are ideal. Let $w(k)$ be the assumed white noise process. For the sine component we obtain,

$$z_s = 2N_p A + \sum_{k=0}^{N_p-1} w(k) \tag{30.8}$$

The mean of z_s is $2N_p A$ and the standard deviation of z_s is $\sqrt{N_p}\sigma_w$ assuming $w(l)$ to be a white noise process. For the moment let us consider a demod SNR, SNR_s based on just the sine component only,

$$SNR_s = 20\log 10\left(\frac{\text{mean}(z_s)}{\sigma_{z_s}}\right) \tag{30.9}$$

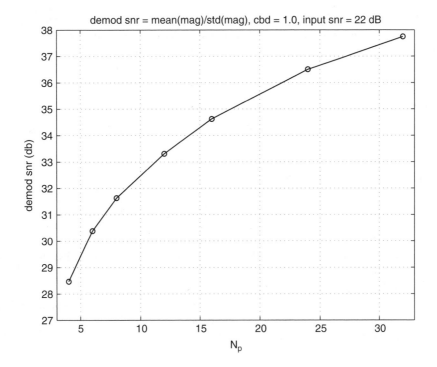

FIGURE 30.11 Demodulation SNR vs. N_p for white noise input SNR of 22 dB, $cbd = 1.0$.

The ratio inside the log term is $\frac{2\sqrt{N_p}A}{\sigma_w}$. Thus the SNR defined as such increases with $\sqrt{N_p}$ that is, for every doubling of N_p the SNR_s improves by 3 dB. It can also be shown that the overall magnitude SNR, SNR_m also improves by 3 dB for every doubling of N_p. The improvement in burst demod SNR, of course, come with format penalty in the servo sector. Values of $N_p = 8$ to 16 have been typically used in recent years.

Figure 30.11 shows simulation results of SNR_m as a function of the number of periods, N_p, of the burst signal. The SNR improves by 3 dB for every doubling of N_p. The simulations are run with the simulation environment of Figure 30.9 using a servo signal at a cbd of 1.0 and input white noise SNR of 22 dB.

30.6.3 Some Other Issues

In order for the servo channel to maintain good demod SNR and ultimately produce a good PES signal it is important that other parts of the servo channel do not compromise the burst demod performance. Specifically the equalization provided by the combination of CTF and FIR for good SAM/Gray detection performance should also result in good demod SNRs. The adaptation noise in the gain recovery and timing recovery algorithms, including any sources of timing jitter, should be small enough to minimize any burst demodulation performance loss.

30.7 Conclusions

We have presented an overview of servo channel signal processing using examples of techniques employed in practical servo channels. We reviewed how PRML coding and detection techniques can be used to improve the detection of the servo data with a time varying Viterbi detector trellis matched to the code. We presented simulation results demonstrating the coding gains and showed that the performance can be further improved by incorporating noise prediction into the detection. Results have been shown for white as well as colored noise. We presented a model for radial incoherence to characterize servo signal impairments during seek mode. We demonstrated that during seek mode, on track coding gain can also translate into improved performance when RI degrades the detection performance. Finally, we reviewed signal processing operations for burst demodulation and presented simulation results of the burst demod SNR.

References

[1] R. Cideciyan et al., A PRML system for digital magnetic recording, *IEEE J. Select. Areas in Commun.*, pp. 38–55 January, 1992.

[2] W. Bliss, An 8/9 rate time varying trellis codes for high density magnetic recording, *IEEE Trans. Magn.*, pp. 2746–2748, September, 1997.

[3] E. Eleftheriou and W. Hirt, Noise-predictive maximum likelihood (NPML) detection for the magnetic recording channel, *Proc. IEEE Intl. Conf. Commn.*, pp. 556–560, 1996.

[4] P. Aziz, Rate M/N code, encoder, detector, decoder for control data, U.S. Patents, 6,751,744, June 15, 2004; 6,606,728, Aug 12, 2003; 6,480,984, Nov 12, 2002.

[5] A. Patapoutian, Wide biphase digital servo information and detection for disk drive using servo viterbi detector, U.S. Patent, August, 1997.

[6] J. Sonntag, Private communication, 1996.

[7] A. Sacks, Position Signal Generation in Magnetic Disk Drives, Ph.D. Thesis, September 5, 1995.

[8] A. Patapoutian, Head position sensing in disk drives *CRC Computer Engineering Handbook*, CRC Press, Boca Raton, U.S. Patent 5,661,760, 2002.

31

Evaluation of Thermal Asperity in Magnetic Recording

M. Fatih Erden

Erozan M. Kurtas

Michael J. Link
Seagate Technology
Pittsburgh, PA

31.1 Introduction **31**-1
31.2 Modeling Thermal Asperities **31**-2
31.3 Effect of TA on System Performance **31**-4
      System Model • System Parameter Settings • Simulation
      Results
31.4 TA Detection and Cancelation Techniques **31**-9
31.5 Conclusion **31**-16

31.1 Introduction

The magneto-resistive (MR) sensors used in today's disk drives are very sensitive to changes in magnetic field, and also to thermal changes. When a surface roughness or asperity comes into contact with the slider on the hard disk, both the surface of the disk and the tip of the asperity are heated, and the resistive properties of MR sensors change. This thermal property of MR sensors was identified shortly after their discovery [1], and is known as thermal asperity (TA).

The change in resistance of MR sensors causes transient jumps in the read back signal as sensed by the read channel. This transient behavior affects the system performance, and thus should be canceled. The system gets affected in two ways:

1. The magnitude of the read-back signal might increase so much that the system may reach its saturation limits. Then, the information about the data at times where baseline was already larger than this saturation limit will be lost.

2. Even if the system does not reach its saturation limits, the sudden significant change of baseline level affects the performance of the whole design.

In this chapter, we will assume no saturation and consider only the second item above. The TA effects of concern in today's drive systems are the so called grown TAs. These TAs develop after the drive is manufactured and are not mapped out as to be the hard TAs which are identified during the manufacturing and build process.

The following section of this chapter provides a model of the effects of a TA as seen at the input of the preamplifier based on TA events captured from known defects on disk drives. The next section illustrates the effect of TA on system performance if not dealt with, and the following one gives an overview of current and past methods for detecting and suppressing TAs. Finally, we conclude the chapter.

31.2 Modeling Thermal Asperities

Figure 31.1 and Figure 31.2 illustrate simulated readback signals without and with a TA, respectively. As can be seen from these figures, immediately after contact with an asperity the transient TA effect quickly and significantly changes the baseline of the read-back signal. The slider and asperity then cool so the baseline of the signal decays to its original level. This behavior is due to the MR element and can be observed both in perpendicular and longitudinal recording systems.

A simplified model for TA effects is necessary in order to quantify the effiency of TA suppression techniques and to aid in modeling on the fly methods to handle TA events. Figure 31.3 illustrates such a model which is described by Stupp et al. [2]. This model fits captured spin stand data and drive data very well. It shows the linear heating of the MR element, the peak temperature, and the exponential decay that characterizes a TA event. According to the model the decay immediately after the TA event is approximated by an exponential function, but later the decay slows down. However, this slow decay is masked by the high pass filter in the preamplifier and read channel for longitudinal recording systems [2]. Thus, we can completely specify the effect of TA with this model together with four parameters. These are:

1. START-TIME: This sets where the TA effect starts.
2. MAX-AMPLITUDE: Whenever there is TA, the value of the signal baseline changes from 0 to the value specified in this parameter times the 0 to peak signal voltage.
3. RISE-TIME: This specifies the rise time required for its base line to rise from 0 to its maximum value (specified at MAX-AMPLITUDE).
4. DECAY-CONSTANT: The TA effect is assumed to decay exponentially, and this parameter specifies the decay constant of the exponential function.

Using this model and measured data, we can model several TA scenarios that typify the conditions that are expected and have been observed in product testing. These scenarios differ in terms of the MAX-AMPLITUDE, RISE-TIME, and DECAY-CONSTANT of the TA effect. MAX-AMPLITUDE from 2 to 7 times the amplitude of the 0 to peak signal voltage, RISE-TIME from 14×10^{-9} to 60×10^{-9} seconds,

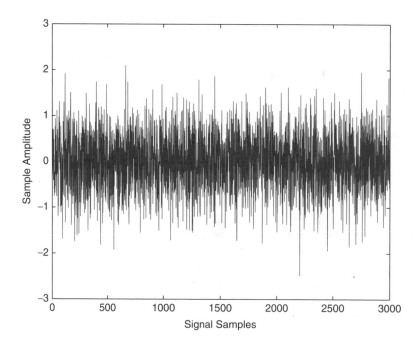

FIGURE 31.1   A typical read-back signal without TA.

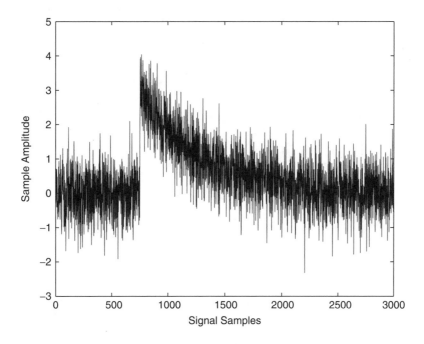

FIGURE 31.2 The same read-back signal in Figure 31.1, this time with TA.

and DECAY-CONSTANT from 25×10^{-9} to 45×10^{-9} seconds cover the range of TA events currently seen in GMR heads (see Figure 31.4). For example, one TA scenario might have a high MAX-AMPLITUDE, a small RISE-TIME, and low DECAY-CONSTANT. Since the amplitude of this TA is very high and has a fast rise time, it is easy to detect. However due to the low exponential decay this type of TA corrupts a large number of written bits. Another TA type might have a low MAX-AMPLITUDE, a large RISE-TIME, and high DECAY-

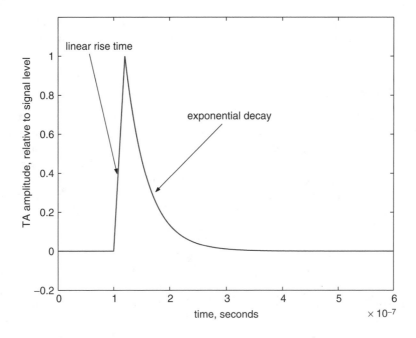

FIGURE 31.3 The simplified model which describes the TA event.

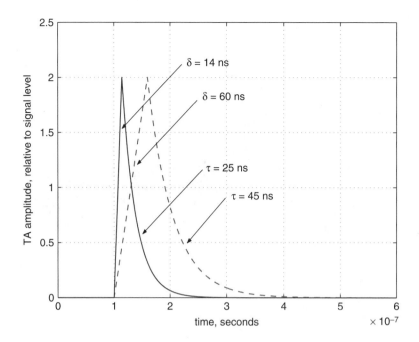

FIGURE 31.4 Typical TA scenarios.

CONSTANT. This TA is more difficult to detect but corrupts significantly fewer written data bits. Note that similar models are currently being developed for tunneling giant magnetoresistive (TGMR) heads.

31.3 Effect of TA on System Performance

Before focusing on TA detection and suppression algorithms, we believe that analyzing the effect of TAs on system performance will motivate the reader for the need of a TA algorithm. To do this, in this section we first explain the simulation environment where we quantify TA effects on system performance, then we focus on the simulation results.

31.3.1 System Model

The channel model used in our simulations is shown in Figure 31.5. We assume the only noise is due to additive white Gaussian noise (AWGN) and the only other system impairment is due to a TA. The real disk drive system is much more complicated, and effects due to media noise, tracking errors, and imperfections in timing recovery and AGC loops also must be considered. However, we will limit ourselves to the simple model in this figure to analyze TA effects with no other distortions or complications.

In Figure 31.5, the binary input sequence, $c_k \in \{\pm 1\}$, is first filtered by an ideal differentiator $(1 - D)/2$, where D is a unit delay operator, to form a transition sequence, $d_k \in \{-1, 0, 1\}$ where -1 and $+1$ represent

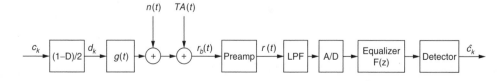

FIGURE 31.5 Channel model used in simulations.

negative and positive transitions while 0 stands for no transition. The sequence d_k then passes through the channel represented by the transition response.

The transition response for a longitudinal recording channel (usually known as a Lorentzian pulse) is given by

$$g(t) = \frac{K}{1 + \left(\frac{2t}{PW_{50}}\right)^2} \tag{31.1}$$

where K is a scaling constant and PW_{50} indicates the width of the Lorentzian pulse at half of its peak value. Similarly, the transition response for a perpendicular recording channel is given by

$$g(t) = erf\left(\frac{2t\sqrt{\ln 2}}{PW_{50}}\right) \tag{31.2}$$

where $erf(\cdot)$ is an error function which is defined by $erf(x) = \frac{2}{\sqrt{\pi}} \int_0^x e^{-t^2} dt$, and PW_{50} determines the width of the derivative of $g(t)$ at half its maximum. The ratio $ND = PW_{50}/T$ represents the normalized recording density which defines how many data bits can be packed within the resolution unit PW_{50}, and the dibit response is defined as $h(t) = g(t) - g(t - T)$.

After convolving the transition sequence d_k with the transition response $g(t)$, we add electronic noise in the system through the signal-to-noise ratio (SNR) value definition given as

$$SNR = 10 \log \frac{E_i}{N_0} \tag{31.3}$$

where E_i is the energy of the normalized impulse response of the recording channel, and N_0 is the power spectral density height of the electronic noise. For convenience, we normalize the impulse response of the recording channel so that E_i becomes unity.

Then, we get the read-back signal $r_b(t)$ before the preamp block by adding the TA effect. In our simulations, we used the model explained in Section 31.2, and chose the four TA parameters accordingly to control the TA effects. Finally, we obtain the read-back signal $r(t)$ after the preamp block by filtering the signal $r_b(t)$ with the preamp transfer function.

After obtaining the read-back signal, it is first low pass filtered, then uniformly sampled at baud rate (i.e., at time instants $t = kT$) assuming perfect synchronization. Following sampling, the baud rate samples of the read-back signal are then equalized by an FIR equalizer $F(z)$ to a target response $G(z)$. The design of generalized partial response (GPR) targets, and their corresponding equalizers can be found in [3]. Finally, the bit stream c_k is detected using the Viterbi detector which is proven to be a way to achieve maximum likelihood sequence detection (MLSD) under white Gaussian channel noise assumption [4].

31.3.2 System Parameter Settings

We believe that it is necessary to explain how we choose the parameters of the simulation environment, and we dedicate this subsection for that purpose.

- Two recording system are considered. These are:
  - Perpendicular recording system
  - Longitudinal recording system
- We set ND to be 1.5 for both recording systems. This low value is chosen to ensure that the system does not have much intersymbol interference (ISI) effect, and we will observe TA effect more clearly.
- We did not consider any of the media jitter and pulse broadening effects in our simulations because we wanted to single out the effect of TA in the system and quantify its effect in a system free from media noise.
- We chose the amount of the electronic noise such that the bit error rate (BER) of the system in absence of TA effect ranges roughly from 10^{-1} to 10^{-4}.

- We simulated the TA effect using the model as explained in the previous section.
- Here, we model the preamp by a simple high-pass filter whose cut-off frequency is equal to $1/1000$ of the channel clock.
- Perpendicular target response: PTar $= [0.8010\ 0.5987\ 0.0025]$
- Longitudinal target response: LTar $= [0.8733\ -0.1983\ -0.4450]$
  - These target responses are general partial response (GPR) targets desined using the monic constraint [3]. We chose the GPR target instead of a fixed targets with integer coefficients because we wanted to do our best to get rid of any misequalization effects in the system.
  - The length of the targets are 3. We have chosen a low ND value (ND $= 1.5$) for both recording systems. Thus, a target response with length 3 is sufficient.
  - These targets are designed at SNR where BER to be less than 10^{-4}, and then fixed for all SNR range of interest.
  - The above target responses are designed in absence of TA effect.
- The detector is fixed as Viterbi detector [4].
- Uniform sampling and perfect synchronization is employed. The reason is to get rid of any timing noise in the system to single out the TA effect.
- Performance of the system is chosen as "BER given TA." This means that, we calculate the BER figure of the system assuming that there is a constant TA effect in the system. We simulate a TA effect within every 32768 bits long block. Then, we calculate the BER of the system, and call that number as "BER given TA."

As can be seen from the previous itemized simulation settings, we fix most of the simulation parameters and vary only a few selected ones. The main motivation is to single out the TA effect in the system, and to observe its effect much more clearly.

31.3.3   Simulation Results

We are now ready to quantify the effect of TA on system performance. For this purpose, we use the model explained in Section 31.2, and vary only the parameter DECAY-CONSTANT, fixing RISE-TIME to 5 symbols, MAX-AMPLITUDE to 3. We choose three specific DECAY-CONSTANT values, and form TA signatures for those. We call them as

1. Severe TA: Affects around 10000 bit duration
2. Medium TA: Affects around 1000 bit duration
3. Minimum TA: Affects around 100 bit duration

before the preamp in the system (see Figure 31.5). Here, we model the preamp by a simple high-pass filter whose cut-off frequency is equal to $1/1000$ of the channel clock. After the preamp block the severe, medium, and minimum TA waveforms corrupt around 1500, 700, and 70 samples, respectively. These numbers are in good agreement with what is observed in real signals.

Figure 31.6 shows the BER given TA versus SNR curves in perpendicular recording system for the cases where we have severe, medium, minimum TA effects, together with no TA in the system. As an be seen from this figure, we do not observe the TA effect that much at low SNR ranges. However, at high SNR values we might see more than a decade difference in BER performance of the system, even with minimum TA signature. Thus, we have to cancel the effect of TA within the perpendicular recording system.

Similarly, Figure 31.7 shows the corresponding results for the longitudinal recording case. Unlike the perpendicular case, in longitudinal recording we do not see that much difference between the severe, medium, and minimum TA cases. Moreover, the difference between them and the "no TA" case reduces, and becomes apparent only at high SNRs.

The main reason of this performance difference between the perpendicular and longitudinal recording systems comes from the difference in their channel responses. Figure 31.8 shows the frequency domain representations of the channel impulse responses. As can be seen from this figure, perpendicular channel

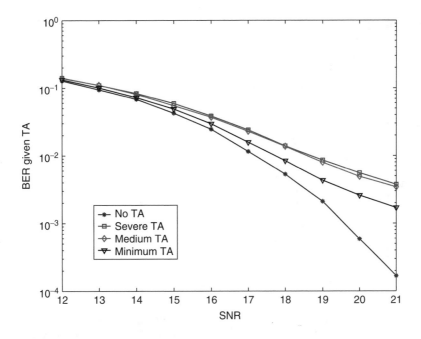

FIGURE 31.6 BER given TA versus SNR curves in perpendicular recording system in presence of "Severe TA," "Medium TA," "Minimum TA," and in absence of TA (i.e., "no TA") ND = 1.5, Target = PTar = [0.8010 0.5987 0.0025].

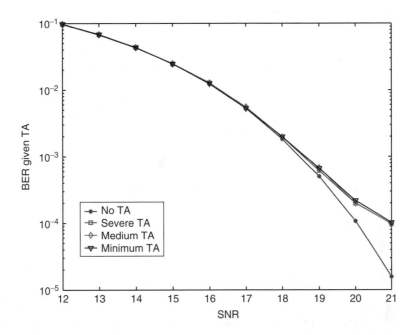

FIGURE 31.7 BER given TA versus SNR curves in longitudinal recording system in presence of "Severe TA," "Medium TA," "Minimum TA," and in absence of TA (i.e., "no TA") ND = 1.5, Target = LTar = [0.8733 −0.1983 −0.4450].

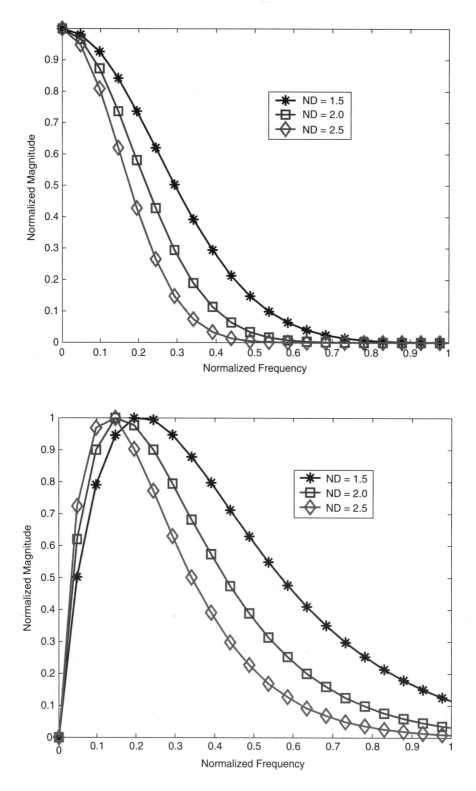

FIGURE 31.8 Impulse response in frequency domain at different NDs for perpendicular recording (top) and for longitudinal recording (bottom).

has considerable amount of energy at the low frequency band while the longitudinal one has very little. As the target responses are trying to match the channel response in the system, they also show similar properties. This can be seen from Figure 31.9 where while the target response for the longitudinal case filters out the low frequency contents of the signal, the one for perpendicular case does not.

As most of the energy of the TA effect is located at the low frequency contents of the signal, the target response for the longitudinal system gets rid of most of the TA effect in the system, while the target response of the perpendicular system preserves the TA effect. This is also shown in Figures 31.10 through 31.12. In Figure 31.10 we show a realization of a read-back signal with TA effect. We then filter this signal with the longitudinal and perpendicular target responses and show those at Figure 31.11 and Figure 31.12. As can be seen from these figures, the longitudinal target response gets rid of most of the TA effect in the system, while, that particular TA effect is preserved in perpendicular one.

However, although the longitudinal recording equalizer target response gets rid of most of the TA effect in the system, from Figure 31.7 we still see loss due to the residual TA at high SNR regions in longitudinal recording. Thus, from Figure 31.6 and Figure 31.7 we conclude that TA affects system performance and should be dealt with. The next section focuses on the conventional methods on TA detection and cancellation algorithms.

31.4   TA Detection and Cancelation Techniques

Sudden transient change of readback signal amplitude level helps develop simple TA detection algorithms. For example, one common method of detecting a thermal asperity in read channels is to compare the envelope of the readback signal with the average read signal envelope, or establish a baseline of the low frequency content of the read back signal [5]. A second method to detect TA events is to low pass filter the output of the variable gain amplifier in the automatic gain control circuit (AGC). The output of the low pass filter is compared to a preset threshold, if the signal is above the threshold a TA event is declared and the suppression of the TA effect is attempted. Klaassen provides an overview of other detection methods, and interested readers may refer to [5].

Once the TA is detected, its effect should be suppressed. A common means of lessening the effects of the TA is to implement a high-pass filter (HPF) whenever TA is detected [5]. The motivation of this method comes from the spectrum of the TA effect. As most of the energy of the TA effect is located at the low frequencies, the HPF filters out most of the TA effect in read-back signal. The high pass corner frequency is moved as high as permitted by low frequency content of the longer data patterns. In addition to the cut-off frequency of the HPF in the system, in [6] the gain of adaptive gain controller (AGC) is lowered when TA is detected to increase the dynamic range of the AGC and timing recovery loops.

Instead of implementing a HPF, Dorfman and Wolf proposes a method for TA suppression in equalizer design [7, 8]. Once the TA effect is detected, their method convolves the channel target response with a digital $(1 - D)$ prefilter (where D is the delay operator whose z-domain representation is z^{-1}). Since the target response is convolved with the $1 - D$ factor, the trellis of the Viterbi is also changed to match the new target. Dorfman demonstrates this technique using an EPR4 target, $(1 - D)(1 + D)^2$ which is changed to $(1 - D)^2(1 + D)^2$. The number of bits corrupted by a TA event is dramatically reduced which can help the overall system reduce the number of ECC symbols required [7]. However this improvement in TA operations reduces the overall performance of the system which is addressed in [8].

The digital prefilter $(1 - D)$ can be considered as a specific high-pass filter, and introduces a DC null at the target shape whenever TA effect is detected. This method is implicitly very similar with the one implementing a HPF. Both introduce a high-pass pole in the system whenever a TA is detected. In the first one, this is done by a high pass filter, in the second one the high-pass pole is inserted into the equalizer through changing its target response.

Once the bulk of the TA effect is cancelled, then the remaining portion can be suppressed by designing appropriate error correcting codes (ECC) within the system. In this section rather than the ECC methods, we focus on the signal processing suppression techniques mentioned above. We first look at the effect of HPF, then consider the equalizer target response together with the HPF.

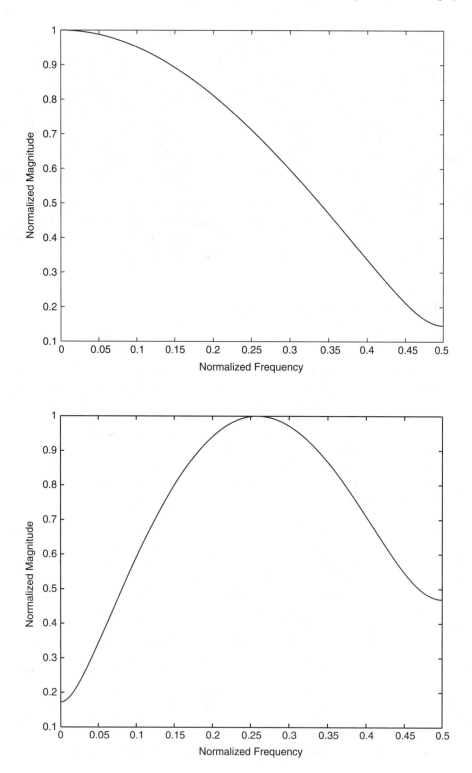

FIGURE 31.9 Normalized frequency response of the perpendicular (top) and longitudinal (bottom) targets.

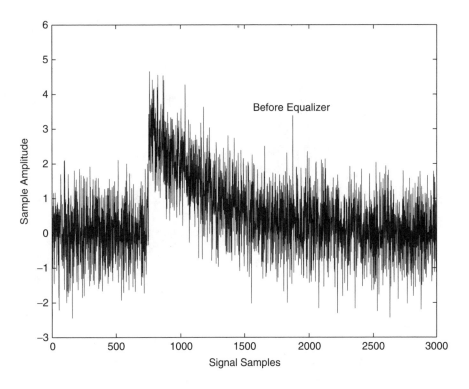

FIGURE 31.10 A realization of a read-back signal with TA effect.

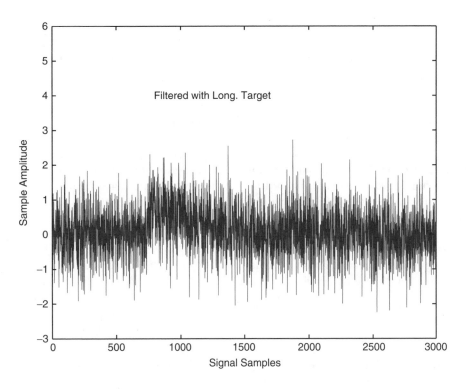

FIGURE 31.11 Read-back signal in Figure 31.10 filtered with the longitudinal target LTar.

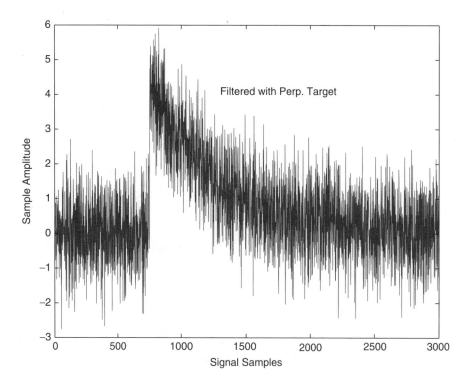

FIGURE 31.12   Read-back signal in Figure 31.10 filtered with the perpendicular target PTar.

Here, we use the system architecture in Figure 31.13. The only difference between this architecture and the one in Figure 31.5 is the "TA Detect Cancel" block which implements a high-pass filter at the analog domain to cancel the TA effect. We chose the z-transform representation of the HPF in the following form

$$H(z) = 1 - \frac{k}{1 - (1 - k)z^{-1}} \tag{31.4}$$

where k is the parameter of the filter, and also determines its cut-off frequency. Smaller the k smaller the cut-off frequency of the filter.

Figure 31.14 shows the effect of the cut-off frequency of the HPF. As we increase the k parameter of the HPF from $1/1024$ to $1/512$, $1/256$, and $1/32$ (which results in increasing the cut-off frequency of the HPF), the TA effect is filtered out more so that the length of the effect reduces. This, intuitively gives us an idea that increasing the HPF cut-off frequency might help in improving the performance of the conventional TA cancelation algorithm.

Next, we obtain the performance of both the longitudinal and perpendicular magnetic recording systems as a function of the cut-off frequency of the HPF. Figure 31.15 shows the SNR versus BER performance plots of longitudinal channel for different k values of the HPF (i.e., different high-pass cut-off frequencies). As can be seen from this figure, changing the k value from $1/1024$ to $1/32$ does not change the performance

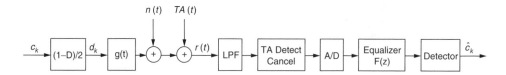

FIGURE 31.13   Channel model with conventional HPF method for TA cancelation.

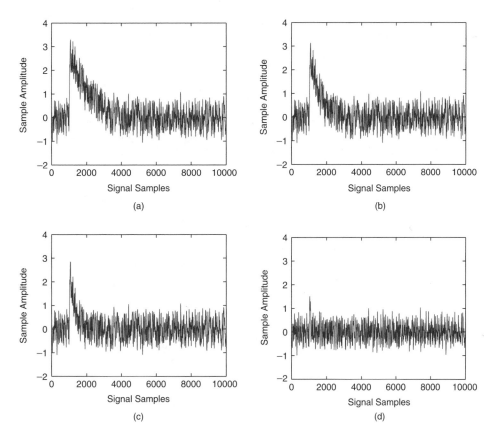

FIGURE 31.14   Readback signal with TA effect after the HPF with (a) $k = 1/1024$, (b) $k = 1/512$, (c) $k = 1/256$, (d) $k = 1/32$.

of the system much. The reason is, the high-pass cut-off frequency implicitly introduced by the channel equalizer target response (here, it is LTar = [0.8733 −0.1983 −0.4450]) is larger than the one introduced by the HPF, even for k equal to 1/32.

We then applied the HPF with different k values to perpendicular recording system. Figure 31.16 shows the results. As we see from these figures, if we choose the k to be 1/512, we obtain slight gain in system performance. However, in perpendicular recording systems increasing the cut-off frequency, for example increasing k from 1/1024 to 1/32, actually deteriorates the performance of the system. This can be explained by the difference at low frequency contents of the longitudinal and perpendicular read-back signals. Longitudinal read-back signals have very little energy content at low frequencies, thus by filtering the low frequencies using the HPF we lose very little information about the written data. However, for perpendicular recording systems, the read-back signal contains considerable amount of energy at low frequencies, which also means considerable amount of information about the written data on recording media. If we increase the cut-off frequency of the HPF, we filter more low frequency content which reduces the TA effect, however, we also lose the useful signal information. That is why, we get worse system performance.

Finally, we took k equal to 1/32 and 1/512 for longitudinal and perpendicular cases, respectively, and designed the optimum equalizer target for that specific k value whenever we have the "Severe TA" signature in the system. Figure 31.17 and Figure 31.18 show the results. As can be seen from those plots, optimizing the target gives very little gain.

Thus, the target response designed for No TA case deals with the TA effects very well once designed accordingly. The effect of HPF cut-off frequency will have very little effect on top of that. As for

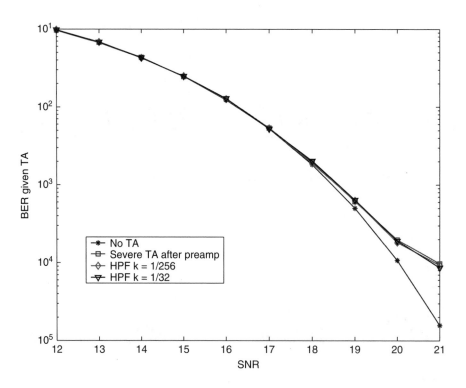

FIGURE 31.15 BER given TA versus SNR curves in longitudinal recording system in presence of "Severe TA," in absence of TA (i.e., "no TA"), and output of conventional TA detection algorithm with HPF of $k = 1/256$, and $k = 1/32$.

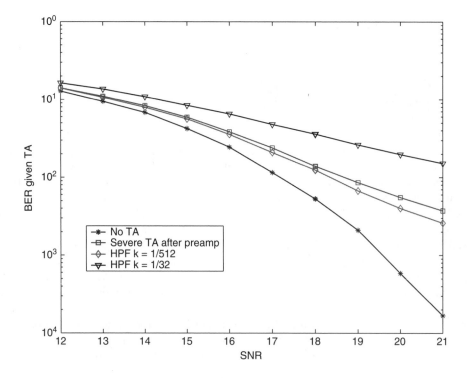

FIGURE 31.16 BER given TA versus SNR curves in perpendicular recording system in presence of "Severe TA," in absence of TA (i.e., "no TA"), and output of conventional TA detection algorithm with HPF of $k = 1/512$, and $k = 1/32$.

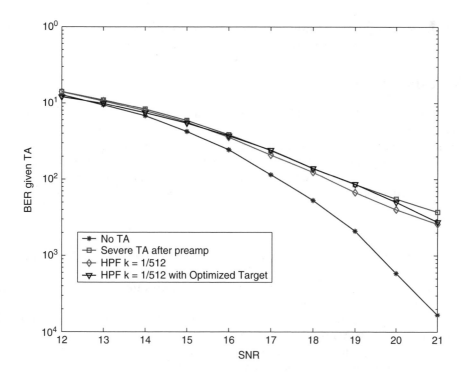

FIGURE 31.17 BER given TA versus SNR curves in perpendicular recording system in presence of "Severe TA," in absence of TA (i.e., "no TA"), with HPF of $k = 1/512$, and with HPF of $k = 1/512$ together with optimized target at every SNR.

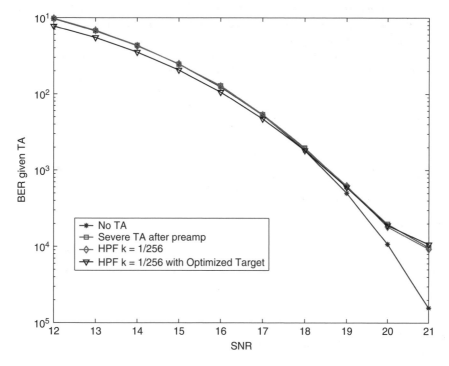

FIGURE 31.18 BER given TA versus SNR curves in longitudinal recording system in presence of "Severe TA," in absence of TA (i.e., "no TA"), with HPF of $k = 1/256$, and with HPF of $k = 1/256$ together with optimized target at every SNR.

perpendicular recording, there is room between the performance of the system with and without TA effects, and playing with the amount of high-pass pole in the system does not help much.

31.5 Conclusion

In this chapter, we analyzed the effect of TA to the performance of read-channel architectures for magnetic recording systems. We first, explained a TA model which fits well with the spin stand and drive data TA signatures, and using this model showed how severe the TA effect can be on system performance. Then, we reviewed the already existing methods. We see that, although the methods in the literature work fine with longitudinal recording channels, for perpendicular recording architectures there is still some room to improve the performance of the system in presence of TA effects.

References

[1] R.D. Hempstead, Thermally induced pulses in magnetoresistive heads, *IBM J. Res. Develop.* vol. 18, pp. 547, 1974.
[2] S.E. Stupp, M. A. Baldwinson, and P. McEwen, Thermal asperity trends, *IEEE Trans. Magn.* vol. 35, No. 2, pp. 752–757, March 1999.
[3] J. Moon and W. Zeng, Equalization for maximum likelihood detector, *IEEE Trans. Magn.*, vol. 31, no. 2, pp. 1083–1088, March 1995.
[4] G. D. Forney Jr., Maximum-likelihood sequence estimation of digital sequences in the presence of intersymbol interference, *IEEE Trans. Inform. Theory*, vol. IT-18, pp. 363–378, March 1972.
[5] K.B. Klaassen and J.C.L. van Peppen, Electronic abatement of thermal interference in (G)MR head output signals, *IEEE Trans. Magn.* vol. 33, no. 5, pp. 2611–2616, September 1997.
[6] R.L. Galbraith, G.J. Kerwin, and J.M. Poss, Magneto-resistive head thermal asperity digital compensation, *IEEE Trans. Magn.* vol. 33, no. 5, pp. 2731–2732, September 1992.
[7] V. Dorfman and J.K. Wolf, A method for reducing the effects of thermal asperities, *IEEE J. Select. areas Commun.* vol. 19, no. 4, pp. 662–667, April 2001.
[8] V. Dorfman and J.K. Wolf, Viterbi detection for partial response channels with colored noise, *IEEE Trans. Magn.* vol. 38, no. 5, pp. 2316–2318, September 2002.

32
Data Detection

32.1	Introduction	32-1
32.2	Partial Response Equalization	32-2
32.3	Decision Feedback Equalization	32-9
32.4	RAM-Based DFE Detection	32-10
	Detection in a Trellis	
32.5	Basic Breadth-First Algorithms	32-13
	The Viterbi Algorithm	
32.6	Noise Predictive Maximum Likelihood Detectors	32-17
32.7	Postprocessor	32-19
32.8	Advanced Algorithms and Algorithms Under Investigation	32-20
	Other Breadth-First Algorithms • Metric—First Algorithms • Bidirectional Algorithms	

Miroslav Despotović

Vojin Šenk
University of Novi Sad
Novi Sad, Yugoslavia

32.1 Introduction

Digital magnetic recording systems transport information from one time to another. In communication society jargon, it is said that recording and reading information back from a (magnetic) medium is equivalent to sending it through a time channel. There are differences between such channels. Namely, in communication systems, the goal is a user error rate of 10^{-5} or 10^{-6}. Storage systems, however, often require error rates of 10^{-12} or better. On the other hand, the common goal is to send the greatest possible amount of information through the channel used. For storage systems, this is tantamount to increasing recording density, keeping the amount redundancy as low as possible, that is, keeping the bit rate per recorded pulse as high as possible. The perpetual push for higher bit rates and higher storage densities spurs a steady increment of the amplitude distortion of many types of transmission and storage channels.

When recording density is low, each transition written on the magnetic medium results in a relatively isolated peak of voltage and peak detection method is used to recover written information. However, when PW_{50} (pulse width at half maximum response) becomes comparable with the channel bit period, the peak detection channel cannot provide reliable data detection, due to intersymbol interference. This interference arises because the effects of one readback pulse are not allowed to die away completely before the transmission of the next. This is an example of a so-called baseband transmission system, that is, no carrier modulation is used to send data. Impulse dispersion and different types of induced noise at the receiver end of the system introduce combination of several techniques (equalization, detection and timing recovery) to restore data. The present chapter gives a survey of most important detection techniques in use today assuming ideal synchronization.

Increasing recording density in new magnetic recording products necessarily demands enhanced detection techniques. First detectors operated at densities at which pulses were clearly separated, so that very simple, symbol-by-symbol detection technique was applied, the so-called *peak detector* [26]. With increased density, the overlap of neighboring dispersed pulses becomes so severe (i.e., large intersymbol interference — ISI) that peak detector could not combat with such heavy pulse shape degradation. To accomplish this task, it was necessary to master signal processing technology to be able to implement more powerful *sequence detection* techniques.

32.2   Partial Response Equalization

In the classical peak detection scheme, an equalizer is inserted whose task is just to remove all the ISI so that an isolated pulse is acquired. However, the equalization will also enhance and colorize the noise (from readback process) due to spectral mismatch. The noise enhancement obtained in this manner will increase with recording density and eventually become intolerable. Namely, since such a full equalization is aimed at slimming the individual pulse, so that it does not overlap with adjacent pulses, it is usually too aggressive, and ends up with huge noise power.

Let us now review the question of recording density, also known as packing density. It is often used to specify how close two adjacent pulses stay to each other, and is defined as PW_{50}/T. At PW_{50}/T ratios above 1 there is no viable solution that includes peak detection systems.

In this chapter we discuss two other receiver types that successfully combat this problem. These are the partial — response equalizer (PRE) and the decision — feedback equalizer (DFE). Both are rooted in old telegraph practice and, just as is the case with peak detector, they take instantaneous decisions with respect to the incoming data. In this chapter we will focus mainly on these issues, together with sequence detection algorithms that accompany PR equalization.

Partial response (PR) [27] (also called *correlative level coding* [2]) is a short small integer-valued channel response modeled as a FIR filter. PR equalization is the act of shaping the readback magnetic recording signal to look like the target signal specified by PR. After equalization, the data is detected using a sequence detector. Of course, quantization by an analog-to-digital converter (ADC) occurs at some point before the sequence detector.

The common readback structure consists of a linear filter, called a whitened matched filter, a symbol-rate sampler (ADC), a PRE and a sequence detector, Figure 32.1. The PRE in this scheme can also be put before the sampler, meaning that it is an analog, not a digital equalizer. Sometimes part of the equalizer is implemented in the analog, the other part in the digital domain. In all cases, analog signal, coming from the magnetic head, should have a certain and constant level of amplification. This is done in a variable gain amplifier (VGA). To keep a signal level, VGA gets a control signal from a clock and gain recovery system. In the sequel, we will assume that VGA is already (optimally) performed. In the design of equalizers and detectors, low power dissipation and high speed are both required. The error performances need to be maintained as well. So far, most systems seek for the implementations in the digital domain, as is the case in Figure 32.1, but it has been shown that ADC may contribute to the high frequency noise during the partial-response target equalization, causing a long settling time in clock recovery loop, as well as degrading performance [29]. In addition, the ADC is also usually the bottleneck for the low-power, high-speed applications. On the other hand, the biggest problem for an analog system is the imperfection

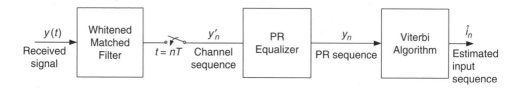

FIGURE 32.1   Maximum-likelihood sequence detector.

of circuit elements. The problems encountered with analog systems include nonideal gain, mismatch, nonlinear hold step, offset etc.

Let us now turn to the blocks shown in Figure 32.1. The first of them, the whitened matched filter, has the following properties [6].

Simplicity: A single filter producing single sample at the output is all that is needed. The response of the filter is either chosen to be causal and hence realizable, or noncausal, meaning some delay has to be introduced, yielding better performance.

Sufficiency: The filter is information lossless, in the sense that its sampled outputs are a set of sufficient statistics for estimation of the input sequence.

Whiteness: The noise components of the sampled outputs are independent identically distributed Gaussian random variables.

The whiteness and sufficiency property follow from the fact that the set of waveforms at the output of the matched filter is an orthonormal basis for the signal space.

The next block is PRE. Shortness of PR yields relatively simple sequence detectors. Kobayashi [8] suggested in 1971 that this type of channels can be treated as a linear finite state machine, and thus can be represented by the state diagram and its time instant labeled counterpart, *trellis diagram*. Consequently, its input is best inferred using some trellis search technique, the best of them (if we neglect complexity issues) being the Viterbi algorithm [1] (if we are interested in maximizing the likelihood of the whole sequence; otherwise, a symbol-by-symbol detector is needed). Kobayashi also indicated that the magnetic recording channel could be regarded as the partial response channel due to the inherent differentiation property in the readback process [7]. This is both present in inductive heads and in magnetoresistive (MR) heads, though the latter are directly sensitive to magnetization and not to its change (this is due to the fact that the head has to be shielded). In other words, the pulse will be read only when the transition of opposite magnet polarities is sensed.

Basic to the PR equalization is the fact that a controlled amount of ISI is not suppressed by the equalizer but rather left for a sequence detector to handle. The nature of the controlled ISI is defined by a partial response. A proper match of this response to the channel permits noise enhancement to remain small even when amplitude distortion is severe. In other words, PR equalization can provide both well-controlled ISI and spectral match.

PR equalization is based on two assumptions:

- The shape of readback signal from an isolated transition is exactly known and determined.
- The superposition of signals from adjacent transitions is linear.

Furthermore, we assume that the channel characteristics are fixed and known, so that equalization need not be adaptive. The resulting partial response channel can be characterized using D-transform of the sequences that occur,

$$X(D) = I(D) \cdot H(D)$$

[6] where $H(D) = \sum_{n=0}^{M-1} h_n D^n$, D represents the delay factor in D-transform and M denotes the order of the PR signals. When modeling differentiation, $H(D) = 1 - D$. The finite state machine (FSM) of this partial response channel is known as the *dicode* system, since there are only two states in the transition diagram.

The most unclear signal transformation in Figure 32.1 is equalization. What does it mean that the pulse of voltage should look like the target signal specified by the partial response (the so-called PR target)? To answer this question let us consider the popular Class IV partial response, or PR4 system.

For magnetic recording systems with PW_{50}/T approximately equal to 2, comparatively little equalization is required to force the equalized channel to match a class-4 partial-response (PR4) channel where $H(D) = (1 - D)(1 + D) = 1 - D^2$. Comparing to the Lorentzian model, PR4 channel shows more emphasis in the high frequency domain. The equalizer with the PR4 as the equalization target thus suppresses the low frequency components and enhances the high frequency ones, degrading the performance of all-digital detectors since the quantization noise, that is mainly placed at higher frequencies, is boosted up.

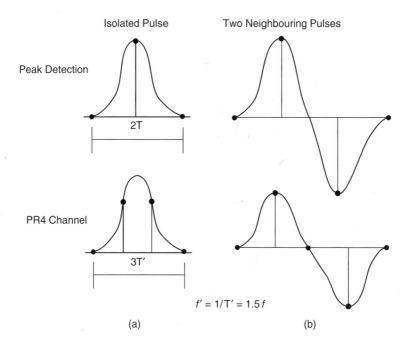

FIGURE 32.2   Capacity of PR4 channel.

The isolated pulse shape in a PR4 system is shown in Figure 32.2. The transition is written at time instant $t = 0$, where T is the channel bit period. The shape is oscillating and the pulse values at integer number of bit periods before the transition are exactly zeros. Obviously, it is this latter feature that should give us future advantage. However, at $t = 0$ and at $t = T$, that is, one bit period later, the values of the pulse are equal to "1." The pulse of voltage reaches its peak amplitude of 1.273 at one half of the bit period.

Assume that an isolated transition is written on the medium and the pulse of voltage shown in Figure 32.2(a) comes to the PRML system. The PR4 system requires that the samples of this pulse should correspond to the bit periods. Therefore, samples of the isolated PR4 pulse will be: $00\ldots011000\ldots$ (of course, "1" is used for convenience, and in reality it corresponds to some ADC level).

Since the isolated transition has two nonzero samples, when the next transition is written, the pulses will interfere. Thus, writing two pulses adjacent to each other will introduce superposition between them, usually called a *dipulse* response as shown in Figure 32.2(b). Here, the samples are $[\ldots, 0, 0, 1, 0, -1, 0, 0, \ldots]$, resulting from

$$
\begin{aligned}
&0\,0\,0\,1 \quad 1 \quad 0\,0\,0 - \text{from the first transition} \\
&\underline{+\,0\,0\,0\,0 -1 -1\,0\,0 - \text{from the second transition}} \\
&= 0\,0\,0\,1 \quad\; 0 -1\,0\,0
\end{aligned}
$$

Once the pulses can be reduced to the predetermined simple shape, the data pattern is easily recovered because superposition of signals from adjacent transitions is known. In the above example, we see that sample "1" is suppressed by "−1" from the next transition. It is a simple matter to check that all possible linear combinations of the samples result in only three possible values: $\{-1, 0, +1\}$ (naturally, we suppose that all parts of the system are working properly, that is, equalization, gain and timing recovery, and that the signal is noise-free). A positive pulse of voltage is always followed by a negative pulse and vice versa, so that the system can be regarded as an AMI (Alternative Mark Inversion) code.

The higher bit capacity of the PR4 channel can best be understood from Figure 32.2. It is observed that PR4 channel provides a 50% enhancement in the recording density as compared with the peak detection

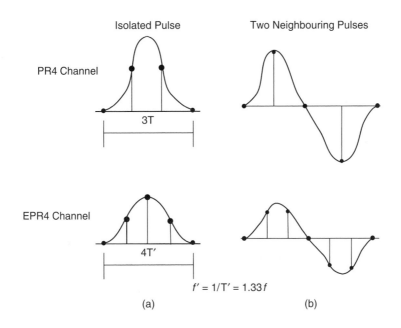

FIGURE 32.3   Capacity of EPR4 channel.

(fully equalized) one, since the latter requires isolation of single bits from each other. In Figure 32.3, we see that the EPR4 channel (explained later) adds another 33% to this packing density.

PR4 has another advantage over all the other PR systems: since $H(D) = 1 - D^2$, the current symbol is correlated to the second previous one, allowing the system to be modeled as two interleaved dicode channels, implying the use of simple dicode detectors for even and odd readback samples. RLL coding is necessary in this case, since nonideal tracking and timing errors result in a residual intermediate term (linear in D) that induces correlation between two interleaved sequences, and thus degrades systems that rely on decoupled detection of each of them.

RLL codes are widely used in conjunction with PR equalization in order to eliminate certain data strings that would render tracking and synchronization difficult. If PR4 target is used, a special type of RLL coding is used, characterized by $(0, G/I)$. Here, G and I denote the maximum number of consecutive 0's in the overall data string, and in the odd/even substrings, respectively. The latter parameter ensures proper functioning of the clock recovery mechanism if deinterleaving of the PR4 channel into two independent dicode channels is performed. The most popular is the $(0, 4/4)$ code, whose code rate is $7/8$, that is, whose rate loss is limited to 12.5%.

There are certainly other PR targets than PR4. The criterion of how to select the appropriate PR target is based on spectral matching, avoiding introducing too much equalization noise, and at the same time keeping trellis complexity as low as possible. For instance, for PW_{50}/T approximately equal to 2.25, it is better to model the ISI pattern as the so-called EPR4 (i.e., extended class-4 partial response) channel with $H(D) = (1 + D)^2 \cdot (1 - D) = 1 + D - D^2 - D^3$. As the packing density goes up, more low frequency components are being introduced (low compared to $1/T$, that also increases as T is shortened; in reality those frequencies are higher than those met for lower recording densities, respectively greater T). This is the consequence of the fact that intersymbol interference blurs the boundary between individual pulses, flattening the overall response (in time domain). The additional $1 + D$ term in the target PR effectively suppresses the unwanted high frequencies. EPR4 enables even higher capacities of the magnetic recording systems than PRIV, observing the difference of 33% in the recording density displayed in Figure 32.3.

However, a practical implementation of EPR4 is much more complex than is the case with PR4. First, the deinterleaving idea used for PR4 cannot be implemented. Secondly, the corresponding state diagram (and consequently trellis) now has 8 states instead of 4 (2 if deinterleaving is used). Furthermore, its output

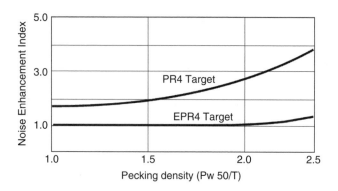

FIGURE 32.4   Equalization noise enhancement in PR channels.

is 5-leveled, instead of ternary for the PR4 and the dicode channel, so that a 4.4 dB degradation is to be expected with a threshold detector. Naturally, if sequence detector is used, such as VA, this loss does not exist, but its elimination is obtained at the expense of a significantly increased complexity of the detector. Furthermore, if such a detector can be used, EPR4 has a performance advantage over PR4 due to less equalization noise enhancement, Figure 32.4.

Let us reconsider the PR equalizer shown in Figure 32.1. Following the approach from [11], its aim is to transform the input spectrum $Y'(e^{j2\pi\Omega})$ into a spectrum $Y(e^{j2\pi\Omega}) = Y'(e^{j2\pi\Omega})|C(e^{j2\pi\Omega})|^2$, where $C(e^{j2\pi\Omega})$ is the transfer function of the equalizer. The spectrum $Y(e^{j2\pi\Omega}) = I(e^{j2\pi\Omega})|H(e^{j2\pi\Omega})|^2 + N(e^{j2\pi\Omega})$ where $H(D)$ is the PR target.

For instance, duobinary PR target $(H(D) = 1 + D)$ enhances low frequencies and suppresses those near the Nyquist frequency $\Omega = 0, 5$, whereas dicode $H(D) = (1 - D)$ does the opposite: it suppresses low frequencies and enhances those near $\Omega = 0, 5$.

In principle, the spectral zeros of $H(e^{j2\pi\Omega})$ can be undone via a linear (recursive) filter, but this would excessively enhance any noise components added. The schemes for tracking the input sequence to the system based on the PR target equalized one will be reviewed later in this chapter For instance, for a PR4 system, a second-order recursive filter can in principle be used to transform its input into an estimate of the information sequence, Figure 32.5.

Unfortunately, if an erroneous estimate is produced at any moment, then all subsequent estimates will be in error (in fact, they will no longer be in the alphabet $\{-1, 0, 1\}$, enabling error monitoring and simple forms of error correction [27]). To avoid this catastrophic error propagation, resort can be taken to a precoder.

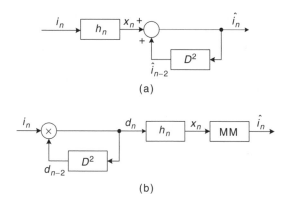

FIGURE 32.5   (a) PR4 recursive restoration of information sequence; (b) Precoder derived from it.

Let us analyze the functioning of this precoder in the case of the PR4 channel (generalization to other PR channels is trivial). Its function is to transform i_n into a binary sequence $d_n = i_n d_{n-2}$ to which the PR transformation is applied, Figure 32.5(b). This produces a ternary sequence

$$x_n = (d * h)_n = d_n - d_{n-2} = i_n d_{n-2} - d_{n-2} = (i_n - 1)d_{n-2}$$

Since d_{n-2} cannot be zero, x_n is zero iff $i_n - 1 = 0$, that is, $i_n = 1$. Thus we can form an estimate $\hat{i}_n$ of i_n by means of the memoryless mapping (MM)

$$\hat{i}_n = \begin{cases} 1, x_n = 0 \\ 0, \text{else} \end{cases}$$

This decoding rule does not rely on past data estimates and thereby avoids error propagation altogether. In practice, the sequences i_n and d_n are in the alphabet $\{0, 1\}$ rather than $\{-1, 1\}$, and the multiplication in Figure 32.5(b) becomes a modulo-2 addition (where 0 corresponds to 1, and 1 to -1).

The precoder does not affect the spectral characteristics of an uncorrelated data sequence. For correlated data, however, precoding need not be spectrally neutral.

It is instructive to think of the precoder as a first-order recursive filter with a pole placed so as to cancel the zero of the partial response. The filter uses a modulo-2 addition instead of a normal addition and as a result the cascade of filter and partial response, while memoryless, has a nonbinary output. The MM serves to repair this "deficiency."

Catastrophic error propagation can be avoided without precoder by forcing the output of the recursive filter of Figure 32.6. to be binary (Figure 32.6(a)). An erroneous estimate $\hat{i}_{n-2} = -i_{n-2}$ leads to a digit

$$\hat{i}_n = x_k + \hat{i}_{n-2} = i_n - i_{n-2} + \hat{i}_{n-2} = i_n - 2i_{n-2}$$

whose polarity is obviously determined by i_{n-2}. Thus the decision $\hat{i}_n$ that is taken by the slicer in Figure 32.6(a) will be correct if i_n happens to be the opposite of i_{n-2}. If data is uncorrelated, this will happen with probability 0.5, and error propagation will soon cease, since the average number of errors in a burst is $1 + 0.5 + (0.5)^2 + \cdots = 2$. Error propagation is thus not a serious problem.

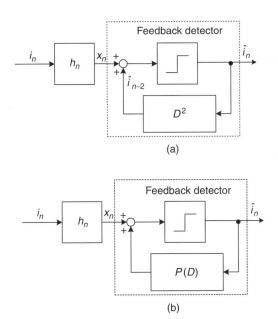

FIGURE 32.6   Feedback detector.

The feedback detector of Figure 32.6. is easily generalized to arbitrary partial response $H(D)$. For purposes of normalization, $H(D)$ is assumed to be causal and monic (i.e., $h_n = 0$ for $n < 0$ and $h_0 = 1$). The nontrivial taps $h_1, h_2, \ldots$, together form the "tail" of $H(D)$. This tail can be collected in $P(D)$, with $p_n = 0$ for $n \le 0$ and $p_n = h_n$ for $n \ge 1$. Hence $h_n = \delta_n + p_n$, where the Kronecker delta function δ_n represents the component $h_0 = 1$. Hence

$$x_n = (i * h)_n = (i * (\delta + p))_n = i_n + (i * p)_n$$

The term $(i * p)_n$ depends exclusively on past digits $i_{n-1}, i_{n-2}, \ldots$ that can be replaced by decisions $\hat{\imath}_{n-1}, \hat{\imath}_{n-2}, \ldots$. We can, therefore, form an estimate $\hat{\imath}_n$ of the current digit i_n according to $\hat{\imath}_n = x_k(\hat{\imath} * p)_n$ as in Figure 32.6(b). As before, a slicer quantizes $\hat{\imath}_n$ into binary decisions $\hat{\imath}_n$ so as to avoid catastrophic error propagation. The average length of bursts of errors, unfortunately, increases with the memory order of $H(D)$. Even so, error propagation is not normally a serious problem [19]. In essence, the feedback detector avoids noise enhancement by exploiting past decisions. This viewpoint is also central to decision-feedback equalization, to be explained later.

Naturally, all this can be generalized to nonbinary data, but in magnetic recording, so far, only binary data are used (the so-called saturation recording). The reasons for this are elimination of hysteresis and the stability of the recorded sequence in time.

Let us consider now the way the PR equalizer from Figure 32.1 is constructed. In Figure 32.7, a discrete-time channel with transfer function $F(e^{j2\pi\Omega})$ transforms i_n into a sequence $y_n = (i * f)_n + u_n$, where u_n is the additive noise with power spectral density $U(e^{j2\pi\Omega})$, and y_n represents the sampled output of a whitened matched filter. We might interpret $F(e^{j2\pi\Omega})$ as comprising of two parts: a transfer function $H(e^{j2\pi\Omega})$ that captures most of the amplitude distortion of the channel (the PR target) and a function $F_r(e^{j2\pi\Omega}) = F(e^{j2\pi\Omega})/H(e^{j2\pi\Omega})$ that accounts for the remaining distortion. The latter distortion has only a small amplitude component and can thus be undone without much noise enhancement by a linear equalizer with transfer function

$$C(e^{j2\pi\Omega}) = \frac{1}{F_r(e^{j2\pi\Omega})} = \frac{H(e^{j2\pi\Omega})}{F(e^{j2\pi\Omega})}$$

This is precisely the PR equalizer we sought for. It should be stressed that the subdivision in Figure 32.7. is only conceptual. The equalizer output is a noisy version of the "output" of the first filter in Figure 32.7, and is applied to the feedback detector of Figure 32.6. to obtain decision variables $\hat{\imath}'_n$ and $\hat{\imath}_n$. The precoder and MM of Figure 32.5. are, of course, also applicable and yield essentially the same performance.

The choice of the coefficients of the PR equalizer (PRE) in Figure 32.1 is the same as for full response equalization, and is explained in the chapter "adaptive equalization and timing recovery." Interestingly, zero-forcing here is not as bad as is the case with full-response signaling, and yields approximately the same result as minimum mean-square equalization. To evaluate the performance of the PRE, let us assume that all past decisions that affect $\hat{\imath}_n$ are correct and that the equalizer is zero forcing (see Chapter "adaptive equalization and timing recovery" for details). The only difference between $\hat{\imath}'_n$ and $\hat{\imath}_n$ is now the filtered

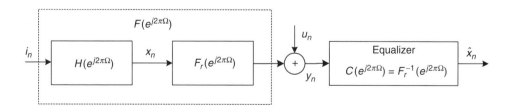

FIGURE 32.7 Interpretation of PR equalization.

noise component $(u * c)_n$ with variance

$$\sigma^2_{ZFPRE} = \int_{-0,5}^{0,5} U(e^{J 2\pi\Omega}) |C(e^{J 2\pi\Omega})|^2 \, d\Omega = \int_{-0,5}^{0,5} \frac{U(e^{J 2\pi\Omega}) |H(e^{J 2\pi\Omega})|^2}{|F(e^{J 2\pi\Omega})|^2} \, d\Omega$$

Since $|H(e^{j 2\pi\Omega})|$ was selected to be small wherever $|F(e^{j 2\pi\Omega})|$ is small, the integrand never becomes very large, and the variance will be small. This is in marked contrast with full-response equalization. There, $H(e^{j 2\pi\Omega}) = 1$ for all Ω, and the integrand in the above formula can become large at frequencies where $|F(e^{j 2\pi\Omega})|$ is small. Obviously, the smallest possible noise enhancement occurs if $H(e^{j 2\pi\Omega})$ is selected so that the integrand is independent of frequency, implying that the noise at the output of the PRE is white. This is, in general, not possible if $H(e^{j 2\pi\Omega})$ is restricted to be PR (i.e., small memory-order, integer-valued). The generalized feedback detector of Figure 32.6, on the other hand, allows a wide variety of causal responses to be used and here $|H(e^{j 2\pi\Omega})|$ can be chosen at liberty. Exploitation of this freedom leads to *Decision Feedback Equalizer* (DFE).

32.3 Decision Feedback Equalization

In this section we will review the basics of decision feedback detection. We will again assume that the channel characteristics are fixed and known, so that the structure of this detector need not be adaptive. Generalizing to variable channel characteristics and adaptive detector structure is tedious, but straightforward.

A DFE detector shown in Figure 32.8, utilizes the noiseless decision to help remove the ISI. There are two types of ISI: *precursor* ISI (ahead of the detection time), and *postcursor* (behind detection time). Feed-Forward Equalization (FFE) is needed to eliminate the precursor ISI, pushing its energy into the postcursor domain. Supposing all the decisions made in the past are correct, DFE reproduces exactly the modified postcursor ISI (with extra postcursor ISI produced by the FFE during the elimination of precursor ISI), thus eliminating it completely, Figure 32.9. If the length of the FFE can be made infinitely long, it should be able to completely suppress the precursor ISI, redistributing its energy into the postcursor region, where it is finally cancelled by feedback decision part. Since no spectrum inverse is needed for this process, noise boosting is much less than is the case with linear equalizers.

The final decision of the detector is made by the memoryless slicer, Figure 32.8. The reason why a slicer can perform efficient sequence detection can be explained with the fact that memory of the DFE system is located in two equalizers, so that only symbol-by-symbol detection can suffice. In terms of performance the DFE is typically much closer to the maximum — likelihood sequence detector than to the LE.

If the equalization target is not the main cursor, but a PR system, a sequence detection algorithm can be used afterwards. A feasible way to implement this with minimum additional effort is the tree search algorithm used instead of VA [5].

The simple detection circuitry of a DFE, consisting of two equalizers and one slicer, makes implementation possible.

The DFE may be regarded as a generalization of the PRE. In the DFE, the trailing portion of the ISI is not suppressed by a forward equalizer but rather canceled by a feedback filter that is excited by past decisions.

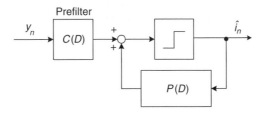

FIGURE 32.8 Decision feedback equalizer.

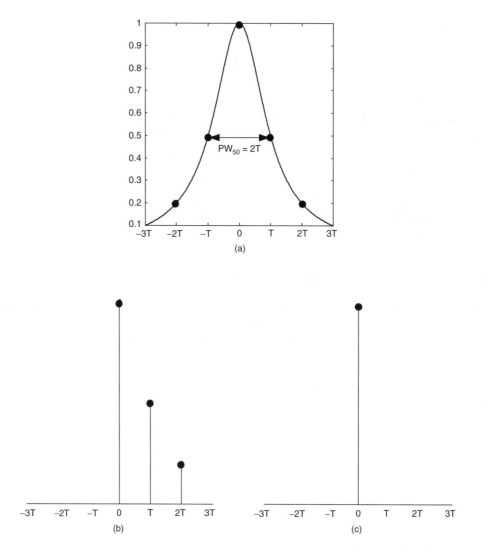

FIGURE 32.9   Precursor and postcursor ISI elimination with DFE: (a) Sampled channel response; (b) After feed-forward filter; (c) Slicer output.

Fortunately, error propagation is typically only a minor problem and it can, in fact, be altogether avoided through a technique that is called Tomlinson/Harashima precoding.

Performance differences between zero-forcing and minimum mean-square equalizers tend to be considerably smaller in the DFE case than for the LE, and as a result it becomes more difficult to reap SNR benefits from the modulation code. It can be proved that DFE is the optimum receiver with no detection delay. If delay is allowed, it is better to use trellis-based detection algorithms.

32.4   RAM-Based DFE Detection

Decision Feedback Equalization or RAM-based DFE is the most frequent alternative to PRML detection. Increase of bit density lead to significant nonlinear ISI in the magnetic recording channel. Both the linear DFE [10, 22] and PRML detectors does not compensate for the nonlinear ISI. Furthermore, the implementation complexity of a Viterbi detector matched to the PR channel grows exponentially with

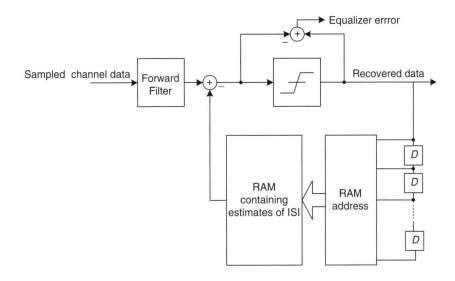

FIGURE 32.10   Block diagram of a RAM-based DFE.

the degree of channel polynomial. Actually, in order to meet requirements for a high data transfer rate, high-speed analog-to-digital (A/D) converter is also needed. In the RAM-based DFE [17, 21], the linear feedback section of the linear DFE is replaced with a look-up table. In this way, detector decisions make up a RAM address pointing to the memory location that contains an estimate of the post cursor ISI for the particular symbol sequence. This estimate is subtracted from the output of the forward filter forming the equalizer output. Look-up table size is manageable and typically is less than 256 locations. The major disadvantage of this approach is that it requires complicated architecture and control to recursively update ISI estimates based on equalizer error.

32.4.1   Detection in a Trellis

A trellis-based system can be simply described as a finite-state machine (FSM) whose structure may be displayed with the aid of a graph, tree or trellis diagram. A FSM maps input sequences (vectors) into output sequences (vectors), not necessarily of the same length. Although the system is generally nonlinear and time-varying, linear fixed trellis based systems are usually met. For them,

$$F\left(a \cdot \mathbf{i}_{[0,\infty)}\right) = a \cdot F\left(\mathbf{i}_{[0,\infty)}\right)$$

$$F\left(\mathbf{i}'_{[0,\infty)} + \mathbf{i}''_{[0,\infty)}\right) = F\left(\mathbf{i}'_{[0,\infty)}\right) + F\left(\mathbf{i}''_{[0,\infty)}\right)$$

where a is a constant, $\mathbf{i}_{[0,\infty)}$ is any input sequence and $F(\mathbf{i}_{[0,\infty)})$ is the corresponding output sequence. It is assumed that input and output symbols belong to a subset of a field. Also, for any $d > 0$, if $\mathbf{x}_{[0,\infty)} = F(\mathbf{i}_{[0,\infty)})$ and $\mathbf{i}'_l = \mathbf{i}_{l-d}$, $\mathbf{i}'_{[0,d)} = \mathbf{0}_{[0,d)}$ then $F(\mathbf{i}'_{[0,\infty)}) = \mathbf{x}'_{[0,\infty)}$, where $\mathbf{x}'_l = \mathbf{x}_{l-d}$, $\mathbf{x}'_{[0,d)} = \mathbf{0}_{[0,d)}$. It is easily verified that $F(\cdot)$ can be represented by the convolution, so that $f\mathbf{x} = \mathbf{i} * \mathbf{h}$, where $\mathbf{h}$ is the system impulse response (this is also valid for different lengths of $\mathbf{x}$ and $\mathbf{i}$ with a suitable definition of $\mathbf{h}$). If $\mathbf{h}$ is of finite duration, then M denotes the system memory length.

Let us now consider a feedforward FSM with memory length M. At any time instant (depth or level) l, the FSM output $\mathbf{x}_l$ depends on the current input $\mathbf{i}_l$, and M previous inputs $\mathbf{i}_{l-1}, \ldots, \mathbf{i}_{l-M}$. The overall functioning of the system can be mapped on a trellis diagram, whereon a node represents one of q^M encoder states (q is the cardinality of the input alphabet including the case when the input symbol is actually a subsequence), while a branch connecting two nodes represents the FSM output associated to the transition between the corresponding system states.

A trellis, which is a visualization of the state transition diagram with a time element incorporated, is characterized by q branches stemming from and entering each state, except in the first and last M branches (respectively called head and tail of the trellis). The branches at the lth time instant are labeled by sequences $\mathbf{x}_l \in X$. A sequence of l information symbols, $\mathbf{i}_{[0,l)}$ specifies a path from the root node to a node at the lth level and, in turn, this path specifies the output sequence $\mathbf{x}_{[0,l)} = \mathbf{x}_0 \bullet \mathbf{x}_1 \bullet \ldots \bullet \mathbf{x}_{l-1}$, where $\bullet$ denotes concatenation of two sequences.

The input can, but need not be, separated in frames of some length. For framed data, where the length of each input frame equals L branches (thus L q-ary symbols) the length of the output frame is $L + M$ branches ($L + M$ output symbols), where the M known symbols (usually all zeros) are added at the end of the sequence to force the system into the desired terminal state. It is said that such systems suffer a fractional rate loss by $L/(L + M)$. Clearly, this rate loss has no asymptotic significance.

In the sequel we will analyze detection of the input sequence, $\mathbf{i}_{(0,\infty)}$, based on the corrupted output sequence $\mathbf{y}_{[0,\infty)} = \mathbf{x}_{[0,\infty)} + \mathbf{u}_{[0,\infty)}$. Suppose there is no feedback from the output to the input, so that

$$P[y_n \mid x_0, \ldots, x_{n-1}, x_n, y_0, \ldots, y_{n-1}] = P[y_n \mid x_n]$$

and

$$P[y_1, \ldots, y_N \mid x_1, \ldots, x_N] = \prod_{n=1}^{N} P[y_n \mid x_n]$$

Usually, $\mathbf{u}_{(0,\infty)}$ is a sequence that represents additive white Gaussian noise sampled and quantized to enable digital processing.

The task of the detector that minimizes the sequence error probability is to find a sequence which maximizes the joint probability of input and output channel sequences

$$P\left[\mathbf{y}_{[0,L+M)}, \mathbf{x}_{[0,L+M)}\right] = P\left[\mathbf{y}_{[0,L+M)} \mid \mathbf{x}_{[0,L+M)}\right] \cdot P\left[\mathbf{x}_{[0,L+M)}\right]$$

Since usually the set of all probabilities $P[\mathbf{x}_{[0,L+M)}]$ is equal, it is sufficient to find a procedure that maximizes $P[\mathbf{y}_{[0,L+M)} \mid \mathbf{x}_{[0,L+M)}]$, and a decoder that always chooses as its estimate one of the sequences that maximize it or

$$\mu\left(\mathbf{y}_{[0,L+M)} \mid \mathbf{x}_{[0,L+M)}\right) = A \cdot (\log_2 P\left[\mathbf{y}_{[0,L+M)} \mid \mathbf{x}_{[0,L+M)}\right]$$

$$-f\left(\mathbf{y}_{[0,L+M)}\right) = A \cdot \sum_{l=0}^{L+M} \log_2 \left(P[\mathbf{y}_l \mid \mathbf{x}_l] - f(\mathbf{y}_l)\right)$$

(where $A \geq 0$ is a suitably chosen constant, and $f(\cdot)$ is any function) is called a maximum-likelihood decoder (MLD). This quantity is called a metric, μ. This type of metric suffers one significant disadvantage because it is suited only for comparison between paths of the same length. Some algorithms, however, employ a strategy of comparing paths of different length or assessing likelihood of such paths with the aid of some thresholds. The metric that enables comparison for this type of algorithms is called the Fano metric. It is defined as

$$\mu_F\left(\mathbf{y}_{[0,l)} \mid \mathbf{x}_{[0,l)}\right) = A \cdot \log_2 \frac{P\left[\mathbf{y}_{[0,l)}, \mathbf{x}_{[0,l)}\right]}{P[\mathbf{y}_{[0,l)}]}$$

$$= A \cdot \sum_{n=0}^{l} \left(\log_2 \frac{P[\mathbf{y}_n \mid \mathbf{x}_n]}{P[\mathbf{y}_n]} - R\right)$$

If the noise is additive, white and Gaussian (an assumption that is not entirely true, but that usually yields systems of good performances), the probability distribution of its sample is

$$p[\mathbf{y}_n \mid \mathbf{x}_n] = \frac{1}{\sqrt{2\pi\sigma^2}} \exp\left(-\frac{(\mathbf{y}_n - \mathbf{x}_n)^2}{2\sigma^2}\right)$$

The ML metric to be used in conjunction with such a noise is the logarithm of this density, and thus proportional to $-(\mathbf{y}_n - \mathbf{x}_n)^2$, that is, to the negative squared Euclidean distance of the readback and supposed written signal. Thus, maximizing likelihood amounts to minimizing the squared Euclidean distance of the two sequences, leading to minimizing the squared Euclidean distance between two sampled sequences given by $\sum_n (y_n - x_n)^2$.

The performance of a trellis-based system, as is the case with PR systems, depends on the detection algorithm employed and on the properties of the system itself. The distance spectrum is the property of the system that constitutes the main factor of the event error probability of a ML (optimum) detector, if the distance is appropriately chosen for the coding channel used [23]. For PR channels with additive white Gaussian noise, it is the squared Euclidean distance that has to be dealt with. Naturally, since the noise encountered is neither white, nor entirely Gaussisan, this is but an approximation to the properly chosen distance measure.

As stated above, the aim of the search procedure is to find a path with the highest possible likelihood, that is, metric. There are several possible classifications of detecting procedures. This classification is in-line with systematization made in coding theory due to fact that algorithms developed for decoding in a trellis are general so that it could be applied to problem of detection in any trellis based system as well. According to detector's strategies in extending the most promising path candidates we classify them into *breadth-first*, *metric-first* and *depth-first* and bidirectional algorithms and into *sorting* and *nonsorting* depending on wheather the procedure performs any kind of path comparison (sifting or sorting) or not. Moreover, detecting algorithms can be classified into searches that *minimize* the *sequence* or *symbol* error rate.

The usual measure of algorithm efficiency is its complexity (arithmetic and storage) for a given probability of error. In the strict sense, arithmetic or computational complexity is the number of arithmetic operations per detected symbol, branch, or frame. However, it is a usual practice to track only the number of node computations, which makes sense because all such computations require approximately the same number of basic machine instructions. A node computation (or simply computation) is defined as the total number of nodes extended (sometimes it is the number of metrics computed) per detected branch or information frame $\mathbf{i}_{[0,L+M]}$. One single computation consists of determining the state in which the node is and computing the metrics of all its successors. For most practical applications with finite frame length, it is usually sufficient to observe node computations since a good prediction of search duration can be precisely predicted. Nevertheless, for asymptotic behavior it is necessary to track the sorting requirements too. Another important aspect of complexity is storage (memory or space), which is the amount of auxiliary storage that is required for detecting (memory, processors working in parallel etc.). Thus, space complexity of an algorithm is the size (or number) of resources that must be reserved for its use, while the computational, or more precisely time complexity, reflects the number of accesses to this resources taking into account that any two operations done in parallel by the spatially separated processors should be counted as one. The product of these two, the time-space complexity is possibly the best measure of the algorithm cost for it is insensitive to time-space tradeoff such as parallelization or the use of precomputed tables, although it also makes sense to keep the separate track of these two. Finally, for selecting which algorithm to use one must consider additional details that we omit here but which can sometimes cause unexpected overall performance or complicate the design of a real-time detector. They include complexity of the required data structure, buffering needs and applicability of available hardware components.

32.5   Basic Breadth-First Algorithms

32.5.1   The Viterbi Algorithm

The Viterbi algorithm was introduced in 1967 as a method of decoding convolutional codes. Forney showed in 1972 [6] that the Viterbi algorithm solves the *maximum-likelihood* sequence detection (MLSD) problem in the presence of intersymbol interference (ISI) and additive white noise. Kobayashi and Tang recognized [7] that this algorithm is possible to apply in magnetic recording systems for detection purposes. Strategy to combine Viterbi detector with partial response (PR) equalization in magnetic recording channel resulted with many commercial products.

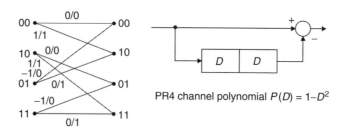

FIGURE 32.11   PR4 channel trellis.

The Viterbi algorithm (VA) is an optimal decoding algorithm in the sense that it always finds the nearest path to the noisy modification of the FSM output sequence $\mathbf{x}_{[0,L+M)}$, and it is quite useful when FSM has a short memory. The key to Viterbi (maximum-likelihood, ML) decoding lies in the *Principle of Nonoptimality* [15]: If the paths $\mathbf{i}'_{[0,l)}$ and $\mathbf{i}''_{[0,l)}$ terminate at the same state of the trellis and

$$\mu\left(\mathbf{y}_{[0,l)}, \mathbf{x}'_{[0,l)}\right) > \mu\left(\mathbf{y}_{[0,l)}, \mathbf{x}''_{[0,l)}\right)$$

then $\mathbf{i}''_{[0,l)}$ cannot be the first l branches of one of the paths $\mathbf{i}_{[0,L+M)}$ that maximize the overall sequence metric. This principle which some authors call the *Principle of Optimality* literally specifies the most efficient MLD procedure for decoding/detecting in the trellis.

To apply Viterbi algorithm as a ML sequence detector for a PR channel, we need to define the channel trellis describing the amount of controlled ISI. Once we define the partial response channel polynomial it is an easy task. An example of such trellis for PR4 channel with $P(D) = 1 - D^2$ is depicted in Figure 32.11. The trellis for this channel consists of four states according to the fact that channel input is binary and channel memory is 2 so there are four possible state values (00, 10, 01, 11). Generally, if the channel input sequence can take q values, and the partial response channel forms the intersymbol interference from the past M input symbols, then the partial response channel can be described by a trellis with q^M states. Branches joining adjacent states are labeled with the pair of expected noiseless symbols in the form *channel_output/channel_ input*. Equalization to $P(D) = 1 - D^2$ results in ternary channel output, taking values $\{0, \pm1\}$. Each noiseless output channel sequence is obtained by reading the sequence of labels along some path through the trellis.

Now the task of detecting $\mathbf{i}_{[0,\infty)}$ is to find $\mathbf{x}_{[0,\infty)}$ that is closest to $\mathbf{y}_{[0,\infty)}$ in the Euclidean sense. Recall that we stated as an assumption that channel noise is AWGN, while in magnetic recording systems after equalization the noise is colored so that the minimum-distance detector is not an optimal one and additional postprocessing is necessary which will be addressed later in this chapter.

The Viterbi algorithm is a classical application of dynamic programming. Structurally, the algorithm contains q^M lists, one for each state, where the paths whose states correspond to the label indices are stored, compared, and the best one of them retained. The algorithm can be described recursively as follows:

1. *Initial Condition*: Initialize the starting list with the root node (the known initial state) and set its metric to zero, $l = 0$.
2. *Path Extension*: Extend all the paths (nodes) by one branch to yield new candidates, $l = l + 1$, and find the sum of the metric of the predecessor node and the branch metric of the connecting branch (ADD). Classify these candidates into corresponding q^M lists (or less for $l < M$). Each list (except in the head of the trellis) contains q paths.
3. *Path Selection*: For each end-node of extended paths determine the maximum/minimum[1] of these sums (COMPARE) and assign it to the node. Label the node with the best path metric to it selecting

---

[1]It depends on whether the metric or the distance is accumulated.

(SELECT) that path for the next step of the algorithm (discard others). If two or more paths have the same metric, that is, if they are equally likely, choose the best one at random. Find the best of all the survivor paths, $\mathbf{x}'_{[0,l)}$, and its corresponding information sequence $i'_{[0,l)}$ and release the bit $i'_{l-\delta}$. Go to Step 2.

In the description of the algorithm we emphasized three Viterbi-characteristic operations add-compare-select (ADC) that are performed in every recursion of the algorithm, so today's specialized signal processors have this operation embedded optimizing its execution time. Consider now the amount of "processing" done at each depth l, where all of the q^M states of the trellis code are present. For each state it is necessary to compare q paths that merge in that state, discard all but the best path, and then compute and send the metrics of q of its successors to the depth $l + 1$.

Consequently, the computational complexity of the VA exponentially increases with M. These operations can be easily parallelized, but then the number of parallel processors rises as the number of node computations decreases. The total time-space complexity of the algorithm is fixed, and increases exponentially with the memory length.

The sliding window VA decodes infinite sequences with delay of δ branches from the last received one. In order to minimize its memory requirements ($\delta + 1$ trellis levels), and achieve bit error rate only insignificantly higher than with finite sequence VA, δ is chosen as $\delta \approx 4M$. In this way the Viterbi detector introduces a fixed decision delay.

Example 32.1

Assume that a recorded, channel input sequence $\mathbf{x}$, consisting of L equally likely binary symbols from the alphabet $\{0, 1\}$, is "transmitted" over PR4 channel. The channel is characterized by the trellis of Figure 32.11, that is, all admissible symbol sequences correspond to the paths traversing the trellis from $l = 0$ to $l = L$, with one symbol labeling each branch, Figure 32.12. Suppose that the noisy sequence of samples at the channel output is $\mathbf{y} = 0.9, 0.2, -0.6, -0.3, 0.6, 0.9, 1.2, 0.3, \ldots$ If we apply a simple symbol-by-symbol detector to this sequence, the fifth symbol will be erroneous due to the hard quantization rule

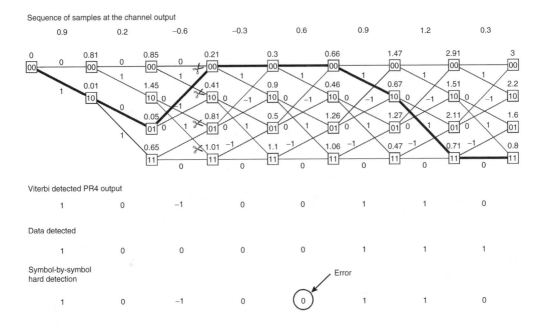

FIGURE 32.12   Viterbi algorithm detection on the PR4 trellis.

for noiseless channel output estimate

$$\hat{y}_k = \begin{cases} -1 & y_k < -0.5 \\ 1 & y_k > 0.5 \\ 0 & \text{otherwise} \end{cases}$$

The Viterbi detector will start to search the trellis accumulating branch distance from sequence **y**. In the first recursion of the algorithm, there are two paths of length 1 at the distance

$$d(\mathbf{y}, 0) = (0.9 - 0)^2 = 0.81$$
$$d(\mathbf{y}, 1) = (0.9 - 1)^2 = 0.01$$

from **y**. Next, each of the two paths of length 1 are extended in two ways forming four paths of length 2 at squared Euclidean distance from the sequence **y**

$$d(\mathbf{y}, (0,0)) = 0.81 + (0.2 - 0)^2 = 0.85$$
$$d(\mathbf{y}, (0,1)) = 0.81 + (0.2 - 1)^2 = 1.45$$
$$d(\mathbf{y}, (1,0)) = 0.01 + (0.2 - 0)^2 = 0.05$$
$$d(\mathbf{y}, (1,1)) = 0.01 + (0.2 - 1)^2 = 0.65$$

and this accumulated distance of four paths labels the four trellis states. In the next loop of the algorithm each of the paths are again extended in two ways to form eight paths of length 3, two paths to each node at level (depth) 3.

Node 00

$$d(\mathbf{y}, (0, 0, 0) = 0.85 + (-0.6 - 0)^2 = 1.21$$
$$d(\mathbf{y}, (1, 0, -1) = 0.05 + (-0.6 + 1)^2 = 0.21 \quad \text{surviving path}$$

Node 10

$$d(\mathbf{y}, (0, 0, 1) = 0.85 + (-0.6 - 1)^2 = 3.41$$
$$d(\mathbf{y}, (1, 0, 0) = 0.05 + (-0.6 - 0)^2 = 0.41 \quad \text{surviving path}$$

Node 01

$$d(\mathbf{y}, (0, 1, 0) = 1.45 + (-0.6 - 0)^2 = 1.81$$
$$d(\mathbf{y}, (1, 1, -1) = 0.65 + (-0.6 + 1)^2 = 0.81 \quad \text{surviving path}$$

Node 11

$$d(\mathbf{y}, (0, 1, 1) = 1.45 + (-0.6 - 1)^2 = 4.01$$
$$d(\mathbf{y}, (1, 1, 0) = 0.65 + (-0.6 - 0)^2 = 1.01 \quad \text{surviving path}$$

Four paths of length 3 are selected as the surviving, most likely paths to the four trellis nodes. The procedure is repeated and the detected sequence is produced after a delay of $4M = 8$ trellis sections. Note, Figure 32.12, that the symbol-by-symbol detector error is now corrected. Contrary to this example, a 4-state PR4ML detector is implemented with two interleaved 2-state dicode, $(1 - D)$, detectors each operating at one-half the symbol rate of one full-rate PR4 detector [31]. The sequence is interleaved, such that the even samples go to the first and the odd to the second dicode detector, Figure 32.13, so the delay D in the interleaved detectors is actually twice the delay of the PR4 detector. A switch at the output re-samples the data to get them out in the correct order.

For other PR channels this type of decomposition is not possible, so that their complexity can become great for real-time processing. In order to suppress some of the states in the corresponding trellis diagram of those PR systems, thus simplifying the sequence detection process, some data loss has to be introduced.

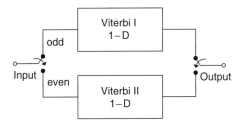

FIGURE 32.13 Implementation of $1 - D^2$ Viterbi detector with two half-rate $1 - D$ detectors.

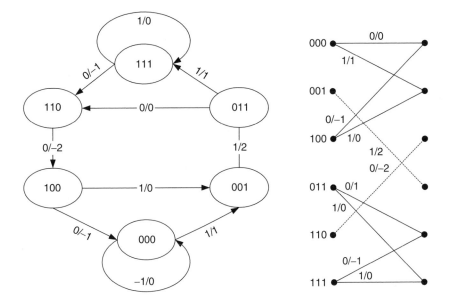

FIGURE 32.14 (1,7) coded EPR4 channel.

For instance, in conjunction with precoding, (1,7) code prohibits two states in EPR4 trellis: [101] and [010]. This can be used to reduce the 8-state EPR4 trellis to 6-state trellis depicted in Figure 32.14., and the number of add-compare-select units in the VA detector to 4. The data rate loss is 33% in this case. Using the (2,7) code eliminates two more states, paying the complexity gain by a 50% data rate loss.

Since VA involves addition, multiplication, compare and select functions, which require complex circuitry at the read side, simplifications of the receiver for certain partial responses were sought. One of them is the dynamic threshold technique [20]. This technique implies generating a series of thresholds. The readback samples are compared with them, just as for the threshold detector, and are subsequently included in their modification. While preserving the full function of the ML detector, this technique saves a substantial fraction of the necessary hardware. Examples of dynamic threshold detectors are given in [5, 26].

32.6 Noise Predictive Maximum Likelihood Detectors

Maximum likelihood detection combined with partial response equalization is a dominant type of detection electronics in today's digital magnetic recording devices. As described above, in order to simplify hardware realization of the receiver, the degree of the target PR polynomial is chosen to be small with integer coefficients to restrict complexity of Viterbi detection trellis. On the other hand, if we increase recording density producing longer ISI, equalization to the same PR target will result in substantial noise enhancement

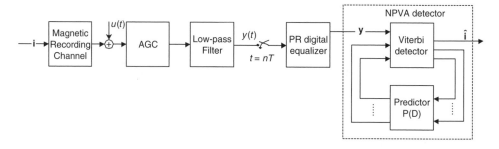

FIGURE 32.15   Block diagram of NPVA detector.

and detector performance degradation. Straightforward solution is to increase the duration of the target PR polynomial decreasing the mismatch between channel and equalization target. Note that this approach leads to undesirable increase in detector complexity fixing the detector structure in a sense that its target polynomial cannot be adapted to changing channel density.

The Noise Predictive Maximum Likelihood (NPML) detector [18, 28] is an alternative data detection method that improves reliability of the PRML detector. This is achieved by embedding a noise prediction/whitening process into the branch metric computation of a Viterbi detector. Using reduced-state sequence-estimation [35] (see also the description of the generalized VA in this chapter), which limits the number of states in the detector trellis, compensates for added detector complexity.

A block diagram of a NPML system is shown in Figure 32.15. The input to the channel is binary sequence, **i**, which is written on the disk at a rate of $1/T$. In the readback process data is recovered via a low-pass filter as an analog signal $y(t)$, which can be expressed as $y(t) = \sum_n i_n h(t - nT) + u(t)$ where $h(t)$ denotes the pulse response and $u(t)$ is the additive white Gaussian noise. The signal $y(t)$ is sampled periodically at times $t = nT$ and shaped into the PR target response by the digital equalizer. The NPML detector then performs sequence detection on the PR equalized sequence **y** and provides an estimate of the binary information sequence **i**. Digital equalization is performed to fit the overall system transfer function to some PR target, for example, the PR4 channel.

The output of the equalizer $y_n = i_n + \sum_{i=1}^{M} f_i x_{n-i} + w_n$ consists of the desired response and an additive total distortion component w_n, that is, the colored noise and residual interference. In conventional PRML detector, an estimate of the recorded sequence is done by the minimum-distance criteria as described for the Viterbi detector. If the mismatch between channel and PR target is significant, the power of distortion component w_n can degrade the detector performance. The only additional component compared to the Viterbi detector, NPML noise-predictor, reduces the power of the total distortion by whitening the noise prior to the Viterbi detector. The whitened total distortion component of the PR equalized output y_n is

$$w_n - \hat{w}_n = w_n - \sum_{i=1}^{N} w_{n-i} p_i$$

where the N-coefficient MMSE predictor transfer polynomial is $P(D) = p_1 D^1 + p_2 D^2 + \cdots + p_N D^N$. Note that an estimate of the current noise sample $\hat{w}_n$ is formed based on estimates of previous N noise samples. Assuming the PR4 equalization of sequence **y** we can modify the metric of the Viterbi detector in order to compensate for distortion component. In this case, the equalizer output is $y_n = x_n - x_{n-2} + w_n$ and the NPML distance is

$$\left[\left(y_n - \sum_{i=1}^{N} w_{n-i} p_i\right) - (x_n(S_k) - x_{n-2}(S_j))\right]^2$$

$$= \left[\left(y_n - \sum_{i=1}^{N} (y_{n-i} - (\hat{x}_{n-i}(S_j) - \hat{x}_{n-i-2}(S_j)) p_i\right) - (x_n(S_k) - x_{n-2}(S_j))\right]^2$$

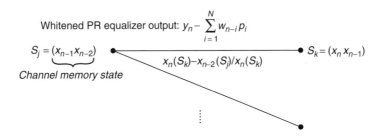

FIGURE 32.16 NPML metric computation for PR4 trellis.

where $\hat{x}_{n-i}(S_j), \hat{x}_{n-i-2}(S_j)$ represent past decisions taken from the Vitrebi survivor path memory associated with state S_j. The last expression gives the flavor of this technique but it is not suitable for implementation so that the interested reader can find details in [18] how to modify this equation for RAM look-up realization. Furthermore, in the same paper, a description of the general procedure to compute the predictor coefficients based on the autocorrelation of the total distortion w_n at the output of a finite-length PR equalizer is given.

32.7 Postprocessor

As we explained, Viterbi detector improves the performance of a read channel by tracing the correct path through the channel trellis [7]. Further performance improvement can be achieved by using soft output Viterbi algorithm (SOVA) [13]. Along with the bit decisions, SOVA produces the likelihood of these decisions, that combined create *soft information*. In principle, soft information can be passed to hard drive controller and used in RS decoder that resides there, but at the present time soft decoding of RS codes is still too complex to be implemented at 1Gb/s speeds. Alternatively, much shorter inner code is used. Because of the nonlinear operations on bits performed by the modulation decoder logic, the inner code is used in inverse concatenation with modulation encoder in order to simplify calculation of bit likelihood. Due to the channel memory and noise coloration, Viterbi detector produces some error patterns more often than others [4], and the inner code is designed to correct these so-called *dominant error sequences* or *error events*. The major obstacle for using soft information is the speed limitations and hardware complexity required to implement SOVA. Viterbi detector is already a bottleneck and the most complex block in a read channel chip, occupying most of the chip area, and the architectural challenges in implementing even more complex SOVA would be prohibitive. Therefore, a postprocessor architecture is used [16]. The postprocessor is a block that resides after Viterbi detector and comprises the block for calculating *error event likelihood* and an inner-*soft error event correcting decoder*.

The postprocessor is designed by using the knowledge on the set of dominant error sequences $E = \{e_i\}_{1 \le i \le I}$ and their occurrence probabilities $p = (p_i)_{1 \le i \le I}$. The index i is referred to as an *error type*, while the position of the *error event end* within a codeword is referred as an *error position*. The relative frequencies of error events will strongly depend on recording density [32]. The detection is based on the fact that we can calculate the likelihoods of each of dominant error sequences at each point in time. The parity bits detect the errors, and provide localization in error type and time. The likelihoods are then used to choose the most likely error events for corrections.

The error event likelihoods are calculated as the difference in the squared Euclidean distances between the signal and the convolution of maximum likelihood sequence estimate and the channel partial response, versus that between the signal and the convolution of an alternative data pattern and the channel partial response. During each clock cycle, the best M of them are chosen, and the syndromes for these error events are calculated. Throughout the processing of each block, a list is maintained of the N most likely error events, along with their associated error types, positions and syndromes. At the end of the block, when the list of candidate error events is finalized, the likelihoods and syndromes are calculated for each of $\binom{N}{L}$ combinations of L-set candidate error events that are possible. After disqualifying those L-sets of candidates which overlap in the time domain, and those candidates and L-sets of candidates which produced a syndrome which does not match the actual syndrome, the candidate or L-set of candidates which remains and which has the highest likelihood is chosen for correction. Finding the error event position and type completes decoding.

The decoder can make two types of errors: it fails to correct if the syndrome is zero, or it makes a wrong correction if the syndrome is nonzero but the most likely error event or combination of error events does not produce the right syndrome. A code must be able to detect a single error from the list of dominant error events, and should minimize the probability of producing zero syndrome when more than one error event occurs in a codeword. Consider a linear code given by an $(n - k) \times n$ parity check matrix H. We are interested in H capable of correcting or detecting dominant errors. If all errors from a list were contiguous and shorter than m, then a cyclic $n - k = m$ parity bit code could be used to correct a single error event [14]. However, in reality, the error sequences are more complex, and occurrence probabilities of error events of lengths 6, 7, 8 or more are not negligible. Furthermore, practical reasons (such as decoding delay, thermal asperities etc.) dictate using short codes, and consequently, in order to keep the code rate high, only a relatively small number of parity bits is allowed, making the design of error event detection codes nontrivial. The code redundancy must be used carefully so that the code is optimal for a given E.

The parity check matrix of a code can be created by a recursive algorithm that adds one column of H at a time using the criterion that after adding each new column, the code error-event-detection capabilities are still satisfied. The algorithm can be described as a process of building a directed graph whose vertices are labeled by the portions of parity check matrix long enough to capture the longest error event, and whose edges are labeled by column vectors that can be appended to the parity check matrix without violating the error event detection capability [3]. To formalize code construction requirements, for each error event from E, denote by $s_{i,l}$ a syndrome of error vector $\sigma_l(e_i)$ ($s_{i,l} = \sigma_l(e_i) \cdot H^T$), where $\sigma_l(e_i)$ is an l-time shifted version of error event e_i. The code should be designed in such a way that any shift of any dominant error sequence produces a nonzero syndrome, that is, that $s_{i,l} \neq 0$ for any $1 \leq i \leq I$ and $1 \leq I \leq n$. In this way we are able to detect a single error event (we rely on error event likelihoods to localize the error event). The correctable shifts must include negative shifts as well as shifts larger than n in order to cover those error events that straddle adjacent codewords, because the failure to correct straddling events significantly affects the performance. A stronger code could have a parity check matrix that guarantees that syndromes of any two-error event-error position pairs $((i_1, l_1), (i_2, l_2))$ are different, that is, $s_{i_1, l_1} \neq s_{i_2, l_2}$. This condition would result in a single error event correction capability. The codes capable of correcting multiple error events can be defined analogously. We can even strengthen this property and require that for any two shifts and any two dominant error events, the Hamming distance between any pair of syndromes is larger than δ, however by strengthening any of these requirements the code rate decreases.

If L_i is a length of the ith error event, and if L is the length of the longest error event from E, ($L = \max_{1 \leq i \leq I}\{L_i\}$), then it is easy to see that for a code capable of detecting an error event from E that ends at position j, the linear combination of error events and the columns of H from $j - L + 1$ to j has to be nonzero. More precisely, for any i and any j (ignoring the codeword boundary effects)

$$\sum_{1 \leq m \leq L_i} e_{i,m} \cdot h_{j-L_i+m}^T \neq 0$$

where $e_{i,m}$ is the mth element of the error event e_i, and h_j is the jth column of H.

32.8 Advanced Algorithms and Algorithms Under Investigation

In this section we give a brief overview of less complex procedures for searching the trellis. It is intended to give background information that can be used in future development if it shows up that NPVA detectors and postprocessing are not capable of coping with ever-increasing storage densities and longer partial responses needed for them. In such cases, a resort has to be made to some sort of reduced complexity suboptimal algorithms, whose performance is close to optimal. Explained algorithms are not yet implemented in commercial products but all of them are a natural extension of already described procedures for searching the trellis.

32.8.1 Other Breadth-First Algorithms

32.8.1.1 The *M*-Algorithm

Since most survivors in the VA usually possess much smaller metrics than does the best one, all the states or nodes kept are not equally important. It is intuitively reasonable to assume that unpromising survivors can be omitted with a negligible probability of discarding the best one. The *M*-algorithm [9] is one such modification of the Viterbi algorithm; all candidates are stored in a single list and the best $M \leq q^M$ survivors are selected from the list in each cycle. The steps of the *M*-algorithm are:

1. *Initial condition*: Initialize the list with the root node and set its metric to zero.
2. *Path extension*: Extend all the paths of length l by one branch and classify all contenders (paths of length $l + 1$) into the list. If two or more paths enter the same state keep the best one.
3. *Path selection*: From the remaining paths find the best M candidates and delete the others. If $l = L + M$ take the only survivor and transfer its corresponding information sequence to the output (terminated case, otherwise use the sliding window variation). Otherwise, go to Step 2.

Defined in this way, the *M*-algorithm performs trellis search, while, when the state comparison in Step 2 is omitted, it searches the tree, saving much time on comparisons but with slightly increased error probability. When applied to decoding/detecting infinitely long sequences, it is usual that comparisons performed in Step 2 are substituted with the so-called ambiguity check [9] and a release of one decoded branch. In each step this algorithm performs M node computations, and employing any sifting procedure (since the paths need not be sorted) perform $\sim M \cdot q$ metric comparisons. If performed, the Viterbi-type discarding of Step 2 requests $\sim M^2 \cdot q$ state and metric comparisons. This type of discarding can be performed with $\sim M \cdot \log_2 M$ comparisons (or even linearly) but than additional storage must be provided. The space complexity grows linearly with the information frame length L and parameter M.

32.8.1.2 The Generalized Viterbi Algorithm

In contrast to the Viterbi algorithm, which is a multiple-list single survivor algorithm, the *M*-algorithm is a single-list multiple-survivor algorithm. The natural generalization to a multiple-list multiple-survivor algorithm was first suggested by Hashimoto [34]. Since all the lists are not equally important, this algorithm, originally called the generalized Viterbi algorithm (GVA), utilizes only q^{M_1} lists (labels), where $M_1 \leq M$. In each list from all q^{M-M_1+1} paths it retains the best M_1 candidates. The algorithm can be described as follows.

1. *Initial condition*: Initialize the starting label with the root node and set its metric to zero.
2. *Path extension*: Extend all the paths from each label by one branch and classify all successors into the appropriate label. If two or more paths enter the same state keep the best one.
3. *Path selection*: From the remaining paths of each label find the best M_1 and delete the others. If $l = L + M$ take the only survivor and transfer its information sequence to the output (for the terminated case, otherwise use the sliding window variant); Go to Step 2.

When $M_1 = M$, and $M_1 = 1$, the GVA reduces to the Viterbi algorithm, and for $M_1 = 0$, $M_1 = M$ it reduces to the *M* algorithm. Like the *M*-algorithm, GVA in each step performs M_1 node computations per label, and employing any sifting procedure $\sim M_1 \cdot q$ metric comparisons. If performed, the Viterbi-type discarding of step 2 requests $\sim M_1^2 \cdot q$ or less state and metric comparisons per label.

32.8.2 Metric—First Algorithms

Metric-first and depth-first sequential detection is a name for a class of algorithms that compare paths according to their Fano metric (one against another or with some thresholds) and on that basis decide which node to extend next, which to delete in metric first procedures or whether to proceed with current branch or go back. These algorithms generally extend fewer nodes for the same performance, but have increased sorting requirements.

Sequential detecting algorithms have a variable computation characteristic that results in large buffering requirements, and occasionally large detecting delays and/or incomplete detecting of the received sequence. Sometimes, when almost error-free communication is required or when retransmission is possible, this variable detecting effort can be an advantage. For example, when a detector encounters an excessive number of computations, it indicates that a frame is possibly very corrupted meaning that the communication is insufficiently reliable and can ultimately cause error patterns in detected sequence. In such situations the detector gives up detecting and simply requests retransmission. These situations are commonly called erasures, and detecting incomplete. A complete decoder such as the Viterbi detector/decoder would be forced to make an estimate, which may be wrong. The probability of buffer overflow is several orders of magnitude larger than the probability of incorrect decision when the decoder operates close to the so-called (computational) cutoff rate.

The performance of sequential detecting has traditionally been evaluated in terms of three characteristics: the probability of sequence error, the probability of failure (erasure), and the Pareto exponent associated with detecting effort.

32.8.3   Bidirectional Algorithms

Another class of algorithms are those that exploit bidirectional decoding/detection which is designed for framed data. Almost all unidirectional procedures have their bidirectional supplements since Forney showed that detecting could start from the end of the sequence provided that the trellis contains a tail. All bidirectional algorithms employ two searches from both sides. The forward search is performed using the original trellis code while the backward one employs the reverse code. The reverse trellis code is obtained from the original code by time reversing.

References

[1] A.J. Viterbi, Error bounds for convolutional codes and asymptotically optimum decoding algorithm, *IEEE Trans. Inf. Theory*, vol. IT-13, pp. 260–269, 1967.

[2] A. Lender, Correlative level coding for binary-data transmission, *IEEE Trans. Commun. Technol.* (Concise paper), vol. COM-14, pp. 67–70, 1966.

[3] B. Vasic, "A graph based construction of high-rate soft decodable codes for partial response channels," to be presented at ICC2001, Helsinki, Finland, June 2001, 10–15.

[4] C.L. Barbosa, Maximum likelihood sequence estimators, A geometric view, *IEEE Trans. Inf. Theory*, vol. IT-35, pp. 419–427, March 1989.

[5] C.D. Wei, An analog magnetic storage read channel based on a decision feedback equalizer, *Ph.D. final report*, University of California, Electrical Eng. Department, July 1996.

[6] G.D. Forney, Maximum likelihood sequence estimation of digital sequences in the presence of intersymbol interference, *IEEE Trans. Inf. Theory*, vol. IT-18, pp. 363–378, May 1972.

[7] H. Kobayashi and D.T. Tang, Application of partial response channel coding to magnetic recording systems, *IBM J. Res. Dev.*, vol. 14, pp. 368–375, July 1979.

[8] H. Kobayashi, Correlative level coding and maximum-likelihood decoding, *IEEE Trans. Inf. Theory*, vol. IT-17(5), pp. 586–594, 1971.

[9] J.B. Anderson and S. Mohan, Sequential coding algorithms: a survey and cost analysis, *IEEE Trans. Comm.*, vol. COM-32, no. 2, pp. 169–176, February 1984.

[10] J. Bergmans, Density improvements in digital magnetic recording by decision feedback equalization, *IEEE Trans. Magnetics*, vol. 22, pp.157–162, May 1986.

[11] J. Bergmans, *Digital Baseband Transmission and Recording*, Kluwer, Dordrecht, The Netherlands, 1996.

[12] J. Hagenauer, Applications of error-control coding, *IEEE Trans. Inf. Theory*, vol. IT-44, no. 6, pp. 2531–2560, October 1998.

[13] J. Hagenauer and P. Hoeher, A viterbi algorithm with soft decision outputs and its applications, *Proc. GLOBECOM 89*, Dallas, Texas, pp. 47.1.1–47.1.7, November 1989.

[14] J.K. Wolf and D. Chun, The single burst error detection performance of binary cyclic codes, *IEEE Trans. Commun.*, vol. 42, no. 1, pp. 11–13, January 1994.

[15] J.L. Massey, Coding and Complexity, *CISM courses and lectures No. 216*, Springer-Verlag, Wien, 1976.

[16] J.L. Sonntag and B. Vasic, Implementation and bench characterization of a read channel with parity check post processor, *Digest of TMRC 2000*, Santa Clara, CA, August 2000.

[17] J.M. Cioffi, W.L. Abbott, H.K. Thapar, C.M. Melas, and K.D. Fisher, Adaptive equalization in magnetic disk storage channels, *IEEE Comm. Magazine*, pp. 14–29, February 1990.

[18] J.D. Coker, E. Eleftheriou, R. Galbraith, and W. Hirt, Noise-predictive maximum likelihood (NPML) detection, *IEEE Trans. on Magnetics*, vol. 34, no. 1, pp. 110–117, January 1998.

[19] J.J. O'Reilly, A.M. de Oliveira Duarte, "Error Propagation in Decision Feedback Receivers," *IEE Proc.*, vol. 132, Pt. F, no. 7, pp. 561–566, December 1985.

[20] K. Knudson et al., Dynamic threshold implementation of the maximum likelihood detector for the EPR4 channel, *Conf. Rec. Globecom '91*, pp. 2135–2139, 1991.

[21] K.D. Fisher, J. Cioffi, W. Abbott, P. Bednarz, and C.M. Melas, An adaptive RAM-DFE for storage channels, *IEEE Trans. Comm.*, vol. 39, no.11, pp. 1559–1568, November 1991.

[22] M. Fossorier, Performance evaluation of decision feedback equalization for the Lorentzian channel, *IEEE Trans. Magn.*, vol. 32, no. 2, March 1996.

[23] M. Despotović and V. Šenk, Distance spectrum of channel trellis codes on precoded partial-response 1-D channel, Facta Universitatis (NIŠ), series: *Electronics and Energetics*, vol. 1 (1995), pp. 57–72, http://factaee.elfak.ni.ac.yu.

[24] P.R. Chevillat and D.J. Costello Jr., A multiple stack algorithm for erasure free decoding of convolutional codes, *IEEE Trans. Comm.*, vol. COM-25, pp. 1460–1470, December 1977.

[25] P. Radivojac and V. Šenk, The generalized viterbi-T algorithm, *Proc. of XXXIX Conference on ETRAN*, vol. 2, pp. 13–16, Zlatibor, Yugoslavia, June 1995.

[26] P.H. Siegel and J.K. Wolf, Modulation and coding for information storage, *IEEE Comm. Magazine*, pp. 68–86, December 1991.

[27] P. Kabal and S. Pasupathu, Partial response signaling, *IEEE Trans. Comm.*, vol. COM-23, no. 9, pp. 921–934, September 1975.

[28] P.R. Chevillat, E. Eleftheriou, and D. Maiwald, Noise predictive partial-response equalizers and applications, *IEEE Conf. Records ICC'92*, pp. 942–947, June 1992.

[29] P.K. Pai, A.D. Brewster and A.A. Abidi, Analog front-end architectures for high-speed PRML magnetic read channels, *IEEE Trans. Magn.*, vol. 31, pp. 1103–1108, 1995.

[30] R.M. Fano, A heuristic discussion of probabilistic decoding, *IEEE Trans. Inf. Theory*, vol. IT-9, pp. 64–74, April 1963.

[31] R.D. Cideciyan et al., A PRML system for digital magnetic recording, *IEEE J. Sel. Areas Comm.*, vol. 10, pp. 38–56, January 1992.

[32] S.A. Altekar, M. Berggren, B.E. Moision, P.H. Siegel, and J.K. Wolf, Error-event characterization on partial-response channels, *IEEE Trans. Inf. Theory*, vol. 45, no. 1, pp. 241–247, January 1999.

[33] S.J. Simmons, Breadth-first trellis decoding with adaptive effort, *IEEE Trans. Comm.*, vol. COM-38, no. 1, pp. 3–12, January 1990.

[34] T. Hashimoto, A list-type reduced-constraint generalization of the viterbi algorithm, *IEEE Trans. Inf. Theory*, vol. IT-33, no. 6, pp. 866–876, November 1987.

[35] V.M. Eyuboglu and S.U. Quereshi, Reduced-state sequence estimation with decision feedback and set partitioning, *IEEE Trans. Comm.*, vol. 36, no. 1, pp. 13–20, January 1988.

33

Detection Methods for Data-dependent Noise in Storage Channels

Erozan M. Kurtas

Jongseung Park

Xueshi Yang

William Radich
Seagate Technology
Pittsburgh, PA

Aleksandar Kavčić
Harvard University
Boston, MA

33.1 Introduction **33**-1
33.2 Detection Methods in the AWGN Channel **33**-2
  Maximum Likelihood Sequence Detection (MLSD) over ISI
  Channel • Maximum A Posteriori Probability (MAP) Symbol
  Detection over ISI Channel
33.3 Recording System Model **33**-7
33.4 Detection Methods in Data-Dependent Noise **33**-9
  Data-dependent Correlated Noise • MLSD Based on
  Pattern-Dependent Noise Prediction • BCJR Algorithm for
  Signal-dependent Channel Noise • SOVA Algorithm for
  Signal-dependent Channel Noise
33.5 Simulations **33**-13
33.6 Conclusions **33**-15

33.1   Introduction

In magnetic recording channels, binary user bits $(0, 1)$s are coded and mapped into bipolar channel symbols $(-1, 1)$, which are then used to modulate the current in the windings of the recording head. This current magnetizes the head and causes a flux pattern that follows the head path and fringes from the head core. The fringing head flux magnetizes the recording media to saturation in opposite directions along with (longitudinal recording) or perpendicular to (perpendicular recording) the moving direction of the recording head. To read data from the disk, the read head senses the change in recording media magnetization and generates the read back signal accordingly. During the writing/reading process, the overall system can be viewed as a system with a certain impulse response corresponding to the bipolar (± 1) input. The impulse response often lasts much longer than one bit interval, especially at high linear recording densities. Consequently, the current readback signal sample is affected by its neighboring samples, which is characterized by intersymbol interference (ISI) [1].

In communications, maximum likelihood sequence detectors (MLSD) [2] are often used to perform signal detection in ISI channels. A salient feature of sequence detection is that its complexity grows exponentially as the ISI length increases. To reduce the cost of implementing MLSD, it is thus a common practice to first process the readback signal by an equalizer, which shapes the overall channel response to some desired target function (referred to as Partial-Response) that is much shorter than the natural channel response, before forwarding the signal to the sequence detector [3]. Since the noise incurred in the channel is also filtered by the equalizer, a direct consequence of employing equalization is that the equalized signal becomes corrupted by correlated noise. Since traditional sequence detectors are optimized for white Gaussian noise, it is thus necessary to whiten the noise to achieve optimal performance.

Noise prediction emerges as an efficient technique to implement noise whitening. Briefly, by exploiting the correlation between noise samples, the noise predictive technique uses past noise samples to estimate the current one. Subtracting the predicted/estimated value from the current noise sample results in a "prediction-error" signal with lower noise variance and diminished or negligible correlation (with sufficient noise-prediction taps). Hence, noise prediction is able to accomplish noise whitening and noise reduction. When noise prediction is integrated into the sequence detector, the so-called noise-predictive maximum-likelihood (NPML) detection arises [4], [5].

Noise in magnetic recording channels possesses another unique feature, which is data dependence. For example, for thin-film media operating at ultra-high recording densities, medium noise is the dominant noise source. Medium noise is partially due to random variations in the geometry of a magnetization transition, so that the system response may change from one transition to another [6, 7]. This type of noise depends on the data pattern written on the disk, and represents a significant portion of the total noise. The data-dependence of medium noise thus necessitates a data dependent, or pattern-dependent, noise prediction/whitening procedure, which gives rise to pattern-dependent noise predictive (PDNP) detectors. The idea of PDNP was proposed in [8, 9] and later on was treated more generally in [10].

The chapter is devoted to a comprehensive theoretical and practical treatment of PDNP detection techniques. The rest of the chapter is organized as follows. In the next section, we first review the classical ML and MAP detection methods in AWGN channel. We then describe the magnetic recording channel model in Section 33.3. Using this model, in Section 33.4, we show that noise in recording channels are correlated as well as data dependent. We then investigate detection techniques that are suitable for such systems. We present numerical examples in Section 33.5, before we conclude the chapter in Section 33.6.

33.2   Detection Methods in the AWGN Channel

In this section, we first review the conventional detectors optimized for additive white Gaussian (AWGN) noise, including maximum likelihood (ML) sequence detection and maximum a posteriori probability (MAP) symbol detection. In Section 33.4, we shall show that signal dependent detectors result as straightforward extensions of these conventional detectors, once the noise whitening point of view is taken.

33.2.1   Maximum Likelihood Sequence Detection (MLSD) over ISI Channel

The ML-optimal input sequence is obtained from the noisy output, $\mathbf{y}$, of an ISI channel by finding $\hat{\mathbf{b}}_{ML}$ in the following expression:

$$\hat{\mathbf{b}}_{ML} = \arg \max_{\mathbf{b}} p(\mathbf{y} \mid \mathbf{b}) \tag{33.1}$$

where $p(\mathbf{y} \mid \mathbf{b})$ is the joint probability distribution function (pdf) of $\mathbf{y}$, conditioned on the unknown channel input $\mathbf{b}$. Since there are 2^N possible sequences of $\mathbf{b}$, where N is the length of the sequence $\mathbf{b}$, finding the solution to the expression (Equation 33.1) is, in general, very difficult for large N. In the following derivations, we represent the vectors, $\mathbf{b}$ and $\mathbf{y}$, using their elements, as

$$\mathbf{b} = \{b_0, b_1, b_2, \ldots, b_{N-1}\} \tag{33.2}$$

$$\mathbf{y} = \{y_0, y_1, y_2, \ldots, y_{N-1}\} \tag{33.3}$$

and subvectors as

$$b_k^l = \{b_k, b_{k+1}, \ldots, b_{l-1}, b_l\} \tag{33.4}$$

$$y_m^n = \{y_m, y_{m+1}, \ldots, y_{n-1}, y_n\} \tag{33.5}$$

By applying the chain rule, the pdf $p(\mathbf{y} \mid \mathbf{b})$ becomes

$$p(\mathbf{y} \mid \mathbf{b}) = p(y_0 \mid \mathbf{b}) p(y_1 \mid y_0, \mathbf{b}) p(y_2 \mid y_0, y_1, \mathbf{b}) \cdots p\left(y_{N-1} \mid y_0^{N-2}, \mathbf{b}\right) \tag{33.6}$$

Further assuming that the channel output $\{y_0, y_1, y_2, \ldots, y_{N-1}\}$ conditioned on the channel input $\mathbf{b}$ are mutually independent, then (Equation 33.6) becomes

$$p(\mathbf{y} \mid \mathbf{b}) = p(y_0 \mid \mathbf{b}) p(y_1 \mid \mathbf{b}) p(y_2 \mid \mathbf{b}) \cdots p(y_{N-2} \mid \mathbf{b}) p(y_{N-1} \mid \mathbf{b}). \tag{33.7}$$

In the case where the noisy channel output $\mathbf{y}$ can be modeled as a Markov process $L(.)$, whose output o_k is observed through an additive channel noise (hidden Markov model) [11], we can write

$$y_k = L(b_{k-I}, b_{k-I+1}, \ldots, b_k) + n_k \tag{33.8}$$

$$= L(S_{k-1}, S_k) + n_k \tag{33.9}$$

$$= o_k + n_k \tag{33.10}$$

where I is channel memory length, and $S_{k-1} \triangleq b_{k-I}^{k-1}$ and $S_k \triangleq b_{k-I+1}^k$ are the previous and current states of the Markov model at time k, respectively. Using this model, we can further simplify expression (Equation 33.7) as shown below:

$$p(\mathbf{y} \mid \mathbf{b}) = \prod_{k=0}^{N-1} p(y_k \mid S_{k-1}, S_k) \tag{33.11}$$

$$= \prod_{k=0}^{N-1} p(o_k + n_k \mid S_{k-1}, S_k) \tag{33.12}$$

$$= \prod_{k=0}^{N-1} p(n_k) \tag{33.13}$$

with some initial state S_{-1}. We note that $p(\cdot)$ in Equation 33.12 and in Equation 33.13 actually represents two different pdf's: $p(\cdot)$ in Equation 33.12 is a pdf of signal plus noise while $p(\cdot)$ in Equation 33.13 is a pdf of noise only. To help clarify the symbols, an example of the state transition diagram of the Markov process, also known as trellis diagram, is shown in Figure 33.1. The example in Figure 33.1 has four states and each connection between the states has a label (b, o, y), where b is the input, o is the output and y is the noisy channel output. If the channel noise n_k is Gaussian, then

$$p(n_k) = \frac{1}{\sqrt{2\pi\sigma^2}} e^{\frac{-1}{2\sigma^2} n_k^2} \tag{33.14}$$

$$= \frac{1}{\sqrt{2\pi\sigma^2}} e^{\frac{-1}{2\sigma^2} (y_k - o_k)^2} \tag{33.15}$$

where σ^2 is the variance of n_k. By taking the logarithm on both sides of Equation 33.14 and Equation 33.15, and after removing irrelevant constant terms, Equation 33.14 and Equation 33.15 can be simplified to

$$\mu_k = -n_k^2 \tag{33.16}$$

$$= -(y_k - o_k)^2 \tag{33.17}$$

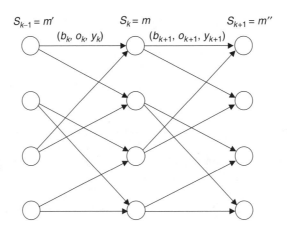

FIGURE 33.1   An example trellis diagram of a Markov model. Circles represent states. b_k and o_k are the input and output of the model, and y_k is the noisy channel output. See related expressions (Equation 33.8 to 33.10).

Left-hand side (or equivalently right-hand side) of expression (Equation 33.17) is known as the branch metric. From Equation 33.13, Equation 33.14 and Equation 33.16, the log-likelihood (logarithm of the likelihood function) of a sequence **b** can be expressed as

$$\log p(\mathbf{y} \mid \mathbf{b}) \propto \sum_{i=0}^{N-1} \mu_k \tag{33.18}$$

Therefore, finding the ML solution to the problem (Equation 33.1) is equivalent to finding the sequence **b** that maximizes $\sum_{i=0}^{N-1} \mu_k$ in Equation 33.18.

Let path metric be the partial sum of the branch metrics

$$M_k = \sum_{i=0}^{k} \mu_i \propto \log p\left(y_0^k \mid b_0^k\right) \tag{33.19}$$

then

$$M_k = M_{k-1} + \mu_k, \quad \text{for all } k \tag{33.20}$$

For each state S_k, we choose a path (survivor) between the two paths which lead to the state S_k based on the following rule:

$$\text{survivor} = \begin{cases} \text{the path with the metric } M_k^0 & \text{if } M_k^0 \geq M_k^1 \\ \text{the path with the metric } M_k^1 & \text{if } M_k^0 < M_k^1 \end{cases} \tag{33.21}$$

where M_k^0 and M_k^1 are the path metrics of the two paths. Expressions (Equation 33.17, Equation 33.20, and Equation 33.21) are recursively applied from $k = 0$ until we find all path metrics for the final states S_{N-1}, and the path with the largest path metric is chosen as the ML solution. Since the number of survivor paths are always the same as the number of states of the Markov model, computational complexity is approximately linearly proportional to the length of input sequence. Outlines of the Viterbi algorithm [2] are provided in Appendix A.

33.2.2 Maximum A Posteriori Probability (MAP) Symbol Detection over ISI Channel

The a posteriori probability (APP) of a symbol, b_k, is defined as

$$APP(b_k) \triangleq P(b_k \mid \mathbf{y}) \tag{33.22}$$

Finding b_k which maximizes the expression (33.22) is known as MAP symbol detection. There are several algorithms to compute or approximate the APP. We present two of them: the BCJR algorithm named after the authors (Bahl, Cocke, Jelinek and Raviv) of the paper [12], which is optimal; and the SOVA (soft output Viterbi algorithm) [13, 14], which is suboptimal.

33.2.2.1 BCJR Algorithm

We assume again that the channels can be modeled as in Equation 33.8, and channel noise is additive white Gaussian. The symbols m', m, and m'' in the following derivations are the past, current and future states of the Markov model respectively. See Figure 33.1 for further clarification of the symbols. To implement BCJR algorithm, we compute $p(b_k, \mathbf{y})$, instead of $P(b_k \mid \mathbf{y})$.

$$p(b_k, \mathbf{y}) = \sum_{m'} \alpha_{k-1}(m') \cdot \gamma_k(m', m) \cdot \beta_k(m) \tag{33.23}$$

where

$$\alpha_{k-1}(m') = p\left(S_{k-1} = m', y_0^{k-1}\right) \tag{33.24}$$

$$\gamma_k(m', m) = p(y_k \mid S_{k-1} = m', S_k = m) \cdot P(b_k) \tag{33.25}$$

and

$$\beta_k(m) = p\left(y_{k+1}^{N-1} \mid S_k = m\right) \tag{33.26}$$

We note that in Equation 33.23, the pair $(S_{k-1} = m', b_k)$ completely determine S_k. Assuming that $P(b_k)$, the a priori probability of b_k, is known, then Equation 33.25 can be evaluated from Equation 33.15. The variables $\alpha_k(m)$ and $\beta_k(m)$ have recursive structures as shown below, and can be computed recursively from the proper initial conditions.

$$\alpha_k(m) = \sum_{m'} \alpha_{k-1}(m') \cdot \gamma_k(m', m) \tag{33.27}$$

and

$$\beta_k(m) = \sum_{m''} \beta_{k+1}(m'') \cdot \gamma_{k+1}(m, m'') \tag{33.28}$$

Typical initial conditions are

$$\alpha_{-1}(m) = \begin{cases} 1, & \text{if } m = 0 \\ 0, & \text{if } m \neq 0 \end{cases} \quad \text{and} \tag{33.29}$$

$$\beta_{N-1}(m) = \begin{cases} 1, & \text{if } m = 0 \\ 0, & \text{if } m \neq 0 \end{cases} \tag{33.30}$$

Equation 33.27 and Equation 33.28 constitute forward and backward recursions, respectively. In implementation, the expressions (Equation 33.23, Equation 33.25, Equation 33.27, and Equation 33.28) are computed in the log domain. We adopt the following convention for notational simplicity:

$$\log p(\cdot) = \bar{p}(\cdot) \tag{33.31}$$

Then, it can be shown that (Equation 33.23, Equation 33.25, Equation 33.27, and Equation 33.28) are expressed in the log domain as follows:

$$\bar{\alpha}_k(m) = \log \sum_{m'} e^{\bar{\alpha}_{k-1}(m') + \bar{\gamma}_k(m',m)} \tag{33.32}$$

$$\bar{\beta}_k(m) = \log \sum_{m''} e^{\bar{\beta}_{k+1}(m'') + \bar{\gamma}_{k+1}(m,m'')} \tag{33.33}$$

$$\bar{\gamma}_k(m, m') = -\frac{1}{2\sigma^2}(y_k - o_k)^2 + \bar{P}(b_k) \tag{33.34}$$

and

$$\bar{p}(b_k, \mathbf{y}) = \log \sum_{m'} e^{\bar{\alpha}_{k-1}(m') + \bar{\gamma}_k(m',m) + \bar{\beta}_k(m)} \tag{33.35}$$

To prevent the numbers from becoming too small, we may do the following operations:

$$\bar{\alpha}_k(m) \leftarrow \{\bar{\alpha}_k(m) - \max_m \bar{\alpha}_k(m)\} \tag{33.36}$$

and

$$\bar{\beta}_k(m) \leftarrow \{\bar{\beta}_k(m) - \max_m \bar{\beta}_k(m)\} \tag{33.37}$$

where $\leftarrow$ means replacement of the left-hand side with the right-hand side. Most applications, such as the Turbo [15] and LDPC decoder [16, 17], require a log-likelihood-ratio which is defined as

$$L(b_k) \overset{\triangle}{=} \log \frac{P(b_k = +1 \mid \mathbf{y})}{P(b_k = -1 \mid \mathbf{y})} \tag{33.38}$$

instead of the APP. The log-likelihood-ratio $L(b_k)$ can be computed using the results from (Equation 33.35),

$$L(b_k) = \bar{p}(b_k = +1, \mathbf{y}) - \bar{p}(b_k = -1, \mathbf{y}) \tag{33.39}$$

Outlines of the BCJR algorithm are provided in Appendix B.

33.2.2.2 SOVA

The high complexity and the long delay in processing (the BCJR algorithm needs an entire sector to begin processing) are the major drawbacks of the BCJR algorithm. Hagenaur et al. developed a suboptimal MAP symbol decoding algorithm, which is known as the soft-output Viterbi algorithm (SOVA) [13]. There are two implementations of SOVA: register exchange mode [13] and trace back mode [14]. SOVA is based on the Viterbi algorithm. Therefore, the decision can be released as soon as a path merge happens, and there is no backward recursion as in the case of BCJR algorithm (SOVA is a forward-only algorithm). We explain the register exchange mode below. The basic idea of SOVA is that we can express the probability of making a wrong decision when we choose the survivor paths on the trellis of the Viterbi algorithm. Specifically, at time k, let the path metric of a survivor path be M_k^s and that of the loser path be $M_k^l(M_k^s \geq M_k^l)$. Then the probability that the wrong survivor was chosen can be expressed as

$$P_k^{sr} = P(\text{choosing wrong survivor}) = \frac{1}{1 + e^{\Delta_k}} \tag{33.40}$$

where

$$\Delta_k = \left(M_k^s - M_k^l \right) / 2\sigma^2 \tag{33.41}$$

Expression 33.40 comes from the following relations.

$$\Delta_k = \log \frac{p\left(y_0^k \mid b_{0,\text{survivor}}^k \right)}{p\left(y_0^k \mid b_{0,\text{loser}}^k \right)} \tag{33.42}$$

$$= \log \frac{P(\text{survivor is correct})}{P(\text{survivor is wrong})} \tag{33.43}$$

$$= \log \frac{1 - P(\text{survivor is wrong})}{P(\text{survivor is wrong})} \tag{33.44}$$

Let us assume that there are g different symbols between the survivor and the loser paths. Let $\hat{P}_j$ ($j = j_1, j_2, \ldots, j_g$) be the probabilities that each of the symbols b_j ($j = j_1, j_2, \ldots, j_g$) on the survivor path is wrong, then we update the probabilities $\hat{P}_j$ based on the following relation:

$$\hat{P}_j \leftarrow \hat{P}_j \left(1 - P_k^{sr} \right) + (1 - \hat{P}_j) P_k^{sr} \tag{33.45}$$

when we choose a survivor path. When the Viterbi algorithm provides the hard output for the symbol b_i, the soft-decision for b_i is

$$P(b_i = +1 \mid \mathbf{y}) = \begin{cases} \hat{P}_i, & \text{if the Viterbi hard decision is 0} \\ 1 - \hat{P}_i, & \text{if the Viterbi hard decision is 1} \end{cases} \tag{33.46}$$

Outlines of the SOVA is provided in Appendix C.

33.3 Recording System Model

In magnetic recording, error correction coding (ECC) and run-length limited (RLL) coding are employed to encode and map user binary bits $\{b_k\}$, $b_k \in \{0, 1\}$ into bipolar channel symbols. The channel symbols are converted to a rectangular current waveform of amplitude -1 or $+1$, determined by b_k, which is subsequently applied to the recording head to magnetize the recording medium. To retrieve the user data, a read head moves along the recording track; it senses the changes in magnetization in the magnetic medium and generates a readback signal which is low-pass filtered, baud-rate sampled, and then processed by the equalizer. The readback signal (prior to low-pass filtering) is given by

$$y(t) = \sum_k x_k h(t - kT) + n_e(t) \tag{33.47}$$

where

$$x_k = b_k - b_{k-1}, \ x_k \in \{-1, 0, 1\} \tag{33.48}$$

is the transition sequence; $h(t)$ is the channel response generated by an isolated transition, that is, from -1 to $+1$, or vice versa ($h(t)$ is often referred to as the transition response); $n_e(t)$ is the electronic noise incurred, assumed to be white Gaussian noise.

We shall note that the transition response $h(t)$ is implicitly time varying and signal dependent, partially due to nonlinearities incurred during the recording and playback process. Such variations are either deterministic or random. While advanced head/medium design and clever writing strategies can reduce these variations, they cannot completely eliminate them. Consequently, transition noise arises due to the

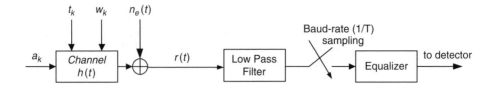

FIGURE 33.2   Transition noise dominated magnetic recording channel.

variations of $h(t)$. More specifically, the transition response $h(t)$ at the kth symbol-interval, denoted by $h_k(t)$, can be written as [18]

$$h_k(t) = h(t - kT - \Delta t_k, w + \Delta w_k) \tag{33.49}$$

where T is the bit interval, and Δt_k and Δw_k are random parameters representing deviations in the position and width of the transition response. The jitter term Δt_k is caused by the fluctuation in the position of written transitions around nominal positions, which varies from one transition to another. The term Δw_k is caused by the fluctuating transition width, resulted from the irregularities of the transition geometry, which become more pronounced as recording densities increase. It is often assumed that $\Delta w_k \geq 0$.

Substituting Equation 33.49 into Equation 33.47, we can write the readback signal as

$$y(t) = \sum_k x_k h(t - kT + \Delta t_k, w + \Delta w_k) + n_e(t) \tag{33.50}$$

The above described system is illustrated in Figure 33.2.

Assuming that Δt_k and Δw_k are small and ignoring the cross-terms, Equation 33.50 can be approximated by its Taylor series expansion, that is.,

$$y(t) \simeq \sum_k x_k h(t - kT, w) + \sum_k x_k \sum_{m=1}^{\infty} \left[\frac{(\Delta t_k)^m}{m!} \frac{\partial^m h(t - kT, w)}{\partial t^m} + \frac{(\Delta w_k)^m}{m!} \frac{\partial^m h(t - kT, w)}{\partial w^m} \right] + n_e(t) \tag{33.51}$$

In Equation 33.51, the first term is the noise-free response and the rest represent the noise. It is clearly shown in Equation 33.51 that the total noise is dependent on the users bits because all factors that involve terms Δt_k and Δw_k are premultiplied by x_k.

The readback signal is low-pass filtered before going through the analog-to-digital (A/D) baud-rate sampler (we assume perfect timing recovery here) and the equalizer. It is not difficult to see that the noise presented at the equalizer output is correlated (colored by the low-pass filter and equalizer) and signal dependent (owing to the second summation term in Equation 33.51), which can lead to significant performance degradation for traditional ML detectors that are optimized for white Gaussian noise.

In this chapter, a Lorentzian pulse is assumed to be the transition response of longitudinal recording. In other words, the transition response for longitudinal recording channel is in the form of

$$h(t) = \frac{1}{1 + \left(\frac{2t}{PW_{50}} \right)^2} \tag{33.52}$$

where PW_{50} denotes the width of the Lorentzian pulse at half of its peak value. The normalized density is defined as

$$D_s = PW_{50}/T \qquad (33.53)$$

which represents the number of bits written in the interval PW_{50}.

33.4 Detection Methods in Data-Dependent Noise

33.4.1 Data-dependent Correlated Noise

In magnetic recording channels, apart from the electronic noise, there are also medium noise and misequalization due to imperfect channel equalization.

As discussed above, medium noise results mainly from transition position/width variation during the magnetization process. Since the positions and shapes of the magnetization transitions are determined by the input channel symbols, medium noise is dependent on the input sequence. The data-dependent property of medium noise is obvious in Equation 33.51, where the second summation term accounts for medium noise and it states that medium noise is determined jointly by channel symbols, jitter/width variations and the recording channel response. Moreover, it is clear that medium noise is colored, and its correlation is determined by the channel characteristics as well as that of the transition position/width variation. In Figure 33.3, we plot the medium noise power corresponding to different data-patterns and

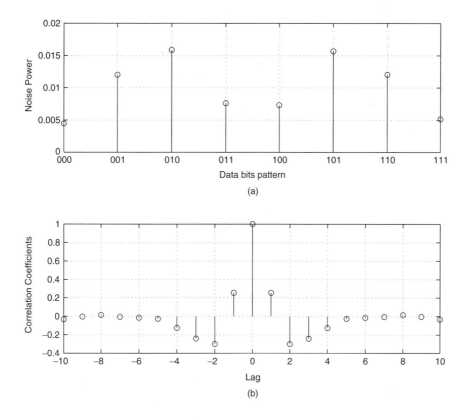

FIGURE 33.3 (a) Transition jitter noise power corresponding to different data patterns; (b) the Correlation structure of transition jitter noise for longitudinal magnetic recording channel at normalized density 2.

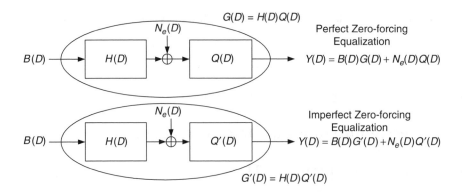

FIGURE 33.4   Imperfect channel equalization causes residual inter-symbol-interference which is data-dependent.

the correlation coefficients of the medium noise in a longitudinal recording channel before channel equalization. The transition jitter Δt_ks are assumed to be i.i.d truncated Gaussian (i.e., $\mid \Delta t_k \mid < T/2$) distributed for different transition. We observe that (1) medium noise power is strongly correlated with data pattern. For example, the noise power corresponding to the 010 and 101 patterns are the largest, since these two patterns have the most transitions (two transitions); (2) significant correlation exists between the transition jitter noise samples.

In the following we show how misequalization is correlated with user bits b_k. Let us represent the cascade of the recording channel, the low-pass filter and the sampler by an equivalent discrete-time linear transversal filter $h_k, k = 0, 1, \ldots, n$. Denote the D-transform of h_k by $H(D)$, that is, $H(D) = \sum_{k=0}^{n} h_k D^k$ and $D = e^{-j\omega T}$ is the delay unit. Denote the partial-response (PR) target by $G(D)$. The zero-forcing equalizer [1] is thus given by $Q(D) \stackrel{\Delta}{=} G(D)/H(D)$, as illustrated in Figure 33.4. In practice, the equalizer is implemented by a linear transversal filter of finite length, whose actual transfer function is different from the ideal $Q(D)$. If we denote the actual transfer function of the equalizer by $Q'(D)$, the effective target realized by the finite-length equalizer is $G'(D) \stackrel{\Delta}{=} H(D)Q'(D)$, which differs from $G(D)$. Referring to Figure 33.4, the equalizer output is given by

$$Y(D) = B(D)G'(D) + N_e(D)Q'(D) \tag{33.54}$$

where $B(D)$ is the input symbol polynomial and $N_e(D)$ denotes the electronic noise polynomial. Now, consider a detector that performs maximum-likelihood sequence detection (MLSD) by assuming a target of $G(D)$ (see Figure 33.5). The total noise seen by the detector is therefore

$$N(D) = Y(D) - B(D)G(D) \tag{33.55}$$

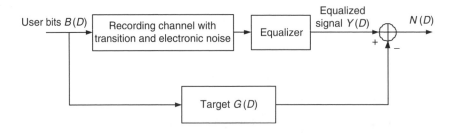

FIGURE 33.5   PRML recording channel with target $G(D)$.

Substituting Equation 33.54 into Equation 33.55, we have

$$N(D) = B(D)[G'(D) - G(D)] + N_e(D)Q'(D) \tag{33.56}$$

where the first term represents the misequalization. Clearly, the misequalization is dependent on the user input data.

33.4.2 MLSD Based on Pattern-Dependent Noise Prediction

It has been shown in the preceding section that the noise seen the by the detector is correlated as well as pattern-dependent. However, to realize maximum-likelihood sequence detection (see Section 33.2.1), the noise samples $\{n_k\}$, $k = 1, 2, \ldots$ must be an independent identically distributed (i.i.d.) sequence. Therefore, in recording channels, it is necessary to *whiten* the noise sequence if conventional MLSD detector is to be used.

Pattern-dependent noise-predictive (PDNP) detection is a technique used to whiten the noise seen by the detector in a signal-dependent fashion. Here we describe PDNP as a generalization of the noise-predictive maximum likelihood (NPML) technique, where pattern-dependent whitening is achieved by exploiting the pattern-dependence of first and second order noise statistics. Note that the resulting detection structure was derived more rigorously in [9], where signal-dependent whitening via an FIR filter was shown to be part of the ML and MAP optimal detectors for sequences over ISI channels in signal-dependent Gauss-Markov noise.

Correspondingly, it is convenient to assume that the noise observed by the detector can be modelled as an auto-regressive Gaussian (also known as a Gauss-Markov) process, that is,

$$n_k(\mathbf{b}) = \sum_{l=1}^{L} a_l(\mathbf{b})n_{k-l}(\mathbf{b}) + e_k(\mathbf{b}) \tag{33.57}$$

where $\mathbf{b}$ denotes the user bit sequence, $\{a_l, l = 1, \ldots, L\}$ are the auto-regressive coefficients, and e_k is a time-uncorrelated Gaussian sequence that depend on the input sequence $\mathbf{b}$. In the above equation, we explicitly state the data dependence of the noise and the associated model coefficients. One should, however, keep in mind that Equation 33.51 will only yield Gaussian noise terms for the case of a first-order Taylor series approximation with Gaussian normal distributed transition width and jitter parameters.

In practice, the coefficient vector $a_l(\mathbf{b})$ are unknown, and can be obtained from the statistics of the noise $n_k(\mathbf{b})$. Specifically, by multiplying both side of Equation 33.57 with the vector $\mathbf{n}(\mathbf{b}) \triangleq [n_{k-1}(\mathbf{b}), n_{k-2}(\mathbf{b}), \ldots, n_{k-L}(\mathbf{b})]^T$ and then taking the expectation of both sides, we get

$$\mathbb{E}\{n_k(\mathbf{b})\mathbf{n}(\mathbf{b})\} = \mathbf{a}(\mathbf{b})^T \mathbb{E}\{\mathbf{n}(\mathbf{b})\mathbf{n}(\mathbf{b})^T\} \tag{33.58}$$

where $\mathbf{a}(\mathbf{b}) = [a_1(\mathbf{b}), a_2(\mathbf{b}), \ldots, a_L(\mathbf{b})]^T$. In the above equation, we have used the fact that $e_k(\mathbf{b})$ is independent of $\mathbf{n}(\mathbf{b})$, that is,

$$\mathbb{E}\{e_k(\mathbf{b})\mathbf{n}(\mathbf{b})\} = 0$$

Thus, we have

$$\mathbf{a}(\mathbf{b})^T = \mathbb{E}\{n_k(\mathbf{b})\mathbf{n}(\mathbf{b})\}R^{-1}(\mathbf{b}) \tag{33.59}$$

and the prediction error variance is

$$\sigma_a^2(\mathbf{b}) = \mathbb{E}\{n_k(\mathbf{b})^2\} - \mathbb{E}\{n_k(\mathbf{b})\mathbf{n}(\mathbf{b})\}^T R^{-1}(\mathbf{b})\mathbb{E}\{n_k(\mathbf{b})\mathbf{n}(\mathbf{b})\} \tag{33.60}$$

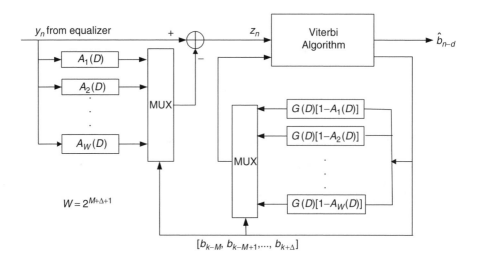

FIGURE 33.6 Pattern-dependent noise predictive maximum-likelihood detection equivalent model. It is shown in the figure that the effective targets are $[1 - A_i(D)]G(D), i = 1, \ldots, 2^{M+\Delta+1}$.

where $R^{-1}(\mathbf{b}) = \mathbb{E}\{\mathbf{n}(\mathbf{b})\mathbf{n}(\mathbf{b})^T\}^{-1}$. It should be noted that although the time index k explicitly appears in the above equations, $\mathbf{a}(\mathbf{b})$ can be independent of k if the system is assumed to be stationary. For nonstationary systems, adaptive algorithms can be used to adjust the predictor coefficients as a function of time.

Now, since $e_k = n_k - \sum_{l=0}^{L} a_l n_{k-l}$ is time-uncorrelated and Gaussian, we can apply Equation 33.14 to compute the likelihoods for sequence detection. In other words, we shall replace n_k with $e_k(\mathbf{b})$ and σ with $\sigma_a(\mathbf{b})$. In practice, the predictor coefficients in $\mathbf{a}$ are restricted to being only dependent on user bits within a finite sliding-window pattern. Using the notation in [10], the window pattern will be denoted by $b_{k-M}^{k+\Delta}$ for nonnegative integers M and Δ. Hence, there are total $2^{M+\Delta+1}$ number of distinct prediction filters. Equivalently, the noise whitening process can be represented by $E_i(D) = [1 - A_i(D)](Y(D) - B(D)G(D))$, where the subscript i corresponds to the ith pattern of $b_{k-M}^{k+\Delta}$, and $E_i(D)$ and $A_i(D)$ denote the whitened sequence and the pattern-dependent noise predictor respectively. Figure 33.6 schematically depicts the noise whitening process.

In view of these discussions, PDNP detection can be implemented as follows:

1. Compute the coefficients of the noise predictors and pattern-dependent variances using Equation 33.59 and Equation 33.60 respectively. If the algorithm is not implemented adaptively, this step is accomplished in the training phase; otherwise, the predictors should be computed adaptively by, for example, the least-mean-square (LMS) algorithm.

2. The Viterbi trellis must be set up such that one is able to determine the signal as well as the predicted noise sample from the transition information. More specifically, if the ISI length is I, to compute the signal and noise samples, it requires the information on $\{b_k, b_{k-1}, \ldots, b_{k-I-L}\}$ for each branch. To compute the predicted noise sample and variance, we need $\{b_{k+\Delta}, \ldots, b_{k-M}\}$. Hence, PDNP requires a trellis with $2^{\max(I+L,\, M)+\Delta}$ states, and state S_k is defined by the bit-sequence $\{b_{k+\Delta}, \ldots, b_{k-\max(I-L,M)+1}\}$.

3. For each transition (or branch) $(\zeta \overset{\Delta}{=} S_{k-1} \to S_k)$, the branch metric is computed as (see Equation 33.14)

$$\lambda_k(S_{k-1}, S_k) = \log p(y_k \mid S_{k-1}, S_k) = -\log \sigma_a(\zeta) - \frac{[y_k - o_k(\zeta) - \hat{n}_k(\zeta)]^2}{2\sigma_a^2(\zeta)} - \frac{1}{2}\log 2\pi \qquad (33.61)$$

where $o_k(\zeta)$ is the signal corresponding to transition ζ, $\hat{n}_k(\zeta)$ is the predicted noise sample for the branch, that is,

$$\hat{n}_k(\zeta) = \sum_{i=1}^{L} a_i(\zeta)[y_{k-i} - o_{k-i}(\zeta)] \tag{33.62}$$

and $\sigma_a(\zeta)$ is the prediction error variance.

4. Proceed as normal Viterbi detection.

33.4.3 BCJR Algorithm for Signal-dependent Channel Noise

We have shown in the above derivations that sequence detection in magnetic recording channels with colored and/or signal-dependent Gaussian noise can be realized with a Viterbi algorithm consisting of modified branch metrics (accordingly, the number of states will need to be expanded to incorporate PDNP). Indeed, this is also the case for the BCJR algorithm (see Section 33.2.2.1). α's and β's can be computed recursively as before using the same formulas: Equation 33.27 and Equation 33.28. However, the γ term should be modified to take into account the pattern-dependent noise whitening process, as shown below :

$$\gamma_k(m', m) = p\left(y_k \mid y_0^{k-1}, S_{k-1} = m', S_k = m\right) \cdot P(b_k) \tag{33.63}$$

Now, incorporating the noise prediction, $p(y_k \mid y_0^{k-1}, S_{k-1} = m', S_k = m)$ should take the form of Equation 33.61, that is,

$$p\left(y_k \mid y_0^{k-1}, S_{k-1} = m', S_k = m\right) = \frac{1}{\sqrt{2\pi \cdot \sigma_a^2(\zeta)}} \exp\left[-\frac{(y_k - o_k(\zeta) - \hat{n}_k(\zeta))^2}{2\sigma_a^2(\zeta)}\right] \tag{33.64}$$

where $\hat{n}_k(\zeta)$ is the same as in Equation 33.62.

33.4.4 SOVA Algorithm for Signal-dependent Channel Noise

SOVA is based on the Viterbi algorithm. Therefore, it is not difficult to modify the SOVA algorithm to be compatible with pattern-dependent Gauss-Markov noise. The only necessary change, like in the Viterbi case, is the branch metric calculation, as given by Equation 33.61 and Equation 33.62.

33.5 Simulations

In this section, we present some simulation results to show that, when medium noise becomes dominant, the detection methods described above can offer performance gains over conventional PRML channels.

The channel model used in the simulation is depicted as in Figure 33.2. In particular, the channel is set up as follows:

1. No precoder is applied.
2. Signal-to-noise ratio (SNR) is defined as

$$SNR = 10 \log_{10} \frac{E_i}{N_0 + M_0} \tag{33.65}$$

where where E_i is the energy of the impulse response of the channel, N_0 is single-sided power-spectrum density height of the electronic noise, and M_0 is the average transition noise energy associated with a single transition.

3. Transition noise is the only medium noise source, and it consists of jitter noise only. The transition jitter Δt_k is truncated Gaussian (bounded by $T/2$, i.e., $|\Delta t_k| < T/2$) with zero mean and variance determined by M_0.

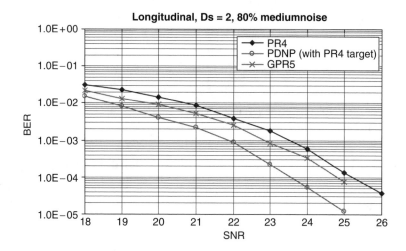

FIGURE 33.7 Performance example of PDNP detection for longitudinal recording channel of normalized density 2 with 80% medium noise. PDNP is based on the PR4 [1 0 −1] target with 32 FIR 4-taps prediction filters. PRML systems based PR4 and GPR5 targets are also shown in the figure for comparison purposes.

4. An 7th-order elliptic low-pass filter (LPF) is assumed at the receiver front end to avoid aliasing.
5. A 21-tap linear transversal minimum-mean-square error (MMSE) equalizer is employed [19].
6. The noise predictive filter length is 4 and the coefficients are optimized in the MMSE sense.
7. Normalized density is fixed at 2.0.

Figure 33.7 shows the bit-error-rate (BER) performance of maximum likelihood sequence detection based on PDNP, when the medium noise accounts for 80% of the total noise. The equalizer is optimized in the MMSE sense for PR4 [1 0 −1] target. The noise predictor length is $L = 4$, and the pattern window is defined by $\Delta = 1$ and $M = 3$. For comparison, in the same figure we also plot the results for classical PRML system for PR4 [1 0 −1] target and the generalized PR (GPR) target of length 5. The GPR target is designed jointly with the equalizer under the monic constraint [19] in the MMSE sense. We can see that when medium noise is significant, PDNP detection can provide large SNR gains over classical PRML systems. However, we should note that the gains observed depend on the comparison target chosen as well as the system parameters. As we observe in Figure 33.7, the gains of PDNP over the GPR5 system is less than the gains seen by comparing to the PR4 system. Figure 33.8 and Figure 33.9 further illustrate such an argument. In Figure 33.8 and Figure 33.9, we give the BER results for the same systems, however, the medium noise level becomes 40% and 20%, respectively. We can see that as medium noise diminishes, although the PDNP detection still outperforms the PRML systems based on PR4 and GPR5 target, the gains decrease with the medium noise level. For example, in Figure 33.9 where the medium noise only accounts for 20% of the total noise, the PDNP detection only provides marginal gains over the PRML system based on GPR5 target. This is due to the fact that when the noise seen by the detector becomes less data dependent, an appropriate designed PRML system where the target well matches the channel can provide near optimal performance.

Complexity wise, the PDNP detection depends on the equalization target, the noise predictor length and the pattern window selected. For example, for the PDNP system shown in Figures 33.7 through 33.9, the number of trellis states required for the PDNP detection is

$$2^{\max(I+L,M)+\Delta} = 2^{\max(2+4,3)+1} = 128 \tag{33.66}$$

By comparison, for the PR4 system, only 4 states are required, and for the GPR5 system, 16 states are required. The complexity of the PDNP detection can be reduced via a variety of ways. For example, by employing decision feedback, the trellis can be significantly downsized. In [10], the choices of the PDNP parameters and various ways for reducing the complexity of PDNP detection are discussed in detail.

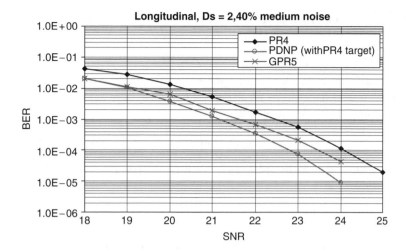

FIGURE 33.8 Performance example of PDNP detection for longitudinal recording channel of normalized density 2 with 40% medium noise. PDNP is based on the PR4 [1 0 −1] target with 32 FIR 4-taps prediction filters. PRML systems based PR4 and GPR5 targets are also shown in the figure for comparison purposes.

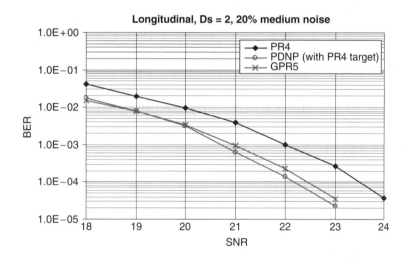

FIGURE 33.9 Performance example of PDNP detection for longitudinal recording channel of normalized density 2 with 20% medium noise. PDNP is based on the PR4 [1 0 −1] target with 32 FIR 4-taps prediction filters. PRML systems based PR4 and GPR5 targets are also shown in the figure for comparison purposes.

33.6 Conclusions

In this chapter, we considered methods for signal detection for recording systems which suffer from data-dependent noise. Specifically, we investigated the pattern-dependent noise prediction (PDNP) detection in detail, which includes the maximum-likelihood (ML) PDNP detection and the maximum a posteriori (MAP) PDNP detection. We also presented numerical simulation results for a longitudinal magnetic recording system. Our results show that the PDNP can offer improved performance for media noise dominated systems over classical PRML detection methods. However, we emphasize that the actual gain numbers highly depend on the operating point (normalized density, media noise percentage) and the reference PRML system (PR target, complexity) one compares the PDNP against.

APPENDIX A: Outlines of the Viterbi Algorithm

A1.1 At time 0

1. Initialize path metrics M_{-1} for all initial states S_{-1}.
2. Compute the branch metrics μ_0 for all possible transitions from S_{-1} to S_0.
3. Compute the path metrics for all states S_0:

$$M_0 = M_{-1} + \mu_0 \tag{33.67}$$

4. Choose the survivor path: Discard the path with the smaller path metric M_0 of the two paths leading to each state S_0.

A1.2 At time k

1. Compute the branch metrics μ_k for all possible transitions between S_{k-1} and S_k.
2. Compute the path metrics for all next states S_k:

$$M_k = M_{k-1} + \mu_k \tag{33.68}$$

3. Choose the survivor path of the two paths leading to each state S_k: Discard a path with the smaller path metric M_k.
4. Check merge: If there is a unanimous decision (merge of the survivors) in the earliest part of the survivor sequences, we make final decision on the bit. Or we force final decision $\hat{b}_{k-D}$ based on majority vote over the survivors with pre-determined decision delay D.
5. Increase k by one and repeat steps $1 \sim 4$ until we find all decisions.

APPENDIX B: Outlines of the BCJR Algorithm

B1.1 Forward Recursion

1. Initialize α's and β's as in Equation 33.29 and Equation 33.30.
2. Beginning from $k = 0$ toward $k = N-1$, as soon as y_k is received, compute γ_k using Equation 33.25.
3. Update α's using Equation 33.27.
4. As soon as reaching the end of forward recursion ($k = N - 1$), begin backward recursion.

B1.2 Backward Recursion

1. Beginning from $k = N - 1$ toward $k = 0$, compute γ_k or use the one computed and saved in the forward recursion.
2. Update β's using Equation 33.28.
3. Compute $p(b_k, \mathbf{y})$ using Equation 33.23 or compute log-likelihood-ratio using Equation 33.35.

APPENDIX C: Outlines of the SOVA

1. Initialize the path metrics of VA for all initial states. Also initialize $\hat{P}_i$, the probability that the decision from the VA is wrong, with zeros.
2. Compute the branch metrics μ_k for all possible transitions between S_{k-1} and S_k.
3. Compute the path metrics for all next states S_k:

$$M_k = M_{k-1} + \mu_k \tag{33.69}$$

4. Choose a survivor of the two paths leading to each state S_k. Discard a path with the smaller path metric. Also compute Δ_k and P_k^{sr} from Equation 33.41 and Equation 33.40. Then update $\hat{P}_j$ using Equation 33.45.

5. Check merge: If there is a unanimous decision in the earliest part of the survivor sequences, we make final decision on the bit. Release soft decision using Equation 33.46.

6. Increase k by one and repeat steps $2 \sim 5$ until we find all decisions.

References

[1] J. G. Proakis, *Digital Communications*, McGraw-Hill, New York, 1995.

[2] G. D. Forney, Maximum-likelihood sequence estimation of digital sequences in the presence of intersymbol interference, *IEEE Trans. Inform. Theory*, vol. IT-18, no. 3, pp. 363–378, 1972.

[3] H. K. Thapar and A. M. Patel, A class of partial response systems for increasing storage density in magnetic recording, *IEEE Trans. Magn.*, vol. 23, no. 5, pp. 3666–3668, 1987.

[4] P. R. Chevillat, E. Eleftheriou, and D. Maiwald, Noise-predictive partial-response equalizers and applications, in *Proceedings of ICC*, pp. 942–947, IEEE, 1992.

[5] J. D. Coker, E. Eleftheriou, R. L. Galbraith, and W. Hirt, Noise-predictive maximum likelihood (npml) detection, *IEEE Trans. Magn.*, vol. 34, no. 1, pp. 110–117, 1998.

[6] H. N. Bertram, *Theory of Magnetic Recording*, Cambridge University Press, England, 1994.

[7] N. R. Belk, P. K. George, and G. S. Mowry, Noise in high performance thin-film longitudinal magnetic recording media," *IEEE Trans. Magn.*, vol. Mag-21, no. 5, pp. 1350–1355, 1985.

[8] J. Caroselli, S. A. Altekar, P. McEwen, and J. K. Wolf, Improved detection for magnetic recording systems with media noise, *IEEE Trans. Magn.*, vol. 33, no. 5, pp. 2779–2781, 1997.

[9] A. Kavcic and J. M. Moura, The viterbi algorithm and markov noise memory, *IEEE Trans. Inform. Theory*, vol. 46, no. 1, pp. 291–301, 2000.

[10] J. Moon and J. Park, Pattern-dependent noise prediction in signal-dependent noise, *IEEE J. Select. Areas on Commn.*, vol. 19, no. 4, pp. 730–743, 2001.

[11] Y. Ephraim and N. Merhav, Hidden markov processes, *IEEE Trans. Inform. Theory*, vol. 48, no. 6, pp. 1518–1569, 2002.

[12] L. R. Bahl, J. Cocke, F. Jelinek, and J. Raviv, Optimal decoding of linear codes for minimizing symbol error rate, *IEEE Trans. Inform. Theory*, vol. 20, pp. 284–287, 1974.

[13] J. Hagenauer and P. Hoeher, A viterbi algorithm with soft-decision outputs and its applications, in *Proceedings IEEE Globecom*, pp. 1683–1686, IEEE, 1989.

[14] J. Hagenauer, Source-controlled channel decoding, *IEEE Trans. Commn.*, vol. 43, no. 9, pp. 2449–2457, 1995.

[15] C. Berrou and A. Glavieux, Near optimum error correcting coding and decoding: Turbo-codes, *IEEE Trans. Commn.*, vol. 44, no. 10, pp. 1261–1271, 1996.

[16] R. G. Gallager, Low-density parity-check codes, *IEEE Trans. Inform. Theory*, vol. 8, no. 1, pp. 21–28, 1962.

[17] D. J. C. Mackay, Good error-correcting codes based on very sparce matrices, *IEEE Trans. Inform. Theory*, vol. 45, no. 3, pp. 399–431, 1999.

[18] J. Moon, Performance comparison of detection methods in magnetic recording, *IEEE Trans. Magn.*, vol. 26, no. 6, pp. 3155–3172, 1990.

[19] J. Moon and W. Zeng, Equalization for maximum likelihood detectors, *IEEE Trans. Magn.*, vol. 31, no. 2, pp. 1083–1088, 1995.

34

Read/Write Channel Implementation

34.1	Introduction	34-1
34.2	Read/Write Channel Architectures	34-2
34.3	Partial Response Equalization with Maximum-Likelihood Detection	34-3
34.4	Magnetic Disk Drive Read/Write Channel Integration	34-4
34.5	Analog Front-Ends	34-5
	Variable Gain Amplifier with Automatic Gain Control • Continuous-Time Filter • Analog-to-Digital Converter	
34.6	Equalizer Architectures in Read Channels	34-9
	FIR Filter Architectures • Review of Representative FIR Filters	
34.7	Viterbi Detector	34-12
	Add-Compare-Select Recursion • Bit-Level Pipelined ACS • Metric Normalization in ACS Design • Survivor Sequence Detection • Data Postprocessing	
34.8	Future Detection: Iterative Decoders	34-19
	Soft-Output Viterbi Decoder • Low-Density Parity-Check Decoders	
34.9	Timing Recovery	34-24
34.10	Write Path	34-27
34.11	Overview of Recently Published Disk-Drive Read/Write Channels	34-28
34.12	Challenges of Further Integration	34-29

Borivoje Nikolić

Engling Yeo
University of California at Berkeley
Berkeley, CA

Michael Leung
Solarflare Communications Inc.
Irvine, CA

Kiyoshi Fukahori
TDK Semiconductor Corp.
Mountain View, CA

34.1 Introduction

The growth in disk drive areal densities in the past has been accomplished by increasing both the linear density of the data bits along a track, and by increasing the track density. The small seek times in disk drives, which signify the time for data retrievals from different location of the disk, are achieved through higher rotational speeds of the disk. These trends, of continuous increase in linear density and rotation speed, increase the data rates of disk-drive read channels. On the other hand, even with the vast advancement in magnetic recording head technology, the reduction in the areal densities is coupled with a decrease in the signal-to-noise ratios at which the data is detected, requiring more sophisticated detection methods to be used. Each process technology generation allows the integration of more complex signal processing schemes to recover the required bit-error rates with lower signal-to-noise ratios. Scaling of the process technology also allows operation at higher data rates, as shown in Figure 34.1. The average annual increase

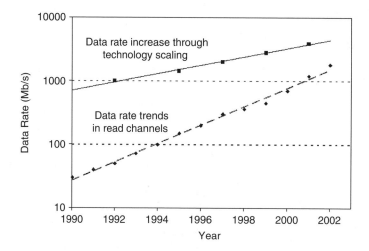

FIGURE 34.1   Data rate trend in magnetic disk drive read channels (Source: IEEE International Solid-State Circuits Conference and manufacturers' web sites).

in disk drive read channel data rates of almost 40% outpaces the rate increase provided by the technology scaling. The demand for high transfer rates, complex signal processing, coupled with stringent cost and power requirements, make the implementation of disk drive signal processing architectures challenging.

This chapter reviews the traditional architectures for signal processing in magnetic read/write channels. Today's dominant architecture, where majority of signal processing is performed in the digital domain, is analyzed in detail. The main challenges and alternatives for implementation of major building blocks are discussed. These include analog front-end signal conditioning (including continuous-time filtering with preequalization), analog-to-digital conversion, adaptive equalization, Viterbi detection, and timing recovery. As the previous chapters have covered the theoretical aspects of operation of these blocks in considerable details, most of the discussion will focus on architectures and techniques that are used in practical read channels. Particular attention is paid to various architectural options in implementing equalizers and detectors with analysis of their impact on performance and power. The techniques for implementing iterative decoders, as possible future detectors are presented and, finally, future integration of the disk-drive electronics is discussed.

34.2 Read/Write Channel Architectures

Advances in integrated circuit technology allow for integration of more complex read/write channels in each generation. Peak detection was used for more than 30 years, and was implemented using multiple chips, usually in bipolar technology. Partial response maximum-likelihood detection, introduced in early 1990s provided performance enhancements at increased complexity.

Disk-drive signal processing using peak detection was initially implemented in analog, continuous-time domain. This was followed by analog sampled-data processing and was finally replaced by digital signal processing (DSP) [1]. As CMOS technology allowed higher integration, with efficient implementation of digital signal processing, it eventually replaced the bipolar and BiCMOS technologies. The development of read channels continues to move the boundary between the analog and digital signal processing closer to the front-end. This trend is also common in most communications receivers.

Analog sampled-data PRML channels dominated the disk-drive market in early- to mid-1990s. Equalization and Viterbi detection were performed in the analog domain, and there was no explicit analog-to-digital conversion [2, 3]. Digital data was output as the decisions from the Viterbi detector [4].

In the late 1990s digital architectures became widespread. The first PRML channels from early 1990s used this architecture [5], but their performance fell short of the analog PRML channels implemented

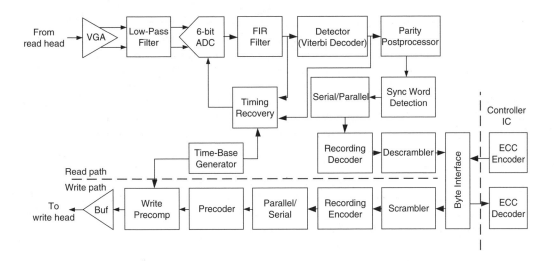

FIGURE 34.2   Digital read/write channel block diagram.

in BiCMOS technology. Consequently, several intermediate architectures appeared; most notably the architectures that replace a complex digital FIR filter with a smaller analog filter [6, 7].

The mainstream digital architecture implements some pre-equalization in the analog domain, but the main equalization is performed in the digital domain using a digital adaptive finite-impulse response (FIR) filter. These channels require a fast, 6-bit analog-to-digital converter (ADC) in the front end.

The digital read/write channel architecture is shown in Figure 34.2. To present the complete information processing in one figure, the outer error-correction code (ECC) block is included in the diagram, although the ECC operation in disk drives resides in the controller IC.

34.3 Partial Response Equalization with Maximum-Likelihood Detection

A contemporary read/write channel is built around partial response equalization with maximum likelihood (PRML) sequence detection. PRML is a suboptimal, but practical solution to sequence detection. The theory of its operation was introduced for digital communications in 1960s [8], proposed for use in magnetic recording systems by Kobayashi and Tang in 1970 [9], and practically implemented in disk drives in 1990 [5, 10]. The equalizer is used to reduce the span of ISI to a few neighboring bits, thus reducing the MLSD complexity.

In general, input and output of the recording channel are modeled using the relationship

$$y(D) = h(D)w(D) \tag{34.1}$$

where the channel transfer function is usually modeled as

$$h(D) = (1 - D)P(D) \tag{34.2}$$

where D denotes a unit sample delay that corresponds to the sampling period T_c. The transfer function has a spectral null at DC, and the polynomial $P(D)$ has a general form of $P(D) = 1 + p_1 D + p_2 D^2 + \cdots + p_N D^N$. Initially the integer coefficients corresponding to the class of polynomials of $P(D) = (1+D)^N$ were widely used. The case $n = 1$ corresponds to a partial-response class 4 (PR4) channel, $n = 2$ corresponds to the (extended-PR4) EPR4 channel, while $n = 3$ corresponds to E^2PR4 channel [11].

The first implemented systems were based on PR4, which allowed interference of one previous sample with the current sample. The increase in storage densities required a shift towards EPR4 channels that are better matched to the channel model for user densities higher than 2.0. Further increase in densities led to generalized partial response polynomials with real coefficients. Longer allowed inter-symbol interference requires the use of more complex Viterbi decoders. Eight-state EPR4 channel detectors were replaced with 16- and 32-state detectors. For example, 16-state targets, such as $(1 - D)(1 + D)(5 + 4D + 2D^2)$ [12] or $(1 - D)(1 + D)(5 + 2D + 2D^2)$ and $(1 - D)(1 + D)(4 + 3D + 2D^2)$ [14], tend to better match the channel response above user densities of 2.5 than the E^2PR4. Various 32-state targets can be used for densities higher than 3.0, and one example is $(1 - D)(1 + D)^2(2 + D + D^2)$ [13]. Further adjustments in matching the channel response lead to diminishing returns in BER performance, while significantly increasing the cost of implementation, because they require more complex detectors.

The equalized sequence is brought to a Viterbi detector that estimates the channel input sequence. The goal of the detector is to resolve which recorded sequence caused a particular response. To resolve N bits of ISI, the detector has to implement a state machine with 2^N states. Therefore, the complexity of the detector increases geometrically with the number of bits that interfere. The exception from the trend is the PR4 detector, where two 2-state $1 - D$ detectors can be interleaved to form one $1 - D^2$ detector.

34.4   Magnetic Disk Drive Read/Write Channel Integration

Typical magnetic disk read/write channel is a mixed-signal system, as shown in Figure 34.2. In the write path, the input data is scrambled and encoded using a recording code. The last block in the write path is the write precompensation, which predistorts write signal to compensate for the anticipated nonlinear effects in the head-medium interface. The digital write signal is sent to a write head driver and recorded onto a disk. To minimize the effects of noise, write current drivers and the read preamplifier are usually placed near recording/reading head.

On the read side, the signal from preamplifier is brought to a read/write channel, where it is first conditioned in the analog domain. In the analog part, read channel has a variable-gain amplifier (VGA), a continuous-time (CT) filter, and an ADC. ADC is followed by the adaptive equalizer, which equalizes the data to a desired target. The adaptive equalizer is of FIR type and is usually implemented in digital domain. The equalized sequence is brought to a Viterbi detector. The estimated sequences from the detector are decoded and descrambled. The equalizer output data, as well as detected data in decision-directed implementations is also used for controlling the VGA, timing recovery, equalizer adaptation and servo control, as shown in Figure 34.2.

Timing recovery is necessary for the read operation of the channel. Its output controls overall chip timing and feeds clock to all other blocks. The time-base generator generates the reference frequency for the write path, and also the reference frequency that is input into the timing recovery. Timing is acquired during the preamble and tracked on detected data. Preamble precedes data in each sector and is also used for adjusting the VGA gain.

Modern read/write channels are implemented in CMOS technology, where exists a tight relationship between the speed, power, and area for an implementation of a given function. Each silicon process technology generation has allowed integration of more complex signal processing schemes into an affordable chip size and has been characterized by exponential trends in data rates, as shown in Figure 34.1. To attain the annual increase of almost 40% in disk drive read channel data rates, new microarchitectural techniques are implemented in succeeding generations. Scaling of the process technology doubles the logic density in each generation, and, coupled with supply voltage reduction, halves the power dissipation of a given function. Stringent cost and power requirements make the implementation of disk drive signal processing architectures challenging, particularly when coupled with the demand for increased transfer rates. The need for low cost dictates small die sizes. Recently reported read/write channel implementations occupy the die size of 10–25 mm^2 and dissipate between 1 and 1.7 W at maximum operating speed [3, 14–18].

Scaling of analog front-ends does not follow the scaling rate of the digital back-ends, because of device matching and noise requirements. The reduction in size of digital back-ends due to sizing does not yield a proportional reduction in overall die size. Hence, instead of reducing the size of read-channel circuits, the technology scaling has been exploited to permit integration of more complex algorithms. As a result, the sizes of digital signal processing blocks have remained approximately constant. The evolution of detectors illustrates this trend: 8-state conventional Viterbi decoders, common in 0.35 μm technology, were replaced with more complex 8-state noise-predictive Viterbi decoders or conventional 16-state decoders in 0.25 μm. This was followed by 16-state noise-predictive decoders in 0.18 μm. Currently, state of the art 0.13 μm detectors incorporate 32-state noise-predictive decoders or 16-state noise predictive detectors with postprocessing [14]. The exception from the trend is the transition from PR4 to EPR4 detectors, which required two technology generations to be feasible, and involved the change from interleaved 2-state detectors to 8-state detectors.

The following sections explore the architectures and implementation tradeoffs of the main building blocks in a read/write channel.

34.5 Analog Front-Ends

The analog front-end of the read channel consists of the VGA, MR asymmetry correction (MRAC), programmable gain amplifier, continuous-time filter (CTF) and an ADC, as shown in Figure 34.3.

34.5.1 Variable Gain Amplifier with Automatic Gain Control

Digital read/write channels employ the VGA with a digital automatic gain control (AGC) loop to maintain a full-scale input to the ADC over the range of input signal amplitudes. Attenuation of the read/write process due to head/media variations, up to the input of the read channel, could be as high as 40 dB, and the VGA has to be able to compensate for it.

VGAs frequently employ differential amplifiers where the gain is controlled through adjusting the biasing tail current and the current to diode connected loads. In bipolar or BiCMOS technology, VGA employs a circuit concept based on Gilbert multiplier cell [19]. To achieve required gain and bandwidth, two differential stages are frequently used.

Amplifiers with linear-gain control have a problem maintaining both the loop stability and the speed of gain acquisition. Linear-gain VGAs are slow with low signal levels and fast, but could be unstable, with high signal levels. This is because the loop gain of the AGC loop and the closed loop bandwidth are proportional to the signal amplitude. A common solution is to use a non-linear-gain amplifier (such as logarithmic) and exponential gain control to achieve the loop gain independent of the signal amplitude as shown in the bipolar realization in Figure 34.4 [20].

The overall transfer characteristic is linear in dB as a function of control voltage ($V_{cp} - V_{cn}$) and the gain ($= V_{out}/(V_{ip} - V_{in})$) is given by $R_L / R_E * exp(kT/q * (V_{cp} - V_{cn}))$. In CMOS technology, the quasi-logarithmic gain function can be accomplished by making use of the square law inherent to the MOS devices.

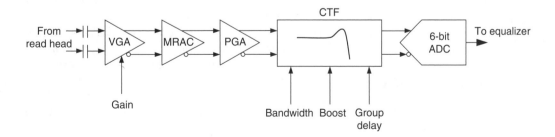

FIGURE 34.3 Block diagram of an analog front-end.

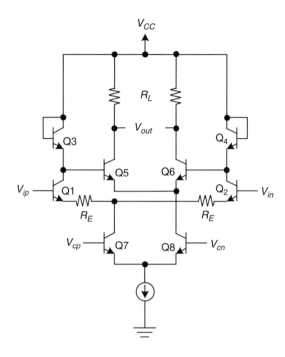

FIGURE 34.4   VGA cell in bipolar technology.

The control for the VGA can be derived in an analog manner by full-wave rectification of the continuous-time filter output [23]. Alternately when the gain control is available in a digital format, the required gain can be accomplished in two steps [22] in which the programmable gain amplifier (PGA) achieves most of the gain required followed by VGA — often an attenuator — thus providing the best signal-to-noise ratio possible.

Digital read channels employ decision-directed AGC. The control is implemented using the least-mean squares algorithm. AGC error derivation should be orthogonal to timing error.

34.5.2   Continuous-Time Filter

The continuous-time (CT) filter employed in the read path has two roles: anti-aliasing for the ADC and some data preequalization in the continuous-time domain, before sampling. The cutoff frequency of the filter is typically set around 20–40% of the data rate, in the example of the EPR4 channels. High frequency boost of 3–7 dB is normally engaged, giving a sharp frequency rolloff of the resulting filter response at around 50% of the data rate. The bandwidth requirement is typically ±10% while its group delay characteristics need to be controlled to a few percent error. The filter pole frequency needs to be varied over a 1:4 range to allow constant magnetizing density over the entire media. The distortion needs to be less than 1%.

The filter is usually implemented as a 5th or 7th order equiripple linear phase filter, and has programmable cut-off, boost and group delay equalization settings [21]. Symmetrical or asymmetrical zeros are used to yield high frequency boost for "pulse slimming" and to provide group delay adjustment. It is implemented using a single-pole section together with 2 or 3 biquad sections to form a 5th or 7th order response.

A biqaud section is typically based on g_m-C architectures shown in Figure 34.5. This simulates an LC low pass filter from V_{in} to V_{out}. Its pole frequency, ω_0, and pole quality factor, Q, are given by $\omega_0 = g_m/C$ and $Q = g_m/g_t$, respectively. Note that g_m and g_t must be individually controlled to yield predictable overall filter response. A simple construction of the g_m stages places its parasitic poles at frequencies much higher

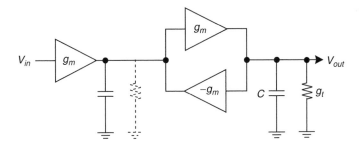

FIGURE 34.5 $g_m - C$ biquad arrangement.

than g_m/C allowing, in principle, a high frequency operation. However, it should be noted that excess phase in the g_m-C integrator due to finite output impedance and parasitic poles and zeros of the g_m stage causes peaking in the quality factor that can be actually realized [28]. Figure 34.6 shows the response of a real integrator, compared to an ideal integrator. Note the excessive phase buildup at ω_0 due to a parasitic pole.

Figure 34.7 shows a commonly used g_m stage in CMOS technology that is made up of a differential pair with a triode degeneration transistor [14]. Important considerations are its tuning range and distortion.

In bipolar or BiCMOS implementations, superior transconductance of the bipolar transistor makes it possible to adjust the filter characteristics by one-time trim at wafer sort [21]. To expand a rather narrow input voltage range of bipolar differential pair, multiple offset differential pairs are summed.

In CMOS implementations, a large variation in g_m-C values over process, temperature and supply voltage makes it mandatory for the filter response be tuned automatically. An on-chip oscillator has been used whose time constant is matched to an accurate off-chip reference tone in a master/slave configuration. [26, 29–31]. Another method involves two control loops, one for the pole frequency accuracy and the other for the quality factor [24]. The pole frequency is tuned by a reference to an external clock and its quality factor by comparing envelope responses of first- and second-order section.

As noted earlier, a good control of the quality factor is more important than that of the pole frequency. Thus, in the design of a filter consisting of several biquad sections, the relative matching of sections that

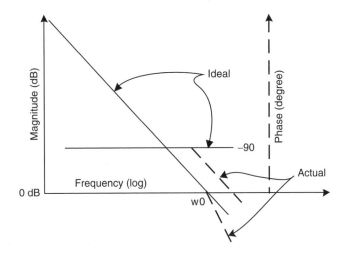

FIGURE 34.6 Amplitude and phase response of ideal and real integrators.

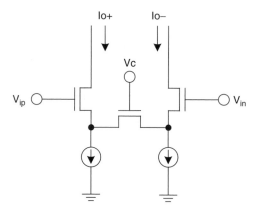

FIGURE 34.7   Commonly used g_m stage with a MOS device biased in triode region.

determines amplitude peaking and group delay variation is of utmost importance. In [25] every section is tuned individually for Q and the pole frequencies are then tuned relative to each other. This is made possible by first making the total integrating capacitance at input and output nodes of a biquad equal, by using dummy transconductance stages. Once the relative matching is accomplished, its absolute bandwidth can be adjusted without affecting the relative shaping of the filter response.

Contemporary analog front-ends for use with magnetoresistive (MR) heads include the thermal asperity (TA) detection and correction circuitry. It is usually implemented as a feedback path from the output of the CTF to the input of the VGA. When an asperity on the disk platter comes into contact with the slider, the slider is being heated up. The change in temperature can affect the MR element's resistance which will, in turn, change its output DC voltage. To detect these sudden changes in voltage, the CTF output is low-pass filtered and compared to the established references. If the signal exceeds the specified upper and lower limits, the TA correction circuitry is engaged.

34.5.3 Analog-to-Digital Converter

Analog-to-digital converter (ADC) converts the band-limited analog signal into the digital domain. Commonly, the ADC is preceded by a track-and-hold (T/H) circuit that relaxes the aperture requirements of the ADC array. To accurately represent the data in the range of operating signal-to-noise ratios, 6-bit ADCs are typically used.

In order to achieve high sampling rates that match disk-drive data throughputs, only parallel ADC architectures can be used. A parallel (or flash) 6-bit ADC uses an array of 64 comparators implemented in parallel. The input signal is compared to the reference voltages, as shown in Figure 34.8. The reference voltages are divided values of the full scale reference, obtained through a string of resistors. Each comparator that has the input voltage greater than its reference outputs a '1,' while those with inputs lower than the reference output a '0.' If there are no mismatches and no offsets between the comparators, the output will be a valid digital word in a "thermometer" code, where a digital value is represented by its equivalent number of ones preceded by zeros. For example a binary value of 000101 (decimal 5) is represented by ... 00011111. If there is a mismatch in comparator input offsets, sequences of alternating zeros and ones, ... 00010111..., (so called "bubbles") in the code may appear. The decoder in Figure 34.8 corrects the bubbles and converts the word from a thermometer code to a binary code.

Area, power, and the input capacitance of the flash converter linearly increases with a number of comparators. Therefore, a 6-bit converter is twice as large as a 5-bit one.

To reduce the power and ADC input capacitance interpolating and folding-and-interpolating ADCs are sometimes used. More details about ADC designs can be found in [33].

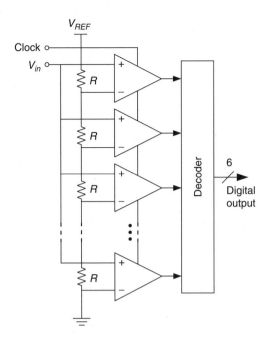

FIGURE 34.8 Block diagram of a flash analog-to-digital converter.

34.6 Equalizer Architectures in Read Channels

Following the analog-to-digital conversion, the data is processed in the digital domain. The equalizer and the detector operate with the serial data, and are the most performance critical. Their throughput frequently can limit the read channel operating speed. Since both the equalizer and the detector are inside the timing recovery loop, their latency is also important. After the sequence detection, the remaining processing is done with parallel, byte-level data.

Transformations of signal processing algorithms lead to tradeoffs in power, area, and throughput/latency of a particular function [34]. Different transformations are suitable for recursive and nonrecursive algorithms. Nonrecursive algorithms, illustrated through FIR filtering in the following section, can be pipelined, retimed, parallelized or transposed to achieve the desired throughput. Pipelining and parallelization of recursive algorithms is more difficult [35], and the throughput can be increased, for example, through loop unrolling and retiming. An example of a recursive algorithm is the Viterbi decoder, analyzed in Section 34.7.

34.6.1 FIR Filter Architectures

Adaptive equalization is frequently implemented through finite-impulse-response (FIR) filtering where the filter taps are updated through the least-mean squares (LMS) algorithm, as described in Chapter 28. Although there does exist a recursion in the filter coefficients update, this feedback loop is not timing critical. The timing constraints of the adaptation loop are relaxed, by implementing the delayed version of the LMS algorithm. As a result, the FIR filter throughput can be increased at the expense of power, area or latency through transformations that include pipelining, parallelism, and transposition.

FIR filtering can be expressed by the equation:

$$y[n] = a_0 x[n] + a_1 x[n-1] + a_2 x[n-2] + \cdots + a_N x[n-N] \tag{34.3}$$

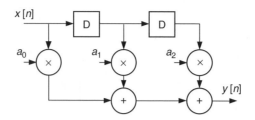

FIGURE 34.9   Direct form of an FIR filter.

Direct implementation of the Equation 34.3 results in what is commonly known as "direct form" FIR filter architecture. An example 3-tap direct-form FIR is shown in Figure 34.9.

The critical path of an N-tap direct-form filter FIR filter implementation consists of one multiplication and $N - 1$ additions together with the register overhead for an N-tap filter. To shorten the critical path the multi-operand addition can be implemented using a tree of carry-save adders [37].

The throughput of a direct-form FIR can be increased by pipelining. This is performed by adding the same number of delay elements to each forward cutset between the input and the output. Pipelining increases the throughput, but also increases the latency of the architecture. The critical path is reduced to one multiplication and one addition plus the register overhead for the example 3-tap FIR from Figure 34.10.

A common method for representing a datapath function is the signal flow graph, where the function is mapped onto a graph. Each node in the graph represents the computation or task and indicates all input signals. Each edge in the graph denotes a linear transformation of the signal. Signal-flow graphs present a convenient way to illustrate some common signal processing transformations. For example, transposition of the graph reverses the direction of all the edges in a signal-flow graph and interchanges the input and output ports without changing the functionality. The direct and transpose graphs of a 3-tap FIR are shown in Figure 34.11(a), and Figure 34.11(b), Block diagram of a transpose form FIR is shown in Figure 34.11(c). The input data are broadcasted to the inputs of all multipliers. The key properties of the transpose form filter are a short critical path, consisting of one multiplication and one addition, with increased input loading [36].

Another convenient representation of the FIR filter uses processing elements (PE), shown in Figure 34.12. Using the PE based representation, derivation of the parallel FIR structure is straightforward. A parallel filter has two parallel paths, each operating at a half the input data rate, as shown in Figure 34.13. The area of the filter doubles, but each processing element has almost twice as much time to finish the computation, without affecting the throughput.

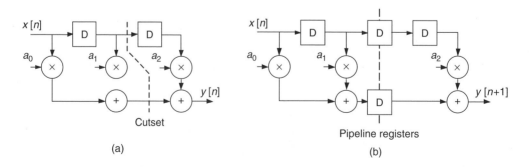

FIGURE 34.10   (a) Forward input-output cutset; (b) pipelined FIR filter.

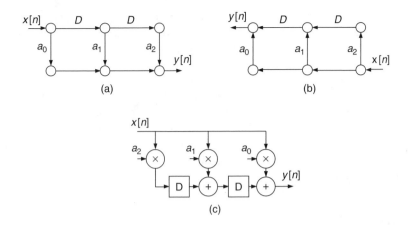

FIGURE 34.11 Signal flow graph representation of a FIR filter: (a) direct form; (b) transpose form; (c) block diagram of a transpose FIR.

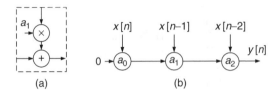

FIGURE 34.12 (a) Processing element (PE) of an FIR filter; (b) PE representation of a direct FIR.

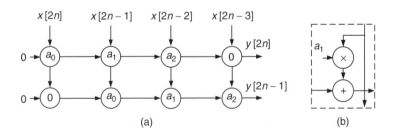

FIGURE 34.13 (a) Parallel FIR filter structure; (b) processing element.

34.6.2 Review of Representative FIR Filters

Architectural transformations lead to throughput-latency-power-area tradeoffs. Chapter 28 surveys the equalization and adaptation architectures implemented in recent disk-drive read channels. This chapter limits the scope to the actual filter architectures.

Digital equalization for PRML channels uses both the direct FIR implementation [38–40], as well as the transposed architecture [43, 44]. An early digital read channel [38] implements a 9-tap adaptive direct FIR filter with 6-bit input data and 10-bit coefficients. The multiplier outputs are added using the carry-save tree and the final carry-lookahead adder in two pipeline stages. Distributed arithmetic approach, [39, 45, 46], precomputes the filter outputs $y(k)$ for an arbitrary set of 1-bit inputs $x(k)$. For m-bit inputs, m identical single bit slices select m outputs from the common precomputed table. To calculate the final outputs, these m outputs are weighted and added together. Two interleaved, parallel paths are used by Mita, et al. [40] and Moloney, et al. [41] to achieve high throughputs in direct FIRs. Wong et al. [41], use four parallel filter paths operating at a quarter of the sampling rate with lower supply voltage to achieve high throughput and low power.

Thon, et al. [43], implement the transpose filter architecture. A recent implementation by Staszewski, et al. [44], is based on two interleaved transposed paths. In this implementation, the data inputs are Booth recoded. The pipeline is level-sensitive-latch based to allow cycle borrowing. Equalizer adaptation is covered in Chapter 28.

In noise-predictive read channels, the PR4 adaptive equalizer, $(1 - D^2)$ is followed by a whitening filter. The read channel from reference [14] uses a noise-whitening filter that can be adjusted between $4 + 3D + 2D^2$ and $5 + 2D + 2D^2$ targets.

34.7 Viterbi Detector

Viterbi detector observes the received and equalized sequence of sample values $y(t)$, and estimates the most likely recorded sequence. Number of possible sequences grows exponentially with the length of the observed data Sequence, making a bank of correlators or matched filters an impractical detector implementation. The Viterbi algorithm performs maximum likelihood detection recursively, and the number of operations grows linearly with the length of the input sequence.

The Viterbi algorithm is commonly expressed in terms of a trellis diagram, which is a time-indexed version of a state diagram. The simplest 2-state trellis is shown in Figure 34.14. Maximum likelihood detection of a digital stream with intersymbol interference can be described in terms of maximizing probabilities of paths through a trellis of state transitions (branches). Each state corresponds to a possible pattern of recently received data bits and each branch of the trellis corresponds to a receipt of the next equalized, noisy, input.

The branch metric is the cost of traversing along a specific branch, as indicated in Figure 34.14. In additive white Gaussian noise (AWGN) they represent the squared difference between the received sample $y(t)$, and corresponding equalization target value s_k:

$$bm_k = (y(t) - s_k)^2 \qquad (34.4)$$

Although branch metric computation is a strictly feed-forward and therefore does not present a speed bottleneck, the implementation of the squaring function can be costly in hardware. In the expanded branch metric expression, $y(t)^2 - 2y(t)s_k + s_k^2$, can be noted that the squared term, $y(t)^2$, is common to all the branch metrics. Since the path metric computation is based on differences of incoming state metrics, this common term can be subtracted from all the branch metrics, simplifying the computation. Therefore, in practice, the branch metric computation is based on multiplication of $y(t)$ by a constant s_k, and addition to a precomputed constant s_k^2.

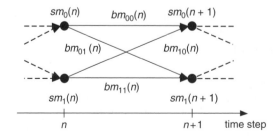

FIGURE 34.14 Two-state trellis.

State metrics, also called path metrics, accumulate the minimum cost of 'arriving' in a specific state. Finally, the path taken through the trellis, represents the survivor sequence, or the most likely sequence of recorded data.

The implementation of the Viterbi detector, a processor that implements the Viterbi algorithm, consists of three major blocks: add-compare-select unit (ACS), branch metrics calculation unit (BMU) and survivor path decoding unit. In the design of the Viterbi detector efficient realization of its central part, the add-compare-select unit, is crucial for the ratio of achievable data throughput and chip area or power.

34.7.1 Add-Compare-Select Recursion

The central part of the Viterbi algorithm is a nonlinear feedback loop, the add-compare-select recursion. ACS unit calculates the sums of the state metrics (sm) with corresponding branch metrics (bm) and selects the maximal (or minimal) to be the new set of state metrics, Equation 34.5.

$$sm_0(n+1) = \min \left\{ \begin{array}{c} sm_0(n) + bm_{00}(n) \\ sm_1(n) + bm_{10}(n) \end{array} \right\} \tag{34.5}$$

The new state metrics at time instant n, $sm_0(n+1)$, $sm_1(n+1)$, are calculated as a minimum of sums of state metrics at time instant n, $sm_0(n)$, $sm_1(n)$, with corresponding branch metrics for transition between the states, $bm_{00}(n)$ and $bm_{10}(n)$, or $bm_{01}(n)$ and $bm_{11}(n)$. These operations are performed in parallel for each state.

The throughput of a Viterbi decoder has traditionally been limited by the difficulty of pipelining the single-step ACS recursion. Figure 34.15 shows the transition trellis of an example 8-state Viterbi decoder, such as the one used in EPR4 channels. The critical-path of a traditional ACS computation extends through the sequential execution of two parallel additions, a comparison and a selection. The new value of state metrics has to be computed in each time instant. This single-cycle recursion prevents straightforward pipelining or parallelization to increase the throughput. Possible alternatives for increasing the throughput include the loop unrolling and retiming of the operations.

The comparison in the ACS recursion is implemented through subtraction, and the most significant bit (MSB) of the result selects the winning path. The ripple-carry implementations of both add and compare operations take advantage of the similarity in carry profiles. The amount of overhead in the critical-path required for executing the subtraction only involves the computation of the MSB of the difference. Fast adder structures such as the carry-select adder will require the subsequent subtraction to follow an abrupt carry profile, which yields small performance gains with large power and area penalties.

An approach to accelerate the ACS computation is by performing the add and compare operations concurrently [49]. The comparison is implemented through a four-input addition of two state metrics

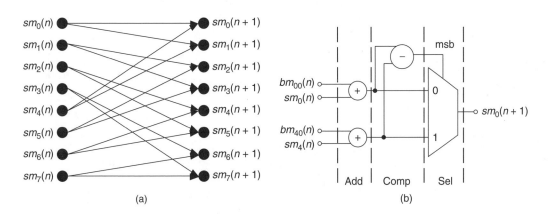

(a) (b)

FIGURE 34.15 (a) Eight state trellis; (b) conventional add-compare-select unit.

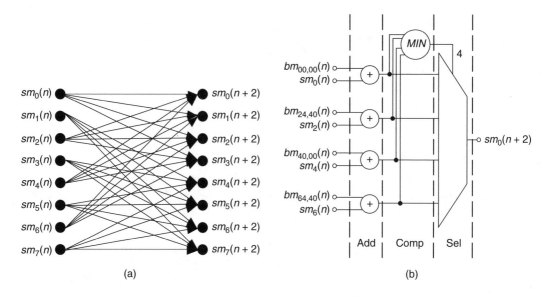

FIGURE 34.16   (a) One step lookahead applied to an eight-state trellis; (b) radix-4 add-compare-select unit.

with two branch metrics. By using two layers of three-to-two carry-save adders, followed by a carry-lookahead adder the four-way addition is faster than the addition followed by a comparison.

Unrolling the ACS recursion to perform two trellis iterations in one clock cycle has been used as a method to increase the throughput [47, 48]. In essence, this lookahead method replaces the original radix-2 trellis with a radix-4 trellis, Figure 34.16(a). A radix-4 ACS computes four sums in parallel followed by a four-way comparison, Figure 34.16(b). In order to minimize the critical-path delay, the comparison is realized using six parallel pair-wise subtractions of the four output sums. The critical-path delay, compared to the radix-2 ACS, is less than doubled, thus increasing the effective throughput.

Finally, an architecture, obtained through retiming and transformation of the ACS unit [50], has a critical path consisting only of a 2-input adder and a multiplexer. The operations are reordered as defined in Equation 34.6. The resulting structure has been labeled as a compare-select-add (CSA) unit, shown in Figure 34.17, resulting in the parallel execution of the compare and add operations. The critical path delay is reduced to the combined delays of the comparator and the multiplexer. Although this modification

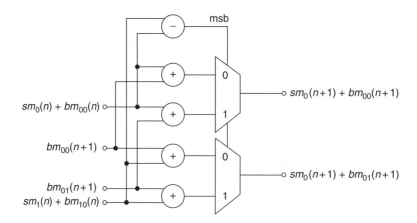

FIGURE 34.17   Parallel compare-select-add unit.

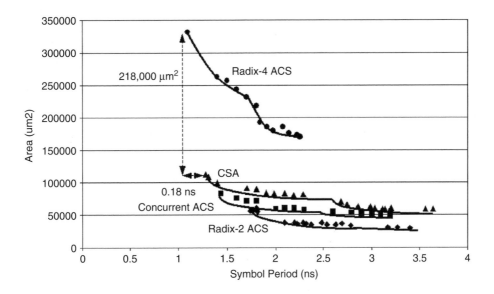

FIGURE 34.18 Area comparison of various ACS structures.

incurs the cost of doubling the number of adders and multiplexers, it is faster than the concurrent ACS.

$$sm_0(n+1) + bm_{00}(n+1)$$
$$= \min \left\{ \begin{array}{l} sm_0(n) + bm_{00}(n) \\ sm_4(n) + bm_{40}(n) \end{array} \right\} + bm_{00}(n+1) \tag{34.6}$$

Various ACS architectures have been compared using standard-cell library in a 0.13 μm technology [51]. The test was conducted on a block of eight ACSs, with interconnect resembling the underlying trellis structure.

The power-throughput and area-throughput comparisons are plotted in Figure 34.18 and Figure 34.19. The synthesis algorithm trades a higher area for delay reduction through sizing and logic transformations.

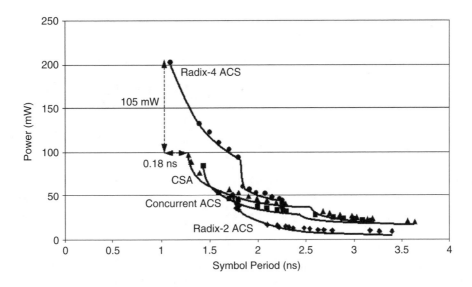

FIGURE 34.19 Power comparison of various ACS structures.

Each curve tracks the same behavior. As the decreasing critical-path constraint approaches a minimum value, the area and power consumption of the synthesized structure increases sharply due to the use of increased gate sizes. The kinks in the curves correspond to points where logic transformations are preferred over increased sizing.

As expected, the radix-4 ACS, which has been accounted for the doubled symbol rate, has the highest throughput. It is faster than the next-fastest structure by a margin of 17%, but requires almost three times the area and two times the power. The radix-4 ACS is consistently larger and consumes more power than any of the other structures. Both the transformed CSA and concurrent ACS are able to improve the throughput with significantly less area and power penalty.

34.7.2   Bit-Level Pipelined ACS

The conventional Viterbi decoding cannot be pipelined because the result of the current ACS operation is needed in the next cycle of recursion. A different approach involves the use of a redundant number system and carry-save addition [52–54]. This approach allows bit-level pipelining of the ACS recursion, allowing high speeds, by adding more computational hardware. A practical realization using pipelined dynamic logic was shown in [48]. State metrics are represented in redundant number system [55], where each binary digit is represented in the base-2 redundant form:

$$S = \sum_{i=0}^{w-1} s_i 2^i \tag{34.7}$$

where $s_i \in \{0, 1, 2\}$, and are 2-bit encoded as $s_i \in \{00, 01, 11\}$.

The key advantage of this approach is that the carry propagates only to the next bit position, which allows bit-level pipelining and very high speeds of operation. However, the comparison is more complex and the redundant representation itself requires more hardware.

34.7.3   Metric Normalization in ACS Design

The path (state) metric is the accumulated sum of branch metrics. Since the branch metrics are positive numbers, the path metric grows continuously with each branch metric added, thus requiring normalization. Several methods of metric normalization have been proposed in the past [56]. Among them, the most attractive method from implementation standpoint is modulo normalization [56, 57].

Modulo arithmetic exploits the fact that the Viterbi algorithm inherently bounds the maximum dynamic range of state metric differences to be [56]:

$$\Delta_{\max} = bm_{\max} \cdot \log_2 N \tag{34.8}$$

where N is the number of states (e.g., 8 in case of EPR4) and $bm_{\max}$ is the maximum value of the single-step branch metrics.

The state metric wordlength is calculated as $\sup[\log_2(\Delta_{\max} + m \cdot bm_{\max})] + 1$, where $m \cdot bm_{\max}$ accounts for the dynamic range increase for one step lookahead implementation ($m = 2$). The length of the state metric is increased by one bit to allow modulo computation and normalization. Therefore, an EPR4 radix-4 ACS with 5-bit branch metrics would require the 8-bit state metrics.

Modulo approach also simplifies the comparison in ACS. The larger among the two numbers is simply found by determining the most significant bit (MSB) of their difference in two's complement representation. If

$$|sm_i - sm_j| < \Delta_{\max} \le 2^{N-1} \tag{34.9}$$

where N is the wordlength of the state metric, and the compared state metrics sm_i and sm_j are in the same half of the number circle in Figure 34.20, and the comparison can be performed by simple subtraction. The most-significant bit (MSB) points to the winning path.

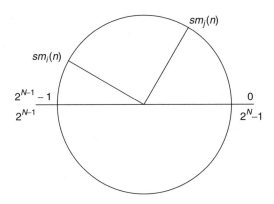

FIGURE 34.20 State metrics sm_i and sm_j represented on a number circle.

34.7.4 Survivor Sequence Detection

In order to decode the input sequence, the survivor path, or the shortest path through the trellis must be traced. The selected minimum metric path from the ACS output points the path from each state to its predecessor. In theory, decoding of the shortest path would require processing of the entire input sequence. In practice, the survivor paths merge after some number of iterations, as shown in a 4-state example in Figure 34.21. From the point they have merged together, the decoding is unique. The trellis depth at which all the survivor paths merge with high probability is referred as the survivor path length. Paths merge with high probability after four to five convolutional code constraint lengths, equivalent to the order of the partial response channel.

A tradeoff exists between two dominant architectures for survivor path decode, the trace-back and the register-exchange [58]. For the detectors with a small number of states, the survivor path is usually determined by using the register exchange. In case of detectors with a larger number of states, RAM based trace-back may have a size advantage. In both cases, the circuit level optimization of memory elements is critical in terms of area and power. A single-step register-exchange structure is shown in Figure 34.22. It consists of flip-flops and two-way multiplexers, controlled by the decisions from the ACS units. Implementation of the register exchange therefore does not affect the critical path of the detector, but does affect its area and power.

By the time they reach the end of survivor registers, all outputs should have converged to the same value. However, this may not always be the case. A common scheme to enhance the register exchange is to employ a global minimum finder to select which survival register output to be used. The global minimum finder consists of a series of pipelined add-select circuits that identify the location of the ACS unit which contains the smallest accumulated path metric. The output of the survival register that corresponds to the minimum metric state is picked as the final output of the Viterbi detector. The use of the global minimum finder

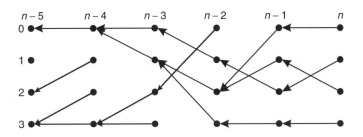

FIGURE 34.21 Survivor paths for a 4-state trellis.

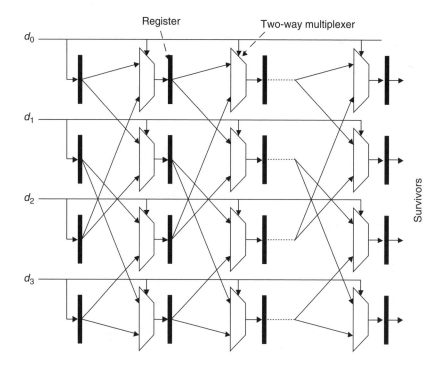

FIGURE 34.22   Register exchange.

can also reduce the survival register length by a factor of 2 [58]. Additionally, fairly accurate preliminary decisions of the Viterbi decoder can be made by picking survival register output of the minimum metric state before the end of complete register exchange. These preliminary decisions can be used to generate timing recovery information with lower latency.

34.7.5 Data Postprocessing

In contemporary read channels, BER at the Viterbi decoder output is further improved through postprocessing. This postprocessing relies on the fact that the errors in Viterbi decoders matched to higher order channels are not uniformly distributed, and some error sequences occur more often than the others. A postprocessor is used to detect these most common errors in the data stream and to correct them [59–61]. A sector of data is divided into smaller sub-frames, of approximately 50 to 100 bits. In each sub-frame, one or more parity bits are added to aid in the detection of dominant errors. The number of parity bits, the location of their insertion and the type of checks being performed depends on the equalization scheme and the recording code being used. It is also specific to the vendors. It is important to point out that adding more parity bits increases the number of error patterns that can be detected, but increases the coding overhead, and is, therefore, carefully optimized [61].

The postprocessor calculates the differences between the inputs to the Viterbi detector and the reconstructed data samples at the output. These differences are fed into various error filters, matched to a list of dominant error patterns for that particular combination of equalization and code. When a dominant-type error occurs in the subframe, it will violate one of the parity constraints and will be detected by the postprocessor. The exact location of the error pattern within the subframe will be indicated by the error filter matched to the error pattern that corresponds to a group of errors that triggers this parity check violation. This postprocessor will correct the dominant errors in the serial data stream before the recording code is decoded.

The postprocessing operation is illustrated on the EPR4 channel example. The postprocessor takes the data stream from the survival sequence register, x'_k, and recreates the expected EPR4 sample values y'_k.

TABLE 34.1 Parity Bits as Indications of Certain Types of Error Patterns

Case	Odd Parity bit	Even Parity bit	Interpretations
A	Correct	Correct	No $\pm\{1\}, \pm\{1-1\}$ or $\pm\{1-11\}$ error in the subframe
B	Correct	Wrong	$\pm\{1\}$ error starting in odd location or $\pm\{1-11\}$ error in even location
C	Wrong	Correct	$\pm\{1\}$ error starting in even location or $\pm\{1-11\}$ error in odd location
D	Wrong	Wrong	$\pm\{1-1\}$ in the subframe

It can be shown that dominant error patterns for EPR4 equalization, reflected to the input data stream are: $e_x = \pm\{1\}$, $e_x = \pm\{1-1\}$, and $e_x = \pm\{1-11\}$. These error patterns can be identified by checking the parity of odd and even bits in the sub-frame. Therefore, two parity bits have to be inserted in each sub-frame. To detect these dominant error patterns, six error filters are used, with coefficients $\pm[1\ 1\ -1\ -1]$, $\pm[1\ 0\ -2\ 0\ 1]$ and $\pm[1\ 0\ -1\ 1\ 0\ -1]$, which are matched to the corresponding error events in the output data stream. The matched filter output will produce a positive maximum at a bit position that corresponds to the position of the error sequence. Constraints on occurrence of each type of error help pin-point the type and position of the error. For example, if x'_k is 1, then the error patterns $+\{1\}, +\{1-1\}$, and $+\{1-1\,1\}$ are cannot start at position k and only error patterns of $-\{1\}, -\{1-1\}$, and $-\{1-1\,1\}$ can start at that bit position. Assuming only single error event occurred in the sub-frame, the errors will be indicated by parity bits as shown in Table 34.1.

In case A, when there are no parity check violations postprocessor will not take any actions. For case D, postprocessor will locate the maximum of the validated outputs from error filters $\pm\{1-1\}$ and correct the sequence accordingly. For cases B and C, postprocessor will locate the maximum of validated odd (or even) outputs of error filters $\pm\{1\}$ and even (or odd) outputs of error filters $\pm\{1-1\,1\}$ and correct the erroneous sequence that corresponds the error filter that gives a higher output value. The postprocessor has to be designed to correctly treat the errors at the subframe boundaries.

The outlined postprocessing technique attempts to perform error correction in decoded data with a rate loss of an error detection code. However, the assumption used in the error correction scheme, that there is a single error event in subframe, that error sequences are confined only to a set of selected dominant error patterns, etc., implies that false correction could occur. As a result of false corrections, the number of error events inside the subframe can increase, making the Reed–Solomon ECC operation in the controller less effective. Furthermore, there is a tradeoff between the code rate loss associated with postprocessing and the rate loss in ECC. Thus, the application of a postprocessing scheme requires an overall system-level consideration.

Implementation of the postprocessor is straightforward. Since the process of error correction is strictly feed-forward and there are no hard latency constraints in the algorithm, the postprocessor can be pipelined and parallelized until it achieves the required throughput.

34.8 Future Detection: Iterative Decoders

As further SNR improvements using advances in PRML technology are becoming more difficult to achieve, a large research effort has been devoted to investigating its possible successors. A promising technique to achieve significant further SNR improvement is the use of iterative decoding 35–36. Initial investigations of iterative detectors for applications in magnetic recording channels created a great deal of interest among researchers in both academia and industry in recent years. Although iterative decoders promise large gains over conventional PRML systems, they have not been used in commercial applications so far. Deployment of iterative decoders, besides impacting the overall system design, would present a substantial increase in complexity of the read/write channels.

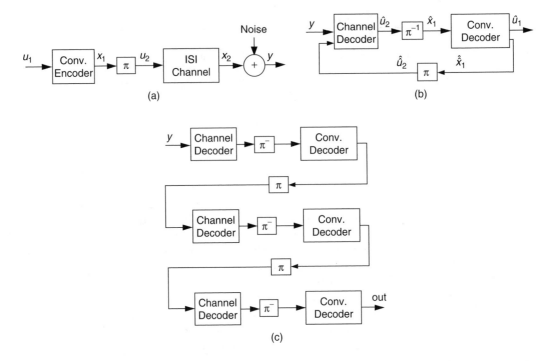

FIGURE 34.23 Serial turbo: (a) encoder; (b) decoder with blocks separated by interleavers/deinterleavers (π/π^{-1}); (c) unrolled decoder for three iterations.

Iterative decoding is performed between the inner decoder, matched to the channel and the outer decoder, matched to either a convolutional or a low-density parity-check (LDPC) code. These decoders exchange soft information (confidence of decoded bits) back and forth between each other.

To implement an iterative decoder it is necessary to replace the conventional, soft-input, hard-output Viterbi decoder with a soft-input, soft-output (SISO) decoder. This soft information can be extracted by using either a maximum-a-posteriori (MAP) decoder, based on a BCJR algorithm [62] or a soft-output Viterbi algorithm (SOVA) [63]. Although BCJR decoder has a small SNR performance advantage, SOVA decoder is a more likely choice for deployment as a channel decoder, because of its smaller complexity [64].

A serial turbo code can be formed by serially concatenating an outer convolutional code with the PR channel acting as the inner code, as shown in Figure 34.23(a). Turbo decoding is performed by iterating between the inner and the outer decoder through the interleavers and deinterleavers, as shown in Figure 34.23(b). Therefore, to maintain the throughput of the iterative decoder comparable to today's read channels, turbo decoder has to be unrolled and pipelined, as shown in Figure 34.23(c). Since the size and power of the SOVA decoder are approximately two times larger than that of the Viterbi decoder [65] unrolling the decoder four times would require more than 20 times increase in decoder area and power, with interleavers included.

34.8.1 Soft-Output Viterbi Decoder

Soft decoding information can be extracted from a Viterbi decoder by modifying its traceback algorithm. The SOVA decoder outputs the log-likelihood of a correctly decoded bit. This value is given by the difference between the path metrics of the two most-likely (ML) paths that trace back to complementary bit decisions, $\hat{x}$ and $\bar{\hat{x}}$. Figure 34.24 shows that the ML path, α, is determined using the Viterbi algorithm with an L-step

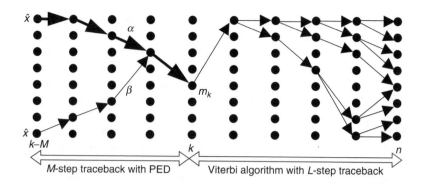

FIGURE 34.24 Two-stage traceback in a SOVA decoder to determine the two most likely paths, α and β.

traceback. This is followed by an M-step traceback that resolves the next ML path, β, based on maximal probability of its deviation from α. Assuming that the absolute values of the path metrics, M_α and M_β, dominate over those of other paths, the probability of selecting β over α (i.e. making the wrong decision) is given by Equation 34.10.

$$P_{\text{err}} = \frac{\exp(-M_\beta)}{\exp(-M_\alpha) + \exp(-M_\beta)}$$

$$= \frac{1}{1 + \exp(\Delta)}; \quad \Delta = M_\beta - M_\alpha \tag{34.10}$$

The log-likelihood of a correct output by the SOVA decoder is given in Equation 34.11.

$$\log\left[P\left(\frac{\text{CorrectDecision}}{\text{WrongDecision}}\right)\right] = \log\left(\frac{1 - P_{\text{err}}}{P_{\text{err}}}\right)$$

$$= \Delta = M_\beta - M_\alpha \tag{34.11}$$

An architecture that implements an 8-state SOVA decoder is shown in Figure 34.25 [66]. The branch metric generator, eight ACS units, and the L-step survivor memory unit (SMU) form the building blocks of a

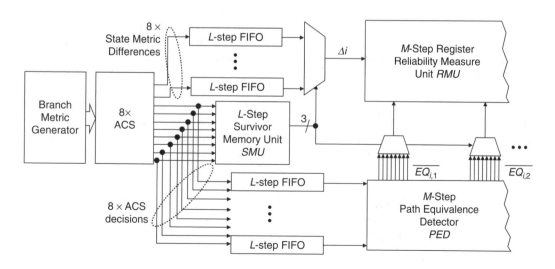

FIGURE 34.25 Architecture of an 8-state SOVA decoder.

conventional Viterbi decoder (e.g., matched to the EPR4 channel). The eight parallel add-compare-select units compute the pairs of cumulative path metrics and select the winning paths in the underlying trellis representation of the convolutional code. Additionally, each ACS also outputs the difference in path metrics between the two competing paths. The path decisions are stored into an array of L-step flip-flop-based first-in, first-out (FIFO) buffers. The delayed signals are used in the M-step path-equivalence detector (PED) to determine the equivalence between each pair of competing decisions obtained through a j−step traceback, $j \in \{1, 2, \ldots, M\}$.

The path metric differences from the eight ACSs are stored in FIFO buffers of depth L. Using the output decision from the SMU as a multiplexer select signal, the delayed metric difference at the most-likely state is sent to a reliability measure unit (RMU). The RMU also receives a list of equivalence test results that are performed on the competing traceback decisions paths that originate from the most-likely state. The selected equivalence results are evaluated in the RMU in order to output the minimum path metric difference reflecting competing traceback paths that result in complementary bit decisions, $\hat{x}$ and $\tilde{x}$.

Throughput of the SOVA decoder, equivalently to a conventional hard-decision Viterbi decoder, is limited by the ACS recursion. Therefore, all the architectural transformations discussed for the Viterbi decoders apply to the SOVA decoders.

34.8.2 Low-Density Parity-Check Decoders

A more practical, but still complex approach to building an iterative decoder is based on using parity checks. Examples of such outer codes are low-density parity-check (LDPC) codes and turbo product (TP) codes. The iterative decoding algorithm, based on the message-passing algorithm, is similar for the LDPC and TP codes. The soft information is extracted from a partial response channel using the SOVA (or, alternatively, a MAP) decoder, and then used in the message-passing algorithm (MPA) for the LDPC or TPC decoding, Figure 34.26 [67]. The major architectural difference is that construction of LDPC codes does not require iterating with the channel and does not require the use of an interleaver [68].

Implementation of an LDPC or a TPC decoder consists of two main parts: computation of message a-posteriori probabilities and passing of messages. Similarly to decoding convolutional codes, LDPC decoders evaluate a-posteriori probability in the log-probability domain, Figure 34.27. The simple even-parity check constraint evaluates $\prod_{n} (1 - 2p_n)$, where p_n represents the probability that a bit x_n equals to 1. The use of log-probability domain simplifies the evaluation of the product, but also requires the table-lookup evaluation of $\log[\tanh(\frac{LLR(x_n)}{2})]$, where $LLR(x_n)$ is defined as the log-likelihood ratio, $\log[\frac{p_n}{1-p_n}]$.

Low-density parity-check decoders operating in the log-probability domain can frequently achieve good error performance with arithmetic precision of just three to five bits. This implies that the lookup tables can be efficiently implemented with simple combinatorial logic that directly implements the required function.

In addition to the calculation of marginal posterior functions, practical decoder implementations can lower the energy consumption per decoded bit by applying stopping criteria to the decoding iterations. This is done, for instance, in an LDPC decoder when all parity check constraints have been met.

An LDPC decoder is required to provide a network for messages to be passed between a large number of nodes. A direct mapping of the network using hard-wired routes leads to congestion in the interconnect fabric for the same reasons [64, 69]. The congestion can be circumvented through the use of memory. However, unlike the interleavers used in turbo codes, which have a one-to-one connectivity, LDPC graphs

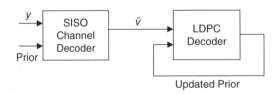

FIGURE 34.26 Iterative decoder consisting of SISO channel decoder and an LDPC decoder.

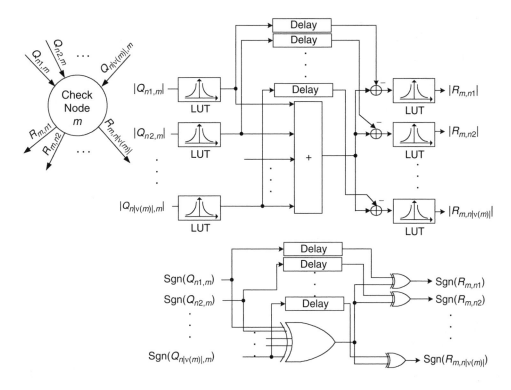

FIGURE 34.27 A posteriori probability computation in LDPC decoders.

have at least a few edges emanating from each variable node. The number of edges is several times larger than that in an interleaver network, and hence requires a larger memory bandwidth.

Practical implementation of a message-passing network is dependent on the structural properties of the graph. In general, the construction of good iterative codes require large numbers of nodes whose interconnections are defined by graphs which are expanders and have an absence of short cycles. These graphs tend to have a disorganized structure, which complicates the implementation of the message-passing network by requiring long, global interconnects or memories accessed through an unstructured addressing pattern. More recently however, graphs with structured patterns have emerged, [70–72] and they simplify the implementation of the decoders.

The order of message computations and their exchange distinguishes two main classes of decoders: parallel and serial. Parallel decoder architectures directly map the nodes of a factor graph onto processing elements, and the edges of the graph onto a network of interconnect. The parallel computation of messages requires the same number of processing elements as the number of nodes in the factor graph. On the other hand, serial decoder architectures distribute the arithmetic requirements sequentially amongst a small number of processing elements. Due to the sparseness of the graph, there is usually a delay between the production and consumption of the messages. Hence, this technique requires additional memory elements to store the intermediate messages.

A parallel realization of any algorithm will frequently be throughput and power efficient, at the expense of increased area. On the other hand, serial realizations require fewer arithmetic units, and make use of memory elements in place of complex interconnect.

In LDPC decoding, there is no interdependency between simultaneous variable-to-check and check-to-variable computation. Parallel LDPC decoders will benefit from throughput and power efficiency, but will require the implementation of large numbers of processing elements and message passing within a congested routing network. In order to ease the difficulty in routing, a common approach is to partition a design into smaller subsets with minimum overlap. However, due to irregularity in the parity

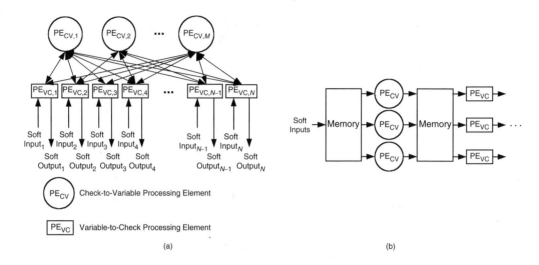

FIGURE 34.28   Parallel (a) and serial (b) architectures for LDPC decoding.

check matrix, design partitioning is difficult and yields little advantage. An example of a parallel LDPC decoder [69], (Figure 34.28(a)) for a 1024-bit rate-$\frac{1}{2}$ code requires 1536 processing elements with an excess of 26,000 interconnect wires to carry the messages between the processing elements. Area of implementation and interconnect routing are the two major issues inhibiting the implementation of parallel architectures.

In a serial LDPC decoder as shown in Figure 34.28(b), the task of message computations is serialized into a small number of processing elements, resulting in a latency of several thousand cycles. Furthermore, the large expansion property of LDPC codes with good asymptotic performance, leads to stalls in processing between the rounds of iteration due to the data-dependencies. In order to capitalize on all hardware resources, it is more efficient to schedule all available processing elements to compute the messages in each round of decoding, storing the output messages temporarily in memory, before proceeding to subsequent round of computations. Hence, although serial architectures result in less area and routing congestion, they lead to dramatically increased memory requirements. The size of the memory required is dependent on total number of edges in the particular code design, which is the product of the average edge degree per bit node, and the number of bits in each block of LDPC code. For example, a serial implementation of a rate-8/9 4608-bit LDPC decoder with variable nodes having an average edge degree of four, will have more than 18,000 edges in the underlying graph. It would have to perform 37,000 memory read or write operations for each iteration of decoding, thus limiting the total throughput. This contrasts with a turbo decoder, whose memory requirement is largely dictated by the size of the interleaver required to store one block of messages. Given the same block size, the memory requirements for the serial implementation of an LDPC decoder is several times larger than that of a turbo decoder.

LDPC decoder architectures with limited level of parallelism require structured code construction, and an example of those is shown in Chapter 38.

34.9 Timing Recovery

As in any communication channel, timing recovery is a key component in the system. The role of timing recovery is to present a perfectly sampled channel to the detector. In a disk-drive read channel, there are two particularly important factors that distinguish its design: the support of multiple data rate for zone recording and servo operation, and the small-overhead timing acquisition. Frequency synthesizers of M/N counter type are used to generate the necessary data rate frequencies. The fundamental requirement for the synthesizer is to supply a stable reference frequency. The synthesizer's PLL bandwidth needs to be balanced between appropriate noise rejection and the need to acquire phase and frequency for zone changes in the actual operation. A separate synthesizer may be needed for servo operations as the transition between servo and non-servo mode can be too fast for the synthesizer to settle down.

The second important factor is the requirement to acquire timing with a very small overhead. A typical read channel will need to acquire timing within 50–100 cycles. To achieve the goal, a few techniques had been developed throughout the years:

- Idle locking: Maintaining the ADC sampling frequency in idle mode to be very close to the frequency of the incoming read data.
- Zero-phase restart (ZPR): Initialization of the ADC sampling clock close to the ideal sampling phase.
- Latency reduction: The stability of the timing recovery loop depends heavily on the latency in the loop. The longer the latency, the less stable is the loop, and thus the loop bandwidth will normally be reduced to maintain stability. In acquisition, the phase detector can use the ADC output directly, instead of the FIR output.
- Gear shifting: In acquisition mode, the bandwidth of the timing loop is higher than that of the tracking mode, and a change in loop bandwidth at the boundary of acquisition and tracking normally takes place. The gear shift can be a single shift or the bandwidth can be shifted down in a series of steps.

The traditional timing recovery is a mixed-signal design based on a phase-locked loop (PLL) structure as shown in Figure 34.29. Two separate voltage-controlled oscillators (VCOs) are used, one for write frequency generation and one for timing recovery. Loop filters are normally implemented in mixed-signal technology, using a digital-to-analog converter (DAC) to drive the analog frequency control signal of the VCO. In idle mode, the timing recovery is locked to the synthesizer, thus keeping the VCO frequency close to the read signal.

An improved scheme using a delay-locked-loop (DLL) structure is shown in Figure 34.30. Clock recovery is carried out by interpolating between cycles of the frequency synthesizer. The frequency synthesizer, usually implemented a ring-oscillator-based VCO, generates initial coarse-resolution phases from the output of the individual delay stages in the VCO. These clock phases will drive a phase interpolator circuit, which will generate fine-resolution phase clock references for the timing-recovery PLL. The loop filter is a digital filter with a digital phase position output. This output is then used to select one of the fine-resolution phases to be used to sample the incoming data. The benefits of this scheme are that there is only one VCO in the system and it is a pure digital scheme. The loop filter is also fully digital and allows a direct implementation of gear shifting and zero-phase restart techniques.

The basic sequencing of the timing recovery loop is described in Figure 34.31. During non-read modes, the timing loop is held in a coast condition with the loop filter coefficients being frozen and the phase multiplexer selection accumulator reset. In this mode, the zero-phase clock drives the system.

In the beginning of the read mode, a ZPR sequence is initiated. The ZPR block calculates the best initial phase. This is accomplished by using the first valid permeable samples to compute the correct phase

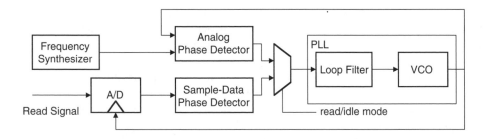

FIGURE 34.29 Timing recovery block diagram.

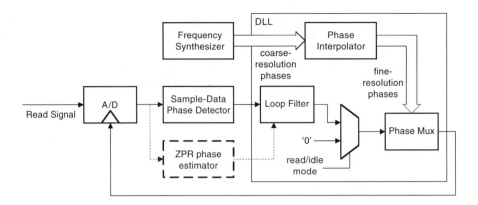

FIGURE 34.30   DLL-based timing recovery.

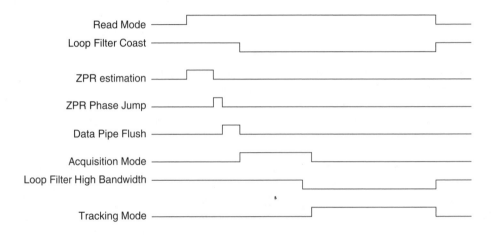

FIGURE 34.31   Sequencing of the timing recovery.

location. Accuracy can be improved by averaging several samples, at the expense of longer ZPR period. The resulting phase is forced into the phase selector and after the pipeline is flushed, acquisition mode will begin as the timing loop exits coast mode and enters acquisition mode.

In the acquisition mode, the loop is closed and the phase detector updates will drive the loop filter. At this time, the phase detector is driven by the acquisition algorithm. Additionally, the loop filter parameters are set to their acquisition values. When the acquisition is complete, the loop will start to transit to the tracking mode. The loop filter is gear shifted and if FIR is bypassed during the acquisition, the FIR output is switched back to drive the phase detector. A few cycles are allowed for the loop to settle down under the new conditions and then the phase detector is switched from the acquisition to the tracking algorithm.

The principles of the timing acquisition and tracking algorithms have been discussed in Chapter 26. Phase error detection algorithm in acquisition mode is designed around the particular preamble pattern chosen. If FIR filter is not bypassed, the acquisition and tracking mode phase error detection can be very similar, except that the tentative decisions of the samples used to compute phase error will either be limited to a few valid levels or be forced to be a predetermined sequence. In the tracking mode, the tentative decisions for phase error detection can be derived from directly slicing the output of FIR to valid levels or by using preliminary survival register output selected by global minimum finder, or even from the final output of the survival registers. The tradeoff is between the accuracy of the decisions and the latency required to form these decisions.

34.10   Write Path

The write data path as shown in Figure 34.2 consists of the scrambler, the recording code encoder, parallel to serial converter, data precoder and the write precompensation block. The write mode sequence of events includes:

1. Preamble pattern is written starting at the beginning of the frame as indicated by an external signal, usually called the write gate (WG).
2. The write path state machine inserts the synchronization mark pattern into the serial data stream, after the preamble pattern is written. Different header formats include the single synchronization pattern or the dual synchronization patterns. In the case of dual synchronization patterns, real data preamble padding will be inserted in between the synchronization patterns.
3. After the synchronization pattern, real channel data, originating from the disk controller via the byte interface, is sent out.

Synchronization patterns are specially designed and are not scrambled. Designated detectors in the read process, as shown in Figure 34.2, are used to detect these patterns. To avoid misalignment of data frames, the requirement of a sync pattern is to have low correlation between the pattern and shifted versions of itself. Thus, a large number of errors has to occur in the detection process to cause a synchronization pattern to be detected at a shifted position, causing a false frame alignment. The byte-interface circuit provides a byte-wide I/O to the controller. The data bytes are then passed to the scrambling function which is implemented normally with linear feedback shift register which implements a prime polynomial, for example, $1 + X^3 + X^{10}$. This scrambler code is XORed with the data byte as it is clocked through the data path. During the read process, a similar linear feedback shift register implementing the same polynomial is used to descramble the data. The scrambled data are then passed to the recording code encoder. The details of the design of various recording codes are covered in Chapters 17 through 21. Encoded data are then passed to the serializer, which normally includes the following functions: generate preamble patterns, generate the sync word, serialize data, and generate the parity bits if postprocessing is used in detection. Data streams are then precoded and finally sent to the write precompensation block.

The write precompensation circuitry is used to precompensate for non-linear bit shift caused in magnetic media/head interface. Nonlinear effects tend to shift one data transition earlier if it is preceded by another transition. The amount of shift is related to how close the preceding transition is to the current transition. The closer the previous transition, the larger is the amount of shift, but the relationship is not linear. The precompensation circuit recognizes specific write data patterns and adds adjustable delays in the time position of write data bits to counteract this nonlinear effect. The nominal requirement of the resolution of delay adjustment is around 1–2% and these delays have to be locked to the data rate. An approach to achieve a two-level write precompensation functionality is outlined in Figure 34.32.

As shown in Figure 34.32, the write precompensation circuit takes in several VCO clock phases, and further interpolates them to produce clock phases of finer resolution. Three phase multiplexers are used to select which one of these fine-resolution clock phases is to be used for the required amount of delay. The phase multiplex for nominal clock would be fixed to always select the earliest phase. A pattern decode circuit will sample the output sequence from the precoder and determine which one of the three clock phases should be used to write out the current data transition. For a two-level write precompensation, the pattern decoder logic is shown in Table 34.2. and the output of the write-precompensation operation

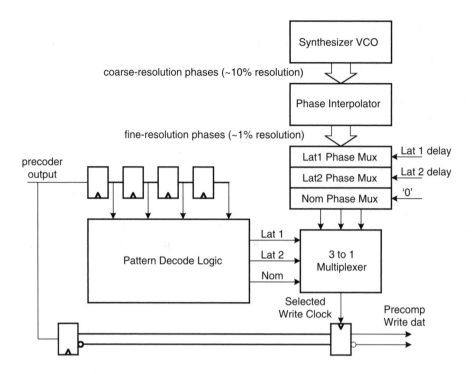

FIGURE 34.32   Block diagram of a two-level write precompensation.

TABLE 34.2   Two-Level Write-Precompensation Scheme

Precomp Level	Precoder Output at Time				Transition between Time k-1 and k
	k-3	k-2	k-1	k	
Level 1	0	0	1	0	Delays by Late1 amount
Level 1	1	1	0	1	Delays by Late1 amount
Level 1	0	1	0	1	Delays by Late1 amount
Level 1	1	0	1	0	Delays by Late1 amount
Level 2	0	1	1	0	Delays by Late2 amount
Level 2	1	0	0	1	Delays by Late2 amount
Level 0	All others				No delay, nominal

is illustrated in Figure 34.33. For example, patterns of [1, 0, 1, 0] and [0, 0, 1, 0] are mapped to Late 1 delay. If more delay levels were available, they these patterns could have been mapped into different delays.

34.11   Overview of Recently Published Disk-Drive Read/Write Channels

Table 34.3 summarizes the performance of recently published or announced disk-drive read/write channels. Conference publications usually provide details of the operation of the devices, while the product announcements present only some of the main facts. The highest throughput up to date has been reported by Marvell's read channel operating at 1.8 Gb/s [73]. The first six entries from the table are collected from the vendors' datasheets and they do not provide details about implementation. The remainder of the data from Table 34.3 is compiled from presentations at major conferences.

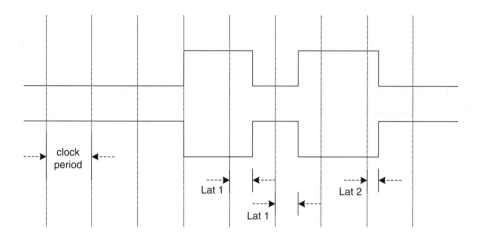

FIGURE 34.33   Write precompensation output.

TABLE 34.3   Performance Summary of Recently Published Detectors used in Disk-Drive Read Channels

Author	Affiliation	Detector	Process (μm)	Code	Speed (Mb/s)	Year
Datasheet	Marvell	N/A	0.13 μm CMOS	N/A	1800	2002. [73]
Datasheet	Marvell	N/A	0.18 μm CMOS	N/A	1200	2000. [74]
Datasheet	Agere	N/A	0.13 μm CMOS	N/A	1200	2003. [75]
Datasheet	ST Micro.	N/A	0.13 μm CMOS	N/A	850	2002. [76]
Datasheet	Marvell	N/A	0.25 μm CMOS	N/A	750	2000. [77]
Nazari	Marvell	Digital 16-state + PP	0.25 μm CMOS	32/34 RLL +parity	500	ISSCC 2000. [78]
Demicheli	ST Micro.	Digital EPR4	0.35 μm BiCMOS	16/17 RLL	450	CICC 1999. [79]
Pan	Datapath	Digital E^2PR4	0.29 μm BiCMOS	8/9 TC	400	ISSCC 1999. [80]
Sutardja	Marvell	Digital EPR4	0.35 μm CMOS	16/17 RLL	360	ISSCC 1999. [81]
Leung/Fukahori	SSi	Analog EPR4	0.8 μm BiCMOS	16/17 RLL	300	ISSCC 1998. [3], [4]
Bishop	Quantum	Digital mE^2PR4	0.35 μm BiCMOS	24/25 RLL	300	ISSCC 1999. [82]

34.12   Challenges of Further Integration

Scaling of CMOS technology presents an opportunity for further integration. Since the introduction of single-chip disk-drive read/write channels, the trend is to integrate increasingly more complex signal processing in the channel. One possible path to continue this trend would be the introduction of iterative decoders. Alternatively, as early integration provided the way to replace multiple read and write functions by a single channel, further cost reduction might be possible by integrating other disk drive functions on the channel die. In the past decade, the main integrated circuits in a disk drive are the line driver (which also includes the preamplifier), the read/write channel, the disk controller, the servo controller, the spindle motor controller and the power conditioning IC. The interface between the read channel and the disk controller has traditionally been a byte (8 bits). To format a disk platter, servo patterns on the disks are written by servo track writers in the initial manufacturing steps of a disk drive.

Spindle motor controller and power conditioning ICs are not good candidates for further integration because they require a high supply voltage. The line driver IC, because of its high linearity and high speed requirements, is generally implemented in bipolar Si/SiGe processes (with a few exceptions where lower performance and lower cost ones were implemented in CMOS) and is physically located very close to the recording head. Therefore, a common approach to further integration is to combine the read channel, the disk controller and the servo controller (typically a DSP processor) together with memory into a single chip [83, 84]. Besides the cost savings this integration provides an opportunity to redesign the recording system.

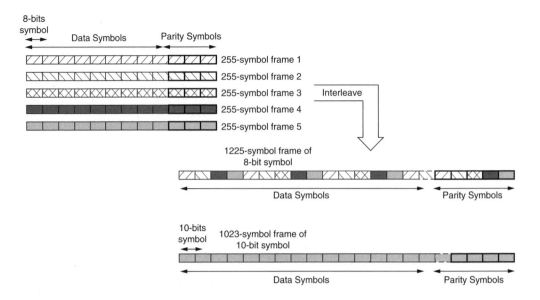

FIGURE 34.34   Construction of 8-bit symbol data frame and 10-bit symbol data frame.

With single-chip implementation of read channel signal processing and ECC, an overall optimization of ECC and channel code can be better carried out.

Another recent advancement in disk drive electronics is the deployment of 10-bit data interface between the read channel and the controller. This results in the use of 10 bits, instead of traditional 8-bit symbols, for the Reed-Solomon ECC encoding [73, 75]. Without covering the details of the ECC design, the advantage of the 10-bit symbol interface is given by the following qualitative example, Figure 34.34.

The use of 8-bit symbol for Reed-Solomon encoding limits the data frame to 255 × 8 bits. A 255-byte frame will suffer a high overall penalty for permeable patterns, frame gaps, etc. Thus larger frames are normally constructed using multiple ECC frames interleaved together, for example, with five 255-byte frame interleaved to form a single 1225 × 8 bits frame. With a 10-bit symbol, the size of the ECC frames is 1023 × 10 bits, similar to the total size from the example above. Assuming that in each case a similar rate loss is chosen, such that the ECC will be able to correct four 10-bit symbol errors with 10-bit encoding, an error will occur with five error symbols in the frame. The ECC of the 8-bit symbol system corrects five 8-bit symbol errors, and thus one 8-bit symbol error in each of its constituent frames. Therefore, an error occurs with only two symbol errors in any of its five interleaves. Thus, for an 8-bit symbol Reed-Solomon encoding to have the same level of data integrity as of 10-bit symbol case, additional rate loss is needed.

Finally, some of the contemporary read/write channel devices are equipped with self-servo-write technology, and are able to write the servo patterns directly, eliminating the use of servo track writers. This is an obvious cost saving in manufacturing.

References

[1] H. Thapar, et al., Hard disk drive read channels: technology and trends, *Proceedings of the 1999 IEEE Custom Integrated Circuits Conference, CICC'99*, San Diego, CA, May 16–19, 1999, pp. 309–316.

[2] R.G. Yamasaki, T. Pan, M. Palmer, and D. Browning, A 72 Mb/s PRML disk-drive channel chip with an analog sampled-data signal processor, *IEEE International Solid-State Circuits Conference, ISSCC'94, Digest of Technical Papers*, San Francisco, CA, February 16–18, 1994, pp. 278–279.

[3] M. Leung et al., A 300 Mb/s BiCMOS EPR4 read channel for magnetic hard disks, *IEEE International Solid-State Circuits Conference, ISSCC'98, Digest of Technical Papers*, San Francisco, CA, February 5–7, 1998, pp. 378–379, 467.

[4] K. Fukahori et al., An analog EPR4 Viterbi detector in read channel IC for magnetic hard disks, *IEEE International Solid-State Circuits Conference, ISSCC'98, Digest of Technical Papers*, San Francisco, CA, February 5–7, 1998, pp. 380–381.

[5] T.J. Schmerbeck, R.A. Richetta, L.D. Smith, L.D, A 27MHz mixed analog/digital magnetic recording channel DSP using partial response signalling with maximum likelihood detection *IEEE International Solid-State Circuits Conference, ISSCC'91, Digest of Technical Papers*, San Francisco, CA, February 13–15, 1991, pp. 136–137, 304.

[6] J. Sonntag, et al., A high speed, low power PRML read channel device, *IEEE Transactions on Magnetics*, vol. 31 no. 2 , pp. 1186–1195, March 1995.

[7] J. Fields et al., A 200 Mb/s CMOS EPRML channel with integrated servo demodulator for magnetic hard disks, *IEEE International Solid-State Circuits Conference, ISSCC'97, Digest of Technical Papers*, San Francisco, CA, February 68, 1997, pp. 314–315, 477.

[8] E.K. Kretzmer, Generalization of a technique for binary data communication, *IEEE Transactions on Communication Technology*, vol. COM-14, pp. 67–68, February 1966.

[9] H. Kobayashi, and D.T. Tang, Application of partial response channel coding to magnetic recording systems, *IBM Journal of Research and Development*, vol. 14, pp. 368–375, July 1970.

[10] R. Cideciyan, F. Dolivo, R. Hermann, W. Hirt, and W. Schott, A PRML system for digital magnetic recording, *IEEE Journal on Selected Areas in Communications*, vol. 10, no. 1, pp. 38–56, January 1992.

[11] H. Thapar and A.M. Patel, A class of partial response systems for increasing storage density in magnetic recording, *IEEE Transactions on Magnetics*, vol. 23, no. 5, pp. 3666–3668, September 1987.

[12] H. Sawaguchi, S. Mita, and J. Wolf, A concatenated coding technique for partial response channels, *IEEE Transactions Magnetics*, vol. 37, no. 2, pp. 695–703, March 2001.

[13] M. Leung, B. Nikolic, L. Fu and T. Jeon, Reduced complexity sequence detection for higher order partial response channels, *IEEE Journal of Selected Areas in Communications*, vol. 19, no. 4, pp. 649–661, April 2001.

[14] N. Nazari, A 500 Mb/s disk drive read channel in 0.25 μm CMOS incorporating programmable noise predictive Viterbi detection and trellis coding, *IEEE International Solid-State Circuits Conference*, San Francisco, CA, February 2000, pp. 78–79.

[15] S. Sutardja, 360 Mb/s (400 MHz) 1.1 W 0.35 μm CMOS PRML read channels with 6 burst 8-20x over-sampling digital servo, *IEEE International Solid-State Circuits Conference*, San Francisco, CA, February 1999, pp. 40–41.

[16] T. Pan et al., A trellis coded E2PRML digital read/write channel IC, 1999 *International Solid-State Circuits Conference, ISSCC'99 Digest of Technical Papers*, 442, San Francisco, CA, February 15–17, 1999, pp. 36–37.

[17] M. Demicheli et al., A 450 Mbit/s EPR4 PRML read/write channel, *Proceedings of the 1999 IEEE Custom Integrated Circuits Conference*, San Diego, CA, May 16–19, 1999, pp. 317–320.

[18] Thapar et al., Hard disk drive read channels: technology and trends, *Proceedings of the 1999 IEEE Custom Integrated Circuits Conference*, San Diego, CA, May 16–19, 1999, pp. 309–316.

[19] B. Gilbert, A precise four-quadrant multiplier with sub-nanosecond response, *IEEE Journal of Solid-State Circuits*, vol. SC-3, pp. 365–373, December 1968.

[20] R.G. Yamasaki, Temperature compensation control circuit for exponetial gain function of an AGC amplifier, U.S. Patent 5 162678, 19912.

[21] G.A. De Veirman and R.G. Yamasaki, Design of a bipolar 10 MHz programmable continuous-time 0.05 equiripple linear phase lowpass filter, *IEEE Journal of Solid-State Circuit*, vol. 27, pp. 324–331, March 1992.

[22] B.E. Bloodworth et al., A 450-Mb/s analog front end for PRML read channels, *IEEE Journal of Solid-State Circuits*, vol. 34, pp. 1661–1675, November, 1999.

[23] D. Choi et al., An analog front-end signal processor for a 64 Mbits/s PRML hard-disk drive channel, *IEEE Journal of Solid-State Circuits*, vol. 29, pp. 1596–1605, December 1994.

[24] J. Silva-Martinez et al., A 10.7-MHz 68-dB SNR CMOS continuous-time filter with on-cChip automatic tuning, *IEEE Journal of Solid-State Circuits*, vol. 27, pp. 1843–1853, December 1992.

[25] W. Dehaene et al., A 50-MHz Standard CMOS pulse Equalizer for hard disk read channels, *IEEE Journal of Solid-State Circuits*, vol.32, pp. 977–988, July 1997.

[26] P. Pai, Equalization and clock recovery for magnetic storage read channels, Ph.D. Dissertation, University of California, Los Angeles, 1996.

[27] R.L. Geiger and E. Sanchez-Sinencio, Active filter design using operational tansconductance amplifiers: a tutorial, *IEEE Circuits and Devices Magazine*, pp. 20–30, March 1985.

[28] H. Khorramabadi and P.R. Gray, High-frequency CMOS continuous-time filters, *IEEE Journal of Solid-State Circuit*, vol. Sc-10, December 1984, pp. 939–948.

[29] C. Petersen et al., A 3-5.5V CMOS 32 Mb/s fully integrated read channel for disk-drives, *Proceedings of IEEE Custom Integrated Circuits Conference — CICC 1993*, San Diego, California.

[30] K.S. Tan and P.R. Gray Fully integrated analog filters using bipolar-JFET technology, *IEEE Journal of Solid-State Circuits*, vol. Sc-13, pp. 814–821, 1978.

[31] Y. Wang, Low-power low-voltage CMOS continuous-time filters for magnetic storage read channel applications, Master of Science Thesis, University of Hawaii, December 1997.

[32] P. Pai et al., A 160 MHz Analog Front-End IC for EPR-IV PRML Magnetic Storage Read Channels, *IEEE Journal of Solid-State Circuits*, vol. 31, pp.1803–1816, November 1996.

[33] R. van de Plassche, *CMOS Integrated Analog-to-Digital and Digital-to-Analog Converters*, 2nd ed., Kluwer, Boston, MA, 2003.

[34] A.P. Chandrakasan and R.W. Brodersen, Low-power CMOS digital design, *IEEE Journal of Solid-State Circuits*, vol. 27, no. 4, pp. 473–484, April 1992.

[35] D.G. Messerschmitt, Breaking the recursive bottleneck, in *Performance Limits in Communication Theory and Practice*, J.K. Skwirzynski, Ed., Kluwer Academic Publishers, Boston, MA, 1988.

[36] R. Jain, P.T. Yang, and T. Yoshino, FIRGEN: a computer-aided design system for high performance FIR filter integrated circuits, *IEEE Transactions on Signal Processing*, vol. 39, no. 7, pp. 1655–1668, July 1991.

[37] R.A. Hawley, B.C. Wong, T.-J. Lin, J. Laskowski, and H. Samueli, Design techniques for silicon compiler implementations of high-speed FIR digital filters, *IEEE Journal of Solid-State Circuits*, vol. 31, no. 5, pp. 656–667, May 1996.

[38] W.L. Abbott et al., A digital chip with adaptive equalizer for PRML detection in hard-disk drives, *IEEE International Solid-State Circuits Conference, Digest of Technical Papers, ISSCC'94*, San Francisco, CA, February 16–18, 1994, pp. 284–285.

[39] D.J. Pearson et al., Digital FIR filters for high speed PRML disk read channels, *IEEE Journal of Solid-State Circuits*, vol. 30, no. 12, pp. 1517–1523, May 1995.

[40] S. Mita et al., A 150 Mb/s PRML chip for magnetic disk drives, *IEEE International Solid-State Circuits Conference, Digest of Technical Papers, ISSCC'96*, San Francisco, CA, February 8–10, 1996, pp. 62–63, 418.

[41] D. Moloney, J. O'Brien, E. O'Rourke and F. Brianti, Low-power 200-Msps, area-efficient, five-tap programmable FIR filter, *IEEE Journal of Solid-State Circuits*, vol. 33, no. 7, pp. 1134–1138, July 1998.

[42] C.S.H. Wong, J.C. Rudell, G.T. Uehara and P.R. Gray, A 50 MHz eight-tap adaptive equalizer for partial-response channels, *IEEE Journal of Solid-State Circuits*, vol. 30, no. 3, pp. 228–234, March 1995.

[43] L.E. Thon, P. Sutardja, F.-S. Lai and G. Coleman, A 240 MHz 8-tap programmable FIR filter for disk-drive read channes, *1995 IEEE International Solid-State Circuits Conference, Digest of Technical Papers, ISSCC '95*, pp. 82–83, 343, San Francisco, CA, 15–17 February 1995.

[44] R.B. Staszewski, K. Muhammad and P. Balsara, A 550-MSample/s 8-tap FIR digital filter for magnetic recording read channels, *IEEE Journal of Solid-State Circuits*, vol. 35, no. 8, pp. 1205–1210, August 2000.

[45] S. Rylov et al., A 2.3 GSample/s 10-tap digital FIR filter for magnetic recording read channels, *IEEE International Solid-State Circuits Conference, Digest of Technical Papers, ISSCC'01*, San Francisco, CA, February 5–7, 2001, pp. 190–191.

[46] J. Tierno et at., A 1.3 GSample/s 10-tap full-rate variable-latency self-timed FIR filter with clocked interfaces, *IEEE International Solid-State Circuits Conference, Digest of Technical Papers, ISSCC'02*, San Francisco, CA, February 3–7, 2002, pp. 60–61, 444.

[47] P. Black and T. Meng, A 140 MB/s 32-state radix-4 Viterbi decoder, *IEEE Journal of Solid-State Circuits*, vol. 27, no. 12, pp. 1877–1885, December 1992.

[48] A.K. Yeung and J.M. Rabaey, A 210 Mb/s radix-4 bit-level pipelined Viterbi decoder, in *IEEE International Solid-State Circuits Conference, ISSCC'95, Digest of Technical Papers*, San Francisco, CA, USA, February 15–17, 1995, pp. 88–89, 344, 440.

[49] S. Sridharan and L. R. Carley, A 110 MHz 350 mW 0.6 mm CMOS 16-state generalized-target Viterbi detector for disk drive read channels, *IEEE Journal Solid-State Circuits*, vol. 35, no. 3, pp. 362–370, March 2000.

[50] G. Fettweis, R. Karabed, P.H. Siegel, and H.K. Thapar, Reduced-complexity viterbi detector architectures for partial response signaling, in *Proceedings IEEE Global Telecommunications Conference*, Singapore, November 13–17, 1995, pp. 559–563.

[51] E. Yeo, S. Augsburger, W.R. Davis, and B. Nikolić, A 500 Mb/s soft-output Viterbi decoder, *IEEE Journal of Solid-State Circuits*, vol. 38, no. 7, pp. 1234–1241, July 2003.

[52] G. Fettweis and H. Meyr, Parallel Viterbi algorithm implementation: breaking the ACS-bottleneck, *IEEE Transactions on Communications*, vol. 37, no. 8, pp. 785–790, August 1989.

[53] G. Fettweis and H. Meyr, High-rate Viterbi processor: a systolic array solution, *IEEE Journal on Selected Areas in Communications*, vol. 8, no. 8, pp. 1520–1534, October 1990.

[54] G. Fettweis and H. Meyer, High-speed parallel Viterbi decoding algorithm and VLSI architecture, *IEEE Communications Magazine*, vol. 29, no. 8, pp. 46–55, May 1991.

[55] A. Avižienis, Signed-digit number representations for fast parallel arithmetic, *IRE Transactions on Computers*, vol. EC-10, pp. 389–400, September 1961.

[56] C.B. Shung, P.H. Siegel, G. Ungerboeck, and H.K. Thapar, VLSI architectures for metric normalization in the Viterbi algorithm, *Proceedings IEEE International Conference on Communications ICC'90*, Atlanta, GA, April 16–19, 1990, pp. 1723–1728.

[57] A.P. Hekstra, An alternative to metric rescaling in Viterbi decoders, *IEEE Transactions on Communications*, vol. 37, no. 11, pp. 1220–1222, November 1989.

[58] P. Black, Algorithms and architectures for high speed Viterbi detection, Ph.D. Thesis, Stanford University, March 1993.

[59] R. Wood, Turbo-PRML: a compromise EPRML detector, *Digest of International Magnetics Conference, INTERMAG '93*, April 13–16, 1993, pp. FD-06.

[60] T. Conway, A new target response with parity coding for high density magnetic recording channels, *IEEE Transactions on Magnetics*, vol. 34, no. 4, pp. 2382–2386, July 1998.

[61] J.L. Sonntag and B. Vasic, Implementation and bench characterization of a read channel with parity check postprocessor, *Digests of the 2000 Magnetic Recording Conference, TMRC'2000*, Santa Clara, CA, August 14–16, 2000, p. B1.

[62] L.R. Bahl, J. Cocke, F. Jelinek and R. Raviv, Optimal decoding of linear codes for minimizing symbol error rate, *IEEE Tranactions on Information Theory*, vol. IT-20, pp. 284–287, March 1974.

[63] J. Hagenauer and P. Hoeher, A Viterbi algorithm with soft-decision outputs and its applications, in *Proceedings IEEE Global Telecommunications Conference*, Dallas, TX, November 27–30, 1989, pp. 47.11–47.17.

[64] E. Yeo, P. Pakzad, B. Nikolić, and V. Anantharam, VLSI architectures for iterative decoders in magnetic recording channels, *IEEE Transactions on Magnetics*, vol. 37, no. 2, part I, pp. 748–755, March 2001.

[65] R. Lynch, E. Kurtas, A. Kuznetsov, E. Yeo, and B. Nikolić, The search for a practical iterative detector for magnetic recording, *Digests of The 2003 Magnetic Recording Conference, TMRC'03*, Santa Clara, CA, August 18–20, 2003, p. B1.

[66] E. Yeo, S. Augsburger, W.R. Davis, and B. Nikolić, A 500 Mb/s soft-output viterbi decoder, *IEEE Journal of Solid-State Circuits*, vol. 38, no. 7, pp. 1234–1241, July 2003.

[67] J. Li, K. R. Narayanan, E. Kurtas, and C.N. Georghiades, On the performance of the high-rate TPC/SPC and LDPC codes over partial response channels, *IEEE Transactions on Communications*, vol. 50, no. 5, pp. 723–735, May 2002.

[68] Z. Wu, *Coding and Iterative Detection for Magnetic Recording Channels*, Kluwer, Norwell, MA, 2000.

[69] A. Blanksby and C.J. Howland, A 690-mW 1-Gb/s 1024-b, rate 1/2 low-density parity-check code decoder, *IEEE Journal of Sold-State Circuits*, vol. 37, no. 3, pp. 404–412, March 2002.

[70] Y. Kou, S. Lin, and M. P.C. Fossorier, Low-density parity-check codes based on finite geometries: a rediscovery and new results, *IEEE Transactions on Information Theory*, vol. 47, no. 7, pp. 2711–2736, November 2001.

[71] J. Rosenthal and P.O Vontobel, Constructions of LDPC codes using Ramanujan graphs and ideas from Margulis, in *Proceedings of the 38th Annual Allerton Conference on Communication, Control, and Computing*, pp. 248–257.

[72] M.M. Mansour and N. Shanbhag, Architecture-aware low-density parity-check codes, *Proceedings of the 2003 International Symposium on Circuits and Systems, ISCAS '03*, vol. 2, pp. II-57–II-60, May 2003.

[73] Marvell 88C7500 Read Channel http://www.marvell.com/products/storage/readchannel/88C7500.jsp

[74] Marvell 88C5520/88C5500 Read Channel http://www.marvell.com/products/storage/readchannel/88C5500.jsp

[75] Agere TrueStore RC6500 Read Channel IC, http://www.agere.com/NEWS/PRESS2003/011403a.html

[76] ST Microelectronics, Leonardo Read/Write Channel, http://www.st.com/stonline/products/promlit/pdf/flcmosleo-1002.pdf

[77] Marvell 88C5200 Read Channel http://www.marvell.com/products/storage/readchannel/88C5200.jsp

[78] N. Nazari, A 500 Mb/s disk drive read channel in 0.25 μm CMOS incorporating programmable noise predictive Viterbi detection and trellis coding, *IEEE International Solid-State Circuits Conference*, San Francisco, CA, February 2000, pp. 78-79.

[79] M. Demicheli et al., A 450 Mbit/s EPR4 PRML read/write channel, *Proceeding of the 1999 IEEE Custom Integrated Circuits Conference*, pp. 317–320, San Diego, CA, May 16–19, 1999.

[80] T. Pan et al., A trellis coded E^2PRML digital read/write channel IC, *1999 International Solid-State Circuits Conference, ISSCC'99 Digest of Technical Papers*, pp. 36–37, 442, San Francisco, CA, February 15–17, 1999.

[81] S. Sutardja, 360 Mb/s (400 MHz) 1.1W 0.35 μm CMOS PRML read channels with 6 burst 8-20x over-sampling digital servo, *IEEE International Solid-State Circuits Conference*, San Francisco, CA, February 1999, pp. 40–41.

[82] A. Bishop et al., A 300 Mb/s BiCMOS disk drive channel with adaptive analog equalizer, *1999 International Solid-State Circuits Conference, ISSCC'99 Digest of Technical Papers*, pp. 46–47, San Francisco, CA, February 15–17, 1999.

[83] S. Nemazie et al., 260 Mb/s mixed-signal single-chip integrated system electronics for magnetic hard disk drives," *1999 International Solid-State Circuits Conference, ISSCC'99 Digest of Technical Papers*, pp. 42–43, San Francisco, CA, February 15–17, 1999.

[84] Marvell 88i5540 System-On-Chip, http://www.marvell.com/products/storage/soc/88i5520.jsp.

VI

Iterative Decoding

35 Turbo Codes   *Mustafa N. Kaynak, Tolga M. Duman, and Erozan M. Kurtas* **35**-1
Principles of Turbo Coding • Iterative Decoding of Turbo Codes • Performance of Turbo
Codes over AWGN Channels • Recording Channels • Turbo Codes for Recording
Channels • Performance of Turbo Codes over Recording Channels • Summary

36 An Introduction to LDPC Codes   *William E. Ryan* **36**-1
Introduction • Representations of LPDC Codes • LDPC Code Design
Approaches • Iterative Decoding Algorithms • Concluding Remarks

**37 Concatenated Single-Parity Check Codes for High-Density
Digital Recording Systems**   *Jing Li, Krishna R. Narayanan,
Erozan M. Kurtas, and Travis R. Oenning* **37**-1
Introduction • System Model • Analysis of Distance Spectrum • Thresholds Analysis
using Density Evolution • Simulation Results • Conclusion

38 Structured Low-Density Parity-Check Codes   *Bane Vasic,
Erozan M. Kurtas, Alexander Kuznetsov, and Olgica Milenkovic* **38**-1
Introduction • Combinatorial Designs and their Bipartite Graphs • LDPC Codes on
Difference Families • Codes on Projective Planes • Lattice Construction of LDPC
Codes • Application in the Partial Response (PR) Channels • Conclusion

39 Turbo Coding for Multitrack Recording Channels   *Zheng Zhang,
Tolga M. Duman, and Erozan M. Kurtas* **39**-1
Introduction • Multitrack Recording Channels • Information Theoretical Limits:
Achievable Information Rates • Turbo Coding for Multitrack Recording
Systems • Discussion

35

Turbo Codes

35.1 Principles of Turbo Coding 35-2
Parallel Concatenated Convolutional Codes • Other
Concatenation Schemes • Other Iteratively Decodable Codes

35.2 Iterative Decoding of Turbo Codes 35-6
Maximum A-Posteriori (MAP) Decoder • The Iterative
Decoder Structure • Iterative Decoding of Serially
Concatenated Convolutional Codes • Other Iterative
Decoding Algorithms

35.3 Performance of Turbo Codes over
AWGN Channels 35-9

35.4 Recording Channels 35-11
Partial Response Channels • Realistic Recording Channel
Models

35.5 Turbo Codes for Recording Channels 35-12
Turbo Coding without Turbo Equalization • Turbo Coding
with Turbo Equalization • Convolutional Coding with an
Interleaver

35.6 Performance of Turbo Codes over Recording
Channels 35-15
Over PR Channels • Over Lorentzian Channels

35.7 Summary 35-18

Mustafa N. Kaynak

Tolga M. Duman
Arizona State University
Tempe, AZ

Erozan M. Kurtas
Seagate Technology
Pittsburgh, PA

Mustafa N. Kaynak: received the B.Sc. degree from Middle East Technical University, Ankara, Turkey, in 1999 and the M.Eng. degree from National University of Singapore, Singapore in 2001, both in electrical engineering, and is currently working toward the Ph.D. degree in electrical engineering at Arizona State University, Tempe, AZ.

His research interests are in digital and wireless communications, magnetic recording channels, iterative decoding algorithms, channel coding, for recording and wireless communication channels, turbo and low density parity check codes.

Tolga M. Duman: received the B.S. degree from Bilkent University in 1993, M.S. and Ph.D. degrees from Northeastern University, Boston, in 1995 and 1998, respectively, all in electrical engineering. Since August 1998, he has been with the Electrical Engineering Department of Arizona State University first as an assistant professor (1998–2004), and currently as an associate professor. Dr. Duman's current research interests are in digital communications, wireless and mobile communications, channel coding, turbo codes, coding for recording channels, and coding for wireless communications.

Dr. Duman is the recipient of the National Science Foundation CAREER Award, IEEE Third Millennium medal, and IEEE Benelux Joint Chapter best paper award (1999). He is a senior member of IEEE, and an editor for IEEE Transactions on Wireless Communications.

Erozan M. Kurtas: received the B.Sc. degree from Bilkent University, Ankara, Turkey, in 1991 and M.Sc. and Ph.D., degrees from Northeastern University, Boston, MA, in 1993 and 1997, respectively.

0-8493-1524-7/05/$0.00+$1.50
© 2005 by CRC Press, LLC

His research interests cover the general field of digital communication and information theory with special emphasis on coding and detection for inter-symbol interference channels. He has published over 75 book chapters, journal and conference papers on the general fields of information theory, digital communications and data storage. He has seven pending patent applications Dr. Kurtas is currently the Research Director of the Channels Department at the research division of Seagate Technology.

Abstract

This chapter describes the parallel and serial concatenated convolutional codes, i.e., turbo codes. First, the use of these codes for additive white Gaussian noise (AWGN) channels is discussed. The iterative decoding algorithm that uses two soft output component decoders as building blocks is presented in detail. Then, the use of concatenated codes over recording channels is reviewed. Partial response channels (e.g., PR4, EPR4) with additive white Gaussian noise are used to illustrate the main ideas. The concepts of turbo decoding without turbo equalization and with turbo equalization as well as serial concatenation of a convolutional code with the channel are explained. As more accurate models, Lorentzian channels are also considered for longitudinal recording. Finally, the effects of precoding and media noise are studied in some depth, and the existence of burst errors is noted.

35.1   Principles of Turbo Coding

In his celebrated work, Shannon proved that, an arbitrarily low probability of error can be obtained in digital transmission over a noisy channel provided that the transmission rate is less than a certain quantity called "channel capacity" [1]. He also proved that, randomly selected codes of very large block lengths can achieve this capacity. However, his proofs were not constructive, that is, he did not give any practical coding/decoding approaches. Since then, coding theorists have been attacking the problem of designing good channel codes with a lot of success (e.g., refer to the standard textbooks [2, 3] for a review of error correcting codes).

Since codes with very long block lengths are expected to perform very well, methods for constructing very long block length codes, that can be encoded and decoded relatively easily, have been investigated intensively. One way of obtaining a large block length code is concatenating two simple codes so that the encoding and the decoding of the overall code is less complex. For instance, Forney concatenated two simple codes in series to obtain a more powerful overall code [4].

Turbo codes proposed in 1993 by Berrou et al., represent a different form of concatenating two simple codes to obtain codes that achieve a near Shannon limit performance [5]. In turbo coding, two systematic recursive constituent convolutional encoders are concatenated in parallel via a long interleaver. For decoding, a suboptimal iterative decoding algorithm is employed. Let us now describe the turbo coding principle in detail.

35.1.1   Parallel Concatenated Convolutional Codes

Turbo codes generated an abundance of literature after their invention in 1993, mainly because of their exceptional performance for very low signal to noise ratios. For example, by using a rate 1/2 turbo code with an interleaver size of 65536 and memory-4 component codes, Berrou et al. [5] demonstrated that a bit error rate of 10^{-5} can be obtained within 0.7 dB of the channel capacity over an AWGN channel. We note that, the capacity for rate 1/2 transmission is at 0 dB[1].

The idea in turbo coding is to concatenate two recursive systematic convolutional codes in parallel via an interleaver. For encoding, the information sequence is divided into blocks of a certain length. The input of the first encoder is the information block and the input of the second encoder is an interleaved version

---

[1]By the statement "the capacity is at 0 dB," we mean that the channel capacity is equal to the transmission rate, that is, 1/2, when the signal to noise ratio is 0 dB.

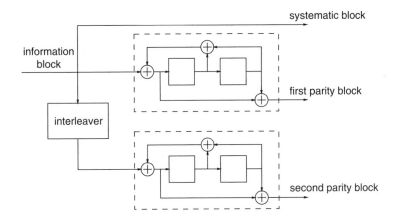

FIGURE 35.1 Rate 1/3 turbo code with 5/7 component codes.

of the information block. The systematic bits and the parity bits of the first and second encoders constitute the codeword corresponding to the information block. As an example, the block diagram of the rate 1/3 turbo code with 5/7 (in octal notation) component convolutional codes[2] is shown in Figure 35.1.

A code rate of 1/3 is not acceptable for many applications. Fortunately, there are means of obtaining higher rate turbo codes that still have a very good performance. To achieve higher code rates, different rate component convolutional codes can be used or certain puncturing schemes can be employed. For example, the rate 1/2 code example provided in the original turbo coding paper [5] is obtained by puncturing half of the parity bits.

For parallel concatenated convolutional codes, the component codes should be chosen as convolutional codes with feedback (hence, the term recursive), on the other hand, they can be nonsystematic. It is also common practice to choose identical component encoders. One can argue that any recursive convolutional encoder has an equivalent feedback-free representation, so the requirement about the recursiveness of the component codes may not be clear. To explain this, we note that, although the codewords generated by the two encoders are the same, they correspond to different input sequences. Therefore, for the purpose of constructing turbo codes, the recursive encoder and its feedback-free version are not equivalent.

The problem of selecting the feedback and the feedforward generator polynomials of the component codes to optimize the performance is studied in [6] and [7], and it is found that, the feedback polynomial should be primitive, whereas we have more flexibility in the selection of the feedforward polynomial. The reason is that, the number of problematic weight 2 information sequences (the sequences that terminate before the end of the block) are lowest if the feedback generator polynomial is primitive.

The other ingredient of the turbo coding scheme, the interleaver, can be chosen pseudo-randomly. When the interleaver is selected very large, in the order of several thousand bits or more, a very good bit error rate performance, usually within 1 dB or so of the Shannon limit is obtained, at the expense of the increased decoding delay. Although pseudo-random interleavers perform well, there are a number of interleaver design techniques that are useful. For instance, the S-random interleaver proposed in [8] provide a significant performance improvement over the pseudo-random interleavers, specifically, for relatively larger signal to noise ratios. For a review of different interleaver design techniques, the reader is referred to [9].

Maximum likelihood decoding of turbo codes is very difficult because of the pseudo-random interleaver used. In general, one has to consider all the possible codewords (there are 2^N possibilities, where N is the

---

[2]The term p/q convolutional code is used to indicate the feedforward and feedback connections in the convolutional code in octal notation. For example, 5/7, or in binary notation 101/111 means that the feedforward link is "connected, not connected and connected" and the feedback link is all connected.

interleaver size), compute the corresponding cost for each one (e.g., Euclidean distance with the received signal), and choose the one with the lowest cost as the transmitted codeword. Even for short interleavers, this is a tedious task. Fortunately, to perform this task practically, an iterative decoding algorithm is proposed in [5]. The iterative decoder offers a near optimal performance, and it is perhaps the most important contribution of the original turbo coding paper [5]. From Shannon's result, we know that codes with large block lengths chosen randomly usually have a very good performance. However, decoding such randomly chosen codes is nearly impossible. In essence, turbo coding is a way to obtain very large block length "random-like" (due to the existence of the interleaver) codes, yet we still have a near-optimal decoding algorithm. Since the iterative decoding algorithm is essential to understand turbo codes, we will describe it in some detail in a separate section.

Let us now give a brief explanation of why turbo codes perform so well. Turbo codes are linear block codes. Therefore, for the purposes of analysis, we can simply assume that the all-zero codeword is transmitted. The possible error sequences corresponding to this transmission are all the nonzero codewords. Consider a codeword with information weight one, that is, a codeword obtained by encoding a weight one information sequence. Since the component encoders are recursive convolutional encoders, the parity sequences corresponding to this information sequence will not terminate until the end of the block is reached, because the component encoders are infinite impulse response filters. With a good selection of the interleaver, if the single "1" occurs towards the end of the information block for one of the component encoders, it will occur towards the beginning of the input block for the other component encoder. Therefore, the codewords with information weight 1, typically, have a large parity weight, hence a large total weight. Furthermore, the interleaver "breaks down" the sequences with information weight greater than one, that is, if the information block results in a lower weight parity sequence corresponding to one of the encoders, it will have a larger weight parity sequence for the other one. Therefore, most of the codewords will have large Hamming weights and they are less likely to be decoded as the correct codeword when the all-zero codeword is transmitted, at least, for an AWGN channel. On the other hand, the interleaver cannot break down all the sequences, and therefore, there will be some codewords with low weights as well, hence the overall free distance of a turbo code is usually small. For a discussion on the effective free distance of turbo codes, see [10]. Since the number of low weight codewords is typically small, although the asymptotic performance of the code (for large signal to noise ratios) is limited by its relatively low free distance, its performance is very good for low signal to noise ratios. We also note that, the most troublesome error sequences for turbo codes are the ones with information weight 2, since those are the most difficult ones for the interleaver to "break down". A more detailed distance spectrum interpretation of the turbo coding scheme can be found in [11].

To predict the performance of turbo codes, one can use Monte Carlo simulation results obtained by the suboptimal iterative decoding algorithm. However, it is also important to develop performance bounds and compare them with simulation results. In [12] and [13] the union bounding technique is applied to derive an average upper bound on the probability of error for turbo codes over an AWGN channel using maximum likelihood decoding. Other more sophisticated bounds on the performance are developed in [14, 15]. Simulation results show that the iterative turbo decoding algorithms perform very well. Particularly, the simulation results and the union bound are very close to each other (in the region where the union bound is tight) [13], which shows that the iterative decoding algorithm is near optimal. In [16], McEliece et al. demonstrated that, the iterative decoding algorithm is not always optimal and may not even converge. However, they also observed that it converges with a high probability.

Terminating the trellis of a recursive convolutional code (i.e., bringing the state of the encoder to the all-zero state) is not possible by appending a number of zeros at the end of the information sequence due to the existence of the interleaver and due the fact that, the states of the two component encoders are in general different from each other. Instead, depending on the current state of the encoder, a nonzero sequence should be appended. In [17], it has been demonstrated that, if the trellis is not terminated, the performance of the turbo code deteriorates. The trellis termination problem is studied in [18] and an algorithm which does not require the knowledge of the current state for either encoders is proposed. The problem of trellis termination for turbo codes is also studied in [19, 20].

FIGURE 35.2 Serial concatenation of convolutional and block codes.

35.1.2 Other Concatenation Schemes

Serial concatenation of convolutional and block codes are proposed in [21]. In this scheme, two component codes are concatenated serially with an interleaver between them as shown in Figure 35.2. The component codes can be chosen as convolutional or block codes. The input of the outer code is the original binary input sequence and the input of the inner code is the scrambled (interleaved) version of the codeword generated by the outer code. Generalization of this scheme to more than two component codes is straightforward. Convolutional codes are usually preferred over block codes because of the existence of simple soft input-soft output decoders and the possibility of greater interleaver gain. In [21], the code selection criteria are studied and it is shown that, for superior bit error rate performance, the inner encoder must be chosen as a recursive convolutional encoder. On the other hand, we have more flexibility to choose the outer code. However, for a better performance, it should have a large free distance.

For decoding, a suboptimal iterative decoding algorithm based on the information exchange between the two component decoders is used. The details of the decoding algorithm for serially concatenated convolutional codes will be given in a separate section.

There are some obvious generalizations of the standard turbo coding scheme. For instance, one can use linear block codes as the component codes instead of the recursive convolutional codes [22]. Or, one can concatenate more than two component encoders in parallel instead of only two [23].

Another concatenation scheme is the hybrid concatenation introduced in [24]. Basically, in this case, a third code is concatenated in parallel to the two serially concatenated component codes as shown in Figure 35.3.

35.1.3 Other Iteratively Decodable Codes

Another family of iteratively decodable capacity approaching codes is the low density parity check (LDPC) codes. These codes were first introduced in 1962 by Gallager [25] and after having been forgotten for almost 30 years, with the extensive research on "turbo-like" codes and iterative decoding algorithms; recently they were rediscovered by MacKay [26]. LDPC codes are linear block codes [2, 27], and they are represented by a large, randomly generated sparse parity check matrix **H**, that is, very few entries of the parity check matrix are ones and the rest are all zeros.

Similar to the turbo codes, LDPC codes can be decoded using a simple, practically implementable iterative decoding algorithm based on the message passing algorithm (or, belief propagation as it is named in the artificial intelligence community). They can be considered as the most serious competitors to turbo codes in terms of the offered performance and complexity. For example, recent results by Richardson et al.

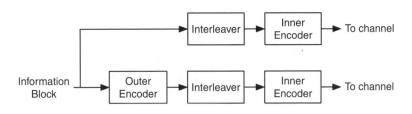

FIGURE 35.3 Hybrid concatenation of convolutional and block codes.

show that, irregular LDPC codes can achieve a performance within 0.13 dB of the channel capacity on AWGN channels [28] and thus outperform any code discovered to date including turbo codes.

35.2   Iterative Decoding of Turbo Codes

In this section, we describe the iterative turbo decoding algorithm in detail based on the presentation in [30]. Consider the standard rate 1/3 turbo coding scheme. Let us denote the input sequence by $d = (d_1, d_2, \ldots, d_N)$ where N is the interleaver length. The encoded sequence consists of three different parts: the systematic bits denoted by $x^s = (x_1^s, x_2^s, \ldots, x_N^s) = d$, the first parity part, that is, the parity bits produced by the first encoder, denoted by $x^{1p} = (x_1^{1p}, x_2^{1p}, \ldots, x_N^{1p})$, and the second parity part denoted by $x^{2p} = (x_1^{2p}, x_2^{2p}, \ldots, x_N^{2p})$, and generated by the second encoder from the interleaved information block.

We assume that the encoded sequence of bits is transmitted using BPSK modulation over an AWGN channel.[3] Therefore, the receiver observes the sequence $R = (R_1, R_2, \ldots, R_N)$ where

$$
\begin{aligned}
R_k &= \left(y_k^s, y_k^{1p}, y_k^{2p} \right) \\
&= \left(\sqrt{E_s} \left(2x_k^s - 1 \right) + \eta_k^s, \sqrt{E_s} \left(2x_k^{1p} - 1 \right) + \eta_k^{1p}, \sqrt{E_s} \left(2x_k^{2p} - 1 \right) + \eta_k^{2p} \right)
\end{aligned}
$$

where E_s is the energy per symbol transmitted. If the overall code rate is r, we have $E_s = r E_b$ where E_b is the energy per information bit. The additive noise terms are distributed according to $\mathcal{N}(0, N_0/2)$, where $N_0/2$ is the noise variance. We refer to E_s/N_0 as the signal to noise ratio (SNR) per symbol, and E_b/N_0 as the signal to noise ratio per information bit.

If some of the parity bits are punctured to obtain a turbo code with a different rate, the decoder operates by inserting "0"s in the observation sequence for the bits that are punctured.

The rest of the section is divided into four parts. In the first part, we present the maximum a-posteriori (MAP) decoding algorithm for convolutional codes [31]. We describe the iterative decoding algorithm of turbo codes that uses the MAP decoders for the component codes as its building blocks in the second part. Then, in the third part, we describe the iterative decoding of serially concatenated convolutional codes. Finally, a brief review of other iterative decoding algorithms, including several simplified turbo decoders, is given.

35.2.1   Maximum A-Posteriori (MAP) Decoder

There are different algorithms for the decoding of convolutional codes. The Viterbi algorithm is the optimal decoder that minimizes the sequence error probability. However, minimizing the sequence error probability does not directly translate into minimizing the bit error probability. In other words, if the performance criteria is the minimum bit error probability, Viterbi algorithm is not optimal. In this case, the MAP decoding algorithm derived in [31] is optimal. Let us now describe the MAP decoder in some detail. Note that, we describe decoder structure with respect to the first component code, the decoder for the second component code is similar.

Let us denote the state of the encoder at time k by S_k, where $k = 0, 1, 2, \ldots, 2^M - 1$, M being the number of memory elements in the encoder. Following a derivation similar to the one in [31], we can show that, the log-likelihood of the information bits can be written as

$$
\begin{aligned}
\Lambda(d_k) &= \log \frac{\Pr[d_k = 1 \mid y^s, y^{1p}]}{\Pr[d_k = 0 \mid y^s, y^{1p}]} \\
&= \log \frac{\sum_m \sum_{m'} \gamma_1(y_k, m', m) \alpha_{k-1}(m') \beta_k(m)}{\sum_m \sum_{m'} \gamma_0(y_k, m', m) \alpha_{k-1}(m') \beta_k(m)}
\end{aligned}
$$

---

[3]The AWGN channel model is not essential, and an arbitrary channel model may also be used.

where $y_k = (y_k^s, y_k^{1p})$, $\alpha_k(m) = p(S_k = m, y_1, \ldots, y_k)$ and $\beta_k(m) = p(y_{k+1}, \ldots, y_N \mid S_k = m)$ are computed using the following forward and backward recursions respectively

$$\alpha_k(m) = \frac{\sum_{m'} \sum_{i=0}^{1} \gamma_i(y_k, m', m)\alpha_{k-1}(m')}{\sum_{m''} \sum_{m'} \sum_{i=0}^{1} \gamma_i(y_k, m', m'')\alpha_{k-1}(m')}$$

$$\beta_k(m) = \frac{\sum_{m'} \sum_{i=0}^{1} \gamma_i(y_{k+1}, m, m')\beta_{k+1}(m')}{\sum_{m''} \sum_{m'} \sum_{i=0}^{1} \gamma_i(y_{k+1}, m'', m')\beta_{k+1}(m')}$$

with the initial values of $\alpha_0(0) = 1$, $\alpha_0(m) = 0$ for $m \neq 0$, $\beta_N(0) = 1$ and $\beta_N(m) = 0$ for $m \neq 0$. For the MAP decoder of the second component code, the initial values are $\alpha_0(0) = 1$, $\alpha_0(m) = 0$ for $m \neq 0$, and $\beta_N(m) = \alpha_N(m)$. Finally, $\gamma_i(y^s, y_k^{1p}, m', m)$ is given by

$$\gamma_i\left(y^s, y_k^{1p}, m', m\right) = p\left(y_k^s \mid d_k = i, S_k = m, S_{k-1} = m'\right) p\left(y_k^{1p} \mid d_k = i, S_k = m, S_{k-1} = m'\right)$$

$$\Pr[d_k = i \mid S_k = m, S_{k-1} = m'] \Pr[S_k = m \mid S_{k-1} = m'].$$

The probabilities $p(y_k^s \mid d_k = i, S_k = m, S_{k-1} = m')$ and $p(y_k^{1p} \mid d_k = i, S_k = m, S_{k-1} = m')$ are directly dependent on the channel characteristics and the probability $\Pr[d_k = i \mid S_k = m, S_{k-1} = m']$ is either zero or one depending on whether the bit i is associated with the transition from state m' to state m or not. The last term, $\Pr[S_k = m \mid S_{k-1} = m']$ uses the a-priori likelihood information on the bit d_k. Assuming that $L(d_k)$ is the a-priori information, we can write

$$\Pr[S_k = m \mid S_{k-1} = m'] = \begin{cases} \frac{e^{L(d_k)}}{1+e^{L(d_k)}} & \text{if } \Pr[d_k = 1 \mid S_k = m, S_{k-1} = m'] = 1 \\ \frac{1}{1+e^{L(d_k)}} & \text{if } \Pr[d_k = 0 \mid S_k = m, S_{k-1} = m'] = 1 \end{cases}.$$

35.2.2 The Iterative Decoder Structure

In this section, we describe the use of the two MAP decoders for the component codes in order to decode the turbo code. First, let us write the log-likelihood of the information bit d_k as the sum of three different terms as

$$\Lambda(d_k) = \log \frac{\sum_m \sum_{m'} \gamma_1'(y_k, m', m)\alpha_{k-1}(m')\beta_k(m)}{\sum_m \sum_{m'} \gamma_0'(y_k, m', m)\alpha_{k-1}(m')\beta_k(m)} + L(d_k) + \log \frac{p\left(y_k^s \mid d_k = 1\right)}{p\left(y_k^s \mid d_k = 0\right)}$$

where $\gamma_i'(y_k^{1p}, m', m) = p(y_k^{1p} \mid d_k = i, S_k = m, S_{k-1} = m')\Pr[d_k = i \mid S_k = m, S_{k-1} = m']$.

In the above equation, the first term is the extrinsic information generated by the current decoder by using the code constraints, the second term is the a-priori information and the last term is the systematic likelihood information.

The iterative decoder works as follows. At each iteration step, one of the decoders takes the systematic information (directly from the observation of the systematic part) and the extrinsic log-likelihood information produced by the other decoder in the previous iteration step to compute its new extrinsic log-likelihood information. Then, this updated extrinsic information is fed into the other decoder for the next iteration step. The extrinsic information of both decoders are initialized to zero before the iterations start. The block diagram of the iterative decoding algorithm is presented in Figure 35.4.

Let us denote the systematic log-likelihood information of the input bit d_k by $L_s(d_k)$, the first extrinsic log-likelihood information by $L_{1e}(d_k)$ and the second one by $L_{2e}(d_k)$, $k = 1, 2, \ldots, N$. $L_s(d_k)$ is directly found from the systematic observation and $L_{1e}(d_k)$ and $L_{2e}(d_k)$ (at each iteration step) can be computed from the code constraints by using the MAP decoding algorithm. In fact, the extrinsic information can be computed by using any other decoding algorithm for systematic codes that accepts log-likelihood values of the information bits and produces updated (independent) log-likelihoods, such as the soft output Viterbi algorithm (SOVA) [32].

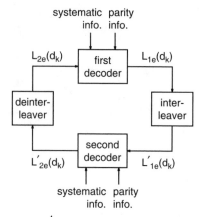

FIGURE 35.4   Iterative (turbo) decoding algorithm.

To summarize, at each iteration step, the log-likelihood of the information bit d_k is given by

$$\Lambda(d_k) = \log \frac{P(d_k = 1 \mid \text{observation})}{P(d_k = 0 \mid \text{observation})}$$
$$= L_s(d_k) + L_{1e}(d_k) + L_{2e}(d_k).$$

The a-priori information (coming from the other decoder) is approximately independent of the extrinsic information generated at the current iteration step. Therefore, further iterations do not deteriorate the performance, and the iterative decoding algorithm converges with high probability. In other words, after a number of iterations, the sum of the three likelihood values converges to the true log-likelihood of the kth information bit, which can be used to make the final decision.

35.2.3   Iterative Decoding of Serially Concatenated Convolutional Codes

Similar to parallel concatenated convolutional codes, the optimal decoding of the serially concatenated convolutional codes is almost impossible due to the interleaver used between the component codes. Therefore, to decode serially concatenated convolutional codes, a practically implementable, suboptimal, iterative decoding algorithm, utilizing soft output MAP component decoders, is used. The block diagram of the decoder is shown in Figure 35.5. At each iteration step, the inner decoder uses the noisy channel observations and the extrinsic log-likelihood ratio (LLR) information of its input block (the codeword generated by the outer encoder) calculated by the outer decoder in the previous iteration and then updates the extrinsic log-likelihood information of its input block using the inner code constraints. This updated extrinsic information, corresponding to the output of the outer code, is used by the outer decoder to

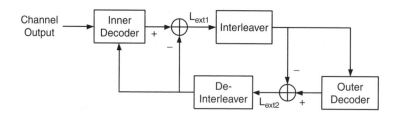

FIGURE 35.5   Iterative decoding of serially concatenated codes.

calculate the log-likelihood information of both the input and the output bits of the outer code based on the code constraints. The LLRs of the output bits of the outer code are sent to the inner decoder for use in the next iteration. After a number of iterations, the log-likelihood information of the input bits generated by the outer decoder is used to recover the transmitted information bits.

There is an important difference between the decoding algorithms of parallel and serial concatenated convolutional codes. For serial concatenation, although the inner decoder calculates only the LLR of its input block, the outer decoder calculates LLRs of both its input and output blocks. However, for parallel concatenation, only the LLRs of the original input block are calculated by both decoders, because in parallel concatenation, for both decoders the noisy channel outputs are available and the extrinsic information exchanged between the decoders are used as the a-priori information by the component decoders to perform the MAP decoding. However, for serial concatenation, the input of the inner code is the codeword generated by the outer code. Therefore, in order for the inner decoder to perform the MAP decoding, it requires the a-priori information, which are the LLRs of the coded bits of the outer code in this case. Likewise for the outer decoder, there is no channel output, so the LLR calculated by the inner decoder for its input is used by the outer decoder instead of the noisy channel outputs as in the parallel concatenation.

In terms of maximum-likelihood performance, serial concatenation is superior to parallel concatenation. In addition, the error floor, caused by the relatively low free distance of the turbo codes, is observed at lower bit error rates for serially concatenated codes.

35.2.4 Other Iterative Decoding Algorithms

The original iterative decoding algorithm for turbo codes [5] is complex. However, there are other simplified iterative decoding algorithms [33–36]. These algorithms provide an appreciable decrease in the complexity of the component decoders at the expense of some performance degradation (typically about 0.5 dB). Jung [37] presents a good comparison of various iterative decoding schemes for turbo codes. Other studies on the iterative decoding algorithm are reported in [38, 39]. Recently, it has also been observed that the original turbo decoding algorithm, among others, can be considered as a special case of a broader class of "belief propagation" algorithms for loopy Bayesian networks that are extensively studied in other fields, especially in artificial intelligence [40].

A general study of iterative decoding of block and convolutional codes is presented in [41]. Using log-likelihood algebra, the authors show that, any decoder that accepts soft inputs (including a-priori values) and delivers soft outputs that can be split into three terms: the soft channel input, the a-priori input, and the extrinsic value, can be used for decoding the component codes. The MAP decoding algorithm derived in [31] and used in [5, 30] and the SOVA algorithm developed in [32, 36] are in this category. Furthermore, the authors provide algorithms for soft decoding of systematic linear block codes, which makes it practical to use block codes (with certain restrictions due to increased complexity in decoding) as component codes in turbo code construction.

35.3 Performance of Turbo Codes over AWGN Channels

In this section, we present a set of results on the performance of turbo codes over AWGN channels. We consider the turbo code with 5/7 component convolutional codes with pseudo-random interleaver, for various block lengths (interleaver lengths). In Figure 35.6, the bit error rate performance is shown with respect to the number of decoding iterations for $N = 1000$, $R_c = 1/2$ turbo code obtained by puncturing half of the parity bits. While the bit error rate (BER) performance improves significantly during the first few iterations; after that, the performance gain is not very significant. Using this particular turbo code, at a BER of 10^{-5}, we can obtain an approximate coding gain of 7 dB over the uncoded system.

In Figure 35.7, the BER performance of $R_c = 1/2$ turbo code is shown for different interleaver lengths after 18 iterations. Larger interleavers improve the BER performance significantly, however the decoding

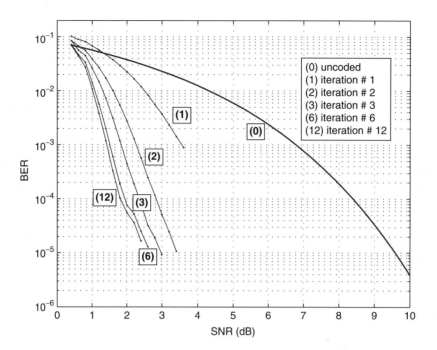

FIGURE 35.6   Performance of the iterative turbo decoder as a function of the number of iterations.

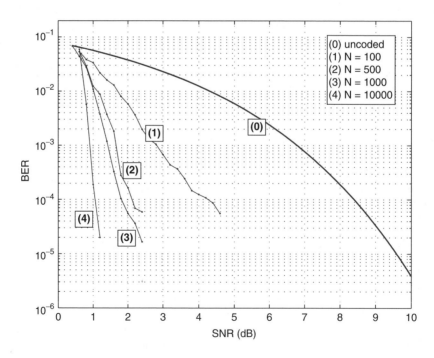

FIGURE 35.7   Performance of the turbo code with 5/7 component codes with different interleaver lengths.

delay increases as well. For example, by choosing $N = 10000$ we obtain a gain of 3 dB over the one using $N = 100$ at a BER of 10^{-4}, and the gain is even more for lower BER values. This improvement is due to the fact that, by using larger interleavers we effectively increase the probability of breaking down the input sequences that will result in low weight codewords, thus we have a larger interleaver gain. We also note that, the capacity for rate 1/2 transmission over an AWGN channel is at 0 dB and by using an interleaver of length 10000, we can obtain a bit error rate of 10^{-5} within 1.2 dB of the channel capacity. Simply by employing larger interleavers we can approach to the channel capacity further [5].

35.4 Recording Channels

Digital magnetic recording channels can be modelled as binary input inter-symbol interference (ISI) channels. In this chapter, we use two specific channel models. The first one is a simplistic partial response (PR) channel representing the recording channel with an equivalent discrete ISI channel with additive white Gaussian noise. This model can be used to represent an "ideal" recording channel. The second one is a Lorentzian channel model (assuming longitudinal recording) which is a more accurate approximation of the realistic magnetic recording channels.

35.4.1 Partial Response Channels

Partial response channels are nothing but ISI channels. For example in magnetic recording, important partial response channels include the PR4 channel (i.e., $(1 - D^2)$ and the EPR4 channel (i.e., $(1 + D - D^2 - D^3)$), where D is the delay operator. In other words, PR4 and EPR4 refer to the equivalent discrete channel models of $y_k = x_k - x_{k-2} + v_k$ and $y_k = x_k + x_{k-1} - x_{k-2} - x_{k-3} + v_k$ respectively, where y_k is the channel output, v_k is additive white Gaussian noise and x_k is the input to the ISI channel.

A PR channel can be considered as rate one nonbinary convolutional code (i.e., the channel outputs are not necessarily 0 or 1). Therefore, it can be modelled as a finite state machine and a trellis can be used to represent its input-output relationship. The optimal method (in terms of minimizing the bit error probability) to recover the information bits from the channel output is to use a MAP decoder (usually called the channel detector) matched to the trellis of the PR channel. This MAP decoder calculates the log-likelihood information of the channel inputs from which the information bits are recovered. Viterbi or soft output Viterbi algorithms can be used to recover the information bits as well.

We define the signal to noise ratio as SNR $= \frac{E_b}{N_0}$, where E_b is the energy per information bit and $N_0/2$ is the two sided additive white Gaussian noise power spectral density. Here $E_s = E_b \cdot R_c$, where R_c is the code rate and E_s is the average energy of the channel outputs.

35.4.2 Realistic Recording Channel Models

The block diagram of a more realistic recording channel model including an outer turbo code is shown in Figure 35.8. Instead of directly recording the uncoded data sequence, we first use an error correcting code, which is the turbo code for our case, to obtain the coded data sequence. This sequence is then interleaved using a pseudo-random interleaver, and may be precoded to obtain another bit sequence. The precoded bit sequence is fed into the write current driver which generates a two-level waveform called write current. The mapping from the binary (precoded) data sequence to the write current is done in such a way that a change in the write current corresponds to a "1", and no change corresponds to a "0". This is called NRZI (non-return-to-zero inverse) recording. Finally, the write head magnetizes the medium in one direction or the other depending on the polarity of the write current.

For longitudinal recording, in the readback process, the output voltage of the readback head corresponding to an isolated positive transition is well modelled as a Lorentzian pulse given by

$$h(t) = \frac{1}{1 + \left(\frac{2t}{T_{50}}\right)^2}$$

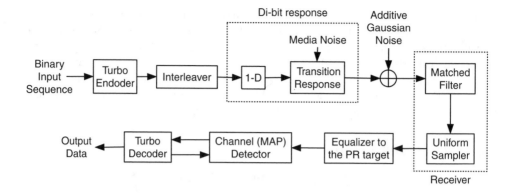

FIGURE 35.8 Block diagram of the magnetic recording system with actual write/read.

where T_{50} is the width of the Lorentzian pulse at 50% amplitude level. The normalized density of the system is defined as $D_n = \frac{T_{50}}{T}$, where T is the duration of each bit interval.

The noiseless readback signal is of the form

$$r(t) = \sum_{k=-\infty}^{\infty} b_k h(t - kT)$$

where $b_k = a_k - a_{k-1}$ is the transition sequence corresponding to the recording channel input a_k. The di-bit response shown in Figure 35.8 corresponds to impulse response of the channel and it is given as $h(t) - h(t - T)$.

The noise in a magnetic recording system is a combination of media noise, head noise and thermal noise generated in the preamplifier. The latter two components can be well modelled as additive Gaussian noise. On the other hand, media noise which may be the result of the stochastic fluctuations on the position of written transitions cannot be generally modelled as additive [42, 43] and it degrades the bit error rate performance significantly. Ignoring the media noise, we define the SNR as

$$\text{SNR} = \frac{E_i}{N_0}$$

where $E_i = \frac{\pi}{2 \times T_{50}}$ is the energy of the impulse response of the recording channel, that is, the derivative of the isolated transition response $h(t)$ and $N_0/2$ is the two sided additive white Gaussian noise power spectral density.

For this channel model the optimum detection consists of a filter matched to $p(t)$, symbol rate sampler and a maximum likelihood sequence detector [44]. In general, the overall noise is not Gaussian, therefore, the use of a matched filter is not optimal. Alternatively, one might use a low pass filter instead of the matched filter as done in most practical systems. After uniform sampling, the receiver output is usually equalized to an appropriate partial response target using a linear equalizer to reduce the computational complexity of the channel detector following the uniform sampler. The function of the channel detector is to compute the log-likelihood values of the transmitted bits for the outer error correction code decoder.

35.5 Turbo Codes for Recording Channels

Turbo codes are applied to digital magnetic recording successfully [45–52]. In particular the simplistic PR channel model assuming Gaussian noise and ideal equalization to the PR target are used [45–51], and large coding gains are obtained. Additionally in [52], it is shown that the performance improvement offered

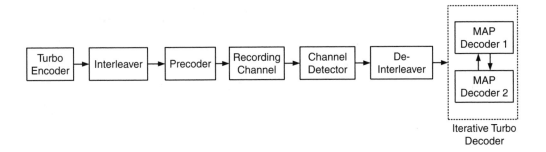

FIGURE 35.9 Block diagram of turbo coding without turbo equalization.

by the turbo coded system is preserved even if the ideal system is replaced by a more realistic Lorentzian channel model (without including the media noise).

We can classify the use of turbo codes for magnetic recording channels in three categories; turbo coding without turbo equalization, turbo coding with turbo equalization and convolutional coding with an interleaver (i.e., serial concatenation of a convolutional code with the ISI channel). These three schemes are detailed in the following subsections.

35.5.1 Turbo Coding without Turbo Equalization

In this scheme, whose block diagram is shown in Figure 35.9, the underlying error correcting code which is used to produce the channel coded bits is a parallel concatenated convolutional code. To obtain the bit sequence that will be stored in the magnetic medium, first the coded data sequence is generated by the turbo encoder. Then, the coded bits are interleaved using a pseudo-random interleaver, and may be precoded to obtain another bit sequence. In this case, the existence of the precoder is not crucial to obtain an interleaving gain, however different precoders will perform differently [53].

The outer decoder is an iterative turbo decoder which uses the log-likelihood values for the channel coded bits (i.e., the systematic and the transmitted parity bits) which are produced by the channel detector. Clearly, appropriate de-interleaving should be employed while passing the computed log-likelihood values to the turbo decoder. The soft information (i.e., the log-likelihoods) about the coded bits are — strictly speaking — correlated. However, since their statistics are not easy to characterize, the decoder assumes that they are the log-likelihoods of the observations from a BPSK transmission over an AWGN channel. To perform the turbo decoding, the variance of the additive Gaussian noise should be specified. Here, we assume that the noise variance is $\frac{N_0}{2}$ though better alternatives may exist. Fortunately, the iterative turbo decoding algorithm is very robust, and although there is not a good reason for assuming an AWGN channel with the specified noise variance, it works properly, that is, errors are corrected in the subsequent iteration steps as it will be illustrated later.

In this scheme, we do not allow the passage of information from the turbo decoder back to the channel detector. Therefore, this scheme is called turbo coding without turbo equalization.

35.5.2 Turbo Coding with Turbo Equalization

This scheme also uses a turbo code as the underlying error correcting code. The only difference is in the decoding algorithm. The extrinsic (new) information about the coded bits produced by the turbo decoder (using the code constraints) is uncorrelated with the original log-likelihoods computed by the channel detector. Therefore, the channel detector can make use of this new information to update the log-likelihoods of the turbo coded bits. By including the channel detector in the iterative decoding algorithm (together with the two MAP decoders), the performance of the decoding algorithm is improved.

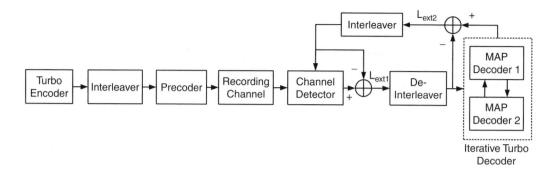

FIGURE 35.10   Block diagram of turbo coding with turbo equalization.

The process of feeding the extrinsic information of the turbo decoder back to the channel detector (and vice versa in the subsequent iteration steps) is named "turbo equalization" [54]. Figure 35.10 illustrates the passage of information between the channel detector and the turbo decoder to perform turbo equalization. Similar to the previous scheme, for turbo coding with turbo equalization, precoding is not essential to obtain an interleaver gain.

35.5.3   Convolutional Coding with an Interleaver

For turbo coding both with and without turbo equalization over recording channels, an outer turbo code is connected "serially" with the recording channel. Since the recording channel acts as a rate 1 inner code, the overall system can be viewed as a serial concatenation scheme [50]. A simpler system can be obtained by replacing the underlying turbo code with a convolutional code. This scheme is called the serial concatenation of a convolutional code with the partial response (ISI) channel [50]. The block diagram of this scheme including the decoder is shown in Figure 35.11.

    The decoding algorithm for this scheme involves the exchange of information between the channel detector and a single MAP decoder for the convolutional code, which is similar to the turbo equalization. In the first iteration step, using the original channel observations, the channel detector computes the log-likelihood of its input by the MAP decoding algorithm and passes this information to the outer decoder. Then, the outer decoder computes its extrinsic information using the code constraints and passes it back to the channel detector. We note that, the outer decoder computes the extrinsic information for the parity bits as well as the information bits of the underlying convolutional code. The iterations are repeated several times to obtain the final likelihoods of the information bits transmitted and bit decisions are made.

    Compared to the case of turbo decoding with turbo equalization, this algorithm is less complex due to the decrease in the number of the MAP decoders from three (two for the turbo decoder and one for the channel detector) to two (one for the outer decoder and one for the channel detector).

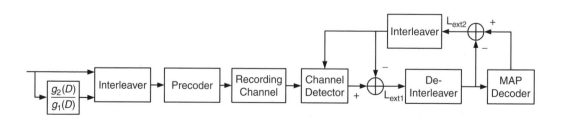

FIGURE 35.11   Block diagram of convolutional coding with an interleaver.

In [21], for serial concatenated codes it is shown that, for superior bit error rate performance, the inner encoder must be chosen as a recursive convolutional encoder. Therefore, unlike the previous two schemes, for the serial concatenation of a convolutional code with the PR channel, the use of a precoder is essential to obtain a large interleaver gain. Precoding makes the PR channel "recursive," thus the precoded channel has good distance properties [53]. As will be shown in the next section, the convolutional coding with an interleaver outperforms turbo coding when precoding is used.

35.6 Performance of Turbo Codes over Recording Channels

In this section, we present a set of results on the performance of turbo codes for magnetic recording channels by using both the simplistic PR channel and the more realistic Lorentzian channel models.

35.6.1 Over PR Channels

For the simulations we consider the $R_c = 4/5$, $N = 10000$, 5/7 turbo code over PR4 and EPR4 channels. After pseudo-random interleaving, the coded bits are precoded with $1/1 \oplus D^2$. Figure 35.12 and Figure 35.13 show the BER for PR4 and EPR4 channels respectively. For comparison purposes, the BER for the uncoded system is also included to the plots.

These results confirm that turbo coding both with and without turbo equalization introduces large coding gains. For instance at 10^{-5} probability of bit error, turbo coding without turbo equalization provides 6 dB and 5 dB coding gains over the uncoded system for PR4 and EPR4 channels respectively. With turbo equalization, the same code introduces an additional 0.5 dB coding gain compared to the system without turbo equalization. However, this additional gain comes at the expense of an increased complexity, because for turbo equalization channel detector is used at every iteration unlike the case without turbo equalization for which the channel detector is used only once.

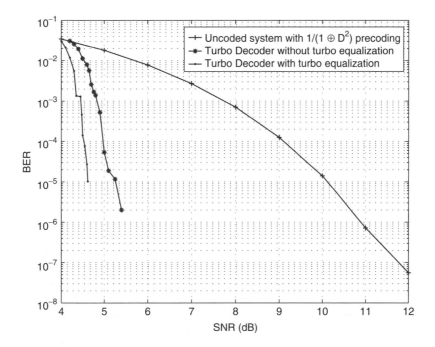

FIGURE 35.12 The BER of the 5/7 turbo code with $R_c = 4/5$ over PR4 channel, $N = 10000$.

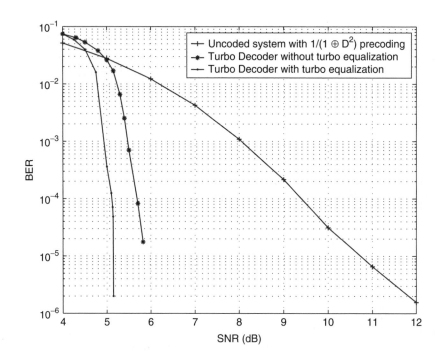

FIGURE 35.13   The BER of the 5/7 turbo code with $R_c = 4/5$ over the EPR4 channel, $N = 10000$.

35.6.2   Over Lorentzian Channels

In this section, we consider the 5/7 and 23/31 turbo codes with $R_c = 16/17$ over the Lorentzian channel with $D_n = 2.5$. In our simulations, we use the (appropriate) iterative decoding algorithm with 15 iterations and equalize the channel to an EPR4 target using a least mean squares (LMS) based linear equalizer with 21 taps.

In Figure 35.14, we present the performance of several codes for an interleaver length of $N = 10016$. We assume that $\frac{1}{1 \oplus D^2}$ precoder is used and there is no media noise. We observe that the high rate coding (turbo or convolutional with an interleaver) provides a significant coding gain of up to 4.5 dB over the uncoded system at 10^{-5} bit error probability. Therefore, the coding gain of turbo codes is mostly preserved when the simplistic PR channel model is replaced by the more realistic Lorentzian channel model. Furthermore, we observe that the convolutional code used together with an interleaver outperforms the parallel concatenated codes, which is in agreement with the observations made for the PR4 equalized ideal magnetic recording channel model of [50].

We believe that the reason why the convolutional code (together with an interleaver) performs better than the turbo code lies in the decoding algorithm. In the case of the convolutional code, the iterative suboptimal decoding algorithm requires the exchange of information between two MAP decoders, whereas in the parallel concatenation case, the exchange of information (when turbo equalization is employed) is between three MAP decoders which results in a "worse" decoder. Performance bounds based on the maximum likelihood decoding computed in [55] support our claim, since the simulation results for the convolutionally coded systems are very close to the bounds based on the optimal (maximum likelihood) decoding. On the other hand, for the turbo codes used in magnetic recording systems, the simulation results (based on the suboptimal iterative decoding algorithm) are worse than the bounds computed using maximum likelihood decoding.

In Figure 35.15, we present the performance of two different codes using different precoders (or, no precoder). We see that the performance of these schemes vary slightly. However the best choice of the precoding scheme is not clear, therefore for code design, one should take the various possibilities into

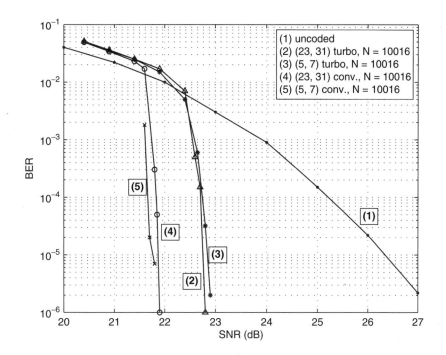

FIGURE 35.14 Performance of several codes and the uncoded system with $D_n = 2.5$ and $1/1 \oplus D^2$ precoder.

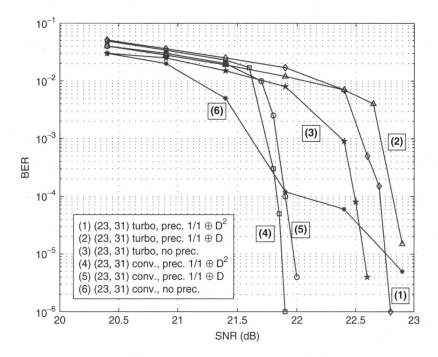

FIGURE 35.15 Performance of two different codes with different precoders with $D_n = 2.5$ and $N = 10016$.

account, as some schemes (including no precoding) may outperform the others. In our example, for the convolutional code the $\frac{1}{1\oplus D^2}$ precoding is the best choice. On the other hand for the turbo code, no precoding is the best choice among the three different precoding schemes used. However as mentioned in [50], at high SNRs, the BER performance is better for both serial and parallel concatenation with precoding. Since we usually do not operate at very low SNRs especially for magnetic recording, we can conclude that precoding improves the performance of the recording channels.

For the simulations, when turbo equalization is performed, improvements of up to 1 dB are observed. It is interesting to note that for both parallel and serial concatenation, the improvement is more if the component code is simpler. This result agrees with the observation that the simple convolutional code outperforms the parallel concatenated code when concatenated serially with the channel [50].

In [51], for the uncoded system it is observed that, at high media noise levels the probability of error cannot be made smaller than a certain value no matter how large the SNR is. On the other hand, the convolutionally coded system is able to tolerate very high media noise levels. This property of turbo based codes is very attractive for use in media noise limited magnetic recording channels.

Although turbo codes provide very low bit error rates for AWGN and magnetic recording channels, they have a "burst" error problem. That is, when there is an error, it is likely that there are many errors within the decoded block [56]. Such errors pose challenging problems for various applications including magnetic recording systems, because the outer error correcting code (ECC) cannot correct the residual errors if they exceed its error correction capability. Therefore, it is standard in magnetic recording literature to study the error bursts and error event distributions to evaluate the performance of the system, see, for instance, [57]. In [58] various burst error identification techniques are introduced for turbo and LDPC coded systems.

35.7   Summary

In this chapter, we studied the parallel concatenated and serial concatenated convolutional codes, that is, turbo codes. First, use of these codes for additive white Gaussian channels is discussed. The encoding procedures and the iterative decoding algorithm based on the two soft output component decoders are presented. Then, the use of concatenated codes over recording channels are discussed. Partial response channels with additive white Gaussian noise are used as simple models to illustrate the main ideas. The concepts of turbo decoding without turbo equalization and with turbo equalization are explained in detail. As more accurate models, Lorentzian channels are also considered for longitudinal recording channels.

To illustrate the capabilities and limitations of the various turbo coding approaches for AWGN and recording channels, we presented a set of results, which verified that, the BER performance of turbo codes is excellent and they provide a large coding gain over the uncoded systems for both channels. Despite this appreciable coding gain, a negative side of turbo codes is the existence of error bursts even at low probability of error values.

References

[1] C.E. Shannon, A mathematical theory of communication, *Bell System Technical Journal*, pp. 1–10, January 1948.

[2] S. Lin and D.J. Costello, Jr., *Error Control Coding: Fundamentals and Applications*. Englewood Cliffs, NJ: Prentice Hall, 1983.

[3] A.M. Michelson and A.H. Levesque, *Error-Control Techniques for Digital Communications*, New York, NY: Wiley, 1985.

[4] G.D. Forney, Jr., *Concatenated Codes*. Cambridge, Massachusetts: M.I.T. Press, 1966.

[5] C. Berrou, A. Glavieux, and P. Thitimajshima, Near Shannon limit error-correcting coding and decoding: Turbo-codes, in *Proceedings of IEEE International Conference on Communications (ICC)*, pp. 1064–1070, 1993.

[6] S. Benedetto and G. Montorsi, Design of parallel concatenated convolutional codes, *IEEE Transactions on Communications*, pp. 591–600, May 1996.

[7] D. Divsalar and F. Pollara, On the design of turbo codes, TDA Progress Report 42-123, JPL, November 1995.

[8] S. Dolinar and D. Divsalar, Weight distributions for turbo codes using random and nonrandom permutations, TDA Progress Report 42-122, JPL, August 1995.

[9] T.M. Duman, *Interleavers for Serial and Parallel Concatenated (Turbo) Codes*, in *Wiley Encyclopedia of Telecommunications*, J.G. Proakis, Ed., Wiley, New York, December 2002, pp. 1141–1151.

[10] D. Divsalar and R.J. McEliece, Effective free distance of turbo codes, *Electronics Letters*, pp. 445–446, February 1996.

[11] L.C. Perez, J. Seghers, and D.J. Costello, Jr., A distance spectrum interpretation of turbo codes, *IEEE Transactions on Information Theory*, pp. 1698–1709, November 1996.

[12] S. Benedetto and G. Montorsi, Unveiling turbo codes: Some results on parallel concatenated coding schemes, *IEEE Transactions on Information Theory*, pp. 409–428, March 1996.

[13] D. Divsalar, S. Dolinar, F. Pollara, and R.J. McEliece, Transfer function bounds on the performance of turbo codes, TDA Progress Report 42-122, JPL, August 1995.

[14] T.M. Duman and M. Salehi, New performance bounds for turbo codes, *IEEE Transactions on Communications*, pp. 717–723, June 1998.

[15] I. Sason and S.S. (Shitz), Improved upper bounds on the ML decoding error probability of parallel and serial concatenated turbo codes via their ensemble distance spectrum, *IEEE Transactions on Information Theory*, vol. 46, pp. 24–47, January 2000.

[16] R.J. McEliece, E.R. Rodemich, and J. Cheng, The turbo decision algorithm, in *Proceedings of Allerton Conference on Communications, Control and Computing*, pp. 366–379, 1995.

[17] S. Benedetto and G. Montorsi, Performance of continuous and blockwise decoded turbo codes, *IEEE Communications Letters*, pp. 77–79, May 1997.

[18] D. Divsalar and F. Pollara, "Turbo codes for deep-space communications," TDA Progress Report 42-120, JPL, February 1995.

[19] W.J. Blackert, E.K. Hall, and S.G. Wilson, Turbo code termination and interleaver conditions, *Electronics Letters*, pp. 2082–2084, November 1995.

[20] O. Joerssen and H. Meyr, Terminating the trellis of turbo codes, *Electronics Letters*, pp. 1285–1286, August 1994.

[21] S. Benedetto, D. Divsalar, G. Montorsi, and F. Pollara, Serial concatenation of interleaved codes: Performance analysis, design, and iterative decoding, TDA Progress Report 42-126, JPL, August 1996.

[22] S. Benedetto and G. Montorsi, Average performance of parallel concatenated block codes, *Electronics Letters*, pp. 156–158, February 1995.

[23] D. Divsalar and F. Pollara, Multiple turbo codes for deep-space communications, TDA Progress Report 42-121, JPL, May 1995.

[24] S. Divsalar and F. Pollara, Hybrid concatenated codes and iterative decoding, TDA Progress Report 42-130, JPL, August 1997.

[25] R.G. Gallager, Low-density parity check codes, *IRE transactions on Information Theory*, vol. IT-8, pp. 21–28, January 1962.

[26] D.J.C. MacKay, Good error-correcting codes based on very sparse matrices, *IEEE Transactions on Information Theory*, vol. 45, pp. 399–431, March 1999.

[27] S.B. Wicker, *Error Control Systems for Digital Communication and Storage*. Upper Saddle River, N.J.: Prentice Hall, 1995.

[28] T. Richardson, A. Shokrollahi, and R. Urbanke, Design of capacity approaching irregular low-density parity check codes, *IEEE Transactions on Information Theory*, vol. 47, pp. 619–637, February 2001.

[29] Z. Wu, *Coding and Iterative Detection for Magnetic Recording Channels*, Norwell, Massachusetts: Kluwer Academic Publishers, 1999.

[30] P. Robertson, Illuminating the structure of code and decoder of parallel concatenated recursive systematic (turbo) codes, in *Proceedings of IEEE Global Communications Conference (GLOBECOM)*, pp. 1298–1303, 1994.

[31] L.R. Bahl, J. Cocke, F. Jelinek, and J. Raviv, Optimal decoding of linear codes for minimizing symbol error rate, *IEEE Transactions on Information Theory*, pp. 284–287, March 1974.

[32] J. Hagenauer and P. Hoeher, A Viterbi algorithm with soft-decision outputs and its applications, in *Proceedings of IEEE Global Communications Conference (GLOBECOM)*, pp. 47.1.1–47.1.7, 1989.

[33] C. Berrou, P. Adde, E. Angui, and S. Faudeil, A low complexity soft-output Viterbi decoding architecture, in *Proceedings of IEEE International Conference on Communications (ICC)*, pp. 737–740, 1994.

[34] P. Jung, Novel low complexity decoder for turbo codes, *Electronics Letters*, pp. 86–87, January 1995.

[35] S.S. Pietrobon and A.S. Barbulescu, Simplification of the modified Bahl decoding algorithm for systematic convolutional codes, in *Proceedings of IEEE International Symposium on Information Theory and Its Applications (ISITA)*, pp. 1073–1077, 1994.

[36] J. Hagenauer, P. Robertson, and L. Papke, Iterative (turbo) decoding of systematic convolutional codes with the MAP and SOVA algorithms, in *Proceedings of ITG Conference, vol. 130, Munich*, pp. 21–29, 1994.

[37] P. Jung, Comparison of turbo code decoders applied to short frame transmission systems, *IEEE Journal of Selected Areas in Communications*, pp. 530–537, April 1996.

[38] S. Benedetto, G. Montorsi, D. Divsalar, and F. Pollara, A soft-input soft-output maximum a posteriori (MAP) module to decode parallel and serial concatenated codes, TDA Progress Report 42-127, JPL, November 1996.

[39] S. Benedetto, D. Divsalar, G. Montorsi, and F. Pollara, Algorithm for continuous decoding of turbo codes, *Electronics Letters*, pp. 314–315, February 1996.

[40] R.J. McEliece, D.J.C. MacKay, and J. Cheng, Turbo decoding as an instance of Pearl's belief propagation algorithm, *IEEE Journal of Selected Areas in Communications*, pp. 140–152, February 1998.

[41] J. Hagenauer, E. Offer, and L. Papke, Iterative decoding of binary block and convolutional codes, *IEEE Transactions on Information Theory*, pp. 429–445, March 1996.

[42] J.G. Proakis, Equalization techniques for high density magnetic recording systems, *IEEE Signal Processing Magazine*, pp. 73–82, July 1998.

[43] H.N. Bertram, *Theory of Magnetic Recording*. Cambridge Press, London 1994.

[44] J.G. Proakis, *Digital Communications*. New York, NY: McGraw-Hill, Inc., 1995.

[45] C. Heegard, Turbo coding in magnetic recording" in *Proceedings of the Information Theory Workshop, San Diego, CA*, pp. 18–19, 1998.

[46] W.E. Ryan, L.L. McPheters, and S. McLaughlin, Combined turbo coding and turbo equalization for PR4-equalized Lorentzian channels, in *Proceedings of Conference on Information Sciences and Systems*, pp. 489–494, March 1998.

[47] W. Pusch, D. Weinrichter, and M. Tafarner, Turbo codes matched to the $1 - D^2$ partial response channel, in *Proceedings of IEEE International Symposium on Information Theory (ISIT)*, p. 62, 1998.

[48] M.C. Reed and C.B. Schelegel, An iterative receiver for the partial response channel, in *Proceedings of IEEE International Symposium on Information Theory (ISIT)*, p. 63, 1998.

[49] W.E. Ryan, Performance of high rate turbo codes on a PR4 equalized magnetic recording channel, in *Proceedings of IEEE International Conference on Communications (ICC)*, pp. 947–951, June 1998.

[50] T. Souvignier, A. Friedman, M. Oberg, P.H. Siegel, R.E. Swanson, and J.K. Wolf, Turbo codes for PR4: Parallel versus serial concatenation, in *Proceedings of IEEE International Conference on Communications (ICC)*, pp. 1638–1642, June 1999.

[51] T.M. Duman and E. Kurtas, Comprehensive performance investigation of turbo codes over high density magnetic recording channels, in *Proceedings of IEEE Global Communications Conference (GLOBECOM)*, vol. 1b, pp. 744–748, December 1999.

[52] T. Souvignier and J.K. Wolf, Turbo decoding for partial response channels with colored noise, *IEEE Transactions on Magnetics*, pp. 2322–2324, September 1999.

[53] K.R. Narayanan, Effect of precoding on the convergence of turbo equalization for partial response channels, in *Proceedings of IEEE Global Communications Conference (GLOBECOM)*, vol. 3, pp. 1865–1871, December 2000.

[54] C. Douillard, M. Jezequel, C. Berrou, A. Picart, P. Didier, and A. Glaivieux, Iterative correction of intersymbol interference: Turbo-equalization, *European Transactions on Telecommunications*, vol. 6, pp. 507–511, September 1995.

[55] T.M. Duman and E. Kurtas, Performance bounds for high rate linear codes over partial response channels, *IEEE Transactions on Information Theory*, vol. 47, pp. 1201–1205, March 2001.

[56] M.N. Kaynak, Turbo and low density parity check codes for AWGN and partial response channels, Technical Report, Department of Electrical Engineering, Arizona State University, December 2002.

[57] T. Souvignier, Turbo decoding for partial response channels. Ph.D. Thesis, Department of Electrical Engineering, University of California, San Diego, 1999.

[58] M.N. Kaynak, T.M. Duman, and E.M. Kurtas, Burst error identification for turbo and LDPC coded systems, in *Proceedings of the 3rd International Symposium on Turbo Codes and Related Topics*, pp. 515–518, September 2003.

36

An Introduction to LDPC Codes

36.1   Introduction 36-1
36.2   Representations of LPDC Codes 36-2
       Matrix Representation • Graphical Representation
36.3   LDPC Code Design Approaches 36-4
       Gallager Codes • MacKay Codes • Irregular LDPC Codes
       • Finite Geometry Codes • RA, IRA, and eIRA Codes
       • Array Codes • Combinatorial LDPC Codes
36.4   Iterative Decoding Algorithms 36-7
       Overview • Probability-Domain SPA Decoder
       • Log-Domain SPA Decoder • Reduced Complexity
       Decoders
36.5   Concluding Remarks 36-17

William E. Ryan
University of Arizona
Tucson, AZ

36.1   Introduction

Low-density parity-check (LDPC) codes are a class of linear block codes which provide near-capacity performance on a large collection of data transmission and storage channels while simultaneously admitting implementable decoders. LDPC codes were first proposed by Gallager in his 1960 doctoral dissertation [1] and were scarcely considered in the 35 years that followed. One notable exception is the important work of Tanner in 1981 [2] in which Tanner generalized LDPC codes and introduced a graphical representation of LDPC codes, now called Tanner graphs. The study of LDPC codes was resurrected in the mid-1990s with the work of MacKay, Luby, and others [3–5] who noticed, apparently independently of the work of Gallager, the advantages of linear block codes which possess sparse (low-density) parity-check matrices.

   This tutorial chapter provides the foundations for the study and practice of LDPC codes. We will start with the fundamental representations of LDPC codes via parity-check matrices and Tanner graphs. Classification of LDPC ensembles via Tanner graph degree distributions will be introduced, but we will only superficially cover the design of LDPC codes with optimal degree distributions via constrained pseudo-random matrix construction. We will also review some of the other LDPC code construction techniques which have appeared in the literature. The encoding problem for such LDPC codes will be presented and certain special classes of LDPC codes which resolve the encoding problem will be introduced. Finally, the iterative message-passing decoding algorithm (and certain simplifications) which provides near-optimal performance will be presented.

36.2 Representations of LPDC Codes

36.2.1 Matrix Representation

Although LDPC codes can be generalized to nonbinary alphabets, we shall consider only binary LDPC codes for the sake of simpicity. Because LDPC codes form a class of linear block codes, they may be described as a certain k-dimensional subspace $\mathcal{C}$ of the vector space $\mathbb{F}_2^n$ of binary n-tuples over the binary field $\mathbb{F}_2$. Given this, we may find a basis $B = \{\mathbf{g}_0, \mathbf{g}_1, \ldots, \mathbf{g}_{k-1}\}$ which spans $\mathcal{C}$ so that each $\mathbf{c} \in \mathcal{C}$ may be written as $\mathbf{c} = u_0 \mathbf{g}_0 + u_1 \mathbf{g}_1 + \cdots + u_{k-1} \mathbf{g}_{k-1}$ for some $\{u_i\}$; more compactly, $\mathbf{c} = \mathbf{u}G$ where $\mathbf{u} = [u_0\ u_1 \ldots u_{k-1}]$ and G is the so-called $k \times n$ generator matrix whose rows are the vectors $\{\mathbf{g}_i\}$ (as is conventional in coding, all vectors are row vectors). The $(n-k)$-dimensional null space $\mathcal{C}^\perp$ of G comprises all vectors $\mathbf{x} \in \mathbb{F}_2^n$ for which $\mathbf{x}G^T = \mathbf{0}$ and is spanned by the basis $B^\perp = \{\mathbf{h}_0, \mathbf{h}_1, \ldots, \mathbf{h}_{n-k-1}\}$. Thus, for each $\mathbf{c} \in \mathcal{C}$, $\mathbf{c}\mathbf{h}_i^T = 0$ for all i or, more compactly, $\mathbf{c}H^T = \mathbf{0}$ where H is the so-called $(n-k) \times n$ parity-check matrix whose rows are the vectors $\{\mathbf{h}_i\}$, and is the generator matrix for the null space $\mathcal{C}^\perp$. The parity-check matrix H is so named because it performs $m = n - k$ separate parity checks on a received word.

A *low-density parity-check code* is a linear block code for which the parity-check matrix H has a low density of 1s. A *regular LDPC code* is a linear block code whose parity-check matrix H contains exactly w_c 1s in each column and exactly $w_r = w_c(n/m)$ 1s in each row, where $w_c \ll m$ (equivalently, $w_r \ll n$). The code rate $R = k/n$ is related to these parameters via $R = 1 - w_c/w_r$ (this assumes H is full rank). If H is low density, but the number of 1s in each column or row is not constant, then the code is an *irregular LDPC code*. It is easiest to see the sense in which an LDPC code is regular or irregular through its graphical representation.

36.2.2 Graphical Representation

Tanner considered LDPC codes (and a generalization) and showed how they may be represented effectively by a so-called bipartite graph, now call a Tanner graph [2]. The Tanner graph of an LDPC code is analogous to the trellis of a convolutional code in that it provides a complete representation of the code and it aids in the description of the decoding algorithm. A *bipartite graph* is a graph (nodes connected by edges) whose nodes may be separated into two types, and edges may only connect two nodes of different types. The two types of nodes in a Tanner graph are the *variable nodes* and the *check nodes* (which we shall call v-nodes and c-nodes, respectively).[1] The Tanner graph of a code is drawn according to the following rule: check node j is connected to variable node i whenever element h_{ji} in H is a 1. One may deduce from this that there are $m = n - k$ check nodes, one for each check equation, and n variable nodes, one for each code bit c_i. Further, the m rows of H specify the m c-node connections, and the n columns of H specify the n v-node connections.

Example 36.1

Consider a $(10, 5)$ linear block code with $w_c = 2$ and $w_r = w_c(n/m) = 4$ with the following H matrix:

$$H = \begin{bmatrix} 1 & 1 & 1 & 1 & 0 & 0 & 0 & 0 & 0 & 0 \\ 1 & 0 & 0 & 0 & 1 & 1 & 1 & 0 & 0 & 0 \\ 0 & 1 & 0 & 0 & 1 & 0 & 0 & 1 & 1 & 0 \\ 0 & 0 & 1 & 0 & 0 & 1 & 0 & 1 & 0 & 1 \\ 0 & 0 & 0 & 1 & 0 & 0 & 1 & 0 & 1 & 1 \end{bmatrix}$$

---

[1] The nomenclature varies in the literature: variable nodes are also called bit or symbol nodes and check nodes are also called function nodes.

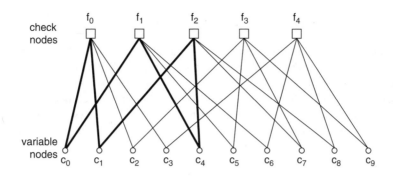

FIGURE 36.1 Tanner graph for example code.

The Tanner graph corresponding to H is depicted in Figure 36.1. Observe that v-nodes c_0, c_1, c_2, and c_3 are connected to c-node f_0 in accordance with the fact that, in the zeroth row of H, $h_{00} = h_{01} = h_{02} = h_{03} = 1$ (all others are zero). Observe that analogous situations holds for c-nodes f_1, f_2, f_3, and f_4 which corresponds to rows 1, 2, 3, and 4 of H, respectively. Note, as follows from the fact that $\mathbf{c}H^T = \mathbf{0}$, the bit values connected to the same check node must sum to zero. We may also proceed along columns to construct the Tanner graph. For example, note that v-node c_0 is connected to c-nodes f_0 and f_1 in accordance with the fact that, in the zeroth column of H, $h_{00} = h_{10} = 1$.

Note that the Tanner graph in this example is regular: each v-node has two edge connections and each c-node has four edge connections (i.e., the *degree* of each v-node is 2 and the degree of each c-node is 4). This is in accordance with the fact that $w_c = 2$ and $w_r = 4$. It is also clear from this example that $mw_r = nw_c$.

For irregular LDPC codes, the parameters w_c and w_r are functions of the column and row numbers and so such notation is not generally adopted in this case. Instead, it is usual in the literature (following [7]) to specify the v-node and c-node *degree distribution polynomials*, denoted by $\lambda(x)$ and $\rho(x)$, respectively. In the polynomial

$$\lambda(x) = \sum_{d=1}^{d_v} \lambda_d x^{d-1},$$

λ_d denotes the fraction of all edges connected to degree-d v-nodes and d_v denotes the maximum v-node degree. Similarly, in the polynomial

$$\rho(x) = \sum_{d=1}^{d_c} \rho_d x^{d-1},$$

ρ_d denotes the fraction of all edges connected to degree-d c-nodes and d_c denotes the maximum c-node degree. Note for the regular code above, for which $w_c = d_v = 2$ and $w_r = d_c = 4$, we have $\lambda(x) = x$ and $\rho(x) = x^3$.

A *cycle* (or *loop*) of length v in a Tanner graph is a path comprising v edges which closes back on itself. The Tanner graph in the above example possesses a length-6 cycle as exemplified by the six bold edges in the figure. The *girth* γ of a Tanner graph is the minimum cycle length of the graph. The shortest possible cycle in a bipartite graph is clearly a length-4 cycle, and such cycles manifest themselves in the H matrix as four 1s that lie on the corners of a submatrix of H. We are interested in cycles, particularly short cycles, because they degrade the performance of the iterative decoding algorithm used for LDPC codes. This fact will be made evident in the discussion of the iterative decoding algorithm.

36.3 LDPC Code Design Approaches

Clearly, the most obvious path to the construction of an LDPC code is via the construction of a low-density parity-check matrix with prescribed properties. A large number of design techniques exist in the literature, and we introduce some of the more prominent ones in this section, albeit at a superficial level. The design approaches target different design criteria, including efficient encoding and decoding, near-capacity performance, or low-error rate floors. (Like turbo codes, LPDC codes often suffer from low-error rate floors, owing both to poor distance spectra and weaknesses in the iterative decoding algorithm.)

36.3.1 Gallager Codes

The original LDPC codes due to Gallager [1] are regular LDPC codes with an H matrix of the form

$$H = \begin{bmatrix} H_1 \\ H_2 \\ \vdots \\ H_{w_c} \end{bmatrix}$$

where the submatrices H_d have the following structure. For any integers μ and w_r greater than 1, each submatrix H_d is $\mu \times \mu w_r$ with row weight w_r and column weight 1. The submatrix H_1 has the following specific form: for $i = 1, 2, \ldots, \mu$, the ith row contains all of its w_r 1s in columns $(i-1)w_r + 1$ to iw_r. The other submatrices are simply column permutations of H_1. It is evident that H is regular, has dimension $\mu w_c \times \mu w_r$, and has row and column weights w_r and w_c, respectively. The absence of length-4 cycles in H is not guaranteed, but they can be avoided via computer design of H. Gallager showed that the ensemble of such codes has excellent distance properties provided $w_c \geq 3$ and $w_r > w_c$. Further, such codes have low-complexity encoders since parity bits can be solved for as a function of the user bits via the parity-check matrix [1].

Gallager codes were generalized by Tanner in 1981 [2] and were studied for application to code-division multiple-access communication channel in [9]. Gallager codes were extended by MacKay and others [3, 4].

36.3.2 MacKay Codes

MacKay had independently discovered the benefits of designing binary codes with sparse H matrices and was the first to show the ability of these codes to perform near capacity limits [3, 4]. MacKay has archived on a web page [10] a large number of LPDC codes that he has designed for application to data communication and storage, most of which are regular. He provided in [4] algorithms to semi-randomly generate sparse H matrices. A few of these are listed below in order of increasing algorithm complexity (but not necessarily improved performance).

1. H is created by randomly generating weight-w_c columns and (as near as possible) uniform row weight.
2. H is created by randomly generating weight-w_c columns, while ensuring weight-w_r rows, and no two columns having overlap greater than one.
3. H is generated as in Step 2, plus short cycles are avoided.
4. H is generated as in Step 3, plus $H = [H_1 \ H_2]$ is constrained so that H_2 is invertible (or at least H is full rank).

One drawback of MacKay codes is that they lack sufficient structure to enable low-complexity encoding. Encoding is performed by putting H in the form $[P^T \ I]$ via Gauss-Jordan elimination, from which the generator matrix can be put in the systematic form $G = [I \ P]$. The problem with encoding via G is that the submatrix P is generally not sparse so that, for codes of length $n = 1000$ or more, encoding complexity is high. An efficient encoding technique employing only the H matrix was proposed in [6].

36.3.3 Irregular LDPC Codes

Richardson et al. [7] and Luby et al. [8] defined ensembles of irregular LDPC codes parameterized by the degree distribution polynomials $\lambda(x)$ and $\rho(x)$ and showed how to optimize these polynomials for a variety of channels. Optimality is in the sense that, assuming message-passing decoding (described below), a typical code in the ensemble was capable of reliable communication in worse channel conditions than codes outside the ensemble are. The worst-case channel condition is called the *decoding threshold* and the optimization of $\lambda(x)$ and $\rho(x)$ is found by a combination of a so-called *density evolution* algorithm and an optimization algorithm. Density evolution refers to the evolution of the probability density functions (pdfs) of the various quantities passed around the decoder's Tanner graph. The decoding threshold for a given $\lambda(x)$-$\rho(x)$ pair is determined by evaluation the pdfs of computed log-likelihood ratios (see the next section) of the code bits. The separate optimization algorithm optimizes over the $\lambda(x)$-$\rho(x)$ pairs.

Using this approach an irregular LDPC code has been designed whose simulated performance was within 0.045 dB of the capacity limit for a binary-input AWGN channel [11]. This code had length $n = 10^7$ and rate $R = 1/2$. It is generally true that designs via density evolution are best applied to codes whose rate is not too high ($R \lesssim 3/4$) and whose length is not too short ($n \gtrsim 5000$). The reason is that the density evolution design algorithm assumes $n \to \infty$ (hence, $m \to \infty$), and so $\lambda(x)$-$\rho(x)$ pairs which are optimal for very long codes, will not be optimal for medium-length and short codes. As discussed in [12–15], such $\lambda(x)$-$\rho(x)$ pairs applied to medium-length and short codes gives rise to a high error-rate floor.

Finally, we remark that, as for the MacKay codes, these irregular codes do not intrinsically lend themselves to efficient encoding. However, as mentioned above, Richardson and Urbanke [6] have proposed algorithms for achieving linear-time encoding for these codes.

36.3.4 Finite Geometry Codes

In [16, 17], regular LDPC codes are designed using techniques based on finite geometries [18]. The resulting LDPC codes fall into the cyclic and quasi-cyclic classes of block codes and lend themselves to simple encoder implementation via shift-register circuits. The cyclic finite geometry codes tend to have relatively large minimum distances, but the quasi-cyclic codes tend to have somewhat small minimum distances. Also, short LDPC codes (n on the order of 200 bits) designed using these techniques are generally better than short LDPC codes designed using pseudo-random H matrices.

The cyclic finite geometry codes have the drawback that the parity-check matrix used in decoding is $n \times n$ instead of $(n - k) \times n$. (It is possible to choose an $(n - k) \times n$ submatrix of the $n \times n$ matrix to decode, but the loss in performance is often non-negligible.) The $n \times n$ matrix is circulant, with its first row equal to a certain *incidence vector* [16]. Another drawback is that the values of w_r and w_c are relatively large which is undesirable since the complexity of the iterative message-passing decoder is proportional to these values. One final drawback is that there is no flexibility in the choice of length and rate, although this issue can be dealt with by code shortening and puncturing.

36.3.5 RA, IRA, and eIRA Codes

A type of code, called a *repeat-accumulate* (RA) code, which has the characteristics of both serial turbo codes and LDPC codes, was proposed in [20]. The encoder for an RA code is shown in Figure 36.2 where it is seen that user bits are repeated (2 or 3 times is typical), permuted, and then sent through an accumulator (differential encoder). These codes have been shown to be capable of operation near capacity limits, but they have the drawback that they are naturally low rate (rate 1/2 or lower).

The RA codes were generalized in such a way that some bits were repeated more than others yielding *irregular repeat-accumulate* (IRA) codes [21]. As shown in Figure 36.2, the IRA encoder comprises a low-density generator matrix, a permuter, and an accumulator. Such codes are capable of operation even closer to theoretical limits than RA codes, and they permit higher code rates. A drawback to IRA codes is that

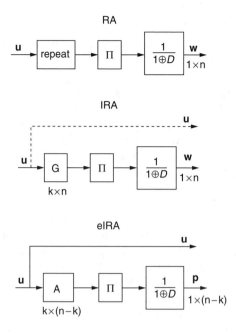

FIGURE 36.2 Encoders for the repeat-acccumulate (RA), the irregular RA (IRA) code, and the extended IRA code (eIRA).

they are nominally non-systematic codes, although they be put in a systematic form, but it is at the expense of greatly lowering the rate as depicted in Figure 36.2.

Yang and Ryan [13–15] have proposed a class of efficiently encodable irregular LDPC codes which are called extended IRA (eIRA) codes. (These codes were independently proposed in [22].) The eIRA encoder is shown in Figure 36.2. Note that the eIRA encoder is systematic and permits both low and high rates. Further, encoding can be efficiently performed directly from the H matrix which possesses an $m \times m$ submatrix which facilitates computation of the parity bits from the user bits [15, 22].

36.3.6 Array Codes

Fan has shown that a certain class of codes called array codes can be viewed as LDPC codes and thus can be decoded with a message passing algorithm [23, 24]. Subsequent to Fan's work, Eleftheriou and Ölçer [25] proposed a modified array code employing the following H matrix format

$$
H = \begin{bmatrix}
I & I & I & \cdots & I & I & \cdots & I \\
0 & I & \alpha & \cdots & \alpha^{j-2} & \alpha^{j-1} & \cdots & \alpha^{k-2} \\
0 & 0 & I & \cdots & \alpha^{2(j-3)} & \alpha^{2(j-2)} & \cdots & \alpha^{2(k-3)} \\
\vdots & \vdots & \vdots & \ddots & \vdots & \vdots & \cdots & \vdots \\
0 & 0 & \cdots & 0 & I & \alpha^{j-1} & \cdots & \alpha^{(j-1)(k-j)}
\end{bmatrix}
\tag{36.1}
$$

where k and j are two integers such that $k, j \leq p$ where p denotes a prime number. I is the $p \times p$ identity matrix, O is the $p \times p$ null matrix, and α is a $p \times p$ permutation matrix representing a single left- or right-cyclic shift. The upper triangular nature of H guarantees encoding linear in the codeword length (encoding is essentially the same as for eIRA codes).

These modified array codes have low error rate floors, and both low- and high-rate codes may be designed, although the high-rate designs perform better (relative to other design techniques). However, as is clear from the description of H above, only selected code lengths and rates are available.

36.3.7 Combinatorial LDPC Codes

In view of the fact that LDPC codes may be designed by constrained random construction of H matrices, it is not difficult to imagine that good LDPC codes may be designed via the application of combinatorial mathematics. That is, design constraints (such as no cycles of length 4) applied to an H matrix of size $(n - k) \times n$ may be cast as a problem in combinatorics. Several researchers have successfully approached this problem via such combinatorial objects as Steiner systems, Kirkman systems, and balanced incomplete block designs [16, 26–29].

36.4 Iterative Decoding Algorithms

36.4.1 Overview

In addition to introducing LDPC codes in his seminal work in 1960 [1], Gallager also provided a decoding algorithm that is typically near optimal. Since that time, other researchers have independently discovered that algorithm and related algorithms, albeit sometimes for different applications [4, 30]. The algorithm iteratively computes the distributions of variables in graph-based models and comes under different names, depending on the context. These names include: the sum-product algorithm (SPA), the belief propagation algorithm (BPA), and the message passing algorithm (MPA). The term "message passing" usually refers to all such iterative algorithms, including the SPA (BPA) and its approximations.

Much like optimal (maximum *a posteriori*, MAP) symbol-by-symbol decoding of trellis codes, we are interested in computing the *a posteriori* probability (APP) that a given bit in the transmitted codeword $\mathbf{c} = [c_0 \, c_1 \ldots c_{n-1}]$ equals 1, given the received word $\mathbf{y} = [y_0 \, y_1 \ldots y_{n-1}]$. Without loss of generality, let us focus on the decoding of bit c_i so that we are interested in computing the APP

$$\Pr(c_i = 1 \mid \mathbf{y})$$

or the APP ratio (also called the likelihood ratio, LR)

$$l(c_i) \triangleq \frac{\Pr(c_i = 0 \mid \mathbf{y})}{\Pr(c_i = 1 \mid \mathbf{y})}$$

Later we will extend this to the more numerically stable computation of the log-APP ratio, also called the log-likelihood ratio (LLR):

$$L(c_i) \triangleq \log \left[\frac{\Pr(c_i = 0 \mid \mathbf{y})}{\Pr(c_i = 1 \mid \mathbf{y})} \right]$$

where here and in the sequel the natural logarithm is assumed.

The MPA for the computation of $\Pr(c_i = 1 \mid \mathbf{y})$, $l(c_i)$, or $L(c_i)$ is an iterative algorithm which is based on the code's Tanner graph. Specifically, we imagine that the v-nodes represent processors of one type, c-nodes represent processors of another type, and the edges represent message paths. In one half-iteration, each v-node processes its input messages and passes its resulting output messages *up* to neighboring c-nodes (two nodes are said to be *neighbors* if they are connected by an edge). This is depicted in Figure 36.3 for the message $m_{\uparrow 02}$ from v-node c_0 to c-node f_2 (the subscripted arrow indicates the direction of the message, keeping in mind that our Tanner graph convention places c-nodes above v-nodes). The information passed concerns $\Pr(c_0 = b \mid \text{input messages})$, $b \in \{0, 1\}$, the ratio of such probabilities, or the logarithm of the ratio of such probabilities. Note in the figure that the information passed to c-node f_2 is all the information available to v-node c_0 from the channel and through its neighbors, excluding c-node f_2; that is, only *extrinsic information* is passed. Such extrinsic information $m_{\uparrow ij}$ is computed for each connected v-node/c-node pair c_i/f_j at each half-iteration.

In the other half-iteration, each c-node processes its input messages and passes its resulting output messages *down* to its neighboring v-nodes. This is depicted in Figure 36.4 for the message $m_{\downarrow 04}$ from

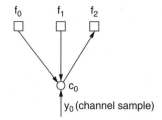

FIGURE 36.3   Subgraph of a Tanner graph corresponding to an H matrix whose zeroth column is $[1\ 1\ 1\ 0\ 0\ \ldots 0]^T$. The arrows indicate message passing from node c_0 to node f_2.

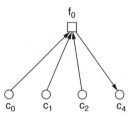

FIGURE 36.4   Subgraph of a Tanner graph corresponding to an H matrix whose zeroth row is $[1\ 1\ 1\ 0\ 1\ 0\ 0\ \ldots 0]^T$. The arrows indicate message passing from node f_0 to node c_4.

c-node f_0 down to v-node c_4. The information passed concerns Pr(check equation f_0 is satisfied | input messages), the ratio of such probabilities, or the logarithm of the ratio of such probabilities. Note, as for the previous case, only extrinsic information is passed to v-node c_4. Such extrinsic information $m_{\downarrow ji}$ is computed for each connected c-node/v-node pair f_j/c_i at each half-iteration.

After a prescribed maximum number of iterations or after some stopping criterion has been met, the decoder computes the APP, the LR or the LLR from which decisions on the bits c_i are made. One example stopping criterion is to stop iterating when $\hat{c}H^T = \mathbf{0}$, where $\hat{c}$ is a tentatively decoded codeword.

The MPA assumes that the messages passed are statistically independent throughout the decoding process. When the y_i are independent, this independence assumption would hold true if the Tanner graph possessed no cycles. Further, the MPA would yield exact APPs (or LRs or LLRs, depending on the version of the algorithm) in this case [30]. However, for a graph of girth γ, the independence assumption is only true up to the $\gamma/2$th iteration, after which messages start to loop back on themselves in the graph's various cycles. Still, simulations have shown that the message passing algorithm is generally very effective provided length-4 cycles are avoided. Lin *et al.* [19] showed that some configurations of length-four cycles are not harmful. It was shown in [31] how the message-passing *schedule* described above and below (the so-called *flooding schedule*) may be modified to mitigate the negative effects of short cycles.

In the remainder of this section we present the "probability domain" version of the SPA (which computes APPs) and its log-domain version, the log-SPA (which computes LLRs), as well as certain approximations. Our treatment considers the special cases of the binary erasure channel (BEC), the binary symmetric channel (BSC), and the binary-input AWGN channel (BI-AWGNC).

36.4.2   Probability-Domain SPA Decoder

We start by introducing the following notation:

- V_j = {v-nodes connected to c-node f_j}.
- $V_j \backslash i$ = {v-nodes connected to c-node f_j}\{v-node c_i}.
- C_i = {c-nodes connected to v-node c_i}.

- $C_i \backslash j = \{$c-nodes connected to v-node $c_i\} \backslash \{$c-node $f_j\}$.
- $M_v(\sim i) = \{$messages from all v-nodes except node $c_i\}$.
- $M_c(\sim j) = \{$ messages from all c-nodes except node $f_j\}$.
- $P_i = \Pr(c_i = 1 \mid y_i)$.
- $S_i =$ event that the check equations involving c_i are satisfied.
- $q_{ij}(b) = \Pr(c_i = b \mid S_i, y_i, M_c(\sim j))$, where $b \in \{0, 1\}$. For the APP algorithm presently under consideration, $m_{\uparrow ij} = q_{ij}(b)$; for the LR algorithm, $m_{\uparrow ij} = q_{ij}(0)/q_{ij}(1)$; and for the LLR algorithm, $m_{\uparrow ij} = \log[q_{ij}(0)/q_{ij}(1)]$.
- $r_{ji}(b) = \Pr($check equation f_j is satisfied $\mid c_i = b, M_v(\sim i))$, where $b \in \{0, 1\}$. For the APP algorithm presently under consideration, $m_{\downarrow ji} = r_{ji}(b)$; for the LR algorithm, $m_{\downarrow ji} = r_{ji}(0)/r_{ji}(1)$; and for the LLR algorithm, $m_{\downarrow ji} = \log[r_{ji}(0)/r_{ji}(1)]$.

Note that the messages $q_{ij}(b)$, while interpreted as probabilities here, are random variables (RVs) as they are functions of the RVs y_i and other messages which are themselves RVs. Similarly, by virtue of the message passing algorithm, the messages $r_{ji}(b)$ are RVs.

Consider now the form of $q_{ij}(0)$ which, given our new notation and the independence assumption, we may express as (see Figure 36.5)

$$q_{ij}(0) = \Pr(c_i = 0 \mid y_i, S_i, M_c(\sim j))$$
$$= (1 - P_i) \Pr(S_i \mid c_i = 0, y_i, M_c(\sim j))/\Pr(S_i)$$
$$= K_{ij}(1 - P_i) \prod_{j' \in C_i \backslash j} r_{j'i}(0) \tag{36.2}$$

where we used Bayes' rule twice to obtain the second line and the independence assumption to obtain the third line. Similarly,

$$q_{ij}(1) = K_{ij} P_i \prod_{j' \in C_i \backslash j} r_{j'i}(1). \tag{36.3}$$

The constants K_{ij} are chosen to ensure that $q_{ij}(0) + q_{ij}(1) = 1$.

To develop an expression for the $r_{ji}(b)$, we need the following result.

Result

(Gallager [1]) Consider a sequence of M independent binary digits a_i for which $\Pr(a_i = 1) = p_i$. Then the probability that $\{a_i\}_{i=1}^M$ contains an *even* number of 1s is

$$\frac{1}{2} + \frac{1}{2} \prod_{l=i}^{M} (1 - 2p_i) \tag{36.4}$$

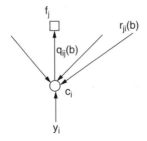

FIGURE 36.5 Illustration of message passing half-iteration for the computation of $q_{ij}(b)$.

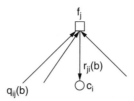

FIGURE 36.6   Illustration of message passing half-iteration for the computation of $r_{ji}(b)$.

Proof 36.1   Induction on M.   □

In view of this result, together with the correspondence $p_i \leftrightarrow q_{ij}(1)$, we have (see Figure 36.6)

$$r_{ji}(0) = \frac{1}{2} + \frac{1}{2} \prod_{i' \in V_j \setminus i} (1 - 2q_{i'j}(1)) \tag{36.5}$$

since, when $c_i = 0$, the bits $\{c_{i'} : i' \in V_j \setminus i\}$ must contain an even number of 1s in order for check equation f_j to be satisfied. Clearly,

$$r_{ji}(1) = 1 - r_{ji}(0) \tag{36.6}$$

The MPA for the computation of the APPs is initialized by setting $q_{ij}(b) = \Pr(c_i = b \mid y_i)$ for all i, j for which $h_{ij} = 1$ (i.e., $q_{ij}(1) = P_i$ and $q_{ij}(0) = 1 - P_i$). Here, y_i represents the channel symbol that was actually received (i.e., it is a not variable here). We consider the following special cases.

BEC. In this case, $y_i \in \{0, 1, E\}$ where E is the erasure symbol, and we define $\delta = \Pr(y_i = E \mid c_i = b)$ to be the erasure probability. Then it is easy to see that

$$\Pr(c_i = b \mid y_i) = \begin{cases} 1 & \text{when } y_i = b \\ 0 & \text{when } y_i = b^c \\ 1/2 & \text{when } y_i = E \end{cases}$$

where b^c represents the complement of b.

BSC. In this case, $y_i \in \{0, 1\}$ and we define $\varepsilon = \Pr(y_i = b^c \mid c_i = b)$ to be the error probability. Then it is obvious that

$$\Pr(c_i = b \mid y_i) = \begin{cases} 1 - \varepsilon & \text{when } y_i = b \\ \varepsilon & \text{when } y_i = b^c \end{cases}$$

BI-AWGNC. We first let $x_i = 1 - 2c_i$ be the ith transmitted binary value; note $x_i = +1(-1)$ when $c_i = 0(1)$. We shall use x_i and c_i interchangeably hereafter. Then the ith received sample is $y_i = x_i + n_i$ where the n_i are independent and normally distributed as $\eta(0, \sigma^2)$. Then it is easy to show that

$$\Pr(x_i = x \mid y_i) = [1 + \exp(-2yx/\sigma^2)]^{-1}$$

where $x \in \{\pm 1\}$.

36.4.2.1   Summary of the Probability-Domain SPA Decoder

1. For $i = 0, 1, \ldots, n - 1$, set $P_i = \Pr(c_i = 1 \mid y_i)$ where y_i is the ith received channel symbol. Then set $q_{ij}(0) = 1 - P_i$ and $q_{ij}(1) = P_i$ for all i, j for which $h_{ij} = 1$.
2. Update $\{r_{ji}(b)\}$ using Equation 36.5 and Equation 36.6.

3. Update $\{q_{ji}(b)\}$ using Equation 36.2 and Equation 36.3. Solve for the constants K_{ij}.
4. For $i = 0, 1, \ldots, n - 1$, compute

$$Q_i(0) = K_i(1 - P_i) \prod_{j \in C_i} r_{ji}(0) \tag{36.7}$$

and

$$Q_i(1) = K_i P_i \prod_{j \in C_i} r_{ji}(1) \tag{36.8}$$

where the constants K_i are chosen to ensure that $Q_i(0) + Q_i(1) = 1$.
5. For $i = 0, 1, \ldots, n - 1$, set

$$\hat{c}_i = \begin{cases} 1, & \text{if } Q_i(1) > Q_i(0) \\ 0, & \text{else} \end{cases}$$

If $\hat{c} H^T = \mathbf{0}$ or the number of iterations equals the maximum limit, stop; else, go to Step 2.

Remark 36.1 This algorithm has been presented for pedagogical clarity, but may be adjusted to optimize the number of computations. For example, Step 4 may be computed before Step 3 and Step 3 may be replaced with the simple division $q_{ij}(b) = K_{ij} Q_i(b)/r_{ji}(b)$. We note also that, for good codes, this algorithm is able to detect an uncorrected codeword with near-unity probability (Step 5), unlike turbo codes [4].

36.4.3 Log-Domain SPA Decoder

As with the probability-domain Viterbi and BCJR algorithms, the probability-domain SPA suffers because multiplications are involved (additions are less costly to implement) and many multiplications of probabilities are involved which could become numerically unstable. Thus, as with the Viterbi and BCJR algorithms, a log-domain version of the SPA is to be preferred. To do so, we first define the following LLRs:

$$L(c_i) = \log \left(\frac{\Pr(c_i = 0 \mid y_i)}{\Pr(c_i = 1 \mid y_i)} \right)$$

$$L(r_{ji}) = \log \left(\frac{r_{ji}(0)}{r_{ji}(1)} \right)$$

$$L(q_{ij}) = \log \left(\frac{q_{ij}(0)}{q_{ij}(1)} \right)$$

$$L(Q_i) = \log \left(\frac{Q_i(0)}{Q_i(1)} \right)$$

The initialization steps for the three channels under consideration thus become:

$$L(q_{ij}) = L(c_i) = \begin{cases} +\infty, & y_i = 0 \\ -\infty, & y_i = 1 \\ 0, & y_i = E \end{cases} \quad \text{(BEC)} \tag{36.9}$$

$$L(q_{ij}) = L(c_i) = (-1)^{y_i} \log \left(\frac{1 - \varepsilon}{\varepsilon} \right) \quad \text{(BSC)}$$

$$L(q_{ij}) = L(c_i) = 2y_i/\sigma^2 \quad \text{(BI-AWGNC)}$$

For Step 1, we first replace $r_{ji}(0)$ with $1 - r_{ji}(1)$ in Equation 36.6 and rearrange it to obtain

$$1 - 2r_{ji}(1) = \prod_{i' \in V_j \setminus i} (1 - 2q_{i'j}(1))$$

Now using the fact that $\tanh[\frac{1}{2} \log(p_0/p_1)] = p_0 - p_1 = 1 - 2p_1$, we may rewrite the equation above as

$$\tanh \left(\frac{1}{2} L(r_{ji}) \right) = \prod_{i' \in V_j \setminus i} \tanh \left(\frac{1}{2} L(q_{i'j}) \right) \tag{36.10}$$

The problem with these expressions is that we are still left with a product and the complex tanh function. We can remedy this as follows [1]. First, factor $L(q_{ij})$ into its sign and magnitude:

$$L(q_{ij}) = \alpha_{ij} \beta_{ij}$$
$$\alpha_{ij} = sign[L(q_{ij})]$$
$$\beta_{ij} = |L(q_{ij})|$$

so that Equation 36.10 may be rewritten as

$$\tanh \left(\frac{1}{2} L(r_{ji}) \right) = \prod_{i' \in V_j \setminus i} \alpha_{i'j} \cdot \prod_{i' \in V_j \setminus i} \tanh \left(\frac{1}{2} \beta_{i'j} \right)$$

We then have

$$L(r_{ji}) = \prod_{i'} \alpha_{i'j} \cdot 2 \tanh^{-1} \left(\prod_{i'} \tanh \left(\frac{1}{2} \beta_{i'j} \right) \right)$$

$$= \prod_{i'} \alpha_{i'j} \cdot 2 \tanh^{-1} \log^{-1} \log \left(\prod_{i'} \tanh \left(\frac{1}{2} \beta_{i'j} \right) \right)$$

$$= \prod_{i'} \alpha_{i'j} \cdot 2 \tanh^{-1} \log^{-1} \sum_{i'} \log \left(\tanh \left(\frac{1}{2} \beta_{i'j} \right) \right)$$

$$= \prod_{i' \in V_j \setminus i} \alpha_{i'j} \cdot \phi \left(\sum_{i' \in V_j \setminus i} \phi \left(\beta_{i'j} \right) \right) \tag{36.11}$$

where we have defined

$$\phi(x) = -\log[\tanh(x/2)] = \log \left(\frac{e^x + 1}{e^x - 1} \right)$$

and used the fact that $\phi^{-1}(x) = \phi(x)$ when $x > 0$. The function is fairly well behaved, as shown in Figure 36.7, and so may be implemented by a look-up table.

For Step 2, we simply divide Equation 36.2 by Equation 36.3 and take the logarithm of both sides to obtain

$$L(q_{ij}) = L(c_i) + \sum_{j' \in C_i \setminus j} L(r_{j'i}) \tag{36.12}$$

Step 3 is similarly modified so that

$$L(Q_i) = L(c_i) + \sum_{j \in C_i} L(r_{ji}) \tag{36.13}$$

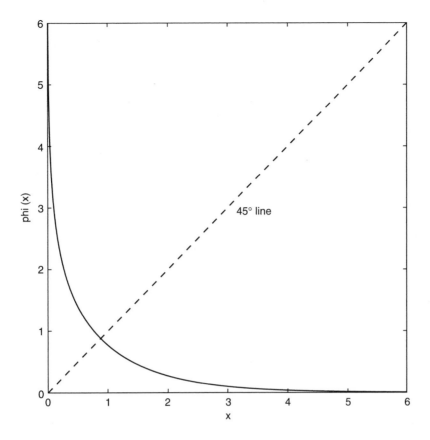

FIGURE 36.7   Plot of the $\phi(x)$ function.

36.4.3.1   Summary of the Log-Domain SPA Decoder

1. For $i = 0, 1, \ldots, n-1$, initialize $L(q_{ij})$ according to Equation 36.9 for all i, j for which $h_{ij} = 1$.
2. Update $\{L(r_{ji})\}$ using Equation 36.11.
3. Update $\{L(q_{ij})\}$ using Equation 36.12.
4. Update $\{L(Q_i)\}$ using Equation 36.13.
5. For $i = 0, 1, \ldots, n-1$, set

$$\hat{c}_i = \begin{cases} 1, & \text{if } L(Q_i) < 0 \\ 0, & \text{else} \end{cases}$$

If $\hat{c} H^T = \mathbf{0}$ or the number of iterations equals the maximum limit, stop; else, go to Step 2.

Remark 36.2   This algorithm can be simplified further for the BEC and BSC channels since the initial LLRs (see Equation 36.9) are ternary in the first case and binary in the second case. See the discussion of the min-sum decoder below.

36.4.4   Reduced Complexity Decoders

It should be clear from the above that the log-domain SPA algorithm has lower complexity and is more numerically stable than the probability-domain SPA algorithm. We now present decoders of lower complexity which often suffer only a little in terms of performance. The degradation is typically on the order of 0.5 dB, but is a function of the code and the channel as demonstrated in the example below.

36.4.4.1   The Min-Sum Decoder [32]

Consider the update Equation 36.11 for $L(r_{ji})$ in the log-domain decoder. Note from the shape of $\phi(x)$ that the term corresponding to the smallest β_{ij} in the summation dominates, so that

$$\phi\left(\sum_{i'}\phi(\beta_{i'j})\right) \simeq \phi\left(\phi\left(\min_{i'}\beta_{i'j}\right)\right)$$

$$= \min_{i'\in V_j\setminus i}\beta_{i'j}$$

Thus, the min-sum algorithm is simply the log-domain SPA with Step 1 replaced by

$$L(r_{ji}) = \prod_{i'\in V_j\setminus i}\alpha_{i'j}\cdot\min_{i'\in V_j\setminus i}\beta_{i'j}$$

It can also be shown that, in the BI-AWGNC case, the initialization $L(q_{ij}) = 2y_i/\sigma^2$ may be replaced by $L(q_{ij}) = y_i$ when the min-sum algorithm is employed. The advantage, of course, is that knowledge of the noise power σ^2 is unnecessary in this case.

36.4.4.2   The Min-Sum-Plus-Correction-Factor Decoder [34]

Note that we can write

$$r_{ji}(b) = \Pr\left(\sum_{i'\in V_j\setminus i}c_{i'} = b\,(\mathrm{mod}\ 2)\mid M_v(\tilde{i})\right)$$

so that $L(r_{ji})$ corresponds to the (conditional) LLR of a sum of binary RVs. Now consider the following general result.

Result

(Hagenauer et al. [33]) Consider two independent binary RVs a_1 and a_2 with probabilities $\Pr(a_i = b) = p_{ib}$, $b \in \{0, 1\}$, and LLR's $L_i = L(a_i) = \log(p_{i0}/p_{i1})$. The LLR of the binary sum $A_2 = a_1 \oplus a_2$, defined as

$$L(A_2) = \log\left[\frac{\Pr(A_2 = 0)}{\Pr(A_2 = 1)}\right]$$

is given by

$$L(A_2) = \log\left(\frac{1 + e^{L_1+L_2}}{e^{L_1} + e^{L_2}}\right) \tag{36.14}$$

Proof 36.2

$$L(A_2) = \log\left(\frac{\Pr(a_1 \oplus a_2 = 0)}{\Pr(a_1 \oplus a_2 = 1)}\right)$$

$$= \log\left(\frac{p_{10}p_{20} + p_{11}p_{21}}{p_{10}p_{21} + p_{11}p_{20}}\right)$$

$$= \log\left(\frac{1 + \frac{p_{10}}{p_{11}}\frac{p_{20}}{p_{21}}}{\frac{p_{10}}{p_{11}} + \frac{p_{20}}{p_{21}}}\right)$$

$$= \log\left(\frac{1 + e^{L_1+L_2}}{e^{L_1} + e^{L_2}}\right)$$

□

If more than two independent binary RVs are involved (as is the case for $r_{ji}(b)$), then the LLR of the sum of these RVs may be computed by repeated application of this result. For example, the LLR of $A_3 = a_1 \oplus a_2 \oplus a_3$ may be computed via $A_3 = A_2 \oplus a_3$ and

$$L(A_3) = \log\left(\frac{1 + e^{L(A_2)+L_3}}{e^{L(A_2)} + e^{L_3}}\right)$$

As a shorthand [33], we will write $L_1 \boxplus L_2$ to denote the computation of $L(A_2) = L(a_1 \oplus a_2)$ from L_1 and L_2; and $L_1 \boxplus L_2 \boxplus L_3$ to denote the computation of $L(A_3) = L(a_1 \oplus a_2 \oplus a_3)$ from L_1, L_2, and L_3; and so on for more variables.

We now define, for any pair of real numbers x, y,

$$\max{}^*(x, y) = \log(e^x + e^y) \tag{36.15}$$

which may be shown [35] to be

$$\max{}^*(x, y) = \max(x, y) + \log\left(1 + e^{-|x-y|}\right) \tag{36.16}$$

Observe from Equation 36.14 and Equation 36.15 that we may write

$$L_1 \boxplus L_2 = \max{}^*(0, L_1 + L_2) - \max{}^*(L_1, L_2) \tag{36.17}$$

so that

$$L_1 \boxplus L_2 = \max(0, L_1 + L_2) - \max(L_1, L_2) + s(L_1, L_2) \tag{36.18}$$

where $s(x, y)$ is a so-called *correction term* given by

$$s(x, y) = \log\left(1 + e^{-|x+y|}\right) - \log\left(1 + e^{-|x-y|}\right)$$

It can be shown [34] that

$$\max(0, L_1 + L_2) - \max(L_1, L_2) = \text{sign}(L_1)\text{sign}(L_2)\min(|L_1|, |L_2|)$$

so that

$$L_1 \boxplus L_2 = \text{sign}(L_1)\text{sign}(L_2)\min(|L_1|, |L_2|) + s(L_1, L_2) \tag{36.19}$$

which may be approximated as

$$L_1 \boxplus L_2 \simeq \text{sign}(L_1)\text{sign}(L_2)\min(|L_1|, |L_2|) \tag{36.20}$$

since $|s(x, y)| \leq 0.693$.

Returning to the computation of $L(r_{ji})$ which we said corresponds to the LLR of a sum of binary RVs, under the independence assumption, we may write

$$L(r_{ji}) = L(q_{1j}) \boxplus \ldots L(q_{i-1,j}) \boxplus L(q_{i+1,j}) \boxplus \ldots L(q_{nj})$$

This expression may be computed via repeated application of Result 2 together with Equation 36.18 (see [34] for an efficient way of doing this). Observe that, if the approximation Equation 36.20 is employed, we have the min-sum algorithm. At the cost of slightly greater complexity, performance can be enhanced by using a slightly tighter approximation, by substituting $\tilde{s}(x, y)$ for $s(x, y)$ in Equation 36.19 where [34]

$$\tilde{s}(x, y) = \begin{cases} c & \text{if } |x + y| < 2 \quad \text{and} \quad |x - y| > 2|x + y| \\ -c & \text{if } |x - y| < 2 \quad \text{and} \quad |x + y| > 2|x - y| \\ 0 & \text{otherwise} \end{cases}$$

and where c on the order of 0.5 is typical.

Example 36.2

We consider two regular Euclidean geometry (EG) LDPC codes and their performance with the three decoders discussed above: the (log-)SPA, the min sum, and the min sum with a correction factor (which we denote by min-sum-c, with c set to 0.5). The first code is a cyclic rate-0.82 (4095, 3367) EG LPDC code with minimum distance bound $d_{min} \geq 65$. Because the code is cyclic, it may be implemented using a shift-register circuit. The H matrix for this code is a 4095 $\times$ 4095 circulant matrix with row and column weight 64. The second code is a (generalized) quasi-cyclic rate-0.875 (8176, 7156) EG LDPC code. Because it is quasi-cyclic, encoding may be performed using several shift-register circuits. The H matrix for this code is 1022 $\times$ 8176 and has column weight 4 and row weight 32. It comprises eight 511 $\times$ 2044 circulant submatrices, each with column weight 2 and row weight 8. These codes are being considered for CCSDS standardization for application to satellite communications [37] (see also [16, 17, 19]).

The performance of these codes for the three decoders on a BI-AWGNC is presented in Figure 36.8 and Figure 36.9. We make the following observations (all measurements are with respect to a BER of 10^{-5}).

- The length-4095 code is 1.6 dB away from the rate-0.82 BI-AWGNC capacity limit. The length-8176 code is closer to its capacity limit, only 0.9 dB away. Regular LDPC codes of these lengths might be designed which are a bit closer to their respective capacity limits, but one would have to resort to irregular LDPC codes to realize substantial gains. Of course, an irregular LDPC code would in general require a more complex encoder.

- For the length-4095 code, the loss relative to the SPA decoder suffered by the min-sum decoder is 1.1 dB and the loss suffered by the min-sum-c decoder is 0.5 dB. For the length-8176 code, these losses are 0.3 and 0.01 dB, respectively. We attribute the large losses for the length-4095 code to the fact that its decoder relies on an H matrix with weight-64 rows. Thus, a minimum among 64 small nonnegative numbers is taken at each check node in the code's Tanner graph, so that a value near zero is usually produced and passed to a neighboring variable node.

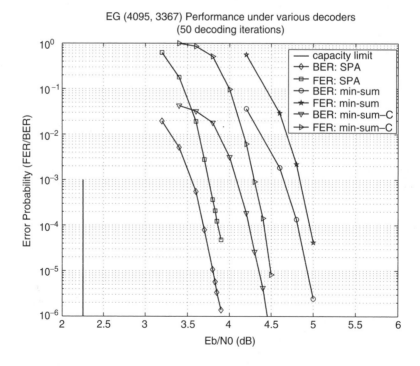

FIGURE 36.8 Performance of a cyclic EG(4095, 3367) LDPC code on a binary-input AWGN channel and three decoding algorithms.

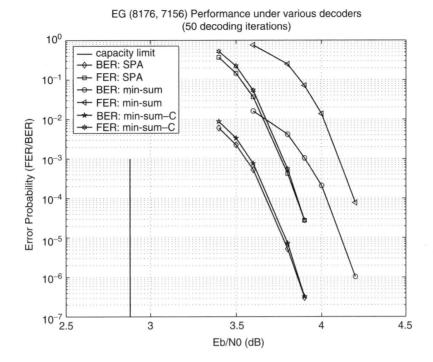

FIGURE 36.9 Performance of a quasi-cyclic EG(8176, 7156) LDPC code on a binary-input AWGN channel and three decoding algorithms.

36.5 Concluding Remarks

Low-density parity-check codes are being studied for a large variety of applications, much as turbo codes, trellis codes, and other codes were when they were first introduced to the coding community. As indicated above, LDPC codes are capable of near-capacity performance while admitting an implementable decoder. Among the advantages LDPC codes have over turbo codes are: (1) They allow a parallelizable decoder; (2) They are more amenable to high code rates; (3) They generally possess a lower error-rate floor (for the same length and rate); (4) They possess superior performance in bursts (due to interference, fading, and so on), (5) They require no interleavers in the encoder and decoder; and (6) A single LDPC code can be universally good over a collection of channels [36]. Among their disadvantages are: (1) Most LDPC codes have somewhat complex encoders, (2) The connectivity among the decoder component processors can be large and unwieldy; and (3) Turbo codes can often perform better when the code length is short. It is easy to find in the literature many of the applications being explored for LPDC codes, including application to deep space and satellite communications, wireless (single and multi-antenna) communications, magnetic storage, and internet packet transmission.

Acknowledgments

The author would like to thank the authors of [19] and [31] for kindly providing preprints. He would also like to thank Yifei Zhang of the University of Arizona for producing Figure 36.7 and Figure 36.8.

References

[1] R. Gallager, Low-density parity-check codes, *IRE Trans. Inf. Theory*, pp. 21–28, January 1962.
[2] R.M. Tanner, A recursive approach to low complexity codes, *IEEE Trans. Inf. Theory*, pp. 533–547, September 1981.

[3] D. MacKay and R. Neal, Good codes based on very sparse matrices, in *Cryptography and Coding, 5th IMA Conf.*, C. Boyd, Ed., *Lecture Notes in Computer Science*, pp. 100–111, Springer, Berlin, Germany, 1995.

[4] D. MacKay, Good error correcting codes based on very sparse matrices, *IEEE Trans. Inf. Theory*, pp. 399–431, March 1999.

[5] N. Alon and M. Luby, A linear time erasure-resilient code with nearly optimal recovery, *IEEE Trans. Inf. Theory*, pp. 1732–1736, November 1996.

[6] T.J. Richardson and R. Urbanke, Efficient encoding of low-density parity-check codes, *IEEE Trans. Inf. Theory*, vol. 47, pp. 638–656, February 2001.

[7] T. Richardson, A. Shokrollahi, and R. Urbanke, Design of capacity-approaching irregular low-density parity-check codes, *IEEE Trans. Inf. Theory*, vol. 47, pp. 619–637, February 2001.

[8] M. Luby, M. Mitzenmacher, M. Shokrollahi, and D. Spielman, Improved low-density parity check codes using irregular graphs, *IEEE Trans. Inf. Theory*, pp. 585–598, February 2001.

[9] V. Sorokine, F.R. Kschischang, and S. Pasupathy, Gallager codes for CDMA applications: Part I, *IEEE Trans. Commn.*, pp. 1660–1668, October 2000 and Gallager codes for CDMA applications: Part II, *IEEE Trans. Commn.*, pp. 1818–1828, November 2000.

[10] D.J.C. Mackay, http://wol.ra.phy. cam.ac.uk/mackay.

[11] S.Y. Chung, G.D. Forney, T.J. Richardson, and R. Urbanke, On the design of low-density parity-check codes within 0.0045 dB of the Shannon limit, *IEEE Commn. Lett.*, vol 5, pp. 58–60, February 2001.

[12] M. Chiani and A. Ventura, Design and performance evaluation of some high-rate irregular low-density parity-check codes," *Proc. 20001 IEEE GlobeCom Conf.*, pp. 990–994, November 2001.

[13] M. Yang, Y. Li, and W.E. Ryan, Design of efficiently-encodable moderate-length high-rate irregular LDPC codes, *Proc. 40th Annual Allerton Conference on Communication, Control, and Computing, Champaign, IL.*, pp. 1415–1424, October 2002.

[14] M. Yang and W.E. Ryan, Lowering the error rate floors of moderate-length high-rate LDPC codes, *Proc. 2003 Int. Symp. on Inf. Theory*, June–July, 2003.

[15] M. Yang, W.E. Ryan, and Y. Li, Design of efficiently encodable moderate-length high-rate LDPC codes, *IEEE Trans. Commn.*, pp. 564–571, April 2004.

[16] Y. Kou, S. Lin, and M. Fossorier, Low-density parity-check codes based on finite geometries: a rediscovery and new results, *IEEE Trans. Inf. Theory*, vol. 47, pp. 2711–2736, November 2001.

[17] R. Lucas, M. Fossorier, Y. Kou, and S. Lin, Iterative decoding of one-step majority-logic decodable codes based on belief propagation, *IEEE Trans. Commn.*, pp. 931–937, June 2000.

[18] S. Lin and D. Costello, *Error-Control Coding: Fundamentals and Applications*, Prentice Hall, New York, 1983.

[19] H. Tang, J. Xu, S. Lin, and K. Abdel-Ghaffar, Codes on finite geometries, submitted to *IEEE Trans. Inf. Theory*, 2002, in review.

[20] D. Divsalar, H. Jin, and R. McEliece, Coding theorems for turbo-like codes, *Proc. 36th Annual Allerton Conf. on Commn., Control, and Computing*ce, pp. 201–210, September 1998.

[21] H. Jin, A. Khandekar, and R. McEliece, Irregular repeat-accumulate codes, *Proc. 2nd. Int. Symp. on Turbo Codes and Related Topics*, Brest, France, pp. 1–8, September 4, 2000.

[22] R. Narayanaswami, Coded Modulation with Low-Density Parity-Check Codes, M.S. thesis, Texas A&M University, 2001, chap. 7.

[23] J. Fan, *Constrained Coding and Soft Iterative Decoding for Storage*, Ph.D. dissertation, Stanford University, December 1999. (See also Fan's Kluwer monograph by the same title.)

[24] J. Fan, Array codes as low-density parity-check codes, *2nd Int. Symp. on Turbo Codes and Related Topics*, Brest, France, pp. 543–546, September 2000.

[25] E. Eleftheriou and S. Ölçer, Low-density parity-check codes for digital subscriber lines, *Proc. 2002 Int. Conf. on Commn.*, pp. 1752–1757, April–May, 2002.

[26] D. MacKay and M. Davey, Evaluation of Gallager codes for short block length and high rate applications, in *Codes, Systems, and Graphical Models: Volume 123 of IMA Volumes in Mathematics and its Applications*, pp. 113–130, Spring-Verlag, New York, 2000.

[27] B. Vasic, Structured iteratively decodable codes based on Steiner systems and their application to magnetic recording, *Proc. 2001 IEEE GlobeCom Conf.*, pp. 2954–2958, November 2001.

[28] B. Vasic, Combinatorial constructions of low-density parity-check codes for iterative decoding, *Proc. 2002 IEEE Int. Symp. Inf. Theory*, p. 312, June/July 2002.

[29] S. Johnson and S. Weller, Construction of low-density parity-check codes from Kirkman triple systems, *Proc. 2001 IEEE GlobeCom Conf.*, pp. 970–974, November 2001.

[30] J. Pearl, *Probabilistic Reasoning in Intelligent Systems*, Morgan Kaufmann, San Mateo, CA, 1988.

[31] H. Xiao and A. Banihashemi, Message-passing schedules for decoding LDPC codes, submitted to *IEEE Trans. Commn.*, May 2003, in revision.

[32] N. Wiberg, Codes and Decoding on General Graphs, Ph.D. dissertation, U. Linköping, Sweden, 1996.

[33] J. Hagenauer, E. Offer, and L. Papke, Iterative decoding of binary block and convolutional codes, *IEEE Trans. Inf Theory*, pp. 429–445, March 1996.

[34] X-Y. Hu, E. Eleftherious, D-M. Arnold, and A. Dholakia, Efficient implementation of the sum-product algorithm for decoding LDPC codes, *Proc. 2001 IEEE GlobeCom Conf.*, pp. 1036–1036E, November 2001.

[35] A. Viterbi, An intuitive justification and a simplified implementation of the MAP decoder for convolutional codes, *IEEE JSAC*, pp. 260–264, February 1998.

[36] C. Jones, A. Matache, T. Tian, J. Villasenor, and R. Wesel, The universality of LDPC codes on wireless channels, *Proc. 2003 IEEE Milcom conf.*, pp. 440–445 October 2003.

[37] W. Fong, White paper for LDPC codes for CCSDS channel Blue Book, NASA GSFC White Paper submitted to the Panel 1B CCSDS Standards Committee, October 2002.

37

Concatenated Single-Parity Check Codes for High-Density Digital Recording Systems

Jing Li
Lehigh University
Bethlehem, PA

Krishna R. Narayanan
Texas A&M University
College Park, TX

Erozan M. Kurtas
Seagate Technology
Pittsburgh, PA

Travis R. Oenning
IBM Corporation
Rochester, MN

37.1 Introduction 37-1
37.2 System Model 37-3
      System Model • Concatenated SPC Codes • Iterative
      Decoding and Equalization
37.3 Analysis of Distance Spectrum 37-5
      Distance Properties of TPC/SPC Coded PR Systems
      • Distance Properties of turbo/SPC Coded PR Systems
37.4 Thresholds Analysis using Density Evolution 37-9
      Introduction to Density Evolution and Gaussian
      Approximation • Problem Formulation • Message Flow
      Within the Inner MAP Decoder • Message Flow Within the
      Outer Code • Thresholds
37.5 Simulation Results 37-13
37.6 Conclusion 37-16

37.1 Introduction

The breakthrough of turbo codes [1] and low-density parity check codes [11, 37] has revolutionized the coding research with several new concepts, among which code concatenation and iterative decoding are being actively exploited both for wireless communications and future digital magnetic recording systems. After being precoded, filtered and equalized to some simple partial response (PR) target, the magnetic recording channel appears much like an intersymbol interference (ISI) channel to an outer code and, hence, many of the techniques used in concatenated coding can be adopted. In particular, the observation that an ISI channel can be effectively viewed as a rate-1 convolutional code leads to the natural format of a serial concatenated system where the ISI channel is considered as the inner code and the error correction code (ECC) as the outer code. With reasonable complexity, iterative decoding and equalization (IDE), or turbo equalization, can be used to obtain good performance gains.

It has been shown by many that turbo codes (based on punctured recursive systematic convolutional codes) and LDPC codes can provide 4-5 dB of coding gain over uncoded systems at bit error rates (BER) of around 10^{-5} or 10^{-6} [2–5], [6–10]. Since magnetic recording applications require BERs in the order of 10^{-15} and since turbo and LDPC codes (as well as many other codes) would have already hit an error floor well before they reach this point, significant coding gains cannot be guaranteed. An effective remedy is to wrap a t-error correcting Reed-Solomon error correction code (RS-ECC) on top of these codes to clear up the residue errors. In this set-up, it is important that the output of the LDPC or turbo decoder (or other ECC codes) do not contain more than t byte errors that may cause the RS-ECC decoder to fail.

Due to the high decoding complexity of turbo codes, current research focuses on lower complexity solutions that are easily implementable in hardware. It has been recognized that very simple (almost useless) component codes can result in an overall powerful code when properly concatenated using random interleavers. Of particular interest is single-parity check codes due to their intrinsic high rates and the availability of a simple and optimal soft decoder. Researchers have applied single-parity check codes directly onto recording channels [32] as well as using them to construct a variety of well-performing codes including LDPC codes, block turbo codes (BTC) (e.g., [14, 20, 29]), multiple-branch turbo codes [17, 18], and concatenated tree (CT) codes [42].

In this work, we propose and study concatenated SPC codes for use in PR recording channels. Two constructions are considered: the first in the form of turbo codes where two branches of SPC codes are concatenated using a random interleaver, and the other in the form of product codes where arrays of SPC codewords are lined up in a multidimensional fashion. We denote the former as turbo/SPC scheme and the latter TPC/SPC scheme, that is, single parity check turbo product codes[1] [7, 19].

We undertake a comprehensive study of the properties of concatenated SPC codes and their applicability to PR recording channels, and highlight new results of this application. The fact that SPC codes are intrinsically very weak codes and that turbo/SPC and TPC/SPC codes are generally worse than LDPC codes on additive white Gaussian noise (AWGN) channels tend to indicate their inferiority on PR channels also. However, our studies show that, when several codewords are combined to form a larger effective block size and when the PR channel is properly precoded, concatenated SPC codes are capable of achieving large coding gains just as LDPC codes, but with less complexity[2].

Theoretical analysis is first conducted to explain the well-known spectral thinning phenomenon and to quantify the interleaving gain. Next, we compute the thresholds for iterative decoding of concatenated SPC codes using density evolution (DE) [21–24] to cast insight into the asymptotic performance limit of such schemes. Finally, we study the distribution of errors at the output of the decoder (i.e., at the input to the RS-ECC decoder) and show that concatenated SPC codes have an error distribution more favorable than that of LDPC codes in the presence of an outer RS-ECC code.

The paper proceeds as follows. Section 37.2 presents the PR system model, followed by a brief introduction to the turbo/SPC and TPC/SPC concatenated schemes. Section 37.3 conducts distance spectrum analysis and Section 37.4 computes the iterative thresholds. Section 37.5 presents the simulation results, which include bit error rate and bit/byte error statistics, and compares them with that of (random) LDPC codes. Finally, Section 37.6 concludes the paper with a discussion of future work in this area.

---

[1] Single parity check turbo product codes are also termed as array codes in [20] and hyper codes in [29].

[2] Randomly constructed LDPC codes typically have quadratic encoding complexity in the length N of the code ($O(N^2)$). It has been shown that several greedy algorithms can be applied to triangulate matrices (preprocessing) to reduce encoding complexity, but the complexity of preprocessing may be as much as $O(N^{3/2})$ [16]. Further, with the exception of a few LDPC codes that have cyclic or quasi-cyclic structures (mostly from combinatorial or geometric designs, see for examples [15]), large memory is generally required (for storage of generator and/or parity check matrices), which is a big concern in hardware implementation.

37.2   System Model

37.2.1   System Model

The digital recording channel is modeled as a L-tap ISI channel, where the channel impulse response is assumed to be a partial response polynomial with AWGN:

$$r_k = \sum_{i=0}^{L-1} h_i x_{k-i} + n_k \tag{37.1}$$

where x_i, y_i, h_i and n_i are transmitted symbols, received symbols, channel response and AWGN, respectively. Specifically, we focus on the PR4 channel with channel response polynomial $H(D) = 1 - D^2$ and the EPR4 channel with $H(D) = (1 - D^2)(1 + D)$.

The overall system model is presented in Figure 37.1. Conforming to the set-up of the current and immediate future recording systems, we use a Reed-Solomon code as the outer wrap (referred to as RS-ECC) to clear up the residue errors. The data is first encoded using an RS-ECC and then the concatenated SPC code (either the turbo/SPC scheme or the TPC/SPC scheme) which we refer to as the outer code. The precoded ISI channel is treated as the inner code. The random interleaver between the outer and inner code works to break the correlation among neighboring bits, to eliminate error bursts, and (in conjunction with the precoder) to improve the overall distance spectrum by mapping low-weight error events to high-weight ones (the so-called spectrum thinning phenomenon). We call this interleaver the "channel interleaver" to differentiate it with the "code interleaver" in the outer code (i.e., the concatenated SPC code). The outer code, the inner code and the channel interleaver together form a serial concatenated system, where turbo equalization can be exploited to iterate soft information between the outer decoder and the inner decoder/equalizer, and then feed the hard decision decoding to the RS-ECC code.

37.2.2   Concatenated SPC Codes

Magnetic recording systems require a high code rate R since for recording systems code rate loss (in dB) is in the order of $10 \log_{10}(R^2)$ rather than $10 \log_{10}(R)$ as in an AWGN channel [27]. Hence, we consider only high-rate codes, that is, 2-branch turbo/SPC codes or 2-dimensional TPC/SPC codes.

Turbo/SPC codes: The turbo/SPC code comprises two parallel branches of $(t + 1, t)$ SPC codes and a random interleaver (Figure 37.2(a)). It should be noted that the interleaver between the parallel branches is of size $K = Pt$, that is, P blocks of data bits are taken and interleaved together before being encoded

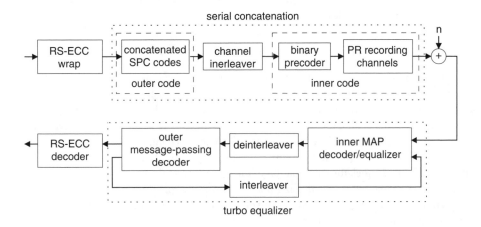

FIGURE 37.1   System model of concatenated SPC codes on (precoded) PR channels.

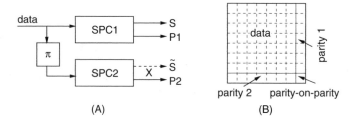

FIGURE 37.2   (a) Structure of a turbo/SPC code. (b) Structure of a 2-dimensional TPC/SPC code.

by the second branch. As we will show later, such combination of P blocks is essential in achieving the interleaving gain. Since the (interleaved) systematic bits are not transmitted in the second branch, the turbo/SPC code we consider is of parameters $(N, K, R) = (P(t + 2), Pt, t/(t + 2))$.

TPC/SPC codes: Another way of concatenating SPC codes is to take the form of product codes. A product code is formed by multidimensional arrays of codewords from linear block codes, where the codeword length, the user data block size, and the minimum distance of the overall code are the product of those of the component codes, respectively [12–14]. When the decoder takes an iterative (i.e. turbo) approach as is typically the case, product codes are also called turbo product codes (TPC) or block turbo codes. We consider 2-dimensional TPC/SPC codes whose code structure is illustrated in Figure 37.2(b). It is worth mentioning that, for the same reason that we combine SPC codewords in the turbo/SPC scheme, several blocks of TPC/SPC codewords will also be combined to form a larger effective block size before passing to the channel interleaver (see Figure 37.1). The resulting TPC/SPC scheme has parameters $(N, K, R) = (Q(t + 1)^2, Qt^2, t^2/(t + 1)^2)$ [7, 9, 19] and it may also appear in the paper as a $Q(t + 1, t)^2$ TPC/SPC code.

It is interesting to note that the two schemes, although differ in lengths and rates, bear structural similarities. Depending on whether it has "parity-on-parity" bits or not, a TPC/SPC code can be either viewed as a serial or parallel concatenation where a linear block interleaver is used between the component codes. It is also worth noting that, from the graph-based point of view, both concatenated SPC schemes can be viewed as special types of LDPC codes where each SPC codeword satisfies a check. The equivalent LDPC code for the turbo/SPC scheme has a variable-node degree profile $\lambda(x) = \frac{1}{t+1} + \frac{t}{t+1}x$ and a check-node degree profile $\rho(x) = x^t$, that for the TPC/SPC code has a variable node degree profile $\lambda(x) = x$ and a check node degree profile $\rho(x) = x^t$, where the coefficient of the term x^j denotes the portion of edges connecting to nodes of degree $j + 1$ [22]. The encoding of these codes involves adding a single parity check bit in each SPC code, and hence is extremely simple and a dense generator matrix need not be explicitly stored as for random LDPC codes.

37.2.3   Iterative Decoding and Equalization

Since an overall maximum likelihood (ML) decoding and equalization of the coded PR system is pro-hibitively complex, the practical yet effective way, is to use turbo equalization to iterate soft extrinsic information (in the form of log-likelihood ratio or LLR) between the inner decoder/equalizer and the outer decoder. From the coding theory, we know that the inner code needs to be recursive in order for a serial concatenated system to achieve interleaving gain. The PR channel alone can be viewed as a rate-1 binary input real-valued output nonrecursive convolutional code. When a binary precoder is properly placed before the PR channel, the combination of the precoder and the channel becomes a recursive code that can be jointly equalized/decoded using maximum a posteriori probability (APP) decoding or the BCJR algorithm [25]. As long as the memory size of the precoder is not larger than that of the PR channel, no additional decoding complexity is introduced by the precoder.

The decoder of the outer concatenated SPC code performs the soft-in soft-out (SISO) message-passing decoding. based on component SPC codes. We use l to denote the number of local iterations within the outer decoder (before LLRs are passed to the inner decoder for the next round of global iteration).

TABLE 37.1 Decoding Complexity in Terms of Number of Operations per Coded bit per Iteration

Operations	Turbo/SPC	TPC/SPC	LDPC	BCJR
Addition	$d\left(3 - \frac{1}{t+1}\right)$	$d\left(3 - \frac{1}{t+1}\right)$	$4s$	$15 \cdot 2^m + 9$
Min/max				$5 \cdot 2^m - 2$
Table lookup	$2d$	$2d$	$2s$	$5 \cdot 2^m - 2$

d-number of branches of the turbo/SPC code or dimensionality of the TPC/SPC code; t-parameter of the component SPC code $(t + 1, t)$; s-average column weight of the LDPC code; m-memory size of the convolutional code.

For a $(t + 1, t)$ SPC code with $a_1 \oplus a_2 \oplus \cdots \oplus a_{t+1} = 0$, where $\oplus$ denotes the binary addition, the soft extrinsic information of bit a_i can be obtained from all other bits in the SPC codes as

$$L_e(a_i) = 2 \tanh^{-1}\left(\prod_{j=1, \, j \neq i}^{t+1} \tanh \frac{L_a(a_j)}{2} \right) \quad i = 1, 2, \ldots, t + 1 \quad (37.2)$$

where L_a stands for the a priori LLR value and L_e the extrinsic LLR value, and $\tanh(\cdot)$ is the hyper tangent function. This is known as the tanh rule decoding of SPC codes.

Since turbo/SPC and TPC/SPC schemes can be viewed as special types of LDPC codes, the same LDPC decoder can be used where all checks are decoded and updated simultaneously. However, below we discuss a slightly different approach, where checks are grouped into two groups (corresponding to upper/lower branches or to row/column component codes) and updated alternatively. This "serial" update is expected to converge a little faster than the "parallel" update.

The turbo/SPC decoder operates much like that of the turbo (convolutional) code. Each branch-decoder takes the extrinsic LLRs from the other branch, adds it to that from the equivalent channel (i.e., the inner code) to form the a priori LLR, and uses Equation 37.2 to compute new extrinsic LLRs to be passed to the other branch for use in the subsequent decoding iterations. After l (local) iterations, the extrinsic LLRs from both branches are added together and passed to the inner code. The TPC/SPC code can be iteratively decoded in a similar manner. Detailed discussion and pseudo-code of the message-passing TPC/SPC decoder can be found in [7, 9].

Table 37.1 compares the complexity of turbo/SPC, TPC/SPC, conventional LDPC (for comparison) and log-domain BCJR decoders (for the channel decoder/equalizer). It is assumed that multiplications are converted to additions in the log-domain [28], and that $\log(\tanh(\frac{x}{2}))$ and its reverse function $2 \tanh^{-1}(e^x)$ are implemented through table lookups. We see that the decoding algorithms for the turbo/SPC and TPC/SPC code require about 2/3 the complexity and about 1/3 the storage space of that for a regular column-weight-3 LDPC code in each decoding iteration.

37.3 Analysis of Distance Spectrum

Despite their similarities in decoding algorithms, turbo/SPC and TPC/SPC codes have very different distance spectra from random LDPC codes. For example, the minimum distance of a randomly-constructed regular LDPC code of column weight ≥ 3 will, with high probability, increase linearly with block length N (for large N). However, the minimum distance of a 2-dimensional TPC/SPC code is fixed to $d_m = 4$ regardless of block lengths and that of turbo/SPC codes in the worst-case scenario is $d_m = 2$. This partially explains why they perform noticeably worse than conventional LDPC codes on AWGN channels. However, on (precoded) ISI channels where iterative decoding and equalization is deployed, these codes perform on par with LDPC codes due to the interleaving gain [7].

Before proceeding to discussion, we first note the results of [30, 31] which stat that, for a serially concatenated system with a recursive inner code, there exists an SNR threshold γ such that for any

$E_b/N_o \geq \gamma$, the asymptotic word error rate (WER) is upper bounded by:

$$P_w^{UB} = O\left(N^{-\left\lfloor \frac{d_m^{(o)}-1}{2} \right\rfloor}\right) \tag{37.3}$$

where N is the interleaver size and $d_m^{(o)}$ the minimum distance of the outer code. This suggests that (i) the outer code needs to have $d_m \geq 3$ in order to achieve the interleaving gain, and (ii) interleaving gain is attainable when $d_m^{(o)} \geq 3$ (and the inner code is recursive). While these serve as a quick and general guideline, caution needs to exercised in extrapolating the result. Specifically, we will discuss two interesting results with the concatenated SPC system[3]. First, although a TPC/SPC system has $d_m^{(o)} = 4 \geq 3$, interleaving gain is not obtainable unless multiple TPC/SPC codewords are combined before passing to the channel interleaver (i.e., $Q \geq 2$). Second, although the ensemble of turbo/SPC systems has $d_m^{(o)} = 2 < 3$ (worst case), interleaving gain still exists so long as multiple SPC codewords are grouped in each branch ($P \geq 2$). Hence, while the concatenated SPC codes we consider in this work are not "good" codes by themselves[4], the combination of these codes with a precoded ISI channel become "good" codes due to spectral thinning.

37.3.1 Distance Properties of TPC/SPC Coded PR Systems

For ease of proposition, we use the precoded PR4 channel with channel response $H(D) = \frac{1-D^2}{1 \oplus D^2}$ as an example to quantify the interleaving gain of the TPC/SPC system. The result can be generalized to any ISI channel.

We first introduce some notations that will be used here and in the analysis of turbo/SPC systems. Let $A_{w,h}$ denote the input-output weight enumerator (IOWE) of a binary code, which enumerates the number of codewords with input Hamming weight w and output Hamming weight h, and A_h the output weight enumerator (OWE) where $A_h = \sum_w A_{w,h}$. Similar notations, A_{w,d_E} and A_{d_E}, are used with the ISI channel, where d_E stands for the (output) Euclidean distance.

We consider an TPC/SPC system with parameters $(N, K, R) = (Q(t+1)^2, Qt^2, t^2/(t+1)^2)$, where each "codeword" of (effective) length N is composed of Q TPC/SPC codewords of length $(t+1)^2$ each. To argue that TPC/SPC codes are capable of interleaving gain on precoded PR channels, we need to show that the average OWE of the TPC/SPC system, $A_{d_E}^{TPC/SPC}$, decreases with interleaver size for small d_E, thus providing a reduction in error rate. Using the ideas in [30, 32, 33] and assuming a uniform interleaver, we have the average OWE given by:

$$A_{d_E}^{TPS/SPC} = \sum_{\substack{l=4 \\ l \text{ even}}}^{N} \frac{A_l^o \times A_{l,d_E}^i}{\binom{N}{l}} \tag{37.4}$$

The sum of the series starts with $l = 4$ because TPC/SPC code has $d_m = 4$ and only even terms are considered because all codewords of the TPC/SPC code have even weights. Since a precoded PR channel is in general a nonregular convolutional code, the all-zeros sequence cannot be treated as the reference codeword. Since it is computationally prohibitive to perform an exact analysis of the full compound of error events pertaining to the inner code (i.e., $A_{l,d_E}^{(i)}$) [35], we take a similar approach developed in [7, 32], and assume that the input to the inner code are independent and identically distributed (i.i.d.) sequences of $\{0,1\}^N$. It then follows that the equivalent trellis corresponding to odd/even bits of the precoded PR4 channel takes the form as in Figure 37.3.

---

[3]We use "system" to denote the concatenation of an outer code with the (precoded) ISI channel, and "scheme" or "code" to denote the outer code only.

[4]A "good" code as defined in [37] is a code that possesses a SNR threshold such that when the channel is better than this threshold, the code can achieve arbitrarily small error probability as the block size goes to infinity.

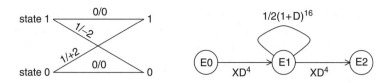

FIGURE 37.3   The equivalent trellis for even/odd bits of a precoded PR4 channel with response $H(D) = \frac{1-D^2}{1\oplus D^2}$. Left: trellis; right: state diagram.

Following similar derivations as in [33], the average error enumerating function with a uniform interleaver is computed as

$$
\begin{aligned}
T(X,Y) &= \frac{X^2 Y^8}{1 - \frac{1}{2}(1+Y^{16})} \\
&= X^2 Y^8 \left[\left(1 + \frac{1}{2} + \frac{1}{2^2} + \cdots \right) + Y^{16} \left(\frac{1}{2} + \frac{2}{2^2} + \frac{3}{2^3} + \cdots \right) + O(Y^{32}) \right] \\
&= X^2 Y^8 [2 + 2Y^{16} + O(Y^{32})]
\end{aligned}
\tag{37.5}
$$

where the exponent of X is the Hamming distance between the error sequence and the i.i.d. reference sequence at the input, and the exponent of Y is the corresponding squared Euclidean distance at the output. The fractional terms in the branch weight enumerator such as $1/2(1 + Y^{16})$ (Figure 37.3) are a direct consequence of the assumption that the input corresponding to that branch can be a 0 or 1 with equal probability [7, 37].

Several things can be observed from the transfer function (Equation 37.5). First, the i.i.d. input error sequence (i.e., without an outer TPC/SPC code) always has input weight 2 for the precoded PR4 channel, since every term in Equation 37.5 corresponds to X^2. Second, any input error sequence of the form $1 + D^{2j}$ results in an error event, and $j = 1$ results in the minimum Euclidean distance (among all such error events) which is 8. Third, every finite weight codeword is the concatenation of k weight-2 input error events for some integer k [7]. For large value of N, let $T_N(X^{2k}, Y)$ denote the truncated weight enumerator truncated to length N, where each error event results from the joint effect of k input error sequences each of weight 2. Hence,

$$
T_N(X^{2k}, Y) \propto \binom{N}{k} X^{2k} Y^{8k} [2 + 2Y^{16} + O(Y^{32})]^k
\tag{37.6}
$$

since there are approximately $\binom{N}{k}$ ways to arrange k error events in a block of length N. For the least nonzero l in the TPC/SPC system, namely $l = 4$ (i.e., $k = 2$ in Equation 37.6), we have $A^i_{l=4, d_E=4} \approx 4\binom{N}{2}$, and $A^o_{l=4} \approx Q[\sqrt{\frac{N/Q}{2}}]^2$, (there are $[(\sqrt{\frac{N/Q}{2}})]^2$ ways in which we can arrange a block of weight 4 within a single TPC/SPC block and there are Q blocks in a codeword of length N.) Substituting them into (Equation 37.4) and using the approximation $\binom{N}{n} \approx N^n/n!$ for large N, we have

$$
A^c_{d_E=4} \propto \frac{\frac{N^2}{Q} \cdot N^2}{N^4} \propto Q^{-1}
\tag{37.7}
$$

Clearly, Equation 37.7 states that the reduction in word error rate is proportional to the number of blocks Q of the TPC/SPC that form a codeword, rather than N as what would be expected from Equation 37.3. This implies that a single TPC/SPC codeword ($Q = 1$), however long it is, does not get additional gain due to the length of the code. This is important for finite-length block sizes, since the achievable interleaving gain is limited to the number of blocks of the outer TPC/SPC codewords that are combined and interleaved. Although we have only discussed the error event corresponding to the least nonzero l (i.e., $l = 4$), it can be shown that for other values of small l, similar arguments hold.

To handle a general ISI channel, it is convenient to consider the precoder separately from the channel. That is, we treat the concatenation of the TPC/SPC and the precoder as a code whose codewords are passed through the ISI channel. Since the interleaving gain is dependent only on the recursive nature of the inner code, an interleaving gain will result regardless of the type of ISI channel. This idea will be further explained in the analysis of turbo/SPC systems.

37.3.2   Distance Properties of turbo/SPC Coded PR Systems

In this subsection, we show that although the minimum distance of the ensemble turbo/SPC codes (with random interleavers[5]) is only 2, an interleaving gain still exists. Here we take a different approach from what we did with the TPC/SPC system, namely, we separate the precoder from the ISI channel, and argue that the combination of the binary precoder (i.e., recursive inner code) and the turbo/SPC code (outer code) results in an interleaving gain, irrespective of what ISI channel follows. This approach obviates the trouble of handling nonregular inner code, and the all-zeros sequence can therefore be used as the reference (for the serial concatenation of the turbo/SPC code and the precoder). For simplicity, we take the precoder $1/(1 \oplus D)$ as an example.

First, it is easy to show, as can also be inferred from Equation 37.3, that outer codewords of weight 3 or more will lead to an interleaver gain. Hence we focus the investigation on weight-2 outer codewords only, and show that the number of these codewords vanishes as P increases, where P is the number of SPC blocks combined in each branch.

Let $A_{w,h}^{(j)}$, $j = 1, 2$, denote the IOWE of the j_{th} SPC branch code that is parallelly concatenated in the outer code. The IOWE of the turbo/SPC codewords, $A_{w,h}^{(o)}$, averaged over the code ensemble is given by:

$$A_{w,h}^{(o)} = \sum_{h_1} \frac{A_{w,h_1}^{(1)} A_{w,h-h_1}^{(2)}}{\binom{K}{w}} \tag{37.8}$$

where $K = Pt$ is the input sequence length.

In each branch where P blocks of $(t + 1, t)$ SPC codewords are combined, the IOWE function is given by (assuming even parity check):

$$A^{SPC}(w, h) = \left(1 + \binom{t}{1} wh^2 + \binom{t}{2} w^2 h^2 + \binom{t}{3} w^3 h^4 + \cdots + \binom{t}{t} w^t h^{2\lceil t/2 \rceil}\right)^P$$

$$= \left(\sum_{j=0}^{t} \binom{t}{j} w^j h^{2\lceil j/2 \rceil}\right)^P \tag{37.9}$$

where the coefficient of the term $w^u h^v$ denotes the number of codewords with input weight u and output weight v. From the coefficients of Equation 37.9, we can obtain the IOWEs of the first SPC branch code: $A_{u,v}^{(1)} = A_{u,v}^{SPC}$. For the second SPC branch, since only parity bits are transmitted, $A_{u,v}^{(2)} = A_{u,v+u}^{(1)}$.

With a little computation, it is easy to show that the number of weight-2 outer codewords is given by

$$A_{h=2}^{(o)} = \sum_{w} A_{w,h=2}^{(o)} = P \binom{t}{2} \left(\frac{P\binom{t}{2}}{\binom{Pt}{2}}\right) = O(t^2) \tag{37.10}$$

where the last equation assumes a large P (i.e., a large block size). Equation 37.10 indicates that the number of weight-2 outer codewords is a function of a single parameter t, which is related only to the rate of SPC codes and not the block length. Now consider the serial concatenation of the outer codewords with the

---

[5]If S-random interleavers are used such that bits within S distance are mapped to at least S distance apart, then the outer codewords are guaranteed to have a minimum distance of at least 3 as long as $S \geq t$.

inner recursive precoder. The ensemble average OWE of the turbo/SPC system, $A_h^{\text{turbo/SPC}}$, can thus be computed as:

$$A_h^{\text{turbo/SPC}} = \sum_{h'} A_{h'}^{(o)} \frac{A_{h',h}^{(i)}}{\binom{N}{h'}} = \sum_{h'} \sum_{w} A_{w,h'}^{(o)} \frac{A_{h',h}^{1/(1 \oplus D)}}{\binom{N}{h'}} \tag{37.11}$$

where the IOWE of the $1/(1 \oplus D)$ precoder is given by [31, 40]:

$$A_{w,h}^{1/(1 \oplus D)} = \binom{N-h}{\lfloor w/2 \rfloor} \binom{h-1}{\lceil w/2 \rceil - 1} \tag{37.12}$$

Substituting Equation 37.10 and Equation 37.12 in Equation 37.11, we get the number of weight-s codewords (for small-s) produced by weight-2 outer codewords (i.e., $h' = 2$), denoted as $A_{h=s}^{\text{turbo/SPC:2}}$, is given as

$$A_{h=s}^{\text{turbo/SPC:2}} = \frac{(t-1)^2}{2} \frac{N-s}{\binom{N}{2}} = O(tP^{-1}) \tag{37.13}$$

where $N = P(t + 2)$ is the length of the overall codeword (or the channel interleaver size). This indicates that the number of small weight s codewords of the overall system due to weight-2 outer codewords (which are caused by weight-2 input data sequences) vanishes as P increases. When the input weight is greater than 2, the outer codeword always has weight greater than 2 and, hence, an interleaving gain can be guaranteed. Hence, an interleaving gain exists for turbo/SPC systems and it is proportional to P.

It is also worth noting that the system model we considered above, namely, the combination of an outer concatenated SPC code and an inner $1/(1 \oplus D)$ code, essentially forms a product accumulate code, which is shown to be a class of high-performance, low-complexity, high-rate "good" codes [38, 40]. Hence, depending on different view-stands, the coded ISI systems we discuss in this work can either be viewed as concatenated SPC codes on precoded ISI channels or product accumulate codes on non-precoded ISI channels.

37.4 Thresholds Analysis using Density Evolution

37.4.1 Introduction to Density Evolution and Gaussian Approximation

Distance spectrum analysis shows that both turbo/SPC and TPC/SPC systems possess good distance spectra and that interleaving gain is achievable for both systems. However, since the analysis assumes a maximum likelihood decoder which differs from the practical iterative decoder, it would be more convincing to account for the suboptimality of iterative decoding in the analysis. Such analysis is possible using the recently developed technique of density evolution [21–24].

Introduced for the analysis of LDPC codes, density evolution unveiled an SNR threshold effect for LDPC codes in that error rate goes to zero when the channel is better than this threshold and that the error rate is bounded away from zero otherwise [21–24]. This threshold clearly marks the performance limit we can expect with the existing suboptimal decoder. Interestingly, the same threshold effect also presents in turbo/SPC and TPC/SPC systems and, hence, the same DE treatment can be used for capacity analysis. However, certain modifications are required.

Due to the space limitation, we go through the critical points in the application of density evolution to turbo/SPC and TPC/SPC systems. Detailed discussion on density evolution and its application to a variety of systems can be found in [21–24] [7].

37.4.2 Problem Formulation

Consider a unified architecture where the precoded PR channel is modeled as an inner rate-1 recursive convolutional code, and the outer code is either a turbo/SPC or TPC/SPC code. A turbo equalizer iterates extrinsic LLR information, denoted as $L_i^{(q)}(a_j)$ and $L_o^{(q)}(a_j)$, between the inner and outer decoders, where subscript i and o denote the quantities associated with the inner and outer codes, respectively. Assuming infinite length and perfect interleaving, the LLR messages are approximated as i.i.d. random variables. During the q_{th} iteration, the outer message-passing decoder generates extrinsic information on the j_{th} coded bit a_j, denoted by $L_o^{(q)}(a_j)$, and passes it to the inner decoder. The inner MAP decoder then uses this extrinsic information (treat as a priori) with the received signal and generates extrinsic information, $L_i^{(q+1)}(a_j)$.

The idea in density evolution is to examine the probability density function (pdf) of $L_o^{(q)}(a_j)$ during the q_{th} iteration, denoted by $f_{L_o^{(q)}}(x \mid a_j)$. For infinite lengths and perfect interleaving, these random variables are i.i.d. Hence, we drop the dependence on j. If the sign of $(\text{sign}(a) \cdot L_o^{(q)}(a))$ is positive, then the decoding algorithm has converged to the correct codeword. The probability that $\text{sign}(a) \cdot L_o^{(q)}(a) < 0$ is

$$
\begin{aligned}
&\Pr(\text{sign}(a) \cdot L_o^{(q)}(a) < 0) \\
&= \Pr(a = +1) \int_{-\infty}^{0} f_{L_o^{(q)}}(x \mid a = +1)\, dx + \Pr(a = -1) \int_{0}^{\infty} f_{L_o^{(q)}}(x \mid a = -1)\, dx \quad (37.14)
\end{aligned}
$$

The key is to find the critical SNR value, or the threshold of the system, such that

$$
\gamma = \inf_{SNR} \left\{ SNR : \lim_{q \to \infty} \lim_{N \to \infty} \Pr\left(\text{sign}(a) \cdot L_o^{(q)}(a) < 0 \right) \to 0 \right\} \quad (37.15)
$$

where N denotes the block size.

Since it is quite difficult to analytically evaluate $f_{L_o^{(q)}}(x)$ for all q, simplification can be made by approximating $f_{L_o^{(q)}}(x)$ to be Gaussian (or Gaussian mixture). This is what is used by Wiberg et al. [34], Chung et al. [22], El Gamal et al. [36], and many others to analyze concatenated codes. Further, Richardson and Urbanke have shown that, for binary input, output symmetric channels, a consistency condition is preserved under density evolution for all messages, such that the pdf's satisfy the condition $f_{L_o^{(q)}}(x) = f_{L_o^{(q)}}(-x) \cdot e^x$ [21]. Imposing this constraint to the approximate Gaussian densities at every step leads to $(\sigma_o^{(q)})^2 = 2m_o^{(q)}$, that is, the variance of the message density equals twice the mean. Under i.i.d. and Gaussian assumptions, the mean of the messages, $m_o^{(q)}$ then serves as the sufficient statistics of the message density. The problem thus reduces to:

$$
\gamma = \inf_{SNR} \left\{ SNR : \lim_{q \to \infty} \lim_{N \to \infty} \text{sign}(a) m_o^{(q)}(x \mid a) \to \infty \right\} \quad (37.16)
$$

To solve the problem formulated in Equation 37.16, we need to examine the message flow within the outer decoder, the inner decoder as well as between the two. In general, we need to evaluate $m_o^{(q)}$ as a function of $m_i^{(q)}$ and vice versa.

37.4.3 Message Flow Within the Inner MAP Decoder

Since it is not straight-forward to derive $m_i^{(q+1)}$ as a function of $m_o^{(q)}$ for the inner MAP decoder, Monte Carlo simulation is used to determine a relationship between $m_i^{(q+1)}$ and $m_o^{(q)}$. This process is denoted as

$$
m_i^{(q)} = \mathcal{F}(m_o^{(q-1)}) \quad (37.17)
$$

where the mean of the message $m_i^{(q)}$ is evaluated at the output of the inner MAP decoder given the input a priori information is independent and Gaussian with mean $\pm m_o^{(q-1)}$ and variance $2m_o^{(q-1)}$.[6] Detailed discussion of Monte Carlo simulation technique for computing γ_i can be found, for example, in [26].

---

[6]Again, due to the nonlinearity of the ISI channel, a sequence of i.i.d. bits are used as the transmitted data.

37.4.4   Message Flow Within the Outer Code

Below we discuss how to compute $m_o^{(q)}$ as a function of $m_i^{(q)}$ for outer concatenated SPC codes. Since both turbo/SPC and TPC/SPC codes can be viewed as special types of LDPC codes, we start with LDPC codes to pinpoint the key steps, and then move onto turbo/SPC and TPC/SPC codes.

Irregular LDPC codes: Both turbo equalization and LDPC decoding are iterative processes. Let us use superscript (q, l) to denote quantities during the qth (global) iteration of turbo equalization and l_{th} (local) iteration within the LDPC decoder. For irregular LDPC codes with bit-node and check-node degree profiles $\lambda(x) = \sum_k \lambda_k x^{k-1}$ and $\rho(x) = \sum_j \rho_j x^{j-1}$, the code rate is given by $R = 1 - \frac{\sum_k \lambda_k/k}{\sum_j \rho_j/j}$. Message flow on the code graph is a two-way procedure that composes of bit updates and check updates, which correspond to summation in the real domain and the so-called check-sum operation or *tanh* rule, respectively. [9, 22, 23]. After L local iterations of message exchange, the message passed over to the inner MAP decoder is the LLR of the bit in the L_{th} iteration after subtracting $L_i^{(q)}$ which was obtained from the inner code and was used as a priori information.

Under the Gaussian assumption, we are interested in tracking the means of $L_b^{(q,l)}$ and $L_c^{(q,l)}$, denoted as $m_b^{(q,l)}$ and $m_c^{(q,l)}$, where subscripts b and c refer to quantities pertaining to bit-nodes and check-nodes, respectively. To handle the irregularity of LDPC codes, we further introduce notations $m_{b,k}^{(q,l)}$ and $m_{c,j}^{(q,l)}$ to denoted message mean associated with bit-nodes of degree k and check-nodes of degree j, respectively. Treating extrinsic information as independent, the means of the extrinsic information at each local iteration l are shown to be [22]

$$\text{bit} - \text{to} - \text{check}: \quad m_{b,k}^{(q,l)} = m_i^{(q)} + (k-1) \cdot m_c^{(q,l-1)} \tag{37.18}$$

$$\text{check} - \text{to} - \text{bit}: \quad m_{c,j}^{(q,l)} = \psi^{-1}\left(1 - \left[1 - \sum_k \lambda_k \psi\left(m_{b,k}^{(q,l)}\right)\right]^{j-1}\right) \tag{37.19}$$

$$m_c^{(q,l)} = \sum_j \rho_j m_{c,j}^{(q,l)} \tag{37.20}$$

where $\psi(x)$ is the expected value of $1 - \tanh(\frac{u}{2})$ where u follows a Gaussian distribution with mean x and variance $2x$. Specifically, $\psi(x)$ is given by:

$$\psi(x) = \begin{cases} 1 - \frac{1}{\sqrt{4\pi x}} \int_{-\infty}^{\infty} \tanh\left(\frac{u}{2}\right) e^{-\frac{(u-x)^2}{4x}} du, & x > 0 \\ 1, & x = 0 \end{cases} \tag{37.21}$$

$\psi(x)$ is continuous and monotonically decreasing on $[0, \infty)$ with $\psi(0) = 1$ and $\psi(\infty) = 0$. The initial condition is $m_b^{(q,0)} = m_c^{(q,0)} = 0$. When x is large (corresponding to low error probability), $\psi(x)$ is shown to be proportional to the error probability [22]. The above derivation is essentially an extension of Chung et al.'s work [22] to the case of turbo equalization. For detailed discussions, readers are directed to [21–24] and the references therein.

After L LDPC decoding iterations, the messages passed from the outer LDPC code to the inner MAP decoder/equalizer follows a mixed Gaussian distribution where λ_k' fraction of bits follow a Gaussian distribution with mean value

$$m_{o,k}^{(q)} = k m_{b,k}^{(q,L)}, \quad k = 1, 2, \ldots \tag{37.22}$$

and variance $2m_{o,k}^{(q)}$. Here λ_k' denotes the percentage of bits that have degree k, and is given by $\lambda_k' = (\lambda_k/k)/(\sum_k \lambda_k/k)$. Hence, we can describe what is passed from the outer decoder to the inner decoder in the qth turbo equalization as $m_o^{(q)} \sim \{<\lambda_k', m_{o,k}^{(q)}>, k = 1, 2, \ldots\}$. This will in turn be used by the inner decoder to generate $m_i^{(q+1)}$.

After q global iterations between the outer and inner decoder (where each iteration contains L local iterations within the LDPC decoder), the threshold can be evaluated as:

$$\gamma_{LDPC} = \inf_{SNR} \left\{ SNR : \lim_{q \to \infty} \cdot m_c^{(q,L)} \to \infty \right\} \qquad (37.23)$$

It is instructive to note that the value of L has a slight impact on the asymptotic threshold, but a quite noticeable impact for finite-length finite-complexity performance. Specifically, it has been shown in [7] that for a given complexity constraint, an optimal value of L can be computed using density evolution for the concatenated system to reach the best performance.

Turbo/SPC codes: Using the degree profiles $\lambda(x) = \frac{1}{t+1} + \frac{t}{t+1}x$ and $\rho(x) = x^t$, the message flow within a Turbo/SPC decoder can be tracked following exactly the same steps as we described above. An alternative procedure stems naturally from the decoding algorithm where checks corresponding to different branches take turns to update. As expected, such a serial scheduling expedites the convergence and improves the performance for finite-length systems, but has little impact on the asymptotic thresholds.

TPC/SPC codes: Although TPC/SPC codes can be viewed as a special type of LDPC codes, the above DE procedure cannot be applied directly. This is because the underlying assumption of a cycle-free code graph does not hold for TPC/SPC codes. Since any rectangular error pattern (or their combination) in the 2-dimensional TPC/SPC bit-array results in a loop in the code graph [7], there always exist cycles of length $4(k+1)$ irrespective of block sizes (k can be any positive integer). Hence, messages being passed in the code graph are not always independent (loop-free operation), and adjustment needs to be made before it is applicable to TPC/SPC codes.

Notice that when the number of local iterations within the TPC/SPC code is restricted to be small, the density evolution method would have operated on cycle-free subgraphs of TPC/SPC codes. In other words, the messages exchanged along each step are statistically independent as long as the cycles have not "closed." Here, we restrict the number of local iterations within TPC/SPC codes to be one row update and one column update, since any more updates in either direction will either pass information to its source or pass duplicate information to the same node, which are unacceptable. On the other side, due to the perfect random interleaver, infinite number of turbo iterations can be performed between the inner and outer decoders if the messages within the outer TPC/SPC code are reset to zero in every new turbo iteration.

For an exact threshold, the density evolution procedure should, in addition to avoiding looping messages, also ensure completeness in the sense that every bit should have fully utilized all the messages (through dependencies) from all the checks. Unfortunately, one row update followed by one column update is not sufficient to exploit all the information contained in all bits and checks [7]. Hence, the resulting threshold is a lower bound[7].

37.4.5   Thresholds

The lower bound on the threshold for TPC/SPC codes, and the threshold for turbo/SPC codes are plotted in Figure 37.4 for precoded PR4 and EPR4 channels. For comparison, we also evaluate LDPC codes on nonprecoded PR channels[8] We consider regular LDPC codes with column weight 3, since regular LDPC codes are shown to be slightly advantageous over irregular ones at short block sizes and high rates such as what will be used in data storage applications [39]. It can be seen that the thresholds of concatenated SPC systems (and their lower bound) are within a few tenths a dB from that of LDPC systems. This indicates they have comparable performance asymptotically.

---

[7]By lower bound, we mean that the exact thresholds of TPC/SPC system should be better than this. Put another way, for a given dB, the achievable code rate (bandwidth efficiency) could be higher, or equivalently, for a given rate, the required SNR could be smaller.

[8]It has been shown that randomly constructed LDPC codes perform better on ISI channels that are not precoded. Hence, the comparison represents the best in both cases and is thus fair.

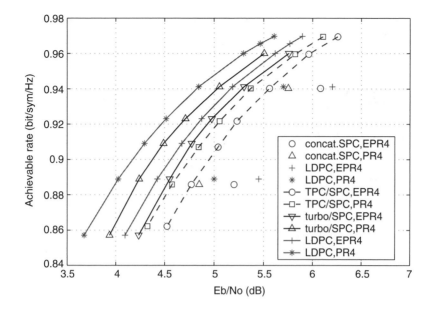

FIGURE 37.4 Iterative thresholds computed using density evolution with a Gaussian approximation. (Solid lines: bounds of LDPC or turbo/SPC systems; dashed lines: lower bounds of TPC/SPC systems.)

Also presented in Figure 37.4 are the corresponding simulation results of a length-4K block evaluated at a BER of 10^{-5}. It is interesting to observe that for practical block sizes concatenated SPC systems may actually (slightly) outperform LDPC systems. Considering the small block size used in the simulation, the $0.5 \sim 1$ dB gap between performance points and threshold curves indicates a good agreement between simulation and analysis.

37.5 Simulation Results

To be applicable to high-density recording systems, the concatenated SPC codes we consider have high rates of 0.89 and 0.94 which are formed from (17,16) and (33,32) TPC codes, respectively. Several codewords are combined and interleaved together to form an effective data block size of (around) 4K bits, the size of a block in a hard-disk drive. At such high code rates and short block sizes, the two concatenated SPC schemes differ very little in lengths, rates and performance. Hence, we do not differentiate them in discussing the simulation results. For comparison, also presented are the results of regular LDPC codes with column weight 3 and similar rates and lengths. In all the simulations presented, 2 iterations are performed within the concatenated SPC decoders and 4 iterations within the LDPC decoder. This makes the decoding complexity of LDPC codes slightly higher than that of concatenated SPC codes, but they are the most efficient scheduling schemes we have found in my experiments.

Bit error rate: Figure 37.5 shows the performance of LDPC codes and concatenated SPC codes over PR4 and EPR4 channels. Gains of 4.4 to 5 dB over uncoded partial response maximum likelihood (PRML) systems are obtained for concatenated SPC codes at a BER of 10^{-5}, which are comparable to those of LDPC codes. All concatenated SPC systems use a binary precoder with polynomial $1/(1 \oplus D^2)$ which is the best-performing precoder for PR4/EPR4 channels, as shown in Figure 37.6. The ISI channels are not precoded in LDPC systems, since as discussed in [26] nonprecoding leads to a better performance than otherwise. This is also confirmed by our simulations. Hence, the comparison is fair as it represents the best cases for both systems.

Error bursts: Although both concatenated SPC and LDPC codes can offer significant coding gains at a BER level of 10^{-7}, it is unclear whether and when they will have an error floor. Therefore, the conventional use of RS-ECC is still necessary to reduce the BER to 10^{-15} as is targeted for recording systems. The RS-ECC

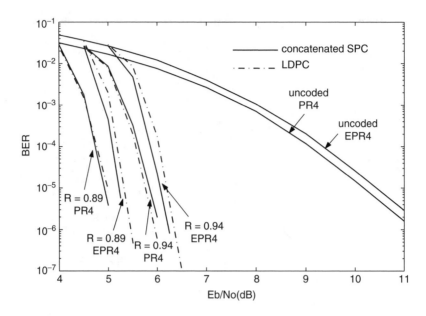

FIGURE 37.5   BER performance of concatenated SPC codes with comparison to LDPC codes on PR recording channels. (Code rate: 0.89 and 0.94; channel model: PR4 and EPR4; evaluated after 10 turbo iterations.

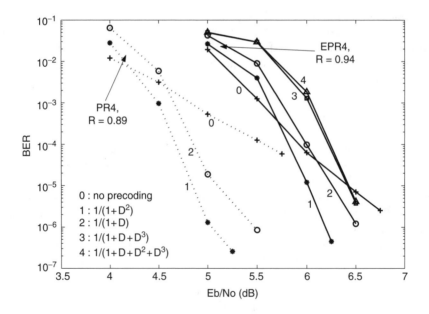

FIGURE 37.6   Effect of binary precoding on the bit error rate of concatenated SPC codes on PR4 and EPR4 channels.

code works on the byte level, capable of correcting up to t byte errors in each data block of size 4K bits or 512 bytes (t is usually around 10 to 20). Hence, the maximum number of uncorrected errors left in each block after the turbo decoding/equalization of concatenated SPC codes and LDPC codes has to be small to guarantee the proper functioning of the RS-ECC code. In other words, block error statistics is crucial to the overall system performance. Unfortunately, this has been largely neglected in most of the previous work.

Figure 37.7 and Figure 37.8 plot the histograms of the number of bit/byte errors for LDPC codes and TPC/SPC codes on EPR4 channels, respectively. The effective block size is 4K and the code rate is 0.94.

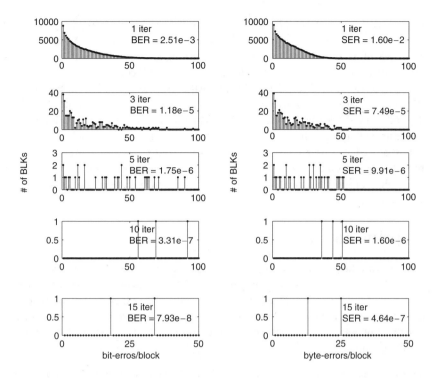

FIGURE 37.7 Statistics of error bursts for LDPC codes on EPR4 channels. (Code rate 0.94, block size $K = 4k$ bits, nonprecoded EPR4 channel, 6.5 dB SNR; X-axis: number of bit/byte errors in each block, Y-axis: number of erroneous blocks; one byte consists of eight consecutive bits; statistics are collected after 1, 3, 5, 10, and 15 iterations over more than 100,000 blocks.)

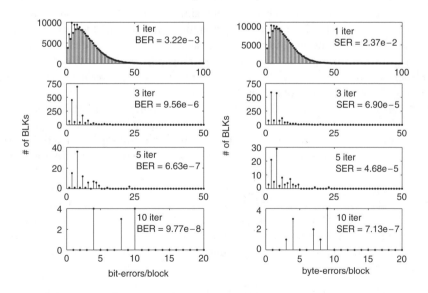

FIGURE 37.8 Statistics of error bursts for concatenated SPC codes on EPR4 channels. (Code rate 0.94, block size $K = 4k$ bits, EPR4 channel with precoder $1/(1 \oplus D^2)$, 6.5 dB SNR; X-axis: number of bit/byte errors in each block, Y-axis: number of erroneous blocks; one byte consists of eight consecutive bits; statistics are collected after 1, 3, 5, and 10 iterations over more than 165,000 blocks.)

The left column in each figure plots bit error histograms and the right byte error histograms, where a byte composes of eight consecutive bits. The statistics are collected over more than $100,000$ blocks of data size 4K bits. At an SNR of 6.5 dB and after the 10_{th} iteration (big loop), the maximum number of symbol errors observed in a single block is less than 10 for concatenated SPC codes (which would be corrected by the RS-ECC code), but around 50 for LDPC codes. If further iterations are allowed, error bursts in LDPC codes are alleviated. Nevertheless a block containing 25 symbol errors is observed after 15 turbo iterations and this may still cause the RS-ECC code to fail. Unless a more powerful RS-ECC is employed, LDPC codes are prone to cause block failure, where all data in that block are presumed lost. It should be noted that although what we have observed suggests that concatenated SPC codes may be more compatible to magnetic recording systems than LDPC codes, the statistics are nonetheless insufficient. Due to the very long simulation time in software, hardware tests over millions or billions of blocks are needed before a convincing argument can be made.

37.6   Conclusion

This paper investigates the potential of applying concatenated single-parity check codes on PR magnetic recording channels, with LDPC codes as a comparison study. Two ways of concatenation are studied, one in the form of parallel turbo codes and the other product codes. While they have very different structures, analysis and simulations show that the two schemes exhibit similar properties in terms of interleaving gain, iterative thresholds and finite-length performances (especially at high rates). Below summarizes the main results of this study.

1. Despite their relatively small minimum distances on AWGN channels, both concatenating schemes are capable of achieving interleaving gain on precoded ISI channels. The key is to have several codeword blocks combined and interleaved together.

2. Density evolution is an effective tool in the analysis of iterative decoding processes by accounting for both the code structure (i.e, codeword space and the mapping of data space to codeword space) and the iterative nature of the decoding algorithms. Thresholds (or their lower bounds) computed using density evolution with a Gaussian approximation indicate these codes have comparable performance asymptotically, and finite-length simulations agree with the analysis.

3. Simulations on high-rate and short-length concatenated SPC systems demonstrate considerable coding gains over uncoded PRML systems. In particular, gains of more than 4.4 dB are observed for length 4K and rate > 0.88 codes on PR4 and EPR4 channels, revealing a performance similar to or slightly better than that of random LDPC codes.

4. In addition to the slightly smaller complexity than that of random LDPC codes, concatenated SPC codes are linear time encodable. Further, they do not require large storage for parity check and generator matrices. The interleaving pattern needs be stored, but algebraic interleavers which can be pseudo-randomly generated "on the fly" can be used to save space in hardware implementation [40, 41].

5. In contrast to LDPC codes whose large error bursts within a block tend to exceed the capacity of RS-ECC wraps, our experiments indicate that the number of errors in each block is typically small with concatenated SPC codes (although there may be more blocks in error). Such a spread error distribution is not only desirable, but also crucial to ensure an overall low error probability in the recording system. However, extensive experiments on massive data needs to be conducted before a firm conclusion can be reached.

To summarize, our study indicates that concatenated SPC codes can be a promising candidate for future high-density magnetic recording systems. However, further experiments need to be carried out on more realistic channel models, like the Lorentzian channel model and (eventually) the real data set collected in the lab.

There are many other interesting problems regarding future digital data recording systems. For example, how to efficiently incorporate the run-length limit (RLL) constraint into the SISO decoding/equalization scheme without affecting much of the overall code rate, performance and complexity? Interleaving is necessary for coded PR systems that use turbo equalizations. How to implement interleaving and deinterleaving schemes that are both space-efficient and time-efficient but still exhibiting good randomness? Most of all, how to achieve a good trade-off among the many competing factors like performance, complexity, delay and cost in a practical setting? These are a few of the many interesting problems that await to be addressed.

References

[1] C. Berrou, A. Glavieux, and P. Thitimasjshima, Near Shannon limit error-correcting coding and decoding: turbo-codes (1), *Proc. IEEE Intl. Conf. Commun.*, pp. 1064–1070, Geneva, Switzerland, May 1993.

[2] T.M. Duman and E. Kurtas, Performance of turbo codes over EPR4 equalized high density magnetic recording channels, *Proc. Global Telecom. Conf.* 1999.

[3] T. Souvignier, A. Friedmann, M. Oberg, P. Siegel, R. Swanson, and J. Wolf, Turbo decoding for PR4: parallel vs. serial concatenation, *Proc. Intl. Conf. Comm.*, pp. 1638–1642, June 1999.

[4] W. Ryan, L. McPheters, and S. McLaughlin, Combined turbo coding and turbo equalization for PR4-equalized Lorentzian channels, *Proc. Conf. Intl. Sci. and Sys.*, 1998.

[5] M. Oberg and P. H. Siegel, Performance analysis of turbo-equalized dicode partial-response channel, *Proc. of 36th Annual Allerton Confer. Commun., Control and Computing*, pp. 230–239, September 1998.

[6] J.L. Fan, A. Friedmann, E. Kurtas, and S. McLaughlin, Low density parity check codes for magnetic recording, *37-th Allerton Conf.*, 1999.

[7] J. Li, K.R. Narayanan, E. Kurtas, and C.N. Georghiades, On the performance of high-rate TPC/SPC and LDPC codes over partial response channels, *IEEE Trans. Commun.*, vol. 50, no. 5, pp. 723–734, May 2002.

[8] T.R. oenning and J. Moon, Low density parity check coding for magnetic recording channels with media noise, *Proc. Intl. Conf. Commun.*, pp. 2189–2193, June 2001.

[9] J. Li, E. Kurtas, K.R. Narayanan, and C.N. Georghiades, On the performance of turbo product codes and LDPC codes over partial response channels, *Proc. Intl. Conf. Commun.*, Helsinki, Finland, June 2001.

[10] H. Song, R.M. Todd, and J.R. Cruz, Performance of low-density parity-check codes on magnetic recording channels, *Proc. 2nd Intl. Symp. on Turbo Codes and Related Topics*, France, September 2000.

[11] R.G. Gallager, Low-density parity check codes, *IRE Trans. Inform. Theory*, pp. 21–28, 1962.

[12] P. Elias, Error-free coding, *IRE Trans. Inform. Theory*, vol. IT-4, pp. 29–37, September 1954.

[13] J. Hagenauer, E. Offer, and L. Papke, Iterative decoding of binary block and convolutional codes, *IEEE Trans. Inform. Theory*, vol. 42, no. 2, March 1996.

[14] R.M. Pyndiah, Near-optimum decoding of product codes: block turbo codes, *IEEE Trans. Commun.*, vol. 46, no. 8, August 1998.

[15] Y. Kou, S. Lin, and M. Fossorier, LDPC codes based on finite geometries, a rediscovery and more, submitted to *IEEE Trans. Inform. Theory*, 1999

[16] T.J. Richardson and R.L. Urbanke, Efficient encoding of low-density parity-check codes, *IEEE Trans. Inform. Theory*, vol. 47, no. 2, pp. 638–656, February 2001.

[17] L. Ping, S. Chan, and K.L. Yeung, Iterative decoding of multidimensional concatenated single parity check codes, *Proc. IEEE Intl. Conf. Commun.*, 1998.

[18] T.R. Oenning and J. Moon, A low-density generator matrix interpretation of parallel concatenated single bit parity codes, *IEEE Trans. Magn.*, pp. 737–741, March 2001.

[19] T. Souvignier, C. Argon, S.W. McLaughlin, and K. Thamvichai, Turbo product codes for partial response channels, *Proc. Intl. Conf. Commun.*, vol. 7, pp. 2184–2188, 2001.

[20] M. Blaum, P.G. Farrell, and H.C.A. van Tilborg, *Array Codes, Handbook of Coding Theory*, V.S. Pless, W.C. Huffman, and R.A. Brualdi, Eds., North-Holland, Amsterdam, pp. 1855–1909, November 1998.

[21] T.J. Richardson, A. Shokrollahi, and R. Urbanke, Design of capacity-approaching irregular low-density parity-check codes, *IEEE Trans. Inform.*, vol. 47, pp. 619–637, February 2001.

[22] S.-Y. Chung, R. Urbanke, and T.J. Richardson, Analysis of sum-product decoding of low-density parity-check codes using a Gaussian approximation, *IEEE Trans. Inform. Theory*, vol. 47, pp. 657–670, February 2001.

[23] J. Li, E. Kurtas, K.R. Narayanan, and C.N. Georghiades, Thresholds for iterative equalization of partial response channels using density evolution, *Proc. Intl. Symp. on Inform. Theory*, Washington D.C., June 2001.

[24] M.G. Luby, M. Mitzenmacher, M.A. Shokrollahi, and D.A. Spielman, Analysis of low density codes and improved designs using irregular graphs, available at http://www.icsi.berkeley.edu/~luby/.

[25] L.R. Bahl, I. Cocke, F. Jelinek, and J. Ravivi, Optimal decoding of linear codes for minimizing symbol error rate, *IEEE Trans. Inform. Theory*, pp. 284–287, March 1974.

[26] K.R. Narayanan, Effect of precoding on the convergence of turbo equalization for partial response channels, *IEEE J. Select. Area. Commun.*, vol 19, pp. 686–698, April 2001.

[27] K. Immink, Coding techniques for the noisy magnetic recording channel: a state-of-the-art report, *IEEE Trans. Commun.*, pp 413–419, May 1989.

[28] P. Robertson, E. Villebrun, and P. Hoeher, A comparison of optimal and sub-optimal MAP decoding algorithms operating in the log domain, *Proc. Intl. Conf. Commun.*, vol. 2, pp. 1009–1013, June 1995.

[29] P.-P. Sauvé, A. Hunt, S. Crozier, and P. Guinand, Hyper-codes: high-performance, low-complexity codes, *Proc. 2nd Intl. Symp. on Turbo Codes and Related Topics*, France, September 2000.

[30] S. Benedetto, D. Divsalar, G. Montorsi, and F. Pollara, Serial concatenation of interleaved codes: Design and performance analysis, *IEEE Trans. Inform. Theory*, vol. 44, pp. 909–926, May 1998.

[31] D. Divsalar, H. Jin, and R. J. McEliece, Coding theorems for "turbo-like" codes, *Proc. Allerton Conf. Commun. and Control*, pp. 201–210, September 1998.

[32] M. Öberg and P. H. Siegel, Parity check codes for partial response channels, *Proc. Global Telecom. Conf.*, vol. 1b, pp. 717–722, December 1999.

[33] L.L. McPheters, S.W. McLaughlin, and K.R. Narayanan, Precoded PRML, serial concatenation, and iterative (turbo) decoding for digital magnetic recording, *IEEE Trans. Magn.*, vol. 35, no. 5, September 1999.

[34] N. Wiberg, Codes and Decoding on General Graphs, Doctoral dissertation, Linköping University, 1996.

[35] J. Li, K.R. Narayanan, and C.N. Georghiades, An efficient algorithm to compute the Euclidean distance spectrum of a general inter-symbol interference channel and its applications, *IEEE Trans. Commun.* preprint (in press).

[36] H. El Gamal, A.R. Hammons, Jr., and E. Geraniotis, Analyzing the turbo decoder using the Gaussian approximation, *Proc. Intl. Symp. Inform. Theory*, 2000.

[37] D.J.C. MacKay, Good error-correcting codes based on very sparse matrices, *IEEE Trans. Info. Theory*, vol. 45, no. 2, March 1999.

[38] J. Li, K.R. Narayanan, and C.N. Georghiades, A Class of linear-complexity, soft-decodable, high-rate, 'good' codes: construction, properties and performance, *Proc. IEEE Intl. Symp. Inform. Theory*, pp. 122–122, Washington D.C., June 2001.

[39] D.J.C. MacKay and M.C. Davey, Evaluation of Gallager codes for short block length and high rate applications, available at http://www.keck.ucsf.edu/~mackay/seagate.ps.gz.

[40] J. Li, K.R. Narayanan, and C.N. Georghiades, Product accumulate codes: a class of capacity-approaching, low complexity codes, *IEEE Trans. Inform. Theory*, January 2004, in preprint (in press).

[41] G.C. Clark, Jr. and J.B. Cain, *Error-Correction Coding for Digital Communications*, Plenum Press, New York, 1981.

[42] L. Ping and K.Y. Wu, Concatenated tree codes: a low-complexity, high-performance approach, *IEEE Trans. Inform. Theory*, vol. 47, pp. 791–799, February 2001.

38

Structured Low-Density Parity-Check Codes

Bane Vasic
University of Arizona
Tucson, AZ

Erozan M. Kurtas

Alexander Kuznetsov
Seagate Technology
Pittsburgh, PA

Olgica Milenkovic
University of Colorado
Boulder, CO

38.1 Introduction **38**-1
38.2 Combinatorial Designs and their Bipartite Graphs... **38**-4
38.3 LDPC Codes on Difference Families **38**-5
38.4 Codes on Projective Planes **38**-7
38.5 Lattice Construction of LDPC Codes **38**-10
  Codes on a Rectangular Subset of an Integer Lattice
38.6 Application in the Partial Response (PR) Channels .. **38**-12
  Channel Model and Signal Generation • Bit Error Rate (BER)
  of the structured LDPC codes. Simulation results
38.7 Conclusion **38**-16

38.1 Introduction

Iterative coding techniques that improve the reliability of input-constrained, intersymbol interference (ISI) channels have recently attracted considerable attention for data storage applications. Inspired by the success of turbo codes [6], several authors have considered iterative decoding architectures for coding schemes comprised of a concatenation of an outer block, convolutional or turbo encoder with a rate one code representing the channel. Such an architecture is equivalent to a serial concatenation of codes [4], with the inner code being the ISI channel. Application of this concatenated scheme in magnetic and optical recording systems is considered in [23, 24].

Theory and practice of the soft iterative detection were facilitated by using the concept of codes on graphs. As shown in the chapter on message passing algorithms, a graph of a code representing the ISI channel is trivial. Therefore, we will focus only on the design and graphical description of the outer code. The prime examples of codes on graphs are low-density parity check codes (LDPC). One of the key results in codes on graphs comes from Frey and Kschischang [36, 45] who observed that iterative decoding algorithms developed for these codes are instances of probability propagation algorithms operating on a graphical model of the code. The belief propagation algorithms and graphical models have been developed in the expert systems literature by Pearl [65] and Lauritzen and Spiegelhalter [45]. MacKay and Neal [48, 58], and McEliece et al. [59] showed that Gallager's algorithm [31] for decoding low-density parity-check codes proposed in the early 1960s is essentially an instance of Pearl's algorithm. Extensive simulation results of MacKay and Neal showed that Gallager codes could perform nearly as well as earlier developed turbo codes [6]. The same authors also observed that turbo decoding is an instance of "belief" propagation and provided a description of Pearl's algorithm, and made explicit its connection to the basic turbo decoding

0-8493-1524-7/05/$0.00+$1.50

algorithm described in [6]. The origins of the algorithm can be found in the work of Battail [3], Hartmann and Rudolph [35], Gallager [32] and Tanner [87].

Application of the LDPC codes and the MPA for their decoding in magnetic and optical recording systems is considered in [51, 52, 82, 83]. In fact, hard iterative decoding of the LDPC codes and their application in storage systems was considered earlier in [46, 47, 86]. A detailed asymptotic analysis of the minimum distance of the LDPC codes and the BER achieved by low complexity decoders can be found in [109].

More recently, Wiberg et al. [107] showed that graphs introduced by Tanner [87] more than two decades ago to describe a generalization of Gallager codes provide a natural setting in which to describe and study iterative soft-decision decoding techniques, in the same way as the code trellis [29] is an appropriate model for describing and studying conventional maximum likelihood soft-decision decoding using Viterbi's algorithm. Forney [27] generalized Wilberg's results and explained connections of various two-way propagation algorithms with coding theory. Frey and Kschischang [30, 36, 45] showed that various graphical models, such as Markov random fields, Tanner graphs, and Bayesian networks all support the basic probability propagation algorithm in factor graphs, similarly as a trellis diagram supports Bahl, Cocke, Jelinek, Raviv's (BCJR) algorithm [3]. Frey and Kschischang also derived a general distributed marginalization algorithm for functions described by factor graphs. From this general algorithm, Pearl's belief propagation algorithm as well as its instances: turbo decoding and message-passing can be easily derived as special cases of probability propagation in a graphical model of the code. A good tutorial on iterative decoding of block and convolutional codes is due to Hagenauer, Offer and Papke [33].

The theory of codes on graphs has not only improved the error performance, but it has also opened new research avenues for investigating alternative suboptimal decoding schemes. It seems that almost any proposed concatenated coding configuration has good performance provided that the used codewords are long and iterative decoding is employed. The iterative decoding algorithms employed in the current research literature are suboptimal, although simulations have demonstrated their performance to be near optimal (e.g., near maximum-likelihood). Although suboptimal, these decoders still have very high complexity and are incapable of operating in the high data rate regimes. The high complexity of the proposed schemes is a direct consequence of the fact that for random codes a large amount of information is necessary to specify positions of the nonzero elements in a parity-check matrix. In this chapter we will introduce well-structured LDPC codes, a concept opposed to the prevalent practice of using random code constructions. Our main focus will be on low-complexity coding schemes and structured LDPC codes: their construction and performance in ISI magnetic recording channels.

In the past few years several random low-density parity-check (LDPC) codes have been designed with performances very close to the Shannon limit [80], see for example, Richardson, Shokrollahi, and Urbanke [75], MacKay [59], and Luby, Mitzenmacher, Shokrollahi, and Spielman [57]. At the same time, significant progress has been made in designing structured LDPC codes. Examples of structured LDPC codes include Kou, Lin and Fossorier's [43] finite geometry codes, Tanner, Sridhara and Fuja's [88] codes constructed from groups, Rosenthal and Vontobel's codes on regular graphs [77], and Johnson and Weller's [37] Steiner system codes. MacKay and Davey [60] also used Steiner systems (a subclass of BIBDs) to construct Gallager codes. Rosenthal and Vontobel [77] constructed some short high-girth codes using a technique by Margulis [62] based on $k-$regular Caley graphs of $SL_2(GF(q))$, the special linear group, and $PGL_2(GF(q))$, the projective general linear group of dimension two over $GF(q)$, the finite field with q elements. Jon-Lark Kim et. al [39] gave another explicit construction of LDPC codes using Lazebnik and Ustimenko's [49] method based on regular graphs. In a series of articles Vasic [91–93, 95, 96, 100], and Vasic, Kurtas and Kuznetsov [94, 97, 98, 100], and [103] introduced several new classes of combinatorially constructed regular LDPC codes, and analyzed their performance in longitudinal and perpendicular recording systems. Sankaranarayanan, Vasic, and Kurtas [79] also showed how to construct irregular codes with a desired degree distribution starting from a combinatorial design. In what follows, we will present a short overview of some of these code constructions.

Although iterative decoding of LDPC codes is based on the concept of codes on graphs, a LDPC code itself can be substantially simplified if it is based on combinatorial objects known as *designs*. In fact, the construction considered in this chapter is purely combinatorial, and is based on *balanced incomplete block designs* (BIBD) [8]. Such combinatorial objects were extensively studied in connection with a large number of problems in applied mathematics and communication theory. Combinatorial designs and codes are very closely connected combinatorial entities, since one can be used to construct the other. For example, codewords of fixed weight in many codes, including the Golay code and the class of quadratic residue (QR) codes, support designs (see, [61, 85]).

As will be shown below, the parity-check matrix of the combinatorially constructed codes can be defined as the incidence matrix of a 2-$(v, c, 1)$ BIBD [17, 55], where v represents the number of parity bits, and c represents the column weight of the parity-check matrix. We are interested in very high rate and relatively short (less than 5000 bits) codes. High rates are necessary to control the equalization loss that is unavoidable in *partial response* (PR) channels [14, 38, 41], while short block lengths are required to maintain compatibility with existing data formats and enable simpler system architecture [91].

The first construction presented uses difference families such as the Bose [10, 11], Netto [66] and Buratti [13] difference families over the group Z_v, while the second construction is related to rectangular integer lattices. The class of codes based on difference families offers the best tradeoff between code rate and code length, but does not produce many high-rate, short-length codes, especially for large column weights of the parity-check matrix. The second construction gives a much larger family of codes, at the expense of the code rate. A third class of structured codes to be discussed in this chapter is the class of codes on projective planes, such as codes on projective and affine geometries, codes on Hermitian unitals and codes constructed from oval designs.

The encoding complexity of combinatorially constructed codes is very low and determined either by the size of the cyclic difference family upon which a block design is based, or by the "vertical" dimension of the lattice for the case of lattice constructions.

We start by introducing BIBDs in Section 38.2 and describing their relation to bipartite graphs and parity-check matrices of regular Gallager codes. In Section 38.3 we introduce several constructions for 2-$(v, 3, 1)$ systems (so called Steiner triple systems), based on cyclic difference families. We present three constructions of cyclic difference families that result in regular codes with column weight $c = 3, 4$ and 5, and give a list of known infinite families. We also give an overview of finite geometry codes in Section 38.4, and a construction based on integer lattices in Section 38.5. The last section of this chapter, Section 38.6, focuses on the performance evaluation of combinatorially constructed LDPC codes over PR channels. The first attempt to apply iterative decoding for ISI channels is due to Douillard et al. [22]. Douillard and his co-authors presented an iterative receiver structure, the so-called "turbo-equalizer," capable of combating ISI due to multipath effects in Gaussian and Rayleigh transmission channels. Soft-output decisions from the channel detector and from the convolutional decoder were used in an iterative fashion to generate bit estimates. Motivated largely by the potential applications to magnetic recording systems, several authors have explored turbo coding methods for some $(1 - D)(1 + D)^N$ channels. Heegard [36] and Pusch, et al. [71] illustrated the design and iterative decoding process of turbo codes for the $(1 - D)$ channel, using codes of rates 1/2 and lower. Ryan, et al. [72, 73] demonstrated that by using a parallel-concatenated turbo code as an outer code, punctured to achieve rates 4/5, 8/9, and 16/17 it is possible to reduce the bit error rate relative to previously known high-rate trellis-coding techniques on a precoded $1 - D$ or $1 - D^2$ channel. Souvignier et al. [84] and McPheters et al. [65] considered serial concatenated systems with an outer code that is a high-rate convolutional code, rather than a turbo code. They found that these convolutional codes perform as well as turbo codes. Additionally, they showed that removal of the channel precoder improves the performance of the turbo-coded system at low SNR, and degrades the performance of the convolutionally-coded system. Oberg and Siegel [67] found the error rates of precoded and nonprecoded serial concatenated systems with high rate block codes. Recently, the performance of LDPC codes in magnetic recording systems was analyzed by Fan, Kurtas, Friedmann and McLauglin [27] and by Dholakia and Eleftheriou [21]. LDPC codes combined with noise prediction are discussed in [74] and [20]. Ryan, McLaughlin et al. [72] combined an iterative decoding scheme with a maximum run length constrained code. Vasic and Pedagani [104] introduced a coding scheme for which the modulation code is completely removed and a channel constraint is imposed by structured (deliberate) error insertion. The scheme uses the power of a LDPC code to correct both random and structured errors. Here, we also should note that the simple turbo product codes with single parity checks (TPC/SPC) can be considered as a subclass of the structured LDPC codes. The TPC with single and multiple parities in columns and rows are considered in [51, 52].

Lattice based LDPC codes in partial response channels were analyzed in [94], and iteratively decodable codes based on Netto CDFs were introduced and studied in [95]. The application of these codes in PR channels assuming decoding with the min-sum algorithm and limited number of iterations was also discussed in [94]. Affine geometry codes that do not contain so-called Pasch configurations and hence have increased minimum distance were introduced in [96]. In [96] LDPC codes based on mutually orthogonal Latin rectangles were constructed and analyzed with respect to their performance in longitudinal magnetic recording channels. The performance of LDPC codes based on Kirkman systems and low-density generator matrix codes in perpendicular magnetic recording were extensively discussed in [98] and [79].

38.2   Combinatorial Designs and their Bipartite Graphs

A balanced incomplete block design (BIBD) with parameters (v, c, l) is an ordered pair (V, B), where V is a v-element set and B is a collection of b c-subsets of V, called blocks, such that each element of V is contained in exactly r blocks and any 2-subset of V is contained in exactly l blocks. Notice that $c \cdot b = R \cdot v$, so that r is uniquely determined by the remaining parameters of the design. We consider designs for which every block contains exactly c points, and every point is contained in exactly r blocks. The notation BIBD(v, c, l) is used for a BIBD on v points, block size c, and index l. A BIBD with block size $c = 3$ is called *a Steiner triple system*. A design is called *resolvable* if there exists a nontrivial partition of its block set B into disjoint subsets each of which partitions the set V. Each of these subsets is referred to as *a parallel class*. Resolvable Steiner triple systems with index $l = 1$ are called *Kirkman systems*. These combinatorial objects have been intensively studied in the combinatorial literature, and some construction methods for them are described in [17, 18, 54]. A BIBD is called symmetric if $b = v$ and $r = c$. A symmetric BIBD(v, c, l) with $c \geq 3$ is equivalent to a finite projective plane [55]. In addition to resolvable designs, one can also use l-configurations for the design of LDPC codes. An *l-configuration* is a combinatorial structure comprised of v points and b blocks such that: (a) each block contains c points; (b) each point is incident with r blocks; (c) two different points are contained in *at most* l blocks.

We define the point-block incidence matrix of a (V, B) design as a $v \times b$ matrix $A = (a_{ij})$, in which $a_{ij} = 1$ if the ith element of V occurs in the jth block of B, and $a_{ij} = 0$ otherwise. If one thinks of points of the design as parity-check equations and of blocks of the design as bits of a linear block code, then it is possible to define the parity-check matrix of a LDPC code as the block-point incidence matrix of the design. Since in a BIBD each block contains the same number of points c, and every point is contained in the same number r of blocks, A defines a parity-check matrix H of a regular LDPC (Gallager) code [32].

For example, the collection $B = \{B_1, B_2, \dots, B_{12}\}$ of blocks $B_1 = \{1, 2, 3\}$, $B_2 = \{1, 5, 8\}$, $B_3 = \{1, 4, 7\}$, $B_4 = \{1, 6, 9\}$, $B_5 = \{4, 8, 9\}$, $B_6 = \{3, 4, 6\}$, $B_7 = \{2, 6, 8\}$, $B_8 = \{2, 4, 5\}$, $B_9 = \{5, 6, 7\}$, $B_{10} = \{2, 7, 9\}$, $B_{11} = \{3, 5, 9\}$, $B_{12} = \{3, 7, 8\}$ is a resolvable BIBD$(9, 3, 1)$ system or Steiner system with $v = 9$, and $b = 12$. The resolvability classes are $\{B_1, B_5, B_9\}$, $\{B_2, B_6, B_{10}\}$, $\{B_3, B_7, B_{11}\}$ and $\{B_4, B_8, B_{12}\}$. The point-block incidence matrix is of the form:

$$A = H = \begin{bmatrix} 1 & 1 & 1 & 1 & 0 & 0 & 0 & 0 & 0 & 0 & 0 & 0 \\ 1 & 0 & 0 & 0 & 0 & 0 & 1 & 1 & 0 & 1 & 0 & 0 \\ 1 & 0 & 0 & 0 & 0 & 1 & 0 & 0 & 0 & 0 & 1 & 1 \\ 0 & 0 & 1 & 0 & 1 & 1 & 0 & 1 & 0 & 0 & 0 & 0 \\ 0 & 1 & 0 & 0 & 0 & 0 & 0 & 1 & 1 & 0 & 1 & 0 \\ 0 & 0 & 0 & 1 & 0 & 1 & 1 & 0 & 1 & 0 & 0 & 0 \\ 0 & 0 & 1 & 0 & 0 & 0 & 0 & 0 & 1 & 1 & 0 & 1 \\ 0 & 1 & 0 & 0 & 1 & 0 & 1 & 0 & 0 & 0 & 0 & 1 \\ 0 & 0 & 0 & 1 & 1 & 0 & 0 & 0 & 0 & 1 & 1 & 0 \end{bmatrix} \tag{38.1}$$

The rate of the code is $R = (b - rank(H))/b$. In general, the rank of H is hard to find, but the following simple formula can be used to bound the rate of a LDPC code based on the parameters of a 2-(v, c, l)

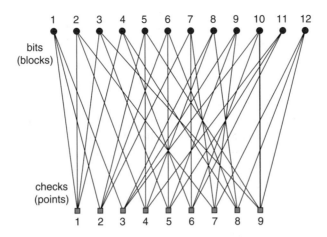

FIGURE 38.1 The bipartite graph representation of the Steiner (9,3,1) system.

design:

$$R \geq \frac{l \cdot x - v}{l \cdot x} \quad \text{where} \quad x = \frac{v(v-1)}{c(c-1)} \tag{38.2}$$

A more precise characterization of the rank (and "p-rank") of the incidence matrix of 2-designs was given by Hamada in [34]. It should be noticed that the above bound is generally loose. For example, for the case of codes constructed from projective planes, the bound is trivially equal to zero, while the actual rate of the codes is quite high (see, [43]). Therefore, for a given code rate, many BIBD codes will have a larger dimension than predicted by the above bound. The construction of maximum rate BIBD codes with $c = 2$ is trivial end reduces to finding K_v, the complete graph [55].

The parity check matrix of a linear block code can be represented as a bipartite graph [63, 87, 107]. The first vertex subset (B) contains bits or variables, while the second subset is comprised of parity-check equations (V). An edge between a bit and an equation exists if the bit is involved in the check. For example, the bipartite graph representation of the Steiner (9,3,1) system whose incidence matrix is given in Equation 38.1 is shown in Figure 38.1.

In order to be properly protected, each bit should be involved in as many equations as possible. On the other hand, iterative decoding algorithms require that the bipartite graph does not contain short cycles [45]. In other words, the *girth* of the graph (i.e., the length of the shortest cycle) must be large. Additionally, in order to allow for efficient iterative decoding, the out-degree of check nodes must be limited too. By using the incidence matrix of a design to define a code, one can observe that the constraint $l = 1$ imposed for BIBDs automatically implies that there are no cycles of length four in the bipartite graph of the code.

38.3 LDPC Codes on Difference Families

Let V be a finite additive Abelian group of order v. Then t c-element subsets of V, $B_i = \{b_{i,1}, \ldots, b_{i,c}\}$, $1 \leq i \leq t$, form a (v, c, l) *difference family* (DF) if every nonzero element of V can be represented in exactly l ways as a difference of two elements within the same member of a family. In other words, every nonzero element of V occurs l times among the differences $b_{i,m} - b_{i,n}$, $1 \leq i \leq t, 1 \leq m, n \leq c$. The sets B_i are called base blocks. If V is isomorphic to Z_v, the additive group of integers modulo v, then the corresponding (v, c, l) DF is called a *cyclic difference family* (CDF).

For example the blocks $B_1 = \{3, 13, 15\}$, $B_2 = \{9, 8, 14\}$, $B_3 = \{27, 24, 11\}$, $B_4 = \{19, 10, 2\}$, and $B_5 = \{26, 30, 6\}$ are the base block of a (31,3,1) CDF of a group $V = Z_{31}$ since the nonzero elements of the difference arrays

$$D^{(1)} = \begin{bmatrix} 0 & 10 & 12 \\ 21 & 0 & 2 \\ 19 & 29 & 0 \end{bmatrix} \quad D^{(2)} = \begin{bmatrix} 0 & 30 & 5 \\ 1 & 0 & 6 \\ 26 & 25 & 0 \end{bmatrix} \quad D^{(3)} = \begin{bmatrix} 0 & 28 & 15 \\ 3 & 0 & 18 \\ 16 & 13 & 0 \end{bmatrix} \quad D^{(4)} = \begin{bmatrix} 0 & 22 & 14 \\ 9 & 0 & 23 \\ 17 & 8 & 0 \end{bmatrix} \quad D^{(5)} = \begin{bmatrix} 0 & 4 & 11 \\ 27 & 0 & 7 \\ 20 & 24 & 0 \end{bmatrix}$$

formed as $D_{i,j}^{(k)} = b_{2,i} - b_{2,j}, 1 \leq k \leq 5)$ are all different.

If G is a group that acts on a set X, then the set $\{gx : g \in G\}, x \in X$, is called the orbit of x. For the case that $G = V$ and $X = B$, where B is the set of all base blocks of a CDF, a BIBD can be defined as the union of orbits of B. If the number of base blocks is s, the number of blocks in BIBD is $b = sv$. For a given constraint (v, c, l), the CDF construction maximizes the code rate, because for a given v the number of blocks is maximized. The parity check matrix of a (v, c, l) CDF LDPC code can be written in the form

$$H = [H_1 H_2, \ldots, H_t] \tag{38.3}$$

where each submatrix is of dimension $v \times v$ and each of the base blocks $B_i = \{b_{i,1}, \ldots, b_{i,c}\}, 1 \leq i \leq t$, specifies positions of nonzero elements in the first column of H_i. The CDF codes have a quasi-cyclic structure similar to Townsend and Weldon's [90] self-orthogonal quasi-cyclic codes and Weldon's difference set codes [105].

The blocks $B_1 = \{3, 13, 15\}$, $B_2 = \{9, 8, 14\}$, $B_3 = \{27, 24, 11\}$, $B_4 = \{19, 10, 2\}$, and $B_5 = \{26, 30, 6\}$ are the base block of a $(31,3,1)$ CDF of the group $V = Z_{13}$. The orbits are given in Table 38.1.

A bound for $d_{\min}$ of Gallager codes with column weight $c = 3$ was first derived in [60]. Another lower bound for $d_{\min}$ is due to Tanner [88], and can be applied to an arbitrary linear code with matrix of parity checks H, represented by a bipartite graph. Using purely combinatorial arguments, lower bounds for the minimum distance were derived in [93]. A general, nontrivial lower bound on $d_{\min}$ for codes based on BIBDs with block size c can easily be obtained by using the fact that these codes are majority-logic decodable. A code is one-step majority-logic decodable if for every bit there exists a set of L parity-check equations that are orthogonal on that bit. In this context, the orthogonality condition imposes the requirement that each of the check equations include the bit under consideration, and that no other bit is checked more than once by any of the equations. If a code is one-step majority decodable, then the minimum distance of the code is at least $L + 1$. From the described construction of LDPC codes based on BIBDs, it follows that $L = c$ and $d_{\min} \geq c + 1$.

As explained in the previous section, once a CDF is known, it is straightforward to construct a BIBD. Constructing a CDF is a complex problem and it is solved only for certain values of v, c and l. One of the first constructions of a difference set is due to Bose [10, 11] and Singer [81]. Most constructions of CDFs

TABLE 38.1   The Orbits of Base Blocks in a (31,3,1) BIBD

B_1 orbit			B_2 orbit			B_3 orbit			B_4 orbit			B_5 orbit			B_1 orbit			B_2 orbit			B_3 orbit			B_4 orbit			B_5 orbit		
3	13	15	9	8	14	27	24	11	19	10	2	26	30	6	19	29	0	25	24	30	12	9	27	4	26	18	11	15	22
4	14	16	10	9	15	28	25	12	20	11	3	27	0	7	20	30	1	26	25	0	13	10	28	5	27	19	12	16	23
5	15	17	11	10	16	29	26	13	21	12	4	28	1	8	21	0	2	27	26	1	14	11	29	6	28	20	13	17	24
6	16	8	12	11	17	30	27	4	22	13	5	29	2	9	22	1	3	28	27	2	15	12	30	7	29	21	14	18	25
7	17	19	13	12	18	0	28	15	23	14	6	30	3	10	23	2	4	29	28	3	16	13	0	8	30	22	15	19	26
8	18	20	14	13	19	1	29	16	24	15	7	0	4	11	24	3	5	30	29	4	17	14	1	9	0	23	16	20	27
9	19	21	15	14	20	2	30	17	25	16	8	1	5	12	25	4	6	0	30	5	18	15	2	10	1	24	17	21	28
10	20	22	16	15	21	3	0	18	26	17	9	2	6	13	26	5	7	1	0	6	19	16	3	11	2	25	18	22	29
11	21	23	17	16	22	4	1	19	27	18	10	3	7	14	27	6	8	2	1	7	20	17	4	12	3	26	19	23	30
12	22	24	18	17	23	5	2	20	28	19	11	4	8	15	28	7	9	3	2	8	21	18	5	13	4	27	20	24	0
13	23	25	19	18	24	6	3	21	29	20	12	5	9	16	29	8	10	4	3	9	22	19	6	14	5	28	21	25	1
14	24	26	20	19	25	7	4	22	30	21	13	6	10	17	30	9	11	5	4	10	23	20	7	15	6	29	22	26	2
15	25	27	21	20	26	8	5	23	0	22	14	7	11	18	0	10	12	6	5	11	24	21	8	16	7	30	23	27	3
16	26	28	22	21	27	9	6	24	1	23	15	8	12	19	1	11	13	7	6	12	25	22	9	17	8	0	24	28	4
17	27	29	23	22	28	10	7	25	2	24	16	9	13	20	2	12	14	8	7	13	26	23	10	18	9	1	25	29	5
18	28	30	24	23	29	11	8	26	3	25	17	10	14	21															

that followed Bose's work are based on the same idea of using finite fields. For example, if one defines the set of integers S as $\{i : 0 \leq i \leq q^2 - 1, \omega^i + \omega \in GF(q)\}$, then S consists of q elements and represents a cyclic difference set modulo $q^2 - 1$, with $l = 1$.

The first construction, considered in [103], is due to Netto [13, 16]. It applies for $c = 3$, and v a power of a prime such that $v \equiv 1 \mod 6$. Let ω be a primitive element of the field [64]. If $v = 6t + 1, t \geq 1$, for $d \mid v - 1$ let Ψ^d be the group of d-th powers in $GF(v)$. Let $\omega^i \Psi^d$ be a coset of d-th powers of the field. Then the set $\{\omega^i \Psi^{2t} \mid 1 \leq i \leq t\}$ defines a Steiner triple system difference family [66, 108] with parameters $(6t + 1, 3, 1)$. The base blocks of this family are typically given in the form $\{0, \omega^i (\omega^{2t} - 1), \omega^i (\omega^{4t} - 1)\}$ or less frequently in the form $\{\omega^i, \omega^{i+2t}, \omega^{i+4t}\}$. Similarly, for $v \equiv 7 \mod 12$ one can show that the set $\{\omega^{2i} \Psi^{2t} \mid 1 \leq i \leq t\}$ defines the base blocks of a so-called Netto triple systems [66].

The second construction is due to Burratti, and is applicable for $c = 4$ and $c = 5$ [13]. For $c = 4$, Burratti's method gives CDFs with v points, provided that v is a prime of the form $v \equiv 1 \mod 12$. The CDF is a set of the form $\{\omega^{6i} B : 1 \leq i \leq t\}$, where base blocks have the form $B = \{0, 1, b, b^2\}$, where again w is a primitive element of $GF(v)$. The numbers $b \in GF(v)$ for several values of v are given in [13]. Similarly, for $c = 5$, the CDF is given by $\{\omega^{10i} B : 1 \leq i \leq t\}$, where $B = \{0, 1, b, b^2, b^3\}$, and $b \in GF(20t + 1)$.

The third construction, also due to Bose [11], is based on a mixed difference system. The sets $\{0.1, 0.2, 0.3\}$, $\{1.i, (2 \cdot u).i, 0.(i + 1)\}$, $\{2.i, (2 \cdot u - 1).i, 0.(i + 1)\}, \ldots, \{u.i, (u + 1).i, 0.(i + 1)\}, 1 \leq i \leq t$, where the elements are all taken $\mod(2 \cdot u + 1)$ and the suffices are taken modulo three, form a mixed difference system. The notation $0.1, 1.i$ etc. used above means, for example, that the symbols 1 and i appearing after the decimal point are indices of 0 and 1, respectively. The mixed difference system uses several copies of the original point set that can be distinguished based on the second index as described above. For more information about this construction, see [1]. It can be shown that a Steiner triple system (where $t = 3$) of order $6 \cdot u + 3$ exists for all $u \geq 0$. In this case $v = 3 \cdot (2 \cdot u + 1), c = 3, l = 1$ are the parameters of BIBD design. Some other interesting constructions based on Latin squares can be found in [17, 18].

38.4 Codes on Projective Planes

The above constructions give a very small set of design with parameters of practical interests. However, so called *infinite families* [16] give an infinite range of block sizes. Infinite families of BIBDs include finite projective geometries, finite Euclidean (affine) geometries, unitals, ovals, Denniston designs as well as certain geometric equivalents of 2-designs (see [16, 62]). For certain choices of the parameters involved, they reduce to finite Euclidean and finite projective geometries. In the following paragraph, we will introduce some basic definitions and concepts regarding the above listed geometric entities. For more details on codes on finite geometries, the reader is referred to [61] and [62].

The parity check matrix of a projective or an affine geometry code is defined as a line-point incidence matrix of the geometry in which points are nonzero m-tuples (vectors) with elements in $GF(p^s)$, where $s > 0$, and p is a prime. The points and lines have the following set of properties: (a) every line consists of the same number of points; (b) any two points are connected by one line only; (c) two lines intersect in at most one point; and (d) each point is an intersection of a fixed number of lines. A finite projective geometry $PG(m, p^s)$ is constructed by using $(m+1)$-tuples of elements x_i from $GF(p^s)$, not all simultaneously equal to zero, called points. Two $(m+1)$-tuples $\mathbf{x} = (x_0, x_1, \ldots, x_m)$ and $\mathbf{y} = (y_0, y_1, \ldots, y_m)$ represent the same point if $\mathbf{x} = \mu\mathbf{x}$, for some nonzero $\mu \in GF(p^s)$. Hence, each point can be represented in $p^s - 1$ ways. All such representations of a point are referred to as its equivalence class. It is straightforward to see that the number of points in $PG(m, p^s)$ is $v = (p^{(m+1)s} - 1)/(p^s - 1)$. The equivalence classes, or points, can be represented by $[\alpha^i] = \{\alpha^i, \beta\alpha^i, \beta^2\alpha^i, \ldots, \beta^{p^s-2}\alpha^i\}$, where $0 \leq i \leq v$ and $\beta = \alpha^v$. Let $[\alpha^i]$ and $[\alpha^j]$ be two distinct points in $PG(m, p^s)$. Then the line passing through them consists of points of the form $[\lambda_1\alpha^i + \lambda_2\alpha^j]$, where $\lambda_1, \lambda_2 \in GF(p^s)$ are not both equal to zero. Since $[\lambda_1\alpha^i + \lambda_2\alpha^j]$ and $[\beta^k\lambda_1\alpha^i + \beta^k\lambda_2\alpha^j]$ are the same point, each line in $PG(m, p^s)$ consists of $p^s + 1$ points. The number of lines intersecting at a given point is $(p^{ms} - 1)/(p^s - 1)$, and the number of lines in projective geometry is

$$b = \left(p^{s(m+1)} - 1\right)\big/(p^{sm} - 1)/(p^{2s} - 1)/(p^s - 1) \tag{38.4}$$

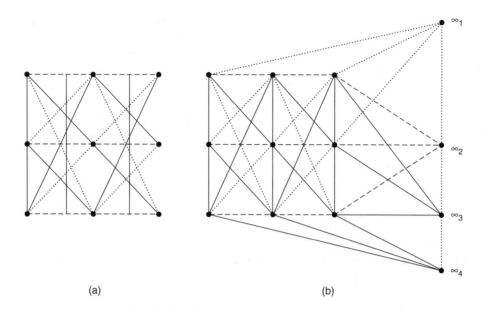

FIGURE 38.2 Obtaining a projective plane from an affine plane by adding points: (a) affine plane; (b) projective plane.

The main difference between projective and affine geometries is that projective planes do not contain parallel lines. Affine planes may be considered as special case of projective planes, since they can be obtained by removing the points belonging to the line at infinity (say $x_0 = 0$). Conversely, a projective plane can be obtained from an affine plane by adding an extra point to each line (a point at infinity), and connecting these points by a line. Figure 38.2 illustrates this idea for a geometry with nine points. Each line in the projective geometry contains one more point, and the points at infinity $\infty_1, \infty_2, \infty_3, \infty_4$ create a line so that no parallel classes exist.

In the parity check matrix each column has weight $p^s + 1$ and each row has weight $(p^{ms} - 1)/(p^s - 1)$. As an illustration of the above discussion, the parity-check matrix of a $PG(2, 3)$ LDPC code is given below.

$$
H = \begin{bmatrix}
1 & 1 & 1 & 1 & 0 & 0 & 0 & 0 & 0 & 0 & 0 & 0 & 0 \\
1 & 0 & 0 & 0 & 1 & 1 & 1 & 0 & 0 & 0 & 0 & 0 & 0 \\
1 & 0 & 0 & 0 & 0 & 0 & 0 & 1 & 1 & 1 & 0 & 0 & 0 \\
1 & 0 & 0 & 0 & 0 & 0 & 0 & 0 & 0 & 0 & 1 & 1 & 1 \\
0 & 1 & 0 & 0 & 1 & 0 & 0 & 1 & 0 & 0 & 1 & 0 & 0 \\
0 & 1 & 0 & 0 & 0 & 1 & 0 & 0 & 1 & 0 & 0 & 1 & 0 \\
0 & 1 & 0 & 0 & 0 & 0 & 1 & 0 & 0 & 1 & 0 & 0 & 1 \\
0 & 0 & 1 & 0 & 1 & 0 & 0 & 0 & 0 & 1 & 0 & 1 & 0 \\
0 & 0 & 1 & 0 & 0 & 1 & 0 & 1 & 0 & 0 & 0 & 0 & 1 \\
0 & 0 & 1 & 0 & 0 & 0 & 1 & 0 & 1 & 0 & 1 & 0 & 0 \\
0 & 0 & 0 & 1 & 1 & 0 & 0 & 0 & 1 & 0 & 0 & 0 & 1 \\
0 & 0 & 0 & 1 & 0 & 1 & 0 & 0 & 0 & 1 & 1 & 0 & 0 \\
0 & 0 & 0 & 1 & 0 & 0 & 1 & 1 & 0 & 0 & 0 & 1 & 0
\end{bmatrix}
\tag{38.5}
$$

Finite geometry LDPC codes were first considered by Kou, Lin and Fossorier [43], for the case $p = 2$.

Since in a finite geometry two lines cannot be incident with the same pair of points, the corresponding parity checks are orthogonal. One advantage of these codes is that the code parameters, such as the minimum distance and the girth of a bipartite graph, are easily controllable (see [43]). These features are the result of the highly regular structures of these codes.

The second class of finite geometry LDPC can be obtained from *algebraic curves* in a projective plane. In a projective plane, an algebraic curve is a collection of points that satisfy a fixed homogeneous algebraic equation of order n, for example, $f(x_0, x_1, x_2) = 0$. An algebraic curve is irreducible if $f(x_0, x_1, x_2)$ is irreducible over the ground field $GF(q)$. The curve meets a line at most in n points. A *conic* is an algebraic curve of order two defined by an equation of the form:

$$f(x_0, x_1, x_2) = ax_0^2 + bx_1^2 + cx_2^2 + fx_1x_2 + gx_2x_0 + hx_0x_2 = 0$$

A conic is irreducible if $f(x_0, x_1, x_2)$ is irreducible over the ground field $GF(q)$. For $k > m$, a $\{k; m\}$-arc in $PG(2, q)$ is a set of k points such that no $m + 1$ points lie on a line. A c-arc in $PG(2, q)$ is a set of c points such that no three points lie on the same line. A c-arc is complete if it is not properly contained in any $(c + 1)$-arc. A line of the plane is said to be a secant, a tangent or an exterior line with respect to the oval, if the number of common points of the line with the oval is 2,1, and 0, respectively. For a given value of q, $(q + 1)$-arcs of $PG(2, q$ odd) are called ovals, and $(q + 1)$-arcs of $PG(2, q$ even) together with a nucleus point (a point for which every line incident to it is a tangent of the oval) are called hyperovals. In $PG(2, 2^m)$, an oval design is an incidence structure with points comprised from the lines exterior to the oval and blocks formed from points not on the oval. A block contains a point if and only if the corresponding exterior point lies on the exterior line. It is a resolvable 2-$(s(s-1)/2, s/2, 1)$ Steiner design where $s = 2^m$. The rank of the incidence matrix of an oval design is $3^m - 2^m$. In the section containing the simulation results, we will present the performance of a code constructed from the nondegenerate conic $x_0 x_2 = x_1^2$. Figure 38.3 shows an oval and its corresponding lines. Codes on ovals were first described in [106].

Unitals or Hermitian arcs are defined as follows. In $PG(2, q)$, with q a perfect square, a Hermitian arc is a $\{q\sqrt{q} + 1, \sqrt{q} + 1\}$-arc. The arc is constructed from an algebraic curve of order $\sqrt{q} + 1$ given by the equation:

$$x_0^{\sqrt{q}+1} + x_1^{\sqrt{q}+1} + x_2^{\sqrt{q}+1} = 0 \tag{38.6}$$

The arc intersects any line of the plane in one or $\sqrt{q} + 1$ points. A code based on a unital can be obtained as the arc constructed from an algebraic curve of order $\sqrt{q} + 1$, where the unital is a $\{q\sqrt{q} + 1, \sqrt{q} + 1, 1\}$ Steiner system. For q a power of 2, the rank of the incidence matrix is $\{q\sqrt{q}$, and for q a power of an odd prime, the rank of the incidence matrix is $(q - \sqrt{q} + 1)\sqrt{q}$. Such designs are treated in a great detail by Assmus and Key and in great detail in [2], but so far they have not been used for the construction of LDPC

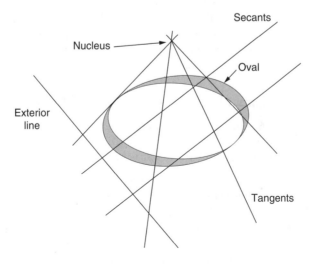

FIGURE 38.3 Visualization of an oval.

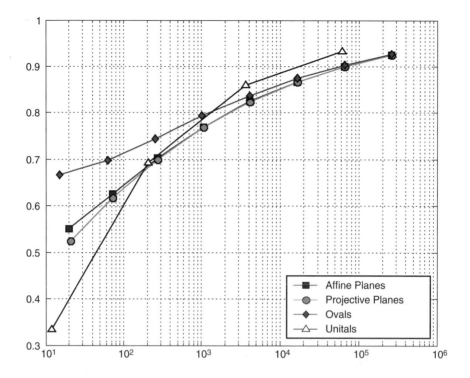

FIGURE 38.4   Achievable length-rate pairs for LDPC codes on projective planes.

codes. Figure 38.4 shows achievable length-rate pairs for LDPC codes on projective planes. Notice that in [43] it was shown that affine and projective geometry codes can be modified in different ways so as to produce a large set of code parameters and to achieve a good trade-off between length and code rate.

38.5   Lattice Construction of LDPC Codes

In this section we address the problem of constructing LDPC codes of large block lengths. The number of parity bits is $m \times c$, where m is a prime, and the blocks are defined as lines of different slopes connecting points of an $m \times c$ integer lattice. The number of blocks is equal to m^2. Integer lattices define l-congurations with index $l = 1$. Here, each 2-tuple is contained in at most $l = 1$ blocks [92]. The goal of the lattice-based construction is to trade the code rate and number of blocks for the simplicity of the construction and for the flexibility of choosing the design parameters. Additionally, as shown in [99], 1-congurations greatly simplify the construction of codes of large girth.

38.5.1   Codes on a Rectangular Subset of an Integer Lattice

Consider a rectangular subset L of the integer lattice, defined by

$$L = \{(x, y) : 0 \leq x \leq c - 1, 0 \leq y \leq m - 1, \}$$

where $m \geq c$ is a prime. Let $\lambda : L \to V$ be an one-to-one mapping from the set L to the point set V. An example of such mapping is a simple linear mapping $\lambda(x, y) = m \cdot x + y + 1$. The numbers $\lambda(x, y)$ are referred to as point labels.

A line is a set of c points specified by its *slope* $s, 0 \leq s \leq m - 1$ and starting point *starting at the point* $(0, a)$, contains the points $\{(x, a + sx \bmod m) : 0 \leq x \leq c - 1\}$, where $0 \leq a \leq m - 1$. Figure 38.5 depicts a rectangular subset of the integer lattice with, $m = 5$ and $c = 3$. In the same figure, four lines with slopes $s = 0, s = 1, s = 2,$ and $s = 3$ are shown. Notice that in the configuration lines of infinite slope

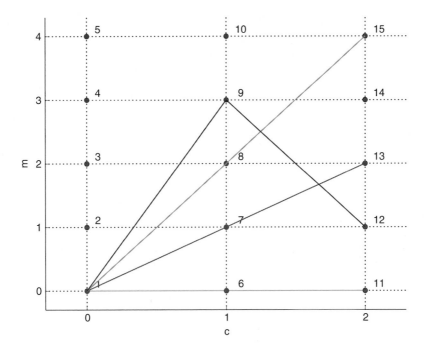

FIGURE 38.5 An example of the rectangular grid for $m = 5$ and $c = 3$.

are not included. In [94] it was shown that the set of blocks B containing all m c-element sets of points in V obtained by taking labels of points along the lines with slopes $s, 0 \le s \le m - 1$ forms a resolvable 1-configuration. Each point in the configuration occurs in exactly m blocks. The lines with slopes $s = 0$, $s = 1, s = 2, s = 3$ and $s = 4$, are $\{\{1, 6, 11\}, \{2, 7, 12\}, \{3, 8, 13\}, \{4, 9, 14\}, \{5, 10, 15\}\}$, $\{\{1, 7, 13\}, \{2, 8, 14\}, \{3, 9, 15\}, \{4, 10, 11\}, \{5, 6, 12\}\}$, $\{\{1, 8, 15\}, \{2, 9, 11\}, \{3, 10, 12\}, \{4, 6, 13\}, \{5, 7, 14\}\}$, $\{\{1, 9, 12\}, \{2, 10, 13\}, \{3, 6, 14\}, \{4, 7, 15\}, \{5, 8, 11\}\}$, and $\{\{1, 10, 14\}, \{2, 6, 15\}, \{3, 7, 11\}, \{4, 8, 12\}, \{5, 9, 13\}\}$. The corresponding parity-check matrix is given by:

$$
H = \begin{bmatrix}
1\ 0\ 0\ 0\ 0 & 1\ 0\ 0\ 0\ 0 & 1\ 0\ 0\ 0\ 0 & 1\ 0\ 0\ 0\ 0 & 1\ 0\ 0\ 0\ 0 \\
0\ 1\ 0\ 0\ 0 & 0\ 1\ 0\ 0\ 0 & 0\ 1\ 0\ 0\ 0 & 0\ 1\ 0\ 0\ 0 & 0\ 1\ 0\ 0\ 0 \\
0\ 0\ 1\ 0\ 0 & 0\ 0\ 1\ 0\ 0 & 0\ 0\ 1\ 0\ 0 & 0\ 0\ 1\ 0\ 0 & 0\ 0\ 1\ 0\ 0 \\
0\ 0\ 0\ 1\ 0 & 0\ 0\ 0\ 1\ 0 & 0\ 0\ 0\ 1\ 0 & 0\ 0\ 0\ 1\ 0 & 0\ 0\ 0\ 1\ 0 \\
0\ 0\ 0\ 0\ 1 & 0\ 0\ 0\ 0\ 1 & 0\ 0\ 0\ 0\ 1 & 0\ 0\ 0\ 0\ 1 & 0\ 0\ 0\ 0\ 1 \\
\hline
1\ 0\ 0\ 0\ 0 & 0\ 0\ 0\ 0\ 1 & 0\ 0\ 0\ 1\ 0 & 0\ 0\ 1\ 0\ 0 & 0\ 1\ 0\ 0\ 0 \\
0\ 1\ 0\ 0\ 0 & 1\ 0\ 0\ 0\ 0 & 0\ 0\ 0\ 0\ 1 & 0\ 0\ 0\ 1\ 0 & 0\ 0\ 1\ 0\ 0 \\
0\ 0\ 1\ 0\ 0 & 0\ 1\ 0\ 0\ 0 & 1\ 0\ 0\ 0\ 0 & 0\ 0\ 0\ 0\ 1 & 0\ 0\ 0\ 1\ 0 \\
0\ 0\ 0\ 1\ 0 & 0\ 0\ 1\ 0\ 0 & 0\ 1\ 0\ 0\ 0 & 1\ 0\ 0\ 0\ 0 & 0\ 0\ 0\ 0\ 1 \\
0\ 0\ 0\ 0\ 1 & 0\ 0\ 0\ 1\ 0 & 0\ 0\ 1\ 0\ 0 & 0\ 1\ 0\ 0\ 0 & 1\ 0\ 0\ 0\ 0 \\
\hline
1\ 0\ 0\ 0\ 0 & 0\ 0\ 0\ 1\ 0 & 0\ 1\ 0\ 0\ 0 & 0\ 0\ 0\ 0\ 1 & 0\ 0\ 1\ 0\ 0 \\
0\ 1\ 0\ 0\ 0 & 0\ 0\ 0\ 0\ 1 & 0\ 0\ 1\ 0\ 0 & 1\ 0\ 0\ 0\ 0 & 0\ 0\ 0\ 1\ 0 \\
0\ 0\ 1\ 0\ 0 & 1\ 0\ 0\ 0\ 0 & 0\ 0\ 0\ 1\ 0 & 0\ 1\ 0\ 0\ 0 & 0\ 0\ 0\ 0\ 1 \\
0\ 0\ 0\ 1\ 0 & 0\ 1\ 0\ 0\ 0 & 0\ 0\ 0\ 0\ 1 & 0\ 0\ 1\ 0\ 0 & 1\ 0\ 0\ 0\ 0 \\
0\ 0\ 0\ 0\ 1 & 0\ 0\ 1\ 0\ 0 & 1\ 0\ 0\ 0\ 0 & 0\ 0\ 0\ 1\ 0 & 0\ 1\ 0\ 0\ 0 \\
\end{bmatrix}
\qquad (38.7)
$$

In general, the parity-check matrix of a lattice code can be written in the form:

$$H = \begin{bmatrix} I & I & I & \cdots & I \\ I & P_{2,2} & P_{2,3} & \cdots & P_{2,m-1} \\ I & P_{3,2} & P_{3,3} & \cdots & P_{3,m-1} \\ \cdots & \cdots & \cdots & \cdots & \cdots \\ I & P_{c-1,2} & P_{c-1,3} & \cdots & P_{c-1,m-1} \end{bmatrix} \tag{38.8}$$

where each submatrix $P_{i,j}$ is a permutation matrix. The power of P which determines $P_{i,j}$ (i.e. the position of the bit 1 the first column of $P_{i,j}$) can be found by using c_i^{j-1}, the ith element of the first base block in the class of blocks corresponding to the jth slope.

Similar parity check matrices have been obtained by several researchers using different approaches. The examples include Tanner's sparse difference codes [89], Blaum, Farrell, and Tilborg's [9] array codes, Eleftheriou and Olcer's [25] array LDPC codes for application in digital subscriber lines, and codes constructed by Kim et al. [39] with girth at least six.

Tanner graphs with large girth can be obtained by a judicious selection of sets of parallel lines included in the integer lattice 1-configuration. The resulting parity-check matrix is of the form of an array of circulant matrices. For example, for $c = 3$ it was shown in [92] that if the slope set represents an "arithmetically constrained" sequence, defined by Odlyzko and Stanley [68], then the resulting codes have girth at least eight. A generalization of this construction for higher girths is also straightforward. Other constructions of codes of large girth include Rosenthal and Vontobel's [77] construction based on an idea by Margulis [62], and the previously mentioned result by Kim et al. [39] based on work of Lazebnik and Ustimenko [49]. The problem of constructing designs with high girth appears to be a very difficult problem in general [5].

38.6   Application in the Partial Response (PR) Channels

As we already mentioned in the introduction for this chapter, the LDPC codes were originally constructed using an ensemble of random sparse parity-check matrices. Such random LDPC codes have been shown to achieve near-optimum performance when decoded by soft iterative decoding algorithms (e.g., message passing algorithm (MPA)), but due to the random structure of their parity check matrix large memory is required for their implementation. The complexity of implementation can be reduced by using structured LDPC codes constructed from algebraic and combinatoric objects described above. The generic channel architecture and decoding algorithms are similar to both random and structured LDPC codes, but due to the special properties of parity check matrices the implementation of the MPA algorithm for structured LDPC codes is much simpler than for the random LDPC codes.

The bit error rate (BER) and other characteristics of the described structured LDPC codes depend on a number of code parameters, such as rate, length, column weight of the parity check matrix, number of short cycles in the bipartite graph of the code, as well as the type of precoder, the type of the PR channel (i.e., equalization), amount of jitter and other factors. We present simulation results describing the effect of some of these factors in a perpendicular magnetic recording channel, but first let us describes the channel model used in simulations.

38.6.1   Channel Model and Signal Generation

Figure 38.6 shows a block diagram of a read/write channel using an LDPC code and an iterative detection scheme for decoding. As we can see from this figure, the user data are encoded by an outer ECC code (usually, Reed-Solomon (RS) code is used), then passed through a run-length limiting (RLL) encoder, and then go to the LDPC encoder. The output of an LDPC encoder can be also precoded before the coded sequence of bits is written on the medium. An analog part of the channel includes read/write heads, the storage medium, different filters, sampling and timing circuits. These components were modeled as described below. Due to electronic and media noise in the channel, the detector can not recover the original user data with arbitrary small error probability, and the ECC decoder is used to detect and correct as many channel errors as possible, decreasing the output BER and SFR of the channel to a level given by the

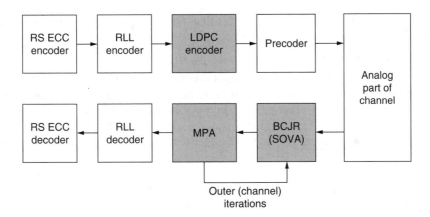

FIGURE 38.6 System view of the encoding and decoding operations in the read/write channel of a disk drive.

channel specifications. Combining RLL and LDPC codes is not a trivial problem which has some technical solutions, for example, described in [72, 104]. Here, we do not consider the effect of an outer ECC code, nor the effect of the RLL code, and in our simulations the user data are directly passed through an LDPC encoder.

Figure 38.7 shows how an input signal of the iterative detection scheme is generated. Formally, the readback signal $V(t)$ of a channel without jitter is defined as follows

$$V(t) = \sum_{k=-\infty}^{\infty} b_k g(t - kT_b) + V_n(t) \tag{38.9}$$

where $g(t)$ is the transition response of the channel, $b_k = \{0, 1, -1\}$ corresponding to no, positive or negative transitions, T_b is the bit interval, and $V_n(t)$ is Additive White Gaussian Noise (AWGN). The transition response of a perpendicular channel is modeled as

$$g(t) = \frac{1}{2} \left(\frac{\pi PW_{50}^2}{2 \ln 2} \right)^{1/4} erf \left(\frac{2\sqrt{2}}{PW_{50}} t \right) \tag{38.10}$$

where PW_{50} is the width of the perpendicular impulse response at 50% of its peak value, and $erf(\cdot)$ is an error function defined as

$$erf(x) = \frac{2}{\sqrt{\pi}} \int_0^x e^{-t^2} dt. \tag{38.11}$$

Here, the amplitude of the transition response is chosen in such a way that the energy of the impulse response is equal to one. In simulations, we defined SNR as $10 \log_{10} E_i / N_0$, where E is the energy of the impulse response which is assumed to be unity, and N_0 is the spectral density hight of the AWGN. The linear density ND is defined as PW_{50} / T_b.

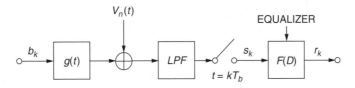

FIGURE 38.7 Signal generation in the magnetic recording channel.

38.6.2   Bit Error Rate (BER) of the structured LDPC codes. Simulation results

In this section, we present the BER performance of the structured LDPC codes in the perpendicular PR2 channel with target [121] at the normalized user density $ND = 2$ (channel ND is adjusted according to the code rate, and is equal to ND/R). The soft output viterbi algorithm (SOVA, [33]) is used on the PRML trellis, and decoding is established by iterating between the inner SOVA decoder and the outer message passing algorithm (MPA, [31,59]). That is, the soft information is extracted from a partial response channel using the SOVA operating on the channel trellis, and then used in the MPA for the LDPC decoding. The bit "likelihoods" obtained from the MPA are passed back to the SOVA as a priori probabilities and so on. The BER results are given below for the scheme in which the compound iteration "SOVA + 4 iterations of MPA" is performed four times before the final hard decision is taken. In other words, the LDPC decoder performs four internal (bit-to-check plus check-to-bit) iterations prior to suppling the inner decoder with extrinsic information. Larger numbers of compound iterations improve the BER, but as we can see from Figure 38.11 four compound iterations is a good compromise between the performance and increased latency and complexity [53].

The BER were evaluated by simulations for two groups of codes. The first group consists of three Kirkman codes with $c = 3$ and length $n = 2420, 4401$, and 5430 (rates $R = 0.95, 0.963$, and 0.966, respectively). The second group consists of lattice codes with $c = 4$ and length $n = 3481, 4489$, and 7921 (rates $R = 0.933$, 0.941, and 0.955, respectively). The BER of the first group of codes and the BER of the random LDPC code with column weight $c = 3$ are shown in Figure 38.8. As we can see from this figure, at the moderate SNR values the random LDPC code has a bit better BER, and in simulations of the size 10^9 bits does not show an error floor or a slope change. At the same time, the Kirkman LDPC codes exhibit a slope change at the BER level $10^{-6} - 10^{-7}$. The BER of lattice LDPC codes with the column weight $c = 4$ and the BER of the random LDPC code with column weight $c = 4$ are shown in Figure 38.9. In this case all BER are almost on the top of each other at the moderate SNR values, but again the lattice LDPC codes exhibit some signs of a slope change at the high SNR values. In Figure 38.10 we compare the BER of the structured LDPC codes with different column weights $c = 3, 4, 5$ and the BER of turbo product code with single parity checks

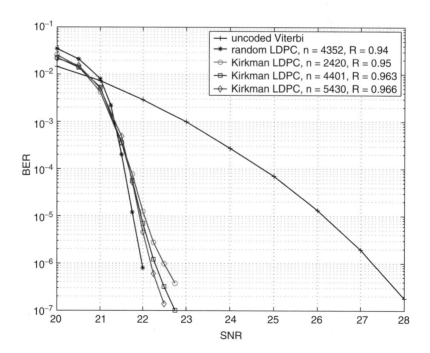

FIGURE 38.8   BER of different Kirkman LDPC codes with column weight 3 in the perpendicular PR2 channel.

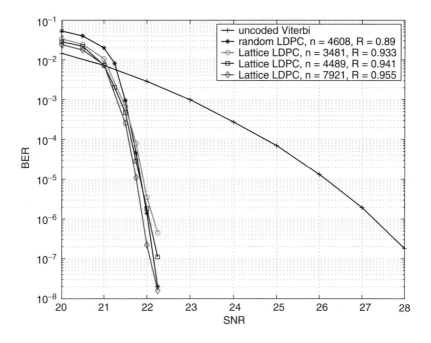

FIGURE 38.9   BER of different lattice LDPC codes with column weight 4 in the perpendicular PR2 channel.

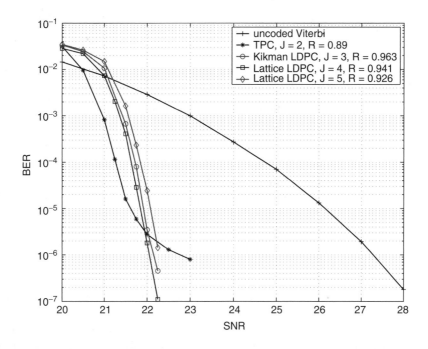

FIGURE 38.10   Comparison of the structured LDPC codes with different column weights of the parity check matrix.

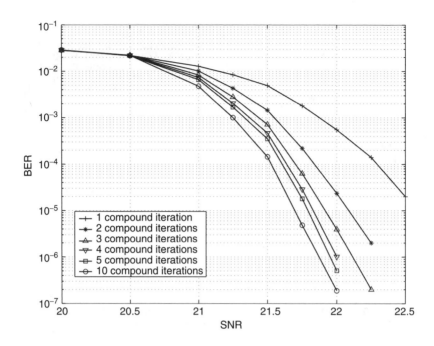

FIGURE 38.11   Effect of the number of iterations on the BER of the lattice LDPC code with $c = 4$.

(TPC/SPC, [51, 52]) considered as an LDPC code with column weight $c = 2$. In this figure, the BER of the TPC code is approximately 0.4 dB better at the BER level 10^{-4}, but has an error floor at the $SNR > 22$ dB, where all structured LDPC codes still are in the "waterfall" region. Finally, in Figure 38.11 we illustrate the typical dependence of the BER from a number of compound iterations G for the lattice and Kirkman LDPC codes. We can summarize the simulation results as follows:

- Kirkman and lattice LDPC codes show the BER that is close to the BER of random LDPC codes, but due to their mathematical structure can lend themselves to low complexity implementations.
- The tested structure LDPC codes with column weights $c = 3, 4$, and 5 exibit a slope change of the BER curves in the high SNR region, but not an error floor in contrast to the TPC/SPC codes.
- Larger numbers of compound iterations improve the BER, and in our simulations four compound iterations were already a good compromise between the performance and increased latency and complexity of the detector scheme.

38.7   Conclusion

We gave an overview of combinatorial constructions of high rate LDPC codes. The emphasis of the exposition was on codes that have column weight not less than three and girth at least six. We discussed BIBD codes obtained using constructions of cyclic difference families due to Bose, Netto and Buratti, and affine integer lattice code. We presented bounds on the minimum distance of the codes and determined the BER performance of these LDPC codes in PR magnetic recording channels by using computer simulations.

Acknowledgement

This work is supported by the NSF under grant CCR-0208597.

References

[1] I. Anderson, *Combinatorial Designs and Tournaments,* Oxford University Press, New York, 1997.

[2] E.F. Assmus and J.D. Key, *Designs and their Codes,* Cambridge University Press, London, 1994.

[3] L. R. Bahl, J. Cocke, F. Jelinek, and J. Raviv, Optimal decoding of linear codes for minimizing symbol error rate, *IEEE Trans. Info. Theory,* Vol. IT-20, pp. 284–287, 1974.

[4] G. Battail, M.C. Decouvelaere, and P. Godlewski, Replication decoding, *IEEE Trans. Info. Theory,* Vol.IT-25, pp. 332–345, May 1979.

[5] R.A. Beezer, The girth of a design, *J. Combinatorial Mathematics and Combinatorial Computing* [in press] (also http://buzzard.ups.edu/pubs.html).

[6] S. Benedetto, G. Montorsi, D. Divsalar, and F. Pollara, Serial Concatenation of Interleaved Codes: Performance Analysis, Design, and Iterative Decoding, *The Telecom. and Data Acquisition Progr. Rep.,* Vol. 42, pp. 1–26, August 1996.

[7] G. Berrou, A. Glavieux, and P. Thitimajshima, Near Shannon Limit Error-Correcting Coding and Decoding: Turbo-Codes, *in Proceedings of IEEE Internatinol Conference on Communications (ICC'93),* Geneva, Switzerland, pp. 2.1064–2.1070, May 1993.

[8] T. Beth, D. Jungnickel, and H. Lenz, *Design Theory,* Cambridge University Press, London, 1986.

[9] M. Blaum, P. Farrell, and H. van Tilborg, Array codes, in *Handbook of Coding Theory,* V. Pless and W. Huffman, Eds., North-Holland, Elsevier, 1998.

[10] R.C. Bose, On the construction of balanced incomplete block designs, *Ann. Eugenics,* Vol. 9, pp. 353–399, 1939.

[11] R.C. Bose, An affine analogue of Singer's theorem, *Ind. Math. Soc.,* Vol. 6, pp. 1–5, 1942.

[12] R.A. Brualdi, *Introductory Combinatorics,* Prentice Hall, Upper Saddle River, New Jersey, 1999.

[13] M. Buratti, Construction of $(q, k, 1)$ difference families with q a prime power and $k = 4, 5$, *Discrete Math.,* 138, pp. 169–175, 1995.

[14] R.D. Cideciyan, F. Dolivo, R. Hermann, W. Hirt, and W. Schott, A PRML system for digital magnetic recording, *IEEE J. Sel. Areas Commn.,* Vol. 10, No. 1, pp. 38–56, January 1992.

[15] M.J. Colbourn and C.J. Colbourn, Recursive construction for cyclic block designs, *J. Stat. Plann. and Inference,* Vol. 10, pp. 97–103, 1984.

[16] C.J. Colbourn and J.H. Dinitz Eds., *The Handbook of Combinatorial Designs,* CRC Press, Boca Raton, 1996.

[17] C.J. Colbourn and A. Rosa, *Triple Systems,* Oxford University Press, Oxford, 1999.

[18] C. Colbourn and A. Rosa, *Steiner Systems,* Oxford University Press (Oxford Mathematical Monographs), London, 1999.

[19] C.J. Colbourn, J.H. Dinitz, and D.R. Stinson, *Applications of Combinatorial Designs to Communications, Cryptography, and Networking,* in "London Math. Soc. Lecture Note Ser.," vol. 267, pp. 37–100, Cambridge Univ. Press, Cambridge, UK, 1999.

[20] M.C. Davey and D.J.C. Mackay, Reliable communication over channels with insertions, deletions, and substitutions, *IEEE Trans. on Inform. Theory,* Vol. 47, No. 2, pp. 687–698, February 2001.

[21] A. Dholakia, E. Eleftheriou, and T. Mittelholzer, On Iterative Decoding for Magnetic Recording Channels, in *Proceedings of Second International Symposium on Turbo Codes,* Brest, France, September 4–7, 2000.

[22] C. Douillard, M. Jaezequel, C. Berrou, A. Picart, P. Didier, and A. Glavieux, Iterative correction of intersymbol interference: turbo equalization, *Eur. Trans. Telecommn.,* Vol. 6, pp. 507–511, September/October 1995.

[23] T.M. Duman and E. Kurtas, Comprehensive Performance Investigation of Turbo Codes over High Density Magnetic Recording Channels, in *Proceedings of IEEE GLOBECOM,* December 1999.

[24] T. M. Duman and E. Kurtas, "Performance Bounds for High Rate Linear Codes over Partial Response Channels," in *Proceedings of IEEE International Symposium on Inform Theory,* Sorrento, Italy, p. 258, *IEEE,* June 2000.

[25] E. Eleftheriou and S. Olcerm, Low-density parity-check codes for digital subscriber lines, *IEEE Int. Conf. on Communications, ICC 2002,* Vol. 3, pp. 1752–1757, 2002.

[26] J.L. Fan, Array Codes as Low-Density Parity-Check Codes, in *Proceedings of 2nd International Symp. on Turbo Codes and Related Topics*, Brest, France, pp. 543–546, September 2000.

[27] J. Fan, A. Friedmann, E. Kurtas, and S.W. McLaughlin, Low Density Parity Check Codes for Partial Response Channels, *Allerton Conference on Communications, Control and Computing*, Urbana, IL, October 1999.

[28] J. Fan, Constrained Coding and Soft Iterative Decoding, in *Proceedings of 2001 IEEE Info. Theory Workshop*, Cairns Australia, pp. 18–20, September 2001.

[29] G.D. Forney, Jr. Codes on graphs: normal realizations, *IEEE Trans. on Info. Theory*, Vol. 47, No. 2, pp. 520–548, February 2001.

[30] G.D. Forney, Maximum-likelihood sequence estimation of digital sequences in the presence of intersymbol interference, *IEEE Trans. Info. Theory*, Vol. 18., No. 3, pp. 363–378, May 1972.

[31] B. J. Frey, *Graphical Models for Machine Learning and Digital Communication*, MIT Press, Cambridge, MA, 1998.

[32] R.G. Gallager, *Low-Density Parity-Check Codes*, MIT Press, Cambridge, MA, 1963.

[33] J. Hagenauer, E. Offer, and L. Papke, Iterative decoding of binary block and convolutional codes, *IEEE Trans. Info. Theory*, Vol 42., No. 2, pp. 439–446, March 1996.

[34] N. Hamada, On the p-rank of the incidence matrix of a balanced or partially balanced incomplete block design and its applications to error correcting codes, *Hiroshima Math. J.*, Vol. 3, pp. 153–226, 1973.

[35] C. Hartmann and L. Rudolph, An optimum symbol-by-symbol decoding rule for linear codes, *IEEE Trans. on Info. Theory*, Vol. 22, No. 5, pp. 514–517, September 1976.

[36] C. Heegard, Turbo Coding for Magnetic Recording, in *Proceedings of IEEE Information Theory Workshop*, San Diego, CA, USA, pp. 18–19, February 1998.

[37] S.J. Johnson and S.R. Weller, Regular Low-Density Parity-Check Codes from Combinatorial Designs, in *Proceedings of 2001 IEEE Information Theory Workshop*, pp. 90–92, 2001.

[38] P. Kabal and S. Pasupathy, Partial-response signaling, *IEEE Trans. Commn.*, Vol. COM-23, pp. 921–934, September 1975.

[39] Jon-Lark Kim, Uri N. Peled, Irina Perepelitsa, and Vera Pless, Explicit Construction of Families of LDPC Codes with Girth at least Six, in *Procedings of 40th Annual Allerton Conference on Communications, Control and Computing*, October 2002.

[40] T.P. Kirkman, Note on an unanswered prize question, *Cambridge and Dublin Mathematics Journal*, 5, pp. 255–262, 1850.

[41] H. Kobayashi, Correlative level coding and maximum-likelihood decoding, *IEEE Trans. Info. Theory*, Vol. IT-17, pp. 586–594, September 1971.

[42] Y. Kou, S. Lin, and M. Fossorier Construction of Low Density Parity-Check Codes: a Geometric Approach, *in Proceedings of Second International Symposium on Turbo Codes*, Brest, France, September 4–7, 2000.

[43] Y. Kou, S. Lin, and M.P.C. Fossorier, Low density parity-check codes based on finite geometries: A rediscovery and new results, *IEEE Trans. Info. Theory*, 2001.

[44] F.R. Kschischang and B.J. Frey, Iterative decoding of compound codes by probability propagation in graphical models, *IEEE J. Select. Areas Commn.*, Vol. 16, pp. 219–230, February 1998.

[45] F.R. Kschischang, B.J. Frey, and H.-A. Loeliger, Factor graphs and the sum-product algorithm, *IEEE Trans. on Info. Theory*, Vol. 47, No. 2, pp. 498–519, February 2001.

[46] A.V. Kuznetsov and B.S. Tsybakov, On unreliable storage designed with unreliable components, *In the book Second International Symposium on Information Theory*, 1971, Tsahkadsor, Armenia. Publishing House of the Hungarian Academy of Sciences, pp. 206–217, 1973.

[47] A.V. Kuznetsov, On the Storage of Information in Memory Constructed from Unreliable components, *Prob. Info. Transmission*, Vol. 9, No. 3, pp. 100–113, 1973 (translated by Plenum from Problemy Peredachi Informatsii).

[48] S. L. Lauritzen, *Graphical Models*, Oxford University Press, Oxford, 1996.

[49] F. Lazebnik and V.A. Ustimenko, Explicit Construction of Graphs with Arbitrary Large Girth and of Large Size, *Discrete Appl. Math.*, Vol. 60, pp. 275–284, 1997.

[50] C.C. Lindner and C.A. Rodger, *Design Theory,* CRC Press, Boca Raton, 1997.

[51] J. Li, E. Kurtas, K.R. Narayanan, and C.N. Georghiades, On the performance of turbo product codes over partial response channels, *IEEE Trans. Magn.*, Vol. 37, No. 4, pp. 1932–1934, July 2001.

[52] J. Li, K.R. Narayanan, E. Kurtas, and C.N. Georghiades, On the performance of the high-rate TPC/SPC and LDPC codes over partial response channels, *IEEE Trans. Commn.*, Vol. 50, No. 5, pp. 723–735, May 2002.

[53] R. Lynch, E. Kurtas, A. Kuznetsov, E. Yeo, and B. Nikolic, The Search for a Practical Iterative Detector for Magnetic Recording, *TMRC 2003*, San Jose, August 2003.

[54] A.C.H. Ling and C.J. Colbourn, *Rosa triple systems,* in *Geometry, Combinatorial Designs and Related Structures* Hirschfeld J.W.P., Magliveras S.S., de Resmini M.J., Eds., Cambridge University Press, London, pp. 149–159, 1997.

[55] J.H. van Lint and R.M. Wilson, *A Course in Combinatorics,* Cambridge University Press, London, 1992.

[56] S. Litsyn and V. Shevelev, On ensembles of low-density parity-check codes: asymptotic distance distributions, *IEEE Trans. Info. Theory*, Vol. 48, No. 4, pp. 887–908, April 2002.

[57] M.G. Luby, M. Mitzenmacher, M.A. Shokrollahi, and D.A. Spielman, Improved low-density parity-check codes using irregular graphs, *IEEE Trans. Info. Theory*, Vol. 47, No. 2, pp. 585–598, February 2001.

[58] D.J.C. MacKay and R.M. Neal, Good Codes Based on Very Sparse Matrices, in *Cryptography and Coding, 5th IMA Conference, in Lecture Notes in Computer Science*, C. Boyd, Ed., 1995, Vol. 1025, pp. 110–111.

[59] D.J.C. MacKay, Good Error-Correcting Codes Based on Very Sparse Matrices, *IEEE Trans. Info. Theory*, Vol. 45, pp. 399–431, March 1999.

[60] D. MacKay and M. Davey, *Evaluation of Gallager Codes for Short Block Length and High Rate Applications,* http://www.cs.toronto.edu/ mackay/CodesRegular.html.

[61] F.J. MacWilliams and N.J. Sloane, *The Theory of Error-Correcting Codes*, North Holland, Amsterdam, 1977.

[62] M.A. Margulis, Explicit group-theoretic constructions for combinatorial designs with applications to expanders and concentrators, *Problemy Peredachi Informatsii*, Vol. 24, No. 1, pp. 51–60, 1988.

[63] R.J. McEliece, D.J.C. MacKay, and J.-F. Cheng, Turbo decoding as an instance of pearl's "Belief Propagation" algorithm, *IEEE J. Select. Areas Commn.*, Vol. 16, pp. 140–152, February 1998.

[64] R.J. McEliece, *Finite Fields for Computer Scientist and Engineers,* Kluwer, Boston, 1987.

[65] L.L. McPheters, S.W. McLaughlin, and K.R. Narayanan, Precoded PRML, Serial Concatenation and Iterative (Turbo) Decoding for Digital Magnetic Recording, in *IEEE International Magnetics Conference*, Kyongju, Korea, May 1999.

[66] E. Netto, "Zur Theorie der Tripelsysteme," *Math. Ann.*, 42, pp. 143–152, 1893.

[67] M. Oberg and P.H. Siegel, Performance analysis of turbo-equalized partial response channels, *IEEE Trans. Comm.*, Vol. 49, No. 3 , pp. 436–444, March 200.

[68] M. Odlyzko and R.P. Stanley, *Some Curious Sequences Constructed with the Greedy Algorithm,* Bell Labs Internal Memorandum, 1978.

[69] J. Pearl, *Probabilistic Reasoning in Intelligent Systems: Networks of Plausible Inference,* Morgan Kaufmann, San Mateo, CA, 1988.

[70] W.W. Peterson and E.J. Weldon, Jr., *Error-Correcting Codes,* MIT Press, Cambridge, MA, 1963.

[71] W. Pusch, D. Weinrichter, and M. Taferner, Turbo-Codes Matched to the 1 - D Partial Response Channel, in *Proceedings of IEEE International Symposium on Inform Theory*, Cambridge, MA, USA, p. 62, August 1998.

[72] W.E. Ryan, S.W. McLaughlin, K. Anim-Appiah, and M. Yang, Turbo, LDPC, and RLL codes in Magnetic Recording, in *Proceedings of Second International Symposium on Turbo Codes*, Brest, France, September 4–7, 2000.

[73] W.E. Ryan, L.L. McPheters, and S.W. McLaughlin, Combined Turbo Coding and Turbo Equalization for PR4-Equalized Lorentzian Channels, in *Proceedings of the Conference on Inform Sciences and Systems*, March 1998.

[74] W.E. Ryan, Performance of High Rate Turbo Codes on a PR4-Equalized Magnetic Recording Channel, in *Proceedings of IEEE International Conference on Communications*. Atlanta, GA, USA, pp. 947–951, June 1998.

[75] T. Richardson, A. Shokrollahi, and R. Urbanke, Design of Provably Good Low-Density Parity-Check Codes, in *International Symposium on Information Theory, 2000. Proceedings*, p. 199, June 2000.

[76] T.J. Richardson and R.L. Urbanke, The capacity of low-density parity-check codes under message-passing decoding, *IEEE Trans. Info. Theory*, Vol. 47, No. 2, pp. 599–618, February 2001.

[77] I.J. Rosenthal and P.O. Vontobel, "Construction of LDPC Codes using Ramanujan Graphs and Ideas from Margulis," in *Proceedings of 2001 IEEE International Symposium on Information Theory*, 2001.

[78] S. Sankaranarayanan, Erozan Kurtas, and Bane Vasic, Performance of Low-Density Generator Matrix Codes on Perpendicular Recording Channels, presented at *IEEE North American Perpendicular Magnetic Recording Conference*, January 6–8, 2003.

[79] S. Sankaranarayanan, B. Vasic, and E. Kurtas, A Systematic Construction of Capacity-Achieving Irregular Low-Density Parity-Check Codes, accepted for presentation in *IEEE International Magnetics Conference*, March 30–April 3, Boston, MA 2003.

[80] C.E. Shannon, A mathematical theory of communication, *Bell Syst. Tech. J.*, pp. 372–423, 623–656, 1948.

[81] J. Singer, a theorem in finite projective geometry and some applications to number theory, *AMS Trans.*, Vol. 43, pp. 377–385, 1938.

[82] H. Song, B.V.K. Kumar, E.M. Kurtas, Y. Yuan, L.L. McPheters, and S.W. McLaughlin, Iterative decoding for partial response (PR) equalized magneto-optical data storage channels, *IEEE J. Sel. Areas Commn.*, Vol. 19, No. 4, April 2001.

[83] H. Song, J. Liu, B.V.K. Kumar, and E.M. Kurtas, Iterative decoding for partial response (PR) equalized magneto-optical data storage channels, *IEEE Trans. Magn.*, Vol. 37, No. 2, March 2001.

[84] T. Souvignier, A. Friedman, M. Oberg, P.H. Siegel, R.E. Swanson, and J.K. Wolf, Turbo Codes for PR4: Parallel Versus Serial Concatenation, in *Proceedings of IEEE International Conference on Communications*, Vancouver, BC, Canada, pp. 1638–1642, June 1999.

[85] E. Spence and V.D. Tonchev, Extremal self-dual codes from symmetric designs, *Discrete Math.*, Vol. 110, pp. 165–268, 1992.

[86] M.G. Taylor, Reliable information storage in memories designed from unreliable components, *Bell Syst. Tech. J.*, Vol. 47, No. 10, 2299–2337, 1968.

[87] R.M. Tanner, A recursive approach to low complexity codes, *IEEE Trans. Infor. Theory*, Vol. IT-27, pp. 533–547, September 1981.

[88] R.M. Tanner, Minimum-distance bounds by graph analysis, *IEEE Trans. Info. Theory*, Vol. 47, No. 2, pp. 808–821, February 2001.

[89] R.M. Tanner, D. Sridhara, and T. Fuja, A Class of Group-Structured LDPC Codes, available ate http://www.cse.ucsc.edu/tanner/pubs.html

[90] R. Townsend and E.J. Weldon, Self-orthogonal quasi-cyclic codes, *IEEE Trans. Info. Theory*, Vol. IT-13, No. 2, pp. 183–195, 1967.

[91] B. Vasic, Low Density Parity-Check Codes: Theory and Practice, National Storage Industry Consortium (NSIC) quarterly meeting, Monterey, CA June 25–28, 2000.

[92] B. Vasic, Combinatorial Constructions of Structured Low-Density Parity-Check Codes for Iterative Decoding, in *Proceedings 2001 IEEE Information Theory Workshop*, pp. 134, 2001.

[93] B. Vasic, Combinatorial Construction of Low-Density Parity-Check Codes, in *Proceedings of 2002 IEEE International Symposium on Information Theory*, p. 312, June 30–July 5, 2002, Lausanne, Switzerland.

[94] B. Vasic, E. Kurtas, and A. Kuznetsov, Lattice Low-Density Parity-Check Codes and Their Application in Partial Response Channels, in Preceedings 2002 *IEEE International Symposium on Information Theory*, p. 453–453, June 30–July 5, 2002, Lausanne, Switzerland.

[95] B. Vasic, Structured Iteratively Decodable Codes Based on Steiner Systems and Their Application in Magnetic Recording, in *Proceedings of Globecom* 2001, Vol. 5 pp. 2954–2960, San Antonio, Texas, November 26–29, 2001

[96] B. Vasic, High-Rate Low-Density Parity-Check Codes Based on Anti-Pasch Affine Geometries, in *Proceedings ICC-2002*, Vol. 3, pp. 1332–1336, New York City, DC, April 28–May 2.

[97] B. Vasic, E. Kurtas, and A. Kuznetsov, LDPC codes based on mutually orthogonal latin rectangles and their application in perpendicular magnetic recording, *IEEE Trans. Magn.*, Vol. 38, No. 5, Part: 1, pp. 2346–2348, September 2002.

[98] B. Vasic, E. Kurtas, and A. Kuznetsov, Kirkman systems and their application in perpendicular magnetic recording, *IEEE Trans. Mag.*, Vol. 38, No. 4, Part: 1, pp. 1705–1710, July 2002.

[99] B. Vasic, K. Pedagani, and M. Ivkovic, "High-rate girth-eight low-density parity check codes on rectangular integer lattices," *IEEE Trans. Commn.* [in press].

[100] B. Vasic, A Class of Codes with Orthogonal Parity Checks and its Application for Partial Response Channels with Iterative Decoding, *Invited paper, 40th Annual Allerton Conference on Communication, Control, and Computing*, October 2–4, 2002.

[101] B. Vasic, E. Kurtas, and A. Kuznetsov, Regular Lattice LDPC Codes in Perpendicular Magnetic Recording, *Intermag 2002 Digest 151, Intermag 2002 Conference*, Amsterdam, The Netherlands, April 28–May 3, 2002.

[102] B. Vasic, High-Rate Low-Density Parity Check Codes Based on Anti-Pasch Affine Geometries, in *Preceegings of ICC-2002*, Vol. 3, pp. 1332–1336, New York City, NY, April 28–May 2.

[103] B. Vasic, E. Kurtas, and A. Kuznetsov, Kirkman Systems and their Application in Perpendicular Magnetic Recording, in *Proceedings 1st North American Perpendicular Magnetic Recording Conference* (NAPMRC), January 7–9th, 2002, Coral Gables, Florida.

[104] B. Vasic and K. Pedagani, Runlength Limited Low-Density Parity Check Codes Based on Deliberate Error Insertion, accepted for presentation at the *IEEE 2003 International Conference on Communications* (ICC 2003) 11–15 May, 2003 Anchorage, AK, USA.

[105] E.J. Weldon, Jr., Difference-set cyclic codes, *Bell Syst. Tech. J.*, 45, pp. 1045–1055, September 1966.

[106] S. Weller and S. Johnson, Iterative Decoding of Codes from Oval Designs, Defense Applications of Signal Processing, 2001 Workshop, Adelaide, Australia, September 2001.

[107] N. Wiberg, H.-A. Loeliger, and R. Ktter, Codes and iterative decoding on general graphs, *Eur. Trans. Telecommn.*, Vol. 6, pp. 513–525, September/October 1995.

[108] R.M. Wilson, Cyclotomy and difference families in elementary abelian groups, *J. Number Theory*, 4, pp. 17–47, 1972.

[109] V.V. Zyablov and M.S. Pinsker, Estimation of the error-correction complexity for Gallager low-density codes, *Probl. info. Transn.*, Vol. 11, pp. 18–28, 1975 (translated by Plenum from Problemy Peredachi Informatsii).

39

Turbo Coding for Multitrack Recording Channels

Zheng Zhang

Tolga M. Duman
Arizona State University
Tempe, AZ

Erozan M. Kurtas
Seagate Technology
Pittsburgh, PA

39.1 Introduction **39**-2
39.2 Multitrack Recording Channels **39**-3
39.3 Information Theoretical Limits:
Achievable Information Rates **39**-5
39.4 Turbo Coding for Multitrack Recording Systems **39**-6
Turbo Codes and Iterative Decoding Algorithm
  • Turbo Coding and Turbo Equalization for ISI Channels
  • MAP Detector and Iterative Decoding for Multitrack Systems
  • Examples
39.5 Discussion **39**-15

Zheng Zhang: received the B.E. degree with honors from Nanjing University of Aeronautics and Astronautics, Nanjing, China, in 1997, and the M.S. degree from Tsinghua University, Beijing, China, in 2000, both in electronic engineering. Currently, he is working toward the Ph. D. degree in electrical engineering at Arizona State University, Tempe, AZ.

His current research interests are digital communications, wireless/mobile communications, magnetic recording channels, information theory, channel coding, turbo codes and iterative decoding, MIMO systems and multi-user systems.

Tolga M. Duman: received the B.S. degree from Bilkent University in 1993, M.S. and Ph.D. degrees from Northeastern University, Boston, in 1995 and 1998, respectively, all in electrical engineering. Since August 1998, he has been with the Electrical Engineering Department Arizona State University first as an assistant professor (1998–2004), and currently as an associate professor. Dr. Duman's current research interests are in digital communications, wireless and mobile communications, channel coding, turbo codes, coding for recording channels, and coding for wireless communications.

Dr. Duman is the recipient of the National Science Foundation CAREER Award, IEEE Third Millennium medal, and IEEE Benelux Joint Chapter best paper award (1999). He is a senior member of IEEE, and an editor for IEEE Transactions on Wireless Communications.

Erozan M. Kurtas: received the B.Sc. degree from Bilkent University, Ankara, Turkey, in 1991 and M.Sc. and Ph.D., degrees from Northeastern University, Boston, MA, in 1993 and 1997, respectively.

His research interests cover the general field of digital communication and information theory with special emphasis on coding and detection for inter-symbol interference channels. He has published over 75 book chapters, journal and conference papers on the general fields of information theory, digital communications and data storage. Dr. Kurtas is the co-editor of the book Coding and Signal Processing for Magnetic Recording System (to be published by CRC press in 2004). He has seven pending patent

0-8493-1524-7/05/$0.00+$1.50
© 2005 by CRC Press, LLC

applications. Dr. Kurtas is currently the Research Director of the Channels Department at the research division of Seagate Technology.

Abstract

Intertrack interference (ITI) is considered as one of the major factors that severely degrade the performance of the practical detectors, especially for narrow-track systems of future. As a result, multi-track systems, where data are written in a group of adjacent tracks simultaneously and read back by multiple heads in parallel, have received significant attention in recent years. In this chapter, we study the turbo coding and iterative decoding schemes suitable for the multitrack recording systems. We describe the maximum a posteriori detector for the multitrack channels in both cases of deterministic ITI and random ITI. We provide simulation results showing that this concatenated coding scheme works very well. In particular, its performance is only about 1 dB away from the information theoretical limits for the ideal partial response channels. It is also shown that the performance achieved by the multitrack system is much better than that of its single-track counterpart. Finally, we provide some results to illustrate the effects of the media noise on the system performance.

39.1 Introduction

The areal density of the hard disk drives has been doubled every 18 months over the last decade [1]. We can increase the areal density in two directions; the axial direction and the radial direction. In the axial direction, we write the information bits close to each other, thus decrease the pulse width which increases the *intersymbol interference* (ISI), while in the radial direction, we decrease the track width which increases the *intertrack interference* (ITI). ITI is considered as one of the major factors that severely degrade the performance of the practical detectors and limit the recording density, especially for narrow-track systems of future. In fact, even with wide tracks, ITI can be present due to possible head misalignment [2]. As a result, multitrack channels, where data are written in a group of adjacent tracks simultaneously and read back by multiple heads in parallel, have received great attention due to their capability in combating the problems caused by the ITI [2–6]. Compared to the single-track systems in the presence of ITI, by using a multitrack recording system, a significantly better performance can be achieved, which in turn results in a density increase for a given performance requirement. In [7, 8], the information theoretical results also show that multitrack systems have higher achievable information rates, which is also justified by the fact that multitrack systems are more robust than their single-track counterparts. Furthermore, we will show later in this chapter that the multitrack systems are not only robust against the ITI, but also to the media noise, which is a type of signal-dependent noise that exists in high-density recording channels.

Another motivation for employing the multitrack system lies in that it provides easier timing and gain control, since the timing and gain information can be derived from any track or any subset of tracks [9]. This brings the benefit of relaxing the synchronization constraints, such as the k constraint of the *run-length limited* (RLL) codes [10], which further increases the storage density. In order to make use of this advantage of multitrack systems, some work has already been performed on the design of the two-dimensional modulation or constrained codes [9, 11–13].

With all these advantages, multitrack systems have the potential to play an important role in the future storage industry. On the other hand, the practical difficulties and costs of developing multiple read/write heads, as well as the inevitable increase of the complexity of the detection and coding/decoding schemes, remain as the main obstacles.

Uncoded multitrack recording systems have been studied in [3, 6, 14–18] (see, also the references therein). These works mainly focus on the development of suitable equalization and detection algorithms. Channel coding techniques are used for the multitrack systems as well in [19, 20] to combat both the intersymbol and intertrack interference, where conventional block and convolutional codes are employed.

In 1993, turbo codes were proposed [21], where the message sequence is encoded by two parallel concatenated convolutional codes that are separated by an interleaver, and decoded by a suboptimal iterative decoding algorithm. The performance of the turbo codes is shown to approach the Shannon limit in the *additive white Gaussian noise* (AWGN) channel when the size of the interleaver is selected sufficiently large. Turbo coding schemes have also been proposed for the ISI channels, such as the magnetic recording channels [22, 23], and it is shown that they can achieve an excellent performance in this case as well.

In this chapter, we consider the application of turbo codes to the multitrack systems (see also [8, 24]). We use a turbo code, or just a single convolutional code with an interleaver, as the outer encoder, and map the coded bits to different groups and transmit them through adjacent tracks. At the receiver, we develop a modified *maximum a posteriori* (MAP) detector for the multitrack systems with deterministic or random ITI. The turbo equalization [25] and the iterative decoding can be performed by exchanging the soft information between the channel MAP detector and the outer soft-input soft-output decoder that corresponds to the outer encoder. We show that the resulting system performance is very close to the information theoretical limits obtained in [7, 8]. We also note that, other turbo-like codes, such as *low-density parity-check* (LDPC) codes [26] and the block turbo codes (or, turbo product codes) [27] can be applied to the multitrack recording systems as well.

The chapter is organized as follows. In Section 39.2, the multitrack recording channel model is given which is nothing but a *multi-input multi-output* (MIMO) ISI channel. In Section 39.3, the capacity and the achievable information rates over such channels with binary inputs are reviewed. These information theoretical limits are useful to evaluate the effectiveness of the specific turbo coding schemes to be proposed. In Section 39.4, a MIMO MAP detector is developed, and the turbo coding and iterative decoding scheme are presented. The performance of the turbo coded systems is then compared with the information theoretical limits developed in Section 39.3, for both multitrack and single-track systems when ITI is present. Finally, the conclusions are provided in Section 39.5.

39.2 Multitrack Recording Channels

Mathematically, magnetic recording channels can be modeled as ISI channels. In addition, in the narrow-track systems, which will be more and more popular in the future recording systems, the intertrack interference will also be present. Both the ISI and the ITI are illustrated in Figure 39.1, where the black block denotes the media track that records the desired signal and the gray ones represent the interfering sources. In the axial direction, the interference comes from the adjacent symbols, and in the radial direction, it results due to the adjacent tracks.

To deal with the increased ITI, we can write the signals to multiple tracks and use multiple heads to read them for joint detection. In an (N, M) multitrack system, for which there are N heads reading M tracks

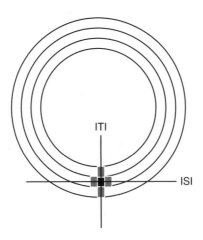

FIGURE 39.1 Multitrack system with intersymbol interference and intertrack interference.

simultaneously, the received signal for the nth head can be expressed as

$$r_n(t) = \sum_{m=1}^{M} a_{n,m} \sum_{k=-\infty}^{\infty} b_{m,k} \cdot p(t - kT) + w_n(t) \quad 1 \leq n \leq N \tag{39.1}$$

where $p(t)$ is the transition response of the channel, T is the symbol duration, $w_n(t)$ is the AWGN with two sided power spectral density of $N_0/2$, and $\{b_{m,k}\}$ is the transition symbol on the mth track with $b_{m,k} = x_{m,k} - x_{m,k-1}$ if $\{x_{m,k}\}$ is the transmitted binary signal.

Equivalently, the received signal is given by

$$r_n(t) = \sum_{m=1}^{M} a_{n,m} \sum_{k=-\infty}^{\infty} x_{m,k} \cdot h(t - kT) + w_n(t) \quad 1 \leq n \leq N \tag{39.2}$$

where $h(t)$ is the pulse response of the channel with $h(t) = p(t) - p(t - T)$. We assume that the pulse responses from different tracks are the same except for the amplitude varying with the distance between the track and the readhead, which is reflected by the coefficients $\{a_{n,m}\}$. This is a simplified model that is justified by the experimental measurements [2]. Other models can also be used, for which the coding/decoding schemes described later are still applicable.

For the sake of comparison, we also consider the single-track channels, which experience interference from adjacent tracks as well, but have only one read head. The receiver detects the desired signal from the corresponding track and considers the others as pure interference.

At the receiver, the output is passed through a matched filter, a sampler at the symbol rate and a noise-whitening filter [10]. The lth received symbol for the nth head is then given by

$$y_{n,l} = \sum_{m=1}^{M} a_{n,m} \sum_{k=0}^{K} x_{m,l-k} \cdot f_k + z_{n,l} \quad 1 \leq n \leq N \tag{39.3}$$

where $\{z_{n,l}\}$ is the white Gaussian noise sequence with variance $\sigma_z^2 = N_0/2$, and $\{f_l\}$ is the set of coefficients of the equivalent discrete-time ISI channel with memory K. We denote the ITI matrix as $A = [a_{n,m}]_{N \times M}$ and consider both cases of deterministic and random ITI.

In practice, the output signal of the magnetic recording channel is usually equalized to an appropriate *partial response* (PR) target using a linear equalizer. Therefore, for simplicity, the ideal PR channels are often used to model the magnetic recording channels, where the received signal can be expressed by Equation 39.3 with $\{f_l\}$ defined by the PR target. For the ideal normalized PR4 channel considered later in this chapter, we have $K = 2$ and $f_0 = 1/\sqrt{2}$, $f_1 = 0$ and $f_2 = -1/\sqrt{2}$.

As a more realistic example, we also use the longitudinal recording channel where the transition response $p(t)$ is modeled by the Lorentzian pulse

$$p(t) = \frac{1}{1 + (2t/PW_{50})^2} \tag{39.4}$$

where PW_{50} is the pulse width at the half height. We define the parameter $D_n = PW_{50}/T$ as the normalized density of the recording system.

In high-density magnetic recording channels, signal-dependent noise, called media noise exists [28, 29]. In this chapter, we only consider the fluctuation of the position of the transition pulse, called jitter noise, which is one of the main sources of the media noise. The received signal for the recording channel with jitter noise is then given by

$$r_n(t) = \sum_{m=1}^{M} a_{n,m} \sum_{k=-\infty}^{\infty} b_{m,k} \cdot p(t + j_{m,k} - kT) + w_n(t) \quad 1 \leq n \leq N \tag{39.5}$$

where $\{j_{m,k}\}$ is the independent zero-mean Gaussian noise term that reflects the amount of position jitter.

The *signal to noise ratio* (SNR) is defined as

$$\text{SNR} = \frac{C_1 \cdot C_2 \cdot E_s}{N_0} \tag{39.6}$$

where E_s is the transmitted symbol energy, and C_1 and C_2 are the normalization factors due to the received ITI and the channel response, respectively. For the deterministic ITI and the random ITI cases, C_1 can be expressed as $\frac{1}{N}\sum_{n=1}^{N}\sum_{m=1}^{M} a_{n,m}^2$ and $\frac{1}{N}\sum_{n=1}^{N}\sum_{m=1}^{M} E[a_{n,m}^2]$, respectively. For the longitudinal channel, $C_2 = \int_{-\infty}^{\infty} h^2(t)\,dt$, and for the normalized PR4 channel, $C_2 = 1$.

39.3 Information Theoretical Limits: Achievable Information Rates

Before we describe the turbo coding/decoding scheme for the multitrack recording channels, we present some results of the achievable information rates, which can be used as the ultimate theoretical limits of the coding/decoding schemes.

For *single-input single-output* (SISO) ISI channels, the capacity is derived in [30]. For multitrack systems with deterministic ITI, a lower bound on the capacity is derived in [31] following a similar approach as in [32]. In both cases, Gaussian inputs are assumed to be used. However, for practical digital communication systems, the inputs are constrained to be selected from a finite alphabet. For example, signals used in the magnetic recording systems are binary. In general, with the use of the constrained inputs, one cannot achieve the "unconstrained" capacity and the gap between the capacity and the achievable information rates under such input constraints is far from ignorable in high SNR regions, especially when the size of the alphabet is small. Therefore, it is also important to determine the information rates achievable under the specific input constraints.

Simulation-based methods have been recently proposed to estimate the achievable information rates for ISI channels with specific binary inputs [33–35]. The main idea is to use a simulation of the channel and employ the BCJR algorithm [36] to estimate the joint probability of the output sequence. Then, this estimate is used to compute the differential entropy of the output sequence and the mutual information between the input and the output, thus the achievable information rates over the noisy channel. By increasing the simulation complexity, one can easily obtain an accurate result. In [37], the maximization of the information rates is preformed for the case of Markov inputs with a certain memory, as apposed to the use of *independent identically distributed* (i.i.d.) inputs.

We can extend these techniques to the case of MIMO ISI channels with deterministic and random ITI. The key step is still the computation of the entropy, and thus the probability of the output sequence for a given channel simulation. To accomplish this for the multitrack systems, we set up the channel trellis based on both the ISI and the ITI, instead of the one that only takes the ISI into account. The details can be found in [8].

To give specific examples in this chapter, we consider the two-track and two-head channel with the ITI matrix

$$A = \begin{bmatrix} 1 & \alpha \\ \alpha & 1 \end{bmatrix} \tag{39.7}$$

where the diagonal elements are set to 1, which represent the amplitudes of the desired signals to each readhead, and off-diagonal elements represent the amplitudes of the ITI from the other track. We also consider a single-track system with $M=2$ and the ITI matrix $[1\ \alpha]$ for comparison. For the case of deterministic ITI, the entries of the matrix are fixed. On the other hand, for the random ITI case, we model α as uniform random variables over $[0, 1]$, which are assumed to be independent spatially and temporally, and known to the receiver.

The information rates per track use for the multitrack and single-track systems with different values of deterministic α are shown in Figure 39.2 for the PR4 channel where the input is constrained to be binary and i.i.d. with equal probability of 1s and 0s. We observe that the improvement in the achievable information

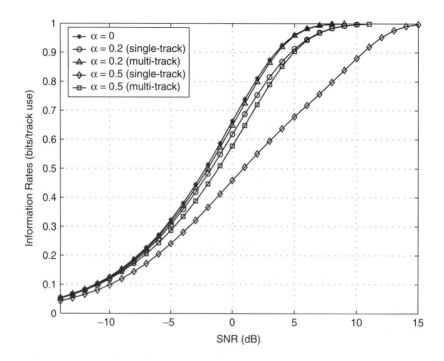

FIGURE 39.2   Information rates for the PR4 channel with deterministic α.

rates by adopting multitrack system is large, especially for high ITI levels. For example, to achieve a transmission rate of 16/17 bits per track use for the single-track system with arbitrarily low probability of error when $\alpha = 0.5$, about 11.6 dB is needed, but only 6.0 dB is enough for the multitrack system.

Figure 39.3 shows the information rates per track use for the PR4 channels with uniform ITI coefficients, which are known to the receiver only. We obtain similar conclusions by comparing the information rates for the multitrack and the single-track systems for this case as well.

In Figure 39.4, we maximize the information rates over all the Markov inputs with a memory $I = 2$ for the $(2, 2)$ multitrack PR4 systems. We observe that in the low-to-medium SNR region, the information rate achieved by the optimized Markov inputs is obviously larger than the one with i.i.d. inputs. Particularly, to achieve a rate of 0.2 bits per track use, there is a gain of 2.6 dB when $\alpha = 0.5$. However, when the transmission rate considered is high, for example, 16/17 bits per track use, there is almost no difference in the required SNR to make the error probability arbitrarily small.

We also present several results for the longitudinal channels with a normalized density of 2.0. Figure 39.5 shows the information rates per track use for the channels with a deterministic α. We also observe the effectiveness of the multitrack recording compared to the single-track one. For example, when α is increased from 0 to 0.5, the SNR loss is about 6.6 dB for the single-track system, whereas it is only 1.7 dB for the multitrack case, for the transmission rate of 16/17 bits per track use.

39.4   Turbo Coding for Multitrack Recording Systems

In 1993, Berrou et al. introduced the turbo codes, which can achieve near Shannon-limit performance over the memoryless AWGN channel [21]. For this reason, turbo coding and decoding techniques have been comprehensively studied and applied to many other channels, such as the ISI channels [22, 23]. In this section, we extend these techniques to the multitrack recording system, which can be viewed as a special case of the general MIMO ISI channels. We first review the turbo codes and iterative decoding algorithm, as well as their application to the ISI channels for completeness.

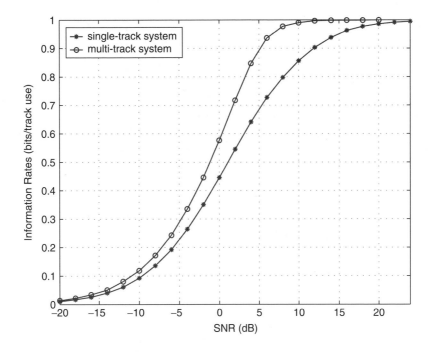

FIGURE 39.3 Information rates for the PR4 channel with uniform α.

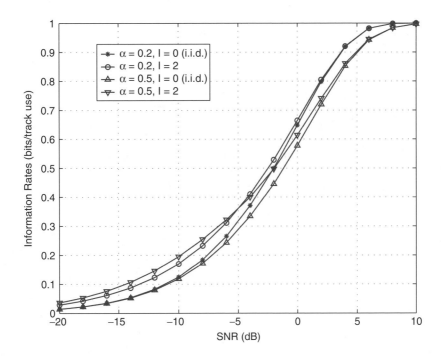

FIGURE 39.4 Maximized information rates of the (2, 2) multitrack PR4 system.

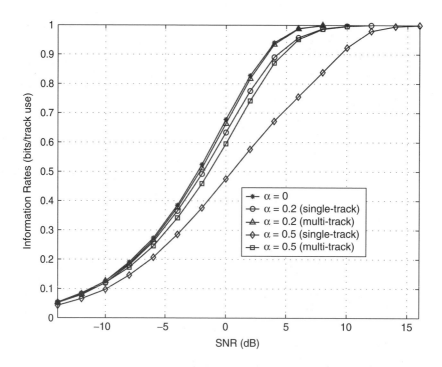

FIGURE 39.5   Information rates for longitudinal recording channels with deterministic $\alpha(D_n = 2.0)$.

39.4.1   Turbo Codes and Iterative Decoding Algorithm

The diagram of a turbo encoder is shown in Figure 39.6, where the uncoded message bits, denoted by **u**, are encoded by two parallel rate-1/2 *recursive systematic convolutional* (RSC) encoders that are separated by an interleaver. The codeword corresponding to the message **u** is then formed by the systematic information bits $\mathbf{x}^s$, and the parity bits, $\mathbf{x}^{1p}$ and $\mathbf{x}^{2p}$, generated by the two RSC encoders. This is an example of a rate 1/3 turbo code, however, by puncturing some of the parity bits, higher rate codes can be easily obtained [21].

The interleaver, which is used to permute the input bits, is very important in the construction of the turbo codes. For example, the average *bit error rate* (BER) over an AWGN channel is shown to be inversely proportional to the interleaver length when a uniform interleaver is used [38] (which is a probabilistic device that takes on each possible permutation with equal probability). This gain is referred as the interleaving gain.

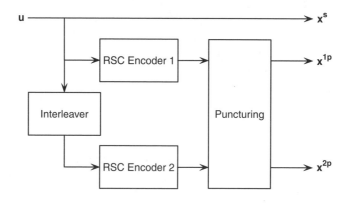

FIGURE 39.6   Turbo code block diagram.

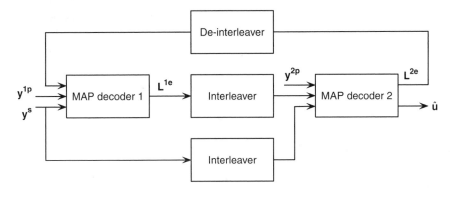

FIGURE 39.7   Block diagram of the turbo decoder.

Pseudo-random interleavers generally offer a good performance, therefore they are often used. However, well designed interleavers can outperform the average (uniform) interleavers significantly, especially in the high SNR region [39, 40].

The optimal decoder for the turbo code in the sense of minimizing the BER is the one that chooses the codeword with the maximum posterior probability. Due to the existence of the interleaver, such a decoder needs to perform an exhaustive search over the entire set of possible codewords. Therefore, its complexity is huge and it is not feasible, especially when the interleaver length is chosen to be relatively large to obtain a significant interleaving gain. A suboptimal decoding strategy is proposed in [21] to solve this problem. For the component convolutional codes, this decoder uses two MAP decoders that iteratively produce soft information about each bit, called extrinsic information, which is exchanged between the decoders. The diagram of the turbo decoder is shown in Figure 39.7, where $\mathbf{y}^s$, $\mathbf{y}^{1p}$ and $\mathbf{y}^{2p}$ are the observations corresponding to the systematic information bit sequence $\mathbf{x}^s$, and the two parity bit sequences, $\mathbf{x}^{1p}$ and $\mathbf{x}^{2p}$, respectively. $\mathbf{L}^{1e}$ and $\mathbf{L}^{2e}$ are the extrinsic information that is computed from the previous MAP decoder and fed to the next one. The turbo decoder works iteratively until the final hard decisions on the uncoded bits are made. For the details of the MAP algorithm and the iterative decoding for the component convolutional codes, the reader is referred to another chapter in this book [41].

39.4.2   Turbo Coding and Turbo Equalization for ISI Channels

After the invention of the turbo codes and the iterative decoding algorithm, many researches have proposed to apply these techniques for the transmission over other channels, including the ISI channels. For the case of ISI channels, instead of the *maximum likelihood sequence detector* (MLSD) [10, 42], it is necessary to use a soft-input soft-output detector or equalizer, which can generate soft reliability information about the coded symbols from the noisy channel observations. This information can then be passed to the turbo decoder to perform the iterative decoding. In addition, the extrinsic information generated by the turbo decoder can also be fed back to the channel detector to update the a priori probabilities. The concept of combining the channel equalizer and the turbo decoder in the iterative decoding scheme is called turbo equalization [25, 43].

It is clear that a MAP detector can be used as the soft-input soft-output detector for the ISI channel, where the soft information is computed based on the *log-likelihood ratio* (LLR) of the coded bit x_l that is transmitted over the ISI channel. We define the LLR as

$$\Lambda(x_l) = \log \frac{p(x_l = 1 \mid \mathbf{y}^n)}{p(x_l = 0 \mid \mathbf{y}^n)} \tag{39.8}$$

where $\mathbf{y}^n = \{y_1, \ldots, y_l, \ldots, y_n\}$ is assumed to be the channel output sequence or the observations with block length n.

To compute the *a posteriori probability* (APP) in Equation 39.8, we can use the BCJR algorithm [36] operating on the trellis of the ISI channel. Suppose the trellis state at the time instance l is S_l, then we have

$$
\Lambda(x_l) = \log \frac{p(x_l = 1, \mathbf{y}^n)/p(\mathbf{y}^n)}{p(x_l = 0, \mathbf{y}^n)/p(\mathbf{y}^n)}
$$
$$
= \log \frac{\sum_{(i,j)\in\Omega(1)} p(S_{l-1} = i, S_l = j, \mathbf{y}^n)}{\sum_{(i,j)\in\Omega(0)} p(S_{l-1} = i, S_l = j, \mathbf{y}^n)} \tag{39.9}
$$

where $\Omega(1)$ and $\Omega(0)$ are the sets of the valid state transitions where $x_l = 1$ and $x_l = 0$, respectively. We can further write $\Lambda(x_l)$ as [41]

$$
\Lambda(x_l) = \log \frac{\sum_{(i,j)\in\Omega(1)} \alpha_{l-1}(i) \cdot \gamma_l(i,j) \cdot \beta_l(j)}{\sum_{(i,j)\in\Omega(0)} \alpha_{l-1}(i) \cdot \gamma_l(i,j) \cdot \beta_l(j)} \tag{39.10}
$$

where $\alpha_l(j)$, $\beta_l(j)$ and $\gamma_l(i,j)$ are defined as

$$
\alpha_l(j) = p(y_1, \ldots, y_l, S_l = j) \tag{39.11}
$$
$$
\beta_l(j) = p(y_{l+1}, \ldots, y_n \mid S_l = j) \tag{39.12}
$$

and

$$
\gamma_l(i,j) = p(S_l = j \mid S_{l-1} = i) \cdot p(y_l \mid S_{l-1} = i, S_l = j)
$$
$$
= p(x_l) \cdot p(y_l \mid S_{l-1} = i, S_l = j) \tag{39.13}
$$

Here $p(x_l)$ is the a priori information about the coded bit x_l, and it can be updated iteratively using the extrinsic information generated by the turbo decoder. We can use the forward and backward recursions to compute $\alpha_l(j)$ and $\beta_l(j)$ in an efficient manner [36]. The details on the generation of the extrinsic information are presented in [41].

In the above discussion, we use the turbo code as the outer encoder, where two *parallel concatenated convolutional codes* (PCCC) separated by an interleaver are employed. As opposed to the PCCC scheme, a *single convolutional code* (SCC) can be employed instead, since the ISI channel can be viewed as a rate-one inner encoder. In this case, a precoder is necessary to make the channel (i.e., the inner code) recursive. Because there is only one component decoder in this case, the computational complexity is smaller than the one for the parallel concatenation, while the performance is shown to be comparable or, even better in some cases [23].

In addition, other turbo-like codes, including the LDPC codes and the block turbo codes, can also be used in place of the SCC or PCCC for the ISI channels [44, 45].

39.4.3 MAP Detector and Iterative Decoding for Multitrack Systems

We now extend the use of the code concatenation to the case of multitrack recording by designing the corresponding MAP detectors for different ITI scenarios. The block diagrams of the transmitter and receiver are shown in Figure 39.8. At the transmitter, we first encode the message bit sequence, denoted by $\mathbf{u}$, by using an outer encoder, such as the PCCC or SCC. After being passed through a random interleaver, the coded bits, represented by $\mathbf{x}$, are divided evenly into M groups, which are sent through the M transmitters or tracks. The N output sequences of the MIMO ISI channels, corrupted by the additive white Gaussian noise, constitute the received signal $\mathbf{y}$. At the receiver, the turbo equalization is used where a modified channel MAP detector takes the channel outputs and the extrinsic information fed back from the outer decoder as its inputs, and generates the soft information about the coded bits. This soft information about the inputs from different transmitters or tracks is deinterleaved and passed to the outer decoder. The outer decoder, that is, a turbo decoder or a MAP decoder for a single convolutional code, generates the extrinsic information, which is then fed back to the channel MAP detector after appropriate processing for the next iteration step. The LLR of the message bits are used to make hard decisions after a number of iterations.

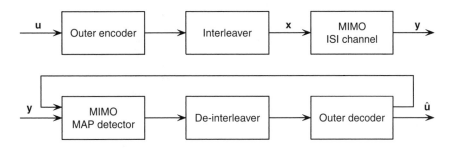

FIGURE 39.8 Block diagram of the turbo coding scheme for the multitrack system.

We use a $(2, 2)$ multitrack system with the ISI memory of K as an example to illustrate the necessary modifications. First we should set up the channel trellis with 2^{2K} states according to the multiple inputs and multiple outputs, which has 2^{2K} states. Assume that, at time instance l, the two coded bits transmitted over the two tracks are $x_{1,l}$ and $x_{2,l}$, and the two output sequences with length n are $\mathbf{y}_1^n = \{y_{1,1}, \ldots, y_{1,l}, \ldots, y_{1,n}\}$ and $\mathbf{y}_2^n = \{y_{2,1}, \ldots, y_{2,l}, \ldots, y_{2,n}\}$. In order to compute the log-likelihood ratios for $x_{1,l}$ and $x_{2,l}$, we compute four probabilities: $P(x_{1,l} = 0, x_{2,l} = 0 \mid \mathbf{y}_1^n, \mathbf{y}_2^n)$, $P(x_{1,l} = 0, x_{2,l} = 1 \mid \mathbf{y}_1^n, \mathbf{y}_2^n)$, $P(x_{1,l} = 1, x_{2,l} = 0 \mid \mathbf{y}_1^n, \mathbf{y}_2^n)$ and $P(x_{1,l} = 1, x_{2,l} = 1 \mid \mathbf{y}_1^n, \mathbf{y}_2^n)$. Thus, the log-likelihood ratios of the two bits can be computed as

$$\Lambda(x_{1,l}) = \log \frac{P\left(x_{1,l} = 1, x_{2,l} = 0 \mid \mathbf{y}_1^n, \mathbf{y}_2^n\right) + P\left(x_{1,l} = 1, x_{2,l} = 1 \mid \mathbf{y}_1^n, \mathbf{y}_2^n\right)}{P\left(x_{1,l} = 0, x_{2,l} = 0 \mid \mathbf{y}_1^n, \mathbf{y}_2^n\right) + P\left(x_{1,l} = 0, x_{2,l} = 1 \mid \mathbf{y}_1^n, \mathbf{y}_2^n\right)} \tag{39.14}$$

$$\Lambda(x_{2,l}) = \log \frac{P\left(x_{1,l} = 0, x_{2,l} = 1 \mid \mathbf{y}_1^n, \mathbf{y}_2^n\right) + P\left(x_{1,l} = 1, x_{2,l} = 1 \mid \mathbf{y}_1^n, \mathbf{y}_2^n\right)}{P\left(x_{1,l} = 0, x_{2,l} = 0 \mid \mathbf{y}_1^n, \mathbf{y}_2^n\right) + P\left(x_{1,l} = 1, x_{2,l} = 0 \mid \mathbf{y}_1^n, \mathbf{y}_2^n\right)} \tag{39.15}$$

Similar to the single-input single-output case, we can employ the BCJR algorithm after a minor modification to compute the APPs. Suppose the trellis state at time instance l is S_l, then

$$\gamma_l(i, j) = P(S_l = j \mid S_{l-1} = i) \cdot P(y_{1,l}, y_{2,l} \mid S_{l-1} = i, S_l = j) \tag{39.16}$$

$$= P(x_{1,l}, x_{2,l}) \cdot P(y_{1,l} \mid S_{l-1} = i, S_l = j) P(y_{2,l} \mid S_{l-1} = i, S_l = j) \tag{39.17}$$

$$\approx P(x_{1,l}) P(x_{2,l}) \cdot P(y_{1,l} \mid S_{l-1} = i, S_l = j) P(y_{2,l} \mid S_{l-1} = i, S_l = j) \tag{39.18}$$

where Equation 39.17 holds because the two observations are independent for a given state transition and the approximation in Equation 39.18 follows due to the use of the interleaver. When the ITI coefficients are random and known to the receiver, we just use them in the decoding as if they are constant for each symbol duration.

We can use a similar MAP detector and coding/decoding scheme for the single-track system, where there is only one received signal and one desired transmitted signal. Compared to the decoding for the multitrack systems, no detection is made for the other transmitted signal, which is considered as pure interference.

39.4.4 Examples

Figure 39.9 shows the performance of this turbo coding/decoding scheme for the $(2, 2)$ PR4 channels with deterministic ITI where α is set to be 0.5. The component convolutional code used for the outer encoder is a $(31, 33)$ code in octal form, the block length of the input to the outer encoder is 10016, the code rate is 16/17 and the iterative decoding algorithm with 15 iterations is used. We employ both SCC and PCCC and consider the use of a precoder, defined by $1/(1 \oplus D^2)$ where $\oplus$ indicates modulo-2 addition, as well. The performance of the uncoded system is obtained by passing the uncoded and unprecoded information bits through the channel and using the channel MAP decoder to detect them. The results show that SCC achieves a better performance compared to the PCCC, and its complexity is lower. The performance is only 1.2 dB away from the theoretical limit computed in Figure 39.2 for a code rate of 16/17 bits per track

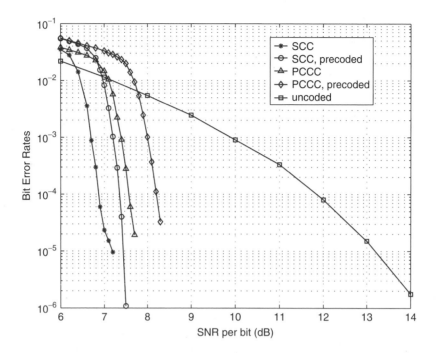

FIGURE 39.9   Performance of turbo coding for (2, 2) multitrack PR4 channels with deterministic α.

use if we consider a BER of 10^{-5} as reliable transmission (where the rate loss of 0.26 dB is considered). This represents a coding gain of about 6 dB relative to the uncoded system. We also notice that, when the SCC is used, the BER curve has lower error floor with the appropriate use of precoding. Therefore, we only consider the combination of the SCC and the precoder in the following examples.

In Figure 39.10, the bit error rates of the iterative decoding with SCC for the multitrack and single-track systems with different values of α are shown. Same parameters are used as the previous example and a precoder of $1/(1 \oplus D^2)$ is still employed. Compared with the information rates shown in Figure 39.2, the distances from the theoretical limits are from 0.8 dB to 1.2 dB (where the rate loss is considered), respectively. To achieve a BER of 10^{-5} for the multitrack systems, there is an SNR gain of 1.6 dB and 5.3 dB over the single-track systems when $\alpha = 0.2$ and 0.5, respectively, which is in line with our expectations from the information theoretical results.

We consider the case with uniform α (known to the receiver) in Figure 39.11 by using the same coding/decoding scheme. The performance is about 1.5 dB away from the limits computed in Figure 39.3 for both multitrack and single-track cases. There is an SNR gain of about 8.2 dB for the multitrack system over the single-track one, which is also expected by comparing the achievable information rates for both systems.

In Figure 39.12, we show some results for the longitudinal channel with normalized density $D_n = 2.0$ and deterministic ITI, which is equalized to a (2, 2) EPR4 channel $(1 + D - D^2 - D^3)$ with the same value of α. We use a single convolutional code as the outer encoder, together with the precoder $1/(1 \oplus D^2)$, and the code generators, the code rate and the block length are the same as in the earlier examples. We observe that, to achieve a BER of 10^{-5}, the required SNR per bit is 2.2 dB higher than the theoretical limit for $\alpha = 0.2$ and 3.0 dB higher for $\alpha = 0.5$. Although the performance is still very good, it is inferior to the previous examples. This may be due to two major factors. First, the equalizer target we used is based on the one-dimensional design and there should be better alternatives which take both the axial and radial interference into consideration. Also, the noise correlation due to the use of the equalizer is not taken into account in the decoding process.

We consider the effects of the jitter noise on the performance in Figure 39.13, where the detector ignores the existence of the jitter. We observe that the jitter degrades the system performance significantly. For

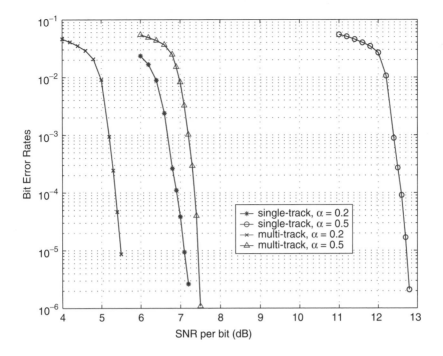

FIGURE 39.10 Performance of the SCC scheme for multitrack and single-track PR4 channels.

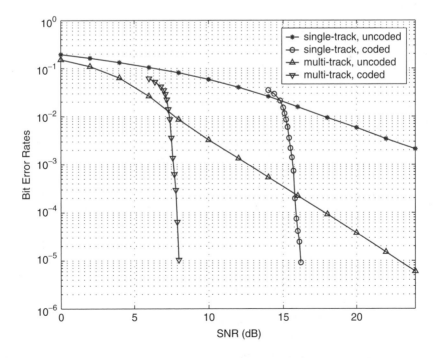

FIGURE 39.11 Performance of the SCC scheme for multitrack and single-track PR4 channels with uniform α.

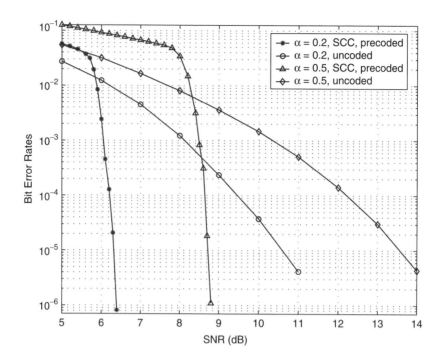

FIGURE 39.12   Performance of the SCC scheme for (2, 2) multitrack longitudinal channels.

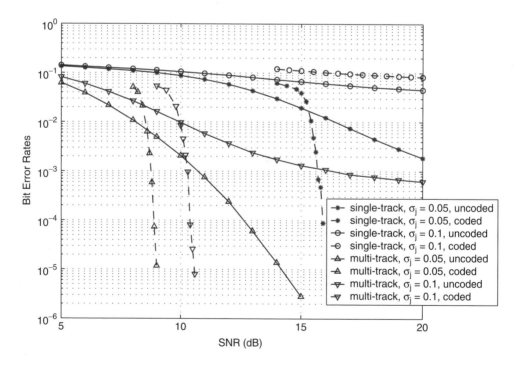

FIGURE 39.13   Performance of the SCC scheme for multitrack and single-track longitudinal channels with jitter noise.

the uncoded system, an error floor is observed, especially for the single-track case. Particularly, when $\sigma_j = 0.1$, the media noise is too high for the iterative decoding algorithm to correctly exchange the soft information in the single-track system, therefore the decoding algorithm almost fails. However, for the multitrack system, the turbo coding scheme still works well and very high coding gain can be obtained compared to the uncoded case.

39.5   Discussion

In order to increase the areal density of the magnetic recording systems, narrow tracks are used, which inevitably causes intertrack interference. We can employ multiple write and read heads to reduce the effects of the ITI. In this chapter, we considered the application of turbo codes to the multitrack recording systems, which are shown to achieve a very good performance (i.e., close to the information theoretical limits), for both cases with deterministic and random ITI. When compared with the single-track systems, the multitrack system has significant advantages in terms of the achievable information rates and the detection performance, which indicates that a multitrack system is a good solution to increase the areal densities of the future recording systems. In addition, we consider the effects of the media noise on the performance of the turbo coding scheme, and show that the multitrack systems are also more robust compared to the single-track ones even in the presence of media noise.

Since the design of the MAP detector for the multitrack systems is independent of the choice of the outer decoder, in addition to the turbo code (or, a single convolutional code), we can apply other turbo-like codes, such as the LDPC codes and block turbo codes, as well.

The MAP detector for the multitrack system can be considered as an optimal soft-output detector for the MIMO ISI channels. However, if a multitrack system with more tracks is considered, the computational complexity for this algorithm is high, as it is exponential in the number of tracks. Therefore, simplified schemes should be used instead, at the cost of some performance degradation. For example, we can give up the turbo equalization if the iterative decoding can be performed within the outer decoder (for the case of PCCC). In addition, we can use other soft-input soft-output detectors with lower complexities, such as the SOVA detector, or even other suboptimal detectors where the complexity increases linearly with the number of tracks.

We also note that, we did not consider the constrained modulation codes, such as the run-length limited codes for multitrack recording in this chapter. To make multitrack recording practical, new schemes combining both constrained coding and channel coding, as well as the iterative decoding algorithm should be employed [46].

References

[1] J. Moon, H. Thapar, B.V.K.V. Kumar, and K.A.S. Immink, (editorial) Signal processing for high density storage channels, *IEEE J. Select. Areas Commun.*, vol. 19, no. 4, pp. 577–581, April 2001.

[2] P.A. Voois and J.M. Cioffi, A decision feedback equalizer for multiple-head digital magnetic recording, in *Proceedings of IEEE International Conference on Communications (ICC)*, vol. 2, June 1991, pp. 815–819.

[3] L.C. Barbosa, Simultaneous detection of readback signals from interfering magnetic recording tracks using array heads, *IEEE Trans. Magn.*, vol. 26, no. 5, pp. 2163–2165, September 1990.

[4] P.A. Voois and J.M. Cioffi, Multichannel signal processing for multiple-head digital magnetic recording, *IEEE Trans. Magn.*, vol. 30, no. 6, pp. 5100–5114, November 1994.

[5] E. Soljanin and C.N. Georghiades, Multihead detection for multitrack recording channels, *IEEE Trans. Inform. Theory*, vol. 44, no. 7, pp. 2988–2997, November 1998.

[6] E.M. Kurtas, J.G. Proakis, and M. Salehi, Reduced complexity maximum likelihood sequence estimation for multitrack high-density magnetic recording channels, *IEEE Trans. Magn.*, vol. 35, no. 4, pp. 2187–2193, July 1999.

[7] Z. Zhang, T.M. Duman, and E.M. Kurtas, On information rates of single-track and multitrack recording channels with intertrack interference, in *Proceedings of IEEE International Symposium on Information Theory (ISIT)*, June-July 2002, p. 163.

[8] Z. Zhang and T.M. Duman, Information rates and coding for multitrack recording channels with deterministic and random intertrack interference, in *Proceedings of 40th Allerton Conference on Communications, Control and Computing (invited)*, Urbana, Illinois, October 2002, pp. 766–775.

[9] M.W. Marcellin and H.J. Weber, Two-dimensional modulation codes, *IEEE J. Select. Areas Commun.*, vol. 10, no. 1, pp. 254–266, January 1992.

[10] J.G. Proakis, *Digital Communications*, McGraw-Hill, Inc. New York, NY, 2001.

[11] J. Lee and V.K. Madisetti, Constrained multitrack RLL codes for the storage channel, *IEEE Trans. Magn.*, vol. 31, no. 3, pp. 2355–2364, May 1995.

[12] E. Soljanin and C.N. Georghiades, Coding for two-head recording systems, *IEEE Trans. Inform. Theory*, vol. 41, no. 3, pp. 747–755, May 1995.

[13] E.M. Kurtas, J.G. Proakis, and M. Salehi, Coding for multitrack magnetic recording systems, *IEEE Trans. Inform. Theory*, vol. 43, no. 6, pp. 2020–2023, November 1997.

[14] W.L. Abbott, J.M. Cioffi, and H.K. Thapar, Performance of digital magnetic recording with equalization and offtrack interference, *IEEE Trans. Magn.*, vol. 27, no. 1, pp. 705–716, January 1991.

[15] P.A. Voois and J.M. Cioffi, Multichannel digital magnetic recording, in *Proceedings of IEEE International Conference on Communications (ICC)*, vol. 1, June 1992, pp. 125–130.

[16] P.S. Kumar and S. Roy, Two-dimensional equalization: Theory and applications to high density magnetic recording, *IEEE Trans. Commun.*, vol. 42, no. 2/3/4, pp. 386–395, February/March/April 1994.

[17] M.P. Vea and J.M.F. Moura, Multichannel equalization for high track density magnetic recording, in *Proceedings of IEEE International Conference on Communications (ICC)*, vol. 2, May 1994, pp. 1221–1225.

[18] M.Z. Ahmed, P.J. Davey, T. Donnelly, and W.W. Clegg, Track squeeze using adaptive intertrack interference equalization, *IEEE Trans. Magn.*, vol. 38, no. 5, pp. 2331–2333, September 2002.

[19] A.M. Patel, Adaptive cross parity code for a high density magnetic tape subsystem, *IBM J. Res. Dev.*, vol. 29, pp. 546–562, November 1985.

[20] N. Kogo, N. Hirano, and R. Kohno, Convolutional code and 2-dimensional PRML class IV for multitrack magnetic recording system, in *Proceedings of IEEE International Symposium on Information Theory (ISIT)*, June 2000, p. 309.

[21] C. Berrou, A. Glavieux, and P. Thitimajshima, Near Shannon limit error-correcting coding and decoding: Turbo-codes, in *Proceedings of IEEE International Conference on Communications (ICC)*, vol. 2, May 1993, pp. 1064–1070.

[22] W.E. Ryan, Performance of high rate turbo codes on a PR4-equalized magnetic recording channel, in *Proceedings of IEEE International Conference on Communications (ICC)*, vol. 2, June 1998, pp. 947–951.

[23] T.V. Souvignier, M. Öberg, P.H. Siegel, R.E. Swanson, and J.K. Wolf, Turbo decoding for partial response channels, *IEEE Trans. Commun.*, vol. 48, no. 8, pp. 1297–1308, August 2000.

[24] E.M. Kurtas and T.M. Duman, Iterative decoders for multiuser ISI channels, in *Proceedings of Military Communications International Symposium (MILCOM)*, vol. 2, October-November 1999, pp. 1036–1040.

[25] D. Raphaeli and Y. Zarai, Combined turbo equalization and turbo decoding, in *Proceedings of IEEE Global Communications Conference (GLOBECOM)*, vol. 2, November 1997, pp. 639–643.

[26] R.G. Gallager, Low-density parity check codes, *IRE Trans. Inform. Theory*, vol. IT-8, pp. 21–28, January 1962.

[27] R. Pyndiah, A. Glavieux, A. Picart, and S. Jacq, Near optimum decoding of product codes, in *Proceedings of IEEE Global Communications Conference (GLOBECOM)*, vol. 1, November-December 1994, pp. 339–343.

[28] J.G. Zhu and H. Wang, Noise characteristics of interacting transitions in longitudinal thin film media, *IEEE Trans. Magn.*, vol. 31, no. 2, pp. 1065–1070, March 1995.

[29] T.R. Oenning and J. Moon, Modeling the Lorentzian magnetic recording channel with transition noise, *IEEE Trans. Magn.*, vol. 37, no. 1, pp. 583–591, January 2001.

[30] W. Hirt and J.L. Massey, Capacity of the discrete-time Gaussian channel with intersymbol interference, *IEEE Trans. Inform. Theory*, vol. 34, no. 3, pp. 380–388, May 1988.

[31] Z. Zhang, T.M. Duman, and E.M. Kurtas, Achievable information rates and coding for MIMO systems over ISI channels and frequency-selective fading channels, accepted for publication in the *IEEE Transactions on Communications*, 2004.

[32] H.E. Gamal, A.R. Hammons, Y. Liu, M.P. Fitz, and O.Y. Takeshita, On the design of space-time and space-frequency codes for MIMO frequency selective fading channels, *IEEE Trans. Inform. Theory*, vol. 49, no. 9, pp. 2277–2292, September 2003.

[33] D. Arnold and H.-A. Loeliger, On the information rate of binary-input channels with memory, in *Proceedings of IEEE International Conference on Communications (ICC)*, vol. 9, June 2001, pp. 2692–2695.

[34] H.D. Pfister, J.B. Soriaga, and P.H. Siegel, On the achievable information rates of finite state ISI channels, in *Proceedings of IEEE Global Communications Conference (GLOBECOM)*, vol. 5, November 2001, pp. 2992–2996.

[35] V. Sharma and S.K. Singh, Entropy and channel capacity in the regenerative setup with applications to Markov channels, in *Proceedings of IEEE International Symposium on Information Theory (ISIT)*, June 2001, p. 283.

[36] L.R. Bahl, J. Cocke, F. Jelinek, and J. Raviv, Optimal decoding of linear codes for minimizing symbol error rate, *IEEE Trans. Inform. Theory*, vol. 20, pp. 284–287, March 1974.

[37] A. Kavčić, On the capacity of Markov sources over noisy channels, in *Proceedings of IEEE Global Communications Conference (GLOBECOM)*, vol. 5, November 2001, pp. 2997–3001.

[38] S. Benedetto and G. Montorsi, Design of parallel concatenated convolutional codes, *IEEE Trans. Commun.*, vol. 44, no. 5, pp. 591–600, May 1996.

[39] T.M. Duman, Interleavers for serial and parallel concatenated (turbo) codes, *Wiley Encyclopedia of Telecommunications*, J.G. Proakis, Ed., December 2002, pp. 1141–1151.

[40] S. Dolinar and D. Divsalar, Weight distributions for turbo codes using random and non-random permutations, *TDA Progress Report*, vol. 42-122, pp. 56–65, August 1995.

[41] M.N. Kaynak, T.M. Duman, and E.M. Kurtas, "Turbo codes," *Coding and Recording Systems Handbook*, B. Vasic and E.M. Kurtas, Eds., CRC Press, Boca Raton, 2003.

[42] G.D. Forney, Maximum-likelihood sequence estimation of digital sequences in the presence of intersymbol interference, *IEEE Trans. Inform. Theory*, vol. 18, no. 3, pp. 363–378, May 1972.

[43] W.E. Ryan, L.L. McPheters, and S.W. McLaughlin, Combined turbo coding and turbo equalization for PR4-equalized Lorentzian channels, in *Proceedings of Conference on Information Sciences and Systems*, Princeton, NJ, March 1998, pp. 489–493.

[44] J.L. Fan, E.M. Kurtas, A. Friedmann, and S.W. McLaughlin, Low density parity check codes for partial response channels, in *Proceedings of Allerton Conference on Communications, Control and Computing*, Urbana, IL, October 1999.

[45] H. Song and J.R. Cruz, Block turbo codes for magnetic recording channels, in *Proceedings of IEEE International Conference on Communications (ICC)*, vol. 1, June 2000, pp. 85–88.

[46] J.L. Fan, Constrained Coding and Soft Iterative Decoding for Storage, Ph.D. dissertation, Stanford University, Stanford, CA, 1999.

Index

A

Adaptive equalization architectures, partial-response channels, **28**-1–13
   continuous time filter configurations, **28**-3–4
   equalizer architectures, **28**-11–12
   finite impulse response filter, least mean square algorithm, **28**-5
   gain equalization, **28**-5–6
   optimization procedure, **28**-6–8
   performance characterization, **28**-6–11
   simulation environment, **28**-6–8
   strategies, **28**-2
Adaptive timing recovery, partial-response channels, **26**-1–27
   jitter, bit error simulation results, **26**-12–14
   lock rate detection, loss of, **26**-14–15
   lock rate loss, **26**-14–15
   lock rate simulations, loss of, **26**-15
   minimum mean square error slope lookup table timing recovery, **26**-5–6
   Mueller and Muller timing loop, **26**-8–9
   noise jitter analysis of timing loop, **26**-10–12
   performance analysis, symbol rate timing loops, **26**-9–12
   phase detector properties, **26**-6–7
   qualitative loop filter description, **26**-10
   quantized slope lookup table phase detector, **26**-8
   signal gain of phase detector, **26**-7
   symbol rate timing recovery schemes, **26**-4–9
   symbol rate VCO, *vs.* interpolative timing recovery, **26**-2–3
   timing loop modes, **26**-3–4
Additive white Gaussian noise channel
   detection methods in, **33**-2–7
   performance of turbo codes over, **35**-9–11
Air bearings, head to media spacing, **4**-4–5
Analog front-end
   analog-to-digital converter, **34**-8–9
   continuous-time filter, **34**-6–8
   read channels, hard drives, **15**-3
   read/write channel implementation, **34**-5–9
   variable gain amplifier, with automatic gain control, **34**-5–6

Applications, modeling recording channel, **5**-9
Asperity, thermal, magnetic recording, **31**-1–16
   modeling, **31**-2–4
   simulation results, **31**-6–9
   system model, **31**-4–5
   system parameter settings, **31**-5–6
   system performance, effect on, **31**-4–9
   thermal asperity detection, cancellation, **31**-9–16
Asymptotic information rate, **17**-2–3
   capacity, **17**-3–4
   counting of sequences, **17**-2–3
Autocorrelation, digital signals, systems, statistical analysis, **7**-5–8
   phase randomising, **7**-7–8
   statistical autocorrelation, **7**-7
   time average autocorrelation, **7**-5–6
Averages, digital signals, systems, statistical analysis, **7**-3
   statistical average, **7**-4–5
   time average, **7**-3–4

B

Bahl, Cocke, Jelinek, Raviv algorithm
   intersymbol interference channel, maximum A posteriori probability symbol detection, **33**-5–6
Beamsplitter, optical recording systems, **3**-16–17
Bidirectional algorithms, data detection, **32**-22
Binary erasure channel, iterative decoding on, message-passing algorithm, **10**-2–7
Bipartite graphs, combinatorial designs, structured low-density parity-check codes, **38**-4–5
Bi-phase code, **1**-11
Birefringence, optical recording media, substrate, **3**-12–13
Bit error, **38**-14–16
Block, trellis code constructions, **20**-7–8
Breadth-first algorithms, **32**-13–17, **32**-21
Buffer, **16**-21–24
Burst, random error correction, **9**-18–21
Burst field, head position estimation, **29**-7–11
Bytes, codes over, error-correcting codes, **9**-8–10

C

Capacity-approaching codes, partial-response channels, **23**-1–21
  branch connections, choosing, **23**-11–13
  branch input-bit assignment, outer low-density parity-check codes, choosing, **23**-15
  branch type numbers in superchannel, choosing, **23**-11
  channel capacity, **23**-3
  channel model, **23**-2–3
  coding theorems for superchannels, **23**-8–9
  dicode channel, optimization results, **23**-16–18
  encoding/decoding system, outer low-density parity-check codes, **23**-13–14
  extended channel trellis, superchannel code rate, choosing, **23**-9–10
  information rates, **23**-6–13
  low-density parity-check decoding, outer low-density parity-check codes, **23**-16
  Markov channel capacity, **23**-4–5, **23**-6
  matched information rate trellis codes, **23**-9–13
  outer code optimization, outer low-density parity-check codes, **23**-15–16
  outer low-density parity-check codes, **23**-13–16
  spectral nulls, channel without, **23**-18
  subcode optimization, outer low-density parity-check codes, **23**-15–16
  trellis representations, **23**-3–4
Capacity of channel, *see Channel capacity*
Channel capacity
  error-correcting codes, **9**-6–8
  information rates, magnetic recording channels, **12**-3–12, **12**-4–5, **12**-5–12, **12**-10–12
  partial-response, **13**-3–4
  upper bound, delayed feedback capacity, partial-response channel capacity, **13**-11–15
    computation, feedback capacity, **13**-13–14
    delayed feedback capacity computation, **13**-14–15
    Markov sources, feedback capacity, **13**-11–12
    stochastic control formulation, **13**-12–13
Channel impediments, modeling recording channel, **5**-9
Codes/coding
  capacity-approaching, for partial-response channels, **23**-1–21
  concatenated single-parity, check, high-density digital recording systems, **37**-1–18
  constrained, **21**-1–10
  constrained binary, with error correction capability, **20**-1–17
  convolutional, partial-response channels, **22**-1–18
  error-correcting, **9**-1–21, **21**-1–10
  inter-track-interference reducing codes, multitrack system, **24**-3–4
  low-complexity encoder, decoder implementations, multitrack system, **24**-4–5
  low-density parity-check, **9**-7–14, **23**-13–18, **34**-22–24, **36**-1–19, **38**-1–21
  maximum transition run, **18**-1–13
  modulation, for storage system, **11**-1–11
  multitrack recording channels, turbo, **39**-1–17
  multitrack systems, **24**-1–29

  spectrum shaping, **19**-1–19
  structured low-density parity-check, **38**-1–21
  turbo, **35**-1–21
Colored noise, modulation codes, storage system, intertrack interference, channels with, **11**-6–7
Combinatorial low-density parity-check codes, **36**-7
  low-density parity-check code, **36**-7
Communication, information theory, magnetic recording channels
  digital signals, systems, statistical analysis, **7**-1–15
  error-correcting codes, **9**-1–21, **21**-1–10
  information theory, **12**-1–10
  message-passing algorithm, **10**-1–18
  modeling recording channel, **5**-1–10
  modulation codes, storage system, **11**-1–11
  partial-response channels, capacity of, **13**-1–19
  partial-response equalization, application to high-density magnetic recording channels, **8**-1–23
  signal, noise generation, magnetic recording channel simulations, **6**-1–20
Communication channel model, magnetic storage, **5**-1–4
Composite constrained combined encoding, spectrum shaping, **19**-15–17
Computer simulation, performance results, partial-response equalization, high-density magnetic recording channels, **8**-20–21
Concatenated convolutional codes, serially, iterative decoding of, **35**-8–9
Concatenated single-parity check codes for high-density digital recording systems, **37**-1–18
  density evolution, thresholds analysis using, **37**-9–13
  distance spectrum, analysis, **37**-5–9
  Gaussian approximation, **37**-9
  iterative decoding, equalization, **37**-4–5
  TPC/single-parity check coded partial-response systems, distance properties of, **37**-6–8
  turbo/single-parity check coded partial-response systems, distance properties of, **37**-8–9
Concatenation
  error-control coding, constrained coding, **21**-3–4
    reverse, **21**-4–6
  modulation codes, storage system, reverse, **11**-9–10
  turbo codes, **35**-5
Constrained binary codes with error correction capability, **20**-1–17; *see also Constrained coding*
  block, trellis code constructions, **20**-7–8
  bounds, **20**-2–4
  channel models, **20**-7
  code constructions, overview, **20**-7–10
  error correcting codes, constrained codes matched to, **20**-9
  error types, **20**-7
  input constraints, **20**-7
  Lee metrics, constructions using, **20**-10
  Levenshtein metrics, constructions using, **20**-10
  linear error correcting codes, combined codes directly derived from, **20**-8–9
  multilevel constructions, **20**-9
  spectral shaping constraints, error correction, **20**-10

standard constrained codes, maximum likelihood decoding of, **20**-10
trellis code construction, **20**-4–6
Constrained coding, error-control coding, **21**-1–10
bit insertion, **21**-6
configurations, **21**-2–8
definitions, **21**-2–3
lossless compression, **21**-6–8
reverse concatenation, **21**-4–6
soft iterative decoding, reverse concatenation, **21**-8–9
standard concatenation, **21**-3–4
Constrained systems, modulation codes, storage system, codes, **11**-2–4
Constraints, intersymbol interference channels, modulation codes, storage system, **11**-4–6
definitions, **11**-5–6
requirements, **11**-4–5
Convolutional codes, for partial-response channels, **22**-1–18
encoding system description, **22**-2–4
partial-response channels, trellis-matched codes, **22**-7–10, **22**-15
run-length limited trellis-matched codes, **22**-10–13, **22**-12–13
trellis codes distance spectrum criterion for, **22**-15
trellis codes for partial-response channels, based upon Hamming metric, **22**-4–7
Cyclic error-correcting codes, **9**-10–11
Cyclic redundancy check, **1**-8

D

Daniel, E.D., **1**-1
Data-dependent noise in storage channels, detection methods, **33**-1–17
additive white Gaussian noise channel, detection methods in, **33**-2–7
Bahl, Cocke, Jelinek, Raviv algorithm, signal-dependent channel noise, **33**-13
correlated noise, **33**-9–11
maximum A posteriori probability symbol detection over, intersymbol interference channel, **33**-5–7
maximum likelihood sequence detection over, intersymbol interference channel, **33**-2–4
pattern-dependent noise prediction, maximum likelihood sequence detection based on, **33**-11–13
pseudo-random bit sequence, playback signal power spectrum of, **33**-2–4
recording system model, **33**-7–9
soft-output Viterbi algorithm, **33**-13, **33**-16–17
Viterbi algorithm, outlines, **33**-16
Data detection, **32**-1–23
advanced algorithms, **32**-20–22
algorithms under investigation, **32**-20–22
bidirectional algorithms, **32**-22
breadth-first algorithms, **32**-13–17, **32**-21
decision feedback equalization, **32**-9–10
M-algorithm, **32**-21
noise predictive maximum likelihood detectors, **32**-17–19

partial-response equalization, **32**-2–9
postprocessor, **32**-19–20
RAM-based decision feedback equalization detection, **32**-10–13, **32**-11–13
trellis detection, **32**-11–13
Viterbi algorithm, **32**-13–17, **32**-21
Dc-free codes, spectrum shaping, **19**-2–9
capacity, dc-free constraint, **19**-4–5
dc-free constraint, **19**-3, **19**-5–6
encoding, decoding of, dc-free constraints, **19**-7–9
state dependent encoding, **19**-9
d-constraint, **1**-11
Decision feedback equalization, data detection, **32**-9–10
Density, recording
increasing, **14**-8
physical limits on, **14**-8–9
Dielectrics, optical recording media, **3**-9–10
Digital magnetic recording system, model of, partial-response equalization, high-density magnetic recording channels, **8**-10–13
Digital signals, systems, statistical analysis, **7**-1–15, **7**-2
autocorrelation, **7**-5–6, **7**-5–8
averages, **7**-3
Markov chains, **7**-14–15
phase randomising autocorrelation, **7**-7–8
power density spectrum, digital signals, **7**-8–14
signals, **7**-2
statistical autocorrelation, **7**-7
statistical average, **7**-4–5
time average, **7**-3–4
Direct overwrite, optical recording media, **3**-14
Disk-drive read/write channels, **34**-28–29
Disk drives, history of development of, **1**-9–15
Disk format control subsystem, hard drive controller functionality, **16**-6–19
channel control signals, **16**-8
data sector format fields, **16**-6–7
data sector read/write sequences, **16**-12–14
error correction code, **16**-16–19
master timing reference, **16**-8–9
read/write fault checking, **16**-14–15
servo spoke format fields, **16**-7
servo spoke read sequence, **16**-9–11
spoke sync detection, **16**-8–9
sync detection windowing, **16**-15
timing data sector R/W operations, **16**-11–12
Dye, organic, optical recoding media, **3**-3–6

E

Electrical World, **1**-2
Electronics, transition noise, simulating recording channel with, **6**-9–20
direct approach, **6**-10–12
indirect approach, **6**-13–16, **6**-19–20
numerical examples, **6**-10, **6**-18–19
simulation methods, **6**-16–18
Electronics noise only, simulating recording channel with, **6**-4–9
communications channel analogy, **6**-4–6

generation of noisy readback signal for given
  signal-to-noise ratio, **6**-6–8
numerical examples, **6**-8–9
EPRML, *see Extended partial-response equalization shifts*
Equalization architectures, adaptive, partial-response
  channels, **28**-1–13
    continuous time filter configurations, **28**-3–4
    finite impulse response filter, least mean square
      algorithm, **28**-5
    gain equalization, **28**-5–6
    optimization procedure, **28**-6–8
    performance characterization, **28**-6–11
Error-control coding
    bit insertion, **21**-6
    constrained coding, **21**-1–10
    definitions, **21**-2–3
    read channels, hard drives, **15**-9
    reverse concatenation, **21**-4–6
    soft iterative decoding, **21**-8–9
    standard concatenation, **21**-3–4
Error-correcting codes, **9**-1–21
    applications, **9**-18–21
    burst, random error correction, **9**-18–21
    bytes, codes over, **9**-8–10
    capacity of channel, **9**-6–8
    cyclic codes, **9**-10–11
    Euclid's algorithm, decoding RS codes with, **9**-16–18
    finite fields, codes over, **9**-8–10
    Hamming codes, **9**-6–8
    key equation, decoding of RS codes, **9**-13–16
    linear codes, **9**-3–6
    Reed Solomon codes, **9**-11–13
    syndrome decoding, **9**-6–8
Error correction capability, constrained binary codes with,
  20-1–17; *see also Error-correcting codes*
    block, trellis code constructions, **20**-7–8
    bounds, **20**-2–4
    channel models, **20**-7
    code constructions, overview, **20**-7–10
    error correcting codes, constrained codes matched to,
      20-9
    error types, **20**-7
    input constraints, **20**-7
    Lee metrics, **20**-10
    Levenshtein metrics, **20**-10
    linear error correcting codes, **20**-8–9
    maximum likelihood decoding of, **20**-10
    multilevel constructions, **20**-9
    post combined coding system architectures, **20**-11–12
    spectral shaping constraints, error correction, **20**-10
    trellis code construction, **20**-4–6
Error performance measures, read channels, hard drives,
  15-9–10
Euclid's algorithm, decoding RS codes with, **9**-16–18
Extended irregular repeat-accumulation code, **36**-5–6

F

Film stack design, optical recording media, **3**-11–12
    optical figure of merit, **3**-11–12
    thermal properties, **3**-12

Finite fields, codes over, error-correcting, **9**-8–10
Finite geometry codes, **36**-5
    low-density parity-check code, **36**-5
Finite impulse response filter
    architectures, read channels, **34**-9–11
    read channels, architectures, **34**-11–12
Finite impulse response-interpolated timing recovery loop
    structure, **27**-9–10
Focus, optical recording system, **3**-17–18
Frequency modulation code, **1**-11

G

Gallager codes, **36**-4
    low-density parity-check code, **36**-4
Gallager decoding scheme, probabilistic iterative, **10**-8–13
    min-sign approximation, **10**-11–13
Gaussian approximation, concatenated single-parity
    check codes, **37**-9
Giant magnetoresistive head, **1**-15
Global controller strategies, hard drive controller
    functionality, **16**-24–26
    data integrity strategies, **16**-24–26
    drive caching, **16**-26

H

Hamming codes, error-correcting, **9**-6–8
Hard drive controller functionality, **16**-1–27, **16**-6–24
    block diagram, **16**-6
    buffer memory subsystem, **16**-21–24
    channel control signals, **16**-8
    component overview, hard drive, **16**-1–3
    controller, role of, **16**-3–4
    data integrity strategies, **16**-24–26
    data sector format fields, **16**-6–7
    data sector read/write sequences, **16**-12–14
    disk format control subsystem, **16**-6–19
    drive caching, **16**-26
    error correction code decoding, data correction,
      16-16–19
    error correction code encoding, **16**-16
    error correction code subsystem, **16**-16
    global controller strategies, **16**-24–26
    hardware subsystems, **16**-6–24
    historical perspective, **16**-4–5
    host interface subsystem, **16**-19–20
    master timing reference, **16**-8–9
    processor subsystem, **16**-20–24
    read/write fault checking, **16**-14–15
    servo spoke format fields, **16**-7
    servo spoke read sequence, **16**-9–11
    servo subsystem, **16**-24
    spoke sync detection, **16**-8–9
    sync detection windowing, **16**-15
    timing data sector R/W operations, **16**-11–12
Hard drive read channels, **15**-1–11
Head design techniques, recording devices, **4**-1–17
    air bearings, head to media spacing, **4**-4–5
    CIP spin valve sensors, **4**-10–13
    CPP sensor designs, **4**-13–15

longitudinal write heads, **4**-7–8
magnetic recording transducers, history of, **4**-1–4
perpendicular heads, **4**-8–10
read head, perpendicular, **4**-10
read/write heads, future developments, **4**-15–16
write heads, **4**-6–7
Head position estimation, **29**-1–12
burst field, **29**-7–11
digital field, **29**-4–7
format efficiency, **29**-5
formatting strategies, **29**-9–10
impairments, **29**-8–9
offtrack detection, **29**-5–7
position estimators, **29**-10–11
servo writing, **29**-3–4
Heat sink/reflector, optical recording media, **3**-10–11
High-density digital recording systems, concatenated
  single-parity check codes for, **37**-1–18
density evolution, thresholds analysis using,
  37-9–13
distance spectrum, analysis, **37**-5–9
Gaussian approximation, **37**-9
iterative decoding, equalization, **37**-4–5
simulation results, **37**-13–16
system model, **37**-3–5
TPC/single-parity check coded partial-response
  systems, distance properties of, **37**-6–8
turbo/single-parity check coded partial-response
  systems, distance properties of, **37**-8–9
High-density magnetic recording channels,
  partial-response equalization, **8**-1–23
computer simulation, performance results, **8**-20–21
detection, partial-response signals, **8**-9–10
digital magnetic recording system, model of,
  8-10–13
intersymbol interference, digital communication
  systems, **8**-3–6
linear equalizer, partial-response targets, **8**-17–18
maximum-likelihood sequence detector, **8**-19–20
optimum detection, AWGN channel, **8**-13–17
partial-response signals, **8**-6–9
symbol-by-symbol detector, **8**-19
Higher order spectral zeros, codes with, spectrum shaping
  codes, **19**-9–17
K-RDSf sequences, **19**-11
partial-response channels, **19**-14–15
zeros at $f == 0$, **19**-10–11
History of magnetic storage, **1**-1–16
disk drives, **1**-9–15
early days, **1**-1–5
tape drives, **1**-5–9

I

Impulse responses, partial-response channel capacity,
  13-18
Information theory, magnetic recording channels, **12**-1–10
bounds on achievable information rates, **12**-6–8
channel capacity, information rates, magnetic
  recording channels, **12**-3–12
estimation of achievable information rates, **12**-8–10

information rates, correlated inputs, **12**-10–12
ISI channels with AWGN, **12**-4–5, **12**-5–12
media noise, realistic magnetic recording channels
  with, achievable information rates, **12**-12–13,
  12-14–17
Integer lattice, codes on rectangular subset of, structured
  low-density parity-check codes, **38**-10–12
Interleaver, convolutional coding with turbo codes,
  35-14–15
Interpolated timing recovery, **27**-1–16
bit error performance, **27**-11–16
conventional architecture, **27**-2
digital timing recovery, **27**-3–4
equalizer, implementation of, **27**-9
finite impulse response-interpolated timing recovery
  loop structure, **27**-9–10
implementation, **27**-7–10
initial sampling phase, estimate of, **27**-7–8
interpolated architecture, **27**-3–7
interpolation filter, implementation of, **27**-9
longitudinal recording, **27**-12–15
magnetic recording applications, **27**-10–16
minimum mean square error interpolation filter,
  27-4–5, **27**-5–7
oversampling system, effect of employing, **27**-9
perpendicular recording, **27**-15–16
signal-to-noise ratio performance, **27**-11
system model, **27**-10
Intersymbol interference, digital communication systems,
  partial-response equalization, high-density
  magnetic recording channels, **8**-3–6
Intersymbol interference channel
Bahl, Cocke, Jelinek, Raviv algorithm, **33**-5–6
maximum A posteriori probability symbol detection
  over, **33**-5–7
maximum likelihood sequence detection over, **33**-2–4
soft-output Viterbi algorithm, **33**-6–7
turbo coding for multitrack recording channels,
  39-9–10
Irregular low-density parity-check codes, **36**-5
low-density parity-check code, **36**-5
Irregular repeat-accumulation code, **36**-5–6
Iterative decoding, **34**-19–24, **35**-6–9
algorithm, **35**-9
equalization, concatenated single-parity check codes,
  37-4–5
high-density digital recording systems, concatenated
  single-parity check codes for, **37**-1–18
log-domain sum-product algorithm decoder,
  36-11–13
low-density parity-check codes, **36**-1–19, **36**-7–17,
  38-1–21
maximum A posteriori decoder, **35**-6–7
multitrack system turbo coding, **39**-10–11
probability-domain sum-product algorithm decoder,
  36-8–11
serially concatenated convolutional codes, iterative
  decoding of, **35**-8–9
structure, **35**-7–8
turbo coding, **35**-1–21, **39**-8–9
Iteratively decodable codes, **35**-5–6

J

Jitter, **26**-10–14

K

k-constraint control, **1**-11

L

Lattice construction, structured low-density parity-check codes, **38**-10–12
Lee metrics, error correction capability, constrained binary codes with, constructions using, **20**-10
Levenshtein metrics, error correction capability, constrained binary codes with, constructions using, **20**-10
Light source, optical recording systems, **3**-15–16
Linear equalizer, partial-response targets, partial-response equalization, high-density magnetic recording channels, **8**-17–18
Linear error-correcting codes, **9**-3–6
Linear minimum mean-square-error equalization, two-dimensional data detection, **25**-5–6
Logical organization, data on disk, **14**-7
Longitudinal, perpendicular recording, physics of, **2**-1–26
    DC remanent noise spectrum, **2**-14–15
    magnetization transition, **2**-14–16
    media noise spectra, **2**-13–16
    noise spectrum, pseudo-random bit sequence, **2**-17
    nonlinear transition shift, **2**-21–22
    playback signal, **2**-8–11, **2**-11–13
    pseudo-random bit sequence, playback signal power spectrum of, **2**-13
    pseudo-random bit sequence in signal-to-noise ratio calculations, **2**-19–21
    signal-to-noise ratio calculations, **2**-17–21
    square-wave pattern, **2**-17
    square-wave pattern in signal-to-noise ratio calculations, **2**-17–19
    transition noise spectrum, **2**-16
    transition parameter, isolated transition, **2**-3–6
Longitudinal write heads, **4**-7–8
Lorentzian channels, turbo codes, performance over, **35**-16–18
Low-density parity-check codes, **36**-1–19
    array codes, **36**-6
    bit error, **38**-14–16
    combinatorial designs, bipartite graphs, **38**-4–5
    combinatorial low-density parity-check codes, **36**-7
    design approaches, **36**-4–7
    difference families, **38**-5–7
    extended irregular repeat-accumulation code, **36**-5–6
    finite geometry codes, **36**-5
    Gallager codes, **36**-4
    Gallager's probabilistic iterative decoding scheme, **10**-8–13, **10**-11–13
    graphical representation, **36**-2–3
    integer lattice, codes on rectangular subset of, **38**-10–12
    irregular low-density parity-check codes, **36**-5
    irregular repeat-accumulation code, **36**-5–6
    iterative decoding algorithms, **36**-7–17
    iterative decoding schemes for, message-passing algorithm, **10**-7–14
    lattice construction, low-density parity-check codes, **38**-10–12
    log-domain sum-product algorithm decoder, **36**-11–13
    low-density parity-check codes, difference families, **38**-5–7
    MacKay codes, **36**-4
    matrix representation, **36**-2
    partial-response channels, application in, **38**-12–16
    probability-domain sum-product algorithm decoder, **36**-8–11
    projective planes, codes on, **38**-7–10
    reduced complexity decoders, **36**-13–17
    repeat-accumulation code, **36**-5–6
    representations of, **36**-2–3
    soft-decision iterative decoding, **10**-13–14
    structured, **38**-1–21
Low-density parity-check decoders, **34**-22–24

M

Magnetic disk drive read/write channel integration, **34**-4–5
Magnetic multilayers, **3**-15
Magnetic recording, basics of, **14**-1–6
Magnetic recording channel capacity, information rates, magnetic recording channels
    bounds on, **12**-6–8
    estimation of, **12**-8–10
    ISI channels with AWGN, **12**-4–5, **12**-5–12
Magnetic recording channel information theory, **12**-1–10
    achievable information rates with media noise, **12**-14–17
    channel capacity, information rates, **12**-3–12, **12**-10–12
    digital signals, systems, statistical analysis, **7**-1–15
    error-correcting codes, **9**-1–21
    information theory, **12**-1–10
    media noise, realistic magnetic recording channels with, achievable information rates, **12**-12–13
    message-passing algorithm, **10**-1–18
    modeling recording channel, **5**-1–10
    modulation codes, storage system, **11**-1–11
    partial-response channels, capacity of, **13**-1–19
    partial-response equalization, application to high-density magnetic recording channels, **8**-1–23
    signal, noise generation, magnetic recording channel simulations, **6**-1–20
Magnetic recording channel simulations, signal, noise generation for, **6**-1–20
    communications channel analogy, electronics noise, **6**-4–6
    numerical examples, **6**-8–9, **6**-10, **6**-18–19
    simulation methods, simulating recording channel, **6**-16–18
Magnetic recording transducers, history of, **4**-1–4
Magnetic storage, history of, **1**-1–16
    disk drives, **1**-9–15
    early days, **1**-1–5
    tape drives, **1**-5–9

Magnetization transition, playback signal of, **2**-8–11

Magneto-optic media, **3**-6–9

    Co/Pt, Co/Pd multilayers, **3**-9

    RE-TM alloys, **3**-7–9

Magnetophon, development of, **1**-3–4

M-algorithm, data detectioin, **32**-21

Manchester code, **1**-11

Marginal function

    computation of, message-passing algorithm, **10**-15–17

    message-passing schedule to compute, **10**-17

Markov chains, **7**-14–15

Markov channel capacity, partial-response channels, capacity-approaching codes, computing, **23**-6

Markov source, computing information rate for, partial-response channel capacity, **13**-4–7

    Monte Carlo method, **13**-6–7

    reformulating expression of mutual information rate, **13**-5–6

    stationary Markov source, **13**-4–5

Maximum A posteriori detector, turbo coding for, multitrack recording channels, **39**-10–11

Maximum A posteriori probability symbol detection, over intersymbol interference channel, Bahl, Cocke, Jelinek, Raviv algorithm, **33**-5–6

Maximum-likelihood sequence detector, partial-response equalization, high-density magnetic recording channels, **8**-19–20

Maximum transition run coding, **18**-1–13

    codes, **18**-4–10

    detector design, constraints, **18**-10–11

    error event characterization, **18**-2–4

    simulation results, **18**-11–12

Media noise

    magnetic recording channels with, **12**-12–13, **12**-14–17

    partial-response channel capacity, **13**-18

    realistic magnetic recording channels with, achievable information rates, **12**-12

Message-passing algorithm, **10**-1–18

    binary erasure channel, iterative decoding on, **10**-2–7

    Gallager's probabilistic iterative decoding scheme, **10**-8–13, **10**-11–13

    low-density parity-check codes, iterative decoding schemes for, **10**-7–14

    marginal function, **10**-15–17

    message-passing schedule to compute, **10**-17

    soft-decision iterative decoding, **10**-13–14

Metal-in-gap heads, **1**-11

Metric-first algorithms, data detection, **32**-21–22

Minimum mean square error interpolation filter

    with channel knowledge, **27**-5–7

    without channel knowledge, **27**-4–5

Minimum mean square error slope lookup table timing recovery, partial-response channels, adaptive timing recovery, **26**-5–6

MM timing loop, *see Mueller and Muller timing loop*

Modeling recording channel, **5**-1–10

    applications, **5**-9

    channel impediments, **5**-9

    communication channel model, magnetic storage, **5**-1–4

    nonlinearity characterization, **5**-8–9

signal-dependent medium noise, **5**-4–5

signal-dependent noise, signal-to-noise ratio definition, **5**-5–8

Modified-frequency modulation code, **1**-11

Modulation codes, storage system, **11**-1–11

    colored noise, intertrack interference, channels with, **11**-6–7

    constrained systems, codes, **11**-2–4

    constraints, ISI channels, **11**-4–5, **11**-4–6, **11**-5–6

    example, **11**-7–8

    reversed concatenation, **11**-9–10

    soft-output decoding, modulation codes, **11**-9

Modulation coding, read channels, hard drives, **15**-8–9

Monte Carlo method to compute, Markov source, partial-response channels, **13**-6–7

Mueller and Muller timing loop, partial-response channels, adaptive timing recovery, **26**-8–9

Multitrack channel model, coding, detection, multitrack systems, **24**-2–3

Multitrack constrained codes, coding, detection, multitrack systems, **24**-3–5

Multitrack recording channels, turbo coding for, **39**-1–17

    information theoretical limits, **39**-5–6

    intersymbol interference channels, turbo equalization, **39**-9–10

    iterative decoding for multitrack systems, **39**-10–11

    maximum A posteriori detector, **39**-10–11

    turbo codes, iterative decoding algorithm, **39**-8–9

Multitrack systems, coding, detection for, **24**-1–29

    ITI reducing codes, **24**-3–4

    low-complexity encoder, decoder implementations, **24**-4–5

    multitrack channel model, **24**-2–3

    multitrack constrained codes, **24**-3–5

    multitrack soft error-event correcting scheme, **24**-5–7

    research, current state of, multitrack codes, **24**-2

    synchronization, constrained coding for, **24**-4

N

Noise jitter analysis of timing loop, partial-response channels, adaptive timing recovery, **26**-10–12

Noiseless channel, codes for, runlength limited sequences, **17**-6–8

Noise prediction

    maximum likelihood detectors, **1**-15, **32**-17–19

    servo detection, **30**-4

Nonlinearity characterization, modeling recording channel, **5**-8–9

Nonlinear transition shift, **2**-21–22

Non-Return to Zero Inverted code, **1**-6–8

NRZI code, *see Non-Return to Zero Inverted code*

O

Offtrack detection, head position estimation, **29**-5–7

$O, G/I$ sequences, runlength limited sequences, **17**-4–5

Optical recording

    beamsplitter, **3**-16–17

    birefringence, **3**-12–13

    Co/Pt, Co/Pd multilayers, **3**-9

dielectrics, 3-9–10
direct overwrite, 3-14
exchange-coupled, field-coupled multilayers, 3-12
focus, tracking servo, 3-17–18
formatting, 3-13–14
GeSbTe, AgInSbTe phase change, 3-3–4
heat sink/reflector, 3-10–11
light source, 3-15–16
magnetic multilayers, 3-15
magnetic properties, 3-12
magneto-optic media, 3-6–9
mechanical properties, 3-13
media, 3-3–15
media noise sources, 3-5–6
noise, 3-18–19
noise sources, 3-14
organic dye, 3-3–6, 3-5, 3-12
phase change, 3-3–6
properties, optical, 3-12–13
read signal, 3-18
RE-TM alloys, 3-7–9
super-resolution, 3-14–15
systems, 3-15–19
thermal properties, 3-12, 3-13
transparency, 3-12–13
Optical recording physics, 3-1–22
Optimum detection, AWGN channel, partial-response
    equalization, high-density magnetic recording
    channels, 8-13–17
Organic dye, optical recording media, 3-3–6

P

Parallel concatenated convolutional codes
    turbo, 35-2–4
Parity-check codes, low-density, 36-1–19
    array codes, 36-6
    combinatorial low-density parity-check codes, 36-7
    design approaches, 36-4–7
    extended irregular repeat-accumulation code, 36-5–6
    finite geometry codes, 36-5
    Gallager codes, 36-4
    graphical representation, 36-2–3
    irregular low-density parity-check codes, 36-5
    irregular repeat-accumulation code, 36-5–6
    iterative decoding algorithms, 36-7–17
    log-domain sum-product algorithm decoder,
        36-11–13
    matrix representation, 36-2
    probability-domain sum-product algorithm decoder,
        36-8–11
    reduced complexity decoders, 36-13–17
    repeat-accumulation code, 36-5–6
Partial-response channel adaptive equalization
    architectures, 28-1–13
    actual equalizer architectures, 28-11–12
    continuous time filter, 28-3–4
    equalization architectures, strategies, 28-2
    finite impulse response filter, least mean square
        algorithm, 28-5
    gain equalization, 28-5–6

optimization procedure, 28-6–8
simulation environment, 28-6–8
Partial-response channel adaptive timing recovery,
    26-1–27
    jitter, bit error simulation results, 26-12–14
    lock rate, loss of, 26-14–15
    lock rate detection, loss of, 26-14–15
    lock rate simulations, loss of, 26-15
    minimum mean square error slope lookup table timing
        recovery, 26-5–6, 26-6–7
    Mueller and Muller timing loop, 26-8–9
    noise jitter analysis of timing loop, 26-10–12
    performance analysis, symbol rate timing loops,
        26-9–12
    signal gain of phase detector, 26-7
    symbol rate timing recovery schemes, 26-4–9
    symbol rate VCO, *vs.* interpolative timing recovery,
        26-2–3
    timing loop modes, 26-3–4
Partial-response channel capacity, 13-1–19
    channel capacity, 13-3–4
    channel capacity upper bound, delayed feedback
        capacity, 13-11–15, 13-14–15
    channel model, 13-2
    feedback capacity, 13-13–14
        Markov sources, 13-11–12
    information rate, channel capacity, 13-3–4
    iterative Markov source optimization algorithm, 13-9
    Markov source, computing information rate for, 13-4–7
    media noise, 13-18
    mutual information rate, reformulating expression of,
        13-8–9
    stationary Markov source, 13-4–5
    tight channel capacity lower bounds, maximal
        information rates, Markov sources, 13-7–11
Partial-response channel capacity-approaching codes,
    23-1–21
    branch connections, choosing, 23-11–13
    branch input-bit assignment, outer low-density
        parity-check codes, choosing, 23-15
    channel capacity, 23-3
    channel model, 23-2–3
    dicode channel, optimization results, 23-16–18
    encoding/decoding system, outer low-density
        parity-check codes, 23-13–14
    extended channel trellis, superchannel code rate,
        choosing, 23-9–10
    information rates, 23-6–13
    low-density parity-check decoding, outer low-density
        parity-check codes, 23-16
    Markov channel capacity, 23-4–5, 23-6
    matched information rate trellis codes, 23-9–13
    number of states in superchannel, choosing, 23-10–11
    outer low-density parity-check codes, 23-13–16
    outer subcode rates, outer low-density parity-check
        codes, 23-15
    spectral nulls, channel without, 23-18
    trellis representations, 23-3–4
Partial-response channel convolutional codes, 22-1–18
    bit stuffing, run-length limited trellis-matched codes,
        22-12–13

distance spectrum criterion, trellis-matched codes, **22**-15
encoding system description, **22**-2–4
partial-response channels, trellis-matched codes, **22**-7–10
run-length limited trellis-matched codes, **22**-10–13
trellis codes for partial-response channels, based upon Hamming metric, **22**-4–7
Partial-response channels
  Monte Carlo method to compute Markov source, **13**-6–7
  structured low-density parity-check codes, application in, **38**-12–16
Partial-response channel turbo codes
  recording, **35**-11
  recording channels, performance over, **35**-15–16
Partial-response equalization
  data detection, **32**-2–9
  detection, partial-response signal, **8**-9–10
  digital magnetic recording system, model, **8**-10–13
  high-density magnetic recording channels, **8**-1–23, **8**-20–21
  intersymbol interference, digital communication systems, **8**-3–6
  linear equalizer, partial-response targets, **8**-17–18
  with maximum-likelihood detection, read/write channel implementation, **34**-3–4
  maximum-likelihood sequence detector, **8**-19–20
  optimum detection, AWGN channel, **8**-13–17
  partial-response signals, **8**-6–9
  symbol-by-symbol detector, **8**-19
Partial-response maximum likelihood channel, **1**-9–10, **1**-13, **15**-3–4, **34**-3–4
Partial-response signals
  high-density magnetic recording channels, detection of, **8**-9–10
  maximum likelihood sequence estimation, read channels, hard drives, **15**-3–4
  partial-response equalization, high-density magnetic recording channels, **8**-6–9
Pattern-dependent noise prediction, maximum likelihood sequence detection based on, **33**-11–13
Perpendicular, longitudinal recording, physics of, **2**-1–26
  DC remanent noise spectrum, **2**-14–15
  isolated magnetization transition, **2**-14–16
  media noise spectra, **2**-13–16
  noise spectrum, pseudo-random bit sequence, **2**-17
  nonlinear transition shift, **2**-21–22
  playback signal of, **2**-8–11
  playback signal power spectrum, **2**-13
  pseudo-random bit sequence, **2**-19–21
  reciprocity principle, playback signal from, **2**-7–13
  signal-to-noise ratio calculations, **2**-17–21
  square-wave magnetization pattern, playback signal of, **2**-11–13
  square-wave pattern, **2**-17–19
  transition noise spectrum, **2**-16
  transition parameter, isolated transition, **2**-3–6
Perpendicular heads, **4**-8–10
Phase change, optical recording media, **3**-3–6

GeSbTe, AgInSbTe phase change, **3**-3–4
magnetic properties, **3**-12
media noise sources, **3**-5–6
organic dye, **3**-5
Phase locked loop, **1**-11
Physical organization, data on disk, **14**-6–7
Physics of longitudinal, perpendicular recording, **2**-1–26
  DC remanent noise spectrum, **2**-14–15
  isolated magnetization transition, **2**-8–11, **2**-14–16
  noise spectrum, pseudo-random bit sequence, **2**-17
  nonlinear transition shift, **2**-21–22
  playback media noise spectra, **2**-13–16
  playback signal power spectrum, **2**-13
  pseudo-random bit sequence, **2**-19–21
  reciprocity principle, playback signal from, **2**-7–13
  signal-to-noise ratio calculations, **2**-17–21
  square-wave pattern, **2**-11–13, **2**-17–19
  transition noise spectrum, **2**-16
  transition parameter, isolated transition, **2**-3–6
Physics of optical recording, **3**-1–22
  beamsplitter, **3**-16–17
  birefringence, **3**-12–13
  dielectrics, **3**-9–10
  direct overwrite, **3**-14
  exchange-coupled, field-coupled multilayers, **3**-12
  film stack design, **3**-11–12
  focus, tracking servo, **3**-17–18
  formatting, **3**-13–14
  heat sink/reflector, **3**-10–11
  light source, **3**-15–16
  magnetic multilayers, **3**-15
  magnetic properties, **3**-12
  magneto-optic media, **3**-6–9
  mechanical properties, **3**-13
  media, **3**-3–15
  noise, **3**-18–19
  noise sources, **3**-5–6, **3**-14
  optical figure of merit, **3**-11–12
  optical properties, **3**-12–13
  optical recording systems, **3**-15–19
  organic dye, **3**-3–6, **3**-5, **3**-12
  phase change, **3**-3–6
  read signal, **3**-18
  RE-TM alloys, **3**-7–9
  substrate, **3**-12–14
  super-resolution, **3**-14–15
  thermal properties, **3**-12, **3**-13
  transparency, **3**-12–13
  waveplate, **3**-16–17
Physics of recording, **14**-1–9
  logical organization, data on disk, **14**-7
  magnetic recording, basics of, **14**-1–6
  physical limits on recording density, **14**-8–9
  physical organization, data on disk, **14**-6–7
Playback media noise spectra, **2**-13–16
  DC remanent noise spectrum, isolated magnetization transition, **2**-14–15
  isolated magnetization transition, **2**-14–16
  noise spectrum, pseudo-random bit sequence, **2**-17
  transition noise spectrum, isolated magnetization transition, **2**-16

Playback signal power spectrum, pseudo-random bit
        sequence, 33-2–4
Postprocessor, read channels, hard drives, 15-7–8
Postprocessor data detection, 32-19–20
Power density spectrum, digital signals, systems, statistical
        analysis, digital signals, 7-8–14
Precompensation, read channels, hard drives, 15-7
Projective planes, structured low-density parity-check
        codes on, 38-7–10
Pseudo-random bit sequence, playback signal power
        spectrum of, 2-13, 33-2–4

Q

Quantized slope lookup table phase detector,
        partial-response channels, adaptive timing
        recovery, 26-8

R

RAMAC, *see Random Access Method of Accounting and
        Control*
RAM-based decision feedback equalization detection,
        32-10–13
    trellis, detection in, 32-11–13
Random Access Method of Accounting and Control,
        1-10–11
Read channel
    organization of data on disk, 14-1–9
    physics, recording, 14-1–9
Read channel coding
    capacity-approaching codes for partial-response
        channels, 23-1–21
    constrained code error-control coding, 21-1–10
    constrained code error correction capability, 20-1–17
    convolutional codes for partial-response channels,
        22-1–18
    maximum transition run coding, 18-1–13
    multitrack systems, coding, detection for, 24-1–29
    runlength limited sequences, 17-1–10
    spectrum shaping codes, 19-1–19
    two-dimensional data detection, error control, 25-1–16
Read channel equalizer architectures, 34-9–12
    finite impulse response filter architectures, 34-9–11
    representative finite impulse response filters, 34-11–12
Read channel hard drive, 15-1–11
    adaptive equalization, 15-4
    analog front end, 15-3
    controller functionality, 16-1–27
    error control coding, 15-9
    error performance measures, 15-9–10
    modulation coding, 15-8–9
    partial-response signaling, maximum likelihood
        sequence estimation, 15-3–4
    postprocessor, 15-7–8
    precompensation, 15-7
    read channel servo information detection, 15-6–7
    thermal asperites, effect of, 15-7
    timing recovery, 15-5–6
    Viterbi detection, 15-4–5

Read channel servo information detection, read channels,
        hard drives, 15-6–7
Read channel signal processing
    adaptive equalization architectures, partial-response
        channels, 28-1–13
    adaptive timing recovery, partial-response channels,
        26-1–27
    data-dependent noise in storage channels, detection
        methods, 33-1–17
    data detection, 32-1–23
    head position estimation, 29-1–12
    interpolated timing recovery, 27-1–16
    read/write channel implementation, 34-1–34
    servo signal processing, 30-1–12
    thermal asperity in magnetic recording, evaluation of,
        31-1–16
Read signal, optical recording system, 3-18
Read/write channel architectures, 34-2–3
Read/write channel implementation, 34-1–34
    analog front-ends, 34-5–9
    analog-to-digital converter, 34-8–9
    architectures, read/write channel, 34-2–3
    continuous-time filter, 34-6–8
    disk-drive read/write channels, 34-28–29
    integration, 34-29–30
    iterative decoders, 34-19–24
    low-density parity-check decoders, 34-22–24
    magnetic disk drive read/write channel integration,
        34-4–5
    partial-response equalization with
        maximum-likelihood detection, 34-3–4
    read channels, equalizer architectures in, 34-9–12
    representative finite impulse response filters, read
        channels, 34-11–12
    timing recovery, 34-24–27
    variable gain amplifier, with automatic gain control,
        34-5–6
    Viterbi decoder, soft-output, 34-20–22
    Viterbi detector, 34-12–19, 34-13–16, 34-16–17,
        34-17–18, 34-18–19
    write path, 34-27–28
Read/write heads, future of, 4-15–16
Reciprocity principle, playback signal from, 2-7–13
Recording density
    physical limits on, 14-8–9
Recording physics, 14-1–9
    increasing recording density, 14-8
    logical organization, data on disk, 14-7
    magnetic recording, basics of, 14-1–6
    physical limits on recording density, 14-8–9
    physical organization, data on disk, 14-6–7
Reduced complexity decoders, low-density parity-check
        codes, 36-13–17
Reed Solomon codes, 9-11–13
Repeat-accumulation code, 36-5–6
Reverse concatenation
    modulation codes, storage system, 11-9–10
    soft iterative decoding, 21-8–9
Runlength limited sequences, 17-1–10
    asymptotic information rate, 17-2–4

counting of sequences, asymptotic information rate, **17**-2–3

maximum transition run constraints, **17**-4

noiseless channel, codes for, **17**-6–8

$O, G/I$ sequences, **17**-4–5

two-dimensional runlength limited sequences constraints, **17**-5–6

weakly constrained sequences, **17**-5

S

Seek mode detection performance, servo signal processing, **30**-6–9

model of radial incoherence, **30**-7

simulation results, **30**-7–9

Serially concatenated convolutional codes, iterative decoding of, **35**-8–9

Servo burst demodulation, **30**-9–11

burst amplitude calculation, **30**-9–10

burst demodulation signal-to-noise ratio, **30**-10–11

Servo coding gain, **30**-3

coding gain/rate tradeoff for servo, **30**-3

example codes, minimum distance, **30**-3

Servo information detection, read channel, **15**-6–7

Servo signal processing, **30**-1–12

burst amplitude calculation, **30**-9–10

burst demodulation signal-to-noise ratio, **30**-10–11

coding gain/rate tradeoff for servo, **30**-3

model of radial incoherence, **30**-7

noise prediction in servo detection, **30**-4

seek mode, **30**-6–9

servo burst demodulation, **30**-9–11

servo coding gain, **30**-3

simulation results, **30**-7–9

track follow mode performance, **30**-4–5

Servo writing, head position estimation, **29**-3–4

Signal, noise generation, magnetic recording channel simulations, **6**-1–20

communications channel analogy, electronics noise only, **6**-4–6

Signal-dependent noise

detection, Bahl, Cocke, Jelinek, Raviv algorithm, **33**-13

signal-to-noise ratio definition, **5**-5–8

Signal processing for read channels

adaptive equalization architectures, partial-response channels, **28**-1–13

adaptive timing recovery, partial-response channels, **26**-1–27

data-dependent noise in storage channels, detection methods, **33**-1–17

data detection, **32**-1–23

head position estimation, **29**-1–12

interpolated timing recovery, **27**-1–16

read/write channel implementation, **34**-1–34

servo signal processing, **30**-1–12

thermal asperity in magnetic recording, **31**-1–16

Signal-to-noise ratio calculations, **2**-17–21

pseudo-random bit sequence, **2**-19–21

square-wave pattern, **2**-17–19

Slope lookup table phase detector, quantized, partial-response channels, adaptive timing recovery, **26**-8

Soft error-event correcting scheme, multitrack, coding, detection, multitrack systems, **24**-5–7

Soft-output decoding, modulation codes, storage system, modulation codes, **11**-9

Soft-output Viterbi algorithm, signal-dependent channel noise detection, **33**-13

outlines, **33**-16–17

Soft-output Viterbi decoder, **34**-20–22

Spectral shaping constraints, constrained binary codes, error correction, **20**-10

Spectrum shaping codes, **19**-1–19

capacity, dc-free constraint, **19**-4–5

composite constrained, combined encoding, **19**-15–17

dc-free codes, **19**-2–9

dc-free constraint, **19**-3, **19**-5–6

encoding, decoding of, dc-free constraints, **19**-7–9

higher order spectral zeros, codes with, **19**-9–17

K-RDS f sequences, **19**-11, **19**-14–15

recording system and, **19**-2

state dependent encoding, **19**-9

zeros at $f == 0$, **19**-10–11

Sperry Univac, **1**-11

Square Tape Cartridge, **1**-8

Square-wave magnetization pattern, playback signal of, **2**-11–13

Statistical analysis, digital signals, systems, **7**-1–15

autocorrelation, **7**-5–8

autocorrelation time average, **7**-5–6

averages, **7**-3

Markov chains terminology, **7**-14–15

phase randomising, **7**-7–8

power density spectrum, digital signals, **7**-8–14

signals, **7**-2

statistical autocorrelation, **7**-7

statistical average, **7**-4–5

time average, **7**-3–4

ST506 disk drive, **1**-12

Stochastic control formulation for partial-response channels, tight channel capacity upper bound, **13**-12–13

Storage, magnetic, history of, **1**-1–16

Storage channels, data-dependent noise in, detection methods, **33**-1–17

additive white Gaussian noise channel, detection methods in, **33**-2–7

Bahl, Cocke, Jelinek, Raviv algorithm, **33**-13

data-dependent correlated noise, **33**-9–11

data-dependent noise, detection methods in, **33**-9–13

maximum A posteriori probability symbol detection over, **33**-5–7

maximum likelihood sequence detection over, **33**-2–4

pattern-dependent noise prediction, maximum likelihood sequence detection based on, **33**-11–13

pseudo-random bit sequence, playback signal power spectrum of, **33**-2–4

recording system model, **33**-7–9

simulations, **33**-13–15

soft-output Viterbi algorithm, **33**-13, **33**-16–17
Viterbi algorithm, outlines, **33**-16
Storage system, modulation codes, **11**-1–11
   colored noise, intertrack interference, channels with,
      11-6–7
   constrained systems, codes, **11**-2–4
   constraints, ISI channels, **11**-4–5, **11**-4–6, **11**-5–6
   example, **11**-7–8
   future developments, **11**-9–10
   reversed concatenation, **11**-9–10
   soft-output decoding, modulation codes, **11**-9
Structured low-density parity-check codes, **38**-1–21
   bit error, **38**-14–16
   combinatorial designs, bipartite graphs, **38**-4–5
   integer lattice, codes on rectangular subset of,
      38-10–12
   lattice construction, low-density parity-check codes,
      38-10–12
   low-density parity-check codes, difference families,
      38-5–7
   partial-response channels, application in, **38**-12–16
   projective planes, codes on, **38**-7–10
Substrates, optical recording media, **3**-12–14
   birefringence, **3**-12–13
   formatting, **3**-13–14
   mechanical properties, **3**-13
   noise sources, **3**-14
   optical properties, **3**-12–13
   roughness, **3**-12–13
   thermal properties, **3**-13
   transparency, **3**-12–13
Super-resolution, optical recording media, **3**-14–15
Symbol-by-symbol detector, partial-response
      equalization, high-density magnetic recording
      channels, **8**-19
Symbol rate timing loops, performance analysis,
      partial-response channels, adaptive timing
      recovery, **26**-9–12
Symbol rate timing recovery schemes, partial-response
      channels, adaptive timing recovery, **26**-4–9
Synchronization, constrained coding, coding, detection,
      multitrack systems, **24**-4
Syndrome decoding, error-correcting codes, **9**-6–8

T

Tape drives, history of development of, **1**-5–9
Thermal asperity, magnetic recording, **31**-1–16
   detection, cancellation, **31**-9–16
   modeling, **31**-2–4
   simulation results, **31**-6–9
   system model, **31**-4–5
   system parameter settings, **31**-5–6
   system performance, effect on, **31**-4–9
3PM code, **1**-11
Tight channel capacity lower bound, maximal
      information rates, Markov sources, **13**-7–11
   iterative Markov source optimization algorithm, **13**-9
   mutual information rate, reformulating expression of,
      13-8–9

Timing loop modes, partial-response channels, adaptive
      timing recovery, **26**-3–4
Timing recovery, interpolated, **27**-1–16
   bit error performance, **27**-11–16
   channel knowledge, minimum mean square error
      interpolation filter, **27**-4–5, **27**-5–7
   conventional architecture, **27**-2
   digital timing recovery, **27**-3–4
   equalizer, implementation of, **27**-9
   finite impulse response-interpolated timing recovery
      loop structure, **27**-9–10
   initial sampling phase, estimate of, **27**-7–8
   interpolated architecture, **27**-3–7
   interpolation filter, implementation of, **27**-9
   longitudinal recording, **27**-12–15
   magnetic recording applications, **27**-10–16
   oversampling system, effect of employing, **27**-9
   perpendicular recording, **27**-15–16
   signal-to-noise ratio performance, **27**-11
   system model, **27**-10
Track follow mode performance, servo codes,
      30-4–5
Tracking servo, optical recording system,
      3-17–18
Transition noise, electronics, simulating recording
      channel with, **6**-9–20
Transition parameter, isolated transition, **2**-3–6
Transition run coding, maximum, **18**-1–13
   codes, **18**-4–10
   detector design, constraints, **18**-10–11
   error event characterization, **18**-2–4
   simulation results, **18**-11–12
Transparency, optical recording media, substrate,
      3-12–13
Trellis codes
   block, **20**-7–8
   constrained binary codes with error correction
      capability, **20**-4–6, **20**-7–8
   construction, **20**-7–8
   error correction capability, constrained binary codes
      with, block, **20**-7–8
   matched information rate, **23**-9–13
   for partial-response channels, convolutional codes,
      based upon Hamming metric, **22**-4–7
   superchannels, information rates, **23**-6–13
Turbo coding, **35**-1–21
   additive white Gaussian noise channels, performance of
      turbo codes over, **35**-9–11
   principles, **35**-2–6
Turbo coding concatenation schemes, **35**-5
Turbo coding for recording channels, **35**-11–12,
      35-12–15
   information theoretical limits, **39**-5–6
   interleaver, convolutional coding with, **35**-14–15
   intersymbol interference channels, turbo equalization,
      39-9–10
   iterative decoding for multitrack systems, **39**-10–11
   Lorentzian channels, **35**-16–18
   maximum A posteriori detector, **39**-10–11
   multitrack, **39**-1–17

partial-response channels, **35**-11, **35**-15–16
turbo codes, iterative decoding algorithm, **39**-8–9
with turbo equalization, **35**-13–14
without turbo equalization, **35**-13
Turbo coding iterative decoding, **35**-6–9
    algorithms, **35**-9
    iterative decoder structure, **35**-7–8
    maximum A posteriori decoder, **35**-6–7
    serially concatenated convolutional codes, iterative
        decoding of, **35**-8–9
Turbo coding parallel concatenated convolutional codes,
    35-2–4
Turbo/single-parity check coded partial-response systems,
    high-density digital recording systems, distance
    properties of, **37**-8–9
Two-dimensional data detection, error control, **25**-1–16
    equalization, two-dimensional, **25**-4–8
    joint detection, decoding, two-dimensional, **25**-13–15
    linear minimum mean-square-error equalization,
        25-5–6
    partial-response equalization, **25**-6–8
    precompensation, two-dimensional, **25**-3–4
    quasi-Viterbi methods, two-dimensional, **25**-8–13
Two-dimensional runlength limited sequences
    constraints, **17**-5–6

V

Viterbi algorithm, **32**-13–17
    generalized, **32**-21
    signal-dependent channel noise detection, outlines,
        33-16
Viterbi decoder, soft-output, **34**-20–22
Viterbi detector, **34**-12–19
    add-compare-select recursion, **34**-13–16
    add-compare-select unit design, metric normalization
        in, **34**-16–17
    bit-level pipelined add-compare-select unit, **34**-16
    data postprocessing, **34**-18–19
    read channels, hard drives, **15**-4–5
    survivor sequence detection, **34**-17–18
    two-dimensional, error control, **25**-8–13

W

Waveplates, optical recording systems, **3**-16–17
Weakly constrained sequences, runlength limited
    sequences, **17**-5
Winchester recording head, **1**-11
Write heads, **4**-6–7; *see also Read/write*
Write path, channel implementation, **34**-27–28